Plant Cell and Tissue Culture

Genetics and Biogenesis of
Mitochondria and Chloroplasts
5–7 September 1974
C. W. Birky, Jr., P. S. Perlman, and T. J. Byers

Regulatory Biology
4–6 September 1975
J. C. Copeland and G. A. Marzluf

Analysis of Ecological Systems
29 April–1 May 1976
D. J. Horn, G. R. Stairs, and R. D. Mitchell

Plant Cell and Tissue Culture
Principles and Applications
6–9 September 1977
W. R. Sharp, P. O. Larsen, E. F. Paddock,
and V. Raghavan

EDITED BY W. R. SHARP,

P. O. LARSEN, E. F. PADDOCK,

AND V. RAGHAVAN

Plant Cell and Tissue Culture

Principles and Applications

Biosciences Colloquium, (4th : 1977,
Ohio State University)

OHIO STATE UNIVERSITY PRESS : COLUMBUS

Library of Congress Cataloging in Publication Data

Biosciences Colloquium, 4th, Ohio State University, 1977.
  Plant cell and tissue culture.

  (Ohio State University biosciences colloquia)
  Includes index.
  1. Plant cell culture—Congresses.   2. Plant tissue culture—
Congresses.   3. Agriculture—Congresses.   4. Plant genetics—
Congresses.   5. Food supply—Congresses.   I. Sharp,
William R., 1936– II. Title.   III. Series: Ohio. State Universi-
ty, Columbus. Ohio State University biosciences colloquia.
QK725.B58   1977   581'.0724   78-13080
ISBN 0-8142-0287-X

Dedicated to the memory of
HERBERT E. STREET
a pioneer in the field of
plant cell and tissue culture
who died unexpectedly on 4 December 1977

Contents

TOTIPOTENCY

Nothing is separate
Nothing's alone
Separation only makes it clear

Wondrous spark of light
Is made of many parts
These aren't apart at any height

Oh plant, I see you as you are
As myself, part of the whole
We together are complete

Oh leaf, I know you as you are
Cell upon cell, part of your whole
Rich green fabric woven so well

Each cell contains the secret within
Each one knows where the other has been

Separation will make the study complete
Which will you be—root, stem, or leaf?

You hold the power deep within
To always return to wherever you've been

Ever-present, power and might
Lulling grace, wondrous and quiet

   Sharon Maraffa

Sharon Maraffa is a graduate student in the Department of Microbiology of the Ohio State University.

Preface

The goals of the Fourth Annual College of Biological Sciences Colloquium were to provide an opportunity in continuing education for the faculty and students of the Ohio State University, to inform international scientists of our current research, and to make a significant contribution to the plant sciences and agriculture. The four-day colloquium attracted over 450 scientists from major institutions and industry, representing 47 states of the United States and 12 foreign countries.

Plant cell and tissue culture techniques have the potential of expediting new breakthroughs in our understanding of fundamental biological principles and providing economic gains when applied to practical problems. Techniques such as somatic and haploid cell cloning, mutagenesis, morphogenesis, embryo culture, and protoplast fusion can be potentially valuable in developing new or genetically modified crops characterized by disease resistance, increased vigor, cold hardiness, and so forth. To cite two examples of economic potential: (1) in forestry, the fusion of haploid protoplasts of slash pine with those of longleaf pine could yield a diploid slash-longleaf hybrid that does not have the "grass" stage of longleaf but retains the straightness and other desirable characteristics of longleaf; and (2) in pathology, phytotoxins have been used as a means of selection for disease resistance in protoplast cultures of tobacco and corn.

The scholarly contributions and enthusiastic dialogue exchanged by participants during the discussion periods were indicative of the impact of this colloquium.

The Ohio State University Office of Academic Affairs is acknowledged for its support in making the publication of these proceedings a reality.

Acknowledgments

Grateful acknowledgment is made to the following institutions and individuals for providing financial support for the Fourth Annual College of Biological Sciences Colloquium:

Baker Company, Inc.
Bellco Glass, Inc.
Brazilian Coffee Institute
California-Florida Plant Corporation
Campbell Institute for Agricultural Research
City National Bank, Columbus, Ohio
Corning Glass Company
Cunningham Gardens
DeKalb Agricultural Research, Inc.
Eli Lilly and Company
Funk Seeds International, Inc.
General Foods
Gilford Instruments
Grand Island Biological Supply Company
International Island Resources
Lakeshore Equipment
Mr. and Mrs. John McCoy, Sr.

Merck, Sharp, and Dohme
Mikkelsens, Inc.
Monsanto Company
Northrup, King and Company
Oglevee Floral Company
OSU College of Agriculture
OSU College of Biological Sciences
OSU Developmental Biology Program
OSU Graduate School
OSU Latin American Studies Program
OSU Office of Academic Affairs
Pan American Plant Company
Pfizer Central Research
Pioneer Hi-Bred International
Rohm and Haas Company
Union Carbide
Weyerhaeuser Company Founddation

Fourth Annual Biosciences Colloquium
College of Biological Sciences
Ohio State University
6–9 September 1977

PLANT CELL AND TISSUE CULTURE
PRINCIPLES AND APPLICATIONS

Organizers

W. R. Sharp, Department of Microbiology, Ohio State University
P. O. Larsen, Department of Plant Pathology, Ohio State University
E. F. Paddock, Department of Genetics, Ohio State University
V. Raghavan, Department of Botany, Ohio State University

Speakers

Trevor G. Arscott	Jeanne Barnhill Jones
Y. P. S. Bajaj	K. K. Kartha
Horst Binding	Roy M. Kottman
J. M. Birkeland	L. R. Krusberg
Donald Boulter	Carl G. Lamm
G. Bruening	Iris A. Mastrangelo-Hough
Raisa G. Butenko	R. L. Meyer
Tsai-Ying Cheng	Knut Norstog
E. C. Cocking	John A. Pino
G. B. Collins	Per Pinstrup-Andersen
Donald K. Dougall	V. Raghavan
David A. Evans	W. R. Sharp
O. L. Gamborg	S. H. Smith
K. L. Giles	Harry E. Sommer
Yury Yu. Gleba	M. R. Sondahl
José Goldemberg	F. C. Steward
Antonio Natal Gonçalves	H. E. Street
James E. Gunckel	N. Sunderland
F. B. Holl	

Part 1

The Current Global Status of Food and Agriculture

ROY M. KOTTMAN

Introductory Remarks

1

The College of Agriculture and Home Economics and the School of Natural Resources at the Ohio State University have been rather heavily involved with a variety of overseas projects sponsored by the United States Agency for International Development and its predecessor agencies throughout the past quarter century. We are concerned not only about the current global status of food and agriculture, but we believe that we have made a substantial contribution to the potential that we now see for an increasing number of nations to help themselves in the task of providing food and fiber for their citizens.

During January of 1977 I was privileged to travel in India and Bangladesh. My earlier visits to India were in 1962 and 1969. On both of those occasions I took the opportunity to visit several countries in both Europe and Asia, with special emphasis on Japan, Taiwan, and the Philippines.

Over the ten-year period from 1965 to 1975, I was able to make rather extensive observations of the agricultural industry in Brazil and somewhat less-extensive observations in several other Latin American countries. During the summer of 1974, it was my good fortune to serve as coleader of a people-to-people tour that took us to Belgium, Holland, Denmark, West Germany, East Germany, Czechoslovakia, and the Soviet Union.

Unfortunately, I have not yet had an opportunity to travel in Africa, and my understanding of the agriculture of that continent is largely based upon

Roy M. Kottman, College of Agriculture and Home Economics, Ohio State University, Columbus, Ohio 43210.

discussions that I have had with members of our faculty who have served in Nigeria, Ethiopia, Uganda, Kenya, and Somalia. I have tried to take advantage of a number of printed materials that I have seen on the subject of African agriculture, but I much prefer firsthand observation.

As a result of my observations and study, I have become convinced that the world's natural resources are quite adequate to meet future world food needs for a population several times that of our present world population. The principal constraints to an adequate food supply are those of unrestrained population growth and unwise governmental policies. Fortunately, population growth has been brought to a condition of near zero population increase in most of the developed countries of the world. Among the developing countries, the People's Republic of China stands as an example of what can be done if a dictator decrees that there shall be zero population growth.

The results of constraint on world population growth are rather encouraging, in that the annual population growth of 66 million that prevailed back in 1965 has now been slowed to about 60 million. Even so, 60 million more people living on planet Earth each year means that every three and one-half years a population equal to that of the United States comes into being. As we think about all that is required to provide food and fiber for that number of people, we can begin to contemplate the magnitude of the world food production challenge that lies ahead of us.

Despite the enormity of that challenge, I for one believe that the future holds great promise for meeting world food needs. My optimism is borne out by the spectacular improvement in world grain yields during the past three decades, when per acre yields of wheat, corn, barley, oats, and several other crops have doubled in just thirty years. Much of this gain was due to the Green Revolution crop varieties that have been utilized quite effectively in several of the developing nations. The so-called Green Revolution was based squarely on research and education backed up by resident instruction programs for young people in a variety of agricultural universities in the developing countries. There are, of course, two major elements in world demand for food. One is based on sheer numbers of people, the other is based on the fact that income levels are rising, and as they rise people in the developing countries are demanding improved levels of nutrition. The result of this dual stress on future world food supplies is simply that the overall supply of food in the world must be doubled by the year 2000.

As we all know, there are basically only two ways to increase the world's food supply: (1) we can cultivate acres that are not now being cultivated, and (2) we can increase per unit yields of crops, livestock, poultry, and fisheries.

At the present time, on a worldwide basis, about 3.6 billion acres are under cultivation. An additional 4.2 billion acres of reasonably good land in Africa, Brazil, and other nations in Latin America, as well as in various other areas throughout the world, can be brought under cultivation.

The greater opportunity for increasing world food production lies in boosting per acre crop yields or per unit livestock, poultry, and fishery yields beyond increases already achieved. I would not be so optimistic as I am about the potential for increasing yields in the developing countries were it not for the fact that our college has been closely associated with what has happened to per unit yields in northwest India.

When our Ohio State University faculty members first went to India under an AID contract in 1955, their mission was that of helping the Indians establish agricultural universities similar to those of our own land grant universities here in the United States. At that time we found average wheat yields of only about 11 bushels per acre being produced in the Punjab, despite the presence of an excellent natural resource base in that state. When we left India in 1972, wheat yields in the Punjab were averaging 60 bushels per acre, and they have continued at that level up to the present time. While I was visiting the Punjab Agricultural University last January, Dr. K. S. Gill, Chairman of the PAU Department of Agronomy, showed me plots of a new variety of wheat (just released to Punjabi farmers in September 1976) that had yielded in test plots throughout the state of Punjab an average of 79 bushels per acre. Dr. Gill showed me other plots where he had a new wheat variety on which one year's data showed yields well beyond 79 bushels per acre.

Another very impressive statistic concerning the Punjab is that 75 percent of their crop acreage is presently under irrigation. This is more than double the irrigated acreage that we found in the Punjab back in 1955. Beyond that, the type of irrigation has changed dramatically from the lifting of water by either human or animal power to the use of internal combustion engines and electric motors, which presently power the tube wells. Another remarkable development has been the number of crops produced on the same piece of land throughout each twelve-month period, which currently stands at 2.6 crops per acre per year in the Punjab.

In the state of Haryana (another of the states where the faculty of our college worked over a seventeen-year period), 47 percent of the land was irrigated as of 1976, and it is my understanding that a new irrigation project has made it possible for 58 percent of the land in Haryana to be irrigated as of this year. In that state, also, 2.4 crops per acre per year make for an abundance of food grains that never would have been achieved had it not been for the development of the Haryana Agricultural University with its integrated agricultural programs of resident instruction, research, and extension.

Those records of 2.6 and 2.4 crops per acre per year provide a marked contrast with the 1.2 crops per acre per year presently produced in West Bengal (the state in which Calcutta is located) or the 1.1 crops per acre per year produced in Bangladesh.

The point I would make is that people in the underdeveloped countries, as is true of people everywhere, possess great potential for responding to research and education and to those incentives and economic developments that will permit a better life for themselves, their children, and their grandchildren. And, like human beings everywhere, they are much better off in being provided an opportunity to help themselves than in being provided charity.

The role of our U.S. universities in helping meet world food supply and demand problems is best discharged in terms of institution building. By that I refer to the providing of opportunity for young people in the developing countries to achieve a meaningful education in the agricultural and biological sciences. Unfortunately, many of the agricultural colleges in the developing countries continue to be woefully inadequate in terms of organization, curriculum, and governmental support of their endeavors.

When one considers that India was producing annually only about 50 million metric tons of food grains back in 1955 but has been producing 118 million metric tons annually in recent years, one almost has to believe that the twenty two agricultural universities that have been developed in India since 1955 have had a great deal to do with that remarkable increase in production. As a matter of fact, India currently has in the neighborhood of a 22 million metric ton surplus of wheat and rice, a part of which reportedly has been utilized to repay a loan from the Soviet Union. This is a truly remarkable accomplishment when one considers that the current population of India is 630 million, with 12 million additional mouths to feed each year. Quite obviously, India's annual population increase must be slowed, but achieving population control is extremely difficult in any nation where there is widespread illiteracy, and where the major source of social security for most citizens resides in their having enough children to take care of them in their old age.

My experiences over the past quarter century have convinced me that future world food supplies can become much more adequate than at present. I am convinced that the deterrents, if any, will not be those of the scientific community but rather those of governments and their interference with or failure to support the scientific community. I was quite shocked when a member of the steering committee for the recently completed National Academy of Sciences study (and president of the Overseas Development Council) stated that "the study appears at a time when the U.S. is looking for more meaningful ways to combat hunger than just giving food away or exporting energy-intensive farming techniques," and my apprehension is not improved when I read it was further stated that "U.S. agricultural research has 'languished' for two decades and needs to be revitalized in areas that will produce new crops. Research should be shared with developing nations, and we should also profit from research done abroad." That such statements could be made in the face of what has taken place over the past twenty five years is truly amazing.

As you might guess, I have had misgivings about that study, as with several others that have been made under the auspices of the National Academy of Sciences in recent years, and when the study to which I have just referred comes out with a statement to the effect that research should emphasize biological productivity—developing crops that do not depend heavily on fertilizers, pesticides, or intensive irrigation—I think it is time for all of us to point out that crops do not grow in a vacuum. They require nitrogen, phosphorus, potash, and various micronutrients, and there is not much prospect that they will ever shed their quaint old habits of requiring these nutrients. Beyond that, when one considers that only 15 percent of the world's cultivated crops are irrigated but that that 15 percent produces 30 percent of our world's production of food and fiber, it becomes quite obvious that intensive irrigation is here to stay. What is most needed are better ways of storing, conserving, and utilizing water, as opposed to so much emphasis on developing new crops that will require greatly reduced supplies of water for their growth. Much the same comment could be made relative to the development of crops that will not require pesticides.

Similarly, I am sure that all of us are well aware that the increased agricultural production that will be needed in the years ahead simply cannot occur without huge inputs of energy other than animal and human energy. The relationship between energy use and economic development here in the United States is simply that for each one billion barrels of oil equivalent that we have consumed we have created $100 billion worth of gross national product and four million new jobs. Whether we obtain our future energy supplies from fossil fuels, from uranium, from solar energy, or from wherever, the point to be remembered is that productivity of any kind utilizes energy. Fortunately, substantial supplies of fossil fuel energy continue to be available in many parts of the globe. Great progress is being made in nuclear and solar energy with the attendant possibility of providing greater material abundance for all mankind on planet Earth.

The use of plant cell and tissue culture shows great promise in several areas of agriculture. The use of haploid plant materials and protoplast fusion provides exciting possibilities for the genetic improvement of plants. Plant tissue culture is currently being used commercially for the eradication of plant pathogens from several plant species that are vegetatively propagated. This use will continue to increase as more technical information is obtained on the regeneration of vegetatively propagated plants from tissue culture. Storage of germplasm and rapid multiplication of plant clones are other areas that show excellent potential for fruitful use of plant tissue culture technology. Many of these areas are discussed elsewhere in this volume. I hope that these papers will provide you with information that will aid you in bridging the technological information barriers that prevent you from achieving your ultimate research and production objectives through the use of plant cell and tissue culture.

Those of us involved with the biological and agricultural sciences have made a tremendous contribution to the well-being of human beings all around the world. I am confident that we are at the point of making scientific breakthroughs of even greater magnitude than those of the past. What we need more than almost anything else is optimism and self-confidence, both individually and collectively.

D. H. GRAHAM, R. L. MEYER, AND N. RASK

The World Food Economy

2

The world food problem is a topic of intense international debate among scientists and governments. Although recent prospects for abundant harvests in many areas of the world hold promise for a partial rebuilding of depleted world reserves of food and feed grains, underlying factors that brought the world food problem into sharper focus in recent years remain largely unchanged. Population growth and increased affluence, increased scarcity and cost of energy, diminished availability of virgin land and water resources, a slowing of productivity growth, and unstable weather patterns—all contribute to uncertainty concerning the world's ability to produce and distribute adequate supplies of food.

In this paper we focus on three general aspects of the world food economy. First, a brief review of recent trends in world production, consumption, and trade is presented. This leads to the conclusion that the hunger problem must be faced and resolved principally in the low income countries (LICs). Problems of energy availability and cost, climatic variability, and adaptability of technology make this a formidable task. In the second section a review of recent agricultural policies and performance in LICs is presented. The third section outlines emerging research priorities and issues important for meeting the world food challenge.

THE WORLD FOOD TRADE AND PRODUCTION SITUATION

In describing the world food picture, it is fashionable to project supply

D. H. Graham, Department of Agricultural Economics and Rural Sociology, Ohio State University, Columbus, Ohio 43210.

and demand conditions into the future. Depending on the assumptions one makes, these projections may indicate optimism or pessimism. Too often the assumptions used are unduly restrictive and overly influenced by the state of food production at the time the projections are made. For example, projections made in the early 1960s assumed continued growth in production at reasonably high levels. However, reacting to the shortages caused by crop failures and increased demand in the 1970s, food production possibilities and attendant problems are now of central concern.

We think it important to identify some of the factors that will shape and direct the supply of, and demand for, food in the future and examine how these factors relate to research. In this context we see some definite constraints being imposed on both the demand for food (principally from low income levels in LICs) and on the use of resources for the production of food (including high energy cost and climate variabilities). It is also probable that these constraints may be significantly different for developing and developed countries, especially in the use of capital- and energy-intensive production technology.

Briefly, the 1960s were a time of abundant, low-cost food. Several factors contributed to this. First, many years of increasing global production had resulted in the buildup of surplus stocks in grain-exporting countries. The United States led the way in the retirement of some cropland in an effort to avoid further buildups and to maintain farm income levels. In some LICs, the Green Revolution provided significant gains in food productivity. Prices for many food and farm products were relatively low, and food aid or imports on concessional terms were available for most countries with supply difficulties. Fertilizer production capacity was more than adequate, leading to low prices for this important input. During this time it seemed that the world had a plentiful supply of inexpensive food and of the inputs necessary for its continued production.

In the early 1970s, however, the picture changed quickly. The United States had been following a practice of gradually reducing its surplus stocks during the late 1960s and early 1970s. This willingness to reduce reserves was associated with an increased demand resulting from a combination of factors including rapid economic growth in 1971–73 in many developed countries, devaluation of the dollar, a change in Russian food import policy and a worldwide food production decline in 1972. These events led to a substantial increase in exports from the United States to Russia in 1972/73 and to developing countries in 1973/74. These increased exports quickly depleted stocks in the United States and other exporting countries. Competition for the remaining food supplies drove prices to exceptionally high levels and ushered in a new period of concern over the world's ability to feed itself. With the depletion of U.S. feed and food reserves, there was also a reduction in the amount of these commodities available for concessional sales to developing countries. Actually, U.S. food aid of various kinds peaked in the early 1960s at a level of $1.6 billion annually. By the early 1970s food aid was cut almost in half and then was offered mostly for dollar credit sales. Production problems and shortages continued through the mid-1970s, and only now are prospects reasonably good for some rebuilding of world stocks.

Looking to the future, it seems likely that the events that have been largely responsible for the shift from a surplus to a precarious world food situation will continue and in some cases intensify. It is also likely that these events will increase rather than lessen the disparity between developing and developed countries. In summary form these include: continued population growth (mostly in LICs), increasing affluence (mostly in developed areas), diminishing productivity increases, and substantial energy price increases (with the attendant depressing affect on agricultural intensification through mechanization and high energy inputs). In addition, food production may have to compete directly with energy produced from biomass in many energy-scarce economies.

The prospects for substantial general economic development in LICs with high population growth rates, resource scarcity, and the need to import petroleum and fertilizer inputs are not good. Economic stagnation will not generate the effective demand for food that will allow the modernization of agriculture for the production of domestic food crops. Thus, agricultural resources in LICs will likely continue to flow more toward the export sector where sufficient demand will be generated.

It is often assumed that the United States and other food-exporting countries, through increased production, can feed the hungry of the world. This is not a likely possibility. First, total U.S. production accounts for only about 12 percent of the food consumed in the world, and U.S. exports in recent years account for only about 3-4 percent. Further, almost all of this exportable surplus is traded to developed countries, principally in Europe and Japan. Actually, in 1972 the LICs exported twice the quantity of farm products that they imported (Barr, 1977). In that year the LICs accounted for 34 percent of world farm product exports, while importing only 17 percent. Developed countries, on the other hand, had a net trade deficit, exporting 55 percent and importing 71 percent of the world trade in farm products.

Furthermore, farm income maintenance in the United States is currently taking precedence over the buildup of world food reserves. As we approach the end of another good wheat production year and a potentially good corn and soybean harvest, the machinery is already in place to bring increased quantities of food and feed grain under government loan (a form of food and feed reserve program) and to mandate set-aside acreages for the next crop year in order to qualify for the crop loan program. Although these programs may result in some temporary build-up of reserves through farmer holdings, the primary motivating force is income maintenance for U.S. farmers; the set-aside acreage provision will surely result in lower production levels, and hence higher world prices, in future years.

Weather and energy are two additional elements that will affect future supplies of food and may lead to even greater disparity between high and low income countries. These factors have not received sufficient attention in supply projections. They will also be important determinants of research direction in production technology.

Most projections for future food supplies assume normal weather. Some attention is paid to recent patterns of highly variable annual fluctuations in weather, but the weather factor is usually dismissed as a determinant factor since it cannot be predicted with any degree of accuracy. Yet it seems quite clear that the normal years (those in our immediate memory), say the period from 1950 to 1970, were highly unusual (and favorable) in terms of annual and seasonal weather stability. Weather extremes, prior to 1940 and since the early 1970s, have been considerably greater.

During this "normal weather" period we witnessed most of the technological revolutions in agriculture, a substantial increase in food output, and a more than doubling of world population. Developing countries, where population levels have risen rapidly, have come to rely on a relatively constant flow of food from domestic production, food aid, or low priced imports. If weather in the future is more variable, we can expect greater year-to-year variability in output within and between regions of the world. Also, total output over an extended period of time will necessarily be less than if more favorable weather were to continue. A potentially more disturbing factor is the possibility that the productivity gains that we have attributed largely to technology must now be partly attributed to good weather. One need only observe average corn yields in the United States since 1955 to appreciate both the stability and growth during the 1950s and 1960s and the contrasting instability and lack of growth exhibited in recent years.

The energy problem will have a significant and persuasive impact on the future world food economy. This impact will affect production technology, levels of output, and also demand, especially in LICs. Energy, like food, has been perceived as abundant and inexpensive. This abundance and low cost has been an essential ingredient in directing economic growth. Modern technology, including agricultural technology, as experienced in developed countries, has focused on maximizing labor productivity through extensive use of mechanization with little concern for economy in energy use. Fertilizer, a key ingredient of the Green Revolution, is energy intensive also. Abundant and inexpensive fossil energy supplies further encouraged this type of growth.

The prospects of diminishing supplies of fossil energy, with production of oil and natural gas peaking within this century, underline the important but transitional role of present forms (and costs) of energy and the agricultural technology that they have fostered. The substantial energy price increases of 1973/74 are only the initial response to fossil energy shortages that will continue to drive up energy prices and affect the production and growth processes.

Changes in energy form, availability, and price are likely to have significant impacts on both world food production and consumption. This will be especially important for LICs, where development aspirations have been fueled by previously low energy prices (and low food prices). The near term (early 1980s) prospects of a substantial further increase in petroleum prices brought about by potential demand in excess of capacity production constraints provides additional urgency to the energy question (CIA, 1977). Given the projected energy supply-price situation, most energy-poor LICs clearly cannot follow the energy-intensive path of development, nor can they rely on substantial imports of energy inputs. Thus, many of these countries must follow a combined path of developing less energy-intensive technology and finding alternative and, it is hoped, renewable sources of energy.

Agriculture, through its biological process of transforming solar energy, is one source of renewable energy. This is especially true in many tropical areas, where year-round growth and abundant solar energy and water are available. From a food production standpoint, however, energy produced from biomass may be detrimental, since use of agricultural resources for energy production may come into direct conflict with the use of these resources for food production. Population growth projections that exceed food production estimates for many of these countries add further emphasis to this possible competition for agricultural resources.

The energy problem may have additional impacts on the composition of demand for domestic food items and thus, indirectly, on agricultural production technology. Historically many LICs have had two general forms of agriculture. One is export oriented. This system is typically capital intensive, highly productive, and provides many of the tropical crops such as coffee, sugar, and bananas sold in markets of developed countries. The more traditional agricultural system, responding to domestic food needs in LICs, is typically labor intensive and registers considerably lower relative yields. Research and policy incentives have generally focused more strongly on the export sector in LICs. In many instances domestic food prices have been held at lower-than-international market prices in an effort to keep the industrial wage bill low and thus favor industrial development in these countries.

Lack of technological adoption in the traditional production system is in part a consequence of the structure of demand for the domestic food crops. The lack of purchasing power and/or low food prices (through policy) does not produce sufficient incentives to stimulate adoption of higher-cost technology by farmers. This lack of effective demand, in the domestic food market of LICs, is strongly responsible for the differences noted in production technology between the domestic and export agricultural sectors in these countries. If one imposes on this system a substantial increase in imported energy costs, it is unlikely that demand for domestic food will be sufficient to attract significant amounts of capital- (and energy-) intensive technology to this kind of agriculture. Thus, labor-intensive agriculture is likely to persist more strongly than anticipated a few years ago under low world energy prices.

To summarize, we see the key elements in the world food picture to be (1) a declining interest on the part of surplus-producing nations to engage in significant food aid activities; (2) a desire in developed economies to maintain farm income levels through production constraints and price supports; (3) a declining number of countries in the surplus food production category; (4) a continued net agricultural trade surplus, but not a food trade surplus, for LICs (i.e., selling more non-food farm products than buying); (5) additional annual variability in food production from less-stable weather conditions; and (6) increased costs of energy-related technologies. We are led to the conclusion that the hungry poor of the world must be fed in large part from food produced with their own resources. Significant international aid is not likely to be forthcoming, nor will they be able to bid effectively for scarce resources or commodities in world trade. This leads us to consider the development possibilities within the LICs themselves.

AGRICULTURE AND THE ISSUES OF DEVELOPMENT IN THE LICS

The foregoing analysis underscores the importance of developing the capacity of local foodstuff production in LICs themselves. Analogous to our earlier discussion of the era of the fifties and early 1960s as the years of surplus and optimism and the post-1970 period as the era of scarcity and pessimism with respect to the world food problem, it is useful to review developments within the LICs during these two distinct periods. To highlight our discussion we will focus on the way in which LICs changed their attitudes and treatment of the agricultural sector during these two periods of recent history in order to place the current LIC food issue in perspective.

Beginning in the early 1950s, many LICs began to consciously design development programs to promote economic growth. Important to these development plans was industrialization. It was felt that industrial growth brought modernization more rapidly than any alternative growth strategy and, as a result, economic policies were designed to promote this sector. Inevitably the agricultural sector was forced to bear the brunt of this effort by being taxed heavily to provide the scarce foreign exchange and domestic capital needed to promote industrial growth. Concessionally priced credit, tariff exemptions, favorable foreign exchange rates, tax holidays, subsidized inputs of energy and transport, and other special favors were granted to develop modern industry.

Rarely, if ever, were these advantages given to agriculture. More importantly, the issue of agricultural productivity was infrequently addressed, and little stress was placed on investment in research or human capital to develop a potentially high rate of return for agriculture. Instead, low productivity levels were taken as given, and the task at hand was one of devising ways of transferring resources from the larger, more traditional agricultural sector to the smaller, more modern industrial sector.

Associated with this sectoral bias was a built-in pessimism concerning the growth potential of agriculture, the presumably weak linkages to promote significant change in the rest of the economy, and finally the negative association of the agricultural sector with the more traditional political structure in the country (especially in Asia and Latin America). The international division of economic activity, with LICs exporting primary products and importing manufactured products, was also attacked by many LIC spokesmen as assigning LICs to permanent poverty. They felt that the international terms of trade were unfavorable to them. In their view the LICs faced an unpromising future for expanding agricultural exports at remunerative prices, while prices for manufactured imports would rise over time. Hence the ideology of import substitution industrialization took firm hold in policymaking circles of most LICs, with the resulting unfavorable treatment of agriculture.

Toward the end of the 1960s, however, the import substitution strategy encountered more problems than promise as one LIC after another experienced a decline and stagnation in economic growth. Numerous problems emerged. The causes of this decline are complex and beyond the scope of this paper, but it is pertinent to point out that the previous neglect of agriculture played an important role in this downturn. The growing penalization of exports led to a drop in the growth rate of agricultural exports and, in some cases, generated an absolute decline in agricultural output. Consequently, the capacity to import industrial capital goods was compromised through the foreign exchange shortage that led to balance-of-payments crises. These developments, in conbination with American grain surpluses, created a large flow of U.S. government P.L. 480 agricultural aid to make up for the shortfall in output. In time these measures of relatively elastic supply of food aid further discouraged attempts to increase local production of agricultural foodstuffs.

At the same time, low to stagnant growth of domestic food output led to supply crises in the growing urban centers of the LICs. Attempts to control price rises merely exacerbated the situation by inducing farmers to hoard stocks or cut back on production. The rapid rate of urban and industrial growth associated with the import substitution industrialization strategy led to growing income disparities between the rural and urban sectors, which in turn increased internal migration into urban centers. Unfortunately, this relative decline in rural manpower was not matched by agricultural productivity increases, given the neglect or outright discrimination against the agricultural sector in the growth strategy of the period. In the end, the agricultural sector became an important bottleneck, adding to inflation and limiting further industrial growth. In this setting P.L. 480 food aid, although helpful in ameliorating certain supply crises, did nothing to resolve the basic problem; namely, reform and modernization of the agricultural sector within the LICs to produce a growing supply of cash crops for export and foodstuffs for the local market at reasonable prices. To accomplish this task, an abrupt shift in priorities and associated policy measures was needed, along with a technological breakthrough creating the possibility for productivity gains within the limited resource setting of poor to semi-industrialized LICs.

From the late sixties through the early seventies, three factors helped remove the earlier bias against agriculture in the LICs. First, technological breakthroughs in new seed varieties for wheat and rice (labeled the Green Revolution) permitted output and productivity increases in some regions of countries such as Mexico, India, Pakistan, the Philippines, and so on. Second, the import bottlenecks and foreign exchange shortages of the early 1960s prompted many LICs to de-emphasize the import substitution and protectionist foreign trade strategy. LIC economies were opened to freer trade and flexible exchange rate systems. Exports were emphasized to earn more foreign exchange. This important shift in priorities aided agriculture (or at least reduced the earlier discriminatory measures), since agricultural products were the most important component of LIC exports. At the same time, concessionally priced credit was extended much more widely to agricultural producers in many countries, and minimum price programs in certain countries reduced some of the risk of price instability. As a result of these measures, the rate of return to agricultural exports rose significantly and output grew substantially.

The third event promoting agriculture in the LICs during this period was the stimulating upturn in the world prices for many tropical and semitropical cash crops up to 1974. World trade in agricultural commodities grew more rapidly during the early 1970s than during any prior period since the Korean War commodity boom. These developments reinforced the LIC domestic drive to stimulate the growth of agricultural output through a battery of credit and other incentives. Also, it should be mentioned that the LICs by this time appreciated the need to invest much more in agricultural research and training institutions and, in the process, to improve the long-run potential for productivity advances in the future.

Finally, it is clear that the shift in emphasis from growth at any cost to growth with equity improved the possibilities for more favorable treatment for the agricultural sector in LICs. By the early 1970s it was recognized that the earlier emphasis on rapid industrialization led to increased income disparities. The earlier hope that industrial growth would generate rapid change and modernization throughout the economy did not materialize. This effort had not created sufficient spread effects (i.e., externalities) to have justified the unusually generous subsidies associated with this growth strategy. Associated with the worsening distribution of income was the growing employment problem, as the capital-intensive technologies invariably drawn upon in the industrialization programs absorbed relatively little employment per unit of invested capital. In summary, these employment and poverty issues underscored the importance of dealing directly with the agricultural sector, since most of the extreme poverty was located within that sector. One could not hope to truly inprove the collective welfare of a representative LIC country unless one improved the income and productivity levels of the least modern and most poverty-stricken sector in the economy, i.e., agriculture. The World Bank and various regional development banks picked up this cue and also promoted and financed programs in rural credit and employment to reinforce efforts undertaken by local governments. In addition, international agricultural research and development centers were established in various parts of the world in the hopes of expanding the potential benefits of the Green Revolution to other commodities and areas.

This shift in strategy and priorities, although clearly needed and long overdue, still fell far short of resolving many of the food-related agricultural sector development dilemmas facing the LICs. Recent experience highlights the difficulties of transforming agriculture, and careful analysis is needed to identify the institutional and technical bottlenecks facing LIC policymakers in their resolve to deal with these issues.

For example, despite the sharp yield and output increases generated by the new seed varieties associated with the Green Revolution, there was a very unequal distribution of the benefits of these technical advances. The benefits of these yields invariably were skewed in favor of the wealthier or larger farmers, who controlled the limited water resources in these countries and who could draw upon a large volume of concessionally priced credit from institutional sources to purchase fertilizer, tractors, and other inputs. In Mexico, India, and Pakistan, regional disparities widened considerably as a result of the very region-specific features of these new varieties. In short, solving the technical question of increased productivity does nothing in itself to resolve the political and institutional issues of disseminating equitably the fruits of these scientific advances within the society. In ignoring these implications, otherwise promising technical change can generate a negative feedback, worsening social and political stability. The post-1973 energy crisis has now added an additional burden for LICs that are attempting to modernize their agricultural sectors, since the recent yield-increasing technologies are also energy intensive. This impels them to engage even more in export activity to pay for the high-cost energy imports required by modern agricultural technology.

Finally, the LIC export emphasis in recent years, although beneficial to agricultural interests, has carried with it two troublesome side effects that condition the LIC capacity to deal with the task of increasing domestic food supplies. First, the recent emphasis on export activity has expanded cash crop production so extensively that lands that had heretofore been reserved for domestic foodstuff production have moved into export activity. As a result, there has been a growing shortage of domestic foodstuff supply for local consumption and a consequent need to import a certain volume of foodstuff to make up for the shortages. Also, the rapid expansion of export crops in LICs has frequently been undertaken with either labor-saving (mechanized) technology on the one hand or, if land-intensive also, capital-intensive technology (fertilizers, insecticides, irrigation, and so on) on the other. The net result has not increased employment opportunities as rapidly as population growth and, in addition, there has been a tendency for land concentration favoring larger farmers and those controlling water resources.

The second major development has been a tendency for the recently established research establishments in LICs to concentrate their investments on export crops and ignore local foodstuffs, with the exception of rice in Southeast Asia. Hence coffee, cotton, cacao, sugar, and so forth have received the major share of research funds and over the years have registered promising gains in productivity. Local foodstuffs have, relatively speaking, been ignored in these LIC research establishments with consequent stagnation or decline in productivity. This development has serious equity and international consequences, in that the higher productivity export crops are generally produced by larger farmers in LICs and used by consumers in developed countries, whereas local foodstuffs are produced by small farmers and form the major part of the diet for the lower income population in the LICs.

In summary, the agricultural sectors in LICs have finally received more attention and policy support in recent years in order to correct the distortions and imbalances generated by the earlier industrial-biased growth strategies. However, in the process the political and institutional framework conditioning policymaking in these countries has frequently compromised the good intentions and initiatives of both the physical and social scientists to resolve these production problems. In the light of these developments, it is instructive to ask to what degree future scientific work can contribute to this increased effort to promote agrucultural development in LICs by focusing its efforts on reducing some of these unfavorable side effects as well as on resolving the basic problem of increased productivity and output.

EMERGING RESEARCH PRIORITIES AND ISSUES

There is an emerging concensus that much more emphasis must be placed on agricultural research, especially in LICs. New technologies are needed that are more appropriate for the resource endowments of the majority of LIC farmers. More research is needed to determine how policies and programs of both developed and developing nations interact to influence patterns of supply, demand, and trade. Major efforts, such as those by the Argicultural Research Policy Advisory Committee (ARPAC) and the National Research Council (NRC), have been made to sort through the myriad problems and interrelations in order to derive key research priorities and issues. These two studies are drawn on heavily in this section of the paper.

Research Priorities

The ARPAC report identified 89 research need areas and 101 problems designated as most important for expanded agriculture research. The NRC report lists 22 priority areas in its list of promising types of research requiring support. From these lists we have identified three general research areas that appear to be among the most important ones in which we are likely to be involved in the coming years. These areas are also related to some of the major uncertainties about future agricultural production described above. The three are (1) plant breeding and genetic manipulation, (2) energy use in agriculture, and (3) weather and climatic influences on crop production.

Plant breeding and genetic manipulation is listed as a specific NRC priority. Research is encouraged along the classic genetic and plant breeding lines as well as the newer approach of producing and transferring genetic changes at the cell level. The objective would be to develop plants more suitable to regions with short growing seasons, high or low temperatures, adverse soil conditions, and nonoptimal rainfall. Lack of appropriate plant strains has limited the expansion of improved varieties in some countries. As agriculture expands into more marginal producing areas, current strains developed for better soils and water conditions may become increasingly inadequate. Furthermore, research in this area may be crucial to successfully meeting the need for improved plant efficiency in energy and water use.

Several research priorities identified relate to energy use in agriculture. As noted earlier, energy is a concern because existing production systems for several crops, including the Green Revolution varieties, have been developed based on expanded use of chemical fertilizer, irrigation water, and mechanization. Farmers in developing countries will be increasingly affected by future petroleum price increases unless some nonprice system is used to ration scarce supplies of fuel and fertilizer. Such a rationing system seems unlikely at this time, however. Research to increase biological nitrogen fixation holds promise as an alternative to chemical nitrogen use. Closely related is research on photosynthesis to improve plant efficiency in use of solar energy from the present conversion of 1 to 3 percent toward the theoretical maximum of 12 percent. Breakthroughs in these two areas would decrease dependency on chemical fertilizers and reduce environmental contamination. But useful results from such research are likely to be several years away and probably will not cover all crops and situations. Therefore, research on chemical fertilizers must proceed at two levels: new sources of supply and efficiencies in fertilizer production are required, and increased efficiency of nutrient use by plants is needed. Fertilizers and fertilization practices need to be designed for the unique crops and the soil and moisture conditions of the tropics.

The relationship of weather and climate to crop production has recently been given increased attention. If weather variability is to be considered the norm rather than the exception, more emphasis is required on measuring and predicting the effects of weather on crops. Plant varieties developed for highly regulated irrigation conditions must be supplemented by others more tolerant to variability. Much research to date, as evidenced by work in the international centers on wheat and rice, has focused on production systems for good soil and water conditions. Since most agriculture in developing countries is in nonirrigated areas, dryland agricultural techniques must be given greater emphasis. Improved knowledge about management of tropical soils is essential. Crop varieties and combinations, planting densities, and water management practices need refinement to guarantee optimum use of available moisture. Aspects of traditional intercropping practices may prove desirable when evaluated in the perspective of weather-induced risks for producers.

The list of priorities could be extended further. For example, biological alternatives for pest control would have high payoff, as would development of crops and varieties tolerant to high soil acidity or nutrient deficiency. But the three research areas listed are likely to pose some of the most important challenges to this agricultural research community and absorb a large share of available resources.

We would be remiss as social scientists if we did not call attention to an issue largely ignored in most analyses of research priorities. That issue is the increasing concern for the distribution of costs and benefits of development as mentioned above. Some physical scientists argue that the distribution question is not their concern. Others conclude that production must be increased before improved distribution can be addressed. The problem with this argument is that improved consumption and nutrition in many countries is intrinsically tied to production. Increasing production on a small subsistence farm may at once contribute to societal goals of both production and nutrition, whereas that same production increase on a large farm may only indirectly affect nutrition. Furthermore, recall that large farms frequently produce for export rather than for the local domestic market.

The point is that breakthroughs in agronomic research, which eventually result in new varieties, crop combinations, production technologies, and so on, affect various groups of producers and consumers in different ways. The dilemma for us as researchers is not only to identify that research with potentially high payoff in terms of production, but to identify and conduct research that makes a contribution to a large part of society, including groups that currently may have limited access to influence the activities of our colleagues in developing countries. Agricultural research is a social enterprise; thus it should benefit society, not just a few select groups. Ideally new crops, varieties, and techniques should be scale neutral with respect to size of farm, and the crops found most frequently only on small farms should receive more attention in our research programs.

Research Organization

The challenge of efficiently conducting agricultural research presents other issues to the research community. In another paper in this volume, Dr. Pinstrup-Andersen discusses important economic issues associated with research. Economic issues can be expected to become increasingly important as research competes with other socially desired expenditures in rapidly inflating economies.

The increased need for LICs to produce for expanding populations implies increasing their expenditures for staff training and research institution building. These expenditures will be required at the same time as the international centers demand more resources for their programs. U.S. institutions likewise need additional funds to maintain and strengthen their capabilities to contribute to international problems. What is not clear is the appropriate division of labor between U.S., international, and LIC institutions. The international centers have been employing unique methods for transferring research results and training LIC researchers. But the quantity of resources absorbed by the centers may have reduced support for U.S. and LIC institutions. What is the appropriate division of labor in research? What are the magnitudes of funding appropriate for each type of institution? What criteria should be used and who should decide on the allocation of funds? Under what conditions does the systems approach to agricultural research advocated by the centers yield superior results to the more laissez-faire approach followed by most U.S. institutions? In what areas should LICs expect to develop their own basic research rather than rely on adaptation of research from developed countries? These and other questions regarding research organization and funding must be faced in coming years.

A disquieting development has been the increasing difficulty of the U.S. government in supporting international research by U.S. institutions. Part of the problem lies with the U.S. public's diminishing support for such activities. Title XII of the 1975 Foreign Assistance Act was designed to provide a stronger legislative base of support, but its slow start has cooled the enthusiasm of some early supporters. Part of the problem is also associated with recent developments such as affirmative action, disclosure, and minority concerns, which may be essential to American society, but which also bog down bureaucratic processes.

A third problem is the lamentable decline in the number and competency of Agency for International Development (AID) staff working in agriculture. AID is the natural vehicle for stronger support for U.S. involvement in international agricultural research. But due to staff reductions and other factors, the agency has suffered a decline in its ability to deal with agriculture. If U.S. resources are to be mobilized to work on the world food problem, AID must rebuild its past competence in agriculture and assume its logical leadership role. Multilateral agencies will draw some U.S. resources into international research, but bilateral programs offer some unique, although perhaps less visible, advantages for long-term programs.

Conclusion

In spite of considerable uncertainties, the overriding concensus of several studies on the future world food economy is that developed countries will reaccumulate surpluses disposed of during the past few years, while several LICs will continue to experience large deficits. Lack of foreign exchange will prevent the deficit countries from importing sufficient foods at the probable higher prices to meet their requirements. Food aid will not close the gap. Thus LICs must look increasingly to themselves to produce more foodstuffs.

During much of the 1950s and 1960s, many LICs pursued strategies to accelerate industrialization. Agriculture was partly penalized, partly ignored in this strategy. Recently many of these countries have placed more emphasis on agriculture and recognize the challenge they face in producing more food.

Several emerging research priorities can be identified in the area of food production. U.S., international, and LIC institutions can and should divide the research task according to their respective comparative advantages. It is not yet clear how this will be done. Nor is it clear how and when AID will rebuild its competency in agriculture in order to mobilize U.S. resources for additional research.

Some have argued that problems of world hunger could be largely resolved in a decade if sufficient resources were assigned to the task. Others are less sanguine about the prospects. We hope that the research discussed in this volume will mark another step toward solving the problems of the world food economy.

REFERENCES

Agricultural Research Policy Advisory Committee. 1975. Research to meet U.S. and world food needs. Report of a working conference. Kansas City, Mo.

Barr, W. 1977. Trade in farm products. ESO-421. Department of Agricultural Economics and Rural Sociology, Ohio State University, Columbus.

Carter, H. O., et al. 1974. A hungry world: The challenge of agriculture. University of California, Berkeley.

Central Intelligence Agency. 1977. The international energy situation outlook to 1985. ER 77-10240 U. Washington, D.C.

Griffin, K. 1974. The political economy of agrarian change. Harvard University Press, Cambridge.

Hayami, Y., and V. W. Ruttan. 1971. Agricultural development: An international perspective. The Johns Hopkins Press, Baltimore.

Little, I., T. Scitovski, and M. Scott. 1970. Industry and trade in some developing countries: A comparative study. Oxford University Press, New York.

National Research Council. 1977. World food and nutrition study: The potential contributions of research. Washington, D.C.

TREVOR G. ARSCOTT

Soil Resources and Their Potential for Food Production in Lesser-Developed Countries

3

INTRODUCTION

Soils of the tropical regions and of several of the lesser-developed regions of the world are predominantly classified as oxisols, ultisols, and alfisols, with the oxisols making up the largest group in the African and American continents. These so-called tropical soils are relatively fragile systems that have historically been used for shifting cultivation. Shifting cultivation by itself is perhaps a rather efficient and low-input method for the use of these soils. It involves the clearing, usually by slash-and-burn methods, of the natural vegetation, and the planting of a crop for one to three years, after which the cultivated area is left to revert to its natural state. The subsequent regrowth of the vegetation, if permitted to persist long enough, results in the rejuvenation of the land.

This relatively simple and efficient system of cultivation breaks down when population pressures and food demands dictate that the originally cleared land be reused without allowing sufficient time for the rejuvenation process to be complete.

In addition, when shifting cultivation is practiced on excessively sloping terrain and under high intensity rainfall conditions, erosion is maximized and the land surface may be damaged beyond repair. Here again population demands sometimes force the practicing cultivator to open up such land areas for crop production.

Trevor G. Arscott, Department of Agronomy, Ohio State University, Columbus, Ohio 43210.

DESCRIPTION OF THE SOILS

Many of the soils that occur in the temperate regions can also be found in the tropical regions, even though they may be limited in extent. However, because of the condition under which they are formed, the soils of the tropics tend to possess certain characteristics that are common to many areas and are somewhat unique for the tropics.

Although classified predominantly as oxisols, ultisols, and alfisols, these tropical soils have certain general characteristics that have a marked influence upon their crop-producing ability: (1) relatively low contents of organic matter in their profiles, and what is there is usually concentrated in the surface of the soil; (2) high levels of aluminum and manganese in the soil profiles, which can be toxic to the growth of certain crop plants; (3) high levels of iron, which usually results in the bright red and yellow coloration of these soils and which may, under certain conditions, impart to soil the ability to harden; and (4) low crop availability of the essential nutrient phosphorus, usually because the high levels of aluminum and iron tie up this element and prevent its assimilation by crop plants. In addition, most tropical soils may be considered to be relatively low in most nutrients considered necessary for crop growth.

POTENTIAL FOR FOOD PRODUCTION

The potential of these soils of the tropics for food production is highly contingent upon the use of proper management practices, which in turn depends upon sound research findings and proven agronomic practices. However, in many cases these practices are not new, but there exists a large communications gap in getting the operators to adopt them.

RESEARCH FINDINGS THAT HOLD PROMISE

Many of the recent research and production findings dealing with the soils of the tropics can be attributed to the work that is being done and the communication linkages that have been formed by the University Consortium on Soils of the Tropics and sponsored by by the U.S. Agency for International Development. Many of these findings are not new but have been exhaustively tested and proven by workers at the International Institutes.

Examples of some of the research findings that have a significant bearing upon the food production potential of the soils of the tropics are erosion control and mulching, aluminum and liming, and classification and productivity.

Erosion Control and Mulching

Surface soil conservation. To the untrained eye a red colored tropical soil may seem to be homogeneous in appearance from the surface to a depth of several meters. The careful observer, on the other hand, will invariably recognize a distinct layer in the upper region of the profile. Invariably it is within this surface layer that the major portion of the natural fertility of the soil is to be found. Studies have shown that if this surface layer is lost through erosion, crop yields are drastically reduced. In many cases it may mean the difference between a crop and no crop.

Prevention of hardening. Some soils of the tropical regions possess the ability to form a rock-hard material called plinthite (Alexander and Cady, 1962) when exposed to the elements through the removal of the natural vegetative cover and the surface layer. The hardening process is many times irreversible and results in a wasteland for crop production. Erosion control and mulching practices prevent the formation of plinthite by maintaining the soil surface, reducing moisture loss, and holding down soil temperatures.

Soil moisture and temperature. Some of the more recent work on the effects of mulching on soil temperatures and moisture retention has been performed by M. K. Wade and P. A. Sanchez (1975) in Peru and by Rattan Lal (1975) at the International Institute for Tropical Agriculture (IITA) in Nigeria. Consistently, mulching has been shown to increase yields by reducing moisture loss, weed infestations, and surface soil temperatures. Results from IITA have shown that when root zone temperatures for corn exceed about 35°C, yield was severely reduced. As is to be expected, the moisture effects of mulching are more pronounced during the dry seasons, whereas the effect on weed control is more pronounced during the wetter months. Also, different mulching materials and their mode of application (incorporated or not incorporated in the soil surface, for example) have different effects on yield, but most mulches will surpass a no-mulch practice in crop growth and production.

Nutrient cycling. Traditionally, the replacement of a tropical forest vegetation with a tree-type crop such as coffee, cacao, peppers, fruit trees, and so on has resulted in more sustained production over a long period of time. Replacing the forest vegetation with crops such as corn, sorghum, millets, and cassava, to name a few, invariably results in a rapid depletion of fertility and yield.

It has long been accepted that the large tropical forest vegetation is maintained by the thrifty cycling of nutrients through living vegetation, decomposing vegetation, and efficient root absorption. By replacing the forest with a corn crop as opposed to a crop of mangoes, for example, the nutrient cycle is broken and rapid nutrient depletion results. On the other hand, a tree crop like mangoes tends to more closely duplicate the forest growth, and relatively small quantities of added nutrients are sufficient to establish and maintain the tree crop. The corn crop, meanwhile, is doomed to failure unless large doses of nutrients are added and other management practices like mulching and erosion control are adopted.

Any mulching practice that will work toward maintaining the environment of the tropical forest and the efficient cycling of nutrients should tend to maintain crop yields. The discovery that fungi and their hyphae play an essential role in the nutrient cycling process (Went and Stark, 1968) of the tropical forest should be further explored by agronomists in order to bypass plant nutrition problems such as the tying up of phosphorus by soil iron and aluminum.

Aluminum and Liming

Liming soils of the tropics to neutralize aluminum. Historically, temperate zone acid soils have been limed to attain a specific pH, usually about 6.5–7.0. The liming of many soils of the tropics to attain such a pH reading has often produced disastrous results due to nutritional imbalances caused by excessive calcium from the liming material.

Recent studies (Kamprath, 1970) have shown that lime recommendations for highly weathered tropical soils based on the amount of exchangeable aluminum in the top soil have been successful.

The use of this method has reduced substantially the amounts of liming materials needed for good crop growth. In the past, based on liming to a given pH, it was not uncommon to use rates of 10 to 30 tons of calcium carbonate per hectare, with mixed results. Based on the concept of liming to neutralize exchangeable aluminum, rates of calcium carbonate are between 2 and 6 tons per hectare, with consistently good results.

Another important consideration relative to aluminum and liming is the knowledge that different crops, and varieties within different crops, tolerate different levels of soil-exchangeable aluminum.

Deep placement of lime. In many cases aluminum toxicity in the soils of the tropics increases with depth, and any liming program should attempt to neutralize the exchangeable aluminum to as great a rooting depth as is possible. It has been shown that where climates are characterized by a pronounced dry season, deep incorporation of lime provides for a greater rooting volume, more available moisture during the dry season, and consequently greater yields (North Carolina State University, 1973).

The adoption of a program for deep lime placement must by necessity depend upon availability of equipment and would seem to be limited to soils of lighter textures and good structure; but where the aluminum saturation of the soil is high, deep placement may mean the difference between a crop and no crop.

Classification and Productivity

Since the introduction of the seventh approximation as a system for soil classification (Soil Survey Staff, 1960) there have been several revisions, culminating in the U.S. soil taxonomy that was published in 1975. With each revision more emphasis was placed on the soils of the tropics, but for the first time a classification system goes so far as to include the soil properties that are important to the growth of plants.

The ultimate value of a classification system would be the ability to predict the yield of a given crop on a specific soil as dictated by that soil's properties. Hopefully, any soil with certain defined properties would produce the same as all identical soils, no matter where in the world they occurred, provided the same management practices, crop, and climate prevailed.

Work is now being performed to test the feasibility of extrapolating results from one country to the other on soils with the same classification, and problems associated with some of the parameters used for the productivity evaluations are constantly being solved.

CONCLUDING REMARKS

A few of the significant studies and findings relative to the soils of the tropical regions and their food-producing potential are discussed. Those mentioned above are by no means the only important studies being performed. Several other investigations are being conducted, such as multiple and continuous cropping experiments, secondary and micronutrient studies, variety and plant population studies, and crop adaptation studies, to mention a few.

With every new finding, and with the interest and effort being placed on the study of the soils of the tropics, the potential increases for these soils to produce significant quantities of food for the earth's ever-increasing population. Naturally, political, sociological, and other constraints, such as communication and marketing problems, also need to be solved to maximize the use of food produced on these soils.

LITERATURE CITED

Alexander, L. T., and J. G. Cady. 1962. Genesis and hardening of laterites in soils. U.S. Dept. Agr. Tech. Bull. 1282.

Kamprath, E. J. 1970. Exchangeable aluminum as a criterion for liming leached mineral soils. Soil Sci. Soc. Amer. Proc. 34:252–54.

Lal, R., B. T. Kang, F. R. Moormann, A. S. R. Juo, and J. C. Moomaw, 1975. Soil management problems and possible solutions in western Nigeria. *In* E. Bornemisza and A. Alvarado (eds.), Soil management in tropical America. Pp. 372–408. North Carolina State University, Raleigh.

North Carolina State University. 1973, 1975. Agronomic-economic research on tropical soils. Annual Reports, 1973 and 1975, Soil Science Department, North Carolina State University, Raleigh.

Soil Survey Staff, 1960. Soil classification—A comprehensive system. Seventh approximation. U.S. Dept. Agr., Washington, D.C.

Went, F. W., and N. M. Stark. 1968. The biological and mechanical role of soil fungi. Proc. Nat. Acad. Sci. USA 60:497–504.

RAISA G. BUTENKO

Food and Energy Research Priorities

4

The most difficult and important problem humanity now faces is to utilize effectively the resources of the earth and at the same time to protect the planet and its mineral wealth, water, and plant and animal life for future generations. The task is not only to provide food for hundreds of millions of people suffering from starvation and malnutrition, but also to procure additional foodstuffs for the rapidly growing human population. It is hardly realistic to think that the population of the planet will decrease in the observable future. During the twenty-five to thirty years to come we shall have to feed, dress, and house many more people than have ever existed during the history of mankind. One may hope that the unwise strife for boundless consumption, so typical of the present-day affluent society, will in the future give way to sensible and voluntary self-restriction. This can only be beneficial for the physical and spiritual health of the human race.

However, the most urgent problem now is increasing the food production with concurrent reduction of nonrenewable energy losses. Although the problem of food and energy has many facets, including social, political, and administrative, we scientists should worry about the problems that can be resolved by the scientific methods at our disposal.

The Green Revolution has resulted in a much higher productivity of major agricultural plants, i.e. wheat, corn, rice, and barley. Farmers may choose from thousands of excellent sorts of vegetables, fruit, cotton, coffee, and tobacco. But the natural and economic resources are not inexhaustible

Raisa G. Butenko, Institute of Plant Physiology, Moscow, USSR.

and the present-day productivity of agriculture should be drastically increased. The productivity of agricultural crops can be raised to a certain extent if the achievements of science are applied to agricultural technology. If plants are selected that are most suited to a given climate and soil, and if losses caused by diseases and pests are reduced and those occurring during transportation and processing of the agricultural products are eliminated, the total amount of food can be increased. But all these measures are not sufficient, and an increase in the productivity of agricultural crops is called for. This higher level of productivity can only be attained if we know more about the nature of higher productivity. This requires more intensive fundamental biological studies on the genetic, molecular, and physiological issues of harvest. What we know today about this matter is obviously insufficient. We must find out how to increase the productivity per hectare, per unit of energy, and per unit of time. Moreover, we must see to it that the beauty of the earth's landscape is not sacrificed and that the environment is improved instead of fouled.

International meetings on food resources and productivity of agricultural plants in Rome in November 1974 and in Harbor Springs in October 1975 named the most important items of research pertaining to higher productivity of agricultural plants. The Harbor Springs Conference, organized by the Michigan State University Agricultural Experimental Station and the Charles F. Kettering Foundation, elaborated, in the course of discussion in six sections, crop productivity research imperatives. These imperatives include the studies of nitrogen and carbon nutrition, water supply, protection from diseases and pests, protection from the action of stress, and the regulation of growth and development of plants. Out of these areas of investigation important for increasing the productivity of agricultural plants, one can single out the most urgent ones: the study of two biochemical processes determining the creation of food and energy on the planet, i.e. photosynthesis and biological fixation of atmospheric nitrogen. Stimulation of these processes may increase the energy influx–to–energy efflux ratio in food production. The types of plants and the technology of their cultivation should be chosen so that the solar energy capture is maximal. Solar energy capture, which is about 1–3 percent, should reach the theoretical level of about 12 percent. Increase in the solar energy capture and productivity of photosynthesis may give life to projects such as that in Brazil, where ethyl alcohol produced from sugarcane and other plants is used as fuel (Goldemberg, 1977; Calvin, 1976).

Stimulation of assimilation of solar energy by plants will make possible its conversion into sufficient quantities of electric energy and production of hydrogen (Calvin, 1977, 1976). Reduction of fuel consumption by agriculture can be achieved by cutting down the utilization of nitrogen fertilizers. Up until the present time the yield of crop plants has directly depended on the amount of nitrogen fertilizers applied to the soil. The consumption of nitrogen fertilizers must increase fourfold by the year 2000. This will not only entail much higher energy expenses but will contaminate the environment and will, possibly, deteriorate the ozone shield of the atmosphere. It is difficult to predict the consequence of the latter. The alternative is the use of biological fixation of nitrogen.

The studies of the genetics, molecular biology, and physiology of nitrogen fixation has become the first priority task. We do not know enough about the fundamental processes associated with nitrogen fixation. This is true of both free-living nitrogen-fixing organisms and, especially, of legume-rhizobia nitro-fixing associations. Stimulation of photosynthesis and increasing its productivity, increasing the productivity of nitrogen fixation by legume plants, and also, possibly, endowing the major grain crops with ability to fix nitrogen reguire the use of new genetic methods. One new technique is revealed in cell and tissue culture. I specialize in this field and shall therefore discuss some aspects of application of cell and tissue cultures for raising the productivity of plants and increasing the food resources of the planet.

Cell cultures can be used as a model system for investigating symbiotic nitrogen fixation, elucidating the factors supplied by plant cells, and triggering the nitrogenase activity of bacteroids (Holsten et al., 1971; Child and LaRue, 1974; Child, 1975; Scowcroft and Gibson, 1975). The study of the photosynthetic activity of cell cultures and alternative interactions between proliferation and photosynthesis have added a great deal to the knowledge of the regulation of photosynthesis (Berlyn and Zelitch, 1975; Smolov et al., 1975). There is no doubt that cell cultures may be used for acceleration and facilitation of plant breeding, the aim of which is to develop plants of higher productivity. Preservation of germplasm of plants existing on the earth and enrichment of the gene pool of agricultural plants by genes of their wild relatives and, possibly, by bacterial genes can also be achieved, now or in the future, with the help of cell culture techniques.

Cell cultures may be used not only as a model or mode for plant improvement, but they are also of interest as one of the promising ways for production of some economically important substances. Figure 1 shows a scheme of the possible application of cell and tissue cultures at present and maybe in the near future.

Even now, the technique of accelerated large-scale propagation of plants by *in vitro* methods and preparation of virus-free plants from meristem may help increase the productivity of vegetatively reproducing plants and

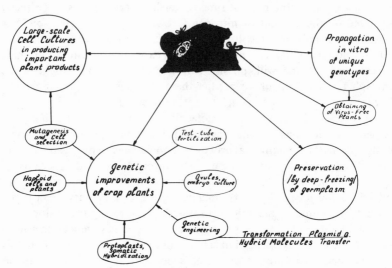

Fig. 1.   Applied aspects of plant cell and tissue cultures.

decrease the energy and time expended for their production (Murashige, 1974; Popov et al., 1976; Quak, 1977). The technique of test-tube fertilization (Kanta et al., 1962; Zenkteler et al., 1975; Shinkareva, 1974), growing on culture media of ovules, inviable hybrid embryoids, and immature seeds (Kursakov, 1970; Raghavan, 1977), accelerates and facilitates distant hybridization and contributes to the genetic heterogeneity of economical plants. *In vitro* methods exist only for some plants. Grain crops and leguminous plants are unlikely to be associated with the cell culture technique as they are now. But considering what has been done for coffee (Monaco et al., 1977), citrus (Button and Kochba, 1977), ornamental plants (Holdgate, 1977), and forest plants (Bonga, 1977), we may be optimistic. To succeed with legumes and grain crops, we lack the fundamental knowledge of the process of morphogenesis, the nature of inductors that bring it about, and the conditions securing the competence of the cells and inductor reception. Still not high is the number of crops (barley, rice, wheat, triticale) for which it is possible to obtain haploids and homozygotic dihaploids via anthers and microspore cultures (Nitsch, 1977; Clapham, 1973; Reinert and Bajaj, 1977; Lukijanuk and Ignatova, 1977; Kucherenko et al., 1977). This technique is important for obtaining the required combination of recessive genes, pure lines for heterosis, and higher production of mutants. The situation here would have been much better if we had enough fundamental studies on the processes involved in the switching of the normally developing male gamete over toward the callus or formation of embryoid structures. It is clear that such a switchover is determined by both the genetic (sort, species) characteristics of a plant and the conditions of cultivation of anthers and microspores. Moreover, if we knew the conditions of the transition of multicellular proembryos to the phase of differentiation and growth of plantlets, we could use the haploid technique more extensively.

The experimental data available now do not allow one to give a precise answer to the question of what can be achieved in the near future for increasing the food crop productivity by hybridization of somatic cells or by attempts to introduce foreign or recombinant molecules of DNA, plasmids, or cell organelles into a higher plant. It is still technically difficult to protect the donor DNA or organelles from being degraded by the enzymes of the recipient cell. Although polyethyleneglycol (PEG) and other agents have been helpful in accomplishing fusion of plant organelles and cells, they also injure both. Little or nothing is as yet known about the possibility of the genetic integration of a donor and a recipient or about replication and transcription of the foreign information. Moreover, genetic modification of the cell with the aim of changing or improving the whole plant is only realistic if we can achieve morphogenesis and regeneration of a plant in a culture of cells and tissues arising from isolated protoplasts. Potrycus et al. (1976) tested several modes of isolation and cultivation of protoplasts from leaf mesophyll of wheat, corn, barley, and rye; in no case was the division of the cells formed from isolated protoplasts achieved. No plants were obtained, either, from the cells formed from isolated protoplasts of leguminous plants (Constabel et al., 1973). In the laboratory of the Institute of Plant Physiology (Moscow), promising results have been obtained by cultivating isolated protoplasts from leaf mesophyll of two kinds of potato species, wild *Solanum chacoense* and cultured *S. tuberosum*. We have succeeded in selecting conditions for effective isolation of protoplasts using leaves of test tube-grown plants. The efficiency of isolation of viable protoplasts reached 90 percent (fig. 2); the plating efficiency relative to the colonies formed was about 60 percent (fig. 3). Obtained callus tissues transferred to morphogenesis-inductive medium produced buds and stems (fig. 4); when rooted they were planted into soil and grew very well, producing tubers and flowers (fig. 4). The number of plant species from which isolated protoplasts can again be grown constantly increases, and this gives hope that genetic manipulations with the cell (cell engineering) will with time become an important approach to attaining a higher productivity of agricultural plants.

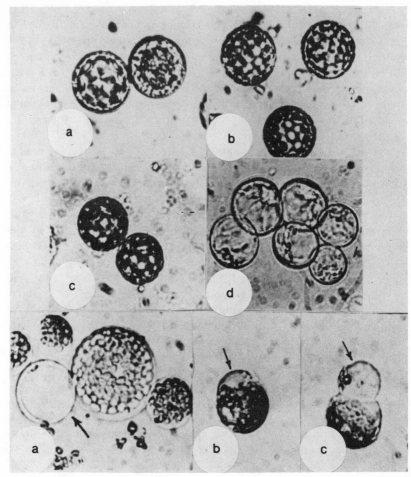

Fig. 2. Isolated protoplasts from leaf mesophyll. Above: (a) *Solanum tuberosum*; (b) *Solanum chacoense*; (c) *S. tuberosum* (dihaploid plant); (d) protoplasts from cell suspension culture of *S. tuberosum*. Below: (a) adhesion; (b, c) fusion.

This is no panacea but a promising line for further research. However, to be able to plan introduction of the Nif-operon into the cells of grain crops we must intensify fundamental research of the organization of genetic material in a eukaryotic cell and regulation of gene activity in plant and bacterial cells. Solving the problem of implantation of symbiotic nitrogen fixation into grain crops requires extensive studies of the genetics and physiology of this process. The results of experiments where *Rhizobium* sp. and *Azotobacter vinelandi* were grown together with tissue culture cells of nonleguminous plants (Davey, 1977; Carlson and Chaleff, 1974) allow one to believe that the search in this direction is worth the effort.

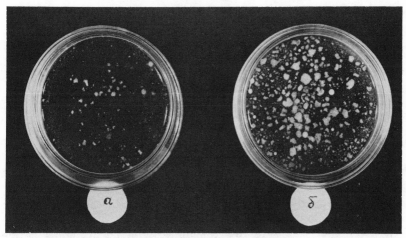

Fig. 3.   Formation of cell colonies after plating of *S. tuberosum* isolated protoplasts on (a) Murashige-Skoog medium and (b) the same with feeder layer (-ray-inactivated tobacco cells).

Fig. 4.   Plant regeneration in callus tissue obtained from isolated potato protoplasts. (Left) Organogenesis in callus tissue of *S. chacoense* potato plants; (center) control plant from tuber of *S. tuberosum*; (right) *S. tuberosum* plant from leaf mesophyll protoplast.

Interspecific and more distant hybridization of somatic cells (Kao et al., 1974; Constabel et al, 1977; Jones et al., 1976) will hardly be more fruitful than sexual hybridization with respect to expanding the limits of incompatibility in distant hybridizations.

If somatic hybridization of distantly related taxons do not produce true hybrids, one can still hope that at least some information exchange will take place in somatic crossing-over. There are indications that this process occurs in tissue cultures (Mariachina and Butenko, 1974).

Also, cybrids present many interesting possibilities (Cocking, 1976; Burgutin et al., 1977; Gleba et al., 1977). Cybrid cell lines (hybrid cytoplasm with a nucleus of one of the parents)—and on this basis whole plants—will yield more knowledge about the regulatory role of the cytoplasm and about the conditions at which cytoplasmic male sterility occurs.

In the problem of application of hybridization and modification of isolated protoplasts, elucidation of the conditions for selection of hybrid or changed product is also a difficult matter (Power and Cocking, 1977). Hybridization of somatic cells, especially distant hybridization and genetic modification of isolated protoplasts, are promising lines of investigation, but one can hardly predict the results now. For the near future more realistic is the application of mutagenesis and selection at the cellular level. Studies of the genetics of the somatic cells of plants should be the theoretical basis for growing and selecting variant cell lines and, further on, genetically altered plants. Such investigations have been started now in some laboratories (Lescure, 1969; Kovaleva et al., 1971; Widholm, 1974; Maliga, 1976).

Table 1 shows variant cell lines reported in tissue and cell cultures. Some of them were capable of regenerating whole plants. This allowed establishment of nucleic or cytoplasmic nature mutations and application of genetic analysis of the progeny. This is the only way to prove the truly mutant nature of the chosen variant. When regeneration of the plant is impossible, the following facts can signify that a genetic change (mutation) has taken place:

1. Sudden appearance of the change, or its emergence only after the application of a mutagenic agent. Low frequency of appearance.

2. Stability of the trait with the selective conditions being absent.

3. Appearance of the altered product as a result of the activity of the mutant gene.

In the latter case some difficulty arises due to complex interactions of the given structural gene with the modifier. This results in the fact that the activity of the mutant gene is phenotypically expressed only in a certain phase of cell differentiation or ontogenesis of the plant. Stability of the changes in a variant line throughout long-term cultivation, including growth under nonselective conditions, lends weighty support to the genetic nature of the changes obtained. Such results have been reported by Kovaleva et al. (1971), Karanova et al. (1977), and Tumanov et al. (1977).

Table 2 shows the differences between the initial strains of *Picea excelsa* and *Triticum aestivum* cell lines selected by resistance to low temperatures

TABLE I

VARIANT CELL LINES REPORTED IN TISSUE CULTURE

Phenotype	Species	Selection[a]	Frequency	Plant[b]	References
Resistance					
NaCl	C. capillaris	cc, ap		+	Sobco et al., 1977
	N. tabacum	sc	—	—	Dix and Street, 1975
	C. annuum	sc	—	—	
Low temperature	N. sylvestris	ap	—	—	Dix and Street, 1976
	C. annuum	ap	—	—	
	P. excelsa	cc	—	—	Tumanov et al., 1977
	Tr. aestivum	cc	—	—	
KClO$_3$	N. tabacum	cc	—	—	Muller and Grafe, 1975
Helminthosporium toxin	Z. mays	cc	—	—	Gengenbach and Green, 1975
2,4-D	N. sylvestris	sc	—	—	Zenk, 1974
Bromodesoxyuridine	N. tabacum	cc	10^{-6}	+	Maliga, 1976
Prototrophy					
Auxin	D. deltoidea	ap	10^{-6}	—	
	N. tabacum	ap	10^{-6}	—	Karanova et al., 1974
Kinetin	D. deltoidea	ap	10^{-6}	—	
Auxotrophy					
(Biotin, arginine, lysine, proline)	N. tabacum	cc	—	+-	Chaleff and Carlson, 1975
Productivity	R. serpentina	cc	—	—	Shamina et al., 1975; Vollosovich et al., 1976

[a] Methods of selection: cc = callus cloning, ap = agar plates, sc = suspension culture.

[b] Plant regeneration from variant line.

TABLE 2

VARIANT CELL LINES RESISTANT TO LOW TEMPERATURE

SPECIES	T°	% OF VIABLE CELLS	
		Wild Strain	Strain after Selection by Low Temperature
Picea exelsa	−10	100	100
	−15	75	100
	−20	0	50
	−25	0	50
	−30	0	50
	−35	0	50
Triticum aestivum cul. Kaukaz	−10	100	100
	−12	75	100
	−14	50	100
	−16	25	100
	−18	0	75
	−20	0	50
	−22	0	50
	−24	0	0

SOURCE: Tumanov et al., 1977.

(Tumanov et al., 1977). Selection of the cells that survived freezing down to −16° C (wheat) and −25° C (white spruce) showed a 6–10° C increase in their resistance on repeated cold exposure, as compared to the initial strain. Selection of spontaneous or induced mutations at the cellular level is only possible on the basis of the principles that are inherent in the cell (resistance to extreme temperatures, salts, bacterial toxins, anomalous metabolite analogues).

We have shown (in the tissue culture laboratory of the Institute of Plant Physiology) the selection of a cell line of *Crepis capillaris* resistant to a high concentration of NaCl in a nutrient medium and regeneration of the plants from these resistant cells. To apply cell selection at the cellular level for obtaining highly productive plants, correlations should be discovered between the integral process in the whole organism and the properties of the cells in the *in vitro* culture. The reverse approach may prove feasible, i.e. obtaining cell cultures from plants with genotypically different productivity and then analyzing the differences between the cells. This was the case when the object of study was the resistance of the cells to low temperatures (Tumanov et al., 1970).

Large-scale cell culture masses for commercial purposes, i.e. for isolating the products of their biosynthesis, can be of great help for agriculture in the near future. Substances that are important in medicine and pharmacy, perfumery and the food industry, antibiotics and compounds possessing antiphytovirus activity, dyes, vitamins, amino acids, enzymes and hormones, and many others can be produced by cultivated cells (Butenko,

1964). Table 3 shows some of the substances synthesized by cell and tissue cultures for the production of which they will be grown on the industrial scale. Microorganisms and microalgae have for a long time been utilized in biosynthetic industry for producing amino acids, vitamins, enzymes, antibiotics, edible protein, and so on. High yields of microbial by-products and the short time within which the products are obtained are the result of their rapid multiplication. The reproduction rate can be increased further by optimization of the growth conditions. In the experiments conducted in a "Biostand" device where yeast cells were grown under optimal growth conditions, the number of cell divisions increased from 5 to 75–80 percent and the generation time (doubling time) reduced from 20 to 5 minutes (Pechurkin and Terskov, 1975).

Utilization of food industry wastes can decrease the cost of microbiological production but the energy expenses are considerable. Autotrophic microalgae are preferable. But, for example, *Chlorella* sp. has cell walls that are hard to break down and contain toxic substances, which prevents it from being used for food and livestock feeding. The use of microalgae for production of amino acids, fats, and physiologically active substances (Semenenko, 1975) is very promising. Microalgae grown in solar light can be used as a system for generating hydrogen, owing to the activity of hydrogenase. However, taking into consideration the inertia of the plant cell and the dependence of the intensity of the process of the phase of the cell cycle and the physiological state of the cell, we may say that it will be unprofitable; in the future the photochemical production of great amounts of hydrogen with a powerful energy source will be preferable.

Highly productive strains of microorganisms used in industry have been created by geneticists. Mutant strain *Penicillium*, used for industrial production of penicillin, synthesizes one thousand times as much of the antibiotic as the starting wild strain.

Cells of higher plants that can be grown in fermentors routinely used in microbiology can be regarded as wild strains to be worked at by geneticists. The technique and the logic used in the microbiological work should be extended to cell cultures of higher plants. This consists of treatment with mutagenic agents for obtaining a wide spectrum of mutants, the plating of the treated cells into the petri dishes where varient lines will be grown and selected, and the use of the selective media and selected conditions.

The resulting variant lines checked with respect to the persistence of the newly formed alterations are studied to find out the biochemical manifestation of the mutant gene. Such studies with cell cultures of higher plants are not numerous. Expansion of these studies is essential. Genetics of somatic cells should be developed as an important theoretical

TABLE 3

SOME COMMERCIALLY IMPORTANT PRODUCTS
IN CELL AND TISSUE CULTURES

Compound	Plant Source	Type of Culture[*]	Concentration	References
Panaxosides (Ginsenosides)	Panax ginseng	s	0.4% purified saponins	Slepyan et al., 1968; Slepyan, 1970
	P. quinquefolium	s	"	Butenko et al., 1970
	P. japonicus	s	"	Furuya et al., 1970
	P. ginseng	c,s		Furuya and Ishii, 1973
Diosgenin, saponins	Dioscorea sp.	c,s	0.5–1.8% (dry weight)	Kaul and Staba, 1969; Sariskova, 1974; Lipski et al., 1977
				Stohs et al., 1974
	Trigonella	c,s	1.8%	Khanna and Jain, 1973
Indolic alkaloids (aimalin, perakin, serpentaine)	Rauwolfia sp.	c,s	~ 2% (total)	Vollosovich et al., 1970, 1976
Coumarins, carotinoids, alkaloids	Ruta graveolens	c,s		Reinhard et al., 1968; Steck et al., 1971; Kuzovkina, 1971, 1973
Tropane alkaloids	Datura stramonium Atropa belladonna Hyosciamus	c,s	1–35 [2]/%	Chan and Staba, 1965; Bereznegovskaja et al., 1975
Volatile oils	Rosa sp.	c	. . . . . .	Kireeva et al., 1977
	Iris sp.	c,s	. . . . . .	Zargarjan et al., 1976
	Pelargonium sp.		. . . . . .	
Antitumors alkaloid	Phytolacca americana	c,s		Liu, 1974

[*] c = callus; s = suspension

foundation for application of cell cultures and improvement of agricultural plants and increase of cell productivity in industrial production. Selection of cell lines for high productivity is performed now on the basis of growth intensity (Shamina and Karanova, 1974). Selection by the content of the initial product has not yet been elaborated. Chaleff and Carlson (1975) showed that mutant cell lines resistant to amino acid analogues contain higher quantities of amino acids, including essential ones. Similar approaches can prove feasible in other cases.

Selection of tissue culture lines of the medicinal plant *Rauwolfia serpentina Benth* growing well on media without phytohormones yielded strains with a doubled amount of alkaloids and with quantitative changes in the spectrum of those produced (Vollosovich et al., 1976; Tuchtasinov, 1977). The strain of *Dioskorea deltoideae Wall* grown on a liquid medium contains 3 to 3.5 times as much diosgenin as the initial strain (Sarkisova et al, 1977; Lipski et al., 1978). Promotion of the industrial growth of cell cultures for the valuable products of their biosynthesis can liberate areas for agricultural plants or preserve natural landscapes. Another important task for the future is the preservation in cell culture of the biosynthetic potential of valuable wild plants that are exterminated during the development of land for agriculture.

The main advantage of industrial cultivation of cell cultures is that this mode of production of valuable plant products does not depend on weather climate. Bryson has called growth of plants adapted to the planet's climate shooting at a moving target (Bryson, 1975), since the earth's climate has changed many times and will be changing, and it is difficult to predict in what direction. So creation of artificial systems manufacturing the products required by man is a necessary step for protection from the consequences of climatic changes. It is common laboratory practice now to grow plant cell suspension cultures in bioreactors and microbiological fermentors. Figure 5 shows a suspension of *Dioskorea deltoidea* grown in a 7-liter MF-107 fermentor. The graph (fig. 6) and table 4 show the main characteristics of the culture, i.e. the growth kenetics determined by the number of cells and dry weight and the content of diosgenin at different

TABLE 4

Diosgenin and Sterins in Suspension Culture of
D. deltoidea (Fermentor MF-107)

Time of Cultivation (days)	% Steroids (dry weight)			Dry Weight of Cells (g/l)	Content of Diosgenin (g/l)
	Diosgenin	Sytosterin	Stigmosterin		
12	0.42	0.01	0.01	9.6	0.040
21	0.74	0.01	0.01	8.3	0.062
37	1.37	0.10	0.10	8.5	0.117

Fig. 5.   Cell suspension culture of *Dioskorea deltoidea Wall* in fermentor MF-107 (batch culture on fifteenth day).

growth phases. Diosgenin is an important product as it is converted, microbiologically or chemically, into steroid hormonal preparations widely used in medicine. The content of diosgenin produced by growing cell suspensions in a fermentor is comparable with that obtained in shaker flasks (fig. 7) and is 1.4 percent by dry weight. The content of diosgenin in the rhizome of tropical plant *D. deltoideae Wall* is about 4–5 percent by dry weight. The rhizome can be used only after a plant has been growing for five years; besides the yield of plants is not high. Therefore, production of diosgenin in fermentors can become commercial.

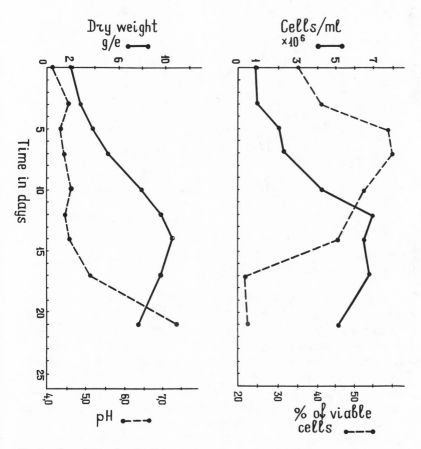

Fig. 6.   Growth kinetics of cell suspension culture of *D. deltoidea* in fermentor MF-107.

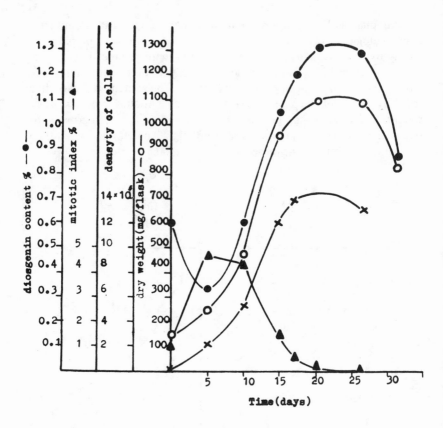

Fig. 7.   Growth kinetics and diosgenin content in cell suspension culture of *D. deltoida* in 500 ml flask on the shaker (100 rpm).

But we can still hardly predict which cultures and which products, and when and where they will be grown and prepared in large-scale industrial installations (fermentors) of the capacity of several thousands of liters. It is also an important research imperative to study the biology of higher plant cells grown in a batch or continuous cultures in large-volume fermentors. We should study autoselection in a cell population, gas regime optimization, pH, and nutrition with the aim of increasing specific growth rate and reducing the time of cell generation. Conditions facilitating the biosynthesis of important products and maintenance of the cultures in the state of stable high synthesis during continuous cultivation are important issues for study.

Commercial success for product biosynthesis from plant cell cultures depends on the cost of the medium components and on the energy consumed by the whole process. The use of molasses, lactose, cotton seed meal, and brewing wastes can drastically reduce the cost of the medium.

It is feasible to select mutants that are productive when grown on cheap media. It is still too optimistic to expect cell cultures capable of photosynthesizing at the same time to intensively synthesize secondary products. This kind of differentation is rather difficult for a plant cell in the *in vitro* culture, as in a whole plant the functions of photosynthesis and biosynthesis of secondary products are usually separated. In this connection it is important to study the problem of energy supply for required synthesis in industrial growth of heterotrophic cells of higher plants. It is not impossible that the way out with respect to carbon nutrition will be to grow together autotrophic and heterotrophic cells. One can imagine the creation of symbiotic relationships between higher plant cells and photosynthesizing algae or bacteria. Mixed cell cultures of carrot and the nitrogen-fixing microorganism *Azotobacter vinelandi*, for example, can grow under conditions that determine the necessity of association (Chaleff and Carlson, 1975).

It is possible that the cell bioengineers of the future will discover other ways for solving the above problems. Maybe they will pose and resolve many other problems that are as yet beyond the reach of today's human intellect.

CONCLUSIONS

The most urgent problems for science research in the area of food and energy in the near future should be:

1. The study of the conditions of increasing the photosynthetic activity of plants; improvement of the influx of products of assimilation to the organs determining the growth and the productivity of plants.

2. The development of the cybernetic approach to the study of a plant; the elaboration of heuristic and, later, mathematical models of the processes resulting in high productivity.

3. Further search for the modes of conversion of the solar energy stored in photosynthesis into fuel (ethyl alcohol, hydrogen) or electrical energy.

4. The study of the physiology and genetics of nitrogen fixation in free-living or symbiotic nitrogen-fixing organisms.

5. Large-scale propagation *in vitro* and obtaining of virus- and mycoplasma-free ornamental and forest plants, vegetables, and potatoes.

6. Application of tissue and cell cultures for acceleration and facilitation of hybridization and plant breeding (fertilization *in vitro*, embryo culture, the use of haploids for developing pure lines, and selection of required mutants and hybrids).

7. Maximal intensification of the studies of embryogenesis and morphogenesis in cell and tissue cultures; search for the inducers of the processes and elucidation of the reasons for cell incompetence to inductive action; application of hybridization of somatic cells for elucidation of the cytoplasmic factors responsible for the reception of inducers and activation of the nuclear morphogenesis program.

8. The study of the genetics, physiology, and biochemistry of the cell grown in a large-scale culture for commercial production of the products of biosynthesis.

9. Search for the modes of acceleration of cell reproduction (reduction of the time of generation and synchronization of the culture).

10. Application of mutagenesis and selection techniques for increasing the productivity of the cells and creating systems for synthesizing new types of plants that produce valuable products.

Imperatives for research in the more distant future may be:

1. The use of cellular engineering techniques for creating plants with high photosynthesizing activity and with their own or symbiotic ability for nitrogen fixation.

2. Construction of specific purpose cells for commercial application or with the aim of developing from them basically new plants.

REFERENCES

Bereznegovskaja, L. N., I. F. Gusev, G. E. Dmitruk, A. V. Smorodin, V. V. Smorodin, N. A. Trofimova, and N. A. Shmicova. 1975. Alkaloid plant tissue and cell culture, p. 196. Tomsk.

Berlyn, M. V., and I. Zelitch. 1975. Photoautotrophic growth and photosynthesis in Tobacco callus cells. Plant Physiol. 56:752–56.

Bonga, J. M. 1977. Applications of tissue culture in forestry. *In* J. Reinert and Y. P. S. Bajaj (eds.), Applied and fundamental aspects of plant cell, tissue, and organ culture. Pp. 93–107. Springer-Verlag, Berlin.

Bryson, R. A. 1975. Shooting at a moving target. Crop productivity—research imperatives. Proc. Intern. Conf. sponsored by the Michigan Agricultural Station and the Charles F. Kettering Foundation. Pp. 108–32.

Burgutin, A. B., R. G. Butenko, L. V. Frolova, and Yu. Yu. Gleba. 1977. Some peculiarities of plants emerged of fused protoplasts of *Nicotiana tabacum* L. mutants. Physiol. Biochem. Cultured Plants 9(3):291-300. (In Russian.)

Butenko, R. G. 1964. Plant tissue culture and plant morphogenesis, p. 640. Nauka, Moscow.

Butenko, R. G., G. S. Bychkova, V. N. Zholkevich, and B. A. Peisakson. 1974. Synchronization of cell division and some peculiarities in the mitotic cycle in suspension culture of *Panax ginseng* C. A. May. III Intern. Congr. of Plant Tissue and Cell Culture, No. 28 (abstr.). University of Leicester, Leicester.

Button, J., and J. Kochba. 1977. Tissue culture in the citrus industry. *In* J. Reinert and Y. P. S. Bajaj (eds.), Applied and fundamental aspects of plant cell, tissue, and organ culture. Pp. 70-82. Springer-Verlag, Berlin.

Calvin, M. 1976. Photosynthesis as a resource for energy and materials. Amer. Sci. 64:270-78.

Calvin, M. 1977. Photosynthesis as a resource for energy and materials. Kemia e. kemi. 2:46.

Carlson, P. S. 1973. Methionine sulfoximine-resistant mutants of tobacco. Science 180:1366-68.

Carlson, P. S., and R. S. Chaleff. 1974. Forced association between higher plant and bacterial cells *in vitro*. Nature 252:393-95.

Chaleff, R. S., and P. S. Carlson. 1975. *In vitro* selection for mutants of higher plants. *In* Genetic manipulation of plant materials. Pp. 351-63. New York-London.

Chan, N. M., and E. Y. Staba. 1965. Alkaloid production by Datura callus and suspension tissue culture. Lloydia 28:55-62.

Child, Y. Y. 1975. Nitrogen fixation by *Rhizobium* sp. in association with non-leguminous plant cell culture. Nature 253:350-51.

Child, Y. Y., and T. A. LaRue. 1974. A simple technique for the establishment of nitrogenase and soybean callus culture. Plant Physiol. 53:88-90.

Clapham, D. 1973. Haploid Hordeum plants from anthers *in vitro*. Z. Pflanzenzuch. 69:142-45.

Cocking, E. C. 1976. A new procedure for the selection of somatic hybrids in plants. *In* D. Dudits, G. L. Farkas, and P. Maliga (eds.), Cell genetics in higher plants. Pp. 141-48. Akadémiai Kiadó, Budapest.

Constabel, F. O., O. L. Gamborg, and L. C. Fowke. 1977. Hybridization of somatic cells from legumes. Intern. Conf. on Regulation of Developmental Processes in Plants, No. 85 (abstr.). Halle.

Constabel, F. O., J. W. Kirkpatrick, and O. L. Gamborg. 1973. Callus formation from mesophyll protoplasts of Pisum sativum. Canad. J. Bot. 51:2105-6.

Davey, M. R. 1977. Bacterial uptake and nitrogen fixation. *In* J. Reinert and Y. P. S. Bajaj (eds.), Applied and fundamental aspects of plant cell, tissue, and organ culture. Pp. 551-63. Springer-Verlag, Berlin.

Dix, P. Y., and H. E. Street. 1975. Sodium-resistant cultured cell lines from *Nicotiana sylvestris* and *capsicum annuum*. Plant Sci. Lett. 5:231-37.

Dix, P. Y., and H. E. Street. 1976. Selection of plant cell lines with enhanced chilling resistance. Ann Bot. 40:903-10.

Furuya, T., and T. Ishii. 1973. The manufacturing of Panax plant tissue culture containing crude saponins and sapogenins which are identical with those of natural Panax roots. Jap. Patent N48-319117.

Furuya, T., H. Kojima, R. Syono, and T. Ishii. 1970. Isolation of panaxatriol from Panax ginseng callus. Chem. Pharm. Bull. 18:2371-72.

Gengenbach, B. G., and C. E. Green. 1975. Selection of T-cytoplasm maize callus cultures resistant to Helminthosporium maydis race T pathotoxin. Crop Sci. 15:645–49.

Gleba, Yu. Yu., N. M. Piven, U. R. Komarnicki, and K. M. Sitnik. 1977. Parasexual and cytoplasmic hybrids (cybrids) of *Nicotiana tabacum* + *N. debneyi* obtained by protoplasts fusion. Dokl. Acad. Sci. USSR (in press).

Goldemberg, J. 1977. Alcohol from plant products: A Brazilian alternative to the energy shortage. *In* this volume.

Holdgate, D. F. 1977. Propagation of ornamentals by tissue culture. *In* J. Reinert and Y. P. S. Bajaj (eds.), Applied and fundamental aspects of plant cell, tissue, and organ culture. Pp. 375–98. Springer-Verlag, Berlin.

Holsten, R. D., R. C. Burns, R. W. F. Hardy, and R. R. Hebert. 1971. Establishment of symbiosis between Rhizobium and plant cells *in vitro*. Nature 232:173–76.

Jones, C. N., I. A. Mastrangelo, H. H. Smith, and H. Z. Liu. 1976. Interkingdom fusion between human (HeLa) cells and Tobacco hybrid (GGLL) protoplasts. Science 193:4251–56.

Kanta, K., N. S. Rangaswany, and P. Maheshwari. 1962. Test-tube fertilization in a flowering plant. Nature 194:1214–17.

Kao, K. N., F. Constabel, M. R. Michayluk, and O. L. Gamborg. 1974. Plant protoplast fusion and growth of intergenetic hybrid cells. Planta 120:215–17.

Karanova, S. L., Z. B. Shamina, and R. G. Butenko. 1977. Inductive genetic variety of Dioskorea deltoidea somatic cells *in vitro*. Prototrophity to phytohormones. Genetika (in press). (In Russian.)

Kaul, B., and E. J. Staba. 1969. *Dioscorea* tissue cultures. I. Biosynthesis and isolation of diosgenin from *Dioscorea deltoidea* callus and suspension cells. Lloydia 31:171–79.

Khanna, P., and S. C. Jain. 1973. Diosgenin, gitogenin, and tigogenin from Trigonella foenum-graecum tissue cultures. Lloydia 36:96–98.

Kireeva. S. A., P. S. Bugorskii, and S. A. Reznikova. 1977. Obtaining of volatile oils var. of Rose tissue culture. I. Accumulation of terpenoids. Sov. Plant Physiol. (in press).

Kovaleva. T. A., Z. B. Shamina, and R. G. Butenko. 1971. Mutagenesis in Rauvolfia tissue culture. Genetika 5:39–43. (In Russian.)

Kucherenko, L. A., P. N. Kcharchenko, and A. I. Davolan. 1978. Tissue culture and isolated organ method applications in rice breeding. *In* Proc. of Acad. of Agric. Sci. Kolos, Moscow (in press).

Kursakov, G. A. 1970. The application of excised embryo culture in the distant hybridization of plum. *In* Plant cell, tissue, and isolated organ culture. Pp. 1047–51. Nauka, Moscow.

Kuzovkina, I. N., G. A. Kuznetsova, and A. M. Smifnov. 1971. Formation of coumarins in root tissue of Ruta graveoleus L. Acad. News 6:929. (In Russian.)

Kuzovkina, I. N., and K. Sendren. 1973. Formation of phytohinoline alkaloids in root tissue of Ruta graveoleus L. Acad. News 2:275. (In Russian.)

Lescure, A. M. 1969. Mutagenèse et sélection de cellules d'Acer pseudoplatanus L. cultivées *in vitro*. Physiol. Veg. 7:237–50.

Lien, A., and A. San Pietro. 1977. An inquiry into biophotolysis of water to produce hydrogen, p. 49. Department of Plant Science, Indiana University.

Lipski, A. H., R. G. Butenko, and V. N. Paukov. 1978. *Dioskorea deltoidea* cell growth and biosynthesis of steroid compounds in batch culture in fermentor (in press).

Liu, Ming-Chin. 1974. A post-seminar report to the US-ROC cooperative science program on plant tissue and cell cultures. Taiwan Sugar 21:1–6.

Lukijanuk, S. F., and S. L. Ignatova. 1977. Obtaining of haploids and dihaploids from Triticale by anthers culture. *In* Apomixis of plant and animal. Proc. of Biol. Inst. of Siberia, Dept. of Acad. of Sci. of USSR, vol. 35.

Maliga, P. 1976. Isolation of mutants from cultured plant cells. *In* D. Dudits, G. L. Farkas, and P. Maliga (eds.), Cell genetics in higher plants. Pp. 59–76. Akadémiai Kiadó, Budapest.

Mariachina, I. Ja., and R. G. Butenko. 1974. Somatic reduction in cabbage tissue culture. Sov. Cytology and Genetika 9:2.

Monaco, L. C., M. R. Söndahl, A. Carvalho, O, J. Crocomo, and W. R. Sharp. 1977. Applications of tissue culture in the improvement of coffee. *In* J. Reinert and Y. P. S. Bajaj (eds.), Applied and fundamental aspects of plant cell, tissue, and organ culture. Pp. 109–26. Springer-Verlag, Berlin.

Müller, A. I., and R. Grafe. 1975. Mutant cell lines of Nicotiana tabacum deficient in nitrate reductase. XII Bot. Congr. Leningrad, No. 304 (abstr.).

Murashige, T. 1974. Plant propagation through tissue cultures. Ann. Rev. Plant Physiol. 26:135–65.

Nitsch, C. 1977. Culture of isolated microspores. *In* J. Reinert and Y. P. S. Bajaj (eds.), Applied and fundamental aspects of plant cell, tissue, and organ culture. Pp. 277–78. Springer-Verlag, Berlin.

Pechurkin, N. S., and I. A. Terskov. 1975. Analysis of growth kinetic and evolution of microbial cells population in control conditions. P. 215. Nauka, Novosibirsk.

Popov, Yu. G., V. A. Visotskaja, and V. G. Trushechkin. 1976. Culture of isolated cherry stem apices. Sov. Plant Physiol. 23(3):513–18.

Potrycus, J., T. C. Harms, and H. Lorz. 1976. Problems in culturing cereal protoplasts. *In* D. Dudits, G. L. Farkas, and P. Maliga (eds.), Cell genetics in higher plants. Pp. 129–40. Akadémiai Kiadó, Budapest.

Power, J. B., and E. C. Cocking. 1977. Selection systems for somatic hybrids. *In* J. Reinert and Y. P. S. Bajaj (eds.), Applied and fundamental aspects of plant cell, tissue, and organ culture. Pp. 497–504. Springer-Verlag, Berlin.

Quak, F. 1977. Meristem culture and virus-free plants. *In* J. Reinert and Y. P. S. Baja (eds.), Applied and fundamental aspects of plant cell, tissue, and organ culture. Pp. 598–616. Springer-Verlag, Berlin.

Raghavan, V. 1977. Applied aspects of embryo culture. *In* J. Reinert and Y. P. S. Bajaj (eds.), Applied and fundamental aspects of plant cell, tissue, and organ culture. Pp. 375–98. Springer-Verlag, Berlin.

Reinert, J., and Y. P. S. Bajaj. 1977. Anther culture: Haploid production and its significance. *In* J. Reiert and Y. P. S. Bajaj (eds.), Applied and fundamental aspects of plant cell, tissue, and organ culture. Pp. 251–64. Springer-Verlag, Berlin.

Reinhard, E., G. Corduan, and O. H. Volk. 1968. Uber gewebekulturen von Ruta graveoleus. Planta Med. 16:8–16.

Sarkisova, M. A. 1974. Diosgenin and steroid saponins in Dioskorea deltoidea tissue culture. III Intern. Congr. Plant Tissue and Cell Culture, No. 259 (abstr.). University of Leicester, Leicester.

Sarkisova, M. A., N. V. Vinnokova, and V. N. Paukov. 1977. Some peculiarities of biosynthesis of steroid saponins in Dioskorea deltoidea Wall cell suspension cultures (in preparation).

Scowcroft, W. R., and A. H. Gibson. 1975. Nitrogen fixation by Rhizobium association with Tobacco and cowpea cell cultures. Nature 253:351–52.

52   Plant Cell and Tissue Culture

Semenenko, V. E. 1975. Intracellular regulation and management of biosynthesis in microalgae. Doctoral thesis, Moscow.

Shamina, Z. B., and S. L. Karanova. 1974. Mutagenic effect of N-Nitroso-N-methyl-urea on tissue culture cell of *Dioskorea deltoidea* Wall. III Intern. Conf. of Plant Tissue and Cell Culture, No. 284 (abstr.). University of Leicester, Leicester.

Shamina, Z. B., S. L. Karanova, and T. A. Kovaleva. 1975. Chemical mutagenesis in plant cell culture, p. 315. XIII Intern. Bot. Congr., No. 174. Leningrad.

Shamina, Z. B., T. A. Kovaleva, and R. G. Butenko. 1976. Effect of nitrogen mustard on Rauwolfia tissue culture. Phytomorphology 22:260–64.

Shinkareva, I. K. 1974. Cytogenetic properties of Tobacco progeny obtained by tissue culture method. Ph.D. dissertation, Institute of General Genetics, Moscow.

Slepyan, L. I. 1970. The results of medicinal studies on Panax ginseng root tissues grown *in vitro*. *In* Plant cell, tissue, and isolated organ culture. P. 249/53. Nauka, Moscow.

Slepyan, L.I., A. G. Vollosovich, and R. G. Butenko, 1968. Medical plant tissue culture and some perspectives of its use in pharmacology. Rastit. Resursii 4(4):457–67. (In Russian.)

Smolov, A. P., A. R. Ignatov, and V. S. Polevaja. 1975. The formation of photosynthetic apparatus function in tissue culture. Sov. Plant Physiol. 22:428–30.

Sobko, V. G., Z. B. Shamina, and B. P. Strogonov. 1977. Resistance of Crepis capillaris cells to high levels of NaCl. XIV Intern. Genet. Congr. (abstr.) (in press).

Steck, W., B. K. Bailey, J. P. Shyluk, and O. L. Gamborg. 1971. Coumarins and alkaloids from cell cultures of Ruta graveoleus. Phytochemistry 10:191–98.

Stohs, S. L., and H. Rosenberg. 1975. Steroids and steroid metabolism in plant tissue cultures. Lloydia 38:181–94.

Stohs, S. J., J. J. Sabatka, and H. Rosenberg. 1974. Incorporation of 4-^{14}C-22, 23-^{3}H-sitosterol into diosgenin by *Dioscorea deltoidea* tissue suspension cultures. Phytomorphology 13:2145–48.

Tuchtasinov, M. H. 1977. Chemical and pharmacological studies of Rauwolfia serpentina tissue culture. Doctoral thesis, Leningrad.

Tumanov, I. I., R. G. Butenko, I. V. Ogolevetz, and V. V. Smetuk. 1977. The increasing of frost-resistance of Picea callus tissue culture by pretreatment with low temperature. Sov. Plant Physiol. 24:5–895.

Tumanov, I. I., I. V. Ogolevetz, and R. G. Butenko. 1970. The influence of low positive and negative temperatures on the isolated tissue of trees. *In* Plant cell, tissue, and isolated organ culture. Pp. 215–19. Nauka, Moscow.

Vollosovich, A. G., and R. G. Butenko. 1970. Alkaloid content in Rauwolfia sp. *In* Plant cell, tissue, and isolated organ culture. Pp. 253–57. Nauka, Moscow.

Vollosovich, N. E., A. G. Vollosovich, T. A. Kovaleva, Z. B. Shamina, and R. G. Butenko. 1976. Strains isolated from the tissue culture of Rauwolfia serpentina Benth. and their productivity. Rastit. Resursii 12:578–83. (In Russian.)

Widholm, J. M. 1974. Selection and characteristics of biochemical mutants of cultured plant cells. *In* H. E. Street (ed.), Tissue culture and plant science. Pp. 287–300. Academic Press, New York.

Zargarjan, O. N., A. V. Marshavina, A. G. Gevorkjan, and L. K. Aslanjans. 1976. Biochemical characteristics of some volatile oils plant tissue and cell cultures. Biol. J. Armenija 29:52–55.

Zenk, H. M. 1974. Haploids in physiological and biochemical research. *In* K. J. Kasha (ed.), Haploids in higher plants. Pp. 339–52. University of Guelph, Guelph.

Zenkteler, M., E. Misiura, and J. Guzowska. 1975. Studies on obtaining hybrid embryos in test tubes. *In* Form, structure, and function in plants. E. M. Johri commemoration volume. Pp. 180–87. University Press, Delhi.

JOSÉ GOLDEMBERG

Alcohol from Plant Products: A Brazilian Alternative to the Energy Shortage

5

INTRODUCTION

The oil crisis of 1973 has affected Brazil in a very harsh way mainly for two reasons. First, the development patterns of the country in the past have been such that the participation of petroleum in the energy balance has increased from a very modest 9.2% in 1941 to 28.0% in 1952 and to 44.8% in 1972 (Wilberg, 1974) (fig. 1.). These patterns are apparent from indications well known to Americans for many decades: cities and roads clogged with cars, air pollution, and deterioration of the quality of life in urban centers.

Second, Brazil produces only 20% of the petroleum it consumes, despite the efforts of PETROBRAS—the state-owned oil company—to find oil. So far it has met with little success except in the continental platform near Campos, in the state of Rio de Janeiro, in the southern part of the country. The remaining 80% (approximately 700,000 barrels per day) has to be imported at a cost of more than 3 billion dollars a year in 1976. This weighs heavily in the country's foreign trade balance; the country has to make an enormous effort to export products (mainly minerals and agricultural products) to compensate.

The development patterns that modernized the southern part of Brazil— in the sense of introducing the latest gadgets and products of industrialized countries—have so far benefited only approximately 20% of the inhabitants concentrated in large cities, with the unpleasant consequences pointed out above. This fraction of the population corresponds to almost 20 million people, which is in itself a large population.

José Goldemberg, Institute of Physics, University of São Paulo, São Paulo, Brazil.

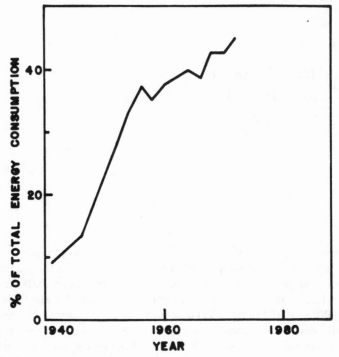

Fig. 1.   Petroleum consumption in Brazil.

An interesting quantitative indication of the "patterns of consumption" adopted in recent decades in Brazil is the number of automobiles, which has grown from a few thousands thirty years ago to about five million in 1975. The country has clearly entered the vicious circle of more cars, more roads, more cars, and so on. The transportation of goods and services has shifted to road transportation, thanks to often disguised subsidies for road construction, with the result that the energy spent in road transportation is approximately 83% of the total (Colombi, 1975), as can be seen in table 1.

Some efforts have been made in the past two years to change this situation, and the bulk of the government actions have been in the direction of reducing gasoline consumption of private cars. An interesting point to stress is that most of the suggestions to that effect originated in the academic community, in particular from the studies of our own group at the University of São Paulo.

The package of measures adopted about one year ago is the following:

1.   Limitation of the maximum speed on roads to 50 mph.

TABLE 1

ENERGY IN TRANSPORTATION OF MERCHANDISE

Sector	Energy Used (%)
Road	82.82
Rail	4.88
Ship	7.14
Air	4.41
Others	0.76
Total	100.00

2. Gradual banning of cars from the center of large cities to discourage people from driving their cars to work or to shopping centers.

3. Improvement of the bus system with the introduction of "executive" buses and exclusive bus lanes.

4. Closing of gasoline stations in the evenings and weekends.

5. Gradual price increase, which makes the gasoline price in Brazil one of the highest in the world.

A precise evaluation of these actions does not yet exist but no great increase in gasoline consumption occurred during the past year, which is encouraging.

However, some of our more drastic suggestions were not adopted, such as allowing the circulation of cars only on alternate days according to the last digit in their license plates (odd digit, odd days; even digit, even days) or simply rationing gasoline according to quotas. Limits to the power of cars produced by the manufacturers (which is a better gasoline-saving measure than velocity limitation) were also not accepted.

Although many industries, buses, and trucks are shifting from gasoline to diesel oil (which is cheaper), prospects for the future are bleak under present conditions. The Brazilian energy balance of 1977 shows that quite clearly (fig. 2).

When projections are made up to 1985, it can be seen that the consumption of liquid combustibles is expected to grow steadily to the equivalent of 1,200,000 barrels a day (a 50% increase in 10 years).

SOLUTIONS

To attenuate this situation a larger emphasis is being put on hydroelectric power production because the country still has many untapped water resources. Unfortunately, most of the remaining large waterfalls are located in the large affluents of the Amazon River (Tocantins, Araguaia, Xingu, Tapajós, Contingo, and Trombetas), which are at least one thousand miles from the large consuming centers in the southern part of the country (Goldemberg, 1976a). There exists here a great challenge to move produc-

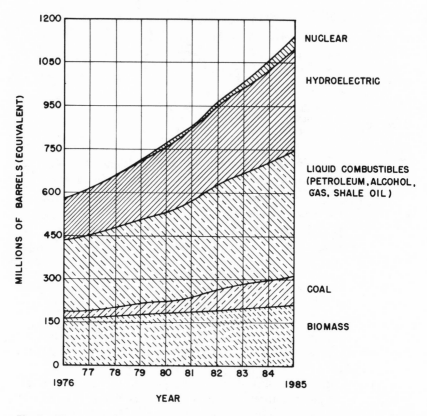

Fig. 2. The Brazilian energy balance. (Source: O Balanco Energetico Brasileiro. 1976. Ministerio de Minas e Energia.)

tion centers near these energy sources or to develop methods of carrying energy very large distances by transmission lines. In addition to that there are hundreds (maybe thousands) of small rivers that can be used for hydroelectric production using the new techniques of bulb type generators. The emphasis on size copied from industrial nations, has relegated these projects to a low priority, with the result that large urban concentrations requiring huge power plants have high pollution and living conditions of cities are deteriorating.

A shift of emphasis in the direction of "small is beautiful" is not in sight, but the pressures for alternative solutions are mounting with a nonnegligible chance of success.

Established habits are, however, hard to change, especially if these habits are supported by strong economic interests such as the car manufacturers. Facing, however, a strong public opinion pressure to save gasoline (and other oil products), a very determined effort by the government was made to "save" the automobile and the consumption habits that go with it.

This is probably the explanation for the adoption of the "alcohol program" by the Brazilian government. Again the suggestions for this program originated in the academic community. The basic idea is to partially (or totally) substitute ethyl alcohol (ethanol) for gasoline in gasoline-driven cars and diesel engines (Goldemberg, 1976b).

Technically the idea is feasible: present internal combustion engines can use or be converted to run on ethanol, and the huge amounts of alcohol needed can be produced in Brazil.

TECHNICAL PROBLEMS IN THE ENGINES

At first glance it might seem that the use of ethanol is not such a good idea because it has a calorific content 39% smaller than gasoline (Stumpf, 1975, 1976, 1977) (table 2). One could conclude from table 2 that the consumption of alcohol is $10,500 / 6,400 = 1.64$, 64% higher than gasoline. That is not so: there are a number of other factors to be taken into account.

TABLE 2

ENERGY CONTENT OF COMBUSTIBLES

Gasoline	10,500 kcal/kg
Anhydrous ethanol	6,400 kcal/kg
Anhydrous methanol	4,700 kcal/kg

The first one is that the power of an engine does not depend on the calorific content of the liquid but on the energy contained in the combustible gas mixture in the cylinders (the combustion energy).

In addition, the power depends also on other facts such as the thermal efficiencies and a cylinder filling factor that has to do with the number of molecules after combustion; this number is larger for alcohol than for gasoline. Taking all these factors into account it turns out that $(\text{Power})_{eth} / (\text{Power})_{gas} = 1.18$. A motor running on ethanol has 18% more power than a motor running on gasoline; this is the reason for using alcohol in racing cars. Consumption, however, is not the same as power. To help in this problem there is first the question of the densities: the density of ethanol is 0.81 and that of gasoline 0.73. Therefore alcohol has more molecules per liter than gasoline.

There are in addition a number of other factors to be taken into consideration; the ratio R of the consumption of the two combustibles is given by $R = $ consumption of ethanol / consumption of gasoline $= (\text{calorific power})_{eth} / (\text{calorific power})_{gas} \times (\text{efficiency})_{eth} / (\text{efficiency})_{gas} \times (\text{density})_{eth} / (\text{density})_{gas}$. The efficiency (liters per hp-hour or grams per kiw-hour) is given by the expression efficiency = thermal equivalent of 1 hp-hour/calorific power $\times$ specific consumption, and represents the energy effectively converted in mechanical energy at the axis of the motor. Putting numbers on this expression one finds (efficiency)$_{gas}$ = 0.27 (27%) and (efficiency)$_{eth}$ = 0.42 (42%), which is quite a surprising result: *a motor running on alcohol has a greater efficiency (42%) than a motor running on gasoline(27%).*

The numbers used in obtaining this result are:

	Ethanol	Gasoline
Thermal equivalent of 1 hp-h	632,000	632,000
Calorific power (kcal/kg)	6,400	10,500
Specific consumption (g/hp-h)	250	220

All together one gets for R the value R = (consumption)$_{eth}$/(consumption)$_{gas}$ = 10,500/6,400 $\times$ 0.27/0.42 $\times$ 0.73/0.81 = 1.014. The consumption of ethanol in liters is only 1.4% greater than the consumption of gasoline.

Alcohol can be mixed with gasoline without any modification in the motors in any amount up to 20%, as shown in figure 3. Above 20% the consumption becomes prohibitively high and is not feasible. In reality one

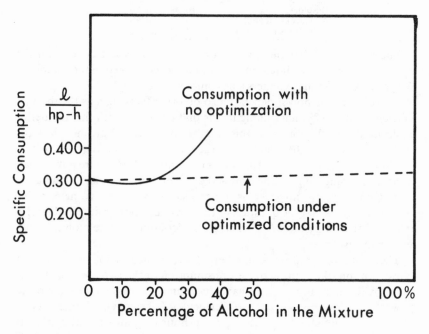

Fig. 3.   Consumption of alcohol mixtures in combustibles.

could use 100% alcohol, but that would require an optimization of the motor, which might be appreciable.

From the point of view of air pollution, alcohol has wider limits of inflammability than gasoline, permitting an air-combustible mixture that burns better, with the consequence that carbon and nitrous monoxide production can be lowered to a large extent. In addition to that, the lead additives can be eliminated.

AGRICULTURAL PROBLEMS IN THE PRODUCTION OF ALCOHOL

The estimated consumption of gasoline in 1980 is 15 billion liters, if no special measures to reduce it are taken. A typical goal of the alcohol program is therefore to replace 20% of the gasoline by alcohol in 1980, which will require 3 billion liters of alcohol. Table 3 summarizes this and other hypotheses (Licio, 1976). The amounts of land needed under some of these hypotheses are large, but not unduly so. Brazil has a total area of 850 million hectares. Of these, approximately 70 million hectares are fertile lands.

TABLE 3

GOALS OF THE ALCOHOL PROGRAM IN BRAZIL

Production (Liters/Year) $\times 10^9$	Cultivated Area Needed in 1,000 ha of Sugarcane[a]
Hypothesis I[b]	1,100
Hypothesis II[c] 16	4,400
Hypothesis III[d] 22	6,000
Hypothesis IV[e] 33	9,000

[a] The average agricultural productivity was taken for simplicity as 60 ton/ha and industrial output as 70 l/ton.
[b] 20% alcohol in gasoline plus 10^9 liters for industry.
[c] 100% alcohol plus 10^9 liters for industry.
[d] 100% alcohol plus 50% of the diesel oil consumption.
[e] 100% alcohol plus 100% of the diesel oil consumption.

One assumes that by the year 2000 the total amount of fertile land in Brazil will increase by 14% (120 million ha). Less than 20% would be necessary to produce all the alcohol needed by then (almost twice hypothesis IV of table 3). This will correspond to only 3% of the Brazilian territory (Gomes da Silva, 1977). This situation is depicted in figure 4.

One should add here that cassava is also a strong candidate for ethanol production. The average agricultural yield of cassava is 29 ton/ha, which is smaller than that of sugarcane; its industrial output, however, is much higher (175 l/ton compared with 70 l/ton for sugarcane). Cassava grows in soils that are poorer than the ones needed for sugarcane, and there are great hopes of using the *cerrado* for this purpose. Almost 12% of the Brazilian territory (100 million ha) is covered by *cerrado*. A complete experiment of

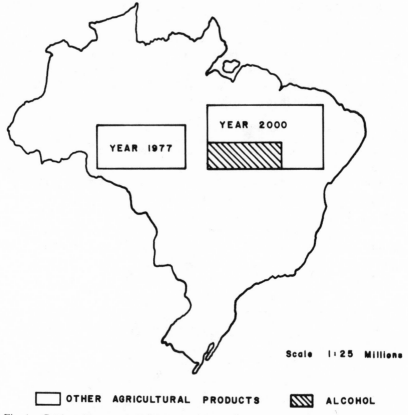

Fig. 4.   Cultivatable areas in Brazil.

alcohol production from cassava is being conducted at present in land of this type in Curvello, State of Minas Gerais, for a production of 100,000 liters of alcohol per day.

REMAINING PROBLEMS

The first important question to ask here is, Since alcohol is being produced for burning in internal combustion engines, what is the energy balance involved in this production? In other words, what is the output energy contained in the alcohol as compared to the total amount of energy needed for its production?

Table 4 shows our most recent results for four cultures: sugarcane, cassava, plant sorghum, and ratoon sorghum (including labor in energy units, machinery, fertilizers, and so on) (Gomes da Silva et al., 1977). They all have a positive balance: sugarcane has a net energy gain of 21,345

TABLE 4
Energy Balance of Ethyl Alcohol Production

| Culture | Agricultural Yield | | Alcohol Production | | | Energy (Mcal/ha/year) | | | | | | | |
| | | | | | | Produced | | | Expended | | | Balance |
	t/ha	t/ha/year	l/t	l/ha	l/ha/year	Alcohol	Residue	Total	Agriculture	Industry	Total	
Sugarcane	72	54	66	4,752	3,564	18,747	17,550	36,297	4,138	10,814	14,952	21,345
Cassava	29[a]	14.5	174	5,046	2,523	13,271	·.··	13,271	2,573	8,883	11,456	1,815
Cassava	29[b]	14.5	174	5,046	2,523	13,271	5,512	18,783	3,868	1,883	12,751	6,032
Sweet sorghum (stems and grains)												
Plant sorghum	[c]	· · · ·	: :	3,775	3,775	19,856	11,830	31,686	4,671	11,883	16,554	15,132
Ratoon sorghum	[d]	· · · ·	: :	2,383	2,383	12,535	7,280	19,815	3,350	7,501	10,851	8,964
Total	: :	: : : :	: :	6,158	6,158	32,391	19,110	51,501	8,021	19,384	27,405	24,096
Sweet sorghum (stems)												
Plant sorghum	32.5	32.5	: :	2,600	2,600	13,676	11,830	25,505	4,671	7,722	12,393	13,113
Ratoon sorghum	20.0	20.0	: :	1,600	1,600	8,416	7,280	15,696	3,350	4,752	8,102	7,594
Total	52.5	52.5	: :	4,200	4,200	22,092	22,092	41,202	8,021	12,474	20,495	20,707

[a] No utilization of stems for steam production.
[b] Utilization of stems for steam production.
[c] 32.5 t of stems and 3.0 t of grains per ha.
[d] 20.0 t of stems and 2.0 t of grains per ha.

Mcal/ha/year and cassava of 6,032 Mcal/ha/year if the stems are utilized for steam production and only 1,315 Mcal/ha/year if they are not used. The superiority of sugarcane from the point of view of energy is evident here, but there are hopes of increasing the agricultural productivity of cassava.

Another problem in the industrial production of alcohol from sugarcane is the handling of very large amounts of liquid (sugarcane syrup) of which only about 7% can be transformed into ethanol. The remaining 93% (13 times the volume of alcohol) still has to be distilled, which means that, for the same size, an alcohol distillery produces 13 times less alcohol than a refinery (in which 100% of the petroleum is distilled into useful oils or combustibles). Typical capacity of a petroleum refinery is 50,000 barrels a day (8,000,000 liters a day), against 100,000 liters of alcohol per day in an alcohol refinery.

Finally the *vinhoto* left in alcohol distilleries is a pollutant, usually thrown in rivers, but it could be used as a fertilizer. The handling of the very large volumes of *vinhoto* is, however, a difficult problem.

CONCLUSIONS

With these qualifications the Brazilian alcohol program is going ahead, and already the city of São Paulo (with its 1.5 million automobiles) is being used as a giant laboratory in which alcohol is being added to the gasoline in the proportion of 20%, with no important problems so far. A fleet of 200 vehicles was also converted to pure alcohol propulsion, and these vehicles are being tested in operation with good results.

If all goes well the program will expand enormously and Brazil might be the first country to run on nonfossil renewable combustibles. This will constitute a major utilization of photosynthesis (and ultimately of solar energy), which is well suited for the tropical climate of Brazil; the experience could possibly be extended to other countries.

LITERATURE CITED

Colombi Neto, J. 1975. O. declínio das ferrovais e o transporte de passageiros de médio e longo percurso. Ph.D. thesis, Faculdade de Economia e Administração, Universidade de São Paulo.

Goldemberg, J. 1976a. Energia no Brasil. Publicaçao ACIESP No. 2. Academia de Ciências do Estado de São Paulo.

Goldemberg, J. 1976b. Problemas de Energia no Brasil. Instituto de Pesquisas. Estudos e Assessoria do Congresso, Brasilia.

Gomes da Silva, J. 1977. O Pro-Alcool e as responsabilidades do setor agricola. 29a. Reunião Anual da Sociedade Brasileira para o Progresso da Ciência, São Paulo.

Gomes da Silva, J., G. E. Serra, J. R Moreira, J. C. Gonçalves, and J. Goldemberg, 1977. Cultural energy balance for ethyl alcohol production. (submitted to Science.)

Licio, A. M. A. 1976. Etanol: combustível e matéria prima. Semana de Tecnologia Industrial. Ministério da Indústria e Comércio. Secretaria de Tecnologia Industrial, Rio de Janeiro.

Stumpf, U. E. 1975. Alcool carburante em mistura de combustível. Açucar e Alcool, 3° Encontro Nacional dos Produtores de Açucar, COPERFLU/APEC, Campos, Rio de Janeiro.

Stumpf, U. E. 1976. Etanol: combustível e matéria prima. Semana de Tecnologia Industrial, Ministério da Indústria e Comércio. Secretaria de Tecnologia Industrial, Rio de Janeiro.

Stumpf, U. E. 1977. Aspectos técnicos de motores a alcool. 29ᵈ Reuniao Anual da Sociedade Brasileira para o Progresso da Ciência, São Paulo.

Wilberg, J. A. 1974. Consumo Brasileiro de Energia. Revista Brasileira de Energia Elétrica 27:17.

PER PINSTRUP-ANDERSEN

Selected Economic Aspects of
Agricultural Research

6

Agricultural research is an essential component of the overall efforts to facilitate expanded agricultural production and productivity. Agricultural research is presumed to seek ways to produce more and/or better food, feed, and fiber at a reduced per unit cost and in such a way as to increase the contribution of agriculture to the achievement of ultimate social and economic goals. Research resources are scarce, whereas research needs are unlimited. Since not all research needs can be attended to, it is important to select the research areas with the highest expected payoff. In order to establish research priorities and allocate available research resources, attempts should be made to estimate the relative contribution and costs of alternative research activities. Such attempts must take into consideration the needs of farmers as well as overall social and economic goals.

Conceptually, optimum allocation of limited research resources is readily defined: available resources must be allocated in such a way as to maximize the contribution of research to established goals. If the cost of each possible research project and the associated contribution to the stated goals were known, research priorities could simply be established by ranking the projects from highest to lowest contribution per research dollar. But research is a search into the unknown. Hence, the outcomes usually cannot be predicted with a great deal of certainty. Furthermore, the value to society of a given research result is not readily predicted. Results

Per Pinstrup-Andersen, Economic Institute, Royal Agricultural University, DK-1871 Copenhagen V, Denmark.

from public research are resources—not consumption goods—and they do not usually carry a market price. Their contribution to society's goals— except for narrow research goals—is realized if, and only if, they are utilized in the production process. The difficulty of evaluating research imputs, particularly the human resource, is another source of uncertainty in research. Thus, no great precision can be expected in the prediction of research payoffs. It is important, however, to utilize whatever knowledge and information can be obtained at a reasonable cost to help assure that scarce research funds are utilized to achieve the objectives to the fullest extent possible.   There seems to be an increasing emphasis, particularly among funding agencies, on quantifying the expected payoffs from alternative lines of research. Such quantitative estimates are convenient for accepting or rejecting any given research project. However, the margin of error associated with the estimates is frequently unacceptable unless an excessive amount of resources is spent on obtaining the data required for the estimation. Furthermore, there may be a tendency to reduce the importance of sound qualitative judgment by researchers and research managers where quantitative estimates—however uncertain—are made. In my opinion, a combination of quantitative estimates, with acceptable error margins, and expert judgment, analyzed within a logical framework, provides the most promising approach to the establishment of research priorities.

The purpose of this paper is to discuss some of the principal economic issues related to research priorities, costs, and benefits. First, worldwide research and extension expenditures are discussed. Then follows a brief analysis of research benefits and their distribution among groups in society. The final section deals with ways to increase net returns from research. Since international agricultural research programs and actual research priorities are treated in the papers by Lamm and Pino, these issues will not be dealt with in any detail in this paper.

AGRICULTURAL RESEARCH AND EXTENSION EXPENDITURES

The current world agricultural research cost is estimated at approximately \$4 billion annually (table 1). One-third of this is spent in North America and Oceania and another 40% in Europe and USSR. Although developing countries occupy about two-thirds of the world's arable land and land used for permanent crops and produce about one-fourth of the world's agricultural product by value, these countries account for only 11 % of all agricultural research expenditures (Evenson and Kislev, 1975).

World agicultural extension costs are about one-third of research costs. Almost one-half of the extension costs occur in developing countries.

TABLE 1

ESTIMATED EXPENDITURES ON AGRICULTURAL RESEARCH AND EXTENSION BY REGION, 1974
(Millions of 1971 Constant U.S. Dollars)

Region	RESEARCH		EXTENSION		Res. Exp./ Ext. Exp.
	Millions of U.S. $	%	Millions of U.S. $	%	
North America and Oceania	1,289.4	33.5	287.6	21.7	4.5
Western Europe	733.4	19.1	183.3	13.8	4.0
Eastern Europe and USSR	860.5	22.4	250.0	18.9	3.4
Latin America	170.3	4.4	121.9	9.2	1.4
Africa	141.1	3.7	224.5	16.9	0.6
Asia	646.0	16.9	258.5	19.5	2.5
World total	3,840.7	100.0	1,325.8	100.0	2.9

SOURCE: Boyce and Evenson, 1975; reprinted with permission.

Hence, the ratio of research to extension costs is highest in developed countries. North America and Oceania spend $4.50 on research for each dollar spent on extension as compared to $0.60 for Africa and $1.40 for Latin America.The principal reasons for this difference in relative emphasis on research versus extension might be (1) that some research results obtained by developed countries are transferable—or believed to be transferable—to developing countries, whereas extension efforts must be carried out where the farmers are; (2) that research is generally considered to be a longer term investment, with a longer period between the initiation of activities and the payoff; (3) that research requires larger investments per man-year in physical facilities, equipment, and personnel; and (4) that the payoff from research may be more difficult to identify, particularly in the short run.

Investments in research and extension have increased rapidly during the last 20 years (figs. 1 and 2). Research costs in 1974 were almost three times greater than in 1959, and extension costs more than doubled during the same period.

In table 2 agricultural research and extension expenditures are presented as a percentage of the value of agricultural production. Latin America, Africa, and Asia spent between 1.2% and 1.8% of the value of the agricultural product on research and from 1% to 2% on extension. North America and Europe spent a higher percentage on research but a lower percentage on extension, compared to developing countries. Research expenditures have increased at a faster rate than the value of the agricultural product in all regions; this is also the case for extension expenditures in the developing world. On the other hand, extension expenditures in North America, Europe, and Oceania have increased at about the same rate as the value of the agricultural product.

Rapid expansions in the network of international agricultural research

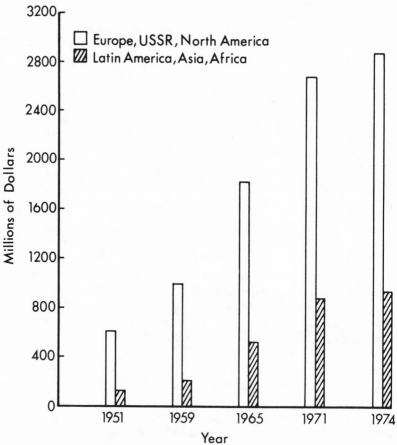

Fig. 1.   Expenditures on agricultural research, 1951–74.

centers during recent years have resulted in research expenditures of considerable magnitude for these centers. The total budget for 1977 is above $80 million, which equals about 7% of research expenditures in Latin America, Africa, and Asia. In figure 3 the increase in international center expenditures is shown. Both operating and capital expenditures are included. Since most of the centers were established recently and only few new centers, if any, are expected to be created in the near future, the rate of growth in expenditures is expected to be reduced considerably.

Among single crops, rice and wheat received the largest research expenditures in developing countries (table 3). About 32% and 19% of all research expenditures for cassava and rice, respectively, came from international research institutes. The proportion of total research expenditures coming from international institutes is smaller for other crops. It is

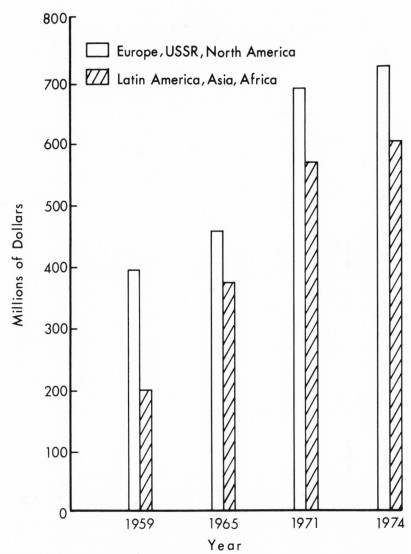

Fig. 2. Expenditures on agricultural extension, 1959–74.

interesting to note that, although in national research there is a high priority on sugarcane, none of the international institutes work on this crop. Sugarcane has not been included in the priority commodities of the international institutes probably because (1) a large part of this crop is grown in large land holdings, (2) a large part of the production is exported to developed countries, and (3) sugar is not generally believed to play a vital

TABLE 2

ESTIMATED EXPENDITURES ON AGRICULTURAL RESEARCH AND EXTENSION IN PERCENT
OF THE VALUE OF AGRICULTURAL PRODUCTION, BY REGION, 1959 AND 1974

	RESEARCH		EXTENSION	
	1959	1974	1959	1974
North America and Oceania	1.5	2.7	0.5	0.6
Western Europe	0.8	2.2	0.5	0.6
Eastern Europe and USSR	1.0	1.8	0.4	0.5
Latin America	0.4	1.2	0.3	0.9
Africa	0.8	1.4	1.2	2.2
Asia	0.6	1.9	0.4	0.9

SOURCE: Boyce and Evenson, 1975; reprinted with permission.

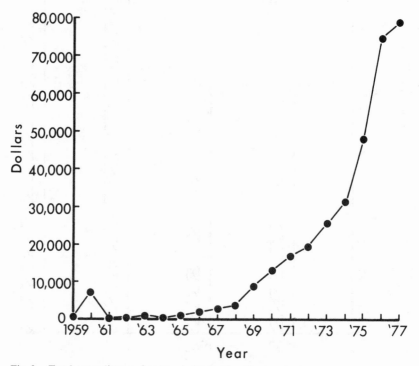

Fig. 3.   Total expenditures of centers in international network.

role in meeting the nutritional requirements of the population of the developing world. The first two premises are probably good reasons why producers themselves or national governments would be in a position to cover the research needs. The third premise, however, is invalid for a number of developing countries where sugar provides a considerable proportion of calorie requirements.

TABLE 3

NATIONAL AND INTERNATIONAL RESEARCH EXPENDITURES FOR THE
MAJOR CROPS IN DEVELOPING COUNTRIES
(Millions of 1971 Dollars per Year)

Crop	International Institute Budget 1976	NATIONAL RESEARCH				
		Latin America	Africa and West Asia	South and Southeast Asia	Total	Grand Total
Rice	7.9	11.7	5.3	17.7	34.7	42.6
Wheat	3.8	11.2	14.8	9.9	35.9	39.7
Sugarcane	0	15.3	8.8	6.8	30.2	30.2
Cassava	1.9	2.0	1.5	0.5	4.0	5.9
Maize	4.1	20.4	6.5	2.7	29.6	33.7

SOURCE: R. E. Evenson, 1976.

Having briefly discussed the research and extension expenditures, we now turn to a discussion of the benefits obtained from these expenditures.

AGRICULTURAL RESEARCH BENEFITS

The benefits from research are difficult to assess for a number of reasons. First, research benefits usually materialize a number of years after the research has been initiated. The research process itself takes considerable time and, even after a certain research output has been obtained, its diffusion and adoption requires considerable time. Second, research benefits may be difficult to separate from the impact of other factors that simultaneously influence agricultural production and prices. Third, the choice of value indicator may be somewhat arbitrary because of the price-quantity interaction; the existence of multiple—and possibly conflicting—goals; and the impact of supply expansions on such goals, e.g., human nutrition, income distribution, and foreign exchange earnings.

However, in spite of these and other difficulties, a number of analyses of the benefits and costs of selected national research programs have been carried out. Some of these studies are listed in table 4. The pioneering work in this area was carried out by Griliches and Schultz at the University of Chicago. Using the concept of economic surplus, the studies estimate the rate of return from investments in agricultural research. Other potential contributions of research, such as improved nutrition, are not considered in these studies.[1] The relative payoff to society from investing in agricultural research versus other activities, such as education, road building, and so on, can then be assessed simply by comparing the respective rates of return. As shown in table 4, annual rates of return to agricultural research are high. Rates above 50% are not uncommon, and only four of the thirty cases reported here provided rates of return of 20% or less. Most governments would have few investment opportunities with higher expected rates of return.

TABLE 4

ESTIMATED ANNUAL INTERNAL RATES OF RETURN FROM AGRICULTURAL
RESEARCH FOR VARIOUS COUNTRIES AND COMMODITIES

Commodity	Country	Time Period	Rate of Return	Study*
Aggregate	U.S.A.	1949–59	35–40	Griliches, 1964
Aggregate	U.S.A.	1949–59	47	Evenson, 1969
Aggregate	U.S.A.	1937–42	50	Peterson and Fitz-Harris, 1975
Aggregate	U.S.A.	1947–52	51	Peterson and Fitz-Harris, 1975
Aggregate	U.S.A.	1957–62	49	Peterson and Fitz-Harris, 1975
Aggregate	U.S.A.	1967–72	34	Peterson and Fitz-Harris, 1975
Aggregate	Japan	1880–1938	35	Tang, 1963
Aggregate	India	. . . . .	40	Jha and Evenson, 1973
Aggregate	India	. . . . .	63	Kahlon et al., 1975
Hybrid corn	U.S.A.	1940–55	35–40	Griliches, 1958
Corn	Peru	1954–67	35–55	Hines, 1972
Corn and sorghum	Mexico	1943–64	26–59	Ardito-Barletta, 1971
Hybrid sorghum	U.S.A.	1940–57	20	Griliches, 1958
Wheat	Mexico	1943–64	69–104	Ardito-Barletta, 1971
Wheat	Colombia	1953–73	11–12	Hertford et al., 1975
Rice	Colombia	1957–72	60–82	Hertford et al., 1975
Rice	Colombia	1957–64	94	Scolie, 1976
Rice	Japan	1915–50	25–27	Akino and Hayami, 1975
Rice	Japan	1930–61	73–75	Akino and Hayami, 1975
Soybean	Colombia	1960–71	79–96	Hertford et al., 1975
Potato	Mexico	1943–64	69	Ardito-Barletta, 1971
Sugarcane	S. Africa	1945–62	40	Evenson, 1969
Sugarcane	Australia	1945–58	50	Evenson, 1969
Sugarcane	India	1945–58	60	Evenson, 1969
Cocoa	Brazil	1923–74	16	Monteiro, 1975
Cocoa	Brazil	1958–74	60	Monteiro, 1975
Cotton	Brazil	1924–67	77+	Ayer, 1970
Cotton	Colombia	1953–72	Negative	Hertford et al., 1975
Poultry	U.S.A.	1915–60	21–25	Peterson, 1966
Pasture	Australia	1948–69	65–80	Duncan, 1972

* All studies from which results were available to the author at the time of preparing this paper are included.

Returns to investment in international research may also be high. Extremely high payoffs were obtained from past wheat research by CIMMYT and collaborating institutions and from past rice research by IRRI and collaborators. Thus, Evenson (1977) estimated that new varieties increased world wheat and rice production by 7 and 13.6 million tons, respectively, during the year 1972/73. Evenson further estimates that during the period 1966/67–1969/70 each research dollar generated an income stream of $303 when invested in wheat research and $142 when invested in rice research. Similar estimations for the period 1970/71–1972/73 resulted in $62 and $108 for wheat and rice, respectively. However, as Evenson points out, these estimates are subject to large errors and should be interpreted as approximate magnitudes only. However, in spite of the difficulties in estimating the exact payoff to this research, there is no doubt that it was extremely high and is likely to exceed the return to most other investment opportunities in the public sector. It is premature to attempt an estimation of the returns to the research carried out by the other international agricultural research institutes, but expectations are high.

Although the above discussion suggests a very handsome return to investment in agricultural research, it is obvious that not all such research results in high returns. In fact, casual observation suggests that a considerable amount of agricultural research has produced very little or no economic return. This is to be expected because of the very nature of research: exploration into the unknown. Furthermore, a certain proportion of the research must be of a more "basic" nature aimed at developing or improving the foundation for more "applied" research. A sound economic evaluation would include both types of research. However, since the payoff from research of a more basic nature is usually indirect and long term, it may easily be underestimated. Even taking these factors into consideration, the full potential payoff from agricultural research is probably not reached in many research institutions, particularly in developing countries. There are many reasons for this. The quality of research resources, whether human or physical, is of critical importance. Many institutions in developing countries suffer from lack of funds for purchasing and/or maintaining needed equipment and for attracting or keeping high-quality personnel. Furthermore, the development of well-qualified agricultural scientists is a slow and expensive process and the number of well-trained scientists is low in many countries despite a considerable increase during recent years.[2] A number of other factors, such as poor research management, lack of well-specified research goals, lack of information, lack of an effective extension service, and so on, may contribute to low research productivity. Some of these factors and how they may be changed to increase research productivity will be discussed in a subsequent section. However, before discussing this, let us briefly look at the distribution of research benefits and costs.

DISTRIBUTION OF RESEARCH BENEFITS AND COSTS

Up to this point we have discussed the research costs and return as a whole. But how are costs and benefits distributed among groups in society? Who gains and who, if anyone, loses? Such knowledge of the distribution of research benefits is critical in helping to decide which groups should pay for the research.

The pattern of distribution of research benefits depends on whether the increased agricultural output replaces food imports, is exported, or is added to the domestic food supply. If a considerable proportion of the additional output is exported or used to replace food imports—the latter being the case for wheat and rice in India and Pakistan during the initial phases of the Green Revolution —farmers are likely to obtain relatively large benefits. If, on the other hand, the additional output is added to prevailing domestic supplies and prices are permitted to fall to a new market equilibrium, consumers will be the principal beneficiaries while the producer sector, as a whole, will lose. Early adopters (larger farmers?) may be worse off than before the research results (new technology) were introduced.

This helps to explain why farmers producing export crops are frequently willing to pay for research on the particular crop, e.g., sugarcane, coffee, and rubber, whereas research on domestically consumed agricultural products usually is financed by the consumer via taxes.

Although there are several other factors that influence the distribution of research benefits and costs, only two will be discussed here. First, which groups in society are most successful in communicating their research needs to the research community? This issue is extremely important and greatly influences research priorities and the resulting knowledge and technology. It has been argued that agribusiness has been very active in communicating its research needs to U.S. universities and experiment stations and, as a result, an unduly large proportion of govenment-funded research effort has been focused on solving the problems of agribusiness rather than those of the farmers who were less capable of communicating their research needs. Furthermore, large farmers generally have easier access to the agricultural research establishment than small farmers. Hence. a disproportional amount of agricultural research may be focused on solving the problems specific to agribusiness and larger farmers. How important this aspect is depends largely on the extent to which new technology is, in fact, biased toward certain farm sizes.

Second, private institutions and individuals are frequently unable to capture the economic benefits from their research activities because the release of research results and related technology cannot be controlled through patents or other arrangements aimed at maintaining exclusive rights. Therefore, although certain research may be highly beneficial to society as a whole or to groups of society such as consumers or farmers, public funds are needed to assure that the research is put into practice. In cases where major economic benefits of research can be captured by private institutions, public funds may not be needed. Thus, 20% to 25% of all agricultural research expenditures in North America originates in private industry. In contrast, only 2% to 5% of agricultural research expenditures in Asia, Africa, and Latin America originate in the private industrial sector (Boyce and Evenson, 1975). Similarly, benefits from publicly funded agricultural research may not be limited to any one country. In such cases, international research institutions may have comparative advantages.

Many of the studies mentioned in table 4 looked at the distribution of research benefits between the producer and consumer sector. In most of these studies the majority of the benefits were obtained by the consumers. Among producers, those who adopt research findings first—in most cases larger farmers—obtain considerable gains. Among consumers, on the other hand, lower income consumers would tend to obtain larger gains from research on basic staples than would higher income consumers. The opposite would be true for higher-cost foods such as meats (Pinstrup-Andersen, 1977).

INCREASING THE RETURNS FROM AGRICULTURAL RESEARCH

Evaluations of past research such as those discussed in the previous section offer certain guidelines for assuring high returns of future research activities. However, the primary contribution of these evaluation studies is in their illustration of the extremely high potential returns to investment in agricultural research, and the resulting impact on governments to allocate more funds to research. But how can we assure that potentially high returns do in fact materialize in future research? In my opinion, four factors are critical: (1) the quality and quantity of research resources with particular reference to the human resources, (2) research organization and management, (3) availability of information regarding potential payoff from alternative research efforts, and (4) the degree of communication between the research group and the final users of the research output, e.g., farmers and agribusiness. These factors are briefly discussed below.

RESEARCH RESOURCES ORGANIZATION AND MANAGEMENT

Well-trained and highly motivated scientists are essential to any sucessful research program. The organization and management must assure that the research efforts are focused on the particular research goals. Ideally, the goals of the research program or institution will coincide with the goals of the individual scientist. To accomplish this, research goals must be clearly specified and the scientists must fully understand how these goals are to be reached and appreciate the contribution of each of the scientists in a particular research program. Therefore, although the overall goals may be partly or fully specified by persons outside the research group, e.g., funding agencies, the scientists must be actively involved in establishing the working objectives and the approach needed to achieve these objectives. A clear specification of the research goals is essential to channel the research efforts toward the desired problems. Unless we specify the desired destination, only by chance will we get there. Although this seems obvious, it is amazing how many research programs do not, in fact, have clearly specified research objectives. Furthermore, goals and means to achieve goals are frequently confused. Finally, the ultimate economic and social goals of research are frequently not translated into working objectives needed by the researchers.

Figure 4 is an outline of the process by which applied agricultural

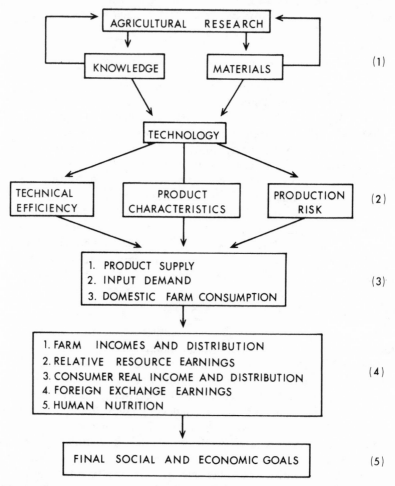

Fig. 4.   Illustration of potential outcomes and implications of agricultural research.

research may contribute to the achievement of social and economic goals to help clarify the distinction between ultimate and immediate goals, on the one hand, and means to reach these goals, on the other. Successful agricultural research produces knowledge and/or improved material, e.g., seed. Such knowledge and improved material may be fed back into the research process for further work, or it may be released to the farmers as new technology. There are three—and only three—potential direct contributions of new technology: (1) increasing technical efficiency, a measure of output per unit of input where both output and input are expressed in physical terms (e.g., production per hectare) of at least one resource; (2) changing the characteristics and composition of products and developing new products (e.g., developing plant types more suited to mechanization and improving the amino acid composition in the protein of a given crop); and (3) reducing production risk. Any other contribution will be indirect; that is, it will come about as a consequence of one or more of the three direct contributions.

There are three potential results of the direct contributions listed above: (1) changing the composition and quantity of the aggregate supply of food, feed, and fiber; (2) changing the composition and quantity of the aggregate resource demand, e.g., increased or decreased employments; and (3) changing the composition and quantity of aggregate domestic farm consumption. Any of these results may contribute to the achievement of national development goals through changes in farm income and its distribution among groups of farmers, relative resource earnings, consumer real income and its distribution among consumer groups, foreign exchange earnings, and human nutrition.

Viewing agricultural research and its potential outcomes and implications as a process reduces the confusion over means and ends. The first level of outcomes (marked by [1] in fig. 4) is clearly a set of means, except when research is carried out for its own sake. The second level represents the working objectives for the agricultural scientist. For research management and for society as a whole, this level expresses alternative apporaches to the goals shown in the fourth level. The third level in figure 4 represents the vehicle by which activities meeting the scientist's working objectives influence the achievement of the final goals. In other words, changes in product supply, input demand, and domestic farm consumption are not themselves goals but are means to reach some final goals.

Two conclusions may be drawn from the discussion above. First, the working objectives for the agricultural scientist must be expressed in terms of technical efficiency, desired product characteristics, and/or production risk. The specific working objectives and the most effective technology to reach these objectives should be determined on the basis of national development goals. Concurrence between the technology specification received and/or established by the scientist and the technology that results in maximum contribution to the achievement of social goals is the joint responsibility of research management and researchers.

Second, research management needs information for resource allocation that is capable of both translating national development goals into working objectives for the agricultural scientist and helping the scientist select the most effective technology to reach the working objectives. The type of information needed and the means to obtain such information will be discussed in a subsequent section.

Having briefly discussed the importance of highly qualified research personnel and a clearly specified set of research goals, let us move to organizational aspects. Several books and articles have been written on efficient research organization, and it would be unrealistic to attempt an exhaustive treatment of this topic here. However, a few points of major relevance to the economics of research should be mentioned. First, research must be purposed-oriented. The working objectives must be specified in such a way that progress toward their achievement can be measured. Furthermore, these working objectives must reflect social and economic goals of the society paying for the research. Research for its own sake is a luxury that few taxpayers would be willing to underwrite. Second, the required research input from the various disciplines must be specified. Although some research, particularly the more basic, may require input from only one discipline, the solution of most agricultural production problems requires a multidisciplinary research input.

Optimal research organization would differ according to the various types of research problems to be solved. Two alternative approaches are discussed below. Both approaches allow for interaction among scientists within individual disciplines and across disciplines. In the first approach the organization is based on one or more multidisciplinary research teams aiming at the solution or one or more specific problems. Administrative responsibility is placed with the team leader and intradisciplinary interaction across teams is informal. Commodity-specific research teams are examples of such an organization. The disciplinary composition of the team depends on the specific problems to be solved for the commodity. The plant breeder usually plays a key role in crop-oriented research teams. It is his responsibility to integrate the desired characteristics into the plant type such as disease or insect resistance, high genetic yield potential, and so on. The breeder receives support for this work of the scientists from the relevant disciplines such as plant pathology, entomology, physiology, and agronomy. The economist should play a dual role in the team: (1) assisting the team in estimating the potential payoffs from alternative research efforts before these efforts are initiated and (2) assisting in estimating the economic viability and potential user acceptability of research results before they are released for commercial application. Both of these roles require the establishment of close linkages between potential users of research results, e.g., farmers, and the research team through the identification of the demand for research output at the user level. This will be further discussed below.

The input-specific research team is another example of multidisciplinary research organization. Again, the composition of the team is determined by the particular input and the problems to be solved. In the case of a chemical input such as chemical fertilizers, the chemical engineer should play a key role if the principal issue is one of developing new fertilizer materials. If, on the other hand, the problem is one of fertilizer management and suboptimal usage at the farm level, the soil scientist and the economist would play the principal role.

Up to this point we have discussed only the multidisciplinary team approach. The second approach is based on disciplinary administrative units, e.g., plant pathology, agronomy, and economics. Research problems of a multidisciplinary nature are approached either by assigning personnel from the disciplinary units to multidisciplinary teams for shorter time periods or by assigning the various parts of the research problem to the appropriate disciplinary units. The international agricultural research institutes with responsibility for more than two commodities were initially organized along disciplinary lines. However, most of them have shifted to a commodity team focus. Such a shift has been occurring only to a very limited extent in national research institutes.

The multidisciplinary team approach is probably the most effective organizational structure in applied agricultural research under most circumstances. However, to be successful, a reward system must be established in such a way that the individual scientist receives his reward from achieving the specific goals established by the team rather than through some discipline-oriented reward system. Furthermore, the team leader must be capable of assuring an integrated research effort without reducing the initiative and motivation of the individual scientist. As opposed to certain other productive processes, successful research is not likely to be accomplished under a rigid pyramid structure in which orders are passed down from above. Successful research depends, more than anything, on the capacity, incentive, and motivation of the individual scientist.

INFORMATION AND COMMUNICATION

In order to establish sound research priorities, information is needed on expected benefits, costs, and time requirements for each of the lines of research considered. Ideally, one would have complete knowledge of research outcomes and their expected contribution to the achievement of established goals as well as the costs and time requirement for each line of research. If this were the case, and assuming a well-defined goal, available research resources could always be allocated in such a way as to maximize the contribution to the achievement of this goal, and no subjective judgment would be needed in the decision-making process.

However, because of the very nature of research, decisions on research priorities will always be subject to uncertainty. But this inherent uncertainty of research outcomes is frequently accompanied by lack of information on issues relevant to the decisions that can in fact be estimated with some certainty. These issues include relative resource scarcity, future output demand, as well as potential production and productivity gains from alternative research outcomes, and their impact on employment, nutrition, and farm revenues. It is argued here that additional information on these and related issues is likely to be highly useful in establishing research priorities.

An effective information system for the allocation of resources in agricultural research must be capable of providing researchers and research management with reliable data that will make possible the establishment and the periodic review of research priorities in such a way as to maximize the expected contribution from research to the achievement of national development goals. The system should also provide a frame of reference within which project priorities can be established and individual projects can be accepted or rejected without great time delays. Extreme care must be taken to avoid a system that imposes heavy bureaucratic procedures on the scientists.

An overall approach for providing the required information for a commodity-oriented research team is suggested elsewhere (Pinstrup-Andersen and Franklin, 1977) and will not be repeated here.

Generally, an effective analytical framework for research resource allocation will include the following activities:

1. Specification of the final goals toward which the research activities should focus.
2. Identification of principal researchable problems believed to significantly limit the achievement of these goals.
3. Estimation of the potential contribution to the achievement of the goals if each of the above problems was eliminated.
4. Determination of the required type of research results (technology or knowledge) to eliminate each of these problems.

5. Estimation of time requirements, costs, and probability of research to produce such results.
6. Estimation of expected contribution to the achievement of the final goals by each research alternative.
7. Consideration of other criteria for choice among research alternatives, e.g., comparative advantage.
8. Specification of immediate research objectives and desired research output.

Quantitative estimates with acceptable error margins would preferably be obtained for items 3, 5, and 6 above, if available at reasonable costs. Acceptable quantitative estimates can sometimes be obtained on item 3 at the local and national level. Items 5 and 6 are much more difficult to quantify because of the uncertainty associated with estimating the probability of success in research and the rate of adoption of research results.

Priorities in applied agricultural research are frequently established on the basis of very limited information about existing problems and their relative economic importance in the production process. The communication between the farm sector and the research institute is often deficient and the demands at the farm level for problem solving research may not be well known by researchers. This is of particular importance in developing countries. Because of lack of such communication, some research in these countries may be irrelevant to the actual farm problems and, thus, research results may not be adopted. On the other hand, large farmers, farm organizations, and agribusiness tend to have easier access to the research institute for the purpose of expressing what they see as the research needs. As a consequence, agricultural research may be focused on the needs of these groups rather than those of smaller farmers (Hightower, 1973).

Low rates of adoption of new technology among small farmers are frequently explained as a result of an ineffective extension service. Whereas the ability of the extension service to assure adoption of new technology may indeed be deficient, one of the primary reasons for low adoption rates may well be that available technology does not meet the most urgent on-farm needs and farmer preferences. A continuous flow of information to the research institute on the potential gains in production, productivity, and risk obtainable from such research activities as (1) developing resistance to existing diseases and insects, (2) changing cultural practices, (3) changing plant types, (4) changing plant response to nutrients, and so on, as well as information on the farmers' preferences with respect to new technology, is likely to be useful to assure that new techology corresponds with the farmers' needs and preferences, hence accelerate adoption and increase research payoff.

Such an information flow may consist of a continuous feedback of information from the farmer through the extension service to the research institutions. Direct contact between researchers and farmers through meetings, farm visits, and such, is another effective vehicle for such information. To complement these, it would be useful to estimate the magnitude of the yield and production losses by each of the principal yield constraints. Such estimates would provide guidelines for the potential returns to research aimed at removing any of these constraints. A procedure to accomplish this is suggested below.

The procedure consists of (1) data collection among a predetermined representative panel of farms, (2) estimation of yield and production losses caused by each factor, and (3) estimation of the market value of these production losses.

Data are needed on the occurrence and severity of each of the factors expected to limit yields in the crop for which the analysis is carried out. These factors vary according to crop and region. In general, the factors include crop diseases, insects, weeds, soil characteristics and fertility, cultural practices, rainfall and water management, plant type, cropping system and use of fertilizers, seed, insecticides, labor, and other inputs.

Data are collected from a representative panel of farms by a trained field team of agronomists and economists. The field team makes periodic visits to each farm throughout a complete growing season. Most data are obtained from direct observation and measurement in the farmer's field, supplemented by subsequent laboratory analyses of soil samples and plant or insect samples for which no immediate diagnosis can be established in the field. Field observations are further supplemented by farmer interviews regarding issues not identifiable in the field, such as input use.

Yield losses are estimated on the basis of a production function analysis, in which observed yields are regressed on the factors expected to influence yields. The area affected by each of the factors is estimated directly from the sample data, the production losses are then estimated as average yield losses multiplied by the area affected.

In order to estimate the market value of the losses caused by each yield-limiting factor, an estimate is made of expected price change due to the production increase that would occur if the factor were removed. This price change is estimated on the basis of existing price elasticities of demand and supply for the particular crop.

The procedure was applied to beans and cassava in a number of regions of Colombia, South America. Figure 5 contains the estimated impact of each of eight factors limiting bean yields in the Cauca Valley region (Pinstrup-Andersen et al., 1976). The eight factors identified as having a

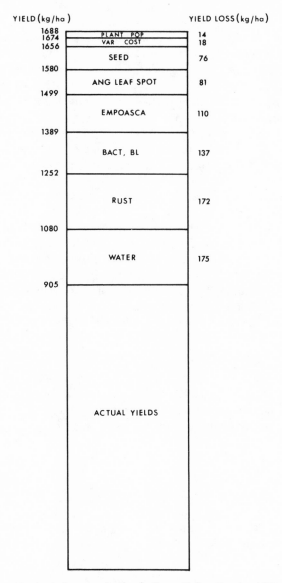

Fig. 5. Loss estimates in bean production in Cauca Valley, October 1974–January 1975.

significant yield-reducing effect were: adverse rainfall, rust, bacterial blight, empoasca, angular leaf spot, use of noncertified seed, suboptimal input use, and suboptimal plant population. The potential payoff from research that would remove any one of these factors was then estimated as the value of the current production loss (the value of the average yield loss multiplied by the total bean acreage). Although such estimates are likely to vary among regions and growing seasons, the procedure provides a first step to obtain guidelines for potential yield increases from developing varieties resistant to selected yield-limiting factors. If, for example, rust resistance could be built into varieties currently used in the region, and all farmers adopted the variety, yields would be expected to increase by 16%. Such a yield increase would add US $1.2 million to the value of the crop at constant prices. Resistance to angular leaf spot, on the other hand, would increase yields by only 8%, and the value of production would increase by about $550,000. Efforts needed to remove yield-limiting factors entail a cost. Hence, the value of each potential production increase less the costs associated with bringing about the production increase provides an estimate of potential net benefits from research and development efforts aimed at removing that specific yield-limiting factor.

The actual net benefits are determined from potential net benefits as well as the probability of developing an appropriate technology to remove a particular yield-limiting factor and the rate of adoption of such technology among farmers. The allocation of research and development funds among these factors is then determined on the basis of relative actual net benefits.

ACKNOWLEDGMENTS

Acknowledgment is due Dr. Eric Craswell, soil scientist; Dr. Mohinder S. Mudahar, economist; and Mr. Chris Dowswell, communications specialist, International Fertilizer Development Center, for valuable comments on an earlier draft of this paper.

1. Hertford and Schmitz (1977) provide a brief discussion of the methodology used. For further details of these methodologies, see the individual reports as listed in table 4.

2. According to data from Boyce and Evenson (1975), the number of agricultural scientists, measured in terms of man-years, in Asia, Africa, and Latin America for 1974 was 3.3 times the number for 1959.

REFERENCES

Akino, M., and Y. Hayami. 1975. Efficiency and equity in public research: Rice breeding in Japan's economic development. Amer. J. Agric. Econ. 57:1–10.

Ardito-Barleta, N. 1971. Costs and social returns of agricultural research in Mexico. Ph.D. dissertation, University of Chicago.

Arndt, T., D. Dalrymple, and V. Ruttan (eds.). 1977. Resource allocation and productivity in national and international agricultural research. University of Minnesota Press, Minneapolis.

Ayer, H. 1970. The costs, returns, and effects of agricultural research in developing countries: The case of cotton seed research in São Paulo, Brazil, Ph.D. dissertation, Purdue University.

Barker, R., and Y. Hayami. 1976. Price support versus input subsidy for food self-sufficiency in developing countries. Amer. J. Agric. Econ. 58.

Boyce, J. K., and R. E. Evenson. 1975. Agricultural research and extension programs. Agricultural Development Council, New York.

De Janvry, A. 1972. Optimal levels of fertilization under risk: The potential for corn and wheat fertilization under alternative price policies in Argentina. Amer. J. Agric. Econ. 54.

Duncan, R. C. 1972. Evaluating returns to research in pasture improvement. Australian J. Agric. Econ. 16.

Evenson, R. 1969a. International transmission of technology in sugarcane production, Mimeographed, Yale University.

Evenson, R. 1969b. The contribution of agricultural research and extension to agricultural production. Ph.D. dissertation, University of Chicago.

Evenson, R. 1976. Notes on agricultural research in the developing nations. Mimeographed, University of the Philippines, Los Banos.

Evenson, R. 1977. Comparative evidence on returns to investment in national and international research institutions. *In* T. Arndt, D. Dalrymple, and V. Ruttan (eds.), Resource allocation and productivity in national and international agricultural research. Pp. 237–64. University of Minnesota Press, Minneapolis.

Evenson, R., and Y. Kislev. 1973. Agricultural research and productivity an international analysis. Paper presented at the 1973 conference of IAAE, São Paulo, Brazil.

Evenson, R. E., and Y. Kislev. 1975. Investment in agricultural research and extension: A survey of international data. Econ. Devel. and Cultural Change 23:507–21.

Fishel, W. (ed). 1971. Resource allocation in agricultural research. University of Minnesota Press, Minneapolis.

Griliches, A. 1958. Research costs and social returns: Hybrid corn and related innovations. J. Pol. Econ. 66:419–31.

Griliches, A. 1964. Research expenditures, education, and the aggregate agricultural production function. Amer. Econ. Rev.

Herford, R., J. Ardila, A. Roches, and C. Trujillo. 1977. Productivity of agricultural research in Colombia. *In* T. Arndt, D. Dalrymple, and V. Ruttan (eds.), Resource allocation and productivity in national and international agricultural research. University of Minnesota Press, Minneapolis.

Hertford, R., and A. Schmitz. 1977. Measuring economic returns to agricultural research. *In* T. Arndt, D. Dalrymple, and V. Ruttan (eds.), Resource allocation and productivity in national and international agricultural research. Pp. 148–67. University of Minnesota Press, Minneapolis.

Hightower, J. 1973. Hard tomatoes, hard times. Schenkman Publishing Co., Cambridge, Mass.

Hines, J. 1972. The utilization of research for development: Two case studies in rural modernization and agriculture in Peru. Ph.D. dissertation, Princeton University.

Kahlon, A. S., H. K. Bal, P. N. Saxena, and D. Jha. 1977. Productivity of agricultural research in India. *In* T. Arndt, D. Dalrymple, and V. Ruttan (eds.), Resource allocation and productivity in national and international agricultural research. University of Minnesota Press, Minneapolis.

Kislev, Y. 1977. The economics of agricultural research—some recent developments. Food Policy, pp. 148–56.

Monteiro, A. 1975. Analiação econômica da pesquisa e extensão agrícola: O caso do cacua no Brasil. M.S. thesis, Universidade Federal de Viçosa, Brazil.

Peterson, W. L. 1966. Returns to poultry research in the U.S. Ph.D. dissertation, University of Chicago.

Peterson, W. L. 1975. The social cost of a cheap food policy: The case of Argentine corn production. Staff paper P75-28, Dept. of Agric. Econ., University of Minnesota.

Peterson, W. L., and J. C. Fitzharris. 1977. Productivity of agricultural research in the United States. *In* T. Arndt, D. Dalrymple, and V. Ruttan (eds.). Resource allocation and productivity in national and international agricultural research. University of Minnesota Press, Minneapolis.

Pinstrup-Andersen, P. 1977. Decision making on food and agricultural research policy: The distribution of benefits from new agricultural technology among consumer income strata. Agric. Admin., no. 4.

Pinstrup-Andersen, P., and F. C. Byrnes (eds). 1975. Methods for allocating resources in applied agricultural research in Latin America. CIAT, Series CE-11.

Pinstrup-Andersen, P., and D. Franklin. 1977. A systems approach to agricultural research resource allocation in developing countries. *In* T. Arndt, D. Dalrymple, and V. Ruttan (eds.), Resource allocation and productivity in national and international agricultural research. University of Minnesota Press, Minneapolis.

Pinstrup-Andersen P., N. de London and M. Infante. 1976. A suggested procedure for estimating yield and production losses in crops. PANS 22(3):359–65.

Scobie, G., and R. Posada. 1976. The impact and political economy of technological change in agriculture: The case of rice in Colombia. Paper presented at IFPRI seminar, Washington, D.C.

Tang, A. 1963. Research and education in Japanese agricultural development. Econ. Stud. Quar.

JORGEN M. BIRKELAND

The Ecology of Science

7

This title, "The Ecology of Science," was chosen for two reasons. First, the term *ecology* is a sort of trash bag into which you can stuff almost anything. Second, the term has taken on a charisma that implies broad perspective and deep insight. Other essays in this volume deal more directly with specific themes. I want to discuss the forces that influence the rate and the direction of the growth of science, and I want to mention problems solved and problems created as a result of solutions.

There is often a tendency to equate science with the scientific method and its tools and techniques. This tendency leads to the mistaken assumption that we can transplant science into the underdeveloped areas by shipping electron microscopes, Geiger counters, supercentrifuges, and such, with, of course, appropriate sheets of instruction in the language of the country. We cannot export science by exporting its tools and techniques. We can set up the finest biological laboratory in an undeveloped area, but it will be of little value unless there are people interested and capable of active participation in scientific work and a cooperative community ready to appreciate and apply their findings.

The deeper we probe into any field, the more refined and precise are the measurements involved, and, incidentally, the greater the cost of research. There was a time when distance could be measured by a "whoop and a holler," length by a foot, and quantity by a handful. The conquest of space or the description of a cell demands precision of measurement beyond the

Jorgen M. Birkeland, Department of Microbiology, Ohio State University, Columbus, Ohio 43210.

imagination of scientists who worked two decades ago. And speaking of precision in instruments, although we do not always think of it in that way, language is the most important tool of science and should be, and usually is, used with precision. Unfortunately, of late, cultists engaged in the propagation of fads have been using terms that mislead or convey half-truths. *Organic, natural,* and *food* have well-accepted meanings. To talk of *organic food* or *natural food* is nonsense. *Organic gardening* apparently implies that, from a nutritional standpoint, there is something mystical or vastly superior in the amino acid or protein from nitrogen in horse manure to that in chemical fertilizers. Is there any evidence that the plant can identify its source of nitrate, oxygen, water, carbon dioxide, or any other chemical? We have enough problems in providing adequate nutrition for an ever-expanding population without deliberately confusing the issue by prostituting our language.

Scientific techniques enable us to get more and more precise answers to certain types of questions. They do not tell us what questions to ask, and this is a matter of crucial and fundamental importance. The answer science gets depends upon the nature of the question asked, and that depends upon our notions of the nature of the universe and of our place in it. These notions or attitudes are influenced by the home, the church, the school, the community, the news media, in fact, by everything in our environment. All affect our outlook and determine not only what we see but the inferences we draw. Our concept of the universe determines what is of greatest consequence and what is of least, and may also suggest how we can cope with our problems. These attitudes are the ecological environment in which science does or does not thrive.

Primitive man lived in a world where he believed the gods responsible for health and the demons for disease. He cajoled and placated the one; he wore charms and fetishes against the other. He reacted logically according to his ideas of the forces governing the world.

The Greek concept of health and disease was based on the balance of the four humors and led quite logically to the practice of bloodletting or phlebotomy.

In the Middle Ages men were preoccupied with problems of salvation and sin. Plagues and pestilences were explained as instances of divine wrath. People had sinned. Whatever happened to them served them right. At times religious man tried to escape his just retribution by showing genuine repentance and flogging or flegellating himself.

Unfortunately, those periods when interest centered almost exclusively on spiritual values were all too often times of distress, poverty, and disease. Too great a preoccupation with life in the afterworld did not provide a favorable mood for developing the knowledge necessary to solve problems of food, shelter, or disease in this life. Action based solely on spiritual values has often been tragically ineffective in dealing with primary problems of survival.

On the other hand, man cannot live by bread alone. As we see evidenced today, too much disregard for moral and spiritual values may lead to a degeneration and disintegration of society that may threaten the very survival of civilization. No one will deny the importance of religion as a cohesive and stabilizing force, and no one would deny that a profound religious faith gives a man more courage and sustains him in times of crisis when all else has failed. Nor could I argue the truth or falsity of any religion. I just want to point out that a people's religion or philosophic belief determines its priorities and has a profound effect on the development of science.

If a man feels that his problems are largely his responsibility and assumes, as most of us do, that natural forces can be understood and explained in physical, chemical, and biological terms, he will value empirical knowledge and rational attitudes and act accordingly. Science is based on empirical knowledge and is limited to things that can be weighed and measured. Most of the progress made in the control of disease and hunger and drudgery has been based on science and technology, and most of it has been made by the Western world.

The Eastern world has traditionally been passive and submissive. Its philosophers encourage the individual to accept his lot, to look inward and attune himself to the universe. Meditation and indifference to material things are outstanding features of his life-style. Where men believe that the test of a man is the things he can do without, it is hardly likely that they will be interested in developing technology with its material benefits. If a man believes that the cow is sacred, it is unlikely that he will contribute to the science of bovine genetics. If he feels that infectious disease is an expression of divine wrath, he will not seek its cure in a laboratory. If he feels that putting his money into temples and statuary will ensure him permanent salvation, he is not going to risk his chances by funding scientific laboratories.

The greatest difficulty we encounter in our relations with each other is in understanding and appreciating attitudes. Meditation and indifference to material things may be valuable in themselves, but they do not contribute to the growth of science.

We are committed to foreign aid. We feel a responsibility to help other countries develop their science so that they may enjoy its benefits. But, while people in most areas of the world seem eager to share in the material goods, the medical knowledge, and the engineering achievements of our technology, they are less eager or find it impossible to adopt the ideas, the attitudes, and the presuppositions that made these achievements possible.

If we are to project the biological research of today to the needs of tomorrow in countries where the needs are most acute, perhaps the greatest obstacle will involve the kinds of questions that come to the minds of poeples whose traditional concept of the universe is so different from ours. It takes time to break with inbred patterns of thought.

The West is dynamic and aggressive. The desire to investigate, to satisfy curiosity, to find answers to certain types of questions, simply cannot be ignored. Such a temper or mood is conducive to scientific inquiry and to its support.

Many of us have had brilliant and capable students from underdeveloped countries and know how frustrated and discouraged they become if, when they return home, they find few opportunities to do what they have trained for because cultural conditions in their countries offer little support for a career in science. It is not that there are deliberate attempts to curb scientific inquiry, but, rather, that long-held beliefs as to what is and what is not important have established different priorities.

History shows that nations have, from time to time, spent their major energies on projects that have had little bearing on the real welfare of the people. The pharaohs built pyramids. The church of the Middle Ages built magnificent cathedrals. Prestige and rivalry are powerful forces in determining how countries spend their wealth. Today the two most powerful nations in the world are competing to see which can build the most effective weapons of destruction and which can be first to conquer space. I cite these two examples because they illustrate so beautifully how national prestige and rivalry may influence the direction and growth of science and technology, not because they illustrate so tragically how two powerful nations can become preoccupied with elements that have so little bearing on the critical problems of the survival of our civilization.

To summarize, science depends upon a method, with its tools and techniques. It depends upon our attitude, our concept of the universe. It depends upon a certain inquisitive mood that drives us to seek answers to questions. These are the factors that have influenced our successes in the control of disease and the development of agriculture.

Because of early and dramatic successes in controlling certain diseases, in improving methods of agriculture, in breeding higher-yielding grains and plants, the scientist developed an easy confidence that amounted at times to smugness in the possibilities of his method. As success followed success, it seemed that all important questions of concern could be answered through science and all of our needs and wants satisfied through technology—that man, indeed, could be master of nature. "Know the truth and the truth shall set you free" was engraved over the archway of many a path to learning, and it seemed that the truth was coming within reach.

This confidence provided an easy optimism that has had a profound effect upon our daily lives and upon our national and foreign policies. We "dreamed the impossible dream." Recently it has taken on some of the aspects of a nightmare. The easy optimism has largely disappeared. Often some of the scientific achievements are viewed with fear and their potentials with alarm. We have become aware that many of today's problems are the result of a solution to earlier ones. Take, for instance, the difficult problems of population and pollution.

Pollution is part of the price we pay for our inventiveness, our ingenuity, and our creativity. It is because of the long-term consequences of having improved our food supply, of having controlled many infectious diseases, and of having allowed ourselves so many of the luxuries of life that we are now in danger of destroying the environment on which we depend. We are criticized for upsetting the balance of nature.

But is this anything new? Man's success in dealing with his problems of survival is the result of his upsetting the balance of nature. What is agriculture if not a deliberate attempt to do this by the application of technology. It might be mentioned here that the invention of the horse collar and barbed wire must be listed along with more recent developments in insecticides, pesticides, and herbicides, with this difference: nature would quickly return to her original state if we removed the fence and plow; the ecological effects of the chemicals are more lasting. It is hard to accept the fact that the consequences of our actions are not determined by our motives.

What is medicine, both preventive and curative, if not a system for thwarting or cheating the whole ecological system? Infectious disease, a special case of parasitism, is an important part of the ecosystem. Like agriculture, disease control may be considered as a highly sophisticated method for upsetting the ecological balance.

Some of the pollution problems can be solved easily by our present technology. Many of our manufactured products are designed for dynamic obsolescence, as, for example, the millions of cars junked each year. All sorts of metal scrap can be salvaged from them and recycled without presenting any really difficult engineering problems. But it will involve some expense. We can scrub the exhaust from factory smokestacks to remove a number of air pollutants. Again, it is expensive. We can process sewage and run it off for irrigation and fertilization, putting back into the land what we have taken from it. We can use the processed sewage, too, as a culture medium for algae, yeasts, and bacteria, potential sources of protein.

Other types of pollution cannot be dealt with so easily. Radiation is one; nondegradable chemicals and plastics are others. These should be kept from entering the ecosystem.

Of even greater concern is nuclear pollution. The question here is how our ever-increasing energy requirements can be met without disrupting our biomass with disastrous results.

When we come to the population problem, we find that a technological fix is not so easy to apply. When we control disease every human instinct in on our side. Everyone wants to live longer and thinks everyone else wants to live longer too. Everyone wants to be free of disease. We all love children in the family, the government needs more individuals to pay taxes, the army needs more men to fight. We are all for reducing the death rate.

When we try to reduce the birth rate, we are fighting every human instinct: the joy of having children, the pride of seeing ourselves reborn, the pleasure of belonging to a large group, the satisfaction of having younger ones to care for you in your declining years.

Most of these attitudes—the attitude in favor of disease control, or, that is, death control, and the attitude opposed to birth control—are influenced by nonrational forces, by strong emotional commitments. We need a rational approach. The fact is that a population doubles every forty years or so; the four billion people inhabiting this globe today will thus have grown to about eight billion in another forty years. We know that; but, as Shelley says, man is peculiar in that he finds it so difficult to imagine what he knows, which is why we are unlikely to be able to convince ourselves that it is urgent to act now.

Our choices should, in fact must, be made in clear recognition of the vast number of conflicting needs and desires. Even if we were omniscient and had complete power, choices would not be easy. You may recall the Lawd in *Green Pastures* saying, "This being God ain't no bed of roses." Neither is a responsible man in a bed of roses. Our options are rarely presented in a language loud and clear. Ability to foresee the long-range consequences is impossible. We cannot know enough about the ultimate effect of our actions. We really cannot write a total impact statement, although we are constantly being asked for one. People want to enjoy the benefits of research, but they also want to be protected from the risks.

Science depends upon freedom of inquiry, a freedom that is always relative. When it comes into conflict with dominant dogmas or beliefs, all kinds of pressures—economic, social, and political—are exerted by the state, by religious institutions, by society.

There is always a question as to whether the scientist or the concerned citizen is the better qualified to judge what types of research should be pursued. Often a question of ethical and moral values is involved, sometimes a question of safety. This would seem to be the crux of the hassle over recombinant DNA research, over abortion and birth control, over the construction of nuclear power plants, over investigations into the nature of intelligence, and many other. How much government funding and how much freedom should these "mad" scientists be allowed?

During my time I have stood with "mankind at the crossroads," lived through "a time for decision" and "an era of rising expectations," and moved into "an age of uncertainty." Never in the history of the world have so many and such a large percentage of its people gone to bed safely at night, awakened to go out in the morning, and returned safely home in the evening. Never have so many been free of the threat of starvation or disease. Yet never has so large a proportion of the people been so restive, so impatient, so disillusioned, so apprehensive, and so troubled as today. The critical question is whether through science and technology we can find the answers to the complex problems that have arisen as a result of our solutions to simpler ones.

CARL G. LAMM

International Agricultural Research Programs of the FAO/IAEA

8

INTRODUCTION

It is my privilege and pleasure to discuss a subject of imminent importance to mankind, namely that of international agricultural research. In this paper I shall limit myself to the joint FAO/IAEA program as an example of international research and training in agriculture.

Let me first spend a few words on the uniqueness of this program, which is the joint responsibility of two United Nations organizations, the Food and Agriculture Organization (FAO) and the International Atomic Energy Agency (IAEA).

The terms of reference, philosophy, and outlook of these two organizations are somewhat different. The FAO is charged with raising levels of nutrition and standards of living, securing improvements in efficiency of production and distribution of food and agricultural products, and bettering the conditions of rural populations. Among its many functions the FAO collects, analyzes, interprets, and disseminates information relating to nutrition, food, and agriculture. Further, the FAO promotes and, where appropriate, recommends national and international action with respect to scientific, technological, social, and economic research relating to nutrition, food, and agriculture. The FAO is primarily concerned with the application of known information to agricultural development and food production, with the emphasis on practical results in the

Carl G. Lamm, Joint FAO/IAEA Division of Atomic Energy in Food and Agriculture, International Atomic Energy Agency, Vienna, Austria.

short term, and with the stimulation of needed research in national and other institutions.

By virtue of its statutes, the IAEA seeks to accelerate and enlarge the contribution of atomic energy to world peace, health, and prosperity, and as one of its functions the IAEA is authorized to encourage and assist research on, and development and practical application of, atomic energy for peaceful uses throughout the world.

THE JOINT FAO/IAEA DIVISION OF ATOMIC ENERGY
IN FOOD AND AGRICULTURE

With the intention of fusing the objectives of the two organizations into one consolidated joint program, and making use of the technical services of both organizations, the Joint FAO/IAEA Division of Atomic Energy in Food and Agriculture was founded in October 1964. This fusion was a logical approach to achieve effective cooperation and coordination in the area of applying nuclear techniques to agricultural sciences and thus represents a marriage between atomic energy and agricultural authorities at an international level.

The Joint FAO/IAEA Division is problem and research oriented and attaches high priority to solving practical problems related to agricultural production and crop, animal, and food protection in the developing world. Thus, the objectives of the Joint Division are to exploit the potential of nuclear techniques in research and development for increasing and stabilizing agricultural production, reducing production costs, improving food quality, protecting agricultural products from spoilage and losses, and minimizing pollution of food and the agricultural environment. It is clear that all activities of the Joint Division are done in the name of both parent organizations.

Since its formation, the Joint Division has been located at the IAEA headquarters in Vienna. It now comprises six sections: soils, irrigation, and crop production; plant breeding and genetics; animal production and health; insect and pest control; chemical residues and pollution; and food preservation. The Joint Division also has technical responsibility for the agricultural section of the IAEA Laboratory outside Vienna. Out of a total of twenty-seven scientists now working for the Joint Division including the laboratory, about two-thirds have fixed short-term contracts and one-third have longer-term contracts. This arrangement, I believe, is important to ensure flexibility and a dynamic approach without losing continuity. To complete the picture, I should add that we have been able to attract many eminent scientists, primarily from this country, to spend their sabbatical year in Vienna, thus injecting their expertise and experience into our joint program.

The Joint Division's budget is allocated from the regular funds of both the FAO and the IAEA. We presently also receive generous external fund allocations on a trust fund basis from the governments of Belgium, Federal Republic of Germany, Sweden, the United Kingdom, and the United States. To this overall picture should be added the fact that over 20 percent of the IAEA technical assistance projects are in agriculture and that, moreover, the Joint Division has technical responsibility for several United Nations Development Program projects.

APPLICATION OF NUCLEAR TECHNIQUES IN
DEVELOPING AGRICULTURE

Let me first point out that alleviating the present serious and world-wide problems of famine, malnutrition, and environmental pollution makes it imperative that all resources and efforts be mobilized. I shall also emphasize here that the isotope and radiation techniques are based on research carried out over a generation ago; their applications have therefore become as "conventional" as many other techniques being routinely used. Whether isotopes (stable or radioactive) are being used as tracers, or the interaction of ionizing radiation with matter is the desired effect, in a given research project the development and application of nuclear methods applied to food and agriculture have gained pace at a tremendous speed since an embryonic start some thirty years ago.

The use of isotopes and radiation in food and agriculture has resulted in faster solutions to a number of practical problems, allowed a more precise measurement of certain phenomena and resultant understanding, and in some cases provided the sole methodology for answering basic and applied problems. This statement is not based on theoretical speculations but on results of profitable applications with benefits to agricultural production in developed and developing countries alike.

It must be well understood that nuclear techniques are tools to be used complementary to, and in conjunction with, other available methods. Isotope and radiation methods can be of value only if they become fully integrated in the overall research efforts to solve specific problems. It is also important to keep in mind that agricultural problems depend heavily on local conditions like climate and soil, and that their solution must be found under the prevailing socioeconomic and ecological conditions.

In conclusion, I therefore wish to stress that as a tool isotope and radiation techniques are neither esoteric nor so sophisticated that developing countries cannot apply them, and they should apply them whenever it is the most economic, the fastest, or the only way to solve a specific problem. Needless to say, however, safe and effective isotope and radiation applications do require specialized experience and facilities, as well as a proper waste management and radiological protection service. Therefore national atomic energy commissions usually join forces with their agricultural authorities in cooperative projects.

THE RESEARCH PRIORITIES OF THE JOINT FAO/IAEA PROGRAM

It is within the above frame that the Joint Division seeks to carry out the duties entrusted to it. We in the Joint Division think of research in terms of developing methods and techniques of immediate and practical applicability to our member states. Our basic aim is that of strengthening the capabilities of national institutes where the actual work is being done; presently over two hundred agricultural institutions around the world cooperate in our programs. But we also interpret the word *research* in a broader sense to include training and other activities designed to facilitate and accelerate an effective and widespread use of improved technology.

The Joint Division actively cooperates with other international, regional, and national organizations. Its objectives and priorities follow from the objectives of the parent organizations. Among the present relevant main priorities of the FAO are the reduction of post harvest losses and control of African Trypanosomiasis, and the Joint Division is actively engaged in both of these areas.

The United States National Academy of Science has just published a report entitled "World Food and Nutrition Study: The Potential Contributions of Research." It represents an excellent assessment of the role of research, it discusses the research spectrum, it emphasizes the importance of more basic research, it attempts to evaluate some gains of research, and it identifies a number of high-priority research areas, all in response to a request from President Ford for recommendations on how U.S. research and development capabilities can best be mobilized to help all countries produce more food and combat malnutrition. I am very happy to note that among the twenty-two selected priority research areas, of which some fifteen fall within the categories of nutrition and food production, the Joint Division is actively engaged in nine (namely, plant breeding and genetic manipulation, biological nitrogen fixation, resistance to environmental stress, pest management, management of tropical soils, irrigation and water management, fertilizer sources, ruminant livestock, and postharvest losses).

THE ACTIVITIES OF THE JOINT FAO/IAEA DIVISION

The activities of the Joint Division can be grouped under three main items: coordination and support of research, technical assistance, and dissemination of information.

Coordination and Support of Research

This activity is furthered through coordinated research programs, where scientists from the developing and developed parts of the world are brought together to cooperate in the solution of a practical agricultural problem relevant to developing countries by means of isotope and radiation techniques. The objectives of each program are initially outlined by specifically convened panels of experts. Competent and interested research institutes are then contacted and invited to take part in the program, which normally lasts for a maximum of five years.

Institutes in developing countries are normally given research contracts with only nominal financial support, ranging from $1,000 to $5,000 annually. Institutes in the developed countries normally participated on a cost-free basis under research agreements. The participants in such a program meet periodically to review the results achieved and to discuss and decide on the future approach.

This mechanism of coordination ensures a high degree of versatility, and in addition to being problem oriented, each program is by nature educative and bears an incentive for further training. The agricultural section of our laboratory supports these programs by training, carrying out irradiation, analyzing samples sent in from the participants (isotopes, mass screening, and so on) distributing isotope-labeled compounds (fertilizers), and devising or testing specific techniques such as mass rearing of insects. However, I wish to point out again that the main bulk of work is done by the participants in their own laboratories and fields.

The Joint Division currently has technical responsibility for twenty-four such coordinated programs, which include over 250 cooperators. Because of the relatively short lifetime of these programs, the objectives can be varied rapidly in response to the changing demands of agriculture in the developing world. Very often, and by necessity, these programs are of a regional nature.

At present, the following coordinated programs are operative or in planning.

In the field of *soil fertility, irrigation, and crop production*, there are three programs dealing with fertilizer and water use efficiency studies on grain legumes, micronutrient studies in rice production with special reference to zinc deficiency, and studies of soil-water regimes.

In the first of these programs the objective is to ascertain which methods of applying fertilizers to grain legumes are most efficient from the point of view of maximizing production without losing the economic benefits associated with the capacity to fix atmospheric nitrogen. Using ^{15}N-labeled urea a method has been developed for obtaining an integrated measure of the total amount of atmospheric nitrogen fixed by a grain legume crop during its growing period.

The micronutrient program aims at developing methods for diagnosing and correcting trace element deficiencies in flooded soils used for growing rice. Besides nitrogen and phosphorus, zinc is the element most often deficient in rice culture. The program has resulted in assessments of the zinc status of different rice growing regions, standardization of analytical techniques, and evaluation of the efficiency of different methods of zinc application.

The soil-water program is designed to develop improved methods for the control of the dynamics of soil water in the field as a basis for the efficient use of soil and water resources. The results obtained have served as a basis for the development of management practices aimed at increasing yields, reducing water and fertilizer losses, and avoiding salt accumulation near the soil surface.

New programs are planned to improve the efficiency of fertilizer management practices in tree crops and to study the role of symbiotic and free-living nitrogen-fixing organisms. We believe the latter program is particularly important, since a limiting factor for increased food production in developing countries is the high cost and shortage of nitrogen fertilizers. In many cases excessive use of nitrogen fertilizers leads to pollution problems. Increased biological fixation of atmospheric nitrogen would have a direct and immediate effect in reducing cost and avoiding pollution. Although the acetylene reduction technique provides a rapid, simple indication of nitrate reductase activity, the use of ^{15}N is the only direct integrated measure of nitrogen fixation and is thus an essential tool in developing agricultural practices that lead to an increased atmospheric nitrogen fixation in agricultural ecosystems. A new program is also planned to develop efficient soil and water management practices under dry-farming conditions in semiarid areas.

In the fields of *plant breeding and genetics*, several coordinated programs are presently being implemented. They all aim at the genetic improvement of crop plants based on concepts, procedures, and methods involving the use of radiation and isotopes. Priority is given to main food crops including fruits and vegetables, as well as to certain industrial crops, and attention is paid to achieving greater yields, better nutritional quality, and higher disease resistance. The programs focus on development of more efficient and economic systems for mutation induction and for the selection and utilization of mutants.

One of these programs aims at increasing the protein content and quality of grains through the induction and selection of mutants. This program has been aided by the development of techniques for mass screening for nitrogen and lysine as well as methods to evaluate mutants genetically, nutritionally, and agronomically. In this work plant breeders, plant physiologists, agronomists, chemists, and nutritionists cooperate; some achievements are barley mutants with a substantially improved amino acid balance, rice mutants with higher protein content, and standardized analytic procedures for mass screening of mutants.

Increasing disease resistance, improving vegetatively propagated plants through somatic mutations, and crossbreeding with cereal mutants are the respective objectives of three individual programs, and improvements have been achieved in respect to characters such as disease and lodging resistance and environmental adaptability (early ripening). This work will continue with stronger emphasis on the transfer of desirable mutant genes into suitable highly productive genotypes for easier use by practical plant breeders.

Finally, programs are designed to gain a deeper insight into the action of mutagens on plant material, to develop appropriate techniques for detecting mutants with improved characteristics, to examine mutants genetically, as well as to design a system for inducing and detecting rare mutations by employing cytogenetic techniques. This work is directed primarily at providing the necessary technology for crop plant improvement through induced mutations.

The use of nuclear techniques for the solution of *animal production and health* problems such as nutrition, reproduction, adaptation, and diseases is the subject of several coordinated programs. In the field of animal nutrition and physiology, one program is designed to discover why certain breeds and species of domestic animals are more adaptable to environmental extremes by studying their water metabolism using tritiated water. Another program aims at early detection of moderate mineral imbalances by identifying any limiting nutrients as animal production becomes more intensive. Various practical techniques have been developed in a program on the utilization of nonprotein nitrogen in ruminants. Here ^{15}N is an indispensable tool for the measurement of microbial protein synthesis in the rumen. Finally, artificial insemination methods for improving the fertility of farm animals in the harsh environment of many developing countries will be sought in a program to evaluate hormonal disfunctions in relation to reproductive capabilities. Radioimmunoassay techniques are used, and the information obtained will help improve reproduction through oestrous synchronization.

A new parasitology program has been established with emphasis on the immunological control of gastrointestinal and tissue parasites using labeled antigenic active fractions and irradiated infectious stages of helminths. Detection and identification of specific antibodies for diagnosing fascioliasis and cysticercosis are progressing, and isolation and maintenance in tissue culture and atypic host animals of the causative agent of East Coast fever, *Theileria parva*, have been achieved. A program to investigate the epizootiology of tick-borne diseases and to devise control measures against ticks is in the planning stage.

Isotope and radiation techniques play an important role in problems of *insect and pest control*. Our current activities deal with insect control; by means of isotope labeling, one objective is to solve specific important pest management problems such as those relating to the optimization of insecticide application, and to target insect ecology, with particular emphasis on rice insects. An international meeting of experts during the fall of 1977 considered the feasibility of developing resistance to insects in crop plants through mutation breeding, possibly leading to a coordinated program.

Several programs emphasize practical aspects and field application of the sterile insect technique for the control of insect pests. Effective control of the Mediterranean fruit fly has been demonstrated in several Mediterranean countries and in the United States, and mass production and release techniques have been optimized. A similar program is under way with the object to control lepidopterous insects attacking tree fruits. Another program aims at developing the sterile insect technique for control of tsetse flies. In this work the laboratory has contributed extensively in devising practical and efficient rearing methods and has pioneered the use of artificial membranes for feeding these bloodsucking insects.

Protecting the environment is the main objective within the realm of *chemical residues and pollution*. Coordinated research programs are designed to minimize undesirable side effects of agrochemical residues; trace contaminants in edible seeds and oil; and improve agrochemical and soil management practices that will minimize fertilizer nitrogen residues as potential pollutants in food, feed, or water and at the same time maximize the utilization of these nitrogen sources by crops. Work is in progress to measure both the soil and fertilizer nitrogen taken up by a crop, the residual fertilizer nitrogen within the rooting zone, and losses of gaseous nitrogen to the atmosphere and through leaching. Nitrogen balances are being established for different soil and crop management practices.

A program on the interaction between chemical residue and soil biota is being initiated, as is a program dealing with microbiological interactions involving chemical residues in aquatic ecosystems, in which the capacity of inland water bodies to absorb contaminants or excessive nutrients without unacceptable effects will be studied.

The final area of activity is *food preservation*. Food preservation is a vital aid to achieving global sufficiency of food supplies, and preserving agricultural products and food by irradiation is receiving increasing recognition. This method for reducing postharvest losses is clean, environmentally attractive, and modest in its energy requirements. The objective of three coordinated research programs is to facilitate the practical application of food irradiation where this produces a wholesome product and offers clear economic and technological advantages.

One of the major problems in the acceptance of irradiated food has been confirmation that such food was in no way harmful to humans. The result of extensive and varied testing has been under review by expert committees appointed jointly by the FAO, the IAEA, and the World Health Organization (WHO). In 1976 such a committee recognized five irradiated food items (potatoes, wheat, chicken, papaya, and strawberries) as "unconditionally safe" for human consumption. Three other foods (rice, fish, and onions) were given "provisional" clearance. It is expected that this evaluation will promote the use of ionizing radiation to preserve these foods, and that it will open the possibilities for international trade with irradiated foods.

The results of the expert meeting, together with the necessary technical data on food irradiation, have been submitted to the FAO/WHO-sponsored Food Standards Programme through the Codex Committee on Food Additives of the Codex Alimentarius Commission for the purpose of standardization of food irradiation at an international level.

Since 1971 the FAO and the IAEA, in collaboration with the Organisation for Economic Cooperation and Development/Nuclear Energy Agency, have sponsored the International Project in the Field of Food Irradiation to execute confirmatory studies on the wholesomeness of irradiated food. At present twenty three countries participate financially in this activity. Another project being planned at an international level is tentatively called the International Facility for Food Irradiation Technology. The purpose of this project would be to assist in international coordination of high-standard independent research, development, and training in the fields of technology, economics, and commercialization of food irradiation. To complete the picture I shall finally mention the Asian Regional Project on Radiation Preservation of Fish and Fishery Products, which was established in 1973 as part of an IAEA-initiated regional cooperative agreement. Presently the activities under this project are part of one coordinated research program in which insect disinfestation of dried and dried-salted fish is the main objective.

Technical Assistance

Currently the Joint Division has technical responsibilities for over fifty small-scale and four large-scale technical assistance projects in developing countries. These projects all aim at applying isotope and radiation techniques as an integrated part of ongoing agricultural training, research, and extension. The small-scale projects usually focus on specific problems, whereas the large-scale projects often embrace many different disciplines.

Each project is designed to support the host institute in terms of training by means of fellowships, as well as by provision of equipment and supplies that cannot be provided locally and of international expertise. Training is indeed of overriding importance, and the Joint Division has technical responsibility for two or three international or interregional training courses each year. In addition, interregional study tours are arranged in specific fields of nuclear applications to agriculture.

Dissemination of Information

International symposia and seminars are important activities in giving scientists in member states the opportunity to exchange views and to keep up to date in highly specialized fields within our terms of reference. An average of two symposia and two seminars annually are part of the Joint Division's program.

Since 1964, over eighty volumes have resulted from the scientific activities of the Joint Division, which also edits a number of regular newsletters to maintain close contact with cooperating and interested scientists.

CONCLUSION

It is my hope that this survey has given you an idea of the philosophy and activities of the joint FAO/IAEA program as an example of international agricultural research.

JOHN A. PINO

Agricultural Research Priorities of the Rockefeller Foundation

9

The U.S. agricultural research community has never before had open to it as it has today the magnificent opportunity to conceptualize and explore the workings of biological and physical phenomena affecting those living forms utilized or utilizable by man for food. There remain few barriers to expanding the parameters of our research to encompass dimensions heretofore impossible to consider.

First of all, in the task of producing and providing adequate food for all people, our perspective has widened to encompass all geographic regions, all climates, all groups and domestic animals under such conditions as are compatible with the laws of the natural system. In a real sense the world is the scientist's oyster. Second, the resource of scientists, disciplines, and institutions is more comprehensive and, when required, can easily be put together into infinite combinations to resolve scientific problems. Third, the advances in knowledge and instrumentation have been so incredibly rapid that practically any research problem that can be defined can be aggressively attacked with reasonable assurance that progress toward its solution can be made.

Although there always may seem to be constraints of one type or another, it is usually the financial limitations that place the real boundaries on research activities. Most of the time, given enough money, other limitations can be overcome. Although the financial resources in this country, both public and private, devoted to agricultural research are

John A. Pino, The Rockefeller Foundation, New York, New York 10036.

considerable—totaling $580 million of federal monies (in fiscal year 1976) for food-related research, $230 million of state funds, and approximately an additional $393 million of private sector funds (ARPAC, 1977)—it is small compared to the value of the agricultural sector. In 1975 the Agricultural Research and Policy Advisory Committee (ARPAC) recommended increasing the amount of public funding by $152 million over a period of four years. The chances that more research money will become available appear reasonably good (Bernstein, 1975). All told this is still less that $4.00 per capita per annum. Support to U.S. institutions for internationally important research projects will become available from USAID funds (now to be covered under the Board for International Food and Agricultural Development Title XII rubric). International funding of agricultural research, except for that channeled through the Consultative Group on International Agricultural Research (CGIAR) system, is largely ad hoc, inconsistent, and fractionated. Other non-U.S. international funding is normally not accessible to U.S. institutions. Given that funds are, and will continue to be, limited, it is essential that priorities be set carefully.

How should these funds be allocated? At first sight one might imagine that it would or should be relatively simple to equate a list of important problems with a corresponding level of funding. In the aggregate that is hardly the way it happens. A major determining factor in this regard is who considers what to be important. In the private sector the priorities for which research funds are allocated is largely determined by what is important to the income of a particular corporation. A company is in business to make money. If its business is manufacturing ketchup, its priority interest may be developing a better canning tomato. Maybe what is good for a company is good for a country, but perhaps not in the same order of priority.

The funds available to state-supported institutions in this country largely support research oriented to resolve problems related to the agricultural interest of the state. The setting of priorities within the state-supported instititional framework is, as it should be, strongly biased toward local interests and pressures. Those interests may be in dairy production, fruit tree research, tobacco, cotton, or perhaps wheat. Often the scope for broad scientific research is limited in such situations.

At the national level U.S. government funding of agricultural research covers a broad spectrum of activities. Although these funds are used largely to support production and product-oriented research, provision is made for support of basic research as well, although the amount and distribution of funds in this latter category are generally conceded as being inadequate.

Research priorities for U.S. agriculture have been set out in a number of reviews of the food problem. One represents the effort of a joint Advisory Committee of the U.S. Department of Agriculture, the National Association of State Universities, and land grant colleges. The approach taken by the team that conducted this study was designed "to produce a report that would reflect the prevalent issues and recommendations for which there appears to be a consensus. . . ." The report lists twelve broad areas of research emphasis that the committee gleaned from numerous reports that it reviewed: basic biological processes; environmental protection and food safety; human nutrition and food quality; pest control; energy conservation and use; production efficiency; processing, storage, distribution, and delivery systems; natural and renewable resources; weather and climate; aquaculture; labor intensive research for the United States and developing countries; and policy analysis and monitoring of systems performance.

The ARPAC report states: "The general feeling expressed in the reports reviewed is that the U.S. must develop a total agricultural research program which considers not only research directed at meeting increased domestic requirements for food and fiber but also recognized the need for research to increase production for export and the urgency of research on how to transfer and adapt new technology to the needs of the developing countries."

The report goes on to say: "Three areas of basic research are singled out in several reports as being particularly in need of intensified effort and having a high probability of significant payoffs in terms of increased food production: (a) photosynthetic carbon dioxide fixation by the green plant through the bioconversion of solar energy; (b) biological nitrogen fixation; (c) genetic splicing and related cell and tissue culture."

A second review of national agricultural research priorities was made in a study begun in June 1975 at the request of President Ford that was addressed to the National Academy of Sciences. Philip Handler, the president of the academy, named Harrison Brown to chair the study. A steering committee of fourteen guided the study in which literally hundreds of persons participated. The study director was Joel Bernstein. The report of the steering committee has recently been released. It is entitled, "World Food and Nutrition Study: The Potential Contributions of Research." This study lists twenty-two priority research areas grouped within four categories as follows: nutrition (4), food production (12), food marketing (2), and policies and organizations (4). Twelve of these directly concern food production, of which seven encompass biological research areas in food production; the others include weather, climate, tropical soils, irrigation and water management, fertilizer sources, and farm production systems. The biological research priorities were given as follows: plant breeding and genetic manipulation; biological nitrogen fixation; photosynthesis; resistance to environmental stress; pest management; ruminant livestock; and aquatic food resources.

Within each of these categories, the report describes the principal lines of research that should be pursued. It is a rather comprehensive analysis of the scope of research activity currently in progress and of those areas that are viewed as needing strengthening. In a sense the report represents the most recent comprehensive review since the publication of the President's Science Advisory Committee report of 1967 (PSAC, 1967).

At a conference on Resource Allocation and Productivity in National and International Agricultural Research, held at Airlie House on 25–29 January 1975, the social scientists concluded: "The capacity of scientists to generate new technology outstrips our understanding of social and economic implications of technical change and our ability to provide guidance for policy makers" (Arndt and Ruttan, 1975).

They signaled four areas relating to research that warrant further research.

1. There is a general need for a more precise understanding of the sources of demand for technical change in agriculture.
2. More analysis of research cost functions and the production process for research is needed.
3. There is need for further understanding of national and international diffusion of agricultural technology and scientific knowledge.
4. There is also need for research on elements of the technical change process that were not well covered at the conference.

One is hard put to react to any conclusions about what are the urgent research priorities based on what has been reported to date. In reviewing most reports that treat the question, one has difficulty identifying what are priorities and what is in reality a total research agenda. Agricultural research priorities as some would view them should include everything relating to food and food products from production to consumption and health. In addition, most reports reflect the various disciplinary, institutional, political, and other biases that are inherent in surveys. Obviously everything cannot be a priority.

We do not disagree with those who argue for the need to apply greater effort to better understand the problems of storage and storage losses, transportation, distribution, processing, and so on, as well as the political and economic phenomena affecting the availability and consumption of food products. The Rockefeller Foundation views the question of research priorities with a strong bias toward the resolution of production problems affecting yields of crop and livestock species in the tropical and subtropical regions of the world. In the broader view, we recognize that this quantitative approach (that is, increasing production) may not resolve the problems of distribution, the balance or imbalance of diets, and ultimately the adequacy or inadequacy of nutrition of all segments of a society. Nevertheless, our major efforts have been, and continue to be, in those areas where yields are low and where food consumption (nutrient intake) is generally lowest.

The Rockefeller Foundation has been concerned with improving the quality and productivity of the basic food crops, that is, those commodities that provide a large percentage of the total nutrient intake of large populations. These crops include the cereals, i.e. wheat, rice, maize, sorghum, millets, and to a lesser extent barley, oats, and rye; the legumes, mainly the common beans (*Phaseolus vulgaris*), chickpeas (*Cicer*), pigeonpeas (*Cajanus cajan*), cowpeas (*Vigna unguiculata*), soybeans *(Glycine max)*, mungbeans (*Vigna radiata*); and tuber crops, including potatoes (*Solanum tuberosum*), cassava *(Manihot esculenta)*, and sweet potatoes (*Ipomoea batatas*). The foundation recognizes the importance of other food crops, including many of the vegetable crops, fruits, nuts, and oil seed crops. However, relatively speaking, our support for research related to these latter groups of crops has been quite small. With respect to animal species utilized for food, our emphasis has been on the ruminant animal mainly because the technology relating to monogastric food animals (poultry and swine) is well advanced and readily transferable, and also because the utilization of vast plant resources not consumed by people can only be harvested and transformed into human food by ruminants.

The foundation further defines its agricultural research priorities in terms of three areas affecting productivity:

First, we must concentrate on previously identified factors that directly affect the yield of important plant species. The barriers to yield are still unknown. Geneticists have made great strides in identifying selections that outyield earlier varieties. Yet we do not know where the ceiling lies nor what factor set that ceiling. Among those genetically determined qualities of whose physiological expression we need more understanding we include (1) photosynthetic efficiency, or, more broadly, the capacity of a plant to assemble and utilize nutrients and energy in growth and/or storage of nutrients; the key seems to be photosynthesis, but other processes are also involved; and (2) wide adaptation, particularly to conditions of stress, including drought tolerance, various conditions of soil toxicity (especially aluminum and magnesium), salt tolerance, and low temperature tolerance.

Second, among the factors indirectly affecting yield, our special interest lies in the area of plant protection. This interest has two principal thrusts:

(1) the understanding and manipulation of the pest or pathogen life cycle so as to disrupt its reproduction efficiency or other behavioral trait, and (2) a better understanding of the defense mechanism of the plant. We believe in the concept of integrating various techniques into a system for plant protection (pest management). However, our concern lies more firmly in the area of basic research that will lead to an understanding of pest biology and the host defense mechanisms.

With regard to livestock our research interest has focused principally on the hemoparasitic diseases, including trypanosomiasis, anaplasmosis, babesiasis and East Coast fever; these diseases are major constraints to livestock production in areas where they are prevalent. Our interest in the human-related aspects of certain protozoan diseases and of the fundamental nature of immunity related to protozoan infections has placed this among our priority areas.

The foundation's third priority research area concerns major ecological systems. As the pressures on natural systems continue to intensify, the need will become more urgent to improve our understanding of major ecological phenomena. Perceptible changes in weather patterns are occurring, increasing desertification, destruction of rain forests, erosion of hillsides, flooding of river deltas, and many similar phenomena. Pressures on land for habitation, agriculture, and resource extraction are causing changes that are irreparable if not irreversible. The foundation places a high priority on the need to study and understand some of the major and more fragile ecological systems of the world. Such studies require comprehensive teams of biological, social, and political scientists working on all facets of the system on a sustained basis.

It is this total system with which we are concerned, and whereas individual scientists may pursue the minutest or perhaps remotest facet of those phenomena involved in the system, in the aggregate we must address the question, "How can we most effectively manipulate the entire process to give greater security and efficiency in meeting the nutrient needs of people?"

Although our focus is problem oriented, relying heavily on applied research, we believe strongly in the support of more fundamental investigations that ultimately will provide the technological basis for new advances in the food producing system.

To view the world's agriculture in one grand sweep and hope to prescribe a set of research priorities that would be valid for all nations equally is unrealistic. The U.S. agricultural research priorities are quite different from those of Peru, Kenya, or New Zealand. These differences are due to the difference in relative importance of various crops or livestock species in each of those countries. The fact of the matter is that many of the lesser-developed countries do not have the resources to sustain even a minimal research effort on their most important food problems. It is paradoxical that nations, though their survival depends upon it, have tended to allow agriculture and food production to follow the unpredictable extractive system, with inadequate scientific, political, or economic interventions to give greater stability. This concern has also led us to place high priority on building and strengthening institutional capacities, especially of developing nations, to carry on country-specific research that demands qualified people and a high degree of scientific sophistication. It has been shown by Evanson that a high correlation exists between a nation's research capacity and its agricultural growth rate (Evenson and Kislev, 1975).

Research must be rapidly translated into production technologies. It is good to remind ourselves occasionally that ultimately it is the farmer who must integrate all of the experience, all of the judgment, weigh the risks, and apply the hard work that delivers the food for people. He must look forward and he must look backward. Yesterday's experience is today's practice, and today's practice is tomorrow's promise. In this country, as well as in many other developed countries, there exists a vast supportive research structure that provides new crops, raises yields, reduces risk, protects the environment, and helps build the productive resource base to sustain future productivity. But that condition does not prevail in much of the world today. The United States is morally committed to provide assistance to poor nations in the resolution of their food and nutrition problems. The preeminence of this nation in agricultural research has placed it in a special role in lending support to other nations' research efforts. Experience of the past thirty years has demonstrated, however, that there must be special efforts geared to the resolution of the peculiar problems of agriculture in the developing nations.

The research programs of the United States and other developed countries can address basic biological problems; for the most part, those of the developing countries cannot. Yet ways must be found to bring the scientific capability of advanced nations to bear on problems of the lesser-developed countries. Closer institutional linkages throughout the international system are necessary, and the strengthening of national research capabilities in the lesser-developed countries should be one of the most important goals of assistance agencies.

REFERENCES

Agricultural Research and Policy Advisory Committee (ARPAC). 1977. Agriculture and food research issues and priorities. U.S. Department of Agriculture, Washington, D.C.

Arndt, T. M., and V. W. Ruttan. 1975. Resource allocation and productivity in national and international agricultural research. Agricultural Development Council, New York.

Bernstein, J. 1975. Communication to the Rockefeller Foundation.

Evenson, R.E., and Y. Kislev. 1975. Agricultural research and productivity. Yale University Press, New Haven.

President's Science Advisory Committee (PSAC). 1967. Report.

W. R. SHARP AND P. O. LARSEN

Plant Cell and Tissue Culture:
Current Applications and Potential

10

During the 1800s the cell theory, which states that the cell is a basic structural unit of all living creatures, was very quick to gain acceptance. However, the second portion of the cell theory, which states that these structural units are distinct and potentially totipotent physiological and development units, failed to gain universal acceptance. The skepticism associated with the latter was because of the inability of scientists such as Schleiden and Schwann to demonstrate totipotency in their laboratories. It was in 1902 that the well-known German plant physiologist, Haberlandt, attempted to cultivate plant tissue culture cells *in vitro* and to, in fact, demonstrate totipotency. However, unfortunately he failed. In the early 1920s workers again attempted to grow plant tissues and organs *in vitro*. Molliard, in 1921, demonstrated limited success with the cultivation of plant embryos, and subsequently Kotte (1922a, 1922b), a student of Haberlandt in Germany, and, independently, Robbins (1922a, 1922b) were successful in the establishment of excised plant root tips *in vitro*. Subsequently Philip White (1931) surveyed the literature in the field and pointed out some of the reasons for earlier failure, as well as suggested areas for future experimentation. He emphasized the suitability of the Kotte-Robbins root tip method for the study of nutrition and critically examined contributions made by other workers. This review was followed by a series of factual papers (1932a, 1932b, 1933a, 1933b) pertaining to experiments

W. R. Sharp, Department of Microbiology, and P. O. Larsen, Department of Plant Pathology, Ohio State University, Columbus, Ohio, 43210.

performed mostly with excised root tips. Finally in 1934 the possibility of growing excised roots of tomato *in vitro* for periods of time without theoretical limits was demonstrated (White, 1934).

Gautheret engaged in experimentation with excised root tips and the cultivation of cambial tissues from trees (Gautheret, 1934, 1935, 1938). These studies were continued, and in 1939 Gautheret, Nobecourt (1937, 1939), and White (1939) published independently descriptions of the successful cultivation for prolonged periods of cambial tissues of carrot and tobacco. These were the first true plant tissue cultures in the strict sense of prolonged cultures of unorganized materials. Philip R. White (1943, 1954, 1963) and Roger J. Gautheret (1959) can be credited with providing a significant impetus to the field with the publication of their authoritative handbooks. Work progressed slowly until 1963, when Philip R. White organized the First International Tissue Culture Congress at Pennsylvania State University. At this congress leaders from around the world in this new area of the botanical sciences presented papers, conducted round tables, held informal discussions, and defined problems to be resolved using the tissue culture approach. It was apparent that plant tissue culture held great promise for the resolution of numerous problems confronting investigators in the disciplines of biochemistry, chemistry, pathology, and physiology. Subsequent congresses have been organized (Strasbourg, 1970; Leicester, 1974; and Calgary, 1978).

Plant cell, tissue culture, and organ culture involve isolating a cell, tissue, or organ and aseptically placing it into a vessel of nutrient medium under controlled environmental conditions (quality and quantity of illumination, temperature, humidity, and so on), with the objective being to obtain rapid asexual multiplication of plant cells or plants. Any given tissue composed of cells with competent nuclei is a suitable explant for the initiation of a plant tissue culture. Vegetative shoot tips, terminal stem tips, axillary buds, stem sections, leaf sections, reproductive parts such as microspores, mega-spores, ovules, embryos, seeds, and spores, as well as isolated cells and protoplasts (wall-less cells) are used for this purpose.

Plant cells and tissues can grow on various formulations of media that are generally classified in two categories, with low salt or high salt. A typical low salt medium is the one developed for tomato root culture (White, 1938), while the most universally used of the high salt media is the one, or a modification thereof, developed by Skoog and his students (Murashige and Skoog, 1962; Linsmaier and Skoog, 1965). In addition to mineral salts, media usually contain an energy source (usually sucrose), vitamins (e.g., thiamine, nicotinic acid, pyridoxine, biotin, and so forth), and growth regulators (e.g., auxins, cytokinins, and gibberellins). Some cells require the addition of complex growth substances to the medium as well as to the growth regulators, e.g., fruit juices, coconut milk, and so on. Growth regulator concentrations in the culture medium are critical to the control of growth and morphogenesis. Generally a high concentration of auxin and a low concentration of cytokinin in the medium promotes abundant cell proliferation with the formation of an unorganized tissue referred to as callus. On the other hand, low auxin and high cytokinin concentrations in the medium result in the induction of shoot morphogenesis. Auxin alone is important in the induction of root primordia. Likewise, it is possible to induce somatic embryogenesis and androgenesis *in vitro* at appropriate growth regulator concentrations. Although explants of all higher plant species are potentially totipotent, only a limited number of species have been observed to undergo high frequency shoot morphogenesis or embryogenesis. Somatic embryogenesis and androgenesis occur at a lower frequency than do shoot and root morphogenesis.

In vitro fertilization, embryo culture, and androgenesis are of special interest to the geneticist and plant breeder. Of special interest has been the possibility for mutation studies with cultured pollen cells and germplasm selection for development of improved cultivars. Moreover, haploids can be used for gene fixation in crops that are outbred and thus provide an alternative to the conventional approach, which often requires as many as ten selfings for fixation. Using the haploid approach, in some species it is possible to go directly to a haploid plant or homozygous diploid plant. The latter involves chemical treatment or selection for homozygous diploids resulting from polyploidy.

Another tool of interest to the geneticist is the fusion of protoplasts from genetically distant species, which allows for the possibility of transferring desired cistron units from protoplasts of one cultivar to the protoplasts of another for the development of new cultivars. Protoplasts can also incorporate foreign genetic material contained in organelles as well as isolated DNA. Protoplast work pertaining to the study of extrachromosomal inheritance, the manipulation of symbiotic nitrogen fixation, and the transfer of the nitrogen-fixing ability to eukaryote cells is promising and will be discussed fully in this volume.

Shoot morphogenesis followed by root morphogenesis or embryogenesis, apical meristem culture, and shoot tip culture have been used for clonal reproduction and vegetative propagation of a number of important cultivars. Apical meristem culture and shoot tip culture involve the culture of explants of ca. 0.1 mm in length and ca. 5.0–10.0 mm in length, respectively, for the induction of cell proliferation and subsequent plantlet proliferation. Both kinds of culture have been effectively used in the eradication of disease from infected plants. Development of adventitious buds from cultured embryos has been an effective mode of culture for propagation of forest trees. Adventitious buds are readily formed from cultured embryos of loblolly pine, pinon pine, white pine, slash pine, and Douglas fir, although rooting is somewhat of a problem.

In tissue culture systems where the objective is vegetative propagation, investigators generally categorize the developmental sequence of events in three or four stages. Stage I occurs following the transfer of an explant onto a culture medium. Enlargement of the explant and/or callus proliferation occurs during this stage. Stage II is characterized by a rapid growth increase of organs and/or the induction of adventitious organs or embryos that occurs sometime following subculture onto medium with or without altered growth factor concentration ratios. Stage III pertains to rooting or hardening to impart some tolerance to moisture stress or confirming a degree of resistance to certain pathogens, but most of all involves the conversion of plant from the heterotrophic to the autotrophic stage. Stage III development may likewise require subculture onto a new medium.

Plant pathologists use the tissue culture approach to eradicate disease and obtain pathogen-free clones, which are characterized by restoration of vigor and increased yield. Furthermore, movement of plants from one country to another or from one region to another can be expedited once pathogen-free material has been isolated. It can also potentially be freeze preserved for storage in a germplasm bank.

The Yoder Brothers Company of Barberton, Ohio, has used meristem culture to eradicate viral disease from chrysanthemums for a number of years. Viral infected plants are subjected to heat therapy in a hot chamber at 100°F; meristems are surgically removed under aseptic conditions and placed into culture on an appropriate medium. Following five weeks of growth, reconstituted plantlets are potted and tested for the presence of various plant pathogenic viruses by grafting explants or rubbing homogenates from the plantlets onto virus indicator plants. If these tests are negative (no indication of viral infection), the reconstituted plantlets obtained from meristem culture are released to a certified area for vegetative propagation.

Another area for plant pathology of current interest is the culture of nematodes. In the past it has been difficult to engage in biochemical studies of nematodes or to study their physiology because of metabolic contaminants from host plants. Now a protocol has been developed whereby monoxenic culture of some nematode species can be achieved. Populations of nematodes are maintained and increased in alfalfa tissue cultures from which they can easily be separated by filtration techniques and then used for experimental purposes.

Other possible applications of plant tissue culture include the synthesis of oil, wood fiber, latex, medicinal products, and so on. Three possible advantages to this approach are: (1) production of compounds under controlled environmental conditions independent of climatic changes and free of microbes; (2) automated control of cell growth and rational regulation of metabolism, possibly contributing to a reduction in labor costs; and (3) the biotransformation of compounds to medicinally useful compounds.

Papers in this volume discuss the physiology of growth and morphogenesis, basic genetic studies, broadening of the genetic base, storage of germplasm, clonal propagation, selection for disease resistance, and secondary product synthesis. It is apparent that we have reached a point in time where we can discuss the application of plant cell and tissue culture.

REFERENCES

Gautheret, R. J. 1934. Culture du tissu cambial. C. R. Acad. Sci. Paris 198:2195-96.

Gautheret, R. J. 1935. Recherches sur la culture des tissus végétaux: Essais de culture de quelques tissus méristématiques. Librarie E. la François, Paris.

Gautheret, R. J. 1938. Sur la repiquage des cultures de tissu cambial de Salix capraea. C. R. Acad. Sci. Paris 206:125-27.

Gautheret, R. J. 1939. Sur la possibilité de réaliser la culture indéfinie des tissus de tubercules de carotte. C. R. Acad. Sci. Paris 208:118-20.

Gautheret, R. J. 1959. La culture des tissus végétaux: Techniques et réalisations. Masson, Paris.

Kotte, W. 1922a. Wurzelmeristem in Gewebekultur. Ber. Deut. Bot. Ges. 40:269-72.

Kotte, W. 1922b. Kulturversuche mit isolierten Wurzels pitzen. Beitr. Allg. Bot. 2:413-34.

Linsmaier, E. M., and F. Skoog. 1965. Organic growth factor requirements of tobacco tissue cultures. Physiol. Plantarum 18:100-127.

Murashige, T., and F. Skoog. 1962. A revised medium for rapid growth and bioassays with tobacco tissue cultures. Physiol. Plantarum 15:473-97.

Nobecourt, P. 1937. Cultures en série de tissus végétaux sur milieu artificiel. C. R. Acad. Sci. Paris 205:521-23.

Nobecourt, P. 1939. Sur la perennité et l'augmentation de volumes des cultures de tissus végétaux. C. R. Soc. Biol. Paris 130:1270.

Robbins, W. J. 1922a. Cultivation of excised root tips under sterile conditions. Bot. Gaz. 73:376-90.

Robbins, W. J. 1922b. Effect of autolyzed yeast and peptone on growth of excised corn root tips in the dark. Bot. Gaz 74:57-79.

White, P. R. 1931. Plant tissue cultures: The history and present status of the problem. Arch. Exp. Zellforsch. 10:501-18.

White, P. R. 1932a. Plant tissue culture: A preliminary report of results obtained in the culturing of certain plant meristems. Arch. Exp. Zellforsch. 12:602-20.

White, P. R. 1932b. Influence of some environmental conditions on the growth of excised root tips of wheat seedlings in liquid media. Plant Physiol. 7:613–28.

White, P. R. 1933a. Plant tissue cultures: Results of preliminary experiments on the culturing of isolated stem-tips of *Stellaria media*. Protoplasma 19:97–116.

White, P.R. 1933b. Concentration of inorganic ions as related to growth of excised root-tips of wheat seedlings. Plant Physiol. 8:489–508.

White, P. R. 1934. Potentially unlimited growth of excised tomato root tips in a liquid medium. Plant Physiol. 9:585–600.

White, P. R. 1938. Accessory salts in the nutrition of excised tomato roots. Plant Physiol. 13:391–98.

White, P. R. 1939. Potentially unlimited growth of excised plant callus in an artificial nutrient. Amer. J. Bot. 26:59–64.

White, P. R. 1943. A handbook of plant tissue culture. Jacques Cattell Press, Tempe, Arizona.

White, P. R. 1954. The cultivation of animal and plant cells. Ronald Press Co., New York.

White, P. R. 1963. The cultivation of animal and plant cells. Ronald Press Co., New York.

Part 2

Physiology of Growth and Morphogenesis

Embryogenesis and Chemically Induced Organogenesis

11

INTRODUCTION

The preparation of a critical review of the chemical control of embryogenesis and organogenesis in tissue and cell cultures faces a number of formidable difficulties. This is highlighted by three important considerations. First, there is the very large number of publications in which some form of morphogenesis is described and details are given of the culture media used and, particularly, of the growth regulators included in such media. However, in almost all such papers the importance of the chemical composition of the media versus other environmental factors, let alone the identification of the critical chemical factors, has not been established. Hence while the culture media may be accepted as permissive of the morphogenic phenomena described, they may, on the evidence provided, in no sense be regarded as determinative, let alone optimal, for the morphogenic expression. Furthermore, the value, for our purpose, of many of these papers is further reduced by lack of quantitative data (the absolute and reproducible frequency of the formation of organized structures) and by failure to consider the possible role of chemical factors derived from the initial explant and operative during the period immediately following excision of the callus from the explant and its independent growth in culture. Second, many publications do not distinguish clearly between chemical factors essential for culture growth and factors playing a

H. E. Street, Botanical Laboratories, University of Leicester, Leicester, LE1 7RH, England.

more specific role in morphogenesis. Since morphogenesis involves growth, all factors essential for growth will be essential for morphogenesis. The peculiar feature of morphogenesis is that the cell divisions are so organized as to result in the emergence of proembryos or organ primordia. Third, and this in the final analysis is the outstanding difficulty, we do not understand for any of the natural plant growth regulatory substances their mechanism of action and interaction at the molecular level. This completely prevents us from distinguishing with any certainty between the permissive and inductive roles of these substances.

The title of this chapter introduces the term *induced*, which implies a phenomenon of induction (an act causing morphogenesis to occur). However, the concept of totipotency—the capacity of plant cells to exhibit the potential for morphogenesis—raises the possibility that we are here concerned with a two-stage process: first, the retention or acquiring in response to appropriate stimuli of morphogenic potential or competence, and then, the expression of this potential in the development of organized structures. The chemical factors of our title may thus possibly play an inductive role (that is, they may be determinants) in both these stages.

HISTORICAL BACKGROUND

Our present approach to morphogenesis has been conditioned by exciting research undertaken some twenty years ago—the classical studies of Skoog and his coworkers on organogenesis in tobacco stem pith explants and those of Steward and others on somatic embryogenesis in carrot cultures. The key paper from Skoog's laboratory appeared in 1956 (Skoog and Miller, 1956). The thesis they developed stemmed from the identification of kinetin (6 furfurylaminopurine) as the growth regulatory factor required in addition to IAA for induction of growth and organogenesis in the callus. In this paper their approach is summarized thus: "These tend to show that quantitative interactions between growth factors, especially between IAA and kinetin (auxin and kinins) and between these and other factors, provide a common mechanism for the regulation of all types of growth investigated, from cell enlargement to organ formation." The subsequent discovery of natural cytokinins and the recognition that they, like auxin, had a range of physiological effects within the whole plant strengthened this concept.

The standpoint developed by Steward is well set out in Steward and Krikorian (1972) and Steward (1976). The work of his group has drawn attention to the essential totipotency of plant cells and led to the recognition that the realization of this totipotency requires that the tissue cells be released from the chemical and physical restrictions imposed upon them by the whole organism, that they be subjected to the chemical stimuli that lift them from the quiescent state into a capacity for active growth, and that they be provided with the essential nutrients that permit them to display their morphogenic potential. Furthermore, Steward has stressed the complexity of the processes involved in this expression, considering that multiple interactions of chemical and physical factors are involved and that they probably have to impinge upon the cells in an appropriate sequence.

From these pioneer studies two contrasted viewpoints have tended to develop, one stressing quantitative interactions between available growth regulatory substances as determinative and the other stressing the seemingly insoluble complexity of the phenomena involved. Inevitably the simpler approach has attracted many concerned with plant regeneration and, although it now has a number of practical successes to record, it has probably resulted in many more nil returns (most of which will have failed to see the light of scientific publication). Furthermore, the pursuance of this "ringing the changes" with the small number of readily available regulatory chemicals has yielded information of practical value rather than resulted in any basic advances in our understanding of the control of morphogenesis. Steward (1976) pinpoints the problem by saying that the need is "to obtain such knowledge that populations of somatic cells, *not only of carrot but of any angiosperm*, could routinely, and in high yield, be caused to develop into plants." Such a statement applies with equal force to the achievement of embryogenesis in the gametophyte cells of angiosperms and of organogenesis in somatic cultures. The present situation is that all cases of experimental morphogenesis in culture, despite their growing number, are in a sense special cases, cases in which we have succeeded without knowing why.

GROWTH REGULATORY SUBSTANCES IN
RELATION TO ORGANOGENESIS

Kohlenbach (1977), in a recent review of plant organogenesis from cell and tissue cultures, has advanced the following generalizations and cited particular examples.

1. Removal of auxin. After a preculture with auxin its removal often leads to root formation. Examples: carrot (Reinert, 1959); *Atropa belladonna* (Thomas and Street, 1970).
2. Ratio of auxin to cytokinin. A high ratio of auxin relative to cytokinin favors root formation, whereas an opposite ratio favors shoot buds. Examples: *Petunia* (Durand et al., 1973); *Macleaya cordata* (Lang and Kohlenbach, 1975); tobacco (Skoog, 1971).
3. Absolute concentration of phytohormones. Example: *Lycopersicon* leaf callus (Padmanabhan et al., 1974), where 2 mg l^{-1} IAA + 2 mg

l^{-1} kinetin (K) causes root initiation and 4 mg l^{-1} IAA + 4 mg l^{-1}K gives shoot formation.

4. Nature of the auxins and cytokinins. Example: *Asparagus* callus (Bui Dang Ha, 1974), where shoots arise in the presence of benzyladenine (BA) in combination with IAA or NAA but not in media containing zeatin (Z) or 2,4-D.

5. Phytohormones other than auxin and cytokinin. Example: tissues derived from sweet potato and potato (Yamaguchi and Nakajima, 1974), where shoot buds may be produced by replacing K with abscisic acid (ABA).

Clearly such generalizations are self-contradictory, and this can be observed even in comparisons between cultures derived from cultivars within the same species. It is certainly true that a number of examples can be quoted where transfer of callus to a medium lacking auxin may result in root emergence, and this may even be enhanced by application of an antiauxin (Thomas and Street, 1970). However, the pioneer studies of Gautheret (1945) not only drew attention to cases where auxin was essential for root formation, but showed that the rhizogenic response to auxin could be critically modified by the sugar supply, illumination, and temperature regime imposed upon the cultures. Auxin may therefore either promote or inhibit rhizogenesis. In some recent work in our laboratory (O'Hara and Street, 1977) on cultures derived from embryos, seedling roots, and nodal meristems of the Maris Ranger cultivar of wheat, we found that in excess of 90 percent of the cultures readily formed roots when transferred from the maintenance medium containing 1 mg l^{-1} 2,4-D to a medium lacking the auxin, and that this capacity persisted during continuous serial subculture of callus. By contrast Chin and Scott (1977), working with the cultivar Mengair, found that callus of embryo origin had a very low capacity for root formation, and that root-derived callus maintained by serial subculture in a medium containing 2 mg l^{-1} 2,4-D + 4 mg l^{-1} p-chlorophenoxyacetic acid (CPA) did not form roots on transfer to medium lacking auxin but only when transferred to media containing 1.0 mg l^{-1} 2,4-D or 0.2 mg l^{-1} CPA or 1.0 mg l^{-1} NAA.

Whereas a capacity for root formation may be long persistent, as in the Maris Ranger wheat callus mentioned above, in sugarcane (Barba and Nickell, 1969) and other monocotyledons (Morel and Wetmore, 1951) it may be lost very quickly during subcultures so rapidly that it can hardly be explained in terms of changes in the nuclear cytology of the cells (Wilson and Street, 1975). Further, there are many cases where cultures fail at any time to show rhizogenesis even when subcultured to a very wide range of growth regulator additions and other cases where roots will arise spontaneously in the presence of quite diverse combinations of growth regulators.

It is true that shoot formation has now been achieved in a number of species by using as auxins either IAA or NAA and as cytokinins K, BA, or N_6-(2-isopentenyl) adenine (2iP). Although in some of these cases an appropriate balance of both types of regulators must be applied, sometimes one or the other (essential or stimulating to callus proliferation) must be omitted to give shoots (Murashige, 1974). Sastri (1963), for instance, with callus of *Armoracea rusticana*, found that auxin additions promoted and kinetin additions suppressed shoot formation. Furthermore, point 5 of Kohlenbach's summary indicates the limited nature of the auxin-cytokinin control of shoot formation. He cites the case of the activity in ABA in sweet potato and potato cultures. To this can be added the observed activity of ABA in promoting shoot bud formation in *Cryptomeria japonica* (Isakawa, 1974) and the instances of GA involvement in shoot bud initiation such as *Ranunculus sceleratus* (Konar and Konar, 1973), *Chrysanthemum* (Earle and Langhams, 1974), and *Limnophila chinensis* (Sangwan et al., 1976).

Further, even the classical case of shoot formation in the tobacco callus system now appears to be less clear cut. Certain phenolic compounds (Lee and Skoog, 1965) and the antiauxin N-1-naphthylphthalamic acid (Feng and Linck, 1970) are promotive of shoot bud formation. The chelating agent, 1,3-diamino-2-hydroxypropane-N,N,N^{1}N^1-tetracetic acid, promotes shoot initiation in microspore-derived tobacco callus (Kochhar et al., 1971), and the polycyclic hydrocarbons like dibenz(a.L)anthracene, benz(a)propane, and chrysene have been reported to be capable of replacing both auxin and cytokinin in promotion of shoot and root formation in haploid callus. The studies of Murashige (1965) on the influence of seven gibberellins on growth and organogenesis in tobacco callus showed that these hormones, at levels promotive or not inhibitory to fresh weight increase, were suppressive of shoot formation, and that this suppression could not be overcome by the simultaneous application of kinetin, auxin, and various antigibberellins. This suppressive action of GA in shoot bud formation has been confirmed by Thorpe and Meier (1973) and Engelke et al. (1973), although the latter workers were able to overcome the GA suppression by addition of a very high level (25 μM) of 2iP.

Again, as with root formation, the importance of light intensity, light quality, and the duration of the daily light period on shoot bud initiation has been amply demonstrated (Murashige, 1974; Seibert, 1973). Since the most frequently observed form of organogenesis observed in tissue cultures is rhizogenesis, there are even more cases where caulogenesis cannot be demonstrated despite the testing of many culture media and the changes in other culture variables. For instance, whereas a precise auxin-cytokinin ratio (BA most effective cytokinin) has proved effective in shoot initiation from cultures of several tree species (Chalupa, 1974; Campbell and Durzan, 1975), de Fossard et al. (1974) could obtain no morphogenic response in callus cultures of *Eucalyptus bancroftii* despite testing 175 different auxin-cytokinin combinations. Again, a capacity for shoot initiation observed during the very early period of culture is frequently, in fact almost always, lost on continuing serial subculture. This has led to the suggestion that in such cases shoot formation involves limited proliferation of meristematic centers directly derived from the primary explant. This interpretation can be advanced in relation to the infrequent and transient formation of shoots by wheat cultures as reported by Shimada et al. (1969) and O'Hara and Street (1977); this shoot formation can also be described as spontaneous insofar as its incidence is insensitive to a wide range of auxin-cytokinin additions to the culture medium.

We are therefore faced with trying to explain (1) the complete failure of most cultured tissues to show somatic embryogenesis, of many to show caulogenesis, and of a very significant number to show any form of morphogenesis; (2) the often rapid loss of a capacity for morphogenesis once the callus or suspension cultures are serially subcultured; (3) the multiplicity of growth regulator additions to culture media that have been reported to be stimulatory or permissive of morphogenesis; and (4) the critical influence that can be exerted by aspects of the culture regime other than the exogenous growth regulators supplied via the culture medium.

These last two questions raise the probability that morphogenic expression is controlled by the levels of *endogenous* growth regulators and other critical metabolites and that these levels determine the response to *exogenous* regulators and other external factors operating in the culture system. To begin to answer this problem we need to consider what is known about such endogenous levels, about the essentiality of exogenous regulators for culture growth, and about the mechanism of action of such applied regulators.

Auxin

Most, if not all, normal plant tissue cultures require for maintenance of their growth in culture the presence in the culture medium of a compound with auxin activity (Gautheret, 1955). IAA, the recognized natural auxin, is sensitive to degradation by oxidation (Kulescha, 1977) and in many cases proves ineffective or less effective than the synthetic auxins, 2,4-D and NAA. Indolebutyric acid alone or in combination with NAA has proved very effective with some tissues (Street and Shillito, 1977); Ingram and Butcher (1972) have reported on the high activity of benazolin (4-chloro-2-oxobenzothiazolin-3yl-acetic acid). Although there is some evidence that

2,4-D at high concentrations may promote cytological instability (and associated loss of morphogenic potential) by interfering with normal spindle formation, this does not seem to be the case at concentrations in the range 2.0–0.2 mg l^{-1} (9.0×10^{-6}–9.0×10^{-7}M) (Bayliss, 1975, 1977a). Its wide use relates to its stability and its enhancement of culture friability.

The fact that tissue cultures require an exogenous auxin does not indicate that they are incapable of endogenous auxin synthesis. Unpublished work by M. Moloney and M. C. Elliot (Leicester Polytechnic) shows that sycamore (*Acer pseudoplatanus*) cells maintained in active growth by 2,4-D contain IAA. Habituation appears to be the result of a reversible, even if relatively stable, activation of endogenous synthesis, or inhibition of the degradation of natural auxin. This raises the important question of whether the activity of substances like 2,4-D is at least in part due to their acting as auxin protectors by suppressing endogenous auxin degradation (Stonier, 1971). Studies involving replacement of 2,4-D by other substituted phenoxyacids in the culture of rose cells has indicated that their suppression of phenol oxidation and polymerization is involved in their growth-promoting activity, but that those that have low (2-chlorophenoxyacetic) or no activity (α-[2-chlorophenoxy]isobutyric acid) in auxin bioassays are less effective in maintaining a high growth rate of the cells (Lam and Street, 1977). The 2,4-D is acting both as an auxin and as a controller of other aspects of cell metabolism.

High levels of auxin and particularly of 2,4-D activate unorganized proliferation (suppress morphogenesis), and the cases where transfer to medium lacking auxin or containing a low level of auxin (and here NAA may be more effective that 2,4-D [Gamborg and Eveleigh, 1968; Walter, 1968]) induces embryogenesis or organogenesis have been interpreted as resulting from a fall in tissue auxin level combined with a reversal of the auxin gradient within the tissue mass (responsible for the development of the necessary polarity?). In line with this, habituated tissues are incapable of morphogenesis. The regeneration of plants from habituated tobacco callus reported by Sacristán and Melchers (1969) and Melchers (1971) was probably a reflection that the cultures involved contained a mixture of habituated and normal cells (Lutz, 1966).

The situation in carrot cultures is that embryogenesis is initiated by transfer to auxin-free medium, although under these conditions callus proliferation ceases and the embryogenic culture cannot be serially propagated. Addition of an antiauxin such as 3,5-dichlorophenoxyacetic acid (3,5-POA) speeds up embryo formation and can enhance the embryo yield. Replacement of 2,4-D by other substituted phenoxyacids shows that the growth rate of the cultures is related to the auxin activity of the acids and that concentrations promoting active callus proliferation suppress embryo formation. However, it is possible, by supplying simultaneously 0.1 mg l^{-i} 2,4-D and 1.0 mg l^{-1} of 3,5-POA, to maintain culture growth and permit a high level of embryo initiation and development up to, but not beyond, the late globular stage (Chandra et al., 1977). The further development of these embryos requires their transfer to auxin-free medium.Thus it seems that the embryos are auxin autotrophic and very sensitive to inhibition by exogenous auxin, and that specific antagonism of the auxin activity of 2,4-D can allow embryo initiation without reduction in callus culture growth rate (not possible with 2,4-D alone), again pointing to a secondary beneficial effect of 2,4-D on the metabolism of cultured cells. This raises the interesting question of whether 2,4-D substitutes for some metabolic regulator (additional to auxin) that is operative *in vivo* and whose action is disrupted by bringing cells into cultures.

In considering responses to auxin, and again particularly to 2,4-D, account should be taken of species differences in uptake and metabolism of the applied regulator. Feung et al. (1976, 1977) have shown that in mono-cotyledonous tissues 2,4-D is predominantly metabolized into a physiologically inactive glycoside derivative, whereas in dicotyledonous tissues this is replaced by the physiologically active amino acid conjugates.

Here, as will be equally apparent in relation to the other growth-regulating substances to be discussed, a better understanding of 2,4-D involvement in morphogenesis and culture growth now requires more intensive study of the levels and gradients of it that are established in tissue and cell cultures, and of its influence on endogenous auxin synthesis/ degradation and other aspects of cellular metabolism.

Ethylene

Morgan and Hall (1962, 1964) showed in work with cotton, that the production of ethylene is closely linked to the level of exogenous auxin applied; and later (Burg and Burg, 1968; Kang et al., 1971) it became clear that rates of ethylene production in tissues are related to their *endogenous* auxin levels. Mackenzie and Street (1970) demonstrated a similar dependence of ethylene production by cultured sycamore cells upon a supply of 2,4-D. The pattern of ethylene production in the cultures showed a sharp peak late in the exponential growth phase at a time when cell aggregation was maximal; immediately following this the cells underwent rapid expansion and the aggregates began to disperse. LaRue and Gamborg (1971) found a similar pattern in cell cultures from a number of species. Although these investigators could not replace the need for an exogenous auxin by supplying ethylene, it is possible, following upon the work of Burg and Burg (1968), that many of the effects of auxin (perhaps aside from promotion of cell division) may be mediated via auxin control of ethylene synthesis. As Osborne (1976) has suggested, it remains possible that gradients of auxin and ethylene set up between cells could, through their interaction, account for many aspects of the control of the rate, extent, and orientation of cell growth. If this proves to be the case, ethylene could be a critical factor in morphogenic expression.

Recent work by Mattoo and Lieberman (1977) suggests that the ethylene synthesizing system could be located at the cell membrane-cell wall complex. They observed that isolated apple protoplasts produced very little or no ethylene compared with whole cell suspensions, that ethylene synthesizing ability was restored as the protoplasts began to regenerate a cell wall, that isolated apple mitochondria and microsomes did not synthesize ethylene, and that ionophore A23187 and difluorobenzene (which cross-links phospholipids with phospholipids and with membrane proteins) both inhibited ethylene synthesis. This evidence for ethylene synthesis at the cell membrane–cell wall interface is of particular interest in view of the parallel evidence that the action of 2,4-D may be similarly located. King (1976) found that the growth response of sycamore cell suspensions to 2,4-D was correlated with the concentration of the regulator in the spent medium and not with its concentration in the intracellular pools. Auxins bind to plasmalemma preparations (Venis, 1977) and influence the activity of enzymes, including the isozymes of peroxidase (Arnison and Boll, 1976) and IAA oxidase (Lee, 1972) that are known to be located in the plasmalemma–cell wall complex.

Although the promotion of adventitious root development in response to exogenous ethylene (Zimmerman and Hitchcock, 1933) and the promotion of apogamy in gametophyte colonies of *Pteridium aquilinum* (Elmore and Whittier, 1973) by endogenously produced or exogenously supplied ethylene involve promotion of organized growth, nevertheless there are a number of instances where the action of ethylene is antagonistic to organization or cytodifferentiation and promotive of unorganized proliferation. At the level of cytodifferentiation we have, for example, the suppressive action of ethylene on chloroplast differentiation in spinach cultures (Dalton and Street, 1976). At the level of promotion of callus initiation and unorganized growth either by ethylene or by 2-CEPA (2-chloroethylphosphonic acid) we have studies with sweet potato (Chalutz and de Vay, 1969), Algerian Ivy and Virginia Gold tobacco (Stoutemyer and Britt, 1970), and *Begonia* × *richmondensis* (Ringe, 1972). 2-CEPA has also been shown to induce additional mitoses in wheat pollen by its application to the donor plants (Bennett and Hughes, 1972) and to promote callus formation from rice anthers (Wang et. al., 1974). Perhaps, however, the most interesting instance is that of the blocking of embryo organization in wild carrot cultures by 2-CEPA (Wochok and Wetherell, 1972) or by ethylene (Huang, 1971).

These observations suggest that the suppression of organized growth by high levels of auxins may be related to their promotion of ethylene synthesis, and that induction of organogenesis or embryogenesis by auxin removal may, at least in part, be due to a falling level of endogenous ethylene in the cells. In particular cases it may therefore be important to study the effect of an experimental reduction in the intracellular level of ethylene, without alteration in the concentrations of the other growth regulators presented. Since ethylene is active at very low concentrations, ethylene absorbents, such as mercuric perchlorate (Young et al., 1951), although effective in reducing the ethylene level in the gas phase of the cultures, may not sufficiently reduce the intracellular level. Dr. O. Reuveni and I are therefore currently establishing conditions whereby cultured cells can be grown at flowing reduced gas pressures (0.1–0.2 atmospheres) while ensuring the availability of an adequate partial pressure of oxygen.

Cytokinins

Tobacco stem pith tissue (Murashige and Skoog, 1962), soybean cotyledonary callus (Miller, 1968), and carrot root tissue (Letham, 1966) can, under appropriate conditions, be used for cytokinin bioassay. These are the special cases of strict dependence upon exogenous cytokinin for growth. Many cultured tissues are strictly cytokinin independent; many are capable of growth in absence of cytokinin, although the growth rate can be increased by an appropriate exogenous supply.

Cytokinin-independent lines have arisen spontaneously (Fox, 1963; Miller, 1969) and can be isolated following appropriate physical (Syõno and Furuya, 1971), chemical (Bednar and Linsmaier-Bednar, 1971; Kaminek and Lustinec, 1974), or tumor-inducing treatments (Braun and White, 1943; Hagen and Marcus, 1973). Such cytokinin-independent lines have been demonstrated to contain endogenous cytokinins (Mackenzie and Street, 1972; Einset and Skoog, 1973; Miller, 1974), the highest levels being detected in cytokinin-independent tumor cells (Miller, 1975). Since cytokinin dependence and independence are reversible, the changes involved have been thought to be under epigenetic control (Meins, 1974).

Cytokinin independence or dependence is never absolute. Sycamore cell suspensions are strictly cytokinin independent as normally propagated. Nevertheless, by holding the cells for a sufficient time in stationary phase and then subculturing at a sufficiently low cell density, it is possible to demonstrate a need for exogenous cytokinin for the reinitiation of growth (Mackenzie et al., 1972). Thus particular growth regulators, for which actively growing cells are independent by virtue of endogenous synthesis, may be needed for the activation of quiescent cells in which biosynthetic pathways are repressed. Similarly, careful standardization of the cultural conditions is essential for the use of the so-called strictly cytokinin-dependent tissues in bioassays. Einset (1977) has shown that a strain of tobacco callus that is cytokinin dependent when grown in presence of less than 3 μM IAA becomes cytokinin independent in presence of 11.4 μM IAA. Soybean callus that is cytokinin dependent in presence of NAA becomes capable of slow growth in absence of cytokinin when supplied with 2,4-D (Witham, 1968). The growth response to cytokinin is therefore critically modified by the level and kind of auxin.

Sycamore cell suspensions, although cytokinin independent, are auxin dependent, and highest growth rates are obtained when 2,4-D is supplied as the auxin. However, in the presence of a high level of cytokinin (10 mg l^{-1} K), the tissue becomes auxin autotrophic, although under this condition the suspension is more aggregated and deeper in color (Simpkins et al., 1970). Einset (1977), in studying the influence of auxin on the cytokinin dependence of tobacco callus, also reported that as the cytokinin level in the medium is increased, the level of IAA required for maximum growth decreases, and that occasionally tissue supplied 2.5 or 5.0 μM 2iP becomes auxin autotrophic. Here, then, we have instances of auxin substituting for a cytokinin and cytokinin for an auxin requirement. Unless we are to regard these as direct substitutions one for another, we must recognize, when considering the action of an externally applied growth regulator, that its activity may reflect not only its direct action or its interaction with exogenously supplied or already established level(s) of a growth regulator (or regulators), but also its modifying influence on the synthesis and/or degradation of other key regulatory substances. This clearly adds an extra dimension to the interpretation of the action of growth regulatory substances on growth and morphogenesis.

Gibberellins

The gibberellins would seem to be of particular interest in the chemical regulation of cytodifferentiation and morphogenesis because of the rather strong evidence that they modify DNA template activity (Jacobsen, 1977) and the possibility for restricted specific activity by individual gibberellins within the large number now known to occur in higher plants. Nevertheless, as described above, we have only the single case of suppression of shoot formation by GA in tobacco callus and so far only three species on which GA stimulation of shoot bud formation is recorded. In relation to cytodifferentiation, only very few instances of GA activity have been reported: inhibiton of xylogenesis in cultured tuber explants of *Helianthus tuberosus* (Minocha and Halperin, 1974) and in *Phaseolus* callus (Haddon and Northcote, 1976), and promotion of fiber differentiation in cultured *Gossypium* ovules (Beasley, 1977).

The pioneer, but in many ways inadequate study of the action of a gibberellin fraction (a mixture containing GA_1, GA_2, and GA_3) by Nickell and Tulecke (1959) upon 49 callus lines from some 25 species showed them to be relatively insensitive to exogenous GA. The only apparent case of a gibberellin-dependent callus is that of staminate cone origin from *Cupressus funebris* (Straus and Epp, 1960).

Nevertheless, there are now several instances of the stimulation of culture growth by GA due mainly or entirely to its promotion of cell expansion: tobacco callus (Murashige, 1965; Helgeson and Upper, 1970; Lance et al., 1976) and suspensions of *Spinacia oleracea* and *Rosa* Paul's Scarlet (S. Fry in our laboratory, unpublished). In *Rosa* culture, 10^{-8}M GA enhances growth in the presence of 2.3 μM K but is inhibitory if K is omitted. Perhaps of greater interest, however, are the studies with sycamore (Stuart and Street, 1969, 1971) and carrot (Syõno and Furuya, 1968) cultures where GA has been shown to be a critical factor for the initiation of cell division in stationary phase cells inoculated at low cell densities. More recent work by J. Hall and M. C. Elliott at the Leicester Polytechnic (unpublished) has shown that the positive response to GA of sycamore cells increases as the cells are maintained in stationary phase, so that when the normal culture period is extended from 21 to 38 days, the cultures require for the induction of cell division GA_3 (at 10^{-3}M) even when the initial density established in subculture is in excess of 1.5×10^5 cells ml^{-1}. This response is specific to GA_3 insofar as GA_{13} at 10^{-3}M was found to be ineffective and a mixture of GA_4 and GA_7 severely inhibitory to cell division. Studies on the methanol extractable GA activity in the cells and in the medium showed that at the onset of stationary phase, cellular GA activity was high but declined to below the detectable level by day 30 of culture, and that as this happened GA appeared in the culture medium. Stationary phase sycamore cells are arrested in G_1 of the cell cycle (Gould and Street, 1975), as are cells arrested by the antigibberellins AMO 1618 and Phosphon D. Hall and Elliott showed that antagonism of these antigibberellins by GA_3 permitted the cells to enter S phase and G_2, suggesting that GA is essential for the transition from G_1 into S phase in stationary phase cells. Here, then, as in the discussion on cytokinins, we see the possible importance of GA for the activation of quiescent cells. In line with this are the reports that GA is stimulatory to callus initiation although not a requirement for subsequent growth as in *Citrus medica* mesocarp (Schroder and Spector, 1957), *Nicotiana tabacum* pith callus (Nitsch, 1968), *Croton bonplandianum* endosperm (Johri and Bhojwani, 1977), and *Gossypium* ovules (Stewart and Hsu, 1977). Later in this chapter reference will be made to the possible importance of conditions operating during callus initiation upon the subsequent morphogenic potential of the resulting cultures. In this connection it may therefore be suggested that more attention should be directed to the influence of gibberellins and cytokinins presented to the cells at this stage, and to how far the presence of particular natural cytokinins or gibberellins may exert a selective action on those cells within the explant that participate in this callus formation.

Abscisic Acid

Reference has already been made to the activity of ABA in promoting shoot buds in sweet potato and potato cultures (Yamaguchi and Nakajima, 1974). There is also evidence that it can act synergistically with auxin in the promotion of callus development from citrus buds (Altman and Goren, 1971) and with a high level of kinetin (5 mg l^{-1}) in promotion of the growth of soybean callus (Blumenfeld and Gazit, 1970). Despite the evidence that ABA secreted by the root cap acts as a growth inhibitor in the geotropic response of roots (Shaw and Wilkins, 1973; Kundu and Audus, 1974) and that it is generally a powerful inhibitor of root growth (Torrey, 1976), we have the observations that at 0.0026 mg l^{-1} it can enhance lateral root initation in cultured tomato roots (Collett, 1970); that at 0.26 mg l^{-1} (1 μM) it can enhance the extension growth of *Pisum* root tip segments (Gaither et al., 1975); and that in the range 0.001–0.01 mg l^{-1} it can enhance linear growth and lateral root development in cultured soybean roots (Yamaguchi and Street, 1977).

These promotive effects of low concentrations of ABA are now sufficiently numerous to suggest that it should be much more generally included in experiments on callus initiation and subsequent morphogenesis. The work of Ammirato (1974) certainly implicates ABA at low concentrations (10^{-6} to 10^{-7}M) in the normalization of embryogenesis in caraway cultures.

NUTRIENTS AND METABOLITES IN RELATION TO
ORGANOGENESIS AND EMBRYOGENESIS

Reduced Nitrogen

The experimental data just do not exist at this time to make an overall critical assessment of the role of mineral ions in the control of morphogenic expression. The media of Murashige and Skoog (1962) and of Linsmaier and Skoog (1965) have found very wide application in the regeneration of plants from tissue and cell cultures, and are characterized by having a high level of potassium and by supplying nitrogen as a mixture of nitrate and ammonium. The inclusion of ammonium is in line with studies that have shown the importance of an appropriate source of reduced nitrogen for the achievement of *in virto* embryogenesis in carrot and belladonna (Thomas and Street, 1972) cultures. The quite extensive work on the nitrogen nutrition of carrot cultures has recently been summarized and extended by Wetherell and Dougall (1976). It now seems clear that growth is reduced and embryogenesis virtually suppressed with nitrate as the sole source of nitrogen, that the addition of a low level of ammonium (0.1 mM NH_4Cl) to such a nitrate medium will permit some embryogenesis, that the optimum level is ca. 10 mM NH_4Cl in media containing 12 to 40 mM KNO_3, that the ammonium requirement can be replaced by glutamine or alanine, and that these latter can support good growth and embryogenesis as sole nitrogen sources. One possible explanation of the role of low levels of ammonium, i.e. nitrate, assimilation as a limiting factor in the availability of amino acids for growth and embryo development, can be inferred from studies with soybean suspension cultures, where a 2 mM NH_4Cl addition to the medium resulted in a marked rise in the level of nitrate reductase in the cells (Bayley et al., 1972). It is well known that young zygotic embryos have exacting requirements for amino acids and are lacking in nitrate reductase. Young somatic embryos will therefore presumably be dependent upon the "nurse" cells of the embryogenic aggregates for an appropriate supply of assimilated nitrogen.

Myo-inositol

Myo-inositol is a standard constituent of bacterial and animal cell culture media. The discovery by Pollard et al. (1961) that the neutral fraction they isolated from coconut milk contained *myo*-inositol, *scyllo*-inositol, and sorbitol, and that *myo*-inositol could fully substitute for the growth-promoting activity of this fraction as assayed by their carrot system, led to its general inclusion in plant tissue culture media (Steinhart et al., 1962). It was subsequently shown that *myo*-inositol was significantly stimulatory to callus growth, usually at concentrations of 50-100 mg l^{-1}, in *Ulmus campestris* (Jacquiot, 1964), *Pelargonium hortorum* (Narayana, 1963), tobacco (Linsmaier and Skoog, 1962), and *Catharanthus roseus* (Wood and Braun, 1961), and that it was essential at 10 mg l^{-1} for the growth of *Fraxinus pennsylvanica* callus (Wolter and Skoog, 1966). *Myo*-inositol has also been reported to promote bud formation in cultures of *Ulmus campestris* (Jacquiot, 1966) and embryogenesis in carrot (Norreel and Nitsch, 1968). The role of *myo*-inositol in tissue cultures, however, remains uncertain, although clearly it is involved in the synthesis of phospholipids and in cell wall pectins (Anderson and Wolter, 1966).

Cyclic AMP

It is tempting to conclude from evolutionary considerations that cyclic AMP must be present and perform regulatory functions in higher plants similar to those it performs in animals, and if this proved to be so then it would become a key metabolite in considering the activity of regulatory substances. However, as Amrhein (1977) has pointed out, the evidence for the natural existence and synthesis of cyclic AMP in higher plants remains suspect and requires us to set aside, at this time, any attempt to interpret the observed responses of higher plants to its external application (these data are well summarized by Sachar et al., 1975). In this connection there is little evidence that the growth regulatory substances (historically termed phyto-hormones) that are discussed above resemble in their mechanism of action the peptide and catecholamine hormones of animals for which cyclic AMP acts as an intracellular mediator (Robison et al., 1971), nor are plant cells normally subjected in their natural environment to "sugar hunger" (Bitensky and Gorman, 1973).

MORPHOGENIC EXPRESSION

The system now being considered is that of callus initiation from a primary organ explant, the growth of the resulting culture as a callus or as a cell suspension, and then the initiation of organs or embryos either spontaneously or after transfer to a morphogenic culture medium. The initiation of the callus involves the induction of unorganized cell division in previously nondividing tissue cells and/or in cells previously undergoing controlled and precisely orientated cell divisions (cambial cells, cells of apical or intercalary meristems). A satisfactory callus induction medium is one that rapidly leads to the appearance of actively proliferating callus. The second stage is the development, in this unorganized callus or suspension, of organized growth leading to organogenesis or embryogenesis, and a satisfactory morphogenic medium is one in which morphogenic expression is rapid and prolific. Such a system is to be distinguished from meristem culture, in which organized growth is maintained throughout, and in which organogenesis occurs from growth centers (incipient primordia) directly derived from those in the primary explant.

Halperin in 1967 introduced, in relation to carrot cultures, the concept that the achievement of embryogenic competence (capacity to express inherent totipotency) occurred during the initiation of the culture from the primary explant, and that the embryos developed exclusively from cell clumps derived from such competent cells. The primary cultures were therefore thought to contain mixed populations of competent and non-competent cells. This offered an explanation of why different cultures (even from the same plant and type of explant) differed in their yield of embryos in morphogenic media and suggested that active proliferation of the competent cells must be essential for retention of the capacity for embryogenesis, whereas conditions preferentially favoring growth of the noncompetent cells would lead to a decline in embryo yield. Strong support for this concept of intercellular competition comes from the work of Smith and Street (1974), who showed that suspension cultures declining in their ability to yield embryos could, by plating, be made to yield cultures of fully restored yield and cultures incapable of yielding embryos. The plating procedure had enabled lines containing only, or predominantly, competent cells to be separated from the noncompetent cells. The basis of the intercellular competition here postulated to underlie changes in embryogenic capacity has to some extent been established by the recent work by Bayliss (1977b). The isolates with fully restored embryo yield, however, again varied in the extent to which they retained this property, and even those that initially showed a stable high yield eventually embarked upon a period during which their embryo yield declined. Examination of the cytology of such cultures showed that their decline in embryogenic capacity was associated with the origin of cells of altered karyotype; aneuploid cells impaired in their capacity to form embryos, but at a selective growth advantage, arose and eventually with repeated subculture became the dominant and ultimately the only kind of cells present. Such cultures were of nil embryogenic capacity and their capacity could not be restored by cloning. The embryogenic capacity of the cultures at any one time is therefore interpreted as reflecting the proportion of competent cells present; this is regarded as being first dependent upon the population composition of the initial culture, subsequently upon competition between competent and noncompetent cells of normal karyotype, and finally upon competition between these and cells of altered karyotype. There are, of course, many cultures from a range of species that lose their morphogenic capacity so rapidly—within the very first few subcultures—that we have in these cases perhaps to make a further postulate that their initial and very transient capacity for morphogenesis depends upon growth regulators and critical metabolites carried over from the initial explant and rapidly diluted out or degraded in culture.

The central concept now being advanced is, then, that morphogenic competence is determined from the time of culture initiation; that the growth regulators and other environmental conditions that are discussed above and that lead to morphogenic expression are permitting such cells to express this competence; and that if such cells are not present no modification of the culture environment will succeed in achieving organogenesis or embryogenesis. This is, I realize, a controversial contention. Against it can be quoted those instances where morphogenic capacity can be restored without cloning or where cultures initially not expressing a particular aspect of morphogenesis come to do so during culture. However, such cases can be accommodated within the present hypothesis if we assume that cultures will appear to have no morphogenic capacity if their proportion of competent cells is sufficiently low and that different aspects of morphogenesis depend upon the presence of cells differing in the nature of their competence (i.e. that the cells capable of initiating rhizogenesis are distinct from those involved in embryogenesis).

It has been convincingly shown using experimentally concocted mixed cell populations in suspension culture, the individual components of which can be separately monitored (cells differing in ploidy or mixtures of competent and noncompetent cells), that the competitive advantage of one of the component populations can be reversed by alteration of the culture medium, the passage length between subcultures, or other cultural conditions (Smith and Street, 1974; Bayliss, 1977c). Whether any particular cell type is at a selective advantage or disadvantage depends upon the culture regime adopted. Furthermore, such changes in cell population composition can be facilitated in work with callus cultures where the progeny of particular cells is stabilized in particular segments of the callus, and these can be selected or used by chance as inoculum for the next culture passage. Hence we would predict alteration in the balance of competent to noncompetent cells or of rhizogenic to embryogenic cells by just the kinds of treatments that have previously been reported to restore or alter qualitatively morphogenic expression.

It is now necessary to consider what evidence there is that the cells responsible for different forms of morphogenesis are distinct and stable in their states of competence at culture initiation. First, there are the many instances where a culture is only rhizogenic or only embryogenic and where cultures persistently lack morphogenic capacity under a wide range of cultural conditions. The first type of culture can be regarded as containing cells with only one state of competence, the second as containing only noncompetent cells. Perhaps of greater interest, however, are cultures initially showing more than one form of morphogenesis but that on continuing culture either show a marked quantitative shift between the two forms of expression or retain only one aspect of morphogenesis. Carrot cultures showing both rhizogenesis and embryogensis can become either solely root-forming or solely embryo-forming. Two carrot cultures were studied in our laboratory by Smith (1973), one predominantly showing, in passage 3, rhizogenesis (R) and the other embryogenesis (E). The morphogenesis displayed by these cultures was then followed in later passages. By passage 15 the initially rhizogenic culture now showed predominantly embryogenesis, whereas the culture showing only occasional roots remained more stable (table 1). The R culture clearly contained two size classes of aggregates, and the use of the 1.5 mm diameter cannula for subculture tended to select for the smaller aggregates which appeared to be those yielding the embryos. The passage 15 R culture was therefore fractionated into three size classes: the larger aggregates, which rapidly sedimented and could be retrieved by decantation; smaller aggregates, present in the decantation fluid and retained by filtration through a layer of surgical gauze; and a fine suspension, which passed the gauze filter. These three fractions were washed with "staled" medium, resuspended, and transferred separately to medium lacking 2,4-D, and embryo and root counts were made after three weeks' incubation (table 2). Clearly this selection resulted in fractionation of the culture into one yielding only

TABLE 1

CHANGES IN MORPHOGENIC EXPRESSION OF TWO CARROT SUSPENSION CULTURES

Culture	Passage No.	Embryo Count (e)	Root Count (r)	e/r
R	3	24	303	0.08
	4	3	187	0.01
	6	177	160	1.1
	15	648	57	11.4
E	3	200	0	∞
	4	930	0	∞
	6	350	3.0	117
	15	2360	12.5	189

NOTE: Both carrot suspension cultures (R and E) were propagated in Murashige and Skoog medium containing 0.1 mg l^{-1} 2,4-D by transfer every 21 days of 0.5 mg culture to 25 ml new medium by automatic pipette fitted with a cannula 1.5 mm in diameter. They were transferred to medium lacking 2,4-D for morphogenic expression. Embryo and root counts include all greater than 0.5 mm in length. (Unpublished data of Smith, 1973.)

TABLE 2

MORPHOGENIC EXPRESSION OF THREE SIZE FRACTIONS

Mean Aggregate Size of Fraction, $mm^3 \times 10^{-2}$	Embryo Count (e)	Root Count (r)	e/r
> 0.3	183	0	
> 7.5	569	18	31.6
>22.0	322	139	2.31

NOTE: Fractions were separated from the fifteenth passage R cultures (table 1) and cultured in medium lacking 2,4-D for 21 days. (Unpublished data of Smith, 1973.)

embryos (fine suspension) and one yielding a higher proportion of roots than the original culture (large aggregates). Plating of the fine suspension yielded clones all of which formed only embryos when tested over several passages. These data strongly support the view that the culture showing two forms of morphogenesis contains two cell populations capable of separation and that, at least in the case of the embryogenic fraction, this then showed no capacity to form roots. The observation by Kessell and Carr (1972) that a carrot suspension culture can be reversibly switched between embryogenesis and rhizogenesis by change in the oxygen tension can be reconciled with this interpretation simply by saying that low oxygen favors expression by the embryogenic cells and high oxygen by the rhizogenic cells. In fact, our limited observations on this phenomenon suggest that the culture shows embryogenesis at both levels of oxygen supply but that the high oxygen condition promotes abnormal development of the root pole of the young embryos.

RETENTION OR ACHIEVEMENT OF MORPHOGENIC COMPETENCE

Narayanaswamy (1977), reviewing regeneration from tissue cultures, makes two statements that can well introduce this section of the review: (1) "The source of the explant cultured is important in determining regeneration potential," and (2) "The physiological age of the explant is another factor which exercises an influence on organ formation." In support of his second point he cites the interesting work of Raju and Mann (1970), which showed that *Echeveria* leaf explants from young leaves initiated only roots, those from older leaves only shoot buds.

Stoutemyer and Britt (1963) have established calluses from adult and juvenile stem segments of *Hedera helix* and have claimed that during many subcultures the calluses from juvenile stems maintain a significantly higher growth rate than those from adult segments and have a higher capacity for root formation. This implies that the cells do not revert to a common physiological state during callus induction but retain respectively juvenile and adult characters. This contention receives some support from work with tree species. Reports of bud formation in cultures from gymnosperm trees refer almost entirely to calluses of seedling or juvenile origin (Isikawa, 1974; Cheng, 1975). Judith Webb, working in our laboratory, has encountered just this problem with cultures of *Pinus contorta* and *Picea sitchensis* (unpublished work). Winton and Shirley (1974) obtained callus but failed to achieve regeneration using explants from 15- to 20-year-old trees of Douglas fir. Morphogenic callus of *Populus* was also obtained from very young (1- to 2-year-old) trees (Wolter, 1968; Chalupa, 1974).

There are also a number of instances where the physiological state of the explants taken from within the same plant seems to influence the morphogenic capacity of the resulting culture. Aghion-Prat (1965), in studies on stem explants and derived callus from a nonphotoperiodic tobacco, could induce only vegetative shoot buds in tissue derived from the basal nodes, whereas flower buds arose from tissue derived from upper segments of stem and particularly from inflorescence explants. Steward et al. (1964) reported the most prolific formation of embryos in cell suspensions derived from carrot embryos. In *Cuscuta reflexa*, embryos were readily obtained from cultures derived from excised immature embryos (Maheshwari and Baldev, 1961). Although there are exceptions, it is still generally true that rhizogenesis in culture is most easily obtained from species that readily form adventitious roots.

Of particular interest here is the observation that plantlets arising in culture via embryogenesis often show spontaneous origin of further embryos at their surface: this has now been observed in *Atropa belladonna* (Rashid and Street, 1973). carrot (McWilliam et al., 1974), *Datura innoxia* (Geier and Kohlenbach, 1973), *Ranunculus sceleratus* (Konar et al., 1972), and *Brassica napus* (Thomas et al., 1976). In *Ranunculus sceleratus* and in *Brassica napus* it has been established that these embryos arose from single cells of the shoot-axis epidermis, cells already well advanced along a pathway of differentiation. Here, as in the natural or rare origin of embryos from synergids and antipodal cells of the embryo sac, we see the retention of competence for embryogenesis in cells of specialized function. This raises the possibility that differentiation—or at least some pathways of differentiation—and competence to embark upon morphogenesis are not necessarily incompatible. Further, since we have presented the evidence that only some of the cells induced to participate in callus formation may be competent, we have the possibility that, whereas all cells whose genome is not impaired by differentiation may be theoretically totipotent, the cells that actually express this totipotency in culture are just those cells in which totipotency has been *physiologically* retained within the differentiating organ. If this is the case, then there may be no such process as achieving competence (dedifferentiation) during the stage of activation of quiescent cells into cells capable of active proliferation, but simply the activation of the mitotic cycle in such competent cells, which then can express their totipotency in a morphogenic medium. The problem of obtaining a morphogenic culture then resolves itself into the need to have an initiation medium that will ensure that the competent cells are involved in callus formation.

To complete this train of reasoning it is necessary to consider some recent work on pollen development in tobacco anthers and its relevance to the origin of pollen embryos when the excised anthers are placed in culture (Horner and Street, 1978). During the ripening of pollen in the tobacco anther, two kinds of pollen grain can be distinguished at the binucleate stage: normal (N) grains characterized by their high frequency, large size, deeply staining cytoplasm, and high starch content; and small (S) grains characterized by their variable and low frequency and weakly staining cytoplasm. From assessment of the frequency of embryo formation in cultured anthers, from correlations between variation in this frequency and variations in embryo yield even between individual pollen sacs of the same anther, and from detailed measurement of the size of pollen grains showing the first stage of embryogenesis, it can be concluded that only the S grains participate in embryogenesis. Studies by Dale (1975) on a similar natural dimorphism in barley anthers have again pointed to the origin of pollen callus only from the infrequent S-type grains. LaCour (1949), in studying a similar dimorphism in the pollen of *Tradescantia bracteata*, proposed that the S grains arose as a result of an abnormal angle between the two spindle axes during the second meiotic division in the pollen mother cells. If this is so, then the competence to embark upon embryogenesis is determined at meiosis and retained. To assess the significance of these observations, we now need to know if natural pollen dimorphism is a consistent feature of anthers yielding pollen embryos, to examine whether such dimorphism can be experimentally induced in species lacking it, and to determine whether the consequence will be that such anthers will then become embryogenic.

Although it is difficult to set these observations aside, the generalization that conversion of noncompetent into competent cells does not occur during culture initiation must be treated with some reservation. There is considerable evidence for the importance of induced division in altering the potential of cells for cytodifferentiation (Street, 1977), and it is possible that bringing tissue cells out of a G_0 state back in G_1 and hence into mitotic cycle may offer an opportunity to establish new patterns of gene transcription. We have earlier referred to the activity of gibberellins and cytokinins in activating arrested cells. The difficulty is to find any clear-cut examples of where an experimental treatment at the time of culture initiation has unambiguously modified the competence of the resulting culture. In our experience (Dix, 1975) callus cultures of *Nicotiana sylvestris* readily showed shoot bud and root initiation but do not yield embryos. Very recently Dr. B. J. Cox (Biosciences Group, Corporate Laboratory of Imperial Chemical Industry, Runcorn), working on regeneration from isolated haploid leaf protoplasts of this species, has observed the resulting small callus colonies to give rise to what clearly appear to be embryos. If this can be confirmed by detailed anatomical study, it would suggest that here G_0 arrested leaf cells had yielded cells with embryogenic competence as a result of the shock of protoplast isolation and the consequent wall regeneration and activation of the division cycle.

SOME GENERAL CONCLUSIONS

Clearly the chemical control of morphogenic expression is a complex

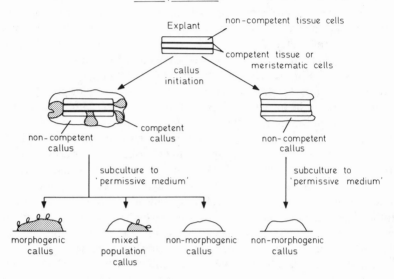

Callus formation involving promotion or failure to promote division
in competent cells

Explant   non-competent tissue cells

                           competent tissue or
           callus   meristematic cells
           initiation

non-competent   competent   non-competent
callus   callus   callus

       subculture to   subculture to
       'permissive medium'   'permissive medium'

morphogenic   mixed   non-morphogenic   non-morphogenic
callus   population   callus   callus
               callus

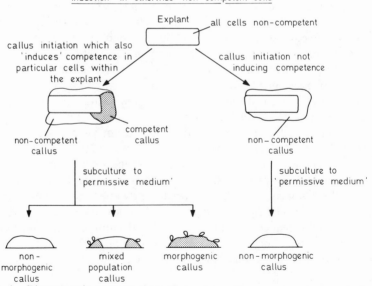

Callus formation involving induction of competence or failure of such
induction in otherwise non competent cells

                    Explant   all cells non-competent

callus initiation which also
   'induces' competence in   callus initiation not
particular cells within   inducing competence
   the explant

non-competent   competent   non-competent
callus   callus   callus

       subculture to   subculture to
       'permissive medium'   'permissive medium'

non-   mixed   morphogenic   non-morphogenic
morphogenic   population   callus   callus
callus   callus

Fig. 1. Diagrammatic representations of morphogenetic expressions in competent and noncompetent cells of a cultured tissue explant.

process involving multiple interactions of growth regulators and critical metabolites. It may be even more complex if there are further natural growth regulators to be discovered (the plant lectins? [Howard et al., 1977]). To interpret the action of morphogenic media, we need to know much more about the endogenous levels of natural regulators established in culture and to understand better how exogenous regulators intervene in the metabolism of cultured cells.

Morphogenic media are interpreted as permitting the cells to express their morphogenic competence. Such morphogenic competence apears to be a stable physiological state only destroyed by karyotype breakdown. The changing morphogenic expression or loss of such expression is explained in terms of competition between competent and noncompetent cells. This competence to express inherent totipotency may be retained during cytodifferentiation, and at this time it is uncertain whether the competence of cells, which confers upon cultures the capacity for morphogenesis, arises by an inductive phenomenon during culture initiation or results from already competent cells present in the primary explant participating in culture formation. Conditions during callus development, including the composition of the initiation medium, may determine the morphogenic potential by selectively promoting division in the competent cells of the explant (fig. 1).

ACKNOWLEDGMENTS

I wish to express my thanks to my research colleagues, N. P. Everett, S. Fry, T. Wang, Judith Webb, and Lyndsey Withers, for drawing my attention to relevant recent references and for helping me locate references regarding which I had imcomplete citations.

I am also indebted to those who have allowed me to draw on their unpublished work and who are acknowledged at the appropriate points in the text.

REFERENCES

Aghion-Prat, D. 1965. Néoformation de fleurs *in vitro* chez *Nicotiana tabacum* L. Physiol. Végét. 3:229–303.

Altman, A., and R. Goren. 1971. Promotion of callus formation by abscisic acid in citrus bud cultures. Plant Physiol. 47:844–46.

Ammirato, P. V. 1974. The effects of abscisic acid on the development of somatic embryos from cells of caraway (*Carum carvi* L.). Bot. Gaz. 135:328–37.

Amrhein, N. 1977. The current status of cyclic AMP in higher plants. Ann. Rev. Plant Physiol. 28:123–32.

Anderson, L., and K. E. Wolter. 1966. Cyclitols in plants: Biochemistry and physiology. Ann. Rev. Plant Physiol. 17:209–22.

Arnison, P. G., and W. G. Boll. 1976. The effect of 2,4-D and kinetin on peroxidase activity and isoenzymic pattern of cotyledon cell suspension cultures of bush bean. Canad. J. Bot. 54:1857–67.

Barba, R., and L. G. Nickell. 1969. Nutrition and organ differentiation in tissue cultures of sugar cane—a monocotyledon. Planta 89:299–302.

Bayley, J. M., J. King, and O. L. Gamborg. 1972. The effect of the source of inorganic nitrogen in growth and enzymes of nitrogen assimilation in soybean and wheat cells in suspension cultures. Planta 105:15–24.

Bayliss, M. W. 1975. The effect of growth *in vitro* on the chromosome complement of *Daucus carota* (L.) suspension cultures. Chromosoma 51:401–11.

Bayliss, M. W. 1977a. The effect of 2,4-D on growth and mitosis in suspension cultures of *Daucus carota*. Plant Sci. Lett. 8:99–103.

Bayliss, M. W. 1977b. The cause of competition in mixtures of two suspension culture lines of *Daucus carota*. Protoplasma 92:117–28.

Bayliss, M. W. 1977c. Factors affecting the frequency of tetraploid cells in a predominantly diploid suspension culture of *Daucus carota*. Protoplasma 92:109–16.

Beasley, C. A. 1977. Ovule culture. Fundamental and pragmatic research for the cotton industry. *In* J. Reinert and Y. P. S. Bajaj (eds.), Applied and fundamental aspects of plant cell, tissue, and organ culture. Pp. 160–78. Springer-Verlag, Berlin.

Bednar, T. W., and E. M. Linsmaier-Bednar. 1971. Induction of cytokinin-independent tobacco tissues by substituted fluorenes. Proc. Nat. Acad. Sci. USA 68:1178–79.

Bennett, M. O., and W. G. Hughes. 1972. Additional mitosis in wheat pollen induced by ethrel. Nature 240:566–68.

Bitensky, N. W., and R. E. Gorman. 1973. Cellular responses to cyclic AMP. Progr. Biophys. Mol. Biol. 26:409–61.

Blumenfeld, A. and S. Gazit. 1970. Interactions of kinetin and abscisic acid in the growth of soybean callus. Plant Physiol. 45:535–36.

Braun, A. C., and P. R. White. 1943. Bacteriological sterility of tissues derived from secondary crown gall tumors. Phytopathology 33:85–100.

Bui Dang Ha, D. 1974. Regeneration of complete plants of *Asparagus* from protoplast cultures. 3rd Intern. Congr. Plant Tissue Cell Culture (abstr. 261). Leicester, England.

Burg, S. P., and E. A. Burg. 1968. Auxin stimulated ethylene formation: Its relationship to auxin inhibited growth, root geotropism, and other plant processes. *In* F. Wightman and G. Setterfield (eds.), Biochemistry and physiology of plant growth substances. Pp. 1275–94. Runge Press, Ottawa.

Campbell, R. A., and D. J. Durzan. 1975. Induction of multiple buds and needles in tissue cultures of *Picea glauca*. Canad. J. Bot. 53:1652–57.

Chalupa, V. 1974. Control of root and shoot formation and production of trees from poplar callus. Biol. Plantarum 16:316–20.

Chalutz, E., and J. E. de Vay. 1969. The production of ethylene *in vitro* and *in vivo* by *Ceratocystis fimbriata* in relation to disease development. Phytopathology 59:750.

Chandra, N., T. H. Lam, and H. E. Street. 1977. The effects of selected aryloxyalkane-carboxylic acids on the growth and embryogenesis of a suspension culture of carrot (*Daucus carota* L.). Z. Pflanzenphysiol. 86:55–60.

Cheng, T. Y. 1975. Adventitious bud formation in cultures of Douglas fir, *Pseudotsuga menziesii*. Plant Sci. Lett. 5:97–102.

Chin, J. C., and K. J. Scott. 1977. Studies on the formation of roots and shoots in wheat callus cultures. Ann.Bot. 41:473–81.

Collett, G. F. 1970. Action de l'acide abscissique sur la rhizogenése. C. R. Acad. Sci. Paris 271D:667–70.

Dale, P. J. 1975. Pollen dimorphism and anther culture in barley. Planta 127:213–20.

Dalton, C. C., and H. E. Street. 1976. The role of the gas phase in the greening and growth of illuminated cell suspension cultures of spinach (*Spinacia oleracea* L). In Vitro 12:485–94.

deFossard, R. A., C. Nitsch, R. Cresswell, and E. C. M. Lee. 1974. Tissue and organ culture of *Eucalyptus*. New Zealand J. For. Sci. 4:267–78.

Dix, P. J. 1975. Plant cell lines resistant to environmental stresses. Ph.D. thesis, University of Leicester, Leicester, England.

Durand, J., I. Potrykus, and G. Donn. 1973. Plantes issuées de protoplastes de *Petunia*. Z. Pflanzenphysiol. 69:26–34.

Earle,. E. D., and R. W. Langhams. 1974. Propagation of *Chrysanthemum in vitro*. II. Production, growth, and flowering of plantlets from tissue cultures. J. Amer. Soc. Hort. Sci. 99:352–58.

Einset, J. W. 1977. Two effects of cytokinin on the auxin requirement of tobacco callus cultures. Plant Physiol. 59:45–47.

Einset, J. W., and F. Skoog. 1973. Biosynthesis of cytokinins in cytokinin-autotrophic tobacco callus. Proc. Nat. Acad. Sci. USA 70:658–60.

Elmore, H. W. and D. P. Whittier. 1973. The role of ethylene in the induction of apogamous buds in *Pteridium* gametophytes. Planta 111:85–90.

Engelke, A. L., H. Q. Hamzi, and F. Skoog. 1973. Cytokinin-gibberellin regulation of shoot development and leaf form in tobacco plantlets. Amer. J. Bot. 60:491–95.

Feng, K. A., and A. J. Linck. 1970. Effects of N-1-naphthylphthalamic acid on the growth and bud formation of tobacco callus grown *in vitro*. Plant Cell Physiol. 11:589–98.

Feung, C., R. H. Hamilton, and R. O. Mumma. 1976. Metabolism of 2,4-D. Identification of metabolites in rice root callus tissue cultures. J. Agric. Food Chem. 24:1013–15.

Feung, C., R. H. Hamilton, and R. O. Mumma. 1977. Metabolism of indole-3-acetic acid. Plant Physiol. 59:91–93.

Fox, J. E. 1963. Growth factor requirement and chromosome number in tobacco tissue cultures. Physiol. Plantarum 16:793–803.

Gaither, D. H., D. H. Lutz, and L. E. Forrence. 1975. Abscisic acid stimulates elongation of excised pea root tips. Plant Physiol. 55:948–49.

Gamborg, O. L., and D. E. Eveleigh. 1968. Culture methods and detection of glucanases in suspension cultures of wheat and barley. Canad. J.Biochem. 46:417–21.

Gautheret, R. J. 1945. La culture des tissus. Gallimard, Paris.

Gautheret, R. J. 1955. The nutrition of plant tissue cultures. Ann. Rev. Plant Physiol. 6:433–84.

Geier, T., and H. W. Kohlenbach. 1973. Entwicklung von Embryonen und embryogenem Kallus aus Pollenkörnern von *Datura metaloides* und *Datura innoxia*. Protoplasma 78:381–96.

Gould, A. R., and H. E. Street. 1975. Kinetic aspects of synchrony in suspension cultures of *Acer pseudoplatanus* L. J.Cell Sci. 17:337–48.

Haddon, L., and D. H. Northcote. 1976. The influence of gibberellic acid and abscisic acid on cell and tissue differentiation of bean callus. J. Cell Sci. 20:47–55.

Hagen, G. L., and A. Marcus. 1973. Cytokinin effects on the growth of quiescent tobacco pith cells. Plant Physiol. 55:90–93.

Halperin, W. 1967. Population density effects on embryogenesis in carrot cell cultures. Exp. Cell Res. 48:170–73.

Helgeson, J. P., and C. D. Upper. 1970. Modification of logarithmic growth rates of tobacco callus by gibberellic acid. Plant Physiol. 46:113–17.

Horner, M., and H. E. Street. 1978. Pollen dimorphism—origin and significance in pollen plant formation by anther culture. Ann. Bot. 42:763–71.

Howard, J., L. Shannon, L. Oki, and T. Murashige. 1977. Soybean agglutinin—a mitogen for soybean callus cells. Exp. Cell Res. 107:448–50.

Huang, S. 1971. M. S. thesis, University of Connecticut (cited from Wochok and Wetherell, 1971).

Ingram, D. S., and D. N. Butcher. 1972. Benazolin (4-chloro-2-oxybenzothiazolin-3yl-acetic acid) as a growth factor for initiating and maintaining plant tissue cultures. Z. Pflanzenphysiol. 66:206–14.

Isakawa, H. 1974. *In vitro* formation of adventitious buds and roots on hypocotyls of *Cryptomeria japonica*. Bot. Mag. Tokyo 87:73–77.

Jacobsen, J. V. 1977. Regulation of ribonucleic acid metabolism by plant hormones. Ann. Rev. Plant Physiol. 28:537–64.

Jacquiot, C. 1964. Application de la technique de culture des tissue végétaux à l'étude de quelques problèmes de la physiologie de l'arbe. Ann. Sci. Forestières 21:317–473.

Jacquiot, C. 1966. Plant tissues and excised organ cultures and their significance in forest research. J. Inst. Wood Sci. 16:22–34.

Johri, B. M., and S. S. Bhojwani. 1977. Triploid plants through endosperm culture. *In* J. Reinert and Y. P. S. Bajaj (eds.), Applied and fundamental aspects of plant cell, tissue, and organ culture. Pp. 398–411. Springer-Verlag, Berlin.

Kaminek, M., and J. Lustinec. 1974. Induction of cytokinin autonomy and cholorphyll deficiency to tobacco callus tissue by streptomycin. Z. Pflanzenphysiol. 73:74–81.

Kang, B. G., W. Newcomb, and S. P. Burg. 1971. Mechanism of auxin-induced ethylene production. Plant Physiol. 47:504–9.

Kessel, R. H. J., and A. H. Carr. 1972. The effects of dissolved oxygen concentration on growth and differentiation of carrot (*Daucus carota*) tissue. J. Exp. Bot. 23:996–1007.

King, P. J. 1976. Studies on the growth in culture of plant cells. XX. Utilisation of 2,4-dichlorophenyoxacetic acid (2,4-D) by steady state cell cultures of *Acer pseudoplatanus*. J. Exp. Bot 27:263–76.

Kochhar, T. S., P. R. Bhalla, and P. S. Sabharwal. 1970. Formation de bourgeons végétatifs par des cals de tabac sous l'influence d'un agent de chelation—l'acide 1-3-diamino-2-hydroxpropane-N,N,N^1,N^1-tetracetique (DHPTA). C. R. Acad. Sci. Paris 271:1619–22.

Kochhar, T. S., P. R. Bhalla, and P. S. Sabharwal. 1971. The effect of tobacco smoke components on organogenesis in plant tissues. Plant Cell Physiol. 12:603–8.

Kohlenbach, H. W. 1977. Basic aspects of differentiation and plant regeneration from cell and tissue cultures. *In* W. Barz, E. Reinhard, and M. H. Zenk (eds.), Plant tissue culture and its biotechnical application. Pp. 355–66. Springer-Verlag, Berlin.

Konar, R. N., and A. Konar. 1973. Interaction of CCC and gibberellic acid in the morphogenesis of *Ranunculus sceleratus* tissue *in vitro*. Phytomorphology 23:105–8.

Konar, R. N., E. Thomas, and H. E. Street. 1972. Origin and structure of embryoids arising from the epidermal cells of *Ranunculus sceleratus* L. J. Cell Sci. 11:77–93.

Kulescha, Z. 1977. Étude de la dégradation, l'absorption et la fixation de l'acide 3-indolyl acétique par le tissu de topinambour. *In* R. J. Gautheret (ed.), La culture des tissus et des cellules des végétaux. Pp. 59–66. Masson, Paris.

Kundu, K. K., and L. J. Audus. 1974. Root growth inhibitors from root tips of *Zea mays*. L. Planta 117:183–86.

LaCour, L. F. 1949. Nuclear differentiation in the pollen grain. Heredity 3:319–37.

Lam, T. H., and H. E. Street. 1977. The effect of selected aryloxyalkanecarboxylic acids on the growth and levels of soluble phenols in cultured cells of *Rosa damescena*. Z. Pflanzenphysiol. 84:121–28.

Lance, B., D. M. Reid, and T. A. Thorpe. 1976. Endogenous gibberellins and growth of tobacco callus cultures. Physiol. Plantarum 36:287–92.

Lang, H., and H. W. Kohlenbach. 1975. Morphogenese in Kulturen isolierter Mesophyllzellen von *Macleaya cordata*. *In* H. Y. Mohan Ram, J.J. Shah, and C. K. Shah (eds.), Form, structure, and function in plants. Pp. 125–33. Sarita Prakashan, Meerut, India.

LaRue, T. A. G., and O. L. Gamborg. 1971. Ethylene production by plant cell cultures. Plant Physiol. 48:394–98.

Lee, T. T. 1972. Changes in IAA oxidase isoenzymes in tobacco tissues after treatment with 2,4-D. Plant Physiol. 49:957–60.

Lee, T. T., and F. Skoog. 1965. Effects of substituted phenols on bud formation and growth of tobacco tissue cultures. Physiol. Plantarum 18:386–402.

Letham, D. S. 1966. A cytokinin in plant extracts: Isolation and interaction with other growth regulators. Phytochemistry 5:269–86.

Linsmaier, E. M., and F. Skoog. 1965. Organic growth factor requirements of tobacco tissue cultures. Physiol. Plantarum 18:100–127.

Lutz, A. 1966. Obtention de plantes de tabac à partir de cultures unicellulaires provenant d'une souche anergiée. C. R. Acad. Sci. Paris 262:1856–58.

Mackenzie, I. A., A. Konar, and H. E. Street. 1972. Cytokinins and the growth of cultured sycamore cells. New Phytol. 71:633–38.

Mackenzie, I. A., and H. E. Street. 1970. Studies on the growth in culture of plant cells. VIII. The production of ethylene by suspension cultures of *Acer pseudoplatanus* L. J. Exp. Bot. 21:824–34.

Mackenzie, I. A., and H. E. Street. 1972. The cytokinins of cultured sycamore cells. New Phytol. 71:621–31.

McWilliam, A. A., S. M. Smith, and H. E. Street. 1974. The origin and development of embryoids in suspension cultures of carrot (*Daucus carota* L.). Ann. Bot. 38:243–50.

Maheshwari, P., and B. Baldev, 1961. Artificial production of buds from embryos of *Cuscuta reflexa*. Nature 191:197–98.

Mattoo, A. K., and M. Lieberman. 1977. Evidence that the ethylene synthesizing enzyme system in plants is associated with a cell wall–cell membrane complex. Fed. Proc. 36:703.

Meins, F., Jr. 1974. Mechanisms underlying the persistence of tumour autonomy in crown-gall disease. *In* H. E. Street (ed.), Tissue culture and plant science 1974. Pp. 233–64. Academic Press, London.

Melchers, G. 1971. Transformation or habituation to autotrophy and tumour growth and recovery. *In* Les cultures de tissus de plantes. Colloques Intern. C.N.R.S. 193:229–34.

Miller, C. O. 1968. Naturally occurring cytokinins. *In* F. Wightman and G. Setterfield (eds.),

Biochemistry and physiology of plant growth substances. Pp. 33–45. Runge Press, Ottawa, Canada.

Miller, C. O. 1969. Cytokinins from a variant strain of cultured soybean cells. Plant Physiol. 44:1035–39.

Miller, C. O. 1974. Ribosyl-*trans*-zeatin, a major cytokinin produced by crown gall tumour tissue. Proc. Nat. Acad. Sci. USA 77:334–38.

Miller, C. O. 1975. Revised methods for purification of ribosyl-*trans*-zeatin from *Vinca rosea* L. crown-gall tumour tissue. Plant Physiol. 55:448–49.

Minocha, S. C., and W. Halperin. 1974. Hormones and metabolites which control tracheid differentiation with and without concomitant effects on growth in cultured tuber tissue of *Helianthus tuberosus*. Planta 116:319–31.

Morel, G., and R. H. Wetmore. 1951. Tissue culture of monocotyledons. Amer. J. Bot. 38:138–40.

Morgan, P. W., and W. C. Hall. 1962. Effect of 2,4-dichlorophenoxyacetic acid in the production of ethylene by cotton and grain sorghum. Physiol. Plantarum 15:420–27.

Morgan, P. W., and W. C. Hall. 1964. Accelerated release of ethylene by cotton following application of indolyl-3-acetic acid. Nature 201:99.

Murashige, T. 1965. Effects of stem-elongation retardants and gibberellin in callus growth and organ formation in tobacco tissue culture. Physiol. Plantarum 18:665–73.

Murashige, T. 1974. Plant propagation through tissue culture. Ann. Rev. Plant Physiol. 25:135–66.

Murashige, T., and F. Skoog. 1962. A revised medium for rapid growth and bioassays with tobacco tissue cultures. Physiol. Plantarum 15:473–97.

Narayana, R. 1963. Growth and differentiation of *Pelargonium hortorum* tissue cultures. Ph.D. thesis, University of Wisconsin.

Narayanaswamy, S. 1977. Regeneration of plants from tissue cultures. In J. Reinert, and Y. P. S. Bajaj (eds.), Applied and fundamental aspects of plant cell, tissue, and organ culture. Pp. 179–206. Springer-Verlag, Berlin.

Nickell, L. G., and W. Tulecke. 1959. Responses of plant tissue cultures to gibberellin. Bot. Gaz. 120:245–50.

Nitsch, J. P. 1968. Studies on the mode of action of auxins, cytokinins, and gibberellins at the subcellular level. In F. Wightman and G. Setterfield (eds.), Biochemistry and physiology of plant growth substances. Pp. 563–80. Runge Press, Ottawa.

Norreel, B., and J. P. Nitsch. 1968. La formation d'embryons végétatifs chez *Daucus carota* L. Bull. Soc. Bot. Fr. 115:501–14.

O'Hara, J., and H. E. Street. 1977. Wheat callus culture: The initiation, growth, and organogenesis of callus derived from various explant sources. Ann. Bot. (in press).

Osborne, D. J. 1976. Control of cell shape and cell size by the dual regulation of auxin and ethylene. In N. Sunderland (ed.), Perspectives in experimental biology. 2:89–102. Pergamon Press, Oxford.

Padmanabhan, V., E. F. Paddock, and W. R. Sharp. 1974. Hormonal control of organogenesis in *Lycopersicon esculentum* Mill. leaf callus. 3rd Intern. Congr. Plant Tissue and Cell Culture (abstr. 66). Leicester, England.

Pollard, J. K., E. M. Shantz, and F. C. Steward. 1961. Hexitols in coconut milk: Their role in nurture of dividing cells. Plant Physiol. 36:492–501.

Raju, M. V. S., and H. E. Mann. 1970. Regenerative studies in the detached leaves of

Echeveria elegans: Anatomy and regeneration of leaves in sterile culture. Canad. J. Bot. 48:1887–91.

Rashid, A., and H. E. Street. 1973. The development of haploid embryoids from anther cultures of *Atropa belladonna*. Planta 113:263–70.

Reinert, J. 1959. Über die Kontrolle der Morphogenese und die Induktion von Adventivembryonen an Gewebekulturen aus Karotten. Planta 53:318–33.

Ringe, F. 1972. Promotion of cell proliferation under the influence of 2-chloroethanephosphonic acid in explants of *Begonia* X *richmondensis* in sterile culture. Z. Pflanzenphysiol. 67:45–48.

Robison, G. A., R. W. Butcher, and E. W. Sutherland. 1971. Cyclic AMP. Academic Press, New York.

Sachar, R., S. R. Taneja, and K. Sachar. 1975. Cyclic AMP—its biological role in higher plants. J. Sci. Industr. Res. 34:54–64.

Sacristán, M. D., and G. Melchers. 1969. The caryological analysis of plants regenerated from tumorous and other callus cultures of tobacco. Molec. Gen. Genet. 105:317–33.

Sangwan, R. S., B. Norreel, and H. Harada. 1976. Effects of kinetin and gibberellin A3 on callus growth and organ formation in *Limnophila chinensis* tissue culture. Biol. Plantarum 18:126–31.

Sastri, R. L. N. 1963. Morphogenesis in plant tissue cultures. *In* P. Maheshwari and N. S. Rangaswamy (eds.), Plant tissue and organ culture—a symposium. Pp. 105–7. Intern. Soc. Plant Morphol. Delhi, India.

Schroeder, C. A., and C. Spector. 1957. Effect of gibberellic acid and indoleacetic acid on growth of excised fruit tissue. Science 126:701–2.

Seibert, M. 1973. Light—the effect of wavelength and intensity on growth and shoot initiation in tobacco callus. In Vitro 8:435.

Shaw, S., and M. B. Wilkins. 1973. The source and lateral transport of growth inhibitors in geotropically stimulated roots of *Zea mays* and *Pisum sativum*. Planta 109:11–26.

Shimada, T., T. Sasakuma, and K. Tsunewaki. 1969. *In vitro* culture of wheat tissues. I. Callus formation, organ redifferentiation, and single cell culture. Canad. J. Genet. Cytol. 11:294–304.

Simpkins, I., H. A. Collin, and H. E. Street. 1970. The growth of *Acer pseudoplatanus* L. cells in a synthetic liquid medium: Response to the carbohydrate, nitrogenous, and growth hormone constituents. Physiol. Plantarum 23:385–96.

Skoog, F. 1971. Aspects of growth factor interactions in morphogenesis of tobacco tissue cultures. *In* Les cultures de tissus de plantes. Colloques Intern. C.N.R.S. 193:115–36.

Skoog, F., and C. O. Miller. 1956. Chemical regulation of growth and organ formation in plant tissue cultures *in vitro*. Symp. Soc. Exp. Biol. 9:118–31.

Smith, S. M. 1973. Embryogenesis in tissue cultures of domestic carrot *Daucus carota* L. Ph.D. thesis, University of Leicester, Leicester, England.

Smith, S. M., and H. E. Street. 1974. The decline of embryogenic potential as callus and suspension cultures of carrot (*Daucus carota* L.) are serially subculted. Ann. Bot. 38:233–41.

Steinhart, C., L. Anderson, and F. Skoog. 1962. Growth promoting effect of cyclitols in spruce tissue cultures. Plant Physiol. 37:60–68.

Steward, F. C. 1976. Multiple interactions between factors that control cells and development. *In* N. Sutherland (ed.), Perspectives in experimental biology. 2:9–23. Pergamon Press, Oxford.

Steward, F. C., and A. D. Krikorian. 1972. Problems of integration and organization. *In* F. C. Steward (ed.), Plant physiology: A treatise, vol. VIC. Pp. 367–419. Academic Press, New York.

Steward, F. C., M. O. Mapes, A. E. Kent, and R. D. Holsten. 1964. Growth and development of cultured plant cells. Science 163:20–27.

Stewart, J. McD., and C. L. Hsu. 1977. Influence of phytohormones on the response of cultured cotton ovules to (2-chlorethyl)-phosphonic acid. Physiol. Plantarum 39:79–85.

Stonier, T. 1971. The role of auxin protectors in autonomous growth. *In* Les cultures de tissus de plantes. Colloques Intern. C.N.R.S. 193:423–35.

Stoutemyer, V. T., and O. R. Britt. 1963. Tissue cultures of juvenile and adult specimens of ivy. Nature 199:397–98.

Stoutemyer, V. T., and O. R. Britt. 1970. Ethrel and plant tissue cultures. Bioscience 29:914 (abstr.).

Straus, J., and R. R. Epp. 1960. Response of *Cupressus funebris* tissue cultures to gibberellins. Science 131:1806–7.

Street, H. E. 1977. Regulation of differentiation in cell and tissue cultures at the molecular-biochemical level. Proceedings of the International Conference on Regulation of Developmental Processes in Plants, Halle (Salle) (in press).

Street, H. E., and R. D. Shillito. 1977. Nutrient media for plant organ, tissue, and cell culture. *In* M. Rechcigl, Jr. (ed.), C.R.C. Handbook Series in Nutrition and Food. 4:305–58. CRC Press, Cleveland.

Stuart, R., and H. E. Street. 1969. Studies on the growth in culture of plant cells. IV. The initiation of division in suspensions of stationary phase cells of *Acer pseudoplatanus* L. J. Exp. Bot. 20:556–71.

Stuart, R., and H. E. Street. 1971. Studies on the growth in culture of plant cells. X. Further studies on the conditioning of culture media by suspension of *Acer pseudoplatanus* L. cells. J. Exp. Bot. 22:96–106.

Syōno, K., and T. Furuya. 1968. Studies on plant tissue cultures. I. Relationship between inocular sizes and growth of calluses in liquid culture. Plant Cell Physiol. 9:103–14.

Syōno, K., and T. Furuya. 1971. Effects of temperature on the cytokinin requirement of tobacco callus. Plant Cell Physiol. 12:61–71.

Thomas, E., F. Hoffmann, I. Potrykus, and G. Wenzel. 1976. Protoplast regeneration and stem embryogenesis of haploid androgenetic rape. Molec. Gen. Genet. 145: 245–47.

Thomas, E., and H. E. Street. 1970. Organogenesis in cell suspension cultures of *Atropa belladonna* L. and *Atropa belladonna* cultivar *lutea* Döll. Ann. Bot. 34:657–69.

Thomas, E., and H. E. Street. 1972. Factors influencing morphogenesis in excised roots and suspension cultures of *Atropa belladonna*. Ann. Bot. 36:239–47.

Thorpe, T. A., and D. D. Meier. 1973. Effects of gibberellic acid and abscisic acid on shoot formation in tobacco callus cultures. Physiol. Plantarum 29:121–24.

Torrey, J. G. 1976. Root hormones and plant growth. Ann. Rev. Plant Physiol. 27:435–59.

Venis, M. 1977. Solubilisation and partial purification of auxin binding sites of corn membranes. Nature 266:268.

Walter, R. E. 1968. Root and shoot initiation in aspen callus cultures. Nature 219:509–10.

Wang, C.-C., C.-S. Sun, and C.-C. Chu. 1974. On the conditions for the induction of rice pollen plantlets and certain factors affecting the frequency of induction. Acta Bot. Sinica 16:43–53.

Wetherell, D. F., and D. K. Dougall. 1976. Sources of nitrogen supporting growth and embryogenesis in cultured wild carrot tissue. Physiol. Plantarum 37:97–103.

Wilson, H. M., and H. E. Street. 1975. The growth, anatomy, and morphogenetic potential of callus and cell suspensions of *Hevea brasiliensis*. Ann. Bot. 39:671–82.

Winton, L. L., and V. Shirley. 1974. Shoots from Douglas fir cultures. Canad. J. Bot. 55:1246–50.

Witham, F. H. 1968. Effect of 2,4-dichlorophenoxyacetic acid on the cytokinin requirement of soybean cotyledonary and tobacco stem pith callus tissue. Plant Physiol. 43:1455–57.

Wochok, Z. S., and D. F. Wetherell. 1971. Suppression of organised growth in cultured wild carrot tissue by 2-chloroethylphosphonic acid. Plant Cell Physiol. 12:771–74.

Wolter, K. E. 1968. Root and shoot initiation in aspen callus cultures. Nature 219:509–10.

Wolter, K. E., and F. Skoog. 1966. Nutritional requirements of *Fraxinus* tissue cultures. Amer. J. Bot. 53:263–69.

Wood, H. N., and A. C. Braun. 1961. Studies on the regulation of certain essential biosynthetic systems in normal and crown-gall tumor. Proc. Nat. Acad. Sci. USA 47:1907–13.

Yamaguchi, T., and T. Nakajima. 1974. Effect of abscisic acid on adventitious bud formation in cultured tissues derived from root tubers of sweet potato and from stem tubers of potato. 3rd Intern. Cong. Plant Tissue and Cell Culture (abstr. 65) Leicester, England.

Yamaguchi, T., and H. E. Street. 1977. Stimulation of the growth of excised cultured roots of soya bean by abscisic acid. Ann. Bot. 41:1129–33.

Young, R. E., H. K. Pratt, and J. B. Biale. 1951. Identification of ethylene as a volatile product of the fungus *Penicillium digitatum*. Plant Physiol. 26:304–10.

Zimmerman, P. W., and A. E. Hitchcock. 1933. Initiation and stimulation of adventitious roots caused by unsaturated hydrocarbon gases. Contr. Boyce Thompson Inst. 5:351–69.

V. RAGHAVAN

Totipotency of Plant Cells: Evidence for the Totipotency of Male Germ Cells from Anther and Pollen Culture Studies

12

INTRODUCTION

The objective of this paper is to reflect current knowledge of the concept of totipotency as applied to plant cells. Particular attention will be paid to the demonstration of totipotency of germ cells of plants—cells of the embryo sac and pollen grain. Since significant new information on the totipotency of the component cells of the pollen grain has become available, this will be dealt with in greater detail.

CONCEPT OF TOTIPOTENCY

The fertilized egg or zygote is a unique cell, distinguished from other cell types by the fact that it passes through a series of divisions whereby the morphological and functional fate of the daughter cells or cell groups becomes recognizably different. The zygote is endowed with the essential genetic blueprint to regenerate a new organism in full multicellularity, sexuality, and structure. Typically, in the ontogeny of a seed plant, the zygote lapses into an embryogenic phase during which the fundamental body plan of the future sporophyte is established. An embryo encased in a seed is sufficiently well developed with morphologically differentiated primordia of the future vegetative organs of the plant that, confronted with

V. Raghavan, Department of Botany, Ohio State University, Columbus, Ohio 43210.

an adequate supply of water, a favorable temperature, and the normal composition of the atmosphere, it can immediately begin to grow into a new plant. Although morphogenetic influences are clearly obvious in the continued division and differentiation of the zygote, they are progressively held in abeyance as the newly formed cells become organized into specialized tissues, which in turn form anatomically recognizable components of organs. As a result, cells born out of the zygote by simple mitotic divisions are unable to express their full genomic potential as parts of a plant, although the competence to do so may persist unchanged throughout their life.

One of the outstanding recent achievements in developmental botany has been the formulation of the concept of totipotency, which essentially implies that all differentiated cells of the plant, except those that have been exclusively programmed into narrow paths of specialization, possess a profound ability to display their full genetic program and embark upon a developmental pathway similar to that of a zygote, leading to the formation of a new plant. Implicit in this definition of totipotency is the fact that in cells that have undergone irreversible differentiation such as those of the tracheids and sieve tubes, no further expression of the genetic potential is possible, and totipotency cannot be demonstrated. Tangible expression of totipotency of a plant cell is its differentiation into a complete plant through embryolike stages.

TOTIPOTENCY OF SOMATIC CELLS

Because experiments leading to the demonstration of totipotency in the diploid somatic cells of plants have been reviewed many times in recent years, it is not necessary to describe here in any detail the mass of well-established facts; lest readers unfamiliar with this field gain an incomplete picture of some of the landmark studies, I shall attempt to describe them only briefly. The ability of undifferentiated somatic cells of plants to recapitulate stages of normal embryogenesis was clearly foreshadowed in the experiments of Haberlandt (1902). However, the definitive experimental demonstration of totipotency was carried out by Steward and co-workers (Steward, Mapes, and Smith, 1958; Steward, Mapes, and Mears, 1958) using cultured carrot root tissue explants. These investigators found that culture of slabs of the secondary phloem of domestic carrot in a solid medium containing coconut milk typically produced proliferating masses of callus made up of simple parenchymatous cells. Transfer of the callus to an agitated liquid medium of the same composition yielded a milky suspension of individual cells and cell groups. The viability of these cells was confirmed by microscopic examination, which revealed intense cytoplasmic streaming. Although the precise origin of the cell aggregates appeared to be a vexing question at first, by examination of a range of cells and cell clumps, it was firmly established that individual cells dissociating from the callus divided, and that cell aggregates arose from cells that, having divided, remained attached. Continued growth of the cell clumps in the liquid medium without subculture led to initiation of root primordia on them. When the rooted aggregates were removed to the surface of a solid medium, shoot initiation and formation of a complete plantlet ensued. In these studies the more arduous task of following a single cell in isolation through its various stages of development into a plantlet was not attempted; it remained for Vasil and Hildebrandt (1965) to show that single cells of a hybrid tobacco grown in isolation from other cells in a defined medium formed completely organized plants capable of flowering.

The morphological entities formed at different stages in the conversion of a single cell into a whole plant possess sufficient internal and external heterogeneity of structure to impress an ardent lover of form in plants; yet they fall short of our definition of totipotency since the single cells do not exactly duplicate the pathway normally followed by a zygote, namely, a lapse into an embryogenic phase before becoming plantlets. From this point of view the work of Reinert (1959), who showed that a strain of callus originating from carrot root, following a succession of changes of nutrient media, underwent differentiation accompanied by formation of normal bipolar embryos, is particularly significant. Without unequivocally demonstrating that the embryos had their origin in single cells of somatic parentage, Reinert showed that when the callus was transferred to a complex medium containing amino acids, vitamins, and hormones, following a long period of subculture in a medium containing coconut milk, it became granular with evident signs of differentiation. Histological examination of the callus showed that the bipolar embryos that appeared during later periods of growth of the callus in the complex medium had their origin from centers of organized development comparable to globular embryos. Transformation of the embryo into plantlets was accomplished by the simple expedient of substituting coconut milk for the hormonal constituents of the medium.

An important milestone along the road of demonstrating complete totipotency of somatic cells of plants occurred a few years later when, almost simultaneously, Steward (1963) and Wetherell and Halperin (1963) reported that free cells of carrot, when spread on an agar plate, gave rise to literally thousands of embryos that faithfully reproduced all known stages of carrot embryogenesis such as globular, heart-shaped, and torpedo-shaped stages. Although the medium composition and other cultural conditions including the starting material for culture employed by these groups of investigators varied somewhat, their work dramatically showed that the overall sequence of development of somatic cells of carrot into a whole plant was remarkably similar to that of a zygote undergoing embryogenesis inside the embryo sac. Finally, the transformation of a single isolated cell into an embryo has also been demonstrated (Backs-Hüsemann and Reinert, 1970), thus reinforcing the conclusion that embryolike structures observed in plated cells suspensions had indeed their origin in single cells. The similarity between normal zygotic embryos and adventitiously formed embryolike structures was emphasized by referring to the latter as embryoids, a term that is now generally accepted.

Since these findings, a number of reports of totipotency of somatic plant cells have appeared; as seen from table 1, plantlets have been shown to arise through an embryogenic pathway from virtually any part of the plant. However, only in a few cases have embryoids been traced to their single-celled origin, but this appears to be a technical problem involved in following embryogenesis of isolated single cells. The key point is that under the influence of propitious stimuli, plant cells that have not embarked upon a pathway of irreversible differentiation are capable of expressing their inherent genetic potential in the same way and to the same extent as a zygote.

TOTIPOTENCY OF GERM CELLS

The egg, the sperm, and the associated haploid cells of the gametophytic generation are generally designated as sexual reproductive cells or germ cells. There is some merit in restricting the use of the term to haploid cells, since germ cells have their origin in simple diploid parenchymatous cells that in other circumstances differentiate into specialized somatic cells. Formation of germ cells involves a complex series of differentiation processes including reduction in the number of chromosomes from the diploid to the haploid level. Embryos that develop from the reduced germ cells in the absence of fertilization are therefore haploids at the cellular level, that is, consist of the single set, the gametic number of chromosomes in their somatic cells.

Egg and Its Component Cells

In angiosperms the egg is formed within the female gametophyte or the embryo sac, which itself is deeply entrenched within the ovule with its multilayered covering of integuments and nucellus. The egg has all the earmarks of a functional cell, with a nucleus surrounded by a blend of cytoplasmic organelles, fully covered by a plasma membrane, and partially covered by a cell wall. In a typical mature embryo sac, besides the egg there

TABLE 1

TOTIPOTENCY AND SOMATIC EMBRYOGENESIS IN PLANT CELLS

Species	Organ	References
Ammi majus	Zygote; from culture of ovulary containing zygote-bearing ovules	Sehgal, 1972
	Hypocotyl	Grewal et al., 1976
	Embryo	Johri and Bajaj, 1964
Amyema pendula	Embryo in cultured ovulary	Johri and Sehgal, 1963
Anethum graveolens	Hypocotyl, cotyledon	Ratnamba and Chopra, 1974
Antirrhinum majus	Stem, leaf mesophyll protoplast	Poirier-Hamon et al., 1974; Sangwan and Harada, 1975
		Reinert et al., 1966
Apium graveolens	Hypocotyl	Williams and Collin, 1976
	Petiole	
Asclepias curassavica	Stem	Prabhudesai and Narayanaswamy, 1974
Asparagus officinalis	Hypocotyl	Wilmar and Hellendoorn, 1968
	Stem	Steward and Mapes, 1971b
	Leaf mesophyll	Jullien, 1974
	Leaf cell protoplast	Bui Dang Ha et al., 1975
	Root	Thomas and Street, 1970
Atropa belladonna	Protoplast from callus originating from stem	Gosch et al., 1975
Begonia semperflorens	Leaf	Sehgal, 1975
Biota orientalis	Embryo cotyledon	Konar and Oberoi, 1965
Bromus inermis	Mesocotyl	Gamborg et al., 1970
Carum carvi	Petiole	Ammirato, 1974
Cheiranthus cheiri	Seedling	Khanna and Staba, 1970
Cichorium endivia	Embryo	Vasil and Hildebrandt, 1966a
Citrus aurantifolia, C. sinensis	Ovulary, ovulary wall, nucellus	Mitra and Chaturvedi, 1972
Citrus sp. (Washington navel)	Nucellus	Button and Bornman, 1971
Coffea canephora	Stem	Staritsky, 1970
Conium maculatum	Hypocotyl	Nétien and Raynaud, 1972
Consolida orientalis	Flower bud	Nataraja, 1971
Coriandrum sativum	Embryo	Steward et al., 1966
Corylus avellana	Embryo	Radojević et al., 1975

TABLE 1—Continued

Species	Organ	References
Crambe maritima	Root	Bowes, 1976
Croton bonplandianum	Endosperm	Bhojwani, 1966
Cucurbita pepo	Fruit pericarp	Schroeder, 1968
	Hypocotyl	Jelaska, 1972
Cuscuta reflexa	Embryo	Maheshwari and Baldev, 1961
Cymbidium sp.	Shoot apex	Steward and Mapes, 1971a
Daucus carota	Embryo	Steward et al., 1964
	Root, petiole, peduncle, seedling	Halperin and Wetherell, 1964
	Epidermal strip of hypocotyl	Kato and Takeuchi, 1966
	Protoplast from root	Kameya and Uchimiya, 1972
	Protoplast from callus of root or petiole origin	Grambow et al., 1972
Dendrophthoe falcata	Embryo	Johri and Bajaj, 1962
Didiscus coerulea	Stem	Ball and Joshi, 1966
Elaeis guineensis	Embryo	Rabéchault et al., 1970
Ephedra foliata	Embryonic root	Sankhla et al., 1967
Eschscholtzia californica	Embryo	Kavathekar and Ganapathy, 1973
Euphorbia pulcherrima	Embryo	Nataraja, 1975
Foeniculum vulgare	Proembryo; from culture of ovulary containing proembryo-bearing ovules	Sehgal, 1964
	Stem	Maheshwari and Gupta, 1965
Gossypium hirsutum	Embryo in cultured ovule	Joshi and Johri, 1972
Hevea brasiliensis	Stem	Wilson and Street, 1975
Hordeum vulgare	Embryo	Norstog, 1970
Ilex aquifolium	Embryo	Hu and Sussex, 1971
Kalanchoe pinnata	Leaf mesophyll	Wadhi and Mohan Ram, 1964
Lemna gibba	Frond	Chang and Chiu, 1976
Macleaya cordata	Leaf mesophyll	Kohlenbach, 1965
Mesembryanthemum floribundum	Leaf, root, stem, shoot tip, hypocotyl, and cotyledons	Mehra and Mehra, 1972

TABLE 1—*Continued*

Species	Organ	References
Nicotiana tabacum	Stem	Haccius and Lakshmanan, 1965
	Hypocotyl, petiole	Prabhudesai and Narayanaswamy, 1973
Nigella damascena	Pedicel, floral bud	Raman and Greyson, 1974
N. sativa	Root	Banerjee and Gupta, 1975
	Leaf	Banerjee and Gupta, 1976
Nuytsia floribunda	Embryo	Nag and Johri, 1969
Panax ginseng	Petiole, leaf, root, and antherophore	Butenko et al., 1968
Pergularia minor	Stem	Prabhudesai and Narayanaswamy, 1974
Petroselinum hortense	Petiole	Vasil and Hildebrandt, 1966b
Petunia hybrida, P. inflata	Leaf, stem	Rao et al., 1973
Pinus palustris	Embryo	Sommer et al., 1975
Pterotheca falconeri	Root, hypocotyl, stem, leaf, petiole, cotyledon, and shoot apex	Mehra and Mehra, 1971
Ranunculus sceleratus	Flower bud	Konar and Nataraja, 1964
	Stem	Konar and Nataraja, 1965a
	Anther	Konar and Nataraja, 1965b
	Stem, petiole, lamina, shoot tip, sepal, and petal	Nataraja and Konar, 1970
	Leaf protoplast	Dorion et al., 1975
Rauvolfia serpentina	Leaf	Mitra and Chaturvedi, 1970
Ricinus communis	Endosperm	Satsangi and Mohan Ram, 1965
Santalum album	Embryo	Rao, 1965
Scurrula pulverulenta	Embryo	Johri and Bhojwani, 1970
Sinapis alba	Hypocotyl	Bajaj and Bopp, 1972
Sium suave	Embryo	Ammirato and Steward, 1971
Solanum melongena	Embryo	Yamada et al., 1967
Tylophora indica	Stem	Rao et al., 1970
Vitis vinifera	Nucellus in cultured ovules	Mullins and Srinivasan, 1976
Zamia integrifolia	Embryo	Norstog, 1965

are seven other nuclei that are organized as naked or partially covered cells according to a characteristic pattern. At the micropylar end of the embryo sac, adjacent to and nearly embracing the egg are two small cells known as synergids. At the chalazal end of the embryo sac, three nuclei differentiate into the antipodal cells. During gametogenesis two nuclei from either pole come to a central position in the embryo sac and fuse, forming a diploid polar fusion nucleus.

There is no unambiguous cytologically supported evidence for the totipotency of the reduced female sexual reproductive cells of angiosperms. However, a number of reports in the literature indicate that under certain conditions the egg or some other haploid component of the embryo sac may regenerate embryos. In interspecific crosses between different species of *Solanum*, Jørgensen (1928) showed that although pollen tubes of *S. luteum* failed to fuse with the egg of *S. nigrum*, the latter, nevertheless, developed into an embryo with a haploid chromosome complement. The role of the egg is less clear among other inter-specific crosses where haploid embryos have been reported to occur. In certain plants, anomalies in the fertilization process during normal reproductive cycle are known to induce parthenogenetic development of the egg; in still other plants, delayed pollination, pollination *in vitro*, and application of temperature shocks, radiation, and chemicals have formed the basis for the induction of parthenogenesis (Battaglia, 1963; Lacadena, 1974). A common feature of these treatments is an interference with the production of normal sperms, although growth of the pollen tube into the vicinity of the egg takes place normally. One could envision that growth of the egg in a milieu conditioned by the presence of a foreign cytoplasm could result in the division of the former.

In serial sections of ovules used to trace the development of the zygote, several investigators have encountered supernumerary embryos, in addition to normal zygotic embryos. Because in some cases the accessory embryos arise close to the egg, they have been thought to originate from the synergids, whereas in other cases their evident position at the chalazal end has given rise to the speculation that they arise from the antipodal cells (Maheshwari and Sachar, 1963). Technical considerations of isolating the delicate female gametophyte of angiosperms have thus far precluded attempts to demonstrate totipotency of their component cells *in vitro*. Recently Norstog (1965) found that culture of the massive parenchymatous female gametophyte of the gymnosperm *Zamia integrifolia* in a high salt medium supplemented with glutamine and alanine induced sporadic formation of typical bipolar embryoids, although their origin was not traced back to single cells.

These observations suggest that by refinements in manipulative tech-

These observations suggest that by refinements in manipulative techniques, it may be possible to demonstrate *in vitro* that even in the absence of fertilization, the female gametophytic cells of angiosperms follow a pathway similar to that of a zygote and regenerate new plants. Although the cells of the female gametophyte contain only a single set of chromosomes, these cells are neither unique nor irreversibly committed to a particular pathway of differentiation. Only their privileged geographical location in the ovule seems to pose the most formidable challenge in demonstrating their totipotency.

The Male Gametophyte

In contrast to the complex female gametophyte of angiosperms, cell differentiation in its almost idealized simple form is seen in the development of the male gametophyte. The formation of microspores following meiosis in the microspore mother cell in the anther is the beginning of a short-lived male gametophytic phase. The microspore matures into the pollen grain, which normally undergoes two mitotic divisions. The first division is unequal with regard to the distribution of the cytoplasm and results in the formation of a small generative cell and a large vegetative cell. The former again divides into two sperms; this second division is normally delayed until the pollen grains are discharged from the anther. The pollen tube grows out of the vegetative cell whose nucleus eventually disintegrates or survives as a vestigial structure in the pollen tube.

Striking differences are discernible between the generative and vegetative cells during early stages of gametogenesis. Difference in size between the two cells was alluded to earlier. The nucleus of the generative cell has a highly condensed and granular chromatin and is surrounded by a markedly basophilic cytoplasm, whereas the vegetative nucleus has diffuse chromatin, a conspicuous nucleolus, and a weakly basophilic cytoplasm. Starch grains that appear in the pollen grain just before or after the first haploid mitosis are generally concentrated in the vegetative cell. At the biochemical level, the vegetative cell has a rapid synthesis of RNA and protein and a negligible synthesis of DNA. In the generative cell DNA content doubles to 2C level, but there is little or no RNA and protein synthetic activity. Differentiation of the vegetative cell thus involves switching off the DNA synthesizing system while RNA synthesis (transcription) proceeds; on the other hand in the generative cell, transcription is slowed down without impairing DNA synthesis.

The problem of totipotency of the male sexual reproductive cells relates to the transformation of the vegetative cell, the generative cell, and the male gametes into functional embryos and plants. The occurrence of plants with paternal characteristics reported in interspecific crosses in *Nicotiana*

(Clausen and Lammerts, 1929; Kostoff, 1929; Kehr, 1951) and *Hordeum* (Davies, 1958) is compatible with the notion that the reduced sperm cells can be activated and can form plants without recourse to fusion with the egg. Cytological studies have led to the view that in these hybrids, following penetration of the egg by the sperm, the nucleus of the former disintegrates, enabling the sperm cell to divide in the female cytoplasm. Plants having dominant male characteristics have also been obtained by X-irradiation of the female flowers and pollination with normal pollen grains (Gerassimowa, 1936; Maly, 1958; Gerlach-Cruse, 1969). It appeared that radiation killed the egg nucleus, and the male gamete alone differentiated into a plant. In certain cases, plants with male characteristics have also been known to arise independently of any breeding programs or treatments, suggesting their origin from a component of a cell of the male gametophyte (Campos and Morgan, 1958, 1960; Goodsell, 1961; Haustein, 1961).

The development of the germ cells through various stages of embryogenic behavior into a plant has not been followed in any of the examples referred to above. Moreover, the potential use of this approach to demonstrate the competence for embryogenesis of the male germ cells is very limited, since it is generally difficult to establish within the confines of the embryo sac whether embryos originate from the vegetative cell, the generative cell, or the sperms.

The clearest demonstration of totipotency of the male sexual reproductive cells of angiosperms, coupled with the evidence that there is no irreversible change in the genome as a result of cell specialization, has come from anther and pollen culture experiments. Guha and Maheshwari (1964) set the theme for a dominant effort in contemporary investigations in developmental botany when they showed that, upon culture of excised anthers of *Datura innoxia* in a mineral salt medium supplemented with coconut milk and other organic additives, embryo-like outgrowths appeared from the sides of the anther in about 6–7 weeks. In later works (Guha and Maheshwari, 1966, 1967) they confirmed the haploid nature of embryoids and their origin from pollen grains. Following upon this work, several investigators have convincingly demonstrated that culture of anthers or occasionally of isolated pollen grains at an appropriate stage of development in a simple mineral salt medium provokes division in the embryogenic pathway in a small percentage of the contained pollen grains (see Vasil and Nitsch, 1975; Raghavan, 1976b, for reviews). As documented in various species of *Nicotiana* and several other plants, the embryogenic pollen grains develop the basic organization of embryos, complete with root and shoot apices, cotyledons, and a vascular system, as they pass through a succession of distinct morphological stages reminiscent of sexually produced zygotic embryos and appear outside the anther wall as plantlets. The transformation of the pollen grain into an embryoid was termed (anther) androgenesis and, aside from its intrinsic interest as a striking developmental process, has been considered as a means of producing isogenic lines of haploid and diploid plants reproducibly and in quantity for breeding purposes.

Early in these studies it was established that androgenesis occurred only if anthers containing fully individualized uninucleate pollen grains were cultured. Nitsch and Nitsch (1970) showed that anthers of *Nicotiana tabacum* cultured at the tetrad stage or at a later stage when starch accumulation had commenced in the binucleate pollen grains failed to form embryoids. The most successful results were obtained when anthers were cultured just before or just after the occurrence of the first microspore mitosis heralding the formation of the generative and vegetative cells. The importance of precise developmental stage for programming pollen grains in the androgenic pathway has been established for a number of plants besides *Nicotiana*, including *Oryza sativa* (Guha et al., 1970), *Atropa belladonna* (Zenkteler, 1971; Narayanaswamy and George, 1972; Rashid and Street, 1973), *Datura metel* (Iyer and Raina, 1972; Narayanaswamy and Chandy, 1971), *Asparagus officinalis* (Pelletier et al., 1972), *Lycium halimifolium* (Zenkteler, 1972), *Capsicum annuum* (George and Narayanaswamy, 1973; Wang et al., 1973; Kuo et al., 1973), *Petunia hybrida* (Sangwan and Norreel, 1975), *Secale cereale* (Thomas et al., 1975) *Saintpaulia ionantha* (Hughes et al., 1975), *Helleborus foetidus, Paeonia lutea, P. suffruticosa, Prunus avium, Bromus inermis, Agropyron repens, Festuca pratensis, Hordeum vulgare* (Zenkteler et al., 1975), apple (Kubicki et al., 1975), and potato (Dunwell and Sunderland, 1973; Sopory and Rogan, 1976). Although embryoids have been obtained only rarely by culturing anthers containing pollen grains at pollen mother cell stage or at pollen meiosis stage, some recent observations indicate that pollen mother cells of certain hybrid clones of *Solanum* in the early meiotic stage have a pronounced tendency to form embryoids naturally *in vivo*. These hybrids are normally male sterile, due to formation of unreduced microspores resulting from failure at telophase I, and embryoids have their origin in the multinucleate pollen mother cells that subsequently undergo regular nuclear divisions (Ramanna, 1974; Ramanna and Hermsen, 1974).

Totipotency of the Vegetative Cell

From an ontogenetic point of view, the crux of the problem in androgenesis is to ascertain whether the entire pollen grain is transformed into an embryoid or whether the pollen grain goes through the gametophytic divisions and the embryoid arises from one of the constituent cells such as the vegetative cell, generative cell, or sperm cell. In such closely investigated systems as *Nicotiana tabacum* (Bernard, 1971; Sunderland and Wicks, 1971) and *Datura metel* (Iyer and Raina, 1972), it appears that in anthers cultured at the uninucleate microspore stage, embryoids are initiated from the vegetative cell arising from an initial asymmetric division of the pollen nucleus. This cell, no larger than many somatic cells, but with a conspicuous, diffuse, lightly staining nucleus and a prominent nucleolus, repeatedly divides by a series of internal divisions without intervening cell enlargement until a compact mass of cells is produced within the confines of the pollen wall. Eventually the pollen wall breaks open and sets free the cellular mass, which, although irregular in outline to begin with, gradually assumes the shape of a typical globular embryo. Continued differentiation of the adventitiously produced globular embryo through successive stages of embryogenesis results in the formation of a seedling plant. In this case, only the vegetative cell is involved in embryogenesis and plantlet formation, whereas the small generative cell disintegrates without dividing. This is thus a clear demonstration of the totipotency of the vegetative cell.

Alternative pathways to embryogenesis involving primarily, but not exclusively, the vegetative cell do exist and have been shown to contribute substantially to the total yield of embryoids in anther and pollen cultures. In anther cultures of *Nicotiana sylvestris, N. tabacum* (Rashid and Street, 1974), *Datura innoxia* (Norreel, 1970; Sopory and Maheshwari, 1972; Sunderland et al., 1974), *Atropa belladonna* (Rashid and Street, 1973), and *Oryza sativa* var. *indica* (Guha-Mukherjee, 1973), and in pollen cultures of *Datura innoxia* (Nitsch and Norreel, 1973), the pollen grains, instead of dividing into microscopically distinct vegetative and generative cells, give rise to two identical cells or nuclei with similar staining properties. These cells subsequently undergo variable planes of divisions and form a typical embryogenic mass of cells. In *N. sylvestris*, where detailed cytological studies have been made, pollen grains with two symmetrical cells have been observed in large numbers during early stages of culture, suggesting that embryoids are formed exclusively through this route; in contrast, pollen grains that differentiate typical vegetative and generative cells become swollen and starch-filled, germinate, or undergo only a limited number of divisions and become nonembryogenic (Rashid and Street, 1974). A third, less frequent route to embryogenesis described in *Datura innoxia* involves division of the microspore nucleus into two nuclei, followed by endoreduplication of the generative nucleus and fusion between chromosome complements of the two nuclei (Sunderland et al., 1974). Divisions in the embryogenic pathway occur only after these complex nuclear events are completed and result in the formation of embryoids and plantlets of variable ploidy levels.

Totipotency of the Generative Cell

Thus far we have accounted for androgenesis in a way that shows that in those pollen grains where the first mitosis results in the formation of distinct vegetative and generative cells, the latter does not participate in the formation of embryoid. Occasionally the generative cell has been reported to undergo a limited number of divisions and form several free nuclei or cells, although the division products are not incorporated into the embryoid and disintegrate with the beginning of growth of the latter (Sunderland and Wicks, 1971; Iyer and Raina, 1972; Rashid and Street, 1974). This observation raised the possibility that the generative cell is not an irreversibly differentiated cell, and that under favorable conditions it might divide in the embryogenic pathway.

In a report of the work on the anther culture of *Nicotiana tabacum*, Devreux et al. (1971) mention the occurrence of one albino plant presumed to have originated from the generative cell. The basis of this claim is genetical, and neither morphological description nor figures are given in support. The first documented proof of totipotency of the generative cell was provided in anther cultures of *Hyoscyamus niger* (Raghavan, 1976a). Cytological observations of pollen grains from anthers of this plant at different times after culture showed that division in the embryogenic pathway occurred as early as 48 hours after culture, delimiting a small generative cell with granular nucleus and a large vegetative cell with diffuse nucleus. Further divisions occurred in the generative cell that initially formed a group of four cells. Variable planes of division in these cells gave rise to a multicellular embryoid enclosed in the exine. Breakage of the exine led to the formation of typical globular and heart-shaped embryoids whose organogenetic part was contributed entirely by the division products of the generative cell. The vegetative cell appeared as a colorless undivided cell at the proximal end of the embryoid, forming a suspensorlike structure, although it is not settled whether the latter is functionally analogous to a similar structure in zygotic embryos. In a slight variation of the segmentation pattern described above, in some pollen grains the vegetative cell was found to divide and form a multicellular suspensorlike structure. Not infrequently, in a small proportion of embryogenic pollen grains, following the first haploid mitosis delimiting a vegetative nucleus and a generative nucleus, both nuclei divided independently, accompanied by wall formation, and contributed to embryoid development (fig. 1 in Raghavan, 1976a).

Additional, and apparently conclusive, evidence for the totipotency of the generative cell and its involvement in embryoid formation in cultured anthers of *H. niger* has come from autoradiographic localization of DNA synthetic patterns in embryogenic pollen grains using ^{3}H-thymidine as a marker (Raghavan, 1977). In line with the cytological observations, embryogenic pollen grains could be separated into groups in which ^{3}H-thymidine incorporation occurred only in cells originating from the generative nucleus and in which incorporation occurred in cells formed from both generative and vegetative nuclei. A feature of embryoids originating exclusively from the generative cell was the complete absence of DNA synthesis in the vegetative cell, which was delineated as a suspensor-like structure. When the suspensorlike region was constituted of two cells, after the first division of the vegetative cell DNA synthesis proceeded in the daughter cells unaccompanied by division and wall formation; in other cases DNA synthesis, mitosis, and cytokinesis were initiated in the vegetative cell, forming a multicellular suspensorlike structure before DNA synthesis occurred in the generative cell. When cells originating from the vegetative nucleus contributed to the formation of the embryoid, grain density due to incorporation of ^{3}H-thymidine was relatively low in these cells in comparison to that found in the division products of the generative nucleus (fig. 1). Why DNA synthesis is generally sluggish in cells formed from the vegetative nucleus is not yet known; it is hoped that a study of the differences in the propensity for division of the vegetative and generative nuclei may provide a clue to an understanding of the mechanism of embryogenic induction in this plant.

The ability of the generative cell of *H. niger* pollen to express its inherent genomic potential in culture and to form embryoids and plantlets is striking when it is realized that in all other plants thus far investigated it is the vegetative cell that functions as the embryo mother cell. Since the generative cell is already programmed for DNA synthesis in advance of the formation of gametes, it probably lapses into a mitotic state with relative ease. The enhanced mitotic activity of the generative cell may in turn be related to the availability of hormones or cell division factors. If this argument is valid, by supplementing the medium with growth hormones it should be possible to reprogram pollen grains of *H. niger* into embryogenesis by a different pathway or to alter the proportion of embryoids formed by the division of the generative cell.

Corduan (1975) found that supplementation of the medium with an auxin such as 2,4-dichlorophenoxyacetic acid (2,4-D) led to the formation of undifferentiated callus rather than normal bipolar embryoids in anther cultures of *H. niger*. By the use of a genetic marker, this investigator

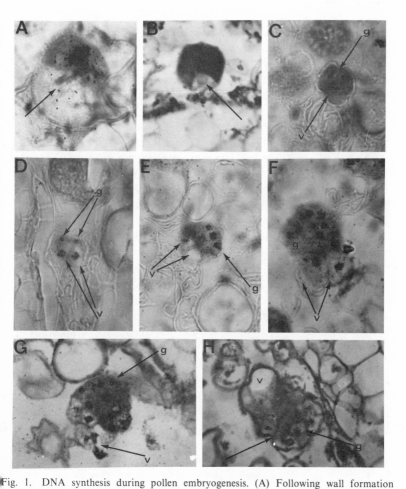

Fig. 1. DNA synthesis during pollen embryogenesis. (A) Following wall formation separating a densely staining generative cell and a lightly staining vegetative cell, ^{3}H-thymidine incorporation occurs in the nucleus of the generative cell; the vegetative cell nucleus (arrow) is unlabeled. (B) A globular embryoid with ^{3}H-thymidine-labeled cells cut off from the generative nucleus; arrow points to the unlabeled nucleus of the vegetative cell. (C–F) Patterns of DNA synthesis in an embryoid in which the generative cell gives rise to the organogenetic part and the vegetative cell forms a two-celled suspensorlike structure. Cells from generative and vegetative nuclei are labeled g and v, respectively in all segments of this figure. (G) An embryoid with several labeled cells cut off from the generative nucleus, and a multicellular suspensorlike structure formed from the vegetative nucleus. Some incorporation of ^{3}H-thymidine is found in the cells of the suspensorlike structure. (H) DNA synthesis in an embryoid formed by the division of both vegetative and generative cells. Arrow points to the separation line observed in such embryoids that demarcates the two groups of cells. (From Raghavan, J. Cell Biol. 73:521–26; reprinted with permission.)

concluded that the pollen callus originated from the *generative cells* probably meaning the entire pollen grain as distinct from the anther wall or what is preferentially known as the somatic tissues of the anther. Since this information is insufficient to distinguish the relative contributions of the two cells of the pollen grain in callus formation, the effect of 2,4-D (2.0 mg/1) on the early divisions of embryogenic pollen grains in cultured anthers of *H. niger* has been investigated. In this study we have developed a technique in which anthers were cut initially into two identical halves and the distal half was discarded. The proximal half (nearer to the filament) was cut longitudinally into two segments, and one segment of each anther was cultured in the basal medium or in a medium containing 2,4-D. By this means it was possible to follow embryogenic development and hormone induced callus formation in pollen grains of the same anther cultured under otherwise identical conditions (Raghavan, 1978).

Pollen grains contained in anther segments cultured in the basal medium followed the same general pathways of embryogenesis described earlier for full anthers. However, there were some differences in the percentages of embryoids formed by different pathways in cultured full anthers and anther segments. Especially striking was the fact that in cultured anther segments there was an appreciable reduction in the percentage of embryoids origi nating exclusively by the division of the generative cell. Compared to the control, in a medium containing auxin, there was a decrease in the number of calluses formed by the division of the generative cell and an increase in the number of calluses formed by the division of both generative and vegetative cells. Culture of anther segments in a medium containing high concentration of auxin (50.0 mg/1) led to a significant reduction in the yield of calluses that were formed almost entirely by the division of both generative and vegetative cells. Transformation of the vegetative cell alone into a callus was not observed in the sampling of anther segments used in this study. Early stages of callus formation from pollen grains by the division of both generative and vegetative cells are shown in figure 2.

From a comparative analysis of the division patterns of embryogenic pollen grains in anther segments cultured in the presence or absence of 2,4-D, one could argue that a gradient of auxin in the mass of pollen grains may be one of the factors that determines the particular pathway of differentia tion followed by an embryogenically determined pollen grain and that callus formation is a secondary effect of auxin. From this point of view, pollen grains that have a high endogenous auxin at the time of culture and that are thus presumably sensitive to added auxin might form embryoids by the division of the generative cell alone. This might account for the

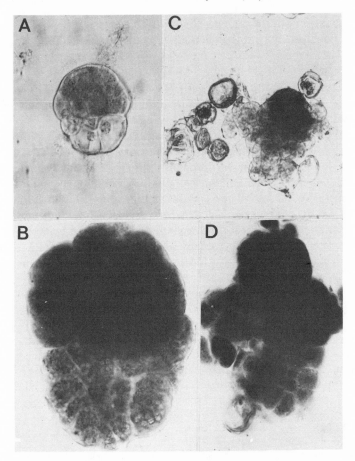

Fig. 2. Ontogeny of pollen calluses formed in cultured anther segments. (A–D) Early stages of callus formed by the division of both vegetative and generative cells. In all segments of this figure, division products of the vegetative cell are lightly colored; those of the generative cell are densely colored.

inhibitory effect of added auxin on the division of embryogenic pollen through this pathway. On the other hand, both generative and vegetative cells might divide in the embryogenic pathway when the endogenous auxin level of the pollen grain is low. The fact that a concentration of added auxin, which inhibits embryogenic division in the generative cell alone, is much less effective in inhibiting division of both vegetative and generative cells in certain pollen grains is consistent with their presumably low auxin content. Thus the critical limitation determining the involvement of the constituent cells of the pollen grain in embryogenesis is the concentration of hormone present in the pollen grain at the time of culture of anthers.

CONCLUSIONS

From an analysis of the ontogeny of pollen embryogenesis explored in some detail above, it appears that the vegetative cell and the generative cell of the angiosperm pollen are not only specialized reproductive units, but also have the innate potential to develop into a multicellular organism. The evidence derived from many species shows fairly convincingly that the vegetative cell has the greatest propensity for embryogenesis *in vitro*. It is probably a peculiar virtue of pollen grains of *Hyoscyamus niger* that allows the generative cell to function as the embryo mother cell. As far as is known, this provides the only system for demonstrating the totipotency of the generative cell. Further technical and conceptual advances in tissue culture and developmental botany should lead to the formulation of experiments to demonstrate totipotency of the single-celled sperms.

REFERENCES

Ammirato, P. V. 1974. The effects of abscisic acid on the development of somatic embryos from cells of caraway (*Carum carvi* L.). Bot. Gaz. 135:328–37.

Ammirato, P. V., and F. C. Steward. 1971. Some effects of environment on the development of embryos from cultured free cells. Bot. Gaz. 132:149–58.

Backs-Hüsemann, D., and J. Reinert. 1970. Embryobildung durch isolierte Einzelzellen aus Gewebekulturen von *Daucus carota*. Protoplasma 70:49–60.

Bajaj, Y. P. S., and M. Bopp. 1972. Growth and organ formation in *Sinapis alba* tissue cultures. Z. Pflanzenphysiol. 66:378–81.

Ball, E., and P. C. Joshi. 1966. Adventive embryos in a callus culture of *Didiscus coerulea*. Amer. J. Bot. 53:612 (abstr.).

Banerjee, S., and S. Gupta. 1975. Embryoid and plantlet formation from stock cultures of *Nigella* tissues. Physiol. Plantarum 34:243–45.

Banerjee, S., and S. Gupta. 1976. Embryogenesis and differentiation in *Nigella sativa* leaf callus *in vitro*. Physiol. Plantarum 38:115–20.

Battaglia, E. 1963. Apomixis. *In* P. Maheshwari (ed.), Recent advances in the embryology of angiosperms. Pp. 221–64. Intern. Soc. Plant Morphol., Delhi.

Bernard, S. 1971. Développement d'embryons haploides à partir d'anthères cultivées *in vitro*. Etude cytologique comparée chez le tabac et le petunia. Rev. Cytol. Biol. Végét. 34:165–88.

Bhojwani, S. S. 1966. Morphogenetic behaviour of mature endosperm of *Croton bonplandiamum* Baill. in culture. Phytomorphology 16:349–53.

Bowes, B. G. 1976. *In vitro* morphogenesis of *Crambe maritima* L. Protoplasma 89:185–88.

Bui Dang Ha, D., B. Norreel, and A. Masset. 1975. Regeneration of *Asparagas officinalis* L. through callus cultures derived from protoplasts. J. Exp. Bot. 26:263–70.

Butenko, R. G., R. V. Grushvitskii, and L. I. Slepyan. 1968. Organogenesis and somatic embryogenesis in a tissue culture of ginseng (*Panax ginseng*) and other *Panax* L. species. Bot. Zh. 53:906–11.

Button, J., and C. H. Bornman. 1971. Development of nucellar plants from unpollinated and unfertilized ovules of the Washington navel orange *in vitro*. J. South Afr. Bot. 37:127–33.

Campos, F. F., and D. T. Morgan, Jr. 1958. Haploid pepper from a sperm. J. Hered. 49:134–37.

Campos, F. F., and D. T. Morgan, Jr. 1960. Genetic control of haploidy in *Capsicum frutescens L.* following crosses with untreated and X-rayed pollen. Cytologia 25:362–72.

Chang, W.-C., and P.-L. Chiu. 1976. Induction of callus from fronds of duckweed (*Lemna gibba* L.). Bot. Bull. Acad. Sinica 17:106–9.

Clausen, R. E., and W. E. Lammerts. 1929. Interspecific hybridization in *Nicotiana*. Haploid and diploid merogony. Amer. Naturl. 63:279–82.

Corduan, G. 1975. Regeneration of anther-derived plants of *Hyoscyamus niger* L. Planta 127:27–36.

Davies, D. R. 1958. Male parthenogenesis in barley. Heredity 12:493–98.

Devreux, N., F. Saccardo, and A. Brunori. 1971. Plantes haploides et lignes isogéniques de *Nicotiana tabacum* obtenues par cultures d'anthères et de tiges *in vitro*. Caryologia 24:141–48.

Dorion, N., Y. Chupeau, and J. P. Bourgin. 1975. Isolation, culture, and regeneration into plants of *Ranunculus sceleratus* L. leaf protoplasts. Plant Sci. Lett. 5:325–31.

Dunwell, J. M., and N. Sunderland. 1973. Anther culture of *Solanum tuberosum* L. Euphytica 22:317–23.

Gamborg, O. L., F. Constabel, and R. A. Miller. 1970. Embryogenesis and production of albino plants from cell cultures of *Bromus inermis*. Planta 95:355–58.

George, L., and S. Narayanaswamy. 1973. Haploid *Capsicum* through experimental androgenesis. Protoplasma 78:467–70.

Gerassimowa, H. 1936. Experimentell Erhaltene haploide Pflanze von *Crepis tectorum* L. Planta 25:696–702.

Gerlach-Cruse, D. 1969. Embryo- und Endospermentwicklung nach einer Röntgenbestrahlung der Fruchtnoten von *Arabidopsis thaliana* (L.) Heynh. Rad. Bot. 9:433–42.

Goodsell, S. F. 1961. Male sterility in corn by androgenesis. Crop Sci. 1:227–28.

Gosch, G., Y. P. S. Bajaj, and J. Reinert. 1975. Isolation, culture, and induction of embryogenesis in protoplasts from cell-suspensions of *Atropa belladonna*. Protoplasma 86:405–10.

Grambow, H. J., K. N. Kao, R. A. Miller, and O. L. Gamborg. 1972. Cell division and plant development from protoplasts of carrot cell suspension cultures. Planta 103:348–55.

Grewal, S., U. Sachdeva, and C. K. Atal. 1976. Regeneration of plants by embryogenesis from hypocotyl cultures of *Ammi majus* L. Indian J. Exp. Biol. 14:716–17.

Guha, S., R. D. Iyer, N. Gupta, and M. S. Swaminathan. 1970. Totipotency of gametic cells and the production of haploids in rice. Curr. Sci. 39:174–76.

Guha, S., and S. C. Maheshwari. 1964. *In vitro* production of embryos from anthers of *Datura*. Nature 204:497.

Guha, S., and S. C. Maheshwari. 1966. Cell division and differentiation of embryos in the pollen grains of *Datura in vitro*. Nature 212:97–98.

Guha, S., and S. C. Maheshwari. 1967. Development of embryoids from pollen grains of *Datura in vitro*. Phytomorphology 17:454–61.

Guha-Mukherjee, S. 1973. Genotypic differences in the *in vitro* formation of embryoids from rice pollen. J. Exp. Bot. 24:139–44.

Haberlandt, G. 1902. Culturversuche mit isolierten Pflanzenzellen. Sitzber Kaiser Akad. Wiss. Berlin, Math. Naturw. Kl., Abt. I. 111:69–92.

Haccius, B., and K. K. Lakshmanan. 1965. Adventiv-Embryonen aus *Nicotiana*-Kallus, der bei hohen Lichtintensitätenkultiviert werde. Planta 65:102–4.

Halperin, W., and D. F. Wetherell. 1964. Adventive embryony in tissue cultures of the wild carrot, *Daucus carota*. Amer. J. Bot. 51:274–83.

Haustein, E. 1961. Eine androgene Halploide von *Oenothera scabra*. Planta 56:475–78.

Hu, C. Y., and I. M. Sussex. 1971. *In vitro* development of embryoids on cotyledons of *Ilex aquifolium*. Phytomorphology 21:103–7.

Hughes, K. W., S. L. Bell, and J. D. Caponetti. 1975. Anther-derived haploids of the African violet. Canad. J. Bot. 53:1442–44.

Iyer, R. D., and S. K. Raina. 1972. The early ontogeny of embryoids and callus from pollen and subsequent organogenesis in anther cultures of *Datura metel* and rice. Planta 104:146–56.

Jelaska, S. 1972. Embryoid formation by fragments of cotyledons and hypocotyls in *Cucurbita pepo*. Planta 103:278–80.

Johri, B. M., and Y. P. S. Bajaj. 1962. Behaviour of mature embryo of *Dendrophthoe falcata* (L.f.) Ettings. *in vitro*. Nature 193:194–95.

Johri, B. M., and Y. P. S. Bajaj. 1964. Growth of embryos of *Amyema, Amylotheca,* and *Scurrula* on synthetic media. Nature 204:1220–21.

Johri, B. M., and S. S. Bhojwani. 1970. Embryo morphogenesis in the stem parasite *Scurrula pulverulenta*. Ann. Bot. N. S. 34:685–90.

Johri, B. M., and C. B. Sehgal. 1963. Chemical induction of polyembryony in *Anethum graveolens* L. Naturwiss. 50:47–48.

Jørgensen, C. A. 1928. The experimental formation of heteroploid plants in the genus *Solanum*. J. Genet. 19:133–211.

Joshi, P. C., and B. M. Johri. 1972. *In vitro* growth of ovules of *Gossypium hirsutum*. Phytomorphology 22:195–209.

Jullien, M. 1974. La culture *in vitro* de cellules du tissu foliaire d'*Asparagus officinalis* L.: Obtention de souches à embryogenèse permanente et régéneration de plantes entières. C. R. Acad. Sci. Paris 279:747–50.

Kameya, T., and H. Uchimiya. 1972. Embryoids derived from isolated protoplasts of carrot. Planta 103:356–60.

Kato, H., and M. Takeuchi. 1966. Embryogenesis from the epidermal cells of carrot hypocotyl. Sci. Papers Coll. Gen. Edu. Univ. Tokyo 16:245–54.

Kavathekar, A. K., and P. S. Ganapathy. 1973. Embryoid differentiation in studies on origin, development, and dormancy of *Eschscholtzia californica*. Curr. Sci. 42:671–73.

Kehr, A. E. 1951. Monoploidy in *Nicotiana*. Heredity 42:107–12.

Khanna, P., and E. J. Staba. 1970. *In vitro* physiology and morphogenesis of *Cheiranthus cheiri* var. Clott of Gold and *C. cheiri* var. Goliath. Bot. Gaz. 131:1–5.

Kohlenbach, H. W. 1965. Über organisierte Bildungen aus *Macleaya cordata* Kallus. Planta 64:37–40.

Konar, R. N., and K. Nataraja. 1964. *In vitro* control of floral morphogenesis in *Ranunculus sceleratus* L. Phytomorphology 14:558–63.

Konar, R. N., and K. Nataraja. 1965a. Production of embryos on the stem of *Ranunculus sceleratus* L. Experientia 21:395.

Konar, R. N., and K. Nataraja. 1965b. Production of embryoids from the anthers of *Ranunculus sceleratus* L. Phytomorphology 15:245–48.

Konar, R. N., and Y. P. Oberoi. 1965. *In vitro* development of embryoids on the cotyledons of *Biota orientalis*. Phytomorphology 15:137–40.

Kostoff, D. 1929. An androgenic *Nicotiana* haploid. Z. Zellforsch. 9:640–42.

Kubicki, B., J. Telezynska, and E. Milewska-Pawliczuk. 1975. Induction of embryoid development from apple pollen grains. Acta Soc. Bot. Poloniae 44:631–35.

Kuo, J.-S., Y.-Y. Wang, N.-F. Chien, S.-J. Ku, M.-L. Kung, and H.-C. Hsu. 1973. Investigations on the anther culture *in vitro* of *Nicotiana tabacum* and *Capsicum annuum* L. Acta Bot. Sinica 15:37–52.

Lacadena, J.-R. 1974. Spontaneous and induced parthenogenesis and androgenesis. *In* K. J. Kasha (ed.), Haploids in higher plants. Advances and potential. Pp. 13–32. University of Guelph Press, Guelph, Ontario, Canada.

Maheshwari, P., and B. Baldev. 1961. Artificial production of buds from the embryos of *Cuscuta reflexa*. Nature 191:197–98.

Maheshwari, P., and R. C. Sachar. 1963. Polyembryony. *In* P. Maheshwari (ed.), Recent advances in the embryology of angiosperms. Pp. 265–96. Intern. Soc. Plant Morphol., Delhi.

Maheshwari, S. C., and G. R. P. Gupta. 1965. Production of adventitious embryoids *in vitro* from stem callus of *Foeniculum vulgare*. Planta 67:384–86.

Maly, R. 1958. Die Mutabilität der Plastiden von *Antirrhinum majus* L. Sippe 50. Z. Vererbungslehre 89:692–96.

Mehra, A., and P. N. Mehra. 1972. Differentiation in callus cultures of *Mesembryanthemum floribundum*. Phytomorphology 22:171–76.

Mehra, P. N., and A. Mehra. 1971. Morphogenetic studies in *Pterotheca falconeri*. Phytomorphology 21:174–91.

Mitra, G. C., and H. C. Chaturvedi. 1970. Fruiting plants from *in vitro* grown tissue of *Rauvolfia serpentina*. Benth. Curr. Sci. 39:128–29.

Mitra, G. C., and H. C. Chaturvedi. 1972. Embryoids and complete plants from unpollinated ovaries and from ovules of *in vivo*–grown emasculated flower buds of *Citrus* spp. Bull. Torrey Bot. Club 99:184–89.

Mullins, M. G., and C. Srinivasan. 1976. Somatic embryos and plantlets from an ancient clone of the grapevine (cv. Cabernet-Sauvignon) by apomixis *in vitro*. J. Exp. Bot. 27:1022–30.

Nag, K. K., and B. M. Johri. 1969. Organogenesis and chromosomal constitution in embryo callus of *Nuytsia floribunda*. Phytomorphology 19:405–8.

Narayanaswamy, S., and L. P. Chandy. 1971. *In vitro* induction of haploid, diploid, and triploid androgenic embryoids and plantlets in *Datura metel* L. Ann. Bot. N. S. 35:535–42.

Narayanaswamy, S., and L. George (*née* Chandy). 1972. Morphogenesis of belladonna (*Atropa belladonna* L.) plantlets from pollen in culture. Indian J. Exp. Biol. 10:382–84.

Nataraja, K. 1971. Morphogenic variations in callus cultures derived from floral buds and anthers of some members of Ranunculaceae. Phytomorphology 21:290–96.

Nataraja, K. 1975. Morphogenesis in embryonal callus of *Euphorbia pulcherrima in vitro*. Curr. Sci. 44:136–37.

Nataraja, K., and R. N. Konar, 1970. Induction of embryoids in reproductive and vegetative tissues of *Ranunculus sceleraturs* L. *in vitro*. Acta Bot. Neerl, 19:707–16.

Nétien, G., and J. Raynaud. 1972. Formation d'embryons dans la culture *in vitro* de tissus de *Conium maculatum* L. Bull. Mensl. Soc. Linne. Lyon 41:49–51.

Nitsch,. C., and B. Norreel. 1973. Effet d'un choc thermique sur le pouvour embryogène du pollen de *Datura innoxia* cultivé dans l'anthère ou isolé de l'anthère. C. R. Acad. Sci. Paris 275:303–6.

Nitsch, J. P., and C. Nitsch, 1970. Obtention de plantes haploides à partir de pollen. Bull. Soc. Bot. Fr. 117:339–59.

Norreel, B. 1970. Etude cytologique de l'androgenèse expérimentale chez *Nicotiana tabacum* et *Datura innoxia*. Bull. Soc Bot. Fr. 117:461–78.

Norstog, K. 1965. Induction of apogamy in megagametophytes of *Zamia integrifolia*. Amer. J. Bot. 52:993–99.

Norstog, K. 1970. Induction of embryolike structures by kinetin in cultured barley embryos. Devel. Biol. 23:665–70.

Pelletier, G. C. Raquin, and G. Simon. 1972. La culture *in vitro* d'anthères d'asperge *(Asparagus officinalis)*. C. R. Acad. Sci. Paris 272:848–51.

Poirier-Hamon, S., P. S. Rao, and H. Harada. 1974. Culture of mesophyll protoplasts and stem segments of *Antirrhinum majus* (snapdragon): Growth and organization of embryoids. J. Exp. Bot. 25:752–60.

Prabhudesai, V. R., and S. Narayanaswamy. 1973. Differentiation of cytokinin-induced shoot buds on excised petioles of *Nicotiana tabacum*. Phytomorphology 23:133–37.

Prabhudesai, V. R., and S. Narayanaswamy. 1974. Organogenesis in tissue cultures of certain Asclepiads. Z. Pflanzenphysiol. 71:181–85.

Rabéchault, H., J. Ahée, and G. Guénin. 1970. Colonies cellulaires et formes embryoides obtenues *in vitro* à partir des cultures d'embryons de palmier à huile *(Elaeis guineensis* Jacq. var. *dura* Becc.). C. R. Acad. Sci. Paris. 270:3067–70.

Radojević, L., R. Vujičić, and M. Nešković. 1975. Embryogenesis in tissue culture of *Corylus avellana* L. Z. Pflanzenphysiol. 77:33–41.

Raghavan, V. 1976a. Role of the generative cell in androgenesis in henbane. Science 191:388–89.

Raghavan, V. 1976b. Experimental embryogenesis in vascular plants. Academic Press, London.

Raghavan, V. 1977. Patterns of DNA synthesis during pollen embryogenesis in henbane. J. Cell Biol. 73:521–26.

Raghavan, V. 1978. Origin and development of pollen embryoids and pollen calluses in cultured anther segments of *Hyoscyamus niger* (henbane). Amer. J. Bot. 65:984–1002.

Raman, K., and R. I. Greyson. 1974. *In vitro* induction of embryoids in tissue cultures of *Nigella damascena*. Canad. J. Bot. 52:1988–89.

Ramanna, M. S. 1974. The origin and *in vivo* development of embryoids in the anthers of *Solanum* hybrids. Euphytica 23:623–32.

Ramanna, M. S., and J. G. T. Hermsen. 1974. Embryoid formation in the anthers of some interspecific hybrids in *Solanum*. Euphytica 23:423–27.

Rao, P. S. 1965. *In vitro* induction of embryonal proliferation in *Santalum album* L. Phytomorphology 15:175–79.

Rao, P. S., W. Handro, and H. Harada. 1973. Bud formation and embryo differentiation in *in vitro* cultures of *Petunia*. Z. Pflanzenphysiol. 69:87–90.

Rao, P. S., S. Narayanaswamy, and B. D. Benjamin. 1970. Differentiation *ex ovulo* of embryos and plantlets in stem tissue cultures of *Tylophora indica.* Physiol. Plantarum 23:140–44.

Rashid, A., and H. E. Street. 1973. The development of haploid embryoids from anther cultures of *Atropa belladonna* L. Planta 113:263–70.

Rashid, A., and H. E. Street. 1974. Segmentations in microspores of *Nicotiana sylvestris* and *Nicotiana tabacum* which lead to embryoid formation in anther cultures. Protoplasma 80:323–34.

Ratnamba, S. P., and R. N. Chopra. 1974. *In vitro* induction of embryoids from hypocotyls and cotyledons of *Anethum graveolens* seedlings. Z. Pflanzenphysiol. 73:452–55.

Reinert, J. 1959. Über die Kontrolle der Morphogenese und die Induktion von Adventivembryonen an Gewebekulturen aus Karotten. Planta 53:318–33.

Reinert, J., D. Backs, and M. Krosing. 1966. Faktoren der Embryogenese in Gewebekulturen aus Kulturformen von Umbelliferen. Planta 68:375–78.

Sangwan, R. S., and H. Harada. 1975. Chemical regulation of callus growth, organogenesis, plant regeneration, and somatic embryogenesis in *Antirrhinum majus* tissue and cell cultures. J. Exp. Bot. 26:868–81.

Sangwan, R. S., and B. Norreel. 1975, Induction of plants from pollen grains of *Petunia* cultured *in vitro.* Nature 257:222–24.

Sankhla, N., D. Sankhla, and U. N. Chatterji. 1967. Production of plantlets from callus derived from root tip of excised embryos of *Ephedra foliata* Boiss. Naturwiss. 54:349.

Satsangi, A., and H. Y. Mohan Ram. 1965. A continuously growing tissue culture from the mature endosperm of *Ricinus communis* L. Phytomorphology 15:26–30.

Schroeder, C. A. 1968. Adventive embryogenesis in fruit pericarp tissue *in vitro.* Bot Gaz. 129:374–76.

Sehgal, C. B. 1964. Artificial induction of polyembryony in *Foeniculum vulgare* Mill. Curr. Sci. 33:373–75.

Sehgal, C. B. 1972. *In vitro* induction of polyembryony in *Ammi majus* L. Curr. Sci. 41:263–64.

Sehgal, C. B. 1975. *In vitro* differentiation of foliar embryos and adventitious buds from leaves of *Begonia semperflorens* Link and Otto. Indian J. Exp. Biol. 13:486–88.

Sommer, H. E., C. L. Brown, and P. P. Kormanik, 1975. Differentiation of plantlets in longleaf pine *(Pinus palustris* Mill.) tissue cultured *in vitro.* Bot. Gaz. 136:196–200.

Sopory, S. K., and S. C. Maheshwari. 1972. Production of haploid embryos by anther culture technique in *Datura innoxia*—A further study. Phytomorphology 22:87–90.

Sopory, S. K., and P. G. Rogan. 1976. Induction of pollen divisions and embryoid formation in anther cultures of some dihaploid clones of *Solanum tuberosum.* Z. Pflanzenphysiol. 80:77–80.

Staritsky, G. 1970. Embryoid formation in callus tissues of coffee. Acta Bot. Neerl. 19:509–14.

Steward, F. C. 1963. The control of growth in plant cells. Sci. Amer. 209(4):104–13.

Steward, F. C., A. E. Kent, and M. O. Mapes, 1966. The culture of free plant cells and its significance for embryology and morphogenesis. *In* A. A. Moscona and A. Monroy (eds), Current topics in developmental biology, 1:113–54. Academic Press, New York.

Steward, F. C., and M. O. Mapes. 1971a. Morphogenesis in aseptic cell cultures of *Cymbidium.* Bot. Gaz. 132:65–70.

Steward, F. C., and M. O. Mapes. 1971b. Morphogenesis and plant propagation in aseptic cultures of *Asparagus.* Bot. Gaz. 132:70–79.

Steward, F. C., M. O. Mapes, A. E. Kent, and R. D. Holsten. 1964. Growth and development of cultured plant cells. Science 143:20-27.

Steward, F. C., M. O. Mapes, and K. Mears. 1958. Growth and organized development of cultured cells. II. Organization in culture⁻ grown from freely suspended cells. Amer. J. Bot. 45:705-8.

Steward, F. C., M. O. Mapes, and J. Smith. 1958. Growth and organized development of cultured cells. I. Growth and division of freely suspended cells. Amer. J. Bot. 45:693-703.

Sunderland, N., G. B. Collins, and J. M. Dunwell. 1974. The role of nuclear fusion in pollen embryogenesis of *Datura innoxia* Mill. Planta 117:227-41.

Sunderland, N., and F. M. Wicks. 1971. Embryoid formation in pollen grains of *Nicotiana tabacum.* J. Exp. Bot. 22:213-16.

Thomas, E., F. Hoffman, and G. Wenzel. 1975. Haploid plants from microspores of rye. Z. Pflanzenzüch. 75:106-13.

Thomas, E., and H. E. Street. 1970. Organogenesis in cell suspension cultures of *Atropa belladonna* cultivar *lutea* Döll. Ann. Bot. N. S. 34:657-69.

Vasil, I. K., and A. C. Hildebrandt. 1966a. Variations of morphogenetic behavior in plant tissue cultures. I. *Cichorium endivia.* Amer. J. Bot. 53:860-69.

Vasil, I. K., and A. C. Hildebrandt. 1966b. Variations of morphogenetic behavior in plant tissue cultures. II. *Petroselinum hortense.* Amer. J. Bot. 53:869-74.

Vasil, I. K., and C. Nitsch. 1975. Experimental production of pollen haploids and their use. Z. Pflanzenphysiol. 76:191-212.

Vasil, V., and A. C. Hildebrandt. 1965. Differentiation of tobacco plants from single isolated cells in microculture. Science 150:889-92.

Wadhi, M., and H. Y. Mohan Ram. 1964. Morphogenesis in the leaf callus of *Kalanchoe pinnata* Pers. Phyton 21:143-47.

Wang, Y.-, C.-S. Sun, C.-C. Wang, and N.-F. Chien. 1973. The induction of the pollen plantlets of *Triticale* and *Capsicum annuum* from anther culture. Scient. Sinica 16:147-51.

Wetherell, D. F., and W. Halperin. 1963. Embryos derived from callus tissue cultures of the wild carrot. Nature 200:1336-37.

Williams, L., and H. A. Collin. 1976. Embryogenesis and plantlet formation in tissue cultures of celery. Ann. Bot. N. S. 40:325-32.

Wilmar, C., and M. Hellendoorn. 1968. Growth and morphogenesis of *Asparagus* cells cultured *in vitro.* Nature 217:369-70.

Wilson, H. M., and H. E. Street. 1975. The growth, anatomy, and morphogenetic potential of callus and cell suspension cultures of *Hevea brasiliensis.* Ann Bot. N.S. 39:672-82.

Yamada, T., H. Nakagawa, and Y. Sinotô. 1967. Studies on the differentiation in cultured cells. I. Embryogenesis in three strains of *Solanum* callus. Bot. Mag. Tokyo 80:68-74.

Zenkteler, M. 1971. *In vitro* production of haploid plants from pollen grains of *Atropa belladonna.* L. Experientia 27:1087.

Zenkteler, M. 1972. Development of embryos and seedlings from pollen grains in *Lycium halimifolium* Mill. in the *in vitro* culture. Biol. Plantarum 14:420-22.

Zenkteler, M., E. Misiura, and A. Ponitka,. 1975. Induction of androgenetic embryoids in the *in vitro* cultured anthers of several species. Experientia 31:289-91.

Embryo Culture as a Tool
in the Study of Comparative
and Developmental Morphology

13

Plant embryo culture has a relatively long history as an experimental science, dating back to the beginning years of the present century. Among its originators were E. Hännig, who first explored the prospect that excised premature embryos might be grown into seedling plants, and H. T. Brown, who studied the effects of various nitrogen compounds upon the development of full-term barley embryos in liquid culture media (Hännig, 1904; Brown, 1906). Hännig's work is of particular interest in the present context, for in his attempts to grow excised premature embryos of several species of crucifers, he observed that they did not continue to develop as embryos in culture, but instead developed directly into small, weak plantlets. He termed this "kunstliche Frühgeburt," which in translation is referred to as precocious germination. Precocious germination of cultured embryos was repeatedly observed by early workers with plant embryo culture (see Narayanaswami and Norstog, 1964; Raghavan, 1966, for partial reviews of the phenomenon). Outstanding among pioneer investigators in embryo culture were C. D. LaRue, who attempted to grow embryos of a number of species and encountered precocious germination in all (LaRue, 1936), and H. B. Tukey, who also reported precocious germination in his attempts to grow the nonviable embryos produced by certain early-ripening varieties of *Prunus* (Tukey, 1933, 1934, 1938). A further observation by both these

Knut Norstog, Department of Biological Sciences, Northern Illinois University, DeKalb, Illinois, 60115.

investigators was that embryos smaller than 0.5 mm in length could not be cultured successfully.

A major breakthrough in embryo culture was the discovery of van Overbeek et al. (1942) that very small (less than 200 μm) hybrid embryos of *Datura*, normally the victims of endosperm-embryo incompatibilities, could be grown to some semblance of maturity, provided the media contained coconut milk. Many subsequent investigations of plant embryogenesis *in vitro* have taken this study as a departure point, and research in both somatic and zygotic embryogenesis can be traced to this important pioneering work.

PRECOCIOUS GERMINATION AND ITS CONTROL

Although coconut milk and other additives and extracts have greatly aided in the culture of excised plant embryos, and have proven very useful in tissue culture as well, encounters with precocious germination have persisted well into the present era of embryo culture. It is at once a troublesome problem in embryogenesis *in vitro* and a valuable tool and aid in understanding plant morphogenesis and its underlying principles. An early example of both the difficulty and the utility inherent in precocious germination is seen in a study of embryogenesis of wild rice (*Zizania aquatica*) done by LaRue and Avery (1938). The wild rice embryo is characterized by an unusually well developed scutellum and epiblast (fig. 1), and it was the hope of the two investigators that these structures might serve them as markers of the developmental potential of the embryo *in vitro*. Their objectives were twofold: (1) to see if embryological development would be continued *in vitro*, and (2) to see if the embryo was determinate or indeterminate (i.e., would it continue to grow beyond the state of differentiation attained *in vivo?*). As it turned out, neither question was answered very satisfactorily. The embryo did not develop *as an embryo* in culture, hence the question of determination could not be answered. Nevertheless, an important conclusion was reached concerning the fundamental nature of embryo development in plants. Plant embryologists (Johansen, 1950; Haccius and Bhandari, 1975) have considered the zygotic plant embryo developmentally to be a preformed mosaic. "Laws" of embryogeny stress a partitioning of developmental potentials as a result of very regular, early cleavage divisions. In the normal embryogenesis of many species, the disposition of embryonic cells is so orderly that it has been suggested that deviations from normal patterns of cell division might result in morphologically aberrant embryos (Johansen, 1950). The work of LaRue and Avery strongly suggested that the plant embryo is epigenetic in development; under conditions leading to precocious germination, the

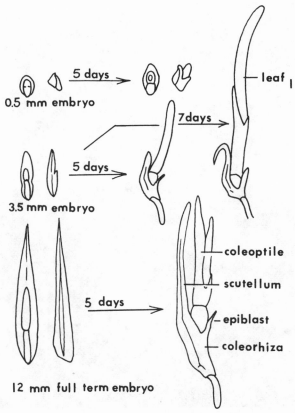

Fig. 1. Development of the embryo of *Zizania aquatica* in culture. Precocious germination occurred in 3.5 mm and 12 mm embryos, whereas 0.5 mm embryos did not develop. (Redrawn from LaRue and Avery, Bull. Torrey Bot. Club 65:11–21.)

youngest embryos capable of growth *in vitro* develop into simple rudimentary plantlets with few leaves and roots in correspondence to those few primordia present at the time of excision (fig. 1). Regarding the determination of embryos, barley embryos growing on a synthetic medium may become larger than natural embryos (4 or more mm long versus 3.2 mm) and form more leaf primordia (4 versus 3), but it has not been shown that the plant embryo *in vitro* is indeterminate (Norstog and Smith, 1963) (fig. 2).

It might, however, be argued that a predisposition for complete development is present in excised premature embryos, provided the culture media are sufficiently sophisticated as to permit the expression of the total developmental potential of the embryo. The media employed by LaRue

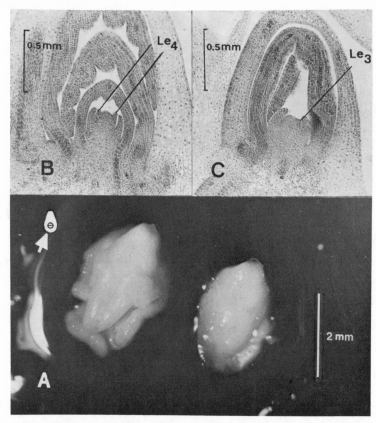

Fig. 2.   Development of cultured barley embryos. (A) Comparison between cultured embryo (left) and excised full-term embryo (right). Relative size of explanted embryo giving rise to embryo at left indicated at arrow. (B) Longitudinal section of apex of embryo cultured 10 days. Note the presence of four leaf primordia (Le₄). (C) Longitudinal section of normal embryo grown *in vivo*. Three leaf primordia are present (Le₃). (From Norstog and Smith, Science 142:1655–56. Copyright 1963 by the American Association for the Advancement of Science. Reprinted with permission.)

and Avery were simple by today's standards and no doubt permitted only the further growth of primordia already present in the embryo, but not the further transition of initials, if any were present, into primordia. It is a moot point; however, the point of totipotency versus predisposition has been clarified by the experiments of Steward (Steward et al., 1964), as well as studies of Reinert (1959, 1967, 1973; Reinert and Backs, 1968), Guha and Maheshwari (1964), and Halperin and Wetherell (1964, 1965; Halperin, 1966). In particular, adventive embryos are characterized by irregular cell arrangements and delayed histogenesis, but nevertheless form bipolar embryos of normal morphology (Haccius and Bhandari, 1975).

Causes of Precocious Germination

Most plant embryos become dehydrated and are metabolically and developmentally inactive upon reaching their full-term development. Seldom do they enter the seedling phase of growth without an intervening period of metabolic inactivity. An exception is the embryo of the red mangrove (*Rhizophora mangle*), which grows directly into a seedling while the fruit hangs on the tree. On occasion, the embryos of cereals also may germinate while the grain is in the inflorescence, usually during extended periods of very high humidity. Nevertheless a desiccated, inactive phase is the normal state of the embryo following the attainment of full-term development. It may be of interest that this phase of development can be duplicated *in vitro* (Norstog, 1966). Excised premature barley embryos have been cultured to the full-term state of development, completely dried down, kept in a desiccator for several weeks, and, finally, rehydrated and grown into green plants.

Depending upon the species or circumstances, the inactive phase of postembryonic development may be quite short, or it may be lengthy. In the latter case, it is true dormancy. Whatever the case, in germination the embryo imbibes water and enters the seedling phase of growth in a few hours. This phase is characterized by protein synthesis as well as water uptake. Protein synthesis during the first few hours of germination appears to be dependent upon preformed, stored mRNA (Gordon and Payne, 1976). Walbot (1971, 1972), in studies utilizing labeled RNA precursors, has shown there is no *de novo* synthesis of RNA in bean embryos during the first twelve hours of germination. Protein synthesis during the imbibition phase of germination is dependent upon mRNA synthesized prior to the desiccation of the embryo during fruit ripening. Ihle and Dure (1970) have demonstrated the presence of a germination inhibitor in premature cotton embryos. Aqueous leaching of embryos released them from this inhibition and permitted precocious germination. Treatment of leached embryos with actinomycin D, which blocks transcription, did not block the formation of a protease, present both in the leached embryo at the time embryogenesis was 68% completed and in the full-term embryo. It was believed that inhibition of precocious germination during embryogenesis resulted from the arrest of mRNA translation by abscisic acid (ABA) or ABA-like molecules.

The embryo of *Taxus baccata* is still immature at the time the seed is fully formed, and an after-ripening period is required for the completion of embryogenesis (LePage-Degivry, 1973a). During this period the embryo is not capable of germination but may be induced to germinate precociously by culturing it for a brief period in liquid nutrient medium (LePage-Degivry, 1970, 1973b; LePage-Degivry and Garello, 1973). After eight days in liquid medium, during which time the embryo does not germinate, it may be transferred to an agar medium, whereupon germination ensues. It has been found that a water-soluble germination inhibitor is leached from the embryo during the liquid culture phase, provided sucrose is present in the medium (LePage-Degivry and Garello, 1973). The inhibitor of dormancy in *Taxus* embryos appears identical to ABA (LePage-Degivry, 1973b, 1973c). King (1976) has reported that high concentrations of ABA in developing wheat grains coincide with maximum growth of embryo and endosperm; he suggests that its role may be that of suppression of precocious germination *in vivo* during grain development. Possibly, leaching out of such ABA or other germination inhibitors may account for precocious germination phenomenon in premature embryos of plants in general.

Precocious germination may be induced in cultures of barley embryos under conditions normally capable of suppressing it, such as high osmolarity, light, and elevated temperature (Norstog, 1972). Particularly effective were concentrations of gibberellic acid (GA) of about 0.1 μg/ml. However, the degree of GA-induced precocious germination was directly proportional to the GA concentration and inversely proportional to the sugar concentration of the medium. For example, precocious germination occurred in embryos cultured in light on media containing 0.1 M to 0.2 M sucrose and GA, but not on media containing GA and 0.3 M and 0.35 M sucrose (fig. 3). Higher concentrations of GA (1.0 μg/ml) produced precocious germination on media containing 0.3 M and 0.35 M sucrose. The effect was mimicked by kinetin, which also produced symptoms of precocious germination in light-grown embryos on media containing 0.2 M sucrose. Abscisic acid counteracted the action of GA with respect to precocious germination.

The relationship between these three classes of plant growth regulators to an extent duplicates their action in controlling germination of dormant strains of barley (Khan, 1975) and in germination of Grand Rapids lettuce (Poggi-Pellegrin and Bulard, 1976). In barley, GA is considered to play an inductive role in germination, whereas ABA acts to inhibit it and cytokinins release GA from ABA inhibition. A similar interaction is believed to occur in lettuce seeds.

Control of Precocious Germination

Although precocious germination was noted rather early in embryo

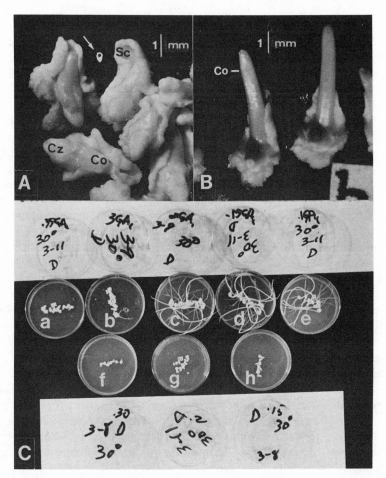

Fig. 3. Responses of barley embryos grown 10 days in media containing several concentrations of gibberellic acid and sucrose. Co=coleoptile, Cz=coleorhiza, Sc=scutellum.
(A) Control culture on media lacking gibberellic acid. Arrow indicates size of initial 0.5
mm explant. (B) Embryos cultured upon media containing gibberellic acid. Note coleoptile elongation symptomatic of precocious germination. (C) Series of cultures upon
media containing 0.1 μg/ml of gibberellic acid and the following sucrose concentrations:
a=0.35 M, b=0.3 M, c=0.2 M, d=0.15 M, e=0.1 M. Culture dishes f, g, and h contain
0.3 M, 0.2 M and 0.15 M sucrose respectively but no gibberellic acid, and represent
controls for cultures b, c, and d. (From Norstog, Phytomorphology 22:134–39; reprinted with
permission.)

culture, the first attempts to control it were made by Dieterich (1924), who
observed that it did not occur in embryos grown in submerged cultures, but
rather such embryos remained poorly developed. Van Overbeek et al.
(1942) employed coconut milk in liquid media to grow *Datura* embryos as

small as 0.15 mm into viable seedlings, apparently without appreciable precocious germination. However, the embryos were malformed and did not develop roots until they were transferred to a coconut milk-free medium. Possibly low oxygen tension in liquid culture is an important factor in suppressing precocious germination. Norstog and Klein (1972) have reported that low O_2 (9%) suppressed precocious germination of barley embryos grown on agar media. Kent and Brink (1947) cultured excised premature embryos of barley on nutrient agar and found that precocious germination occurred on a basic agar medium containing only sucrose and mineral salts, but that precocious germination was delayed if casein hydrolysate, tomato juice, or other natural plant extracts were added. Subsequently (Ziebur et al., 1950) it was discovered that the delay of precocious germination in barley could be obtained in the absence of casein hydrolysate (1%) with elevated concentrations of sucrose (up to 12.8%), although casein hydrolysate was slightly more active in supporting embryonic development than a high concentration of sucrose alone. Rijven (1952) reported essentially the same effect of sucrose in suppressing precocious germination. Sucrose concentrations of 12 and 18% were employed to culture embryos as small as 100 μm. Germination of *Capsella* embryos cultured on high sucrose media was not obtained unless embryos were transferred to media of lower osmolarity (2 to 6% sucrose). An interesting point is that Rijven determined the requirement for high osmolarity by analyses of tonicity values of the embryos, which were isotonic with a solution of 0.375 M sucrose. More recently Smith (1971) has measured the osmolarity of the liquid endosperm of bean, by the freezing-point depression method, and found it to be 0.7 osmolar when embryos are in the heart stage and 0.5 osmolar when they are in the late cotyledon stage. Smith observed that "the 0.7 osmolar value is more than double the osmolarity of any medium reported for the culture *in vitro* of plant embryos." On the basis of analyses of the environment of the developing embryo, one therefore may argue in favor of the use of media of high osmolarity in embryo culture.

In a departure from those who claimed high osmolarity was the key to successful embryo culture and the prevention of precocious germination, Raghavan and Torrey (1963) stated that premature heart-stage embryos of *Capsella* could be grown successfully without precocious germination in liquid or upon semisolid media of low osmolarity (2% sucrose), provided ample nitrogen was present in the medium. Barley embryos also have been grown without precocious germination upon media of relatively low osmolarity (0.2 osmolar) (Norstog and Klein, 1972; Umbeck and Norstog, 1978). In particular, the ability of cultured *Hordeum* embryos to develop without precocious germination appears to be related to relatively high concentrations of potassium (Norstog, 1967), and embryonic differentiation without precocious germination occurred in media containing 470 $\mu g/ml$ KNO$_3$, 700 $\mu g/ml$ of KCl, concentrations of 4.3 mM ammonia (supplied as ammonium malate), and a sucrose concentration of 2% (Umbeck and Norstog, 1978) (fig. 4). The induction of adventive embryogenesis in carrot tissue culture also is stimulated by combined K$^+$ and NH$^+_4$ (Brown et al., 1976), and in this system neither Na$^+$ nor NH$^+_4$ could replace K$^+$.

It is not clear what the role of high osmolarity actually is in the early development of plant embryos *in ovulo*. Since precocious germination *in vivo* probably is prevented by inhibitors (Ihle and Dure, 1970; LePage-Degivry, 1973b) as well as by osmotic pressure and possibly low O$_2$, it may be that the more important role of high osmolarity is to prevent plasmolytic damage to the very delicate proembryo in its early cleavage divisons.

THE CULTURE OF UNDIFFERENTIATED PROEMBRYOS

Whereas the pathways of development of somatic embryos derived in tissue culture seem to have been pretty well worked out (Nitsch, 1969; Sunderland, 1973; Reinert, 1973), the culture of undifferentiated zygotic embryos and the analysis of their early differentiation has not been advanced in so spectacular a fashion. Probably the smallest embryos regularly to be cultured on a production-line basis are those of orchids. The orchid embryo measures about 100 μm in diameter, has little endosperm, and requires either a symbiotic association with mycorrhiza-like fungus or a nutrient medium to support its differentiation and development. Interestingly, in the mature seed the orchid embryo is actually an undifferentiated proembryo and only later develops into an asymmetrical embryo called a protocorm, rather than the bipolar embryo characteristic of plants in general and monocotyledons in particular (Curtis and Nichol, 1948; Raghavan and Torrey, 1964). The protocorm may be described as an irregular, calluslike mass of tissue from which one or more roots and buds are initiated, and there is no embryonic axis per se. Proembryos of other plants than orchids have been reported to form protocorms rather than bipolar embryos *in vitro* (DeMaggio and Wetmore, 1961; Norstog, 1956, 1965) (fig. 5). The smallest proembryo successfully cultured is that of sugarcane, in which an embryo 60 μm in diameter was cultured and grown into a mature plant, but regrettably no data pertaining to its development *in vitro* were given (Warmke et al., 1946). Barley embryos as small as 100 μm have been cultured, and the smallest grew into protocorm-like masses from which shoots and roots regenerated (Norstog, 1956, 1961, 1965).

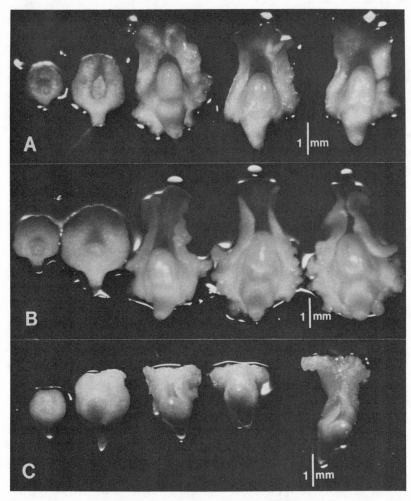

Fig. 4. Sets of representative barley embryos cultured 10 days on media containing combinations of concentrations of ammonia and abscisic acid. (A) Embryos, from left to right, cultured on media containing the following series of concentrations of ammonia (in mM): 0.0, 2.1, 4.3, 6.4, 8.6. (B) Embryos cultured upon same series of ammonia concentrations as in A, but with 0.01 μg/ml of abscisic acid added to each medium. Note that embryos appear more robust than corresponding embryos in (A). (C) Embryos cultured upon same series of ammonia concentrations as in (A) and (B), but 1.0 μg/ml of abscisic acid added to each medium. Note inhibition of axis and hypertrophy of the scutellum.

Globular embryos of *Capsella* less than 80 μm were cultured by Raghavan and Torrey (1963) and cell divisions and differentiation observed. Although the embryos did not attain complete development, differentiation

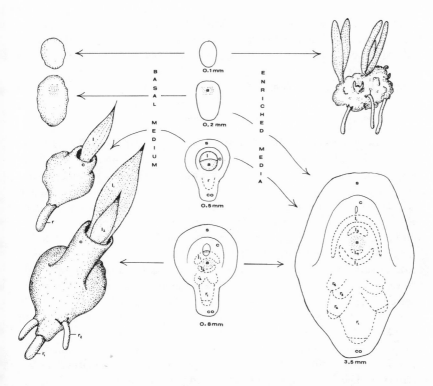

Fig. 5. Diagram illustrating developmental pathways of barley embryos *in vitro*. In left column responses of embryos to a depauperate medium are shown. Smaller embryos do not grow; larger embryos develop by precocious germination. In right column development of embryos upon enriched media is represented. Smaller embryos form protocorms; larger embryos differentiate normally into complete, full-term embryo (lower right). (Based upon Norstog, Amer. J. Bot. 48:876–84.)

of cotyledons and roots occurred. Their development, however, was irregular. Whether the tendency of proembryos to form protocorms *in vitro* is an intrinsic characteristic of most if not all undifferentiated young embryos in which the pathways of development are as yet undetermined, or whether protocorm formation only reflects a failure of media and methods to support early and normal developmental pathways, is not known. However, it should be pointed out that a general tendency in somatic embryogenesis *in vitro* is the formation first of a group of undifferentiated cells, possibly comparable to a protocorm, from which proembryonic buds subsequently arise. These in turn may form regular, symmetrical embryos (Reinert, 1959, 1973; Halperin and Wetherell, 1964; Haccius, 1963; Haccius and Bhandari, 1975). Slightly larger (<200 μm) but externally unorganized barley embryos have been cultured routinely and develop into complete, symmetrical embryos (Norstog, 1965) (fig. 5). Therefore, the prior differentiation of primordia may not be required for normal embryogenesis in cultured embryos. However, an enriched medium is required for this development; otherwise there will be no further differentiation of primordia and organs (fig. 5).

Nutrition of Premature Zygotic Embryos

Although coconut milk has been used in the culture of both zygotic embryos (Van Overbeek et al., 1942; Warmke et al., 1946; Norstog, 1956, 1961) and somatic embryos (Steward et al., 1964; Sussex, 1972), the trend in embryo and tissue culture has been in the direction of synthetic media in which all of the components are capable of chemical definition (see Raghavan, 1966, for a review of major advances in development of embryo culture media up to 1965). In an early attempt in the direction of defining media for embryo culture, we have noted that Rijven (1952) attempted a rational approach toward embryo culture and analyzed such factors as pH of the ovule, tonicity of embryo and medium, and organic components of the endosperm. As a result, he concluded that pH should be on the acid side (ca. 5.0) and, as noted before, the osmolarity of the medium should be high (7.5 M) and a source of reduced nitrogen (preferably glutamine) should be incorporated in the medium. A similar approach was used by Mauney et al. (1967) in determining the culture requirements of premature cotton embryos. The cotton seeds were harvested when 12 to 14 days old, and the small volume of liquid endosperm (ca. 0.02 ml) was analyzed for organic acids using two-dimensional paper chromatography. The results indicated that only malic acid was present in detectable quantities. Its concentration was found to be over 7 mg/ml. Further exploration of salts of malic acid showed that sodium malate inhibited embryo growth but that ammonium malate was strongly growth-promoting. In addition, Mauney et al. (1967) observed that embryos did not require as high an osmolarity with ammonium malate in the medium as they did without it. Raghavan and Torrey (1964) observed that orchid embryos grew better upon media containing ammonium salts rather than nitrate during their early development. However, after completion of the protocorm stage of morphological development and leaf differentiation (60 days in culture) upon an ammonium medium, seedlings were able to assimilate nitrate and continue normal growth. Gamborg and Shyluk (1970) showed that the growth of soybean cells could be enhanced by ammonia, in contrast with nitrate, but only if the ammonium ion were combined with citrate, malate, fumarate, or succinate anions.

Smith (1971, 1973) attempted a comprehensive analysis of bean endosperm in an attempt to develop a medium for the culture of globular bean embryos. He too found a high concentration of ammonium ion and malate in the liquid endosperm, and in fact reported that the level of these ions was much higher than that utilized in any tissue or embryo culture medium.

Umbeck and Norstog (1978) analyzed the responses of premature barley embryos to various concentrations of ammonium malate in media, and concluded that embryos could be grown on media of relatively low sugar (2%) provided sufficient ammonia was provided. Optimum development of premature barley embryos (0.5 mm long, as compared with 3.2 mm size of full-term embryo) was attained with a simple medium containing 6.4 mM of ammonia. A higher concentration (8.6 mM) was not inhibitory, but did not enhance further development. Lower concentrations were growth limiting, and no development occurred on a medium lacking ammonia and containing nitrate as a sole nitrogen source. Thus it appears not only that the high osmolarity of media is not necessary to the development of premature but differentiated embryos *in vitro*, but that the ammonium ion, preferably supplied as the salt of an organic acid (citrate or malate) is indispensable.

HORMONES AND EMBRYO CULTURE

The first investigator to employ a hormone in embryo culture was LaRue (1936), who found that IAA in low concentrations (0.05 μg/ml) promoted the development of embryos of several species including *Zea mays*. Raghavan and Torrey (1963) found that low concentrations of IAA and kinetin were required for continued development *in vitro* of globular *Capsella* embryos.

Although auxin is present in liquid endosperm (Smith, 1971, 1973), which is the natural environment of the developing embryo, and together with cytokinins and adenine has been found to support the differentiation of globular *Capsella* embryos (Raghavan and Torrey, 1963), exogenous auxins generally seem not to be required for the growth *in vitro* of plant embryos (LaRue, 1936; Rijven, 1952; Rietsema et al., 1953; Veen, 1963). This observation seems to be in keeping with reports that the induction of somatic embryos in tissue culture is inhibited by higher levels of exogenous auxin and stimulated by reduced or zero concentrations (Halperin, 1966; Reinert, 1959, 1973; Steward, 1970).

Gibberellic acid has been reported to induce precocious germination in cultured bean embryos (Skene, 1969) and in barley embryos (Schooler, 1960; Norstog, 1972). At very low levels (ca. 0.01 μg/ml) it may promote embryogenesis without precocious germination (Norstog, unpublished report).

Cytokinins alone are either ineffective or slightly growth promoting for very young embryos of *Capsella* (Veen, 1963), but they have been reported to promote growth and differentiation of embryos when combined with IAA (Veen, 1963) or with IAA and adenine sulfate (Raghavan and Torrey, 1963). Kinetin has been reported to induce the formation of adventive embryos by cultured barley embryos (Norstog, 1970) (fig. 6).

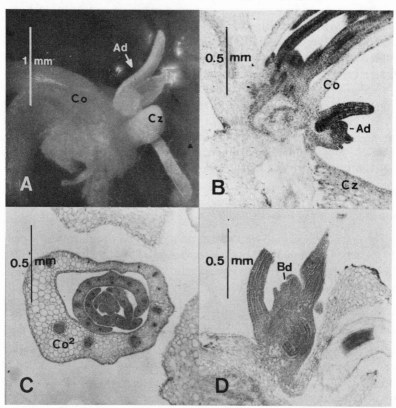

Fig. 6.   Development of adventive embryos upon cultured barley embryos. Ad=adventive embryo, Co=coleoptile of parent embryo, Cz=coleorhiza of parent embryo, Co^2=coleoptile of adventive embryo. (A) Surface view of adventive embryo. (B) Sagittal section of early adventive embryo. (C) Cross section through coleoptile of adventive embryo. (D) Sagittal section of older adventive embryo. Note its bipolarity. (From Norstog, Devel. Biol. 23:665–70; reprinted with permission.)

Khan (1975) has suggested that there is an interaction between gibberellins, cytokinins, and ABA in the control of germination in barley, and a similar interaction of these three growth regulators appears to control germination in lettuce (Poggi-Pellegrin and Bulard, 1976). Khan proposes that in barley germination, gibberellins promote germination, ABA acts as a GA-antagonist, and cytokinins are envisioned as releasing gibberellins from ABA antagonism. Somewhat the same interactions as proposed by Khan influence development and precocious germination of premature barley embryos *in vitro* (Norstog, 1972). Gibberellic acid (GA_3) was reported to promote precocious germination, and ABA counteracted gibberellin-induced precocious germination. Kinetin also promoted precocious germination, possibly by interacting with GA and ABA.

A more promotive role for ABA has been reported in cultures of somatic embryos of caraway (Ammirato, 1973, 1974, 1977), where ABA, in combination with zeatin and GA_3, markedly reduced aberrant embryo development. A similar normalizing influence of ABA has been detected in barley embryo cultures (Umbeck and Norstog, 1978). In these cultures low levels of ABA in combination with NH_4^+ (fig. 3) produced healthy embryos and, in particular, promoted the differentiation of the scutellum.

In general ABA appears to act as a growth inhibitor in tissue cultures (Gamborg and LaRue, 1971; Li et al., 1972), but it also has been reported to promote the growth of callus (Blumenfeld and Gazit, 1970; Altman and Goren, 1971), the rooting of cuttings (Chin et al., 1969), and adventitious bud formation (Heide, 1968). That it may not be a growth inhibitor for developing embryos is indicated by a report alluded to earlier that increasing concentrations of ABA in developing wheat grains coincides with a period of maximum embryo and endosperm growth (King, 1976), and by the observations that relatively low levels of ABA (ca. 0.01 $\mu g/ml$) promote the differentiation of cultured barley embryos (Umbeck and Norstog, 1978) (fig. 3).

Probably a role of ABA in the natural development of embryos is to inhibit the germinative phase of growth and, at the same time, permit embryogenesis. Its apparent normalizing effect in cultures of somatic and zygotic embryos may only consist of the inhibition of endogenous gibberellins and/or the transcription of mRNA-regulating synthesis of germination compounds, as suggested by Ihle and Dure (1970), and not a specific stimulation per se of differentiation and development. Nevertheless, the possibility that low levels of exogenous GA_3 may promote embryogenesis without precocious germination (Norstog, unpublished) casts some doubt upon the interpretation that ABA acts solely as an antigibberellin in embryogenesis. In summary, it appears that the ammonium ion can be employed as a sole supplement to simple tissue culture media as a promoter of embryo differentiation and growth, that media of high osmolarity are not required for normal embryogenesis, and that hormones generally are not required for embryogeny *in vitro*, although combinations of GA, ABA, and cytokinins may be helpful in "regularizing" development.

EMBRYO CULTURE AND COMPARATIVE MORPHOLOGY

Questions about the homology of embryos and other stages in the plant life cycle (as, for example, the developmental potentials of gametophytes and sporophytes) have been the subject of experimentation using tissue and embryo culture techniques. Tulecke (1953) showed that the male gametophyte (pollen) of *Ginkgo* could be cultured and grown as a rapidly dividing tissue. LaRue (1948) discovered that the megagametophytes of cycads were capable of producing sporophyte characteristics including buds, roots, and embryo-like structures. Subsequently investigators have determined that the male gametophyte generation of plants is capable of producing embryos of essentially normal sporophytic form (Guha and Maheshware, 1964; Nitsch, 1969). Thus the essential developmental homology of the sporophytic and gametophytic plant generations have been demonstrated. In a sense, also the direct production of adventive embryos by cultured zygotic embryos without an intervening callus stage, as reported by Konar and Oberoi (1965), Hu and Sussex (1971), and Norstog (1970), is an example of the essential homology of zygotic and somatic embryos. In barley the adventive embryos were reportedly formed from the epidermis and developed into bilateral, bipolar embryos with closed coleoptiles (fig. 6). This development is noteworthy because *de novo* development of grass embryos *in vitro* is not known to occur; rather, in cultures of somatic tissues of grasses, isolated root and shoot initiation occurs in callus (see, for example, Cheng and Smith, 1975).

In embryos of grasses, several structures have been the source of much argument insofar as the interpretation of their homology (refer to fig. 1 for orientation) (Brown, 1960). The scutellum has variously been described as a leaf sheath, a cotyledon, a leaf, a ligule, an aborted axis, a haustorium, and a structure of no homologous significance (Arber, 1934; Brown, 1960; Guignard, 1975). Similar pronouncements have been made about the coleoptile, which has been termed variously a modified leaf (Reeder, 1953), a leaf sheath (Arber, 1934), and the first leaf of an axillary bud (Guignard, 1975). The epiblast has been thought to be a vestigial cotyledon (Roth, 1955), and the coleorhiza has been proposed to be a highly modified radicle (Foard and Haber, 1962). Possibly the responses exhibited by these organs *in vitro* may throw some light upon this subject that descriptive comparisons have not. In theory, at least, a modified or even vestigial structure may retain some potentialities for development exhibited by their normal counterparts. Foard and Haber (1962), for example, have shown that in gamma-irradiated embryos of wheat, hairs resembling root hairs are produced from the coleorhiza even though cell division was completely eliminated by irradiation. In barley cultures the embryo, which normally lacks an epiblast, sometimes forms one (Norstog, 1961). The scutellum of the barley embryo is highly responsive to manipulations of media and embryo. When barley embryos were cultured in light upon GA-containing media, the scutellum developed chlorophyll. Apparent upon the adaxial surface of the scutellum *in vivo* is a trough-like depression in which the coleoptile lies. That this depression is not simply the result of pressure of the pericarp on the coleoptile is revealed in cultures of excised premature embryos. A well-developed scutellar trough, surrounded by a semicircular ridge, forms in embryos grown in media containing NH_4^+ (fig. 7), and this development is accentuated when the medium contains, in addition, low levels of ABA (Umbeck and Norstog, 1978). It is believed that the trough and ridge of the scutellum are homologous with the sheath and leaf of the foliage leaf of barley, in particular, and of grasses in general.

The culture of barley embryos on media containing kinetin induced the formation of trichomes on the scutellum as well as upon the coleoptile (Norstog, 1969). In both instances the trichomes resemble leaf hairs (fig. 8), and bear no resemblance to coleorhizal hairs or root hairs. Foard and Haber (1962) have reported that hairs formed upon the coleorhiza and epiblast during germination of gamma-irradiated wheat embryos resemble root hairs in ontogeny and morphology and are not like trichomes. Neither did these hairs respond similarly to gamma irradiation, which inhibited trichome development completely but permitted development of hairs of root, epiblast, and coleorhiza. Indole acetic acid (IAA) promoted the growth of hairs by epiblast, coleorhiza, and root, but not of leaf trichomes. The growth retardant, CCC (2-chloroethyl-trimethylammonium chloride) inhibited leaf expansion but not that of epiblast, and the epiblast and coleoptile responded differently to GA, which promoted coleoptile elongation but inhibited epiblast expansion. Foard and Haber concluded that the epiblast and coleorhiza and the root were morphologically alike, and that the epiblast is not a leaf homolog. In an arena in which a fertile imagination may give rise to almost any interpretation of ontogenetic and phylogenetic homology, an experimental approach with emphasis upon developmental patterns and potentialities may be a tool of major utility.

Microsurgery and Embryogenesis

Much has been learned about the developmental potentialities of shoot and root apices through the painstaking effort of workers such as Wardlaw (1955), Steeves and Sussex (1957), Sussex and Clutter (1960), and Ball (1960), in which microdissection, sometimes followed by culture of excised apices and primordia, has demonstrated the great plasticity and regenerative potential of plant meristems. In a very preliminary way, these metho-

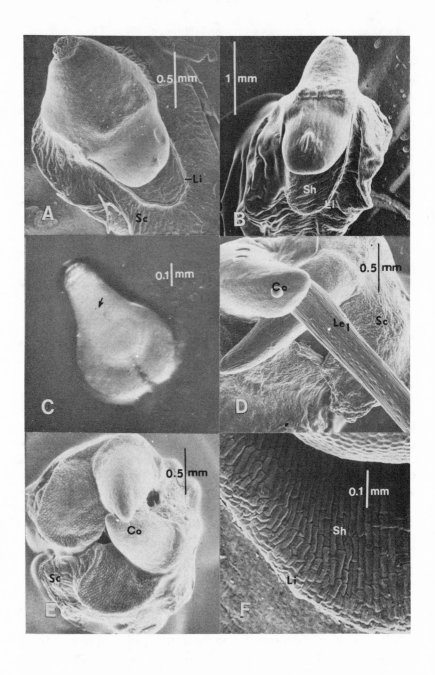

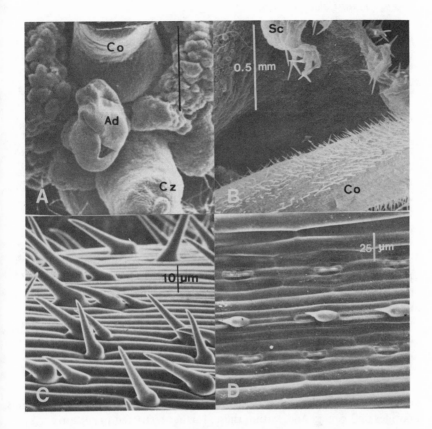

Fig. 7 (opposite). Scanning electronic micrographs of fresh-wet barley embryos. Co=coleoptile, Le₁=first foliage leaf, Li=scutellar ligule, Sc=scutellem, Sh=sheath of scutellum. (A) Natural, full-term embryo. (B) Embryo cultured 10 days on medium containing ammonia and abscisic acid. Note development of scutellar sheath and ligule. (C) Excised embryo wounded by sagittal incision extending to point indicated by arrow. (D) Embryo like that in (C) cultured 8 days upon media containing ammonia but lacking abscisic acid. Note absence of scutellar ligule and sheath. (E) Bisected embryo cultured 8 days upon media containing both ammonia and abscisic acid. Note in particular the development of normal scutellar ligule and sheath by each half embryo. (F) Enlarged view of scutellar sheath. Note elongate epidermal cells characteristic of foliage leaf, in particular of leaf sheath epidermis.

Fig. 8. Scanning electron micrographs for illustrating reponses of cultured barley embryos to hormones. Ad=adventive embryo, Co=coleoptile, Cz=coleorhiza, Sc=scutellum. (A) Induction of adventive embryo by kinetin. (B) Induction of trichomes upon scutellum and coleoptile of barley embryo by kinetin. (C) Enlarged view of kinetin-induced trichomes of barley coleoptile. They develop from short cells, as do normal leaf trichomes shown in (D).

dologies also have been applied to plant embryos (LaRue, 1952) and to our own studies of barley embryos. Barley embryos do not appear to have much regenerative potential *in vitro*, but do have a limited capacity for restoring the superficial integrity of their form (Umbeck and Norstog, 1978). In a series of cultures in which small embryos (0.5–0.7 mm) were bisected sagittally through the apex, coleoptile, and scutellar primordia, it was found that on occasion split apices and primordia were healed and restored the half-embryos to a semblance of the form of the normal embryo (fig. 7). For example, half-coleoptiles often formed hollow, conical structures superficially like normal coleoptiles, and half-scutella formed nearly complete scutella with well-developed sheath and ligule regions. A surprising finding was that ABA greatly promoted the differentiation and healing of such mutilated embryos. In fact, scutellar ligules and sheaths occurred in regenerating scutella only when ABA was present in the medium. Ideally microdissection experiments should be carried out upon undifferentiated proembryos in order to assess the state of compartmentalization of developmental potentialities of such early embryos. Preliminary attempts at such work with barley cultures (Norstog, unpublished) have not met with much success, owing to a lack of both surgical finesse and media capable of supporting regeneration of minute pieces of embryonic tissue. Although there seems to be a general belief among botanists that adventive embryogenesis *in vitro* has settled once and for all questions about embryogenetic potentialities of cells, it must be pointed out that adventive embryogenesis currently is a phenomenon restricted to comparatively few familes and genera and, within these, further restricted to certain cell and tissue types. It is remarkable, for example, that adventive embryogenesis paralleling early zygotic embryogenesis seems largely to be exhibited by pollen grains, whereas the development of embryos from callus follows a different pattern (see, for example, Halperin, 1969; Nitsch, 1969). Thus further experimentation in the growth of premature zygotic embryos seems justified, if only to explore the requirements for the induction of early pathways of embryogeny. No doubt information thus gained will also prove to have application to the culture of somatic tissues and the induction of somatic embryogenesis.

LITERATURE CITED

Altman, A., and R. Goren. 1971. Promotion of callus formation by abscisic acid in citrus bud cultures. Plant Physiol. 47:844–46.

Ammirato, P. V. 1973. Some effects of abscisic acid on the development of embryos from caraway cells in suspension culture. Amer. J. Bot. Suppl. 60:22–23.

Ammirato, P. V. 1974. The effects of abscisic acid on the development of somatic embryos from cells of caraway (*Carum carvi* L.). Bot. Gaz. 135:328–37.

Ammirato, P. V. 1977. Hormonal control of somatic embryo development from cultured cells of caraway. Plant Physiol. 59:579–86.

Arber, A. 1934. The Gramineae: A study of cereal, bamboo, and grass. At the University Press, Cambridge.

Ball, E. 1960. Sterile culture of the shoot apex of *Lupinus albus*. Growth 24:91–110.

Blumenfeld, A., and S. Gazit. 1970. Interaction of kinetin and abscisic acid in the growth of soybean callus. Plant Physiol. 45:535–36.

Brown, H. T. 1906. On the culture of excised embryos of barley on nutrient solutions containing nitrogen in different forms. Trans. Guinness Res. Lab. 1:288–99.

Brown, S., D. F. Wetherell, and D. K. Dougall. 1976. The potassium requirement for growth and embryogenesis in wild carrot suspension cultures. Physiol. Plantarum 37:73–79.

Brown, W. V. 1960. The morphology of the grass embryo. Phytomorphology 10:215–23.

Cheng, T., and H. H. Smith. 1975. Organogenesis from callus culture of *Hordeum vulgare*. Planta 123:307–10.

Chin, T., M. M. Meyer, Jr., and L. Beevers. 1969. Abscisic-acid-stimulated rooting of stem cuttings. Planta 88:192–96.

Curtis, J. T., and M. A. Nichol. 1948. Cultures of proliferating orchid embryos *in vitro*. Bull. Torrey Bot. Club 75:358–73.

DeMaggio, A. E., and R. H. Wetmore. 1961. Morphogenetic studies on the fern *Todea barbara*. III. Experimental embryology. Amer. J. Bot. 45:551–65.

Dieterich, K. 1924. Über die Kultur von Embryonen ausserhalb des Samens. Flora 117:379–417.

Foard, D. E., and A. H. Haber. 1962. Use of growth characteristics in studies of morphologic relations. I. Similarities between epiblast and coleorhiza. Amer. J. Bot. 49:520–23.

Gamborg, O. L., and T. A. G. LaRue. 1971. Ethylene production by plant cell cultures: The effects of auxins, abscisic acid, and kinetin on ethylene production in suspension cultures of rose and *Ruta* cells. Plant Physiol. 48:399–401.

Gamborg, O. L., and J. P. Shyluk. 1970. The culture of plant cells with ammonium salts as the sole nitrogen source. Plant Physiol. 45:598–600.

Gordon, M. E., and P. I. Payne. 1976. *In vitro* translation of the long-lived messenger ribonucleic acid of dry seeds. Planta 130:269–73.

Guha, S., and S. C. Maheshwari. 1964. *In vitro* production of embryos from anthers of *Datura*. Nature 204:497.

Guignard, J. L. 1975. Du cotylédon des monocotylédons. Phytomorphology 25:193–200.

Haccius. B. 1963. Restitution in acidity-damaged plant embryos—regeneration or regulation? Phytomorphology 13:107–15.

Haccius, B., and N. N. Bhandari. 1975. Delayed histogen differentiation as a common primitive character in all types of nonzygotic embryos. Phytomorphology 25:91–94.

Halperin, W. 1966. Alternative morphogenetic events in cell suspensions. Amer. J. Bot. 53:443–53.

Halperin, W. 1969. Morphogenesis in cell cultures. Ann. Rev. Plant Physiol. 20:395–418.

Halperin, W., and D. F. Wetherell. 1964. Adventive embryony in tissue cultures of the wild carrot, *Daucus carota*. Amer. J. Bot. 51:274–83.

Halperin, W., and D. F. Wetherell. 1965. Ontogeny of adventive embryos of wild carrot. Science 147:756–58.

Hännig, E. 1904. Zur Physiologie Pflanzenlicher Embryonen. I. Über die Kultur von Cruciferen-Embryonen assur-halb des Embryosachs. Bot. Zeit. 62:45–80.

Heide, O. 1968. Stimulation of adventitious bud formation in *Begonia* leaves by abscisic acid. Nature 219:960–61.

Hu, C. Y., and I. M. Sussex. 1971. *In vitro* development of embryoids on cotyledons of *Ilex aquifolium*. Phytomorphology 21:103–7.

Ihle, J. N., and L. Dure III. 1970. Hormonal regulation of translation inhibition requiring RNA synthesis. Biochem. Biophys. Res. Comm. 38:995–1001.

Johansen, D. A. 1950. Plant embryology. Chronica Botanica, Waltham.

Kent, N. F., and R. A. Brink. 1947. Growth *in vitro* of immature *Hordeum* embryos. Science 106:547–48.

Khan, A. A. 1975. Primary, preventive, and permissive roles of hormones in plant systems. Bot. Rev. 41:391–420.

King, R. W. 1976. Abscisic acid in developing wheat grains and its relationship to grain growth and maturation. Planta 132:43–51.

Konar, R. N., and Y. P. Oberoi. 1965. *In vitro* development of embryoids on the cotyledons of *Biota orientalis*. Phytomorphology 15:137–40.

LaRue, C. D. 1936. The growth of plant embryos in culture. Bull. Torrey Bot. Club 63:365–82.

LaRue, C. D. 1948. Regeneration in the megagametophyte of *Zamia floridana*. Bull. Torrey Bot. Club 75:597–603.

LaRue, C. D. 1952. Growth of the scutellum of maize in culture. Science 115:315–16.

LaRue, C. D., and G. S. Avery. 1938. The development of the embryos of *Zizania aquatica* in the seed and in artificial culture. Bull. Torrey Bot. Club 65:11–21.

LePage-Degivry, M. T. 1970. Acide abscissique et dormance chez les embryons de *Taxus baccata* L. C. R. Acad. Sci. Paris 271:482–84.

LePage-Degivry, M. T. 1973a. Étude en culture *in vitro* de la dormance embryonnaire chez *Taxus baccata* L. Biol. Plantarum 15:264–69.

LePage-Degivry, M. T. 1973b. Intervention d'un inhibiteur lié dans la dormance embryon-naire de *Taxus baccata* L. C. R. Acad. Sci. Paris 277:177–80.

LePage-Degivry, M. T. 1973c. Influence de l'acide abscissique sur le développement cultivés *in vitro*. Z. Pflanzenphysiol. 70:406–13.

LePage-Degivry, M. T., and G. Garello. 1973. La dormance embryonnaire chez *Taxus baccata*: Influence de la composition du milieu liquide sur l'induction de la germination. Physiol. Plantarum 29:204–7.

Li, H. C., E. L. Rice, L. M. Rohrbaugh, and S. H. Wender. 1972. Effects of abscisic acid on phenolic content and lignin biosynthesis in tobacco tissue culture. Physiol. Plantarum 23:928–36.

Mauney, J. R., J. Chappell, and B. J. Ward. 1967. Effects of malic acid salts on growth of young cotton embryos *in vitro*. Bot. Gaz. 128:198–200.

Narayanaswami, S., and K. Norstog. 1964. Plant embryo culture. Bot. Rev. 30:587–628.

Nitsch, J. P. 1969. Experimental androgenesis in *Nicotiana*. Phytomorphology 19:389–404.

Norstog, K. 1956. The growth of barley embryos on coconut milk media. Bull. Torrey Bot. Club 83:27–29.

Norstog, K. 1961. The growth and differentiation of cultured barley embryos. Amer. J. Bot. 48:876–84.

Norstog, K. 1965. Development of cultured barley embryos. I. Growth of 0.1–0.4 mm embryos. Amer. J. Bot. 52:538–46.

Norstog, K. 1966. Influence of nutritional and physical factors on the growth of cultured barley embryos. Amer. J. Bot. Suppl. 53:613–14.

Norstog, K. 1967. Studies on the survival of very small barley embryos in culture. Bull. Torrey. Bot. Club 94:223–29.

Norstog, K. 1969. Morphology of coleoptile and scutellum in relation to tissue culture responses. Phytomorphology 19:235–41.

Norstog, K. 1970. Induction of embryo-like structures by kinetin in cultured barley embryos. Devel. Biol. 23:665–70.

Norstog, K. 1972. Factors relating to precocious germination in cultured barley embryos. Phytomorphology 22:134–39.

Norstog, K., and R. M. Klein. 1972. Development of cultured barley embryos. II. Precocious germination and dormancy. Canad. J. Bot. 50:1887–94.

Norstog, K., and J. Smith, 1963. Culture of small embryos of barley on defined media. Science 142 (1963): 1655–56.

Poggi-Pellegrin, M., and C. Bulard. 1976. Interactions between abscisic acid, gibberellins, and cytokinins in Grand Rapids lettuce seed germination. Physiol. Plantarum 36:40–46.

Raghavan, V. 1966. Nutrition, growth, and morphogenesis of plant embryos. Biol. Rev. 41:1–58.

Raghavan, V., and J. G. Torrey. 1963. Growth and morphogenesis of globular and older embryos of *Capsella* in culture. Amer. J. Bot. 50:540–51.

Raghavan, V., and J. G. Torrey. 1964. Inorganic nitrogen nutrition of the seedlings of the orchid, *Cattleya*. Amer. J. Bot. 51:264–74.

Reeder, J. 1953. The embryo of *Streptochaeta* and its bearing on the homology of the coleoptile. Amer. J. Bot. 40:77–80.

Reinert, J. 1959. Über die Kontrolle der Morphogenese und die Induction von Adventivembryonen an Gewebekulturen aus Karotten. Planta 53:318–33.

Reinert, J. 1967. Factors of embryo formation in plant tissues cultivated *in vitro*. *In* Sur les cultures de tissus de plantes. Colloques Intern. C.N.R.S. 920:33–40.

Reinert, J. 1973. Aspects of organization—organogenesis and embryogenesis. *In* H. E. Street (ed.), Plant tissue and cell culture. Pp. 338–55. University of California Press, Berkeley.

Reinert, J., and D. Backs. 1968. Control of totipotency in plant cells growing *in vitro*. Nature 220:1340–41.

Rietsema, J., S. Satina, and A. F. Blakeslee. 1953. The effect of indole-3-acetic acid on *Datura* embryos. Proc. Nat. Acad. Sci. USA 39:924–33.

Rijven, A. H. G. C. 1952. *In vitro* studies on the embryo of *Capsella bursa-pastoris*. Acta Bot. Neerl. 2:157–200.

Roth, I. 1955. Zur morphologischen Deutung des Grasembryos and verwandter Embryotypen. Flora 142:564–600.

Schooler, A. B. 1960. The effect of gibrel and gibberellic acid (K salt) in embryo culture media for *Hordeum vulgare*. Argon. J. 52:411.

Skene, K. G. M. 1969. Stimulation of germination of immature bean embryos by gibberellic acid. Planta 87:118–92.

Smith, J. G. 1971. An analytical approach to the culture of globular bean embryos. Ph.D. thesis, University of Michigan, Ann Arbor.

Smith, J. G. 1973. Embryo development in *Phaseolus vulgaris*. II. Analysis of selected organic ions, ammonia, organic acids, amino acids, and sugars in the endosperm liquid. Plant Physiol. 51:454–58.

Steeves, T. A., and I. M. Sussex. 1957. Studies on the development of excised leaves in sterile culture. Amer. J. Bot. 44:665–73.

Steward, F. C. 1970. Totipotency, variation, and clonal development of cultured cells. Endeavour 29:117–24.

Steward, F. C., M. O. Mapes, A. E. Kent, and R. D. Holsten, 1964. Growth and development of cultured plant cells. Science 143:20–27.

Sunderland, N. 1973. Pollen and anther culture. *In* H. E. Street (ed), Plant tissue and cell culture. Pp. 205–39. University of California Press, Berkeley.

Sussex, I. M. 1972. Somatic embryos in long-term carrot tissue cultures: Histology, cytology, and development. Phytomorphology 22:50–59.

Sussex, I. M., and M. E. Clutter. 1960. A study of the effect of externally supplied sucrose on the morphology of excised fern leaves *in vitro*. Phytomorphology 10:87–99.

Tukey, H. B. 1933. Artificial culture of sweet cherry embryos. J. Hered. 24:7–12.

Tukey, H. B. 1934. Artificial culture methods for isolated embryos of deciduous fruits. Proc. Amer. Soc. Hort. Sci. 32:313–22.

Tukey, H. B. 1938. Growth patterns of plants developed from immature embryos in artificial culture. Bot. Gaz. 99:630–65.

Tulecke, W. 1953. A tissue derived from the pollen of *Ginkgo biloba*. Science 117:599–600.

Umbeck, P., and K. Norstog. 1978. Effects of abscisic acid and ammonia on morphogenesis of cultured barley embryos (submitted).

van Overbeek, J., M. E. Conklin, and A. F. Blakeslee. 1942. Cultivation *in vitro* of small *Datura* embryos. Amer. J. Bot. 29:472–77.

Veen, H. 1963. The effect of various growth regulators on embryos of *Capsella bursa-pastoris* growing *in vitro*. Acta Bot. Neerl. 12:129–71.

Walbot, V. 1971. RNA metabolism during embryo development and germination of *Phaseolus vulgaris*. Devel. Biol. 26:369–79.

Walbot, V. 1972. Rate of RNA synthesis and tRNA end-labeling during early development of *Phaseolus*. Planta 108:161–71.

Wardlaw, C. W. 1955. Embryogenesis in plants. Methuen and Co., London.

Warmke, H. E., J. Rivera-Perez, and J. A. Ferrer-Monge. 1946. The culture of sugarcane embryos *in vitro*. Inst. Trop. Agric. Univ. Puerto Rico, 4th Ann. Report, pp. 22–23.

Ziebur, N. K., R. A. Brink, L. H. Graf, and M. A. Stahmann. 1950. The effect of casein hydrolysate on the growth *in vitro* of immature *Hordeum* embryos. Amer. J. Bot. 37:144–48.

Comparative Studies of Anther and Pollen Culture

14

INTRODUCTION

In the past decade great advances have been made in the production of haploid and diploid homozygous plants by anther culture (for review, see Sunderland and Dunwell, 1977). However, low yields are still a serious drawback particularly in food and other economic crops. Various ways of increasing yields have been reported. In *Datura innoxia* and several species of *Nicotiana*, anthers excised and cultured from chilled buds have been shown to be more productive than anthers cultured directly from the plant (Nitsch and Norreel, 1973; Nitsch, 1974). Addition of activated charcoal to agar media has also proved advantageous, as for example in anther culture of tuberous *Solanums* (Irikura, 1975). However, some workers believe that the anther wall constitutes a major obstacle to progress, in that it provides an unfavorable environment for the developing pollen, either by restricting gaseous exchange and/or nutrient availability, or by producing toxic materials. If this is so, greater yields are likely to result, and the pollens of more species induced to divide, by isolating the pollen and culturing it in chemically defined media (Nitsch, 1974, 1977). Although these may be good theoretical reasons favoring pollen culture, there is still little factual evidence to support the claim.

A development that has received too little attention in the past is the use of liquid media in anther culture, and it is this theme that provides the

N. Sunderland, John Innes Institute, Colney Lane, Norwich, England.

background of the present paper. Studies at the John Innes Institute have shown liquid culture to be more reliable and more effective in both the Solanaceae and Gramineae, and to allow a more accurate assessment of results. Moreover, by a slight modification of the procedure used for the Solanaceae, high-yielding pollen cultures can be obtained—an approach that requires no mutilation of the anthers and leaves the anther wall intact (Sunderland and Roberts, 1977b). The new approach relies heavily on a chilling pretreatment of buds, but for longer periods and at slightly higher temperatures than usually recommended. New facts have thus come to light on the nature of the chilling effect.

ANTHER CULTURE IN LIQUID MEDIA

Advantages of liquid media in anther culture were first intimated at the Haploid Conference held in Guelph (1974) when Devreux announced greatly enhanced yields in tobacco by transfer of anthers from agar-to shake-culture at the time of anther-opening (see also Devreux et al., 1975). Shaking dislodged the developing embryos into the medium in which they continued to develop free of the anther tissues. Effects of competitive inhibition imposed by crowding of embryos inside the anther were thus alleviated. Since then, Wernicke and Kohlenback (1975, 1976) have shown that anthers of several Solanaceous plants cultured in liquid media for the whole of the culture period, but kept stationary, are also more productive than anthers plated on either agar or agar-charcoal media. Anthers do in fact float without support on shallow layers of medium (4–5 ml) in plastic petri dishes (5 × 2 cm) and rarely sink if left undisturbed. The pollen develops not only in association with the anther (attached fraction) but also as "free" units in the medium (fig. 1C; fig. 2A,C). Gaseous exchange is apparently sufficient to allow development of the "free" fraction without further aeration. As in agar culture (Pelletier and Ilami, 1972; Mii, 1976), occasional anthers turn brown rapidly, and the enclosed pollen is killed. There is no evidence, however, that such anthers influence the viability of others in the same dish. Rapid browning and death of the pollen also occur if the anthers sink before embryogenesis has commenced. Shaking causes anthers to sink and should therefore be avoided during the early stages of culture (table 1).

Substitution of liquid for solid media has not altered previous findings in respect to anther-staging, chilling, or ploidy levels. Thus, some species respond maximally at the first pollen division (*Datura innoxia, Nicotiana tabacum*), whereas others respond better just before the division (*Hyoscyamus niger*) or just after it (*N. paniculata, N. knightiana, Atropa belladonna* [cf. Rashid and Street, 1973]) (table 2). Chilling of buds for 4 days at 4°C

TABLE 1

Anther Culture of Datura Innoxia in Liquid Medium

	SHAKE		STATIONARY	
	Dark	Light	Dark	Light
Percent cultures with embryos	22	22	77	100
Mean embryo yield per anther	1	1	48	83

Means of 9 cultures.

Four petri dishes (5 × 2 cm) each containing 5 ml A medium and inoculated with one anther from each of two buds having pollen at the late unicellular to early bicellular stage. Dishes incubated for 28 days at 28°C either in darkness or under a high pressure sodium lamp (12 h day) on a rotary shaker (120 rpm). Unchilled buds from greenhouse plants (January–February) given supplementary light.

A medium—major salts (half strength) and iron (full strength) of Murashige and Skoog (1962) plus sucrose 2%, pH 5.5.

TABLE 2

Anther Culture (Liquid Medium) of Various Solanaceous Plants
at Different Developmental Stages

SPECIES		DEVELOPMENTAL STAGE					
		Early to Mid-Unicellular		Late Unicellular to Early Bicellular		Bicellular to Early Starch	
		NC	C	NC	C	NC	C
Datura innoxia	a	96(27)	100(28)	96(22)	100(16)	60(10)	94(16)
	b	46	45	62	75	12	19
Nicotiana tabacum cv.	a	69(36)	87(39)	97(33)	94(67)	89(27)	93(14)
White Burley	b	17	32	75	67	19	36
N. paniculata	a	0(28)	0(14)	5(37)	3(33)	36(31)	52(31)
	b	0	0	<1	<1	2	2
N. knightiana	a	0(24)	0(28)	7(68)	7(71)	28(32)	37(27)
	b		no record—mixture of callus and embryos				
Hyoscyamus niger v.	a	44(41)	38(32)	22(36)	35(20)	21(24)	11(27)
pallidus	b	4	3	4	2	<1	<1
Atropa belladonna v.	a	8(39)	2(49)	29(28)	50(24)	47(17)	50(26)
lutea	b	<1	<1	2	1	2	4

NC—Unchilled; C—buds chilled at 4°C for 4 days in polythene bags, wrapped in aluminum foil to minimize water loss.

a = Percent cultures with pollen embryos or other structures. Number of cultures examined (in parentheses).

b = Mean yields of embryos and other structures per anther. Multiple structures scored as one unit.

Each culture comprised two (*Datura*) or four (all others) anthers from the same bud in 5 ml liquid A medium (table 1). Stationary cultures at 28°C in darkness for 14 days then in light (Grolux 500 klx) for 21 days at 25° C. Greenhouse plants (June–August).

prior to anther excision (Nitsch and Norreel, 1973) slightly increases both the percentage of successful cultures and the yield per anther at certain stages (table 2). Embryos are predominantly haploid in the three *Nicotiana* species, but of varying ploidies in the other species.

Liquid culture has not overcome the wide differences in response between species. However, in the low-yielding species of table 2, sufficient plants to meet most requirements can obtained by scaling up the procedure. Moreover, in these species, initial culture yields in no way reflect the ultimate plant yields attainable. This is due to the production of morphogenic callus, multipolar, and other anomalous structures in addition to normal bipolar embryos (Sunderland and Roberts, 1977a).

POLLEN CULTURE WITHOUT MUTILATION OF ANTHERS

In all the species of table 2, except *Datura innoxia*, the "free" fraction arises from pollen liberated into the medium after dehiscence of the anthers. Such cultures are to all intents and purposes pollen cultures. In fact, the liberated pollen, even though shed in an immature state of embryogenesis, will continue to grow if the anthers are removed from the petri dish. Shedding continues throughout the culture period. Hence, by frequent transfer of anthers to fresh medium, a series of pollen cultures (without anther wall) can be prepared from the same batch of anthers. Anther-stage is the most important criterion for early dehiscence. Thus, in *Nicotiana tabacum* little opening takes place at the early to mid unicellular stage (microspores) over the first 14 days (table 3). Time to anther-opening decreases with increasing anther age, and at the mitotic and early bicellular stages, development of the shed pollen begins after about 6 days. At these two stages, pollen shed between 6 and 14 days accounts for more than half of the total embryo yield, the highest yields being attained at the early bicellular stage (microgametophytes). Bud-chilling is also essential to ensure early anther-dehiscence. After a 6-day chilling in *N. tabacum*, shed pollen starts to develop before 6 days (table 3). Yields in the 0–6 day fraction increase with increased duration of chilling and after a 12-day chill are highly productive (table 4). Chilling for longer that 12 days leads to progressive deterioration of the buds, but successful cultures have been obtained in both *Nicotiana* and *Hyoscyamus* species after chilling for as long as 24 days (fig. 1). In *N. tabacum*, higher yields are attained in dark-grown cultures (table 4).

Because isolated tobacco pollen has specialized nutritional requirements, it will not grow in the simple media used in anther culture (Nitsch, 1974). Hence, successful developement of the shed pollen implies conditioning of the medium by the anther tissues. Tapetum has long been held to have a nutritive function, and it is thus possible that the conditioning stems directly from degenerating tapetal cells. The serial production of pollen cultures from the same batch of anthers (table 3) clearly indicates conditioning to be a continuous process, contact between anthers and medium

TABLE 3

EMBRYO YIELDS IN CULTURES OF TOBACCO POLLEN SHED*
FROM ANTHERS OF DIFFERENT DEVELOPMENTAL STAGES

Pollen Fraction	Days of Anther Culture	Developmental Stage (corolla length mm)							
		Unicellular (12–14)		Mitotic (16–18)		Early Bicellular (19–24)		Bicellular to Early Starch (27–35)	
		NC	C	NC	C	NC	C	NC	C
Shed	0–6	0(3)	0(1)	0(7)	14(4)	0(4)	35(5)	0(6)	1(2)
	6–10	0	0	55	287	197	481	5	1
	10–14	0	0	71	608	77	337	4	5
Attached	14–35	47	6	154	656	76	544	21	10
Mean embryo yield per anther		4	<1	23	139	29	116	3	1

NC—Unchilled; C—buds chilled for 6 days at 7°C.
Number of cultures examined (in parentheses).

Each series started with 12 anthers from three buds of similar stage. Anthers transferred to fresh medium at 6, 10, and 14 days, each dish being resealed and reincubated after removal of anthers. Liquid A medium (see table 1) 5 ml per dish. Stationary cultures at 28°C in darkness. All dishes transferred at 14 days to light (Grolux 500 klx) for 21 days at 25°C. Greenhouse plants (January –March) given supplementary light.

* The term shed pollen is used here to mean "pollen shed from anthers into liquid medium."

TABLE 4

Embryo Yields in Dark- and Light-Grown Cultures of Shed Tobacco Pollen

Pollen Fraction	Days of Anther Culture	Light		Dark	
		NC	C	NC	C
Shed	0–6	0(6)	437(6)	0(6)	1533
	6–12	213	1003	228	1943
Attached	12–35	195	929	197	1367
Mean embryo yield per anther		20	119	22	242

NC—Unchilled; C—buds chilled for 9–12 days at 7°C.

Mean of 6 experiments.

Each series started with 20 anthers from four buds (late unicellular to early bicellular pollen). Anthers transferred to fresh medium at 6 and 12 days, each dish being resealed and reincubated after removal of the anthers. Liquid A medium (table 1) 5 ml per dish. Stationary cultures at 28°C. Illumination by high pressure lamp for 12 hr per day. Greenhouse plants (February–March) given supplementary light.

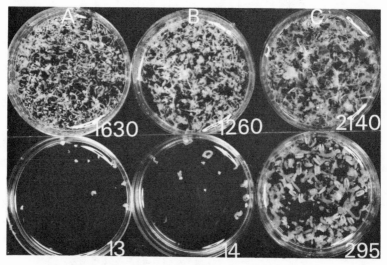

Fig. 1. Cultures of pollen shed from anthers floating in liquid medium. Upper three dishes: *Nicotiana tabacum* cv. White Burley. Lower: *Hyoscyamus albus*. In both species, anthers taken from 6 buds harvested at the late unicellular to early bicellular stages and stored in polythene bags for 24 days at 7–8°C. After 24 days, anthers excised and floated in 5 ml liquid A medium at 28°C (dark). After 7 days of culture, anthers removed, dishes resealed and reincubated for a further 28 days (0–7 day fraction, A). Anthers inoculated into fresh A medium. After a further 7 days of culture, anthers again removed, dish resealed and reincubated for 21 days (7–14 day fraction, B). Residual anthers cultured in fresh A medium for a further 21 days (14–35 day fraction, C). Figures represent total numbers of embryos and other structures per dish.

for as little as 4 days being sufficient to support growth. Nevertheless, tests have shown that the conditioned medium is growth limiting and that the performance of shed pollen is greatly enhanced by use of the three supplements recommended by Nitsch (1974), glutamine, serine, and inositol (table 5). Of these, glutamine is the key component (table 6).

Separation of the developing embryos from the anthers is to some extent dependent upon the number produced. Thus, in relatively low-yielding species like *Hyoscyamus albus*, most of the embryos remain in association with the anther (fig. 1). Such cases call for some form of agitation as soon as the anthers open in order to dislodge the pollen into the medium. Pollen shed from anthers at an early stage of culture clearly has considerable potential for development in large-scale batch cultures.

POLLEN CLUTURE BY MECHANICAL AND SURGICAL
TREATMENT OF ANTHERS

Wernicke and Kohlenbach (1977) have recently reported results on

TABLE 5

GROWTH OF SHED TOBACCO POLLEN IN DIFFERENT MEDIA

DAYS OF CHILL	DAYS OF ANTHER CULTURE	MEDIUM				
		A	B	C	D	E
9	9	8	16	42	0	1223
12	9	10	119	221	0	1720

Mean embryo yields per culture from 2 experiments in each series.

A—A medium (Table 1); B—conditioned A medium; C—conditioned A medium plus fresh A medium (equal parts); D—extract prepared from ten 9-day cultured anthers boiled for 5 min in 10 ml A medium, cooled and filter-sterilized; E—AGSI medium (A medium plus *l*-glutamine 800 mg/l, *l*-serine 100 mg /l and *m*-inositol 5000 mg/l)

Three dishes each with 20 anthers at the mitotic stage floated on A medium for 9 days at 28°C in darkness. Anthers then removed for preparation of medium D. Conditioned medium plus shed pollen centrifuged. Conditioned medium used for B and C. Pollen divided into 5 aliquots and washed twice in A medium. Each aliquot mounted in 5 ml of medium.

TABLE 6

GROWTH OF ISOLATED TOBACCO POLLEN IN MEDIA SUPPLEMENTED
WITH GLUTAMINE, SERINE, AND INOSITOL

	MEDIUM							
	A	AS	AG	AI	AGS	ASI	AGI	AGSI
Embryo yield per culture	0	0	885	11	988	0	997	952

A—A medium (Table 1); S—*l*-serine 100 mg/l; G—*l*-glutamine 800 mg/l; I—*m*-inositol 5000 mg/l.

Four dishes with 20 anthers at the mitotic stage floated on A medium for 7 days at 28°C in darkness. Pollen isolated as described in table 7 and divided into 8 aliquots. Each aliquot mounted in 5 ml of medium.

Nicotiana tabacum and *Hyoscyamus niger* not unlike those given above, except that the anthers are opened by slitting them along the line of dehiscence. In these experiments failure of the anthers to dehisce may be attributed to the use of unchilled material. Surgical operation is appropriate to species like *Datura innoxia* in which "free" embryos develop in liquid culture only after disruption of the anther wall by the embryos themselves. The method is probably limited in application to species having large anthers.

The homogenization procedure of Nitsch (1974) for isolating pollen is advantageous in that it is applicable to all species and gives pollen fractions without anther-preculture. However, in *Nicotiana tabacum*, anther-preculture is a prerequisite for development of the isolated pollen, 4 days being the prescribed period following a 4-day chilling of the buds. In previous studies anther-preculture has generally been carried out on agar media or on agar-charcoal media (Bajaj et al., 1977), the isolated pollen being suspended in liquid media. In my experiments liquid media have been used throughout and effects of longer periods of anther-preculture and of bud-chilling tested against both anther-culture controls and fractions of shed pollen. Anther-preculture for more than 3-4 days proved unsatisfactory largely because of anther-dehiscence, but chilling for 8 instead of 4 days increased the performance of the isolated pollen (table 7). The results confirmed the findings of Nitsch (1974) in all respects, but the isolated pollen consistently gave lower yields that either shed pollen or anther-culture controls. At present, the choice for pollen culture in *N. tabacum* is between relatively poor-yielding fractions isolated after 3 days of anther-preculture (table 7) or higher-yielding fractions shed between 0 and 6 days (table 4).

For *Datura innoxia* and other species whose anthers do not dehisce in liquid culture, the homogenization procedure is the obvious choice. Moreover, isolated pollen of *D. innoxia* will grow without preculture of the anthers (Nitsch and Norreel, 1973). Here again, however, the isolated pollen has proved unreliable, more than half the cultures on average failing to produce embryos in contrast to a 100 percent response in anther-culture controls (table 8). Anther-preculture enhances the performance of the isolated pollen but not to the level of the controls (table 9). However, in this species, occasional experiments have resulted in pollen cultures strikingly more productive than anther-culture controls (fig. 2). The reason for the batch-to-batch variation is not fully understood. Pollen is lost during isolation by mechanical means, by toxicity of the wall homogenate, or by the sudden change in osmotic environment of the pollen. Viability counts by fluorescein diacetate reveal losses often exceeding 50 percent (table 10).

TABLE 7

COMPARISON OF EMBRYO YIELDS IN CULTURES OF ANTHERS, ISOLATED POLLEN, AND SHED POLLEN OF NICOTIANA TABACUM

POLLEN FRACTION	DAYS OF CHILL	DAYS OF PRECULTURE BEFORE ISOLATION OF POLLEN										NO ISOLATION ANTHER-CULTURE CONTROLS	
		1		3		5		7		9		35	
		I	II	I	II	I	II	I	II	I	II	I	II
Isolated	0	0	0	0	0	0	0	0	3	1	5	22	35
	4	0	0	0	3	0	9	0	0	0	30	37	68
	8	0	0	0	71	0	35	2	34	2	63	120	116
Shed	0	0	0	0	0	0	0	0	0	1			
	4	0	0	0	0	0	0	0	0	8			
	8	0	0	0	0	0	0	34	34	90			

Mean yields per anther from two experiments.

I—Pollen cultured in A medium; II—pollen cultured in AGSI medium.

Anthers precultured in liquid A medium and pollen thus shed into A medium. Five dishes each containing 5 ml A medium inoculated with one anther from each of 25 buds of optimal stage. As controls, one anther removed from each dish and floated on 5 ml A medium, one other from each dish floated on 5 ml AGSI medium. At 1, 3, 5, 7, and 9 days one dish harvested from each group, dead anthers rejected, remainder (about 20) removed and dish resealed (shed fraction). The anthers removed were gently ground in a Potter glass homogenizer having a loose-fitting pestle. Homogenates filtered through 100 μ nylon and filtrate centrifuged for 4 min at 100g. Supernatant discarded, pellet resuspended in 10 ml A medium, divided into two aliquots and recentrifuged. One pellet mounted in 5 ml A and the other in 5 ml AGSI (table 5) (isolated pollen fraction). Dishes incubated at 28°C in darkness. Greenhouse plants (November–January) grown under supplementary light. Buds chilled at 7°C.

TABLE 8

COMPARISON OF EMBRYO YIELDS IN CULTURES OF ANTHERS AND
ISOLATED POLLEN OF DATURA INNOXIA

DEVELOPMENTAL STAGE		ANTHER		ISOLATED POLLEN	
		NC	C	NC	C
Early to mid unicellular	a	80(10)	100(8)	20	38
	b	13	51	1	6
Late unicellular to early bicellular	a	86(7)	100(3)	29	67
	b	52	16	6	2

NC—Unchilled; C—buds chilled for 4 days at 7°C.
a = percent cultures with embryos; number examined (in parentheses). b = mean embryo yields per anther.

Each anther culture contained four anthers in 5 ml liquid A medium, two from each of two buds of similar stage. Each pollen culture contained pollen isolated from the other four anthers of the same two buds by the procedure given in table 7. Isolated pollen mounted in A medium. Stationary cultures at 28°C for 14 days in darkness, then in light (Grolux 500 klx) for 21 days at 25°C. Greenhouse plants (March–April) under supplementary light.

TABLE 9

EMBRYO YIELDS IN ANTHER CULTURES OF DATURA INNOXIA AND
IN CULTURES OF POLLEN ISOLATED FROM PRECULTURED ANTHERS

DAYS OF CHILL	DAYS OF PRECULTURE BEFORE ISOLATION OF POLLEN				NO ISOLATION ANTHER CULTURES
	0	1	2	3	35
0	0	0	0	5	469
2	0	355	408	434	1063
3	3	0	491	195	1080
4	1	0	12	319	299
5	0	0	104	0	1133
8	10	0	8	57	571
Mean yield per anther	<1	15	43	42	192

Buds chilled at 7°C.
Each anther culture contained four anthers in 5 ml liquid A medium, two from each of two buds of optimal stage. Pollen isolated from the other four anthers of the same two buds by the procedure given in table 7. Isolated pollen mounted in 5 ml liquid A medium. Incubation conditions as in table 7. Greenhouse plants (March–April) under supplementary light.

Neither inclusion of protective agents (polyvinylpyrrolidone, dextran sulphate) in the isolation medium nor isolation in media of different tonicities at low temperatures has overcome the batch variability.

NATURE OF THE CHILLING EFFECT

In the original experiments on chilling, buds were simply supported in water and kept in a refrigerator at 4–5°C (Nitsch and Norreel, 1973; Nitsch, 1974). The procedure adopted here (see table 2) is space-saving and allows

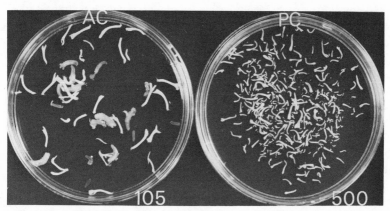

Fig. 2. Anther (AC) and pollen (PC) cultures of *Datura innoxia*. AC: 16 anthers (2 from each of 8 buds harvested at the late unicellular to early bicellular stages and chilled for 4 days at 4°C) (9 cm dishes containing 16 ml liquid A medium). PC: Pollen isolated from 16 anthers (2 from each of the same 8 buds) by the procedure outlined in table 7 and mounted in 16 ml A medium. Cultures incubated at 28°C (dark). Figures represent total numbers of embryos per dish.

TABLE 10

SURVIVAL OF TOBACCO POLLEN DURING ISOLATION AND CULTURE

	Percent Viable Pollen
Before isolation	74
After isolation	28
Washed pollen after 24 hr	5
Unwashed after 24 hr	0.2

Five buds of optimal stage. One anther from each used to assess viability before isolation of the pollen. Pollen isolated from the remaining anthers as in table 7 and divided into two aliquots. After washing, one aliquot mounted in the original isolation medium (unwashed), the other in fresh A medium (washed).

for systematic storage. Buds can be harvested daily as they pass through the critical stage, thus allowing maximal use of the plants available. While in this method of storage the rate of bud deterioration varies between species, buds of *D. innoxia* are the only ones so far encountered that do not last longer than about 6–8 days. The temperature of storage is important. The most successful cultures in the present experiments were carried out on buds stored at 7–8°C.

As already indicated, one effect of prolonged chilling in species like *N. tabacum* is to hasten anther-dehiscence. After 9 days of culture, nearly 50 percent of the pollen has been shed from anthers cultured from buds chilled for 12 days at the early bicellular stage compared with less than 5 percent from anthers cultured without a chilling treatment (table 11). Chilling also leads to a higher frequency of induction. Thus, after 9 days of culture, there are over 10 and 20 times more embryos respectively in attached and shed fractions derived from 12-day chilled material than in the corresponding fractions derived from unchilled material. Moreover, the embryos derived from chilled material are considerably more advanced. Whereas in unchilled material, embryogenic division has only just begun at 9 days, many of the embryos in 12-day chilled material have reached the 64-cell stage and beyond at 9 days. It follows that division begins sooner in cultures derived from chilled buds. Fluorescence counts indicate that there are more viable cells than proembryos in unchilled material at 9 days but these are clearly insufficient to increase the size of the embryo pool to that of the chilled material.

In *N. tabacum* pollen embryos are formed from the presumptive vegetative cell, and in unchilled material division commences after about 6

TABLE 11

EMBRYO SURVIVAL IN CULTURES OF ATTACHED AND SHED FRACTIONS
OF TOBACCO POLLEN

AVERAGE NUMBER OF POLLEN UNITS PER ML AFTER 9 DAYS' CULTURE	ATTACHED		SHED	
	NC	C	NC	C
	49000	34800	2190 (4.3%)	28020 (44.6%)
Percent embryos at 9 days (acetocarmine)	0.5	11.8	0.9	27.4
Average number of embryos per ml at 9 days	245	4110	20	7680
Average number of embryos per culture at 9 days	1225	20550	100	38400
Average number of embryos per culture at 35 days	106	1043	29	465
Percent recovery	8.7	5.0	29	1.2
Percent viable pollen at 9 days (fluorescein diacetate)	2.6	13.9	2.6	26.2
Average number of viable pollen per ml at 9 days	1274	4840	57	7340
Average number of viable pollen per culture at 9 days	6350	24200	285	36700
Percent recovery based on viable pollen at 9 days	1.7	4.3	10.2	1.3
Mean embryo yield per anther	13.3	130	3.6	58

Means of 6 experiments in each series.

NC—Unchilled; C—buds chilled for 12 days at 7–8°C. Shed fractions—pollen shed after 9 days from 8 anthers of optimal stage in 5 ml liquid A medium.

Attached—pollen remaining in the 8 anthers at 9 days and isolated into 5 ml A medium. Embryo yields determined at 35 days in control dishes each containing 8 anthers from the same buds. Conditions of incubation as in table 5.

days (Sunderland and Wicks, 1971). However, in 12-day chilled material, less than 24 hr of culture is required to initiate division. The vegetative cell divides in a plane parallel or perpendicular to that of the first pollen division (fig. 3C,D). The second embryogenic division occurs invariably at right angles to the first to give a typical quadrant of cells. The embryogenic grains are asynchronous and continue to enter division up to about 6 days, when the most-advanced embryos contain 12 or more vegetative-cell derivatives. The generative cell is still attached to the intine when the vegetative cell divides, and generally it remains attached as the embryo develops. As in unchilled material, the generative cell does not participate in embryogenesis but degenerates without dividing or after only limited division. In the chilled material there is probably a greater frequency of single divisions of the cell as it is attached to the intine (fig. 3F). Degenera-

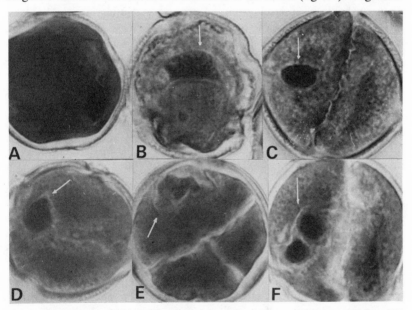

Fig. 3. Embryogenic divisions in pollen of *Nicotiana tabacum* following a 12-day chilling (7–8°C) of buds at the early bicellular pollen stage. (A) Densely stained (acetocarmine) nonembryogenic pollen grain in an anther immediately after excision from the chilled buds. (B) Light-staining embryogenic grain from the same anther as (A). (C) First embryogenic division of the vegetative cell after only 24 hr anther culture at 28°C (dark) in liquid A medium. Plane of division probably parallel to that of the first pollen division. (D) Another grain from the same anther as (C) but showing the plane of division at right angles to that of the first pollen division. (E) Second embryogenic division after 72 hr culture; generative cell already showing signs of degeneration. (F) Another grain from the same anther as (E) but showing division of the generative cell; no further division of the cell observed. Arrows point to the generative cell in all cases.

tion of the generative cell begins after about 3 days (fig. 3E), so that by 9 days the embryo pool comprises embryos of three types, all composed of many vegetative cell derivatives plus either the original generative cell, two generative cell derivatives, or none.

Cytological, cytophotometric, and ultrastructural studies in both *N. tabacum* and *D. innoxia* have all contributed to the view that gametophytic processes are checked in certain grains before the generative cell becomes detached from the intine. In *N. tabacum*, as the vegetative nucleus undergoes DNA replication prior to division, there is a fundamental reorganization of the other cell components (Dunwell and Sunderland, 1974) and loss of RNA and protein (Bhojwani et al., 1973). Other grains in the same anther in which the gametophytic program is not checked synthesize RNA and protein, and two populations of pollen emerge distinguishable by different staining properties (Sunderland and Wicks, 1971). A similar divergence of development occurs in the anthers during chilling (fig. 3A,B). It may be concluded that excision of the bud followed by low-temperature storage at 7–8° C likewise checks gametophytic processes, and the emergence of a pool of light-staining grains poised for division implies a process of regression probably analogous to that previously studied in unchilled material.

Anomalous pollen (B units, Sunderland, 1973) such as have been found to varying extents in different species in culture have not been encountered in anthers of *N. tabacum* excised from buds chilled for 12 days at the early bicellular stage. Although such B units, when formed, do contribute in various species to the embryo pool, and are also known to arise in certain genotypes after low-temperature treatment (Sax, 1935), they are not the source of increased embryo yields in the present experiments. Reorientation of the plane of the microspore division, which can lead to B pollen formation, is not the principal effect of chilling, although this has been repeatedly stated to be so by other workers (e.g., Nitsch, 1974, 1977; Reinert and Bajaj, 1977). The most plausible explanation of the increased embryo pool in chilled material is that the low temperature affords a greater rate of survival of the pollen after excision of buds than occurs in culture at high temperatures. The data of table 11 indicate that after 9 days of culture, without chilling, less than 3 percent of the pollen has survived. During chilling, however, there is only a small reduction, from about 68 to 60 percent viable pollen grains over 12 days, and at anther excision about half the grains exhibit embryogenic properties. At low temperature, therefore, more grains are able to complete the sequence of events that lead to the first embryogenic division.

The question inevitably arises as to why gametophytic processes are checked only in certain grains. It should be possible to divert all the pollen into embryogenesis by chilling buds when the pollen is still unicellular. This, however, is not so. In *N. tabacum* buds at 7-8° C, unicellular pollen slowly divides to give predominantly typical microgametophytes, and divergence of two populations again proceeds. There thus appears to be some restriction, other than the commitment to the usual pollen program, that determines the pool size of embryogenic grains. Such a restriction may be imposed by events preceding excision of the buds as has recently been suggested by Horner and Street (1978). However, the possibility cannot be entirely eliminated that chilling affects specific inducers or inhibitors that may be involved in the switch to embryogenesis.

EMBRYO SURVIVAL AND THE PROBLEM OF EMBRYO YIELDS

The data of the preceding section indicate that conservation of viable pollen inside the anther at low temperature gives much higher induction frequencies than has previously been achieved by any empirical manipulation of the cultural conditions. The numbers of dividing proembryos produced in *N. tabacum* after a 12-day chilling are phenomenally high (table 11). Yet few survive to the plantlet stage. Given the best cultural conditions so far devised in this species, the highest yields attained, on a per-anther basis, do not amount to more than about 200 on average, a figure equivalent to about one plant to every 200 pollen grains. The presence of the anther wall certainly constitutes a hazard once the anthers are in culture, and could well be contributing to early death of proembryos by production of toxic materials. The low recovery values (see table 11) clearly underline the inadequacy of present cultural procedures and media to maintain more than a small percentage of the pollen grains induced. More embryos apparently survive at low- than at high-plating densities, implying exhaustion of an essential nutrient or nutrients.

CONCLUSION

Haploid studies should benefit by the recent diversification of techniques and procedures. Rather than aiming at one procedure, we should use the one best fitted to a species and the particular problem in hand. Agar-charcoal media may be the best milieu in some cases, liquid culture in others. Both anther and pollen culture are laborious for species having small anthers, as in many cereals; the development of a technique for culture of whole barley inflorescences (Wilson, 1977) is thus likely to be of great value. For mutation studies, high-yielding pollen cultures might seem obligatory, but simple anther culture in liquid media could well be just as effective. For biochemical studies of pollen embryogenesis, on the other hand, high-yielding synchronized cultures are essential, and to this end further intensive study is needed especially on isolation procedures that preserve pollen viability. In my opinion the major obstacle to progress in this field is our inability to maintain embryo growth on a large enough scale either in the presence or absence of the anther wall.

REFERENCES

Bajaj, Y. P. S., J. Reinert, and E. Heberle. 1977. Factors enhancing *in vitro* production of haploid plants in anthers and isolated microspores. *In* R. J. Gautheret (ed.), La culture des tissus et des cellules des végétaux. Travaux dédiés à la mémoire de Georges Morel. Pp. 47–58. Masson, Paris.

Bhojwani, S. S., J. M. Dunwell, and N. Sunderland. 1973. Nucleic-acid and protein contents of embryogenic tobacco pollen. J. Exp. Bot. 24:863–69.

Devreux, M., U. Laneri, and P. de Martinis. 1975. Reflexions sur nos cultures d'anthères de plantes cultivées. Giorn. Bot. Ital. 109:335–49.

Dunwell, J. M., and N. Sunderland. 1974. Pollen ultrastructure in anther cultures of *Nicotiana tabacum*. II. Changes associated with embryogenesis. J. Exp. Bot 25:363–73.

Horner, M., and H. E. Street. 1978. Pollen dimorphism—origin and significance in pollen plant formation by anther culture. Ann. Bot. 42:763–71.

Irikura, Y. 1975. Cytogenetic studies on the haploid plants of tuber-bearing *Solanum* species. I. Induction of haploid plants of tuber-bearing *Solanums*. Res. Bull. Hokkaido Nat. Agr. Exp. Sta. No. 112.

Mii, M. 1976. Relationships between anther browning and plantlet formation in anther culture of *Nicotiana tabacum* L. Z. Pflanzenphysiol. 80:206–14.

Murashige, T., and F. Skoog. 1962. A revised medium for rapid growth and bioassays with tobacco tissue cultures. Physiol. Plantarum 15:473–97.

Nitsch, C. 1974. Pollen culture—a new technique for mass production of haploid and homozygous plants. *In* K. J. Kasha (ed.), Haploids in higher plants: Advances and potential. Pp. 123–35. University of Guelph Press, Guelph, Ontario, Canada.

Nitsch, C. 1977. Culture of isolated microspores. *In* J. Reinert and Y. P. S. Bajaj (eds.), Applied and fundamental aspects of plant cell, tissue, and organ culture. Pp. 268–78. Springer-Verlag, Berlin.

Nitsch, C., and B. Norreel. 1973. Effet d'un choc thermique sur le pouvoir embryogène du pollen de *Datura innoxia* cultivé dans l'anthère ou isolé de l'anthère. C. R. Acad. Sci. Paris 276:303–6.

Pelletier, G., and M. Ilami. 1972. Les facteurs de l'androgénèse *in vitro* chez *Nicotiana tabacum*. Z. Pflanzenphysiol. 68:97–114.

Rashid, A., and H. E. Street. 1973. The development of haploid embryoids from anther cultures of *Atropa belladonna* L. Planta 113:263–70.

Reinert, J., and Y. P. S. Bajaj. 1977. Anther culture: Haploid production and its significance. *In* J. Reinert and Y. P. S. Bajaj (eds.), Applied and fundamental aspects of plant cell, tissue, and organ culture. Pp. 251–67. Springer-Verlag, Berlin.

Sax, K. 1935. The effect of temperature on nuclear differentiation in microspore development. J. Arnold Arbor. 16:301–10.

Sunderland, N. 1973. Pollen and anther culture. *In* H. E. Street (ed.), Plant tissue and cell culture. 1st ed. Pp. 205–39. Blackwell, Oxford.

Sunderland, N., and J. M. Dunwell. 1977. Anther and pollen culture. *In* H. E. Street (ed.), Plant tissue and cell culture. 2d ed. Pp. 223–65. Blackwell, Oxford.

Sunderland, N., and M. Roberts. 1977a. Anther and pollen culture. Sixty-seventh Ann. Rep. John Innes Institute. Pp. 60–65.

Sunderland, N., and M. Roberts. 1977b. A new approach to pollen culture. Nature 270:236–38.

Sunderland, N., and F. M. Wicks. 1971. Embryoid formation in pollen grains of *Nicotiana tabacum*. J. Exp. Bot. 22:213–26.

Wernicke, W., and H. W. Kohlenbach. 1975. Antherenkulturen bei *Scopolia*. Z. Pflanzenphysiol. 77:89–93.

Wernicke, W., and H. W. Kohlenbach. 1976. Investigations on liquid culture medium as a means of anther culture in *Nicotiana*. Z. Pflanzenphysiol. 79:189–98.

Wernicke, W., and H. W. Kohlenbach. 1977. Versuche zur Kulture isolierter Mikrosporen von *Nicotiana* und *Hyoscyamus*. Z. Pflanzenphysiol. 81:330–40.

Wilson, M. H. 1977. Culture of whole barley spikes stimulates high frequencies of pollen calluses in individual anthers. Plant Sci. Lett. 9:233–38.

F. C. STEWARD AND A. D. KRIKORIAN

Problems and Potentialities of Cultured Plant Cells in Retrospect and Prospect

15

RETROSPECT

Introduction

Some seventy-five years have elapsed between Haberlandt's prophetic insight in 1902 (Krikorian and Berquam, 1969) into the potentialities of maintaining isolated cells from higher plants and this colloqium. The march of events through the early use of aseptic heterotrophic growth of tissue explants or isolated root tips was slow to focus attention on free cells of angiosperms, although they were seen to survive in such situations as sloughed-off root cap cells. The setting from which modern aseptic cultures, the cloning of plants via cells, somatic embryogenesis, and the use of isolated plant protoplasts all emerged may be recognized in the immediate pre- and post-World War II years. But, in the subsequent thirty to forty years, what was a prophetic vision to Haberlandt has exploded into research activity that is now worldwide, that invades virtually all the botanical and plant physiological journals, has its technical handbooks on methods, as well as its associations of adherents to accepted disciplines and dogmas, and a literature that has proliferated into a succession of newsletters, multiauthored monographs, symposia, and "workshop" discussions on the international scale. No article of this length can, therefore, be comprehensive in content or rationally cover all the data that has emerged.

F. C. Steward and A. D. Krikorian, Department of Biology, State University of New York, Stony Brook, New York 11794.

Therefore, the objective of this summary is to recognize the current status of the subject, to note its theoretical accomplishments on the one hand and the limitations of the current almost frenetic search for technological applications on the other. Although one of us (F.C.S.) has given a personal, illustrated account of his own involvement in this dramatic story, this written version is done jointly because together we relate to the past, are involved in the present, and anticipate the future. Moreover, we have already attempted such a synthesis up to 1973 (Steward and Krikorian, 1975), although since then, whole volumes have appeared (Street, 1974, 1977; Kasha, 1974; Thomas and Davey, 1975; deFossard, 1976; Reinert and Bajaj, 1977; Gautheret, 1977). Also, a very large number of papers and many well-funded industrial schemes to seek profitable applications have been launched (Cell Culture and Tissue Culture Symposium, 1977, and refs. there cited). Our method will be, whenever possible, to avoid repetition by referring to our earlier writings and to supplement them or bring them up to date. By way of preamble, however, the following general observations may be made.

The scientific scene has totally changed over the last forty years. Vastly increased numbers of workers in every scientific field flock to those areas that, often temporarily, have a popular appeal, and plant cell and tissue culture is no exception. All these scientific workers and their output, largely of uncoordinated information, create pressures on all means of communication; from the printed word as in scholarly journals, digests and abstracts, annual reviews and symposia, to the public press and television and electronic media, e.g., as in so-called on-line information services and computer bibliographic services, in which authors may even bypass the forms and restraints of conventional publication. Thus, the former, often lengthy, lapse in time between the laboratory and publication, which fostered understanding, has largely disappeared. Moreover, research, instead of an avocation quietly conducted by the few with limited means and little thought of priority or publicity, must now face the continued realities of funding. The temptation is to exploit ever-narrower objectives and, in the struggle for survival, minutiae are exaggerated, possibilities seen soon masquerade as realities, and, often aided by television and radio, the public or industry becomes involved so that patents and propaganda soon confuse the trail. Therefore, we ask for some indulgence, for we may need to question whether, in the field of plant cell and tissue culture, the scientific advances or the status of the art have always justified the public clamor they arouse or even the current resources they absorb. Finally, we should confess that the rigorously preplanned, circumscribed, investigation on preconceived ground rules, where the end not merely justifies but presupposes the means, rarely, in our view, adds (or has added) those required new dimensions to the understanding of plants. A few signposts *en route* from Haberlandt, to what one may regard as the exponential growth phase of cell and tissue culture research and development, may illustrate this. Such signposts will show that the field of cell and tissue culture research actually responded, whether fortuitously or by design, to understanding drawn from peripheral areas of knowledge. Conversely, it has stagnated when it has relapsed into a self-perpetuating "tissue-cult".

Haberlandt: His Vision

Haberlandt's perspectives of, and comments on, the potentialities of surviving cells (culminating in his dictum that "one could successfully grow artificial embryos from vegetative cells") did not arise from narrowly focusing upon the properties of the cells in question. Moreover they were independent of any preoccupations with the now seemingly sacred aseptic cultural or experimental conditions. On the contrary, Haberlandt saw the cells he examined against the background of his knowledge of, and speculations about, the cell theory and how the plant body was organized into its respective tissue systems and how it developed from its single cell precursor in the zygote or fertilized egg. There can be little doubt that it was Haberlandt's knowledge of cases of apomixis and apomictic development (inherited from his teacher Julius von Wiesner) that alerted him to the reality that certain mature body cells, without playing a part in the sexual reproductive cycle, could in their environment in the plant body be prompted to develop *in situ* like fertilized eggs into embryos. Similarly, Haberlandt was prepared by his own interpretations of the hormone concept, in logical descent from Julius Sachs, and reinforced by his studies on wound healing (Krikorian, 1975, p. 75 et seq.) that chemical substances could induce quiescent mature cells to embark upon growth and cell division as they begin their apomictic development. And if this could occur in species of *Hieracium*, why not in the parenchyma of other plants? Although Haberlandt could not pursue this experimentally because of his lack of understanding of the causes of cell division (which he felt constrained to pursue through wound healing), nevertheless it was in his laboratory that Kotte in 1922 developed the technique for culturing root tips. Thereafter, his influence spread to the French school under Nobécourt and Gautheret. (Almost simultaneously in the United States, W. J. Robbins published on root tip culture, followed by P. R. White in the 1930s; for refs. cf. Krikorian and Berquam, 1969; Krikorian, 1975).

The Intervening Years

Much later than Haberlandt, J. H. Priestley, a keen interpreter in his day of the contrasted organization of shoot and root apices, who was dedicated to the search for causal physico-chemical explanations for the responses of plants, was attracted by the problems and practical importance of vegetative propagation (Priestley, 1926). In this context Priestley restudied the renewed cell divisions that could occur in the cut surface of potato tubers. Whereas Haberlandt attributed the cell division activity to a particular hormone ("leptohormone" derived from the proximity of such vascular elements as phloem), Priestley preferred to see the events as they transpired in terms of the access of oxygen to the cut surface (which became blocked by suberin), the renewed "or wound respiration," the disappearance of starch from the surface cells, and the renewed protoplasmic syntheses as they embark upon a course leading eventually to division as part of a cork cambium. Secure in his belief that the divisions were innate to the cells adjacent to the cut surface, Priestley did not invoke, in this system, *the* growth hormone (prior to the recognition of IAA) as it was then being investigated in oat coleoptiles, if only because in that system the known responses due to auxin did *not* involve cell multiplication and division (Swingle, 1940 and 1952, for historical references). (Nevertheless, it was in large measure due to much later work, in the 1940s and thereafter, on the growth of aseptic tissue explants, free from the then current dogma that *all* the requirements for their rapid growth were already defined, that the study of growth-promoting substances returned to those that induce cell multiplication by division in contrast to the one, IAA, then concerned with cell enlargement.) But the investigations on vegetative propagation in the mid-twenties around Priestley and on storage organs had two consequences that were to influence the later turn of events.

In the study of the wound-healing response on carrot slices a proliferating callus (not a cork periderm as on potato) was noted to form—particularly over the cambium of the storage root (or over successive cambia as in beet root slices). If these calluses were allowed to continue and become large, they spontaneously produced adventitious buds that developed into miniature shoots with adventitious roots. (But here, as in many much later studies of morphogenesis in proliferations on relatively massive tissue explants [especially of tobacco], the nutrients and stimuli were furnished by the tissue of origin acting as a "nurse" system.) Second, the observations of the events as they occurred in potato cells during wound healing provided the essential background for the interpretation of the subsequent work done on potato tuber slices in the study of ion absorption, respiration, protein synthesis, and nitrogen metabolism. All of this later played a real, but indirect, part in the development of systems conducive to a breakthrough in the stalemate in the use of aseptic plant tissue cultures that for too long had been preoccupied with, if not bogged down in, demonstrations of potentially unlimited growth of subcultured callus cultures or of root tips (see Krikorian, 1975, for detailed references).

But one should also note here and give credit to those who, again in the 1920s, noted the distinctive development of cells and tissues from the growing regions of shoot and root apices and from cambia. The challenge of "physiological differentiation" between even contiguous cells of complex tissues (e.g., stomata, guard cells, subsidiary cells, glands, and hair cells all in the epidermis; fibers, parenchyma, sieve tubes, and companion cells, as well as ray cells of the secondary phloem, as in *Tilia*; the astonishing range of form and composition in the parenchyma of a plant like *Pellionia*, as noted by Thoday, 1933, etc.) was well appreciated long before the difficulty of explaining it in terms of cells that initially receive equal "doses" of DNA at mitosis. (The dilemma is now even more acute, since such complex tissues may give rise in culture to free cells that do not show mature characteristics of parenchyma *in situ* as they move in culture toward embryos that can regenerate whole plants [Steward, 1970a].)

The hope that the definitive signals or messages that control the genetic information that determines specialized compositions of cells *in situ* during normal development might be simulated in unorganized cultured cells and tissues is of long standing. But all the early optimistic attempts to test this possibility failed, for until normal form was recapitulated in plantlets developed from cells, control of the desired biochemistry usually remained elusive (Krikorian and Steward, 1969). But why should this not have been expected? For instance, in the case of mint the divergence of chemical composition as between leaf, stem, root, and so on, may be very dramatic; and it responds, equally dramatically, to the interplay of environments (photoperiod and night temperatures) with inorganic nutrition (Steward et al., 1962; Steward, 1963). Moreover, the changes in composition that occur during development are not to be thought of as though regulated by a single gene control or by a single factor, but rather they must involve large blocks of genetic information that, as it were, are simultaneously activated or repressed *in situ* as major control switches are tripped. It is as if the parenchyma of potato tubers, activated to develop in the apices of stolons (in marked contrast to comparable tissue at the tips of the main axis, which may form flowers), responding to specific combinations of environmental stimuli, is simultaneously programmed toward its particular carbohydrate and nitrogen metabolism, its cation (K^+) accumulation, its ability to form periderm, and so on. Therefore, the idea that controls of such complexity affecting such diverse responses in intact organized growing regions could

be simply replaced by external factors affecting unorganized cell proliferations now seems ingenuous (Krikorian and Steward, 1969; Steward, 1970b,
1976).

In short, neither in reviewing the subject of cell and tissue culture nor in
planning research should one lose sight of the fact that the problems of
culturing angiosperm free cells and their subsequent potentialities are
inherently part of the larger problems of interpreting growth, development
metabolism, and morphogenesis. Thus, it is artificial and mistaken to
isolate, conceptually, the problems of cell and tissue culture and to identify
them as a distinctive discipline, and especially so by the mere exploitation
of aseptic techniques. Indeed, the sequence of events as discussed below
fortifies this view, for they arose out of broad interests in devising
alternative methods to attack basic problems of cell physiology and growth
leading to new applications of aseptic procedures. The key events here
mentioned stress that because the experimental objectives never became
too narrow, or the systems too restricted, (in the name of control and
definition) that the responsiveness of fully viable cells that can grow was
maintained and this opened unexpected leads. How else could studies of
ion-uptake in potato or carrot disks have moved logically, and seemingly
inevitably, to totipotency of free carrot cells and somatic embryogenesis?
In broad outline this happened as follows.

From Potato Tuber Slices to Somatic Embryogenesis

Prior to 1940 one of us (F.C.S.) had recognized that the ability of cells at
a cut potato tuber surface to accumulate ions (cations and anions) against a
concentration gradient derived from the driving force of cells able to grow
and divide. Although a system composed of very thin potato-tuber slices
(only a few cells thick) in a very dilute aerated solution revealed the
metabolic background for the ionic accumulations, a better system, more
able to cope with long-term experiments in the presence of organic solutes,
was required in which to compare and contrast the activated cells with their
quiescent counterparts.

After 1946 a tissue culture system in which fully mature quiescent cells
could be contrasted with those most actively growing was sought. To be
most adaptable to quantitative experiments, this system needed to expose
the cells to liquid, not semisolid, media. Among the many tissue sources
tried, the best proved to be very small, standard, explants (2–3 mg) of
secondary phloem (free of cambium or ray tissue) precisely cut from
mature carrot roots. Among many possible sources of growth-inductive
stimuli (natural and synthetic), the liquid endosperm of the coconut (coconut milk, CM, or water) excelled in its effect on the carrot tissue. Other

natural fluids or extracts (from young fruits of *Juglans, Aesculus*, or *Zea* to extracts of female gametophyte of *Ginkgo*) could emulate the role of CM. Curiously, the slices of potato tuber did not then so respond by external growth until, much later (Steward and Caplin, 1951), the effect of CM along with 2,4-D and many other auxin synergist produced an actively proliferating culture (Shantz et al., 1955).

To render the carrot-coconut system suitable for the cell physiological experiments, a new system of growing carrot explants in aseptic liquid culture was developed, and, in this system, the first criterion of their growth was fresh weight; but this was soon supplemented by the number and average sizes of cells. Faced with the obvious problems of (1) the chemical growth induction, whether by the constituents of coconut milk or other means; (2) the changed metabolism induced in the explants as they grew, in contrast to the quiescent tissue; and (3) the recognized speed with which small explants grew under these conditions, the objectives that had prompted the technique in the first place, i.e., understanding the motive forces in growing cells that drive initially the machinery of ion accumulation were deferred—appropriately so because, later, they were taken up when much more was known about the events of growth induction (Steward and Mott, 1970; Mott and Steward, 1972, and refs. there cited).

At first the carrot explants were grown one, two, or three at a time in special "T" tubes with 10 ml of medium so rotated at 1 rpm that the tissue alternated between liquid and air. (The best relation between these periods, dependent on the geometry of the tubes, was determined.) However, this system, devised for the uptake of ions, was also used to assay for natural or synthetic growth factors for carrot, and their inhibitors and to investigate their metabolic manifestations, especially in protein synthesis for endogenous sources of nitrogen (Bidwell et al., 1964). Unexpectedly, however, it also led to the means of (1) obtaining viable free carrot cells in very large numbers, (2) culturing these, and (3) growing them into embryos and plantlets. This occurred as an unforeseen consequence of modifying the experimental vessels to obtain amounts of actively growing tissue large enough for biochemical studies. Fifty or one hundred explants were used in single vessels (made from 1-liter pyrex flasks) containing 250 ml of medium and in which the "ends" of many tubes (albeit with one body) were represented by "nipples" that diverged radially around the flask. Under these circumstances the explants grew very uniformly, but, as they rolled against each other, they gently pushed off from their periphery cells "that can grow into a medium in which they may grow and into a free space in which they do so independently." Thus, the most successful means of obtaining viable free cells and of culturing them in liquid into plants (in 1953 and 1957 respectively) in the Cornell laboratory originated as a "fringe benefit" of technique devised for other purposes (Krikorian, 1975, p. 83 et seq., for references).

Thenceforward, the course of investigation diverged along the following lines:

The responses by quiescent mature cells to complex natural fluids like coconut milk (out of all proportion to those due to IAA alone) not only justified their use, despite the criticism that they were undefined (Gautheret, 1952; White, 1953), but also focused attention, again, on stimuli to *cell multiplication*, and put the role of IAA in cell enlargement in coleoptiles into perspective and led to fractionations of the natural fluids. The now widespread use of inositol in culture media is traceable to its presence as an active component of CM along with other hexitols (Pollard et al., 1961). Meanwhile, synergisms, as between CM and various synthetic auxins (2,4-D, NAA, etc.), and as between parts of CM or *Aesculus* fluid, showed first the responsiveness of the cells to a spectrum of concurrently interacting substances and later to their useful application in sequences. Moreover, the events of "chemical growth induction," like those of normal development, were seen to be susceptible to further interactions with environmental factors (notably photoperiod and day and night temperature). Although both the CM and *Aesculus* systems have yielded *in part* their chemically defined components (Steward and Krikorian, 1971, p. 65, for references), it is still true that their full potency has not yet been accounted for (van Staden, 1976). Even so, it has been useful to speak of growth-promoting systems I and II mediated by inositol and IAA respectively, each with their synergistically active components (AF_1 and AF_2) that elicit their distinctive, but complementary, responses in carrot root phloem explants (fig. 1a and b). The current usefulness of inositol, IAA, and zeatin in culture media should, however, not encourage the belief that *all* the simple soluble naturally occurring stimuli to growth and cell multiplication are thereby known. To the extent that the liquid endosperms or fluids from vesiculate embryo sacs (as from coconut, maize, or *Aesculus*) bring about effects in carrot not easily recapitulated, they still seem to do so as our best approximation to the nourishment of young embryos *in situ* (Steward and Bleichert, 1972).

The explanation of chemical growth induction is, however, not only limited by our incomplete knowledge of the agents concerned, whether auxins or the so-called cytokinins and such, and whether these act singly or in combinations, it is equally restricted by our knowledge of where in the complexity of the cells they act and what are the salient features of metabolism causally concerned in the resultant stimulus. Autoradiographic studies, though still incomplete, suggest that individual substances

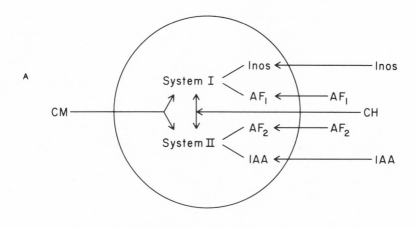

Fig. 1. Component parts, and their interactions, of systems that induce growth in carrot phloem explants. (A) Endogenous and exogenous relations: system I comprises the various growth factors (AF_1) the activity of which is mediated by *myo*-inositol (Inos), and the *Aesculus* active fraction (AF^i_{aesc}) that contained an IAA-rhamnose-glucose complex; system II comprises the various natural and synthetic growth factors (AF_2) the activity of which is mediated by IAA, e.g., the many adenyl compounds of which zeatin (Zeat.) is a naturally occurring example; CH (casein hydrolyzate) extends the range of exogenous components of systems I and II and links them together; CM (coconut milk) represents a complete and balanced system that achieves results over and above what can be attributed solely to exogenous systems I and II, even in the presence of CH. (See also fig. 2 of Steward and Bleichert. 1972.) (B) Further modulation of growth-promoting interactions and systems: iron is essential for the growth induction of small carrot explants as by CM and its component parts; CH extends the range of both systems I and II, the active fractions of which (AF_1; AF_2) include both known and unidentified substances (a, b, c, etc.); light and temperature intervene to modulate the responses to exogenous growth factors; and the over-all response may still be reversibly inhibited by treatments with abscisic acid (ABA) and gibberellin (GA_3). (See also fig. 5.2 of Steward and Krikorian, 1971.)

(like IAA or zeatin) may elicit different responses at sites where they act alone and at others where they act in combination. Others, linked to inositol, may elicit still different consequences at other sites (Steward and Israel, 1972; Israel et al., 1978).

The responses of freely suspended cells in liquid culture and extensive subculture yielded their reward in somatic embryogeny neither immediately nor deliberately. After long-continued subculture in flasks, a strain of freely suspended carrot cells in filtered inocula were transferred to small "T" tubes for certain experiments. But the decisive event, unplanned and unpredicted, was their formation of roots and their continued growth thereafter only as organized roots. But then if roots, why not shoots? In the then mistaken idea that shoots might only arise after growth on semisolid media, in direct contact with air, and after geotropic stimulus, cells were transferred to large Erlenmeyer flasks and grown in a different room; this did produce plants and even some with miniature storage roots. This showed long ago (1957) that even after long-continued growth as free carrot cells, they still retained all the information to be complete plants, but it identified neither the factor(s) nor their combinations or sequences that best elicited this response (Krikorian, 1975, p. 84 et seq., for references). But for any umbellifer it became apparent that this response occurred most easily for cells of embryo or seedling origin, although it could occur from any part of the plant body if their cells could be induced to grow. Even so, the stimuli to cause the growth and to release embryogenesis varied greatly with the cultivar concerned and with the site in the plant body that yielded the explanted material. Refinements of the technique enabled more faithful, consistent, and uniform recapitulations of normal zygotic embryogeny when the cultured somatic cells were most uniform and kept small. In an "art" (rather than a rigorously prescribed procedure where all the many factors were defined absolutely) the conditions of initial stimuli and nutrients, of environments and media "conditioned" by growth of the culture in bulk, of filtration and spreading of cells in agar, and so on, were adjusted empirically to foster embryogenesis (Steward et al., 1975, and ref. there cited).

Over the years carrot plants primarily from root explants via cells were routinely grown, and regrown from resultant crops of roots through successive vegetative "generations" (Steward and Mapes, 1963). In doing all this repetitively and routinely, it was noted that cells that were repeatedly subcultured, without organization into plantlets, for long periods, eventually lost their morphogenetic competence, first in the ability to form shoots and then roots. By contrast, when similar cells formed plantlets, the next "generation" of cells from the phloem of their developed roots always gave rise to embryos with relative ease. Thus, the totipotency of carrot cells may be more easily reexpressed through each morphogenetic cycle of growth than it is retained, unimpaired, during long-continued standardized subculture. Hence, in the indefinite subculture of cells, their morphogenetic propensities may be so deeply repressed that special empirical devices may be needed to restore their expression (Steward et al., 1967).

Another essential feature of this process of somatic embryogenesis awaited a new turn of events for its discovery. For purposes of an experiment in space, to be described later in this paper, the morphogenetic induction of embryos from single cells needed to be rendered uniform and consistent, to occur in agar, and above all to be feasible for study in darkness and within a period as short as twenty days. The surprise was that, after appropriate prior culture of cells to develop their morphogenetic propensity, the initial embryogenetic expression is positively stimulated by exposure for several days to complete darkness (Steward et al., 1975). Thus, although the later development of the first leaf primordia needs light, their induction proceeds readily in the dark. The conditions of the zygote in the embryo sac, replete with specially balanced nutrients and stimuli, nourished by the parent sporphyte in what one may now regard as "physiological darkness," now find their parallel in small somatic cells that express their totipotency in darkness in media adjusted to their needs. Thousands of carrot embryos may now develop on small dishes and can grow thereafter into plants (fig. 2a and b). But the responses that have been, and are, easily possible with carrot and other umbellifers (e.g., *Sium*) were also emulated with cells from a tobacco (*N. rustica*), *Asparagus, Cymbidium,* and others, though with appropriate and empirical variations on the theme of initial stimuli, media, and environments. However, these responses are still far from universal.

Metabolism and Nutrition: Later Integrated Studies

The use of the carrot explant system, in rotated tubes and flasks, for its originally intended purposes, also led to investigations of metabolism and respiration and of solute (organic and inorganic) concentrations and water content of cells whether quiescent, in active growth, or as they mature,. The first observation was that cells *in situ* are very different in form and content from their counterparts that grow and multiply in isolation. The indications were that the stored components of quiescent cells are greatly transformed as growth induction occurs, and that neither the nature nor the content of their soluble substances necessarily reflects their role as intermediates of synthesis or metabolism. The picture that emerged was one of cells that are highly compartmented, as evidenced by their fine structure, so that nitrogen compounds and other metabolites could move through phases of temporary storage to sites of active synthesis or metabolism as the pace of turnover (of carbon or nitrogen) is regulated by

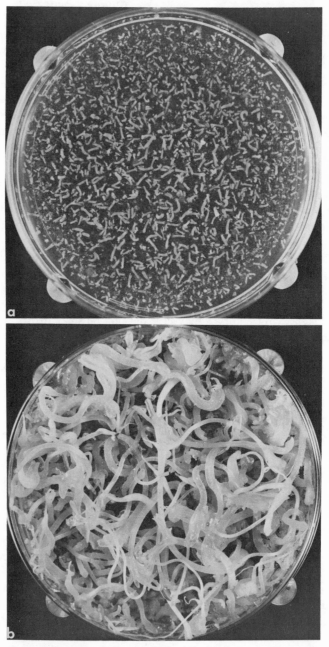

Fig. 2. Embryos from free totipotent carrot cells. (a) Crop of embryos after 20 days' growth in darkness from cells distributed in an agar medium containing basal nutrients and inositol 25 mg/liter; (b) a similar crop after 50 days' growth in darkness.

exogenous stimuli and environments. By the use of radioactively labeled substrates and measurements of the changes in specific activity of internal constituents, light was shed on the compounds that are close to sites of active synthesis and those that are more remote (Bidwell et al., 1964). But studies of the solutes of carrot explants subjected to a wide range of combinations of growth-promoting stimuli (acting singly and in combination) and of various trace elements (also acting singly and in combination) show how great is the range within which these various parameters are subject to external control (Steward and Rao, 1970). In the area of ion uptake and solute content, the results were most illuminating. They showed that the *first* solutes to be accumulated in cells as they multiply in heterotrophic media and begin to vacuolate are organic ones (sugars, nitrogen compounds, organic acids with potassium as the conspicuous inorganic ion that essentially accumulates, but only to a relatively low level, in solutes that are not yet conspicuously high in their osmotic value.) After cell division subsides and external nutrient supplies are depleted, or withdrawn, the cells acquire high osmotic values (i.e., the activity of water is internally reduced) and now cations (K^+) and anions Cl^- or Br^-) replace the osmotically active organic solutes. The cells that create form and substance out of simple molecules, which locally reduce entropy as they create complexity, develop internal reservoirs of organic solutes that increase free energy and locally reduce the activity of water. Subsequently these solutes may be depleted and replaced by mainly K^+ and Cl^- ions accumulated from their ambient media. Hence, the morphological problem of how cells get the message to return to their growing, even embryonic state, how they abandon their quiescent and resting metabolism for the activity compatible with their growth, also finds a parallel in their solute contents. The nutrients and stimuli that result in the recrudescence of growth control, at one and the same time, the number and sizes of cells, their total water content, and organic and inorganic solutes they contain (Mott and Steward, 1972; Steward and Mott, 1970; Steward, 1971).

The cultured carrot system thus became a means to investigate how metabolism is regulated in compartmentalized cells as they grow, to see the finer points of its relations to trace elements and their nutritional interactions with each other and with growth factors, and the impact of all this on the solutes of cells and the ion transports into their vacuoles. This was, and is, so, because by these means one can trace out all these events, more selectively than on the intact plant body or in preformed virtually mature organs, as cells originate by multiplication and as they mature.

THE CONTEMPORARY SCENE

Thus far, progress has been seen in retrospect. One should now succinctly survey the contemporary scene so as to identify the salient problems, and assess their status. This will be done with respect to:

the status of media and the modern means of culturing embryos, growing regions, explanted organs, tissues, cells, protoplasts, and organelles;

totipotency, somatic embryogenesis, clonal and micropropagation;

diversification, physiological differentiation, chemical composition and biochemical potentialities of cultured cells and tissues;

developments involving "genetic engineering" and routes to modify genetic constitution.

Plant Parts: In Situ and in Culture

A mounting dilemma should now be faced. This concerns how the stability of organisms as physical systems is maintained as morphogenesis proceeds from gametes, zygotes, on to embryos and through the increasing complexity of cells in tissues and organs to maturity and reproduction again.

Relevant thoughts here are:

(1) "The price to be paid for morphological complexity is that all the living cells cannot exercise to the full their innate capacities, for this is determined by their position in the plant body."

(2) The property of physical systems that measures their complexity, order and non-randomness (their inprobability) is entropy, and the Boltzman equation relates entropy, by a constant that has dimensions, to a quantity ($\log W$) that is a statistical measure of that order versus disorder ($-\text{entropy} = K \log 1/W$).

(3) We have no similarly rigorous system or quantitative measure to define the internal properties of plants as physical systems that will determine their stability as they develop in complexity. This is so even though, by reduction of their organization, we have learned much about the minutiae of their particular syntheses, processes, and metabolic events. Thus, we can only recognize complementarity at various levels, autotrophic vs. heterotrophic nutrition, anabolism vs. catabolism, endergonic vs. exergonic reactions, cycles of oxidation and reduction, synthesis and breakdown, cell multiplication and enlargement, and in structural terms shoot vs. root, vegetative and reproductive growth, and so on.

(4) Therefore, it should not be surprising if each of the topics to be discussed soon reaches a seeming impasse at which progress depends upon further penetration of the ultimate problems of life; in particular, precisely how the inherited life cycle and the events of differentiation are programmed and whence does it, at the stage of zygote or totipotent cell,

acquire its drive and, though powered by nutrition and metabolism, its persistence through ontogeny.

Organelles. Preparations of cellular organelles, whether aseptic or not, have been used primarily for biochemical purposes. These range from membranes, protoplasmic granules and particles (ribosomes), endoplasmic reticulum (whether "smooth" or "rough"), golgi bodies (dictyosomes), vacuoles, mitochondria, plastids, nuclei and their nucleoli, and so on. Ultimately, each cell, following its origin at division, provides the environment for these organelles either to be maintained or to develop. If, theoretically, it were possible to recapitulate this environment externally, the separate culture of organelles is not inconceivable, although only in the case of the more obviously self-duplicating structures (nuclei, plastids, mitochondria) does this seem even remotely feasible. Even isolated nuclei do not divide. It is one thing, therefore, to maintain preparations of surviving organelles (and plastids are the salient example), but their continued multiplication in culture is another. But the preparation from *aseptically cultured cells and tissues* of aseptic organelles for study or attempted culture is both feasible and desirable (Steward and Krikorian, 1975, p. 164).

Protoplasts. Although free protoplasts, isolated for manipulations and so forth, have long been known (Steward and Krikorian, 1975, p. 153, and refs. there cited), the use of protoplasts from angiosperm cells, divested enzymatically of their walls, has recently burst into a most active field even though the procedures are now at least seventeen years old (Evans and Cocking, 1975, and refs. cited). Although the presence of a cell wall protects plant cells against external osmotic changes, it is a barrier against protoplasmic fusion and certain penetrations; this is normally circumvented in the case of gametes (male and female). The objectives of work with protoplasts are (1) to ensure and study their reformation of wall, (2) to encourage the reconstituted cell to divide and grow, either with or without change to its genetic information, while in the free state, and (3) to be the means of circumventing the sexual process as the means of changing the genetic information that will determine the subsequent course of metabolism or morphogenesis.

Just where does work with isolated protoplasts now stand? First, it has become relatively easy and routine to obtain in large numbers, free, washed, clean surviving protoplasts (Evans and Cocking, 1975; Vasil, 1976, and refs. there cited). Depending upon the source, the preparative and "incubation" procedure, and perhaps even the laboratory, cell wall regeneration occurs with greater or less ease. Melchers (1977) even claims that by use of a phospholipid the problem of fusing protoplasts is "virtually solved."

But, the viability of the cells that have reconstituted walls as shown by the frequency with which they divide and subsequently grow may often be overemphasized.

Although protoplasmic fusions, leading often to grotesque forms or heterocaryons are frequent (Kao et al., 1974), their usefulness may be limited because they usually do not "go anywhere." With few striking exceptions (Melchers and Labib, 1975; Dudits et al., 1976; Melchers and Sacristán, 1977; Smith et al., 1976), fusions between protoplasts with different genotypes (producing hybrids) are still, for most purposes, impractical (Bottino, 1975, and refs. there cited; Melchers, 1977; Heyn et al., 1974). In this context strange observations have been made of the association between the protoplasts of tobacco cells and of human tumor (Hela) cells. The nuclei of the Hela cells were able to survive and even divide in the tobacco cytoplasm (Jones et al., 1976).

In all the above the crucial problem is one of numbers. When the number of reconstituted cells from a population of protoplasts is very low, one should ask "Why?" When the number of reconstituted cells that divide, proliferate, and grow is even lower and is far less frequent than the division and growth of free cells with their original walls intact, as in our experience is usually so, this also poses unanswered questions.

But it has been our hope that free protoplasts of carrot might behave like naked zygotes and lead on in large numbers to somatic embryos. Although cultured free cells of carrot can do this (Steward et al., 1975), disappointingly their protoplasts, derived from the same clone of cultivated totipotent cells, do not readily do so without first forming proliferating calluses which then yield plantlets adventitiously (fig. 3) (Vasil, 1976; Melchers, 1977; Reinert and Bajaj, 1977, and refs. there cited; Krikorian and Steward, 1979).

In other words, and over many years, the work with free protoplasts has produced more propaganda than substance. If we really knew why one naked fertilized egg in its embryo sac develops so infallibly, this might shed light on how to improve the performance of protoplasts from somatic cells, quiescent or in active growth.

Cells. Proliferating tissue cultures transferred as callus masses from agar to liquid with substantial degrees of shaking have been the prime sources of free cells (Torrey and Landgren, 1977, and refs. there cited). In our view the usual error is to agitate the cells too violently. In extreme cases with heavy inocula, the cells may even be said to be growing in a "homogenate of broken cells." In some situations, e.g., for cells from root caps, they will readily slough off into water. But our procedure has been to grow small explants directly in liquid and to allow these gently to abrade their surfaces against each other as their containers revolve about a horizontal axis. Then

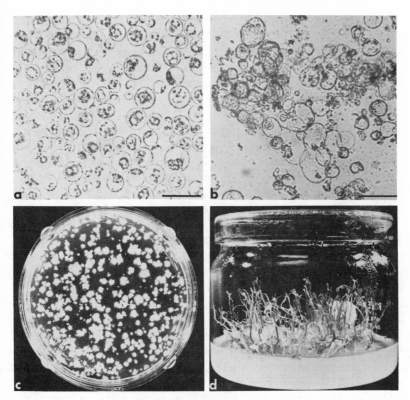

Fig. 3.   Regeneration of plantlets from cultured carrot cells after enzymatic digestion with "Cellulase R_{10}" and "Macerozyme R_{10}" (a pectinase). (a) Protoplasts from totipotent cells grown in suspension culture (scale bar, 100 μm); (b) wall regeneration and cell multiplication in a complex "regeneration medium" (scale bar, 100μm); (c) growth of cell clusters into colonies after distribution and growth in an agar medium (scale, 50 mm diameter dish); (d) plantlets regenerated from totipotent cell colonies taken from (c), distributed in a growth regulator–free medium and grown for 30 days (scale, 100 mm diameter deep petri dish).

after filtration to screen out small clumps, followed by aseptic transfer from one flask to inoculate another, the crops will increase over time (Steward et al., 1975). But it is in the nature of angiosperm cells to adhere and remain in organic contact via plasmodesmata, and the more normal the conditions of culture, the more they will do this (Wallner and Nevins, 1974). It has seemed that many have made a virtue of cultivating absolutely free cells (as, e.g., in the case of the ubiquitous culture, from Cambridge, England, of *Acer pseudoplatanus* cells), which may be contrasted with freshly isolated cultures from *Acer pennsylvanicum* that yielded free cells with obvious tendencies to organize (Steward et al., 1970). But to insist upon cells that remain free is merely to accentuate a feature that removes them even more from their normal growth habit and thus to discourage the first prerequisite for morphogenesis, i.e., the formation of small proembryonic multicellular globules. (In fact, the much used *Acer* cell culture seems to have lost over the years its resemblance to its source, and one wonders why people persist in its use instead of restarting new strains as needed.)

Another feature of long cultivation is the exaggerated respect paid to maintaining cells only under strictly constant conditions, as in chemostats: although that constancy is what may appeal to investigators, it may not reflect an essential requirement of the cells. To express their full potentialities, cells need to be exposed to both variable combinations and sequences of stimuli; therefore, to design experiments that maintain them elaborately and rigidly constant is to condition the cells against responding, and the investigators from expecting them to do so.

Tissues. Although the term *tissue cultures* is ubiquitous, rarely is this strictly true (Bailey, 1943). Many so-called tissue, or proliferating, cultures, randomly proliferated on relatively large explants grown on agar media and subcultured, may be highly variable and very obscure as to their origin. Where large explants are placed on semisolid media, they essentially furnish nutrients and act as "nurse" tissue or organs, so that the relationship of the consequential proliferations to the media as furnished may be very remote. A general rule is that the smaller the original explant and the more precisely it was anatomically located, the more frequently the cultures are renewed or reestablished, the more meaningful the results will be in the understanding of the plants in question.

Although the number of successfully grown callus cultures seems very large, it is not so in terms of the total range of genera, species, and cultivars. Moreover, it is still true that in many particular cases the growth responses to accepted practices may be very meager. (In fact, too much is often made of very slowly and erratically growing cultures.) So long as there are so many recalcitrant systems, these techniques cannot really claim to be universally applicable or useful. A broad case in point is that of the monocotyledons, which are distinguished by their obvious lack of cambium, and from this follows the great difficulty encountered in stimulating mature tissue explants to proliferate. In spite of some success in this direction, it is still true to say that explants from monocotyledons are harder to grow, and especially so in suspension culture, than the most responsive of the dicotyledons. But even well-understood proliferating cultures (like those of carrot explants as stimulated by CM and casein hydrolyzate and growing in liquid) may not yet have achieved their maximum growth potential. (Incidentally, it was clear from actual measurements that cell generation times were, on the average, shorter in the surface cells of explants than in the corresponding free cells either on agar or in liquid [cf. also Yeoman, 1976, for a discussion of this topic].) Also, the best conditions for morphogenesis are not necessarily those that stimulate most rapid multiplication.

Explanted organs. Explantation into aseptic culture of preformed but very immature organs has been practiced. Fruits (Nitsch, 1963) and young leaves of both ferns and angiosperms (Steeves and Sussex, 1972, and ref. there cited) are prominent examples. The motivation here has been to see how far the normal development *in situ* can be continued along the paths already initiated but after the organ in question is excised, and, if so, to see what properties of the medium contribute to this development. Although substantial success has been reported for both fruits and leaves, the conspicuous feature is the difference between what would have transpired on the intact plant and what actually does occur in culture. The gaps between the nutrients and stimuli furnished to the organ *in situ* and what can be furnished and assimilated in culture are still substantial. Moreover the interesting stages in organogenesis have long since happened *before* this kind of experiment begins. Surely it is what prompts a leaf initial by position to become a leaf which is the major problem—not merely how big a leaf finally emerges.

Apices. The culture of isolated root tips severed from the axis was one of the early and prolific applications of aseptic culture (Butcher and Street, 1964). The success in culturing these relatively short apical segments was clearly due to the fact that the organization of the root apex, with its deep-seated origin of lateral organs, relatively far from the tip, enables the entire functional apex to be easily removed into an aquatic medium that roots are "designed" to tolerate. Even the newer knowledge of the compact "quiescent center" of root tips (Feldman and Torrey, 1976) and the subjacent regions of activity (the modern equivalent of Hanstein's histogens) is compatible with this being removed intact in what are relatively short root-tip segments. The work on such cultures enabled the inorganic media to be designed (following familiar culture solutions, Street, 1969) and with the minimum organic nutrients to satisfy the heterotrophic needs of the isolated root tips. The successes of repeatedly excised tips in classical cases (e.g., tomato) should, however, not blind one to the fact that root-tip cultures of many plants, (both dicots and monocots) eventually "peter out." And remarkably little has finally emerged from this sort of work except that root tips, in contrast to shoot tips, are more easily grown with the normal substances translocated being replaced exogenously. Some have resorted to simulations of the organic food supply (aseptically) via the vascular system and of the inorganic nutrients via the external solution

(Raggio and Raggio, 1956) in their attempts to bridge the gap between attached and detached root tips. But the most striking feature of excised root-tip cultures is that they are so completely dominated by the root habit that they virtually never (except in some unusual cases like *Convolvulus*, Bonnett and Torrey, 1965) form buds or regenerate shoots. This stands in marked contrast to the behavior of *cell* cultures, even from roots, which may embark on embryogenesis in which they generate *both* root tips and shoot tips *de novo*. (In fact, had vegetative reproduction of plantlets from root-tip cultures been a really general and feasible prospect, it could have nullified the present emphasis on cloning via cells.)

The shoot apex presents a very different problem. Although the quiescent centers of root apices have their counterparts in shoots, the major growth in shoot tips occurs in localized centers of activity that occur along the flanks of the apical dome, and these arise in rapid succession (Dormer, 1972). But the organization of shoot tips (quite apart from their probably dependence on translocates from roots) with their tunica, rib-meristems, pro-cambial strands, meristems d'attente, and so on, presents a greater degree of complexity in which *the smaller the segment isolated, the greater the difficulty encountered in its independent growth* (Steward and Krikorian, 1975, and refs. there cited). Thus segments that are really small "cuttings" that will develop root initials usually present little problem because the phyllotactic sequence of leaf primordia is initially intact, probably through more than one plastochron. But very small segments that purport to isolate the central dome *without its preformed leaf primordia* became so hard to grow that it is still a moot point whether the central tip of the shoot apex *alone* has been, or can be, cultured (Smith and Murashige, 1970; Shabde and Murashige, 1977).

Embryos. It is logical here to consider the cultivation of entire explanted embryos (as in embryo culture) with their shoot and root apices intact and in organic connection. This has long been practiced in the case or orchids, where the embryos are "primitive" and little more than minute, globular forms (Raghavan, 1976, 1977; Rao, 1977). There have been, and are, useful applications in which by these means, embryos can be removed from natural inhibitors (Randolph and Randolph, 1955). A recent and commercially promising application of this sort is the successful rearing to maturity of excised embryos of "Makapuno" coconut. In this instance embryo development *in situ* is prevented by rotting of the endosperm (de Guzman and Del Rosario, 1964, 1974). But in most of these cases the definitive events of morphogenesis already may have occurred, and the culture medium merely attempts to simulate that which functions or should function *in situ*. It was work of this sort on abortive embryos in *Datura* by

Blakeslee and van Overbeek that first brought coconut milk, or coconut water, into prominence and its use in cell and tissue culture work followed (Raghavan, 1976, for refs).

Nutrient media. A discussion about nutrient media usually includes recipes with more or less complex formulations and claims for their effectiveness (Gamborg et al., 1976). In retrospect, the inorganic requirements of angiosperm isolates need not be different from the mineral salt solutions that suffice for whole plants. There may, however, be subtle points about total concentrations, proportions, and so on, and whether or not nitrogen should be supplied as nitrate or as reduced nitrogen (ammonium or organic-N, such as casein hydrolyzate, glutamine, urea, etc.). All this involves no great nutritional principle. The old arguments about the form of carbohydrate, largely resolved in favor of sucrose, now seem rather sterile; but since the cultures may not become green, their needs for some vitamin supplements are now usually recognized. It is when cultures initiate from minute, relatively unorganized explants, and when dormant cells need to be activated and when subsequent morphogenesis may be sought, that the present range of stimuli involving growth-regulating substances are most called for.

In dicotyledons when the explants from mature organs include cambium, the stimuli over and above general nutrients to induce their reactivated growth may be minimal; it is when the influence of cambium is neither available nor persuasive that more effective combinations of stimuli are needed. Despite the fact that CM to supplement a basal nutrient solution proved so effective on carrot explants, it (or its morphological equivalents) was never regarded as a "panacea" in all such cases. In fact, it is only too obvious that there are many instances in which it alone was either ineffective or even deleterious.

But the view that has emerged is not one of an ideal fixed nutrient solution with a single set of additive stimuli. Rather, it is an array of combinations of inorganic nutrients and of organic stimuli that, by their interactions, modulate not only the growth (e.g., in cell number and size) that ensues but also the metabolites formed and the organic and inorganic content of the cells. To the interactions between such partial systems as IAA and zeatin-like cytokinins (AF_2), and inositol and other substances (AF_1), there should be added the need for, and the interactions between, the trace elements (fig. 1). In all such work the basal nutrient medium needs to be prepared as near trace-element-free as possible, and even its organic stimuli (even CM) should be available for test in a trace-element-free form (Steward and Rao, 1970, 1971). Although the needs for iron and boron are paramount, interactions between Fe and the other trace metals (Cu, Zn,

Mn, and Mo) in various combinations with iron, when they are added either singly or from two to four at a time, have been studied (Steward and Rao, 1970). Polygonal diagrams have been useful to present the data from many such tests made on a single clone of explants, and they have been used to show the variations in composition in response to 24 combinations of the component parts of Systems I and II (Steward and Rao, 1971) as well as in response to as many as 96 combinations of trace elements and growth stimuli! Therefore, it should be no surprise that there is no universally applicable nutrient solution or set of inductive stimuli that is best for all purposes. The responses observed and the required triggers or stimuli will in any one situation depend upon the extent of, and the means by which, the first freely multiplying cells developing *in situ* became repressed as they formed part of the given plant body or organ.

Totipotency, Somatic Embryogenesis, and
Clonal Propagation

All these topics may be considered together.

The theoretical possibility (or principle) of totipotency of living angiosperm cells that have originated by continuous equational divisions, as from a zygote, should now be accepted, even though the evidence that supports it is from a very limited number of plants (Steward et al., 1975; Steward and Krikorian, 1975). Where cells remain alive in mature tissues and organs, the release of their inherent totipotency presupposes:

1. the inductive stimuli that allow them to respond to nutrients and overcome their quiescence or dormancy—this involves the whole gamut from CM or *Aesculus* fluid, and so on, with their component parts, auxins, cytokinins, and their interactions, synthetic growth regulators, and stimuli all of which are probably not yet recognized (Steward and Krikorian, 1971, and refs. there cited);

2. the culture of the cells in small units that escape thereby as completely as possible from the regulatory control of the tissue or organ of their orgin;

3. the provision of environments, media, and chemical stimuli that foster morphogenesis (in contrast to mere multiplication).

Only when isolates from mature tissue of adult plants can produce cells capable of recapitulating embryogenesis in large numbers and in liquid to give mature plants are these techniques technologically inviting. The ideal here (rarely met except in carrot and other umbellifers) is when shoot and root apices develop in a small globular unit and form an axis *at the outset* as in normal development. The compromise often is that random proliferations on large explants may give rise, often slowly and unpredictably, to relatively few, and usually very variable, adventitious buds or plantlets; this may have its technical uses but it should be the challenge to promote a more nearly somatic embryogenesis as from free cells (Murashige, 1977; Hussey, 1978).

One may use here the present status of the carrot system in illustration. After the realization that free cells of carrot root phloem cultured in liquid retain their totipotency and can give rise to plantlets and that this property may be released by different inductive stimuli acting on different tissue sources, and that other members of the Umbelliferae share this property with carrot, one might imagine that this system was too well known to demand much attention. Not so, although it required the challenge of a new situation to significantly add to this knowledge (cf. page 231). This new situation arose from the need to establish whether single cells (e.g., of carrot) could pass through normal embryo development unaffected by zero gravity as in space. The test was made possible by access to the USSR/USA Joint Biological Space Mission Kosmos 782, in which the spacecraft was equipped with a centrifuge to furnish 1 g controls. Morphogenetically competent free cells were prepared and distributed in agar in small plastic petri dishes. All biological activity was reversibly reduced and cell multiplication prevented by maintaining the cultures at 4°C. Once in orbit, the dishes experienced normal temperatures (22°C) for their growth. In anticipation of a later and fuller account (Krikorian and Steward, 1978), one may, therefore, show figure 4 (a–k) and from this state the following conclusions:

1. At least 2000 organized forms developed per dish after 20 days in space at 0 g.

2. These forms were classified, and there was ample evidence that embryonic forms developed in space at 0 g and equally so on a 1 g centrifuge in space (cf. Krikorian and Steward, 1978).

3. Although shoots initiated at 0 g in 20 days of flight, they did not develop in the dark but they did so *very rapidly* when they were brought into the light at 1 g.

4. So far as these tests go, somatic embryogenesis from free cells to embryos can occur equally well at 0 g or 1 g in space, and the resultant plantlets, which were induced in large numbers, developed into normal plants in the greenhouse. (A remote possibility that single cells on agar, although unable to divide at low temperature, could still be "programmed" so that they subsequently "remembered" gravity when they grew at 0 g, may still remain to be tested.)

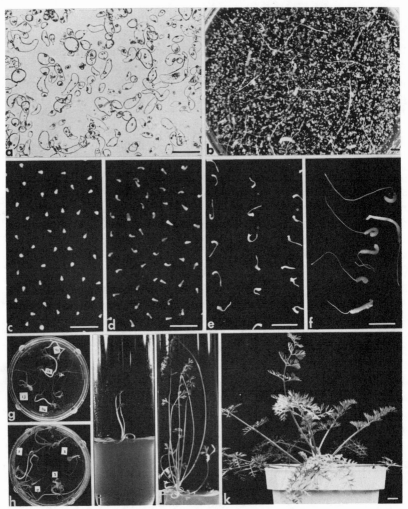

Fig. 4. Development of somatic embryos from totipotent carrot cells under near zero gravity conditions and their subsequent growth into mature plants. (a) General appearance of cells from suspension culture after gradation through sieves with decreasing pore size (scale bar, 100μm). (b) Development of carrot cells into embryonic forms after 20 days in space. The low magnification (scale bar, 1 mm) and the dispersal of the profuse crop of embryos *within* as well as *on* the agar medium make it difficult to distinguish easily all the forms but typical globular, heart (c), and torpedo-shaped (d) (scale bar, 5 mm), as well as mature embryos, (e and f) (scale bars 1 cm) were present and could be aseptically isolated. Embryos at each of these stages could be removed from the dishes and further grown to maturity in successive steps; (g) and (h) show 50 mm diameter dishes each with several representative embryos that arose during flight and were transplanted to fresh medium and grown for approximately 1 month; (i) and (j) show two 35 mm diameter culture tubes with plantlets as from (g) and (h) at time of initial transfer to tube (i) and after 40 days' subsequent growth (j); (k) carrot plant reared from (i) and grown in soil for 2 months (scale bar, 1 cm.).

Nevertheless, there is still a need to define fully the interacting environmental variables and chemically inductive ones that release the morphogenetic potential of cultured cells. But the essential dilemma of embryogenesis, whether somatic or zygotic, remains. What internal forces, or considerations of stability, might intervene in a globular embryo, or a proembryonic cultured globule, to promote root and shoot initials, to endow these with their respective developmental patterns, and, thereafter, allow them to exhibit their evident complementarity and stability? If one knew, for example, what localized event, or consideration, could prompt the first two leaf initials with an apex between, as in most dicotyledonous plants, or the seemingly quite different and more asymmetrical sequence, as in most monocotyledonous plants, to occur and why these events complement the respective development of root tips (and vice versa), the problems of both normal morphogenesis and simulated somatic embryogenesis would then, but not until then, seem to be soluble.

But is all this covered by current practices traceable to Skoog and Miller's report of apparent control of adventitious shoot and root formation on proliferated callus of *Nicotiana tabacum* cv. 'Havana' Wisconsin 38? There are now many instances in which random proliferation has been fostered by a cytokinin plus an auxin; and shoot and root organization is induced thereafter by a controlled auxin/cytokinin ratio. But this device is by no means universally effective. Frequently shoots *or* roots emerge (roots usually more frequently than shoots); and even when this occurs, *uniform* plantlets are not easy to obtain in large numbers (Steward and Krikorian, 1971, p. 82; Hussey, 1978).

From experiences based on carrot, other umbellifers, and other plants in which somatic embryogenesis has been achieved from cells grown in suspension culture, the advantages of this route over adventitious shoot and root formation on proliferated calluses or explants seem manifest. These inherent advantages flow from the large numbers of uniform embryos and normal plantlets that are possible. Although criticisms are made that cytological aberrations are common in cultured cells (D'Amato, 1975; Torrey and Landgren, 1977), these are not always encountered nor do they necessarily persist into the resultant plantlets. Strains of cultivated carrot cells have shown deviations from the normal diploidy of their source, but the propagated plants examined nevertheless were again normal diploids (Mitra et al., 1960). But, at another extreme, *Haplopappus gracilis* rapidly produced from its normal diploid cells with 4 chromosomes, cells from which many distinctive clones could be reared, with various chromosome numbers—even up to 64 (Mitra and Steward, 1961). Also, since haploidy has been seen to occur in cultures of free cells, an opportunity may have been missed to study in these systems the biochemistry to foster the reduction division as a part of meiosis, even as the ability to control other changes in chromosome complements by exogenous treatments could be profitable. Indeed, if plants can be reared from cells, one should not be deterred from doing this by the threats of chromosomal changes, for the incidence of the latter may prove to be as controllable as the morphogenesis. Undoubtedly, however, the propagation of plants from suspended cells involves resources and skills not commonly brought to bear upon the problem. From what has been said, the need for more broadly based studies of, first, the growth induction and, second, the embryogenesis should be apparent. But after embryos are in the course of formation and early development, it requires skills, and for large numbers, a specialized kind of "nursery technique" to bring these to the state in which industry can also use them. The young plantlets are very succulent and need protection in their aseptic state from drying out and the eventual incidence of disease and such.

Against this general background, the following comments on cloning desirable plants via free cells or callus cultures may be appropriate.

1. Obviously these methods (Nickell and Heinz, 1973; Murashige, 1977) are only technologically rewarding where conventional means of vegetative propagation are inadequate (Hartmann and Kester, 1975; Holdgate, 1977; Hussey, 1978). They are particularly valuable when it is desired to transmit the visible properties of *mature plants* to clones of propagules by the thousands.

2. The attainment, routinely, of large numbers of uniform plants being part of the essential problem, the initial material multiplied as cells, capable of subculture, should be brought through the stages of somatic embryogenesis in liquid.

3. The source of cells to be propagated should be somatic tissue from mature plants. Although seedling or embryo sources may often prove easier, it is a delusion to regard this as of more than theoretical interest since it may leave essential practical obstacles to be overcome. The culture of calluses and the regeneration therefrom of plantlets in cases where plants have both juvenile and adult forms holds a clear warning here. In some cases (*Nicotiana*) the regenerated plantlets produce *directly* the morphology of their source, whereas in other (*Passiflora*) they must always go through their juvenile phase (Marcavillaca and Montaldi, 1967).

4. If it should be that isolated protoplasts in *large numbers* could be induced to simulate zygotes and develop as embryos, then this would be very profitable (Vasil, 1976, and refs. there cited). However, the weight of evidence is that free protoplasts that adequately retain

their viability (like those from carrot explants) only give rise to regenerated cells with walls in a very small proportion of the cases, and that plantlets from such protoplasts only arise in such infinitesimally small numbers that their present bearing on feasible mass vegetative propagation seems not to be in sight.

5. Many plant calluses grown on bulb scales, parts of mature organs like floral parts (Bush et al., 1976), conifer needles and cotyledons (Sommer et al., 1975; Cheng, 1977; Mehra-Palta et al., 1977; Mott et al., 1977; Smeltzer et al., 1977; Thomas et al., 1977), and so on, have given rise to plants and are in fact being developed for commercial propagation (Durzan and Campbell, 1974; Winton and Huhtinen, 1976; Bonga, 1977, and refs. there cited). It would seem wiser however to persist with the culture of free cells from such calluses in the hope that a more precisely simulated somatic embryogenesis, analogous to that of carrot and such, may emerge (Radojević et al., 1975; Skolmen and Mapes, 1976).

6. The present over-all effort being expended on attempted propagation of commercially valuable plants by so-called tissue-culture means must be large (Murashige, 1974; Cell Culture and Tissue Culture Symposium, 1977; Reinert and Bajaj, 1977, and refs. there cited). A case in point is that of *Hemerocallis*. In this monocotyledonous plant, explanted tissues, especially those from floral organs, can be induced to proliferate and can be subcultured. Thereafter, very small inocula can be used to develop growths from which adventitious shoots and roots emerge. Although this can be achieved in very large numbers, it has not yet been possible to induce very small cell clusters to undergo somatic embryogenesis (fig. 5). Virtually all the published cases of clonal micropropagation from callus are comparable in principle to that here briefly described for *Hemerocallis*. But this still leaves much to be desired in comparison with true somatic embryogenesis from single cells (see p. 835 below).

Therefore, instead of arousing extravagant hopes based on what seems theoretically possible, it seems that much more conservative effort should be expended to make the possibilities not only feasible but adequately controllable before the hope of commercial exploitation goes too far.

Biochemical Potentialities of Cultured Cells

This problem concerns the ability to control desirable chemical compositions in cultured systems in which cells that retain the genetic information of zygotes merely grow rapidly without creating normal morphology.

Fig. 5.   Organized development in cultured explants of *Hemerocallis*. (a) Close-up of material as grown in liquid, scale bar 1 cm; units such as these may be subdivided and inoculated as tiny inocula (ca 1–2 mm) onto an agar medium in 100 mm diameter plastic petri dishes as shown in (b). After an appropriate period of time (in this case 56 days), these inocula will proliferate and yield shoots and roots adventitiously (c); plantlet (d) derived from material as at (c) by subdivision and further grown, scale bar 1 cm. Such plantlets can be grown to maturity after transfer to soil. (See p. 835 of this volume for details.)

Although these objectives have long been in view, the returns have been meager (Krikorian and Steward, 1969; Teuscher, 1973; Butcher, 1977; Staba, 1977). The reasons are not far to seek. Valuable, so-called secondary metabolites often occur in highly specialized situations. Their biochemistry may require that subtle combinations of many genes be appropriately mobilized. At the time of their accumulation the products in question may occur in organs of perennation the formation of which is dependent upon interacting environmental conditions. But, as mentioned earlier, the crucial question is what, in the factors that determine the complementarity and stability of contrasting organs or even of adjacent cells, decrees that a given location or a given morphology should, as it were, "trigger" its specific metabolites to accumulate. Furthermore, the general feature is that when quiescent cells are reactivated and grow, their secondary metabolic characteristics disappear just as, when grown free, they lose their mature forms and closely resemble each other even as eggs in different embryo sacs are much alike (Steward et al., 1969).

Therefore, the odds are long and weigh heavily against the possibility of easily giving to cells in culture the same signals as those to which they respond during development *in situ*. These odds are the more impressive when one notes that carrot phloem cells in cultured explants, or when free, do not ordinarily accumulate reserves of orange-red carotene until and unless (by the techniques of somatic embryo genesis) they give rise to little plants that begin to initiate storage roots, and cells *in situ* again receive the appropriate signals (Steward, 1970a). Much the same story can be told for starch in potato cells. Two recent and comprehensive reviews (Butcher, 1977; Staba, 1977) evaluate similar problems in detail. But it is noteworthy that they essentially reach the same conclusions that Krikorian and Steward had already expressed in 1969! This conclusion was, and is, that normal biosynthetic pathways characteristic of intact plants and mature organs are not easily recapitulated *accurately and at will* in unorganized cell and tissue cultures. Surprisingly, therefore, a German firm, probably well known to many, has recently advertised for sale at a very high price a report based primarily on a literature survey that purports to show how aseptically cultured plant cells and tissues may be used to achieve the industrial production of various biochemicals. This seems misleading and premature. Also, those who point to the precedent of the microbial fermentation industry lose sight of some essential differences between bacteria and fungi and higher plants. Whereas the former carry out much of their metabolism superficially in free connection with their ambient medium, the latter have cells that are highly complex with an internal fine structure and highly compartmentalized organelles. Characteristically, higher plant cells do not usually release their contents to the ambient medium, although they transfer metabolites to other parts of the plant body. Such interactions between cells, tissues, and organs are often vitally involved in the total biosynthetic outcome.

But the short answer follows from what we currently know, chemically, about how protein is made and how sugar is synthesized. With this great body of information on hand, it might seem laudable to strive to eliminate cells and the plant body as the prime source of protein synthesis in nature, or the green leaf as the ultimate source of sugar by photosynthesis. But for the foreseeable future, the odds are still overwhelmingly in favor of the biochemistry working best in the appropriate morphological setting in plants that grow and bring in, at the appropriate time and place, the genetic information they inherit.

Developments Based on Genetic Engineering

This topical section is here very brief since comprehensive reviews exist (Heyn et al., 1974; Bottino, 1975; Constabel, 1976; Vasil, 1976; Bajaj, 1977; Power and Cocking, 1977; Hess, 1977; Giles, 1977; Davey, 1977, *in* Reinert and Bajaj, 1977; Hollaender, 1977).

In our review published in 1975, we stressed the thinking surrounding expectations on the use of cultured plant cells in various genetic studies that had been, until then, largely confined to microorganisms (Steward and Krikorian, 1975, p. 158 et seq.). The hope was, and still is, that since very large numbers of plant cells can be grown in culture or obtained as protoplasts, rare events such as mutation and somatic recombination of genetic material might be more easily detected. Tissue and cell culture techniques theoretically should indeed permit the selection of certain types of mutants by adjustment of the medium or environmental factors (e.g., the isolation of drug-resistant mutants by plating cells in a medium containing the drug, or isolating auxotrophs by standard methods used with simpler forms of life). Also, fusion of protoplasts involving two protoplast types, each carrying a different trait (e.g., specific resistance to a particular drug), has been suggested as a potential method for selecting hybrid cells. For instance, plating protoplasts so treated to enhance fusion from cells showing individual resistances to specific drugs in a medium containing both drugs and observing the persistence of cells or colonies should again, theoretically, enable one to isolate fused products carrying resistance to both inhibitors. But it appears that drug-resistant mutants are recessive and virtually impossible to detect, so that, to date, no drug-resistant mutants and very few nutritional mutants have been isolated using cultured plant cells or protoplasts (Aviv and Galun, 1977).

Papers outlining the possibilities of "replica plating" of higher plant cells have appeared, however (Scheider, 1976; Schulte and Zenk, 1977). In this work it seems important to note that there is no convincing evidence that single cells have been utilized. This is an absolute prerequisite of the technique. Moreover, as pointed out earlier (Steward and Krikorian, 1975, p. 161), multiple metabolic pathways in higher plant cells (especially those involving nitrogen) make it inherently improbable to select for and demonstrate true auxotrophy—hence the many so-called leaky mutants. It has been a very real problem, therefore, to establish whether a loss or seeming change in a phenotypic or metabolic characteristic is a result of gene mutation or is in response to a cytoplasmic change or to "turning on or off" of unaltered genes by an epigenetic system in the cells (Widholm, 1972a,b,c, 1974; Sung, 1976). In fact, stable changes in phenotype may take place through persistent extrachromosomal phenomena acting on gene expression. In still other instances, the presumptive changes observed are adaptations to a given environment. Moreover, alterations in the structure of chromosomes that may occur in plant cells in culture are also known to change phenotypes without necessarily changing the genes *per se*. Nevertheless, the attempted genetic exploitation of cultured plant cells and protoplasts has expanded rapidly. Protoplast fusion, so called transgenosis, or transformation and nuclear and other organelle transfers have all been attempted with varying degrees of success. But all these operations have been limited by problems of selection, stability, and even viability of the end products.

The report by Carlson et al., (1972) that "parasexual hybrids" identical to the normal genetic cross had been obtained by fusing protoplasts of mesophyll cells from *Nicotiana langsdorfii* and *N. glauca* stimulated much discussion and work (Smith, 1974a, b; Steward and Krikorian, 1975, p. 158 et seq.). Despite early difficulties, this work has since been repeated and extended by Smith et al. (1976) but with interesting differences. Specifically, the later hybrids show a high and variable chromosome number. This indicates that the nuclear events are not yet fully controllable. Analysis of so-called Fraction I protein of the chloroplasts from these hybrids revealed that the mixture of the two chloroplast types from each "parent" was almost exclusively segregated into one "parental" type (Kung et al., 1975; Chen et al., 1977). The significance of this finding for work designed to improve photosynthetic efficiency by introducing "high efficiency chloroplasts" from one group of plants into protoplasts of another, followed by causing these fused products to develop into plants, is clear. It may be too optimistic to expect "foreign" chloroplasts to operate and multiply successfully in a new environment. Intraspecific hybrids of *Nicotiana* and of *Petunia* have also been produced using protoplast fusion (Smith et al., 1976, and refs. there cited; Melchers, 1977). However, to date no plant has been produced by these means that could not have arisen by normal sexual reproduction.

In addition to the above difficulties (selection of mutants, etc.), another problem arises. It is accepted that homozygous diploid plants, achieved by treating haploid cells or plantlets of another origin (via so-called androgenesis) with colchicine or some appropriate treatment (either artificial or natural) would be a great boon to plant genetic research. But as was pointed out in 1973 (Steward and Krikorian, 1975, p. 161 et seq.), relatively few plants, e.g., species of *Nicotiana, Datura, Oryza, Brassica*, and so on, had yielded haploids using the anther culture technique. Despite the subsequent increase in number of species that have yielded haploids (Kasha, 1974; Reinert and Bajaj, 1977, pp. 251 et seq.), the total is still very small, and other species have been tested without success. Indeed, even in genera that respond relatively easily (i.e., *Nicotiana*) there are species and even cultivar differences in their formation of haploids from anthers (Tomes and Collins, 1976; Hlásniková, 1977).

Even so, the availability of haploids obtained by androgenesis have enabled improvements to be made in rice and tobacco via the technique of so-called combination breeding. Since homozygous diploids can be obtained in a single operation, this provides uniform breeding material for conventional crossing that ordinarily would be available only after considerable inbreeding and back-crossing. In a similar way, it has even been suggested that propagation of potato by tubers will one day be replaceable with true seeds that yield plants that are virus-resistant (Melchers, 1977.)

The concept of genetic engineering requires that new or foreign genetic information, either as organelles such as chloroplasts, nuclei, or DNA (as in so-called plasmids, viruses, chromosome fragments, or even intact chromosomes) be introduced into essentially totipotent cells or protoplasts that thereafter must function effectively in the new environment (Heyn et al., 1974; Steward and Krikorian, 1975 p. 163 et seq.). It is also necessary to be able to induce such cells to grow into mature plants that retain and express the desired characteristics of the modified genome. Although some modest progress has been made in the uptake of genetic information by isolated protoplasts (Potrykus and Lörz, 1976; Lörz and Potrykus, 1976; Suzuki and Takebe, 1976; Hess, 1977, *in* Reinert and Bajaj 1977), it is still premature to believe that transferred DNA is really integrated into the functions of the cells (Vasil, 1976; Melchers, 1977). This is also true for the transfer of chloroplasts and nuclei. Viruses, on the other hand, especially RNA viruses, have been introduced and made to function in protoplasts (Reinert and Bajaj, 1977) and there is some evidence that single stranded bacteriophage DNA can be taken up by tobacco protoplasts (Suzuki and Takebe, 1976).

By far the most convincing evidence for stable incorporation of plasmid DNA into higher plant cells, however, derives from work on the transformation of normal cells of tobacco into tumors in a crown gall system utilizing a virulent plasmid of *Agrobacterium tumefaciens* (Tempe et al.,

1976; Chilton et al., 1977). (Whether the plasmid "genes" code for a tumor-producing factor[s] or act indirectly to "turn on" genes of the host plant that then elicit the tumor state has not yet been established however.) It has been said (Chilton et al., 1977, p. 268) that "crown gall tumors . . . can be viewed as a feat of genetic engineering on the part of *A. tumefaciens.*" Since elaborate techniques exist for purifying and selectively replicating small segments of DNA from bacteria (Glover, 1976), it has been suggested that functional genes coding for nitrogen fixation from *Rhizobium* might be successfully transferred into nonleguminous plants and thus minimize the need for exogenous nitrogen (Heyn et al., 1974, p. 53; Child, 1975; Snowcroft and Gibson, 1975; Vasil, 1976 p. 149 et seq.; Davey, 1977, p. 551, *in* Reinert and Bajaj, 1977). An earlier and more simplistic approach conceived of the synthesis of a hybrid nitrogen-fixing organism by means of parasexual hybridization between protoplasts of a legume and a non-fixer. Recently protoplasts of the mycorrhizal fungus *Rhizopogon* incubated with protoplasts of the free-fixing bacterium *Azotobacter vinelandii* have yielded colonies of fungus that can reduce acetylene (Giles and Whitehead, 1977; Hollander, 1977). But it is still true that speculation in this area is far ahead of achievement, and it remains to be shown that a true symbiotic relationship can be achieved between a non-fixer angiosperm and a microorganism that normally fixes symbiotically (Carlson and Chaleff, 1974; Vasil, 1976, p. 149).

In short, if change of genetic makeup is ever to be of practical importance, introduction of the change or new genome will have to be made at the level of spores, haploid protoplasts, zygotes, or somatic totipotent protoplasts or cells. But the "change" or "correction" will be no more effective than the means available to re-create embryos and growing them to mature plants. Hence, the importance of getting unequivocal evidence of changed genetic constitution (via fusion of protoplasts or introduction of new genetic material with meaningful biochemical consequences) coupled with the essential need to rear mature plants in large numbers from the genetically changed products cannot be overemphasizsed.

PROSPECTS

In the broad picture prospects have changed very little since they were assessed in 1973, despite the fact that much has been gained in the interim in the direction of technical accomplishments. Much still remains to be known about the induced control of somatic cell division, especially in situations of seeming recalcitrance toward accepted practices. But it is even more pressing to bring into activated divisions, and eventual direct embryogenesis, the isolated protoplasts that are now so readily producible.

This is the more imperative because the isolated protoplasts are the ultimate targets for the genetic analysis and the genetic engineering approaches to the problems of development or metabolism.

The improvement, for routine use, of haploid cells that will yield haploid plants and protoplasts for use in plant breeding by procedures that bypass sexuality is an obviously important area to be developed.

In the continuing search for means of culturing cells to plants, especially when economically important plants may need to be rapidly multiplied or cloned, emphasis will need to be put on most accurately recapitulating normal embryogenesis somatically, and, for this, liquid culture of the cell seems preferable. In this context, the monocotyledons still seem to present special difficulty and a special challenge, as indeed do many other economically important plants, such as forest trees. When applying present knowledge, emphasis should here be placed on investigating somatic cells from *mature* tissues and organs as the preferred starting point for such studies even though this may present, at the outset, greater difficulty. The current impression is that plans to practice very large-scale clonal propagation of plants (especially conifers) via cultured cells may have often anticipated by too far the precise laboratory techniques on which such plans should be based. Therefore, in the prospects for the future, the rigorous establishment of the conditions under which any given species or cultivar will consistently produce clones from cultured cells via true somatic embryogenesis should have high priority.

But the impressive feature is, how much diversity normally arises in the behavior and ultimate composition of the cultured plant cells from the impact upon them of interacting and nutritional variables (including the trace metals). In fact, one should here recall the extent of the environmental controls over normal development and morphogenesis exerted early in development (e.g., vernalization of cereals while still in the ear; the facts of floral induction by temperature in bulbs and biennials often simulated by gibberellins, etc.) when assessing how much can be expected from, or attributed to, rearrangements of the genetic materials during development of the kind now familiarly studied under the guise of "genetic engineering."

It is, in fact, unsatisfying to conclude this review without asking how the most recent trends in genetic engineering as related to development in animal cells compare with the trends in plants. Briefly, the well-known work on frogs (in which hitherto somatic nuclei to be effective had to be transferred to enucleated oocytes) has had the following sequel. In the attempt to dissect the gene-controlling molecules in oocytes, purified DNA (genes) from nuclei of cells at various stages in development are being injected into eggs. The consequences are being followed by the activity of these controlled assemblies of genes on subsequent protein complements.

The hope is that it will be possible to identify by these means the regulator molecules, presumed to exist in fertilized developing oocytes, that can control development. In other words, this may more elegantly show that the genetic information must be sequentially controlled in development, but it may not immediately show how this control is normally achieved (Gurdon, 1977).

The closest analogy to this work using plants is to be seen in the *Agrobacterium tumefaciens* plasmid that causes tumors. In this system the morphological response to the induced change in the genetic complement is visible because it involves a drastic and sudden departure from the normal developmental trend. Thus the animal system (notably the frog) lends itself to this type of work by virtue of the fact that it is normally progressively arrested or delimited during development, whereas the plant system is here limited by its great asset, namely, its demonstrable totipotency throughout. If, as it seems, the crucial problem still is to understand how the controls of normal zygotic development occur, than botanists may need to decide how best to use effectively the cultured totipotent systems of higher plants to this end.

ACKNOWLEDGMENTS

Many of the views expressed here derive from personal experience that was gained largely from the years spent at Cornell University, Ithaca, N.Y., particularly in the Laboratory for Cell Physiology, Growth, and Development, where the authors' collaborations commenced. Some investigations described arose out of long-continued support under successive grants GM 09609 to one of us (F.C.S.) from the National Institutes of Health, and especially since 1971, the work has been supported by National Aeronautics and Space Administration grants (the most recent of which have been NAS 2-7846 and NSG-7270) and carried out at the State University of New York at Stony Brook by A.D.K.

LITERATURE CITED

Aviv, D., and E. Galun. 1977. An attempt at isolation of nutritional mutants from cultured tobacco protoplasts. Plant Sci. Lett. 8:299–304.

Bailey, I. W. 1943. Some misleading terminologies in the literature of "plant tissue culture." Science 93:539.

Bidwell, R. G. S., R. A. Barr, and F. C. Steward. 1964. Protein synthesis and turn-over in cultured plant tissue: Sources of carbon for synthesis and the fate of the protein breakdown products. Nature 203:367–73.

Bonga, J. M. 1977. Applications of tissue culture in forestry. *In* J. Reinert and Y. P. S. Bajaj (eds.), Applied and fundamental aspects of plant cell, tissue, and organ culture. Pp. 93–108. Springer-Verlag, Berlin.

Bonnett, H. T., and J. G. Torrey, 1965. Chemical control of organ formation in root segments of Convolvulus cultured in vitro. Plant Physiol. 40:1228–36.

Bottino, P. J. 1975. The potential of genetic manipulation in plant cell cultures for plant breeding. Rad. Bot. 15:1–16.

Bush, S. R., E. D. Earle, and R. W. Langhans. 1976. Plantlets from petal segments, petal epidermis, and shoot tips of the periclinal chimera, Chrysanthemum morifolium 'Indianapolis.' Amer. J. Bot. 63:729–37.

Butcher, D. N. 1977. Secondary products in tissue culture. In J. Reinert and Y. P. S. Bajaj (eds.). Applied and fundamental aspects of plant cell, tissue, and organ culture. Pp. 668–93. Springer-Verlag, Berlin.

Butcher, D. N., and H. E. Street. 1964. Excised root culture. Bot. Rev. 30:513–86.

Carlson, P. S., and R. S. Chaleff. 1974. Forced association between higher plant and bacterial cells in vitro. Nature 252:393–94.

Cell Culture and Tissue Culture Symposium. 1977. Hort. Sci. 12:125–52.

Chen, K., S. G. Wildman, and H. H. Smith. 1977. Chloroplast DNA distribution in parasexual hybrids as shown by polypeptide composition of Fraction 1 protein. Proc. Nat. Acad. Sci. USA 74:5109–12.

Cheng, F. Y. 1977. Factors affecting adventitious bud formation of cotyledon culture of Douglas fir. Plant Sci. Lett. 9:179–87.

Child, J. J. 1975. Nitrogen fixation by a Rhizobium sp. in association with non-leguminous plant cell cultures. Nature 253:350–51.

Chilton, M. D., M. H. Drummond, D. J. Merlo, D. Sciaky, A. L. Montoya, M. P. Gordon, and E. W. Nester. 1977. Stable incorporation of plasmid DNA into higher plant cells: The molecular basis of crown gall tumorigenesis. Cell 11:263–71.

Constabel, F. 1976. Somatic hybridization in higher plants. In Vitro 12:743–48.

D'Amato F. 1975. The problem of genetic stability in plant tissue and cell cultures. In Crop genetic resources for today and tomorrow. IBP2. Pp. 335–48. At the University Press, Cambridge.

De Fossard, R. A. 1976. Tissue culture for plant propagators. University of New England, Armidale, New South Wales, Australia.

De Guzman, E. V., and D. A. Del Rosario. 1964. The growth and development of Cocos nucifera L. 'Makapuno' embryo in vitro. Philippine Agric. 48:82–94.

De Guzman, E. V., and D. A. Del Rosario. 1974. The growth and development in soil of makapuno seedlings culture in vitro. Res. Bull. Nat. Res. Coun. 21:1–16.

Dormer, K. J. 1972. Shoot organization in vascular plants. Syracuse University Press, Syracuse.

Dudits, D., K. N. Kao, F. Constabel, and O. L. Gamborg. 1976. Embryogenesis and formation of tetraploid and hexaploid plants from carrot protoplasts. Canad. J. Bot. 54:1063–67.

Durzan, D. J., and R. A. Campbell. 1974. Prospects for the mass production of improved stock of forest trees by cell and tissue culture. Canad. J. For. Res. 4:151–74.

Evans, P. K., and E. C. Cocking. 1975. The techniques of plant cell culture and somatic cell hybridization. In R. H. Pain and B. J. Smith (eds.), New techniques in biophysics and cell biology, 2:127–58. John Wiley & Sons, New York.

Feldman, L. J., and J. G. Torrey. 1976. The isolation and cultura *in vitro* of the quiescent center of *Zea mays*. Amer. J. Bot. 63:345-55.

Gamborg, O. L., T. Murashige, T. A. Thorpe, and I. K. Vasil. 1976. Plant tissue culture media. In Vitro 12:473-78.

Gamborg, O. L., and L. R. Wetter (eds.). 1975. Plant tissue culture methods. National Research Council of Canada, Saskatoon.

Gautheret, R. J. 1952. Remarques sur l'emploi du lait de coco pour la réalisation des cultures de tissus végétaux. C. R. Acad. Sci. Paris 235:1321-24.

Gautheret, R. J. (ed.). 1977. La culture des tissus et des cellules des végétaux. Travaux dédiés à la memoire de Georges Morel. Masson, Paris.

Giles, K. L., and H. Whithead. 1976. Uptake and continued metabolic activity of *Azotobacter* within fungal protoplasts. Science 193:1125-26.

Glover, D. M. 1976. The construction and cloning of hybrid DNA molecules. *In* R. H. Pain and B. J. Smith (eds.), New techniques in biophysics and cell biology. 3:125-45. John Wiley & Sons, New York.

Gurdon, J. 1977. Egg cytoplasm and gene control of development. Proc. Roy. Soc. Lond. 198B:211-47.

Hartmann, H. T., and D. E. Kester. 1975. Plant propagation: Principles and practices. 3d edition. Prentice-Hall, Englewood Cliffs, N.J.

Heyn, R. F., A. Rörsch, and R. A. Schilperoort. 1974. Prospects in genetic engineering of plants. Quart. Rev. Biophys. 7:35-73.

Hlásniková, A. 1977. Androgenesis *in vitro* evaluated from the aspects of genetics. Z. Pflanzenzüchtg. 78:44-56.

Holdgate, D. P. 1977. Propagation of ornamentals by tissue culture. *In* J. Reinert and Y. P. S. Bajaj (eds.), Applied and fundamental aspects of plant cell, tissue, and organ culture. Pp. 18-43. Springer-Verlag, Berlin.

Hollander, A., et al. (eds.). 1977. Genetic engineering for nitrogen fixation. Basic life sciences, vol. 9. Plenum Publishing Corp., New York. 538 pp.

Hussey, G. 1978. The application of tissue culture to the vegetative propagation of plants. Sci. Prog. 65:185-208.

Israel, H. W., H. J. Wilson, and F. C. Steward. 1978. The localization in cultured carrot cells of factors that induce their growth. Ann. Bot. 42 (in press).

Jones, C. W., I. A. Mastrangelo, H. H. Smith, and H. Z. Liu. 1976. Interkingdom fusion between human (Hela) cells and tobacco hybrid (GGLL) protoplasts. Science 193: 401-3.

Kao, K. N., F. Constabel, M. R. Michayluk, and O. L. Gamborg. 1974. Plant protoplast fusion and growth of intergeneric hybrids. Planta 120:215-27.

Kasha, K. J. (ed.). 1974. Haploids in higher plants: Advances and potential. University of Guelph, Guelph, Ontario, Canada.

Krikorian, A. D. 1975. Excerpts from the history of plant physiology and development. *In* P. J. Davies (ed.), Historical and current aspects of plant physiology: A symposium honoring F. C. Steward, Pp. 9-97. College of Agriculture and Life Sciences, Ithaca.

Krikorian, A. D., and D. L. Berquam. 1969. Plant cell and tissue cultures: The role of Haberlandt. Bot. Rev. 35:58-88.

Krikorian, A. D., and F. C. Steward. 1969. Biochemical differentiation: The biosynthetic potentialities of growing and quiescent tissue. *In* F. C. Steward (ed.), Plant physiology: A treatise. 5B:227-326. Academic Press, New York.

Krikorian, A. D., and F. C. Steward. 1978. The morphogenetic responses of cultured totipotent cells of carrot (*Daucus carota* var. *carota*) at zero gravity. Science 200:67–68.

Krikorian, A. D., and F. C. Steward. 1979. Is gravity a morphological determinant in plants at the cellular level? *In* W. R. Holmquist (ed.), Life sciences and space research, vol. 17. Pergamon Press, Oxford (in press).

Kung, S. D., J. C. Gray, S. G. Wildman, and P. S. Carlson. 1975. Polypeptide composition of Fraction I protein from parasexual hybrid plants in the genus *Nicotiana*. Science 187:353–55.

Lôrz, H., and I. Potrykus. 1976. Uptake of nuclei into higher plant protoplasts. *In* D. Dudits, E. L. Farkas, and P. Maliga (eds.), Cell genetics in higher plants. Pp. 239–44. Akadémiai Kiadó, Budapest.

Marcavillaca, M. D., and E. R. Montaldi. 1967. Diferentes formas de lojas producidas por yemas adventicias inducidas experimentalmente en lojas aisladas de *Nicotiana tabacum* L. *y Passiflora coerulea* L. Revista de Investigaciones Agropecuarias, INTA, Buenos Aires, Argentina, ser. 2, 4(1):1–17.

Mehra-Palta, A., R. H. Smeltzer, and R. L. Mott. 1977. Hormonal control of induced organogenesis from excised plant parts of loblolly pine (*Pinus taeda* L.) *In* TAPPI Conference Papers. Pp. 15–20. Madison, Wisconsin.

Melchers, G. 1977. The combination of somatic and conventional genetics in plant breeding. Naturwiss. 64:86–110.

Melchers, G., and G. Labib. 1975. Somatic hybridization of plants by fusion by protoplasts. I. Selection of light resistant hybrids of "haploid" light sensitive varieties of tobacco. Molec. Gen. Genet. 135:277–94.

Melchers, G., and M. D. Sacristán. 1977. Somatic hybridization of plants by fusion of protoplasts. II. The chromosome numbers of somatic hybrid plants of 4 different fusion experiments. *In* R. J. Gautheret (ed.), La culture des tissus et des cellules des végétaux. Travaux dédiés à la mémoire de Georges Morel. Pp. 169–77. Masson, Paris.

Mitra, J., M. O. Mapes, and F. C. Steward. 1960. Growth and organized development of cultured cells. IV. The behavior of the nucleus. Amer. J. Bot. 47:357–68.

Mitra, J., and F. C. Steward, 1961. Growth induction in cultures of *Haplopappus gracilis*. II. The behavior of the nucleus. Amer. J. Bot. 48:358–68.

Mott, R. L., R. H. Smeltzer, and A. Mehra-Palta. 1977. An anatomical and cytological perspective on pine organogenesis *in vitro*. *In* TAPPI Conference Papers. Pp. 9–14. Madison, Wisconsin.

Mott, R. L., and F. C. Steward, 1972. Solute accumulation in plant cells. V. An aspect of nutrition and development. Ann. Bot. 36:915–37.

Murashige, T. 1977. Clonal crops through tissue culture. *In* W. Barz, E. Reinhard, and M. H. Zenk (eds.), Plant tissue culture and its biotechnological application. Pp. 392–403. Springer-Verlag, Berlin.

Nickell, L. G., and D. J. Heinz. 1973. Potential of cell and tissue culture techniques as aids in economic plant improvement. *In* A. M. Srb (ed.) Genes, enzymes, and populations. Pp. 109–29. Plenum Publishing Corp., New York.

Nitsch, J. P. 1965. The *in vitro* culture of flowers and fruits. *In* P. Maheshewari and N. S. Rangaswamy (eds.), Plant tissue and organ culture—a symposium. Pp. 198–214. Intern. Soc. Plant Morphol., Delhi.

Pollard, J. K., E. M., Shantz, and F. C. Steward. 1961. Hexitols in coconut milk: Their role in the nurture of dividing cells. Plant Physiol. 36:492–501.

Potrykus, I., and H. Lörz. 1976. Organelle transfer into isolated protoplasts. *In* D. Dudits, G. L. Farkas, and P. Maliga (eds.), Cell genetics in higher plants. Pp. 183–90. Akadémiai Kiadó, Budapest.

Priestley, J. H. 1926. Problems of vegetative propagation. J. Roy. Hort. Soc. 51 (part 1):1–16.

Radojević, L., R. Vujčić, and M. Nešković. 1975. Embryogenesis in tissue culture of *Corylus avellana* L. Z. Pflanzenphysiol. 73:33–41.

Raggio, M., and N. Raggio. 1956. A new method for the cultivation of isolated roots. Plant Physiol. 9:466–69.

Raghavan, V. 1976. Experimental embryogenesis in vascular plants. Academic Press, London.

Raghavan, V. 1977. Applied aspects of embryo culture. *In* J. Reinert and Y. P. S. Bajaj (eds.), Applied and fundamental aspects of plant cell, tissue, and organ culture. Pp. 375–97. Springer-Verlag, Berlin.

Randolph, L. F., and F. R. Randolph. 1955. Embryo culture in *Iris* seed. Amer. Iris Soc. Bull. 139:2–12.

Rao, A. N. 1977. Tissue culture in the orchid industry. *In* J. Reinert and Y. P. S. Bajaj (eds.), Applied and fundamental aspects of plant cell, tissue, and organ culture. Pp. 44–69. Springer-Verlag, Berlin.

Reinert, J., and Y. P. S. Bajaj (eds.). 1977. Applied and fundamental aspects of plant cell, tissue, and organ culture. Springer-Verlag, Berlin.

Scheider, O. 1976. The spectrum of auxotrophic mutants from the liverwort *Sphaerocarpus donnellii* Aust. Molec. Gen. Genet. 144:63–66.

Schulte, U., and M. H. Zenk. 1977. A replica plating method for plant cells. Physiol. Plantarum 39:139–42.

Shabde, M., and T. Murashige, 1977. Hormonal requirements of excised *Dianthus caryophyllus* L. shoot apical meristems in vitro. Amer. J. Bot. 64:443–48.

Shantz, E. M., F. C. Steward, M. S. Smith, and R. L. Wain. 1955. Investigations on the growth and metabolism of plant cells. VI. Growth of potato tissue in culture: The synergistic action of coconut milk and some synthetic growth-regulting compounds. Ann. Bot. 19:49–58.

Skolmen, R., and M. O. Mapes. 1976. *Acacia Koa* Gray plantlets from somatic callus tissue. J. Hered. 67:114–15.

Smeltzer, R. H., A. Mehra-Palta, and R. L. Mott. 1977. Influence of parental tree genotype on the potential for *in vitro* clonal propagation from loblolly pine embryos. *In* TAPPI Conference Papers. Pp. 5–8. Madison, Wisconsin.

Smith, H. H. 1974a. Model genetic systems for studying mutation, differentiation, and somatic cell hybridization in plants. *In* Polyploidy and induced mutations in plant breeding. Pp. 355–65. International Atomic Energy Agency, Vienna.

Smith, H. H. 1974b. Model systems for somatic cell plant genetics. Bioscience 24:269–76.

Smith, H. H., K. N. Kao, and N. C. Combatti. 1976. Interspecific hybridization by protoplast fusion in *Nicotiana*. J. Hered. 67:123–28.

Smith, R. H. and T. Murashige. 1970. *In vitro* development of the isolated shoot apical meristem of angiosperms. Amer. J. Bot. 57:562–68.

Snowcroft, W. R., and A. H. Gibson. 1975. Nitrogen fixation by *Rhizobium* associated with tobacco and cowpea cell cultures. Nature 253:351–52.

Sommer, H. E., C. L. Brown, and P. P. Kormanik. 1975. Differentiation of plantlets in longleaf pine (*Pinus palustris* Mill.) tissue cultured *in vitro*. Bot. Gaz. 136:196–200.

Staba, E. J. 1977. Tissue culture and pharmacy. *In* J. Reinert and Y. P. S. Bajaj (eds.), Applied and fundamental aspects of plant cell, tissue, and organ culture. Pp. 694–702. Springer-Verlag, Berlin.

Steeves, T. A., and I. M. Sussex. 1972. Patterns in plant development. Prentice-Hall, Englewood Cliffs, N.J.

Steward, F. C. 1963. Effects of environment on metabolic patterns. *In* L. T. Evans (ed.), Environmental control of plant growth. Pp. 195–214. Academic Press, New York.

Steward, F. C. 1970a. Totipotency, variation, and clonal development of cultured cells. Endeavour 29:117–24.

Steward, F. C. 1970b. Cloning cells and controlling the composition of crops. Progress-Unilever Quart. 54:44–51.

Steward, F. C. 1971. Interacting effects of nutrients, growth factors and environment, and metabolism and growth. *In* R. M. Samish (ed.), Recent advances in plant nutrition. 1:595–621. Gordon & Breach Science Publishers, New York.

Steward, F. C. 1976. Multiple interactions between factors that control cells and development. *In* N. Sunderland (ed.), Perspectives in experimental biology. Vol. 2: Botany. Pp. 9–23. Pergamon Press, Oxford.

Steward, F. C., P. V. Ammirato, and M. O. Mapes. 1970. Growth and development of totipotent cells: Some problems, procedures, and perspectives. Ann. Bot. 34:761–87.

Steward, F. C., and E. F. Bleichert. 1972. Partial and complete growth-promoting systems for cultured carrot explants: Synergistic and inhibitory interactions. *In* D. J. Carr (ed.), Plant growth substances. Pp. 668–78. Springer-Verlag, Berlin.

Steward, F. C., and S. M. Caplin. 1951. A tissue culture from potato tuber: The synergistic action of 2,4-D and of coconut milk. Science 113:518–20.

Steward, F. C., K. J. Howe, F. A. Crane, and R. Rabson. 1962. Growth, nutrition, and metabolism of *Mentha piperita* L. Parts I–VII. Cornell Univ. Agric. Expt. Station Memoir 379:1–144.

Steward, F. C., and H. W. Israel. 1972. Multiple interactions between media, growth factors, and the environment of carrot cultures: Effects on growth and morphogenesis. *In* D. J. Carr (ed.), Plant growth substances. Pp. 679–85. Springer-Verlag, Berlin.

Steward, F. C., H. W. Israel, R. L. Mott, H. J. Wilson, and A. D. Krikorian. 1975. Observations on growth and morphogenesis in cultured cells of carrot (*Daucus carota* L.). Phil. Trans. Roy. Soc. 273B:33–53.

Steward, F. C., A. E. Kent, and M. O. Mapes. 1967. Growth and organization in cultured cells: Sequential and synergistic effects of growth-regulating substances. Ann. N.Y. Acad. Sci. 144:326–34.

Steward, F. C., and A. D. Krikorian. 1971. Plants, chemicals, and growth. Academic Press, New York.

Steward, F. C., and A. D. Krikorian. 1975. The culturing of higher plant cells: Its status, problems, and potentialities. *In* H. Y. Mohan Ram, J. J. Shah, and C. K. Shah (eds.), Form, structure, and function in higher plants. Pp. 144–70. Sarita Prakashan, Meerut City.

Steward, F. C., and M. O. Mapes. 1963. The totipotency of cultured carrot cells: Evidence and interpretations from successive cycles of growth from phloem cells. J. Indian Bot. Soc. 42A:237–47.

Steward, F. C., M. O. Mapes, and P. V. Ammirato. 1969. Growth and morphogenesis in tissue and free cell cultures. *In* F. C. Steward (ed.), Plant physiology: A treatise. 5B:329–76. Academic Press, New York.

Steward, F. C., and R. L. Mott. 1970. Cells, solutes, and growth: Salt accumulation in plants re-examined. Intern. Rev. Cytol. 28:275–370.

Steward, F. C., and K. V. N. Rao. 1970. Investigations on the growth and metabolism of cultured explants of *Daucus carota*. III. The range of responses induced in carrot explants by exogenous growth factors and by trace elements. Planta 91:129–45.

Steward, F. C., and K. V. N. Rao. 1971. Investigations on the growth and metabolism of cultured explants of *Daucus carota*. IV. Effects of iron, molybdenum, and the components of growth promoting systems and their interactions. Planta 99:240–64.

Street, H. E. 1969. Growth in organized and unorganized systems. Knowledge gained by culture of organs and tissue explants. *In* F. C. Steward (ed.), Plant physiology: A treatise. 5B:3–244. Academic Press, New York.

Street, H. E. (ed.). 1974. Tissue culture and plant science 1974. Academic Press, London.

Street, H. E. (ed.). 1977. Plant tissue and cell culture. 2d ed. Blackwell Scientific Publications, Oxford.

Sung, Z. R. 1976. Mutagenesis of cultured plant cells. Genetics 84:51–57.

Suzuki, M., and I. Takebe. 1976. Uptake of single-stranded bacteriophage DNA by isolated tobacco protoplasts. Z. Pflanzenphysiol. 78:421–33.

Swingle, C. F. 1940. Regeneration and vegetative propagation. Bot. Rev. 6:301–54.

Swingle, C. F. 1952. Regeneration and vegetative propagation. II. Bot. Rev. 18:1–13.

Tempe, J., A. Petit, M. Holsters, M. van Montagu, and J. Schell. 1977. Thermosensitive step associated with transfer of the Ti plasmid during conjugation: Possible relation to transformation in crown gall. Proc. Nat. Acad. Sci. USA 74:2848–49.

Teuscher, E. 1973. Probleme der Produktion sekundärer Pflanzenstoffe mit Hilfe von Zellkulturen. Pharmazie 28:6–17.

Thoday, D. 1933. Some physiological aspects of differentiation. New Phytol. 32:274–87.

Thomas, E., and M. R. Davey. 1975. From single cells to plants. Springer-Verlag, New York.

Thomas, M. J., E. Duhoux, and J. Vazart. 1977. In vitro organ initiation in tissue cultures of *Biota orientalis* and other species of the Cupressaceae. Plant Sci. Lett. 8:395–400.

Tomes, D. T., and G. B. Collins. 1976. Factors affecting haploid plant production from *in vitro* anther cultures of *Nicotiana* species. Crop Sci. 16:837–40.

Torrey, J. G., and C. R. Landgren. 1977. Mitosis and cell division in cultures of higher plant protoplasts and cells. *In* R. J. Gautheret (ed.), La culture des tissus et des cellules des végétaux. Travaux deśdiés à la mémoire de Georges Morel. Pp. 148–68. Masson, Paris.

Van Staden, J. 1976. The identification of zeatin glucoside from coconut milk. Physiol. Plantarum 36:123–26.

Vasil, I. K. 1976. The progress, problems, and prospects of plant protoplast research. Adv. Agron. 28:119–60.

Wallner, S. J., and D. J. Nevins. 1974. Changes in cell walls associated with cell separation in suspension cultures of Paul's Scarlet Rose. J. Exp. Bot. 25:1020–29.

White, P. R. 1953. A comparison of certain procedures for the maintenance of plant tissue cultures. Amer. J. Bot. 40:517–24.

Widholm, J. M. 1972a. Tryptophan biosynthesis in *Nicotiana tabacum* and *Daucus carota* cell cultures: Site of action of inhibitory tryptophan analogs. Biochim. Biophys. Acta 261:44–51.

Widholm, J. M. 1972b. Cultured *Nicotiana tabacum* cells with an altered anthranilate synthetase which is less sensitive to feedback inhibition. Biochim. Biophys. Acta 261:52–58.

Widholm, J. M. 1972c. Anthranilate synthetase from 5-methyltryptophan-susceptible and resistant cultured *Daucus carota* cells. Biochim. Biophys. Acta 279:48–57.

Widholm, J. M. 1974. Cultured carrot cell mutants: 5-methyltryptophan-resistance trait carried from cell to plant and back. Plant Sci. Lett. 3:323–30.

Winton, L., and O. Huhtinen. 1976. Tissue culture of trees. *In* J. P. Miksche (ed.), Modern methods in forest genetics. Pp. 243–64. Springer-Verlag, Berlin.

Yeoman, M. M. (ed.). 1976). Cell division in higher plants. Academic Press, London.

Part 3

Genetics

HAROLD H. SMITH AND IRIS A. MASTRANGELO-HOUGH

Genetic Variability Available
through Cell Fusion

16

INTRODUCTION

The main use of somatic cell or protoplast fusion is, almost by definition, the enhancement of genetic variability. It involves the bringing together of two, usually genetically diverse, genotypes and thus serves in genetic and plant breeding research essentially the same function as cross-pollination or sexual hybridization. But the methods and sometimes the consequences are different; and the as yet unrealized potentialities offer new avenues of research for the future.

The results to date have been modest. However, much has been accomplished by improving technique, and in this field the method is still the main message. The development of protoplast fusion as a tool in plant genetics has not been through mastery of a single technique, but rather it requires completion of a whole series of procedures, each of which must be successfully executed in order to achieve the desired end.

The main steps are: (1) isolation of protoplasts, (2) fusion followed by selection of heterokaryocytes, (3) culture *in vitro* of hybrid cells, and (4) regeneration to whole plants. These have been accomplished with a few model systems that will be described briefly. The battery of techniques employed is basic, necessary, and part of the routine methodology of fusing protoplasts to enhance genetic variability. Recent results suggest some new directions or emphases in fusion research, and these will be discussed. It is

Harold H. Smith and Iris A. Mastrangelo-Hough, Department of Biology, Brookhaven National Laboratory, Upton, New York.

now clear that plant and animal cells can be fused together, which may open up unusual opportunities for creating novel genetic variability and for making more direct use of the extensive experience in animal somatic cell genetics.

PLANT HYBRIDIZATION THROUGH PROTOPLAST FUSION

Nicotiana glauca + *N. langsdorffii*

The production of a mature interspecific hybrid by fusion of protoplasts, obtained by enzymic digestion of the cell wall, was first reported by Carlson et al. in 1972. The materials used were leaf mesophyll cells of *Nicotiana glauca* (2n = 24) and *N. langsdorffii* (2n = 18). The less-stringent cultural requirements for growth *in vitro* of the hybrid cells, compared to the parental ones, served as a selective screen to recover preferentially the heterospecific fusion products. Under the conditions of this experiment, the parasexually produced hybrids were found to be identical in every way, including chromosome number (2n = 24 + 18 = 42), to the sexually produced amphiploid.

In view of the uniqueness of this result and the improvements in technique that subsequently became available, experiments were carried out aimed at confirming and extending the original finding. Polyethylene glycol (PEG) was used to facilitate adherence of protoplasts (Kao et al., 1974). The results (Smith et al., 1976) confirmed, in general, the earlier work, but differed in that the new parasexual hybrids had high and variable chromosome numbers that ranged from 2n = 56 to 64. Among 23 mature parasexually produced hybrid plants, the average somatic chromosome number was 59.3. At meiosis most chromosomes formed bivalents, and as a consequence pollen fertility and seed set were usually satisfactory. The range in 2n chromosome numbers observed, which was exclusively from the mid-50s to the mid-60s, suggests that the hybrids may have resulted from triple protoplast fusions involving two of *N. langsdorffii* and one of *N. glauca* (60 chromosomes) or two of *N. glauca* and one of *N. langsdorffii* (66 chromosomes). Furthermore, it suggests that this was followed by a loss of chromosomes during the prolonged proliferation of the hybrid cultures (a common phenomenon in plant tissue cultures), and that these hyperaneuploids were more successful than other genotypes in yielding viable and differentiated cultures that produced the mature hybrid plants.

F_2 progeny were obtained (Smith, 1976a) by self-pollinating 13 of the parasexual hybrids. Each F_2 family retained many of the morphological features of the particular parasexual hybrid from which it was derived, and also the approximate same chromosome number (range, 2n = 54 to 62; average 2n = 58.9). We now have mature plants growing of sixteen different F_3 and four different F_4 families. Each of these cultures is distinguished by a unique constellation of features, representing a novel recombination of characteristics of the two parental species. For example, the leaf patterns displayed in figure 1 are from F_3 families each with consistent differences in size or shape of leaves.

Petunia hybrida + *P. parodii*

A second success in interspecific plant hybridization by protoplast fusion was reported for *Petunia hybrida* + *P. parodii* (Power et al., 1976). The hybrid colonies were selected by a dual procedure. One half was based on a difference in growth of parental protoplasts so that those of *P. parodii*

2

Fig. 1. Leaves of (top line, left to right): *N. glauca* (GG), amphiploid (GGLL), and *N. langsdorffii* (LL). Below are rows of leaves from plants in four different parasexual F_3 families: first row, consistently large leaves; second row, consistently misshapen leaves; third row: varied leaf shapes; fourth row, consistently small round leaves.

never grew beyond a small colony on Murashige-Skoog medium (Murashige and Skoog, 1962), while those of *P. hybrida* produced a callus. The other half of the selection procedure was based on a difference in sensitivity to actinomycin D, *P. hybrida* protoplasts being more sensitive than those of *P. parodii*. Complementation occurred in somatic hybrids so that colonies of combined genotype were able to grow beyond the colony stage in Murashige-Skoog medium (like *P. hybrida*) and were insensitive to actinomycin D (like *P. parodii*). The mature somatic hybrids, in common with those produced by cross pollination, could be distinguished from the parental types in flower color, morphology, and peroxidase isozyme pattern. Chromosome numbers in the parasexual hybrids were from 2n = 28 (expected from the addition of the two parental genomes) to a minimum of 2n = 24.

More recently, Izhar and Power (1977) have demonstrated differences among several species, inbred lines, and hybrids of *Petunia* with respect to their potential for growth on media containing specific phytohormone combinations. These differences offer possible means for selecting somatic hybrids, particularly since the sexually produced hybrids grew better than either parent on a range of hormone formulations. For example, protoplasts of neither *P. hybrida* cv Comanche nor *P. parodii* were able to develop into full calli on a medium containing 5.0 mg/l of 2,4-D, yet the sexual hybrid between them was able to do so. The circumstances are reminiscent of the *Nicotiana* system described above, where the hybrid cell colonies are selected out of a mixture containing parental types on the basis of a genetically governed differential response to the phytohormone content of the medium.

Nicotiana tabacum + N. tabacum

Another series of mature hybrid plants has been produced in *Nicotiana tabacum* by protoplast fusion, utilizing other methods of fusion and selection. Melchers and Labib (1974) isolated mesophyll protoplasts from induced haploids (n = 24) of two chlorophyll-deficient, light-sensitive mutant strains of *N. Tabacum* and fused them according to the method of Keller and Melchers (1973) that makes use of high pH and a high calcium ion concentration. The two mutants complement each other and give a green hybrid callus that can be picked out as distinct from the yellow calli of either parental type. The selected calli were regenerated into plants that were green in color and similar to the sexually produced hybrid. Cytological analyses of 108 somatic hybrid plants so produced (Melchers and Sacristán, 1977) revealed that, although many (54%) had 48 chromosomes, a few (12%) had variations around this number (as 46, 47, 49 and 50); others

(23%) were "triploids" with 72 chromosomes and variations around this number (as 67, 69, 70, 71, and 73), and some (12%) were "tetraploids" with 96 chromosomes and variations around this number.

In another series of experiments on parasexual hybrid production in *N. tabacum*, Gleba et al. (1975) have reported success by using a semidominant chlorophyll-deficient mutant to mark the hybrid nuclear genome and, simultaneously, another (which gives variegated leaves in hybrids) to mark the plastids.

The extreme importance of an effective selection system in order to recover hybrids is emphasized by the low rate of recovery reported even for these genome combinations that are known to be stable. Power et al. (1976) found that the frequency of *Petunia hybrida* + *P. parodii* hybrid plants was 1 from $1-8 \times 10^5$ protoplasts. Melchers and Labib (1974) indicated that 12 intraspecific hybrids originated from 2.2×10^6 calli tested.

CHROMOSOME INSTABILITY

Regeneration of true-breeding somatically stable hybrid plants, a major aim of interspecific cell fusions, can be hindered by perturbations in chromosome number in the hybrid cells. This may be caused by conditions *in vitro*, hybridity itself, or a combination of both.

Culture-Induced Chromosome Changes

The chromosome constitution of plants regenerated from somatic fusion experiments is often variable, ranging from diploid and hypodiploid to triploid, tetraploid, and a variety of aneuploids. These aberrant chromosome numbers may result from multiple fusions, in addition to the well-documented phenomena of tissue culture-induced changes. The numerical alterations may be followed by selective growth and accumulation of those cell types that are most well adapted to the culture conditions (Blakely and Steward, 1964; Murashige and Nakano, 1967; Singh et al., 1972). These changes are of considerable importance since aneuploid cells that successfully grow in culture have been reported to lose morphogenetic potential (Murashige and Nakano, 1967).

Various combinations of phytohormones and other constituents of the culture medium may affect the degree of polyploidy (Bennici et al., 1971; Bayliss, 1977) and of aneuploidy (Fox, 1963). Extensive aneuploidy may also develop from other culture conditions, unidentified except that they were unrelated to the phytohormones present in the medium (Ogura, 1976).

Both stable euploid and aneuploid plants have been recovered from some cell fusion experiments. In other plants regenerated from tissue culture, however, the normal mechanisms that retain strict euploidy in plant meristems, thus ensuring a uniform distribution of chromosomes to offspring (D'Amato, 1977), can become disturbed. For example, Ogura (1976) found that first- and second-generation tobacco plants, produced from self-pollinating tissue culture regenerates, were chimeras with highly variable chromosome numbers and phenotypes. At the same time, from one second-generation selfing, stable diploid offspring were obtained. These results can be compared with those establishing relatively stable aneuploid lines among second- and third-generation offspring of *Nicotiana glauca + N. langsdorffii* (Smith et al., 1976).

Chromosome Elimination

In somatic hybrid cells made between species that are sexually incompatible, tissue culture-engendered chromosome changes may be compounded by the process of preferential chromosome elimination. Examples of the latter, whereby the chromosomes of one parent are eliminated, are well known in the literature on sexually produced interspecific hybrids. In crosses between *Hordeum vulgare* and *H. bulbosum*, there is a rapid and complete loss of *bulbosum* chromosomes in the embryo (Subrahmanyam and Kasha, 1973; Bennett et al., 1976). Partial elimination of nonspecific or random chromosomes of only one parent has been observed in a number of *Nicotiana* hybrids including *N. tabacum* × *N. plumbaginifolia* and *N. suaveolens* × *N. glutinosa* (Gupta and Gupta, 1973). Mitotic abnormalities, cell cycle asynchrony, and gene interactions as, e.g., restriction enzyme systems (Davies, 1974) have been suggested as possible causes of the elimination process.

There are two examples of selective elimination of chromosomes in parasexual interspecific hybrid cells. In one (Power et al., 1976), there appears to have been a total loss of *Petunia hybrida* chromosomes following protoplast fusion with *Parthenocissus tricuspidata*. In the other, Kao (1977) reported that six months after fusion, five hybrid cell lines of *Nicotiana glauca +* soybean retained the original soybean chromosome set but had lost a number of *N. glauca* chromosomes, which can be seen to fragment, stick, and form bridges at anaphase.

Although the nuclei of a number of plant heterokaryocytes have been observed to fuse (Constabel et al., 1975; Dudits et al., 1976a) and to enter mitosis together (Kao, 1977; Kao et al., 1974), stabilized hybrid cells have usually not been detected after a period of continued culture. This may not be just a consequence of combining distantly related genomes. For example, from the literature on animal somatic cell fusions, it is known that human HeLa + chicken erythrocyte heterokaryons synthesize DNA synchronously, indicating equal response of the disparate genomes to the same molecular signals (Johnson and Harris, 1969a). In contrast, the more closely related combination of human HeLa + mouse Ehrlich ascites heterokaryons do not have synchronized DNA syntheses (Johnson and Harris, 1969b). Instead, the mouse ascites nucleus dominates, so that most of the Ehrlich nuclei synthesize DNA and most of the HeLa nuclei do not.

It may be possible to control the loss of chromosomes in hybrid cells by selection for genetic markers carried by each species, as was done to recover parasexual hybrids in the experiment described earlier, or by cloning. Kao (1977) found that physically segregating heterokaryons of *N. glauca* + soybean by cloning promoted the continued division of these hybrid cells; otherwise, the *N. glauca* chromosomes were gradually lost. It is likely that other hybrid cell lines can also be maintained by removing the hybrid cell from competition with parental cells.

NEW AREAS OF RESEARCH EMPHASIS

Extending the Range of Hybridization through Protoplast Fusion

So far all of the hybrids that have been produced by fusing protoplasts are obtainable by cross-pollination. There would be little genetic interest in parasexual hybridizaton if, by this process, no more could be achieved than by sexual means. Reviewers of this field envisage the use of protoplast fusion as a means of combining, by addition, the genotypes of two evolutionary lines that have diverged to the extent that they cannot be hybridized sexually. We are attempting to extend the range of hybridization by protoplast fusion through further exploitation of the tobacco genetic tumor selection method described above.

The scheme requires predicting that a hybrid, which has not previously been possible to obtain by cross-pollination, would be expected to be tumorous, and consequently have less-stringent tissue culture requirements, if it could be made by protoplast fusion. The basis for prediction is the observation of Näf (1958) that species of *Nicotiana* can be divided into two groups (arbitrarily called plus and minus) such that intragroup hybrids develop normally, but intergroup hybrids form tumors and their tissues can be cultured on media that do not contain phytohormones.

Based on a survey of the literature on *Nicotiana* species hybrids (East, 1935; Kostoff, 1943; Kehr and Smith, 1952; and Goodspeed, 1954), we have selected for experimentation four different interspecific combinations that would be expected to be tumorous (i.e., combine plus and minus groups) but have failed to yield hybrids by cross-pollination. These combinations are:

> *N. debneyi* (n = 24) + *N. langsdorffii* (n = 9)
> *N. bigelovii* (n = 24) + *N. langsdorffii* (n = 9)
> *N. bigelovii* (n = 24) + *N. longiflora* (n = 10)
> *N. rustica* (n = 24) + *N. longiflora* (n = 10)

To date, through the efforts of Dr. Francesco Salamini and Mr. Nicholas Combatti, we have conducted a series of fusion experiments utilizing mesophyll cells of *N. debneyi* and *N. langsdorffii*. These were carried through essentially the same steps as was done with *N. glauca* + *N. langsdorffii*, namely, enzyme digestion of the cell wall, fusion of protoplasts with the aid of PEG, growth from protoplast to cell colony on fully enriched M3 medium, and selection of those calli that grew on a hormone-less Murashige-Skoog medium (Smith et al., 1976). In order to maximize the chances for successful growth of callus when placed on a medium lacking phytohormones, the suspension cultures were maintained under optimal conditions on M3 for a month or longer before transfer. In all, 8,915 colonies were eventually transferred onto a solid hormoneless medium, where 1,911 (21%) continued to grow. Of these, 441 (23%) differentiated, forming at least rudimentary organs. We have now succeeded in regenerating 75 (17%) mature plants from these calli. It is disappointing to report that the regenerated plants are all *N. debneyi*. From the cell size and the morphology of the leaves and flowers (Smith, 1943), it appears that 46 are ± 2n, 25 are ± 4n, and 4 are higher ploidy. The experiment is still in progress, and it is possible that some of the still undifferentiated and partially differentiated calli may prove to be of hybrid genotype.

The main lesson learned so far from this experiment is that calli of *N. debneyi* (in contrast to those of *N. glauca* and *N. langsdorffii*) become habituated and will grow on a hormoneless medium if the transfer to that medium is preceded by a month or more of culture on M3. Thus, *N. debneyi* was able to escape the selective step that was so effective with *N. glauca* and *N. langsdorffii*. It remains to be seen if *N. debneyi* + *N. langsdorffii* hybrids can be selected out of a mixed culture containing the two parental-type protoplasts.

It is now clear that for further exploitation of the genetic tumor selection system, it will be necessary to identify the specific cell culture requirements of each species separately, as well as the precise conditions under which the parasexual hybrids can be picked out. Both this system and that of Izhar and Power (1977) point up the differentially better growth, or less-exacting *in vitro* needs, of certain hybrids compared with their parents. These observations, together with the widespread occurrence of *in vivo* hybrid vigor among interspecific and intervarietal crosses, suggests the possibility that a generally useful principle of *in vitro* selection might be based on the more rapid growth of hybrid tissue in culture (Smith, 1976b). Is there any basis for this expectation?

To answer this question the library at Brookhaven National Laboratory undertook a literature search, utilizing Lockheed's DIALOG, of BIOSIS Previews (Biological Abstracts/Bioresearch Index) for the period of January 1972 to July 1976. Titles of papers were sought that combined the three topics: plants, hybrid vigor or heterosis, and cell or tissue culture. No references whatsoever were found. We concluded that little, if any, research in this specific area has been reported. It would appear to be a fertile field for investigation. A general correlation between vigor *in vivo* and *in vitro* is not expected, except for the specific case of heterosis. This phenomenon has been attributed (East, 1936) to many heterozygous combinations of nondefective alleles, each with slightly different physiological action, that act in a complementary manner. Thus, a more efficient physiological condition is produced that finds phenotypic expression as hybrid vigor. The increased growth of hybrid over parents, according to this interpretation, might equally well be expressed *in vitro*.

As noted earlier (Carlson et al., 1972): "We are investigating one further method for preferentially recovering parasexually produced hybrids. Protoplasts containing potentially complementing recessive nuclear albino mutations from different species are isolated and fused. Only calli that have regenerated from fused cells should appear as green colonies on the petri plate." In this program at Brookhaven National Laboratory, we are concentrating on the family *Solanaceae* and are seeking complementation with all possible combinations of available chlorophyll-deficient mutants of *Nicotiana tabacum, Datura stramonium, Lycopersicon esculentum, Solanum chacoense,* and *Capsicum frutescens.*

Chloroplast Distribution in Parasexual Hybrids

A priori an expected difference between plant hybrids produced by protoplast fusion vs. cross pollination is that, in the latter, organelles would be of maternal derivation only, whereas parasexual hybrids would be expected to have a mixture of both paternal and maternal organelles. Surely, this would be the condition in an original hybrid cell resulting from complete fusion of two different protoplasts. Only recently has it been possible to trace two genetically marked parental chloroplasts in mature parasexual hybrids (Chen et al., 1977). The materials used were the *N.*

glauca (G) + *N. langsdorffii* (L) fusion hybrids described above. Crystalline Fraction I protein (ribulose, 5-diphosphate carboxylase/oxygenase) was obtained from the leaves and resolved by electrofocusing (Kung et al., 1974) into the large subunit polypeptides, coded by chloroplast DNA, and the small subunit polypeptides, coded by nuclear DNA.

Of sixteen different mature parasexual hybrids analyzed, only one retained a mixture of G and L large subunits, thus forming a genetic variant unique to hybridization by protoplast fusion; and the rest were of one type, either G or L. The reason for the rapid sorting out into a monotypic population of chloroplasts is unknown. It does not appear to be due to random fixation alone. Active selection against cells that contain a mixture of chloroplasts may combine with random elimination to cause rapid fixation of one type. Little is understood about organelle interactions during plant development. Hybridization by protoplast fusion is a new tool that can be used to explore this area of basic biology of plants. Since chloroplasts of either parent can, with *N. glauca* and *N. langsdorffii*, come to dominate in a hybrid cytoplasm, chloroplast uptake into protoplasts may, in general, be a promising method of generating a specific kind of hybridity.

Chromosomal Introgression Following Fusion

Interspecific hybridization by cross-pollination has long been used as a first step in generating genetic variability. This variability is utilized mainly by repeatedly backcrossing the hybrid (or amphiploid) to one parental species with selective retention of a desirable gene or block of genes from the donor species. The donor genes may be carried as homologous chromosomes, rearranged chromosomes, or chromosome fragments translocated to the host species. Fertile hybrids produced by protoplast fusion could also be used in the same way to effect gene introgression. In addition, however, fusion and culture *in vitro* offer further opportunities for generating and utilizing hybridity. These are illustrated in the results reported by Kao (1977), who fused protoplasts of *Glycine max* (soybean) and *Nicotiana glauca*. Up to 39 percent of the protoplasts in the PEG-treated population were heterokaryocytes. When these were isolated and individually cultured, they divided indefinitely and each produced many millions of cells within two to three months. This particular combination is one of some seventeen widely different protoplast fusion products, obtained by the Saskatoon group (Constabel, 1976), that are capable of at least a few cell divisions. The prolonged series of mitoses of the soybean–*Nicotiana glauca* cultures shows that some long-term viable products can be produced by protoplast fusion between phylogenetically more distant taxa than would be possible in conventional hybridization. Soybean is classified in the family Leguminosae, order Rosales and subclass Archichlamydeae (distinct petals), and *Nicotiana glauca* is in the family Solanaceae, order Tubiflorae and subclass Metachlamydeae (sympetalous). Such a wide taxonomic gap could not be bridged by cross-pollination.

A second potentially advantageous feature of hybridization by protoplast fusion is illustrated by the chromosomal behavior and its consequences in the soybean–*Nicotiana glauca* combination. The two genomes did not divide synchronously during the early cell generations, and *N. glauca* chromosomes underwent mitotic disturbances that gradually reduced their size, and in time, their numbers. However, some of the reconstructed *N. glauca* chromosomes were still retained in later cell generations and became synchronized with the full soybean complement. In this way a few gene blocks of *N. glauca* were introgressed into soybean cells during a few months of culture. This amount of genetic elimination would take several sexual generations to accomplish by conventional means. The utility of the method becomes more evident if the *N. glauca* chromosomes can be marked by specific genetic characteristics upon which selection *in vitro* could be applied. In more general terms, this phenomenon might be developed into a technique for introgressing desirable donor genes or chromosomes from distant taxa into a host plant.

PLANT-ANIMAL FUSION

Thus far we have considered the sources of genetic variability available to plants to be confined to the plant kingdom. This may not permanently be so. The initial step toward producing plant + animal hybrid cells has been achieved by PEG-induced fusion to form interkingdom heterokaryons. The first such fusion reported was between yeast protoplasts and hen erythrocytes (Ahkong et al., 1975b). This was rapidly followed by fusions of human-derived HeLa cells with tobacco protoplasts (Jones et al., 1976) and with carrot protoplasts (Dudits et al., 1976b). Ultrastructure studies of fusing tobacco protoplasts with avian erythrocytes indicate that the membranes fuse in a manner similar to plant + plant or animal + animal cell fusions and this is then followed by cytoplasmic mixing (Burgess and Fleming, 1974; Willis et al., 1977).

Ahkong and collaborators suggest that "fusogenic" chemicals including PEG cause protein/glycoprotein particles to aggregate within the membrane, leaving protein-free regions (Ahkong et al., 1975a). When these regions are apposed, it is suggested that the structural lipids interact and intermix, and the membranes fuse. It seems probable that those genetic structures whose successful transfer among animal cells is dependent on membrane agglutination, fusion, and vesicle formation may be potential sources of new genes for introducing into plant protoplasts. The transfer and incorporation of genes by chromosome uptake has been achieved with mammalian systems (Willecke and Ruddle, 1975; McBride and Athwal, 1977). Chromosomes may be ingested by mammalian cells in a manner similar to uptake of chloroplasts by plant protoplasts (Marx, 1977; Bonnett, 1976).

A first step has been made in attempting chromosome-mediated gene transfer from animal cells to plant cells. With Paul V. C. Hough, Brookhaven National Laboratory, *N. glauca* × *N. langsdorffi* protoplasts have been co-cultured with isolated Chinese hamster chromosomes in a ratio of 1:100 for up to six hours in plant protoplast culture medium. The incubation mixture was stained with carbol fuchsin and examined. During this time the mammalian chromosomes maintain their typical metaphase morphology and some appear to adhere to the plasmalemma of the protoplast (fig. 2). Metaphase chromosomes were not seen in the cytoplasm or vacuoles of the protoplasts. However, a small number of protoplast nuclei were seen to contain darkly stained bodies that resembled the metaphase chromosomes (fig. 3).

The question might well be asked, "Is there any reasonable hope that viable higher plant cells can be established that contain introduced genes, chromosomes, or genomes from animal or other widely distant sources?" The following observations may lend some support to an affirmative answer. (1) Chloroplasts have been introduced into host protoplasts of widely different plant origin, as spinach chloroplasts into fungal (*Neurospora*) protoplasts (Vasil and Giles, 1975) and algal (*Vaucheria*) chloroplasts into carrot protoplasts (Bonnett, 1976). The incorporated chloroplasts remained morphologically intact and may have retained some physiological activity. (2) The gastropod mollusc *Elysia atroviridis* ingests chloroplasts of the green siphonaceous seaweed *Codium fragile*. These plant organelles are retained within digestive cells of the marine slug where they undergo sustained active photosynthesis. It seems likely that the symbiotic chloroplasts supply glucose to the animal cells (Trench, 1975).

(3) In the interkingdom fusion of HeLa cells and tobacco protoplasts the HeLa nucleus appears to enter and remain in the tobacco cell. To what extent can a human nucleus survive and divide within a plant cell or at least in a culture medium particularly designed for growth of plant cells? HeLa suspension cultures were incubated in 100 percent HeLa medium and in mixtures containing 25, 30, and 35 percent M3 plant culture medium (fig. 4). With increasing proportions of M3 the HeLa cell counts were reduced, but growth continued for at least 300 hours (and possibly could continue

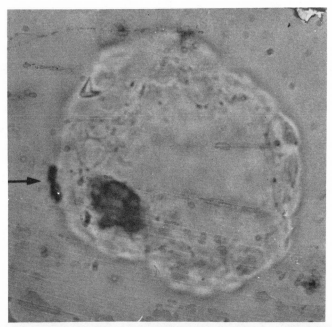

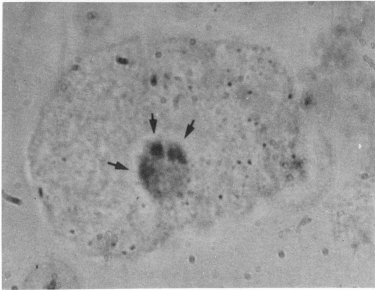

Figs. 2 and 3.　Carbol fuschin–stained protoplasts of amphiploid *N. glauca* × *N. langsdorffii* with isolated metaphase chromosomes from Chinese hamster cells. 1200×. 2 (top). Chromosome appears to adhere to protoplast plasmalemma. 3 (bottom). Arrows point to structures in protoplast interphase nucleus that resemble Chinese hamster chromosomes in morphology.

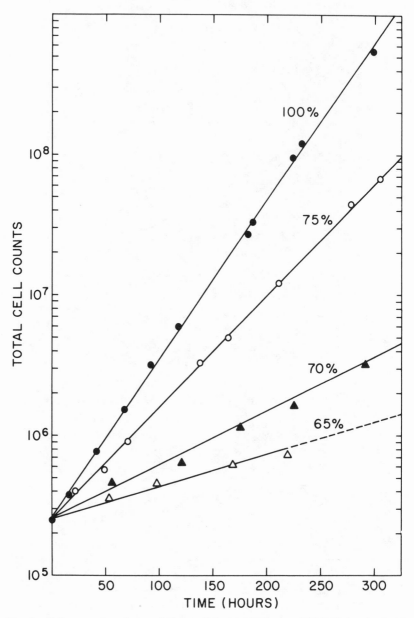

Fig. 4. Cell counts in cultures of HeLa cells suspended in standard Eagle's animal cell culture medium (100%) and mixed with 25, 30, and 35 percent standard (M3) plant cell culture medium.

indefinitely) in as much as 30 percent plant medium. (4) The pattern of DNA sequence organization, i.e., the interspersion of short repetitive sequences, long repetitive sequences, and single-copy sequences, in the genome of *Nicotiana tabacum* is very similar to that found in animal species, and in some protists as well. Zimmerman and Goldberg (1977) concluded that the widespread phylogenetic occurrence of this pattern strongly implies that it "performs an essential, but yet unknown, biological function in eukaryotic organisms." (5) Plant and animal biosynthetic molecules routinely cofunction *in vitro*. Wheat germ translation systems are commonly used to synthesize enzymes from mammalian messenger RNA (Chang and Littlefield, 1976).

(6) According to the theory of evolution of eukaryotic cells by serial symbiosis (Margulis, 1970), there was an early symbiotic combination of aerobic bacteria with a prokaryote host cell, and the incorporated bacteria evolved into mitochondria. At a later stage, it is postulated, blue-green algae were incorporated that gave rise to the symbiotically existing intracellular chloroplasts found in higher plants. Whether or not one agrees with the evolutionary aspects of this interpretation, it is clear that the eukaryotic cell does contain multiple quasi-independent genetic information-carrying structures, these do provide for enhanced genetic variability, and this condition must have a high adaptive value because the eukaryotic cell has been such a successful biological structure. Furthermore, these observations suggest that fusion and incorporation may not be new methods for enhancing genetic variability, but may have been an integral process in the evolution of higher forms. What is important today is that cell fusion can be accomplished in the laboratory between distantly related forms, and its advantages and usefulness can be explored to a degree not previously possible.

LESSONS FROM ANIMAL CELL GENETICS

An examination of the older, more-developed field of animal cell genetics shows three areas in which results with animal systems may increase our understanding of chromosomal events in somatic hybrid plant cells and extend the scope of plant cell genetic experiments. These research areas suggest (1) various factors that may contribute to the evolution of the hybrid cell by chromosome elimination; (2) the use of microcells and chromosomes as gene sources for somatic plant cells; and (3) the possibility of introducing plant genes into sexually reproducing animals.

Chromosome Elimination

Chromosome elimination in animal somatic cell hybrids is an important process that has been used with enormous success to map chromosomes and establish linkage relationships (McKusick and Ruddle, 1977). Cell hybrids have been produced within species and between species from different orders and phyla, e.g., human + mouse, rat kangaroo + Chinese hamster, chicken + mouse, mosquito + human (Ruddle, 1972). Chromosome elimination from one of the parent lines may or may not occur in intraspecific hybrids. In contrast, interspecific cell hybrids usually lose random chromosomes of one parent while conserving the other parental set.

Intraspecific cell hybrids that lose chromosomes do so at two rates: either rapidly or at a slow rate of about 1 chromosome lost per 20 cell generations (Bernhard, 1976). In human + mouse hybrids, the average number of human chromosomes retained is 7, with a range of 1 to 20 after 30 generations (Ruddle and Creagan, 1975). In general, chromosome rejection in interspecific cell hybrids is rapid and occurs immediately after fusion during the first few mitoses (Ruddle, 1972). It then becomes stabilized so that there is slow or little subsequent elimination. Thus, clones of hybrid cells can be isolated that retain specific groupings of chromosomes of the parent cell that is prone to elimination (McKusick and Ruddle, 1977). Bernhard (1976) suggests that different mechanisms operate during rapid and slow chromosome loss.

The causes of chromosome elimination have not been identified, but factors that may contribute to the loss are related to difference in (a) cycle synchrony of parental cells, (b) cell generation times, and (c) species origin of cells.

Cycle synchrony. When one nucleus in a heterokaryon is in late G_2 or M phase, it induces dissolution of the nuclear membrane and chromosome condensation in the other nucleus if the latter is in interphase. A result is incomplete condensation of chromosomes and subsequent unequal distribution to daughter cells. Induced premature condensation of chromosomes and its aftermath in asynchronous heterkaryons does not account for preferential chromosome elimination, since elimination from both parents would be expected when asynchronous cell populations are fused (Ruddle, 1972). Nevertheless, chromosome loss can be minimized and recovery of viable hybrid cells increased by fusing parental cells that have been co-synchronized (Rao and Johnson, 1972).

Cell generation time. Kao and Puck (1970) suggest that relative generation times of parent cells influence their ability to retain chromosomes in hybrids. For example, human cell generation time is longer than in the mouse. In human + mouse cells, the cells divide more rapidly as human chromosomes are rejected.

Chinese hamster cells divide twice as fast as mouse cells at 37° C. This disparity is reduced at 31°C. At 37°, more Chinese hamster chromosomes are retained than at 31° in Chinese hamster + mouse hybrids (Bernhard, 1976). This is not a stable condition, however. After extended cultivation at 31° or 37°, progressive loss of mouse chromosomes was observed. The suggested interpretation is that these new cell types were faster growing and thus became dominant (Bernhard, 1976). Supporting this hypothesis is the retention of high numbers of human chromosomes in human + mouse hybrids when cloning is done immediately after fusion (Ruddle, 1970). This strategy reduces competitive stresses that would be found in more heterogeneous cell populations.

Species origin. An inherent superiority of one species over another is probably not the reason why it maintains genome integrity in hybrid cells (Ruddle, 1972; Ruddle and Creagan, 1975). Pontecorvo (1971) could cause loss of either mouse or Chinese hamster chromosomes in their hybrid cells by treating one parent or the other with X-rays or BUdR prior to fusion.

More recently, attempts have been made to alter the rejection of human chromosomes in human + mouse hybrids. In contrast to the earlier work, Pontecorvo (1974) found that human chromosomes, not mouse chromosomes, continued to be eliminated after mouse chromosomes were irradiated. But more human chromosomes are retained in hybrids with a double set of mouse chromosomes (Bernhard, 1976). Human chromosome loss can be fully reversed, however. Mouse chromosomes are lost in hybrids between human cells transformed by the oncogenic virus, SV-40, and diploid mouse cells (Minna and Coon, 1974). Ruddle and Creagan (1975) suggest that adaptation of hybrid cells to the environment *in vitro*, rather than species-specific differences of the individual parental cells, is playing a significant role in the elimination of chromosomes from hybrid cells.

*Microcell-mediated Chromosome Transfer and
Chromosome-mediated Gene Transfer*

Selected whole chromosomes can be introduced into recipient cells via microcells (Fournier and Ruddle, 1977). Chromosome transfer is a method for introducing one gene or a few genes into recipient cells (McBride and Ozer, 1973; Willecke and Ruddle, 1975).

Microcells are produced from monolayer cultures by treatment with colcemid and cytocholasin B, followed by centrifugation. A microcell has a micronucleus that contains a small fraction of the diploid chromosome set surrounded by a rim of cytoplasm and an intact cell membrane. The chromosomes are protected, and cell fusion technology can be applied. Since microcells do not survive, simplified, one-sided selection for donor cell markers in recipient cells is possible. Fournier and Ruddle (1977) report that 1–5 intact mouse chromosomes are reproduced and function in mouse, Chinese hamster, and human recipient cells.

Several different genes coding for enzymes have been transferred to recipient cells by uptake of donor metaphase chromosomes. In the hybrid cells the genetic material is not visible as chromosomes or chromosome fragments (Degnen et al., 1976). Base pair homology is probably not involved in association of the donor gene with the recipient genome since a particular transferred gene has been located among different-size classes of recipient chromosomes (Spandidos and Siminovitch, 1977). It seems likely that widely divergent taxa may be hybridized by this method. The observed frequency of gene transfer is approximately 10^{-6} and 10^{-7}. The true frequency may be closer to 10^{-4} if it is assumed that the transferred gene is approximately 0.1 percent of the donor genome and unselected genes go undetected (McBride and Athwal, 1977).

Since many mammalian cells carry mutations that allow powerful biochemical selection techniques to be applied for genetic studies, a plant-animal cell genetic system could exploit this existing resource. Drug-resistant mutants, temperature-sensitive mutants, nutritional auxotrophs, and membrane-associated markers are available in animal cells. A number of these may be suitable for transfer to plant cells that, in general, lack well-defined mutants. For example, α-amanitine-resistant hamster cells have a mutant RNA polymerase (Amati et al., 1975); an aminopterin-resistant Chinese hamster cell line carries a dominant gene mutation that confers increased activity to dihydrofolate reductase (Flintoff et al., 1976). Transfer of these markers into plant protoplasts could be attempted by cell fusion, microcell fusion, or chromosome transfer in a plant tissue culture medium supplemented with the appropriate inhibitors.

Where Does It Go from Here?

Until recently, plant somatic cell genetic studies had the unique distinction, when compared with the more sophisticated animal studies, of producing sexual offspring from parasexual parents. The exciting experiments of B. Mintz (Mintz and Illmensee, 1975) with mouse teratocarcinoma cells removes this distinction. By injecting euploid genetically marked totipotent teratocarcinoma cells into mouse blastocysts, she has shown that these tumor cells can normalize and form all somatic tissue; the blastocysts develop into mature adults whose germ line cells differentiated from teratocarcinoma cells. The F_1 progeny from these experiments had one parent, which in the cell lineage sense, had been a tumor!

Teratocarcinoma cells may be a way of transferring genes of other species into adult mice. The introduction of plant genes into a fully differentiated mammal might be accomplished by first making hybrid mouse teratocarcinoma + plant cells and injecting them into the blastocyst. It is now possible to consider joining the two kingdoms on a cellular level and conceivably to recover adult plant/animal chimeras. These new avenues of research offer wider opportunities to enhance genetic variability through fusing cells and combining organelles from the entire spectra of living things. Ultimate limitations may be encountered, not so much in the ability to transfer genetic components, but in physiological incompatibilities or in the function of regulatory systems.

ACKNOWLEDGMENTS

This research was carried out at Brookhaven National Laboratory, under the auspices of the U.S. Energy Research and Development Administration.

LITERATURE CITED

Ahkong, Q. F., D. Fisher, W. Tampion, and J. A. Lucy. 1975a. Mechanisms of cell fusion. Nature 253:194–95.

Ahkong, Q. F., J. I. Howell, J. A. Lucy, F. Safwat, M. R. Davey, and E. C. Cocking. 1975b. Fusion of hen erythrocytes with yeast protoplasts induced by polyethylene glycol. Nature 255:66–67.

Amati, P., F. Blasi, V. DiPorzio, A. Riccio, and C. Traboni. 1975. Hamster α-amanitine resistant RNA polymerase II able to transcribe polyoma virus genome in somatic cell hybrids. Proc. Nat. Acad. Sci. USA 72:753–57.

Bayliss, M. W. 1977. The causes of competition between two cell lines of *Daucus carota* in mixed culture. Protoplasma 92:117–27.

Bennett, M. D., R. A. Finch, and I. R. Barclay. 1976. The time rate and mechanism of chromosome elimination in *Hordeum* hybrids. Chromosoma 54:175–200.

Bennici, A., M. Buiatti, F. D'Amato, and M. Pagliai. 1971. Nuclear behavior in *Haplopappus gracilis* callus grown *in vitro* on different culture media. *In* Les cultures de tissus de plantes. Colloques Intern. C.N.R.S. 193:245–50.

Bernhard, H. P. 1976. The control of gene expression in somatic cell hybrids. Intern. Rev. Cytol. 47:289–325.

Blakely, L. M., and F. C. Steward. 1964. Growth and organized development of cultured cells. VII. Cellular variation. Amer. J. Bot. 51:809–20.

Bonnett, H. T. 1976. On the mechanism of the uptake of *Vaucheria* chloroplasts by carrot protoplasts treated with polyethylene glycol. Planta 131:229–33.

Burgess, J., and E. N. Fleming. 1974. Ultrastructural studies of the aggregation and fusion of plant protoplasts. Planta 118:183–93.

Carlson, P. S., H. H. Smith, and R. D. Dearing. 1972. Parasexual interspecific plant hybridization. Proc. Nat. Acad. Sci. USA 69:2292–94.

Chang, S. E., and J. W. Littlefield. 1976. Elevated dihydrofolate reductase messenger RNA levels in methotrexate-resistant BHK cells. Cell 7:391–96.

Chen, K., S. G. Wildman, and H. H. Smith. 1977. Chloroplast DNA distribution in parasexual hybrids as shown by polypeptide composition of Fraction I protein. Proc. Nat. Acad. Sci. USA 74:5109–12.

Constabel, F. 1976. Somatic hybridization in higher plants. In Vitro 12:743–48.

Constabel, F., D. Dudits, O. L. Gamborg, and K. N. Kao. 1975. Nuclear fusion in intergeneric heterokaryons. Canad. J. Bot. 53:2092–95.

D'Amato, F. 1977. Cytogenetics of differentiation in tissue and cell cultures. *In* J. Reinert and Y. P. S. Bajaj (eds.) Applied and fundamental aspects of plant cell, tissue, and organ culture. Pp. 343–57. Springer-Verlag, Berlin.

Davies, R. D. 1974. Chromosome elimination in inter-specific hybrids. Heredity 32:267–70.

Degnen, G. E., I. L. Miller, J. M. Eisenstadt, and E. A. Adelberg. 1976. Chromosome-mediated gene transfer between closely related strains of cultured mouse cells. Proc. Nat. Acad. Sci. USA 73:2838–42.

Dudits, D., K. N. Kao, F. Constabel, and O. L. Gamborg. 1976a. Fusion of carrot and barley protoplasts and division of heterokaryocytes. Canad. J. Genet. Cytol. 18:263–69.

Dudits, D., I. Rasko, G. Y. Hadlaczky, and A. Lima-de-Faria. 1976b. Fusion of human cells with carrot protoplasts induced by polyethylene glycol. Hereditas 82:121–24.

East, E. M. 1935. Genetic reactions in *Nicotiana*. I. Compatibility. Genetics 20:403–13.

East, E. M. 1936. Heterosis. Genetics 21:375–97.

Flintoff, W. F., S. M. Spindler, and L. Siminovitch. 1976. Genetic characterization of methotrexate-resistant Chinese hamster ovary cells. In Vitro 12:749–57.

Fournier, R. E. K., and F. H. Ruddle. 1977. Microcell-mediated transfer of murine chromosomes into mouse, Chinese hamster, and human somatic cells. Proc. Nat. Acad. Sci. USA 74:319–23.

Fox, J. E. 1963. Growth factor requirements and chromosome number in tobacco tissue cultures. Physiol. Plantarum 16:793–803.

Gleba, Yu. Yu., R. G. Butenko, and K. M. Sytnyak. 1975. Protoplast fusion and parasexual hybridization in *Nicotiana tabacum* L. Genetika 221:1196–98. (In Russian.)

Goodspeed, T. H. 1954. The genus *Nicotiana*. Chronica Botanica, Waltham, Mass.

Gupta, S. B., and P. Gupta. 1973. Selective somatic elimination of *Nicotiana glutinosa* chromosomes in the F_1 hybrids of *N. suaveolens* and *N. glutinosa*. Genetics 73:605–12.

Izhar, S., and J. B. Power. 1977. Genetical studies with *Petunia* leaf protoplasts. I. Genetic variation to specific growth hormones and possible genetic control on stages of protoplast development in culture. Plant Sci. Lett. 8:375–83.

Johnson, R. T., and H. Harris. 1969a. DNA synthesis and mitosis in fused cells. II. HeLa-chick erythrocyte heterokaryons. J. Cell Sci. 5:625–43.

Johnson, R. T., and H. Harris. 1969b. DNA synthesis and mitosis in fused cells. III. HeLa-Ehrlich heterokaryons. J. Cell Sci. 5:645–97.

Jones, C. W., I. A. Mastrangelo, H. H. Smith, H. Z. Liu, and R. A. Meck. 1976. Interkingdom fusion between human (HeLa) cells and tobacco hybrid (GGLL) protoplasts. Science 193:401–3.

Kao, F. T., and T. T. Puck. 1970. Genetics of somatic mammalian cells: Linkage studies with human-Chinese hamster cell hybrids. Nature 228:329–32.

Kao, K. N. 1977. Chromosomal behaviour in somatic hybrids of soybean-*Nicotiana glauca*. Molec. Gen. Genet. 150:225–30.

Kao, K. N., F. Constabel, M. R. Michayluk, and O. L. Gamborg. 1974. Plant protoplast fusion and growth of intergeneric hybrid cells. Planta 120:215–27.

Kehr, A. E., and H. H. Smith. 1952. Multiple genome relationships in *Nicotiana*. Cornell Univ. Agric. Expt. Sta., Memoir 311, Ithaca, New York.

Keller, W. A., and G. Melchers. 1973. The effect of high pH and calcium on tobacco leaf protoplast fusion. Z. Naturforsch. 28C:737–41.

Kostoff, D. 1943. Cytogenetics of the genus *Nicotiana*. States Printing House, Sofia, Bulgaria.

Kung, S. D., K. Sakano, and S. G. Wildman. 1974. Multiple peptide composition of the large and small subunits of *Nicotiana tabacum* Fraction I protein ascertained by fingerprinting and electrofocusing. Biochim. Biophys. Acta 365:138–47.

McBride, O. W., and R. S. Athwal. 1977. Chromosome mediated gene transfer with resultant expression and integration of the transferred genes in eukaryotic cells. Brookhaven Symp. Biol. 29:116–26.

McBride, O. W., and H. L. Ozer. 1973. Transfer of genetic information by purified metaphase chromosomes. Proc. Nat. Acad. Sci. USA 70:1258–62.

McKusick, V. A., and F. H. Ruddle. 1977. The status of the gene map of the human chromosome. Science 196:390–405.

Margulis, L. 1970. Origin of eukaryotic cells. Yale University Press, New Haven, Conn.

Marx, J. L. 1977. Gene transfer in mammalian cells: Mediated by chromosomes. Science 197:146–48.

Melchers, G., and G. Labib. 1974. Somatic hybridization of plants by fusion of protoplasts. I. Selection of light resistant hybrids of "haploid" light sensitive varieties of tobacco. Molec. Gen. Genet. 135:277–94.

Melchers, G., and M. D. Sacristán. 1977. Somatic hybridisation of plants by fusion of protoplasts. II. The chromosome numbers of somatic hybrid plants of 4 different fusion experiments. *In* R. G. Gautheret (ed.), La culture des tissus et des cellules de végéteaux. Travaux dédiés à la mémoire de Georges Morel. Pp. 169–77. Masson, Paris.

Minna, J., and H. G. Coon. 1974. Human × mouse hybrid cells segregating mouse chromosomes and isozymes. Nature 252:401–4.

Mintz, B., and K. Illmensee. 1975. Normal genetically mosaic mice produced from malignant teratocarcinoma cells. Proc. Nat. Acad. Sci. USA 72:3585–89.

Murashige, T., and R. Nakano. 1967. Chromosome complement as a determinant of the morphogenetic potential of tobacco cells. Amer. J. Bot. 54:963–70.

Murashige, T., and F. Skoog. 1962. A revised medium for rapid growth and bioassays with tobacco tissue cultures. Physiol. Plantarum 15:473–97.

Näf, U. 1958. Studies on tumor formation in *Nicotiana* hybrids. I. The classification of parents into two etiologically significant groups. Growth 22:167–80.

Ogura, H. 1976. The cytological chimeras in original regenerates from tobacco tissue cultures and in their offspring. Jap. J. Genet. 51:161–74.

Pontecorvo, G. 1971. Induction of directional chromosome elimination in somatic cell hybrids. Nature 230:367–70.

Pontecorvo, G. 1974. Induced chromosome elimination in hybrid cells. *In* R. L. Davidson and F. F. de la Cruz (eds.), Somatic cell hybridization. Pp. 65–69. Raven Press, New York.

Power, J. B., E. M. Frearson, C. Hayward, D. George, P. K. Evans, S. F. Berry, and E. C.

Cocking. 1976. Somatic hybridisation of *Petunia hybrida* and *P. parodii.* Nature 263:500–502.

Rao, P. N., and R. T. Johnson. 1972. Premature chromosome condensation: A mechanism for the elimination of chromosomes in virus-fused cells. J. Cell Sci. 10:495–513.

Ruddle, F. H. 1970. Utilization of somatic cells for genetic analysis: Possibilities and problems. Symp. Intern. Soc. Cell Biol. 9.233–64.

Ruddle, F. H. 1972. Linkage analysis using somatic cell hybrids. Adv. Hum. Genet. 3:173–235.

Ruddle, F. H., and R. P. Creagan. 1975. Parasexual approaches to the genetics of man. Ann. Rev. Genet. 9:407–86.

Singh, B. D., B. L. Harvey, K. N. Kao, and R. A. Miller. 1972. Selection pressure in cell populations of *Vicia hajastana* cultures *in vitro.* Canad. J. Gen. Cytol. 14:65–70.

Smith, H. H. 1943. Studies on induced heteroploids of *Nicotiana.* Amer. J. Bot. 30:121–30.

Smith, H. H. 1972. Plant genetic tumors. Progr. Expt. Tumor Res. 15:138–64.

Smith, H. H. 1976a. Characterization of somatic hybrid plants and further exploitation of a selective system. *In* D. Dudits, G. L. Farkas, and P. Maliga (eds.), Cell genetics in higher plants. Pp. 163–69. Akadémiai Kiado, Budapest.

Smith, H. H. 1976b. Genetic engineering with tobacco protoplasts. *In* Proc. 6th Internat. Tobacco Sci. Congr. Pp. 75–80. Jap. Tob. & Salt Public Corp., Tokyo.

Smith, H. H., K. N. Kao, and N. C. Combatti. 1976. Interspecific hybridization by protoplast fusion in *Nicotiana.* Confirmation and extension. J. Hered. 67:123–28.

Spandidos, D. A., and L. Siminovitch. 1977. Genetic analysis by chromosome-mediated gene transfer in hamster cells. Brookhaven Symp. Biol. 29:127–34.

Subrahmanyam, N. C., and K. Kasha. 1973. Selective chromosomal elimination during haploid formation in barley following interspecific hybridization. Chromosoma 42:111–25.

Trench, R. K. 1975. Of "leaves that crawl": Functional chloroplasts in animal cells. Symp. Soc. Expt. Biol. 29:229–65.

Vasil, I. K., and K. L. Giles. 1975. Induced transfer of higher plant chloroplasts into fungal protoplasts. Science 190:680.

Willecke, K., and F. H. Ruddle. 1975. Transfer of the human gene for hypoxanthineguanine phosphoribosyltransferase via isolated human metaphase chromosomes into mouse L-cells. Proc. Nat. Acad. Sci. USA 72:1792–96.

Willis, G. E., J. X. Hartman, and E. D. de Lameter. 1977. Electron microscope study of plant-animal cell fusion. Protoplasma 91:1–14.

Zimmerman, J. L., and R. B. Goldberg. 1977. DNA sequence organization in the genome of *Nicotiana tabacum.* Chromosoma 59:227–52.

Manipulating Symbiotic Nitrogen Fixation

17

INTRODUCTION

Much of the attention generated by global food shortages and nutritional deficiencies has focused on protein levels in agricultural crops and, consequently, on the nitrogen input to crop production. To most plants, atmospheric nitrogen is unavailable unless it has been oxidized or reduced to a form of fixed nitrogen. However, legume species, in conjunction with rhizobia, are among those plants that can utilize atmospheric nitrogen. Biological and industrial chemical fixation are the two main sources of fixed nitrogen (fig. 1) in agriculture. The advent of cheap nitrogen fertilizers has reduced the once-important role of legumes in cropping schemes. Recent increases in energy costs have been reflected in the production of industrial fixed nitrogen, both in terms of capital investment and in the costs of production. Nitrogen fertilizer may become an expensive input in farm production. Environmentalists have also raised concerns over the potential long-term pollutant effects of heavy chemical fertilizer use on ground water reserves.

In agricultural research, there is renewed interest in the contribution of biological nitrogen fixation as an alternative technology to continued high use of nitrogen fertilizers. In particular, the legume-*Rhizobium* symbiosis has been reexamined intensively (see Quispel, 1974; Evans, 1975; Newton and Nyman, 1976). The interest generated in nitrogen fixation coincided with many of the recent advances in molecular biology and genetic engineering. Numerous calls for the transfer of the nitrogen fixation (*nif*)

F. B. Holl, Prairie Regional Laboratory, Saskatoon, Saskatchewan, Canada.

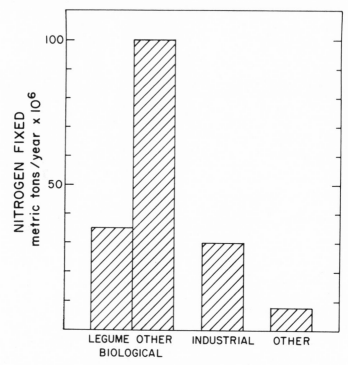

Fig. 1.   Comparative amounts of nitrogen fixed annually. Data are from Quispel (1974), Silver and Hardy (1976), and Evans (1975).

genes to a variety of crops and for the panacea of nitrogen-fixing wheat have resulted (e.g., see Hardy, 1976). Many of the more speculative proposals have been made by individuals outside the field of nitrogen fixation research, and it is reassuring to see attempts being made to make a realistic assessment of the problems and potential for exploiting nitrogen fixation (Evans and Barber, 1977).

The proposition that symbiotic nitrogen fixation should be manipulated makes three implicit assumptions: (1) genetic variation for the character exists and can be detected; (2) alterations or extension of the characteristic are desirable goals; and (3) symbiotic nitrogen fixation can be manipulated with techniques that are, or might become, available.

In this paper I shall outline the present status of research efforts in nitrogen fixation and discuss some of the goals and techniques for manipulation that have been suggested, with particular emphasis on the legume-*Rhizobium* symbiosis.

SYMBIOTIC NITROGEN FIXATION

In order to exploit any biological system, it is first desirable to understand its function. The legume-*Rhizobium* symbiosis is a complex interaction of two organisms, genomes, and physiological systems within a range of environments. In spite of the long history of investigation of fixation, we are still a long way from appreciating or elucidating totally its complexity.

INFECTION

Multiplication

Rhizobia present in the soil as indigenous strains or a specific inoculum are attracted to, and multiply in, the legume rhizosphere. This stimulatory growth response reflects a low-order specificity for legumes in contrast to nonlegumes. Bacterial survival in the soil is influenced by a variety of factors including dessication, pH, calcium levels, temperature, and competition with other microorganisms.

Root Hair Curling

Rhizobial multiplication in the rhizosphere is followed by a specific host-bacterial interaction, resulting in root hair curling (Yao and Vincent, 1969). This initial stage in the invasion process, characterized by a high degree of specificity, is influenced by the genetic constitution of both macro- and microsymbiont. With few exceptions, root hair curling is the prelude to bacterial invasion. The number of root hairs deformed is dependent upon the host species, as is the proportion of infection loci that subsequently result in nodule development (Nutman, 1959).

Infection Thread Formation

Invasion of the plant root via the deformed root hairs is accompanied by the formation of the infection thread having the appearance of an invagination of the hair wall. The infection thread is histochemically similar to the root hair. Thread growth proceeds from the tip as a consequence of plant cell wall deposition to contain the dividing rhizobia. It is not known how the bacterium initiates the infection thread, and not all infection threads result in nodule formation (Nutman, 1959)—abortion occurring at various stages of thread development.

Nodule Development

In regions where nodules will ultimately develop, the root cortical cells enlarge and divide. Dividing cells in the nodule meristem are usually polyploid. The development of the nodule is dependent upon the induction of local meristematic growth in the root cortex. Several hypotheses have been suggested to explain the control of nodule initiation. In a critical discussion of these factors, Libbenga and Bogers (1974) indicate the complexity of the process and the inconclusive nature of the available data. On reaching the meristematic zone, the infection thread becomes more divided and ramifies through the central cells; rhizobia are released into the cytoplasm and continue to multiply. The host cells enlarge and become surrounded by dividing uninvaded cells. Ultimate nodule morphology is a function of the host plant (Dart, 1976). A diagrammatic nodule cross section is shown in figure 2, illustrating the relationship of the various regions.

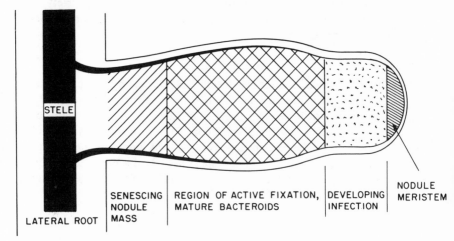

Fig. 2.   Diagrammatic cross section of a mature nodule showing the range of development and association with the lateral root.

Bacteroid Development

Within the infected cells rhizobia cease division and develop into the pleomorphic bacteroids that may superficially resemble the vegetative cells (e.g., cowpea, soybean) or may form greatly enlarged structures, as in clover and alfalfa. The bacteroid (fig. 3) is the site of nitrogen fixation.

Nitrogenase Enzymology

Within the bacteroids, the reduction of N_2 to ammonia is catalyzed by the enzyme nitrogenase. Since the development of asymbiotic rhizobia

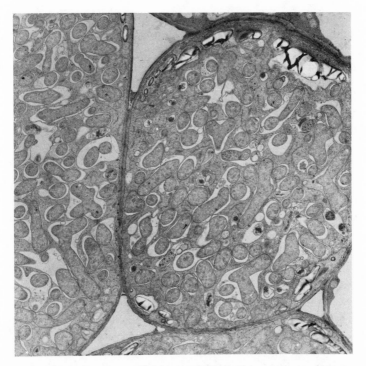

Fig. 3. Electron micrograph of a root nodule of *Pisum sativum* L. cv. Trapper. The specimen was fixed in glutaraldehyde and osmium tetroxide and stained with lead citrate and uranyl acetate. Magnification ($\times$7500). The pleomorphic bacteroids are the site of nitrogen fixation.

capable of reducing acetylene (Child, 1976), it is now accepted that the genetic message for nitrogenase structure is located exclusively in the bacterial genome.

Most nitrogenases investigated have similar, but not necessarily identical, properties. They are complex proteins composed of Fe-protein and Mo-Fe protein subunits. The smaller Fe-protein MW 50,000–70,000) is oxygen-sensitive and in a number of systems including soybean bacteroids has been shown to be cold-labile. The molybdenum-iron component of nitrogenase is large (MW 100,000–300,000) and is somewhat less sensitive to oxygen than the iron protein (Eady et al., 1972). It is not cold-labile.

Nitrogenase catalyzes the following reaction:

$$N_2 + 8H^+ + 6e^- + nATP \rightarrow 2NH_4^+ + nADP + nPi$$

The stoichiometric relationship between ATP hydrolysis and electron transfer is not entirely clear. Available data suggest that under physiological conditions the $ATP:2e^-$ ratio is probably in the range of 4–5 (Ljones, 1974).

Nitrogenase exhibits broad substrate specificity. The enzyme can reduce a variety of substrates; for example, $2H^+ \rightarrow H_2$, $HCN \rightarrow CH_4 + NH_3 + CH_3NH_2$ and $C_2H_2 \rightarrow C_2H_4$. This substrate versatility has had ramifications with regard to assay methodology and estimates of the efficiency of nitrogen fixation.

Ancillary Metabolism

To fuel the nitrogen-fixing reaction, the host plant must provide bacteroids with the products of photosynthesis. The significance of the source of carbon skeletons and energy will be discussed in greater detail in the section on photosynthesis and nitrogen fixation.

Oxygen and Leghaemoglobin

A physiological conflict arises in the nodule tissue over the requirements of the bacteroid for high O_2 and aerobic ATP production versus the low pO_2 environment necessary for nitrogenase function. The presence in the nodule of the hemoprotein, leghaemoglobin, is associated with active nitrogen fixation.Leghaemoglobin is contained within the membrane envelope surrounding the bacteroids and appears to facilitate the diffusion of the O_2 required for aerobic functions, while maintaining a low pO_2 environment within the nodule tissue.

Fixation Products

The primary product of nitrogen fixation, ammonia, undergoes conversion to amides and amino acids. Although the post-fixation reactions of the nodule have not been completely elucidated, a number of enzyme systems have been implicated. In particular, evidence supporting the existence and function of the glutamine synthetase glutamine:2 oxoglutarate aminotransferase (GS:GOGAT) pathway has been reviewed by Miflin and Lea (1976). Robertson et al. (1975a) found that the levels of GS and GOGAT in the plant fraction of the nodule increased during nodule formation. Fixed nitrogen may be excreted from the bacteroid as ammonia and assimilated into glutamine by GS in the nodule cytosol. Conversion of the major portion of the glutamine produced into asparagine has been suggested, since the latter compound is a major export product of the nodules (Robertson, Farnden, Warburton, and Banks, 1975; Robertson, Warburton, and Farnden, 1975; Scott et al., 1976).

The involvement of GS in nitrogen assimilation during fixation is also supported by studies implicating GS in the control of nitrogenase. The repressive action of ammonia on nitrogenase synthesis does not appear to be a direct effect (Brill, 1975). In *Klebsiella*, Streicher et al. (1974) and Tubb (1974) suggested that the deadenylated form of GS mediated a reversal of ammonia repression. Since other fixing systems, e.g., peas, do not appear to show the adenylating control of GS (Kingdon, 1974), it remains to be seen whether GS involvement in nitrogenase regulation will prove to be a general phenomenon.

Assimilation of fixed nitrogen via enzyme systems in the nodule cytosol (i.e., plant origin), and possible links to nitrogenase synthesis regulation, indicate further avenues of potential genetic control through manipulation of the plant genome.

Hydrogen Evolution

Nitrogenase-catalyzed hydrogen evolution in cell-free enzyme preparations has been well established (Burns and Hardy, 1975). The ATP-dependent production of hydrogen *in vivo* had not been resolved until a recent report by Schubert and Evans (1976). They observed hydrogen evolution to be a general phenomenon among nodulated legumes; and the biological energy lost via this activity meant that only 40–60 percent of the electron flow to nitrogenase was transferred to nitrogen. Hydrogen evolution was concluded to be a major factor affecting the efficiency of agronomically important legumes. Variation was observed. Bacteria such as the cowpea rhizobium being more efficient. No evidence in this work was given as to whether the more efficient cowpea strain (which is promiscuous) would give equally high activity over a range of hosts, or whether there is a host influence on the process. Variation in the degree of hydrogen evolution or the ability to recycle the hydrogen via hydrogenase intimates that the prospect of increased efficiency does exist (Evans and Barber, 1977) and that this trait might be incorporated into commercial rhizobial strains for more efficient energy utilization.

WHOLE PLANT BEHAVIOR

Measurement of Nitrogen Fixation

The nitrogenase catalyzed reduction of acetylene (C_2H_2) to ethylene (C_2H_4) described by Dilworth (1966) was rapidly developed into a facile and sensitive assay technique for the estimation of nitrogen fixation (Hardy et al., 1973). There are now a number of variations used for the nondestructive assay of nitrogenase activity, although some reservations have been expressed regarding the numerical relationship between the reduction of acetylene and that of nitrogen (Burris, 1974).

Field Measurements

Using the acetylene reduction technique, estimates of nitrogen fixation by grain legumes in the field have demonstrated the pattern of activity portrayed in figure 4. Under normal cultivation conditions, grain legumes appear to fix symbiotically only 25–50 percent of the total plant nitrogen (Silver and Hardy, 1976; Holl and LaRue, 1976). Grain legumes, especially soybeans, have shown little or no response to applied N. Rates of 134–440 kg N/ha produced no significant yield increases (Hinson, 1975; Johnson et al., 1975; Bezdicek et al., 1974). Recently, Parker and Harris (1977) examined the interaction of N levels and molybdenum on the yields of nodulating and nonnodulating soybeans. In addition to a molybdenum requirement under nitrogen-fixing conditions, they also observed a positive yield response to applied nitrogen. The observation by Parker and

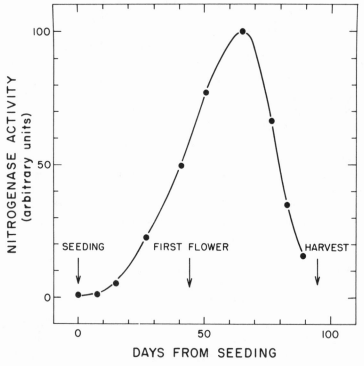

Fig. 4. Seasonal pattern of nitrogen fixation (C_2H_2) in *Pisum sativum* L. grown at Saskatoon, with key phases of the cropping cycle indicated.

Harris (1977) that a differential response was observed between nodulating and nonnodulating cultivars suggests some interesting variations in the assimilatory nitrate reductase and symbiotic pathway of N acquisition by the plant that have not yet been clarified. These differences may be reflected in the relative energy costs to the host plant of the two pathways.

Photosynthesis and Nitrogen Fixation

In the symbiotic fixation of nitrogen, the energy required is provided ultimately by photosynthesis. The significant interaction of photosynthesis and nitrogen fixation has been recognized for some time. Wilson (1940), in his landmark monograph, stated, "Photosynthesis and nitrogen fixation must be considered as inter related processes, and the function of one should not be emphasized at the expense of the other." Earlier, Allison and Ludwig (1934, 1939) had also emphasized the roles of carbohydrate as a source of energy and of carbon skeletons for combination with assimilated nitrogen.

Renewed interest in photosynthesis and nitrogen fixation has accompanied advances in analytical techniques and the search for more-efficient fixation. Silver and Hardy (1976) conclude that photosynthate available to the nodule is the major limitation to nitrogen fixation in soybeans, although this may be an oversimplification of the complex carbohydrate-nitrogen relationship.

Photosynthesis provides a pool of substrates that may provide electrons, reductant, and energy to nitrogenase, as well as carbon skeletons to ammonia assimilation. Metabolic functions such as photorespiration, which affect the size of this pool, may affect nitrogen fixation. Hardy et al. (1976) have proposed that a reduction in photorespiration in C_3 plants would increase nitrogen fixation in soybeans.

As plant functions, photosynthesis and photosynthate supply are regulated through the host plant genome. Attention is now being directed to identifying the genetic basis of the photosynthesis-fixation interaction and to examining the potential for increased photosynthate supply to the nitrogen -fixing process.

Environmental Stress

The influence of the environment on symbiotic nitrogen fixation has been recognized for some time. The balance between host and bacterium may be altered even under conditions where plant growth is apparently unaffected. The effects of environmental variation have been reviewed in some detail (Vincent, 1965; Lie and Mulder, 1971; Lie, 1974).

Symbiosis is known to be affected by the composition of the gas phase in the soil. Variations in the levels of N_2, O_2, and CO_2 in the soil atmosphere reflected in the performance of nitrogen fixation may be affected under waterlogging conditions. Soil moisture conditions leading to waterlogging, or even slight dessication, have been shown to produce significant reductions in nodule activity (Sprent, 1969, 1971, 1972).

Soil acidity may also depress symbiotic nitrogen fixation. Some attempts have been made to determine the mechanism of this depression (Lie, 1969), and although Munns (1968) showed that the acid-sensitive step in nodulation might involve root hair curling, there is no definitive explanation for this phenomenon as yet.

In none of the environmental influences has the mechanism involved been adequately clarified. No genetic basis has been established, with the exception of the temperature-dependent nodulation of a pea cultivar described by Lie (1971). There is considerable circumstantial evidence, however, to implicate host plant controls in many of these functions.

Symbiotic Interactions

In the introduction to this paper, it was pointed out that breeding for, or manipulation of, symbiotic fixation is in part dependent upon the existence of genetic variation for the character. Variation in bacterial strains and in plant hosts has been known for some time (Fred et al., 1932; Nutman, 1956; Holl and LaRue, 1976). Detailed genetic studies have usually dealt with either bacterial or host variation, but seldom the interaction.

Genetic variation may be partitioned into three components: (1) additive effects of the host plant genotype; (2) additive effects of the rhizobial genotype, (3) nonadditive (interaction) effects of plant-rhizobial combinations.

Mytton (1975) investigated the white clover-*Rhizobium trifolii* symbiosis and reported that nonadditive interaction effects accounted for almost one-quarter of the phenotypic variation observed. In more recent data on the *Vicia faba-Rhizobium leguminosarum* symbiosis, Mytton et al. (1977) found that more than 70 percent of the phenotypic variation could be attributed to nonadditive specific host genotype × rhizobium genotype interactions. It is well to recall in any discussion on manipulating the process the sophisticated interrelationship that exists between plant and rhizobium, especially the contribution of interaction effects to overall symbiotic performance.

EXPLOITING THE SYMBIOSIS

Rapid strides in molecular biology and the potential for genetic engineering have led many plant researchers to examine the prospects for applying such techniques to manipulate nitrogen fixation in crop plants. It is evident that both genetic and physiological constraints limit symbiotic nitrogen fixation. It therefore would be appropriate to consider some of these modifications. I intend to examine only those areas where a genetic basis has been demonstrated or is likely to be involved.

MICROSYMBIONT MANIPULATIONS

Asymbiotic Nitrogen Fixation and Rhizobium Breeding

The main experimental interest in rhizobia has always been its ability to interact with legumes in the nitrogen-fixation symbiosis. Historically this function could not be demonstrated in the free-living bacterium, and consequently there was little incentive to undertake extensive genetic analyses. The report in 1971 by Holsten et al. of nitrogen fixation *in vitro* by cultured plant cells and rhizobia provided the primary impetus to efforts in a number of laboratories that culminated first in reports of *in vitro* systems on solid media (Phillips, 1974; Child and LaRue, 1974) and subsequently, asymbiotic fixation by free-living rhizobia (Pagan, et al., 1975; Kurz and LaRue, 1975; McComb et al., 1975).

Some of the initial interest in the *in vitro* system was directed toward its potential as a system to study the infection process. It soon became apparent that the plant cell-bacterium association was not entirely comparable to infection *in vivo* (Child and LaRue, 1976). Moreover, the behavior of plant cells in the *in vitro* systems did not always correspond to the genetic constitution of the host plant from which they were derived. Investigation of the behavior *in vitro* of three cell lines of field pea (*Pisum sativum* L. (Holl, unpublished) derived from two normally nodulated cultivars, Trapper and Century, and a third double recessive nonnodulating non-fixing mutant, 72H1-2 (Holl, 1975b), was carried out. Under conditions described by Child and LaRue (1974), acetylene reduction was observed in all cell cultures (table 1). The most active combinations utilized the cowpea rhizobium strain 32H1 as one partner. It was apparent that the genetic barriers to nodulation and fixation in the host mutant either were not operable or were bypassed *in vitro* and that the scope of genetic studies of the plant-bacterium interaction *in vitro* was limited.

Nevertheless, the ability to grow rhizobia under defined conditions as a free-living organism and to induce and measure acetylene-reducing activity under such conditions now provides researchers with a useful tool for genetic analysis and manipulation of the bacterium. The establishment of artificial growth environments and the development of such systems as the "artificial nodule" described by Shanmugan and Valentine (1976) will assist efforts to elucidate the biochemical functions of fixation and to investigate

TABLE 1

ACETYLENE REDUCTION BY CULTURE IN VITRO OF
PISUM SATIVUM L. AND RHIZOBIUM SPP.

Cultivar	Cell Culture[a]	Bacterial Strain[b]	Specific Activity nmoles C_2H_4 hr^{-1} gm dry wt^{-1}
Century	PC-1271	. . . . . .	0.2[c]
		pea isolate	3.7
		32H1	90.2
Trapper	TR-1073	. . . . . .	0.6
		pea isolate	1.6
		32H1	29.1
72HI-2	PA-1073	. . . . . .	4.7
		pea isolate	2.8
		32H1	65.4

[a] Cell cultures were initiated from root tissue and maintained in suspension culture in medium B5 or B5-C (Gamborg et al., 1968). Cultures were generously provided by Dr. O. L. Gamborg and Mr. J. Shyluk of the Prairie Regional Laboratory, Saskatoon.

[b] The pea isolate was a single colony isolate of *Rhizobium leguminosarum* from nodule tissue on cv. Trapper. 32H1 is a cowpea rhizobium that was kindly provided by Dr. J. J. Child, Prairie Regional Laboratory, Saskatoon.

[c] Activities shown are the mean values of a number of replicates.

the effect of mutation on rhizobial behavior. It should be possible to select, under asymbiotic conditions, bacterial strains that will show altered phenotypes symbiotically. Whether the *in vitro* system will provide a tool to select directly or indirectly for such characteristics as increased competitiveness, soil survival, or infectivity remains to be evaluated.

Nif Genetics

The genes coding for nitrogen fixation have received most attention in the free-living nitrogen-fixing organisms such as *Klebsiella pneumoniae*. The ability to transfer *nif* genes between bacteria utilizing specific bacteriophages has revealed some important features of the molecular biology of fixation. Some of the *nif* genes form a cluster near the histidine operon of the *K. pneumoniae* chromosome map (Streicher et al., 1971, 1972); other phenotypes do not map in this region and may function in the ancillary metabolism of fixation. Mapping of the *Klebsiella* chromosome in a number of laboratories has resulted in the identification of eight *nif* genes distributed in two clusters on either side of the *his* locus of the linkage map. Three of the genes are structural genes for nitrogenase, and the remainder code for ancillary proteins or function as regulators.

Because of the undeveloped state of rhizobial genetics, little progress had been made until recently in genetically mapping the *Rhizobium* chromosome. Some earlier gene transfer experiments have been reported (Raina and Modi, 1975) and a linkage map has been described for an *R. lupini* strain (Heumann et al., 1973). Beringer and Hopwood (1976) used a specific R-factor plasmid (R68.45) to map nodulating strains of *R. leguminosarum*. The authors indicate that the development of this recombination system should provide considerable potential for genetic analysis of mutants of symbiotic nitrogen fixation. Map linkages in *R. meliloti* have also been worked out using the conjugative plasmid RP4 (Meade and Signer, 1976). The development of rhizobial genetics will be of considerable benefit in future investigations of the nitrogen-fixing symbiosis.

MACROSYMBIONT ALTERATIONS

Although the basic genetics of plants, including many legumes, have been generally more advanced than for rhizobia, relatively little effort has been made to identify or to exploit the genetic contribution of the host to the symbiosis. Examples of genetically controlled host modifications of the nodulation/fixation cycle have been reported and reviewed (Nutman, 1956; Holl, 1975b; Holl and LaRue, 1976). Manipulation of the macrosymbiont may provide more-stable genetic changes than those that might be accomplished in rhizobial populations.

Considerable basic genetic and biochemical analysis remains to be carried out in legumes before we have a clear picture of the exploitable variation. Hardy et al. (1971) suggested that the growth cycle and fixation measurements could provide some useful comparative parameters: (1) initiation time (days from seeding and onset of fixation), (2) doubling time (N doubling during exponential portion of growth phase), and (3) termination time (end of exponential phase). Little attempt has been made, however, to characterize available germ plasm in these kinds of terms, or to define the extensive variability that exists even in a self-pollinating crop such as field peas. The potential for classical manipulation of the macrosymbiont has been somewhat overshadowed by the more spectacular prospects of innovative techniques of genetic manipulation.

Combined Nitrogen and Duration of Fixation

The ability to nodulate and fix nitrogen in the presence of high levels of soil nitrogen and to extend the period of exponential activity would contribute significantly to a more efficient symbiosis. Varying responses to nitrogen have been reported, but nitrate-resistant plants selected in the greenhouse have not shown similar behavior in field studies (Holl and LaRue, 1976). In peas (*Pisum sativum* L.) cultivars have been detected in which fixation continued during pod-filling (Holl, unpublished). It remains to be seen whether this variation will be reproducible under field conditions. In neither characteristic has evidence been reported on the genetic basis of these variations from what is considered the typical nodulation/ fixation pattern. Until the genetic regulation of these functions can be elucidated, positive developments are likely to be slow.

Investigation of the acetylene-reducing pattern in *Vicia faba* suggested that this seed legume may continue its period of active fixation beyond the point of termination in peas, soybeans, and peanuts. This behavior has not been characterized genetically, but could provide a source of genetic variability for transfer to other legume species. Some attempts to cross peas with *Vicia faba* have been made (Gritton and Wierzbicka, 1975). Embryonic development was observed, but was halted by abnormal changes occurring in the hybrid endosperm after 6–7 days, and the authors concluded that attempts to culture the embryo were not feasible. Some preliminary experiments in embryo culture of *Pisum* sp. and *Vicia* sp. have been conducted to evaluate the prospects for further investigation of a *Vicia* × *Pisum* combination and the possible consequences on symbiotic nitrogen fixation (Holl, unpublished results).

Photosynthesis and Productivity

Direct evidence to link crop yield with photosynthesis is lacking, yet yield must be dependent upon the net photosynthetic activity of the plant. Although the interaction of photosynthesis and nitrogen fixation in grain legumes has been appreciated for some time (Wilson, 1940), it is only recently that efforts have intensified to define this interaction. The CO_2 feeding experiments reported by Hardy and Havelka (1975) reaffirmed the dependence of nitrogen fixation upon a photosynthetically derived energy source. In C_3 plants the effects of photorespiration in reducing the net efficiency of photosynthesis have also been noted (Zelitch, 1976).

Although photosynthesis is subject to genetic control (Moss, 1976), there is a large genotype × environment interaction for photosynthetic rate that often masks the genetic component. Difficulties in obtaining reproducible measurements of the process have compounded difficulties in providing a firm experimental basis for breeding improvements. Greater progress might be expected from an elucidation of some of the individual stages of photosynthesis and selection for specific markers. Stomatal frequency (Ciha and Brun, 1975) and ribulose diphosphate carboxylase (Frey, 1974) are examples of instances where genotypic differences may be of some utility. The effect of photorespiration on net photosynthetic efficiency in C_3 plants has received considerable attention in recent years. Extensive screening programs of a number of crop species have been undertaken to select C_4-like mutants in C_3 plant populations—no such examples have found (Menz et al., 1969; Chollet and Ogren, 1975). Zelitch (1975) has noted similarities between photorespiration and glycolic acid metabolism. He proposed that inhibiting or altering glycolic acid synthesis or further metabolism could produce net photosynthetic rate increases of at least 50 percent. Zelitch and Day (1973) have described pedigree selection experiments for lower photorespiration/faster net photosynthesis in tobacco. Although such plants could be detected, successive generations of selection failed to establish the characteristic in the plant populations.

The complexity of genetic control over photosynthesis and its sensitivity to environment have led to proposals for more innovative approaches to the problems of improvement. One of the options available is the development of plant tissue culture systems such as those of Zelitch and his coworkers (Zelitch, 1976). They were able to grow haploid tobacco cells under photoautotrophic conditions in the absence of sucrose. The rates obtained are comparable to activity *in vivo* when measured on a chlorophyll basis (300 mol of CO_2/ mg chlorophyll^{-1}/ hr^{-1}). Using such systems, it should be possible to define conditions under which variation in photorespiration and/or net photosynthetic rate may be reproducibly selected. The ability to select cells and regenerate plants using *in vitro* methodology will be central to the ultimate success of such an approach.

From comparative estimates of the energy costs to the plant of nitrogen fixation and nitrate assimilation, it was concluded that there were no significant differences (Gibson, 1976; Minchin and Pate, 1973). A substantial proportion (32 percent) of total photosynthate transported to the roots is used in nodule growth and metabolism (Minchin and Pate, 1973). This energy is not available to the plant, and seedling vigor may be adversely affected (Holl, unpublished results). Silsbury (1977) has recently compared the growth and energy requirements of *Trifolium subterraneum* L. cv. Woogenellup under symbiotic and mineral nitrogen conditions. He concluded that the energy requirement for symbiotic nitrogen fixation is substantially greater than that required for the assimilation of mineral nitrogen from the soil. The direct costs of symbiosis to the plant have not been completely clarified; however, should a substantial difference between symbiosis and nitrate assimilation be established, it will have a significant input to calculations of the overall benefits of nitrogen fixation in legumes.

INTERACTIONS AND NEW SYMBIOSES

It was noted earlier in this paper that a significant component of genetic variation was due to nonadditive or interaction effects. More intensive studies on the interaction of *Rhizobium* and legumes from both a biochemical and genetic perspective are only now beginning to be carried out.

Chemotaxis

An account of early investigations of the attraction of rhizobia to the root hair region (chemotaxis) was given by Wilson (1940). Excretion of a plant product was implicated (Thornton, 1929). Ludwig and Allison (1935) confirmed that chemotactic material did exist, but was nonspecific, since corn and wheat had the same effect as the homologous legume. Wilson (1940) indicated that the attraction of rhizobia to the appropriate legume is likely to exhibit greater specificity than the general bacterial growth stimulation observed using nonlegumes.

Recent investigations of chemotaxis by Currier and Strobel (1976, 1977) showed differential attraction of six strains of *Rhizobium* to root exudates of legumes and nonlegumes. Acid or pronase treatment of a trefoil root exudate destroyed its ability to attract trefoil rhizobia, but the active component was not heat-labile (boiling for 10 min). Purification of the trefoil attractant increased its activity toward trefoil rhizobia, but also resulted in the attraction of other rhizobial species, e.g., sain-foin, cowpea, *R. japonicum,* and *R. meliloti.* Although attraction at this stage does occur, the level of specificity is low and does not appear to be especially conducive to improvement via genetic manipulation.

Phytohaemagglutinins (Lectins)

Although the attraction and multiplication of bacteria in the rhizosphere exhibits a low degree of specificity, the subsequent infection process is highly specific. Criteria regulating this interaction are still largely unknown.

Lectins are a group of proteins found in many plant seeds with the ability to bind sugars and polysaccharides, analogous to the antigen antibody reaction (Lis and Sharon, 1973). Hamblin and Kent (1973) proposed a role for lectin in the specific interaction of *Phaseolus vulgaris* and *Rhizobium phaseoli.* Further study of the possible involvement of lectin in nodulation was reported by Bohlool and Schmidt (1974), who suggested that a legume lectin interaction with bacterial surface polysaccharides could account for host-*Rhizobium* specificity.

Recent work on the *Rhizobium*-clover symbiosis may shed additional light on the biochemical nature of host specificity. Dazzo and Hubbell (1975) described cell surface polysaccharides of *R. trifolii* that are specific- ally cross-reactive with antigens on the outer surface of clover roots. The bacterial antigen bound specifically with a multivalent lectin extracted from seeds of *Trifolium repens.* Further reports (Dazzo et al., 1976; Dazzo and Brill, 1977) conclude that clover roots contain proteins (lectins) that cross-link complementary polysaccharides on the surface of clover root hairs and infective *R. trifolii* through a 2-deoxglucose-sensitive binding site. Some controversy exists regarding the involvement of lectins in nodulation specificity. In studies of several fluorescent-labeled lectins and a variety of rhizobial strains, Law and Strijdom (1977) were unable to show any consistent evidence to support a role for lectins in *Rhizobium* specificity. In view of the rather elegant studies of Dazzo and his coworkers described above, it would appear that, in the *Rhizobium*-clover symbiosis at least, reasonable evidence exists to postulate a role for lectins in the specific attraction of *R. trifolii* to clover root hairs. Should these observations reflect a general phenomenon in legumes, it would be possible to ascertain whether host-controlled genetic variation for this character exists.

Infection Studies In Vitro

Although I have earlier indicated some limitation to the potential for *in vitro* studies of the symbiotic infection process, there are two approaches that may permit elucidation of some of the stages of infection and/or provide a vehicle for genetic analysis and selection.

The first and only successful attempt to nodulate isolated cultured legume roots was reported by Lewis and McCoy (1933) using *Phaseolus vulgaris* L. Raggio et al. (1957) described a technique for the culture and nodulation of isolated black wax bean roots. Surprisingly, this technique has not since been exploited as a vehicle for the investigation of nodulation and nitrogen fixation.

Cell culture techniques developed following the initial report of symbiosis *in vitro* by Holsten et al. (1971) did not require bacterial invasion of the plant cell to synthesize active nitrogenase. Reporter (1976) has developed what he describes as the synergetic transmembrane method to examine the transfer of substances between plant cells and rhizobia necessary to produce active nitrogenase and to elucidate why symbiosis occurs in legumes, but not other crop plants. Hermina and Reporter (1977) have produced soybean root hair cultures to examine the association *in vitro* with *R. japonicum*. These differentiated cultures are easy to infect with rhizobia and can be used as a lectin source. Soybean lectin concentrations comparable to those found in cotyledons have been isolated (Pueppke and Bauer, 1976). These latter culture systems could provide the basis for a biochemical comparison of some of the non-nodulating genetic variants in soybean that are known (Holl and LaRue, 1976).

Techniques for the biochemical analysis of the legume-*Rhizobium* interaction are being developed. If we are to exploit this methodology effectively, considerable attention should be directed to measuring and evaluating the available genetic variation, particularly the significant contribution of the nonadditive component as described by Mytton et al. (1977).

GENETIC ENGINEERING

Nif Gene Transfers

The development of new techniques in molecular genetics and interest in genetic engineering in plants has provoked numerous suggestions regarding the transfer of nitrogen-fixing ability. Streicher et al. (1971, 1972) and Dixon and Postgate (1971, 1972) reported *nif* gene transfer using a transducing phage and conjugation to correct *nif* recipient phenotypes within and between *Klebsiella pneumoniae* and *Escherichia coli*. The *nif* genes were functional in either species. Subsequently, Dunican and Tierney (1974) used an R factor to transfer the nitrogen-fixing genes from *R. trifolii* to a strain of *Klebsiella aerogenes*. Dixon et al. (1976) constructed a P plasmid (RP41) carrying the *K pneumoniae nif* gene that was used in transfers to *Agrobacterium tumefaciens* and *R. meliloti*. In neither of the latter instances was acetylene reduction detected, but cross-reacting material to *Klebsiella* nitrogenase molybdo-protein was present in both cultures, indicating transcription and translation of the introduced *nif* genes. Plasmid RP41 was also used to transfer the *Klebsiella nif* genes to *Azotobacter vinelandii*, where expression was observed (acetylene reduction) (Cannon and Postgate, 1976). These data indicate that the regulatory apparatus of an obligate aerobe (*Azotobacter*) could provide a suitable environment and controls for the function of nitrogenase specified by the genes of an anaerobe (*Klebsiella*). It was not possible in these experiments to ascertain whether hybrid enzymes containing components of both donor and recipient were functioning—although this might account for the lower activities observed in *nif* exconjugants.

Considerable flexibility of *nif* gene transfer and expression can be accomplished in bacteria. Schell et al. (1976) have discussed a possible method of transferring *nif* activity to higher plants involving the Ti plasmids responsible for determining the oncogenicity of *Agrobacterium tumefaciens*. They describe evidence suggesting that a given segment of the Ti plasmid is transposable to plant DNA. Schell and his colleagues have constructed Ti:RP4 cointegrates that have the same wide host range as the original RP4 plasmid (Schell et al., 1976). They propose to use this cointegrate as a vehicle to transfer the *nif* genes (on RP4) to plants to discover whether the *nif* message can be integrated into plant DNA via the transposable element of the Ti plasmid. Model experiments might be carried out by infection of plant tissue in culture under controlled conditions where the material could subsequently be assessed for its ability to fix nitrogen (or to show positive evidence of *nif* gene transfer and maintenance).

Somatic Hybridization

The ability to hybridize cells of unrelated plant species or genera via protoplast fusion and plant regeneration may expand gene pools for disease and pest resistance, tolerance to stress, and quality improvements (Gamborg, this volume). Combinations of legumes and nonlegumes have often been suggested as vehicles to effect the transfer of nitrogen-fixing ability to nonlegumes. Protoplast fusion and culture are now a reality, however, with the exception of species such as tobacco, which may also be sexually crossed (Melchers and Labib, 1974), and moss (*Physcomitrella patens*) (Grimsley et al., 1977), in which vegetative regeneration is a common feature, and reproducible plant regeneration has not been obtained. Hybrid cells of legumes and nonlegumes have been obtained (Gamborg, this volume). If we cannot solve the obstacles to regeneration of hybrid material, such novel technology is of little more than academic interest.

Another fusion approach to the transfer of nitrogen-fixing ability was reported by Giles and Whitehead (1976). Polyethylene glycol-induced fusion of protoplasts of *Rhizopogon* sp. with vegetative cells of *Azotobacter vinelandii* produced fungal cultures capable of acetylene reduction and nitrogen fixation (^{15}N assays). The modified fungus was able to grow on media without a defined nitrogen source. No details of the mechanism of operation were postulated, nor were tests reported as to whether the altered material could form a mycorrhizal relationship with the normal host *Pinus radiata*.

DNA Transformation in Plants

Since the work of Ledoux's group of Mol, Belgium, was first reported in the 1960s, various investigators have considered the prospects of gene transfer using exogenous DNA (Ledoux, 1971; Holl et al., 1974). Controversy has arisen over many of the early results with reports by Kleinhofs et al. (1975) and Redei et al. (1976), which discuss alternative interpretations of the observed phenomenon. Other reports of DNA-mediated plant transformation have been reported using bacteriophage vectors (Doy et al., 1973) and isolated DNA (Turbin et al., 1975; Soyfer et al., 1976, 1977). Analysis of subsequent generations of treated material by Soyfer et al. led them to conclude that alterations observed in initial generations were not stably transmitted.

Holl (1975a, 1977) described preliminary results in which attempts were being made to repair a recessive nonnodulating phenotype of *Pisum sativum* L. by feeding DNA from a wild-type cultivar. Occurrences of repair had been observed in treated plants. Although many of the variations were unstable, as observed by Soyfer, corrected F_2 plants have been selected for further analysis (Holl, unpublished results). No genetic analysis of the corrected types has been completed.

The phenomena that might occur during DNA feeding, whether via isolated DNA, bacteriophages, or plasmids, are presently not understood. The possibility that dry seeds might be an ideal recipient for the introduction of foreign DNA was briefly raised by Holl (1973) in a comparison of the deoxyribonuclease activity of soybean protoplasts, cells, and germinating seedlings. Simon (1974) has shown that dry seeds contain no functional membranes until the onset of imbibition and hydration. Ions, hormones and proteins may traverse the nonfunctional membranes at this stage (Shannon and Francois, 1977). Just as competence is recognized in bacterial transformation, an analogous phase may exist in plant seeds, during which time it is possible to introduce a variety of foreign compounds, including DNA.

PROSPECTS

Proponents of nitrogen fixation research predict that increased biological fixation will increase crop productivity and decrease the reliance of the producer on industrial nitrogen fertilizers. Proposals for the improvement of fixation range from simple alterations of the legume host to the most refined methods of molecular biology.

The available literature on symbiotic fixation is extensive. Nevertheless, we are far from a definite understanding of this complex interaction of plant and bacterium. There is no doubt that a significant contribution will be made by attempts to evaluate the genetic variation in available plant germplasm and to assess the considerable genetic variability of the interaction. The use of tissue culture systems and the development of rhizobial genetics should significantly enhance our knowledge and understanding of the biochemistry and regulation of symbiotic fixation. Our ability to evaluate the effects of variation in photosynthesis and photorespiration will depend upon technological advances in measurement of appropriate parameters, the selection of markers reflecting photosynthetic behavior, and the further development of analytical systems such as the tissue culture approaches of Berlyn and Zelitch (Zelitch, 1976).

Genetic manipulations of symbiosis remain an attractive objective. Traditional approaches to the exploitation of genetic variation in legumes should extend our understanding and create cultivars with improved agronomic potential. The more innovative techniques of protoplast fusion, DNA transformation, and *nif* gene transfer to extend the host range of symbiotic nitrogen fixation require considerable assessment. Concern has been expressed over the possible consequences of recombinant DNA approaches to gene transfer in plants (Marx, 1977). In addition to the moral and ethical questions, we require more precise information on the costs to the plant of nitrogen fixation if we are to evaluate the balance sheet of symbiosis effectively. Symbiotic nitrogen fixation has evolved in the natural environment. Under the "artificial" environment of modern crop production systems, we may be able to exploit our knowledge to improve and/or extend symbiotic fixation to enhance crop productivity.

ACKNOWLEDGMENTS

I thank D. Olson for technical assistance in original work presented in this manuscript, Dr. J. J. Child for the cowpea rhizobial culture (32H1), and Dr. O. L. Gamborg and Mr. J. Shyluk for the cell cultures of *Pisum sativum* L. I thank also Mr. A. Lutzko, Mr. E. Knapp, and Mr. L. Nesbitt for their preparation of slides and figures, and especially Mrs. G. Forrest, Mrs. J. Cooper, and Mrs. J. Yorke for their efforts in producing the finished manuscript under difficult circumstances.

REFERENCES

Allison, F. E., and C. A. Ludwig. 1934. The cause of decreased nodule formation on legumes supplied with abundant combined nitrogen. Soil Sci. 37:431–43.

Allison, F.E., and C. A. Ludwig. 1939. Legume nodule development in relation to available energy supply. J. Amer. Soc. Agron. 31:149–58.

Beringer, J. E., and D. A. Hopwood. 1976. Chromosomal recombination and mapping in *Rhizobium leguminosarum*. Nature 264:291–93.

Bezdicek, D. F., R. F. Mulford, and B. H. Magee. 1974. Influence of organic nitrogen on soil nitrogen, nodulation, nitrogen fixation, and yield of soybeans. Soil Sci. Soc. Am. Proc. 38:268–73.

Bohlool, B. B., and E. L. Schmidt. 1974. Lectins: A possible role for specificity in the *Rhizobium*-legume root nodule symbiosis. Science 185:269–71.

Brill, W. J. 1975. Regulation and genetics of bacterial nitrogen fixation. Ann. Rev. Microbiol. 29:109–29.

Burns, R. C., and R. W. F. Hardy. 1975. Nitrogen fixation in bacteria and higher plants. Springer-Verlag, Berlin. 189 pp.

Burris, R. H. 1974. Methodology. *In* A. Quispel (ed.). The biology of nitrogen fixation. Pp. 9–33. Elsevier Publishing Co., New York.

308   Plant Cell and Tissue Culture

Cannon, F. C., and J. R. Postgate. 1976. Expression of *Klebsiella* nitrogen fixation genes (*nif*) in *Azotobacter*. Nature 260:271–72.

Child, J. J. 1976. New developments in nitrogen fixation research. Bioscience 26:614–17.

Child, J. J., and T. A. LaRue. 1974. A simple technique for the establishment of nitrogenase in soybean callus cultures. Plant Physiol. 53:88–90.

Child, J. J., and T. A. LaRue. 1976. Legume-*Rhizobia* symbiosis in tissue culture: Techniques and application. *In* W. E. Newton and C. J. Nyman (eds.), Proceedings of the international symposium on nitrogen fixation. 2:447–55. Washington State University Press, Pullman.

Chollet, R., and W. L. Ogren. 1975. Regulation of photorespiration in C_3 and C_4 species. Bot. Rev. 41:137–79.

Ciha, A. J., and W. A. Brun. 1975. Stomatal size and frequency in soybeans. Crop Sci. 15:309–13.

Currier, W. W., and G. A. Strobel. 1976. Chemotaxis of *Rhizobium* spp. to plant root exudates. Plant Physiol. 57:820–23.

Currier, W. W., and G. A. Strobel. 1977 Chemotaxis of *Rhizobium* spp. to a glycoprotein produced by birdsfoot trefoil roots. Science 196:434–35.

Dart, P. J. 1976. Infection and development of leguminous nodules. *In* R. W. F. Hardy, and W. S. Silver (eds.), A treatise on dinitrogen fixation, Sect. 111. Biology. Pp. 367–472. Wiley-Interscience, New York.

Dazzo, F. B., and W. J. Brill. 1977. Receptor site on clover and alfalfa roots for *Rhizobium*. Appl. Environ. Microbiol. 33:132–36.

Dazzo, F. B., and D. H. Hubbell. 1975. Cross-reactive antigens and lectin as determinants of symbiotic specificity in the *Rhizobium*-clover association. Appl. Microbiol. 30:1017–33.

Dazzo, F. B., C. A. Napoli, and D. H. Hubbell. 1976. Adsorption of bacteria to roots as related to host specificity in the *Rhizobium*-clover symbiosis. Appl. Environ. Microbiol. 32:166–71.

Dilworth, M. J. 1966. Acetylene reduction by nitrogen-fixing preparations from *Clostridium pasteurianum*. Biochim. Biophys. Acta 127:285–94.

Dixon, R., F. Cannon, and A. Kondorosi. 1976. Construction of a P plasmid carrying nitrogen fixation genes from *Klebsiella pneumoniae*. Nature 260:268–71.

Dixon, R. A., and J. R. Postgate. 1971. Transfer of nitrogen-fixation genes by conjugation in *Klebsiella pneumoniae*. Nature 234:47–48.

Dixon, R. A., and J. R. Postgate. 1972. Genetic transfer of nitrogen fixation from *Klebsiella pneumoniae* to *Escherichia coli*. Nature 237:102–3.

Doy, C. H., P. M. Gresshoff, and B. G. Rolfe. 1973. Biological and molecular evidence for the transgenosis of genes from bacteria to plant cells. Proc. Nat. Acad. Sci. USA 70:723–26.

Dunican, L. K., and A. B. Tierney. 1974. Genetic transfer of nitrogen fixation from *Rhizobium trifolii* to *Klebsiella aerogenes*. Biochem. Biophys. Res. Commun. 57:62–72.

Eady, R. R., B. E. Smith, K. A. Cook, and J. R. Postgate. 1972. Nitrogenase of *Klebsiella pneumoniae*: Purification and properties of the component proteins. Biochem. J. 128:655–75.

Evans, H. J. (ed.) 1975. Enhancing biological nitrogen fixation. National Science Foundation, Washington, D.C.

Evans, H. J., and L. E. Barber. 1977. Biological nitrogen fixation for food and fiber production. Science 197:332–39.

Fred, E. B., I. L. Baldwin, and E. McCoy. 1932. Root nodule bacteria and leguminous plants. University of Wisconsin, Madison.

Frey, N. M. 1974. Ribulose 1,5-diphosphate carboxylase activity of *Hordeum vulgare* L. Ph. D. thesis, University of Minnesota.

Gamborg, O. L., R. A. Miller, and K. L. Ojima. 1968. Nutrient requirements of suspension cultures of soybean root cells. Exp. Cell Res. 50:151–58.

Gibson, A. H. 1976. Somiplan symposium. Kuala Lumpur. Cited in Silsbury, 1977.

Giles, K. L., and H. Whitehead. 1976. Uptake and continued metabolic activity of *Azotobacter* within fungal protoplasts. Science 193:1125–26.

Grimsley, N. H., N. W. Ashton, and D. J. Cove. 1977. The production of somatic hybrids by protoplast fusion in the moss *Physcomitrella patens*. Molec. Gen. Genet. 154:97–100.

Gritton, E. T., and B. Wierzbicka. 1975. Embryological study of a *Pisum sativum* × *Vicia faba* cross. Euphytica 24:277–84.

Hamblin, J., and S. P. Kent. 1973. Possible role of phytohemagglutinin in *Phaseolus vulgaris* L. Nature 245:28–30.

Hardy, R. W. F. 1976. Potential impact of current abiological and biological research on the problem of providing fixed nitrogen. *In* W. E. Newton and C. J. Nyman (eds.), Proceedings of the international symposium on nitrogen fixation. 2:693–717. Washington State University Press, Pullman.

Hardy, R. W. F., R. C. Burns, R. R. Hebert, R. D. Holsten, and E. K. Jackson. 1971. Biological nitrogen fixation: A key to world protein. *In* T. A. Lie, and E. G. Mulder (eds.), Biological nitrogen fixation in natural and agricultural habitats. Plant soil special volume. Pp. 561–90. Martinus Nijhoff, The Hague.

Hardy, R. W. F., R. C. Burns, and R. D. Holsten. 1973. Applications of the acetylene-ethylene assay for measurement of nitrogen fixation. Soil Biol. Biochem. 5:47–81.

Hardy, R. W. F., and W. D. Havelka. 1975. Nitrogen fixation research: A key to world food. Science 188:633–43.

Hardy, R. W. F., W. D. Havelka, and B. Quebedeaux. 1976. Opportunities for improved seed yield and protein production: N_2 fixation, CO_2 fixation, and O_2 control of reproductive growth. *In* Genetic improvement of seed proteins. Pp. 196–230. National Academy of Sciences, Washington, D. C.

Hermina, N., and M. Reporter. 1977. Root hair cell enhancement in tissue cultures from soybean roots: A useful model system. Plant Physiol. 59:97–102.

Heumann, W., A. Puhler, and E. Wagner. 1973. The two transfer regions of the *Rhizobium lupini* conjugation. Molec. Gen. Genet. 126:267–74.

Hinson, K. 1975. Nodulation responses from nitrogen applied to soybean half-root systems. Agron. J. 67:799–804.

Holl, F. B. 1973. Cellular environment and the transfer of genetic information. Colloques Intern. C.N.R.S. 212:509–16.

Holl, F. B. 1975a. Innovative approaches to genetics in agriculture. Canad. J. Genet. Cytol. 17:517–24.

Holl, F. B. 1975b. Host plant control of the inheritance of dinitrogen fixation in the *Pisum-Rhizobium* symbiosis. Euphytica 24:767–70.

Holl, F. B. 1977. Molecular genetic modification in legumes. *In* I. Rubenstein, R. L. Phillips, C. E. Green, and R. J. Desnick (eds.), Molecular genetic modification of eukaryotes. Pp. 149–58. Academic Press, New York.

Holl, F. B., O. L. Gamborg, K. Ohyama, and L. E. Pelcher. 1974. Genetic transformation in plants. *In* H. E. Street (ed.), Tissue culture and plant science. Pp. 301–27. Academic Press, New York.

Holl, F. B., and T. A. LaRue. 1976. Genetics of legume plant hosts. *In* W. E. Newton, and C. J. Nyman (eds.), Proceedings of the international symposium on nitrogen fixation. 2:391–99. Washington State University Press, Pullman.

Holsten, R. D., R. C. Burns, R. W. F. Hardy, and R. R. Hebert. 1971. Establishment of symbiosis between *Rhizobium* and plant cells *in vivo*. Nature 232:173–76.

Johnson, J. W., L. F. Welch, and L. T. Kurtz. 1975. Environmental implication of N fixation by soybeans. J. Environ. Qual. 4:303–6.

Kingdon, H. S. 1974. Feedback inhibition of glutamine synthetase from green pea seeds. Arch. Biochem. Biophys. 163:429–31.

Kleinhofs, A., F. C. Eden, M. D. Chilton, and A. J. Bendich. 1975. On the question of the integration of exogenous bacterial DNA into plant DNA. Proc. Nat. Acad. Sci. USA 72:2748–52.

Kurz, W. G. W., and T. A. LaRue. 1975. Nitrogenase activity in rhizobia in absence of plant host. Nature 256:407–8.

Law, I. J., and B. W. Strijdom. 1977. Some observations on plant lectins and rhizobium specificity. Soil Biol. Biochem. 9:79–84.

Ledoux, L. (ed.). 1971. Informative molecules in biological systems. North-Holland Publishing Co., Amsterdam.

Lewis, K. H., and E. McCoy. 1933. Root nodule formation on the garden bean: Studies by a technique of tissue culture. Bot. Gaz. 95:316–29.

Libbenga, K. R., and R. J. Bogers. 1974. Root-nodule morphogenesis. *In* A. Quispel (ed.), The biology of nitrogen fixation. Pp. 430–72. Elsevier Publishing Co., New York.

Lie, T. A. 1969. The effect of low pH on different phases of nodule formation in pea plants. Plant Soil 31:391–405.

Lie, T. A. 1971. Symbiotic nitrogen fixation under stress conditions. *In* T. A. Lie and E. G. Mulder (eds.), Biological nitrogen fixation in natural and agricultural habitats. Plant soil special volume. Pp. 117–27. Martinus Nijhoff, The Hague.

Lie, T. A. 1974. Environmental effects on nodulation and symbiotic nitrogen fixation. *In* A. Quispel (ed.), The biology of nitrogen fixation. Pp. 555–82. Elsevier Publishing Co., New York.

Lie, T. A., and E. G. Mulder (eds.). 1971. Biological nitrogen fixation in natural and agricultural habitats. Plant soil special volume. Martinus Nijhoff, The Hague. 590 pp.

Lis, H., and N. Sharon. 1973. The biochemistry of plant lectins (phytohemagglutinins). Ann. Rev. Biochem. 42:541–74.

Ljones, T. 1974. The enzyme system. *In* A. Quispel (ed.), The biology of nitrogen fixation. Pp. 617–38. Elsevier Publishing Co., New York.

Ludwig, C. A., and F. E. Allison. 1935. Some factors affecting nodule formation on seedlings of leguminous plants. J. Amer. Soc. Agron. 27:895–902.

McComb, J. A., J. Elliott, and M. J. Dilworth. 1975. Acetylene reduction by *Rhizobium* in pure culture. Nature 256:409–10.

Marx, J. L. 1977. Nitrogen fixation: Prospects for genetic manipulation. Science 196:638–41.

Meade, H., and E. Signer. 1976. Abstr. Am. Soc. Microbiol., p. 106. Cited in Beringer and Hopwood, 1976.

Melchers, G., and G. Labib. 1974. Somatic hybridisation of plants by fusion of protoplasts. I. Selection of light resistant hybrids of "haploid" light-sensitive varieties of tobacco. Molec. Gen. Genet. 135:277–94.

Menz, K. M., D. N. Moss, R. Q. Cannell, and W. A. Brun. 1969. Screening for photosynthetic efficiency. Crop Sci. 9:692–94.

Miflin, B. J., and P. J. Lea. 1976. The pathway of nitrogen assimilation in plants. Phytochemistry 15:873–85.

Minchin, F. R., and J. S. Pate. 1973. Carbon balance of a legume and the functional economy of its root nodules. J. Exp. Bot. 24:259–71.

Moss, D. N. 1976. Studies on increasing photosynthesis in crop plants. *In* R. H. Buris and C. C. Black (eds.), CO_2 metabolism and plant productivity. Pp. 31–41. University Park Press, Baltimore.

Munns, D. N. 1968. Nodulation of *Medicago sativa* in solution culture. I. Acid-sensitive steps. Plant Soil 28:129–46.

Mytton, L. R. 1975. Plant genotype × rhizobium strain interactions in white clover. Ann. Appl. Biol. 80:103–7.

Mytton, L. R., H. H. El-Sherbeeny, and D. A. Laws. 1977. Symbiotic variability in *Vicia faba*. 3. Genetic effects of host plant, *rhizobium* strain, and host x strain interaction. Euphytica 26:785–91.

Newton, W. E., and C. J. Nyman (eds.). 1976. Proceedings of the international symposium on nitrogen fixation. Vols. 1 and 2. Washington State University Press, Pullman.

Nutman, P. S. 1956. The influence of the legume in root nodule symbiosis. Biol. Rev. 31:109–51.

Nutman, P. S. 1959. Some observations on root-hair infection by nodule bacteria. J. Exp. Bot. 10:250–63.

Pagan, J. D., J. J. Child, W. R. Scowcroft, and A. H. Gibson. 1975. Nitrogen fixation by *Rhizobium* cultured on a defined medium. Nature 256:406–7.

Parker, M. B., and H. B. Harris. 1977. Yield and leaf nitrogen of nodulating and non-nodulating soybeans as affected by nitrogen and molybdenum. Agron. J. 69:551–54.

Phillips, D. A. 1974. Factors affecting the reduction of acetylene by *Rhizobium*-soybean cell associations *in vitro*. Plant Physiol. 53:67–72.

Pueppke, S., and D. Bauer. 1976. Distribution and localization of the soybean lectin in *Glycine max*. Plant Physiol. 57 (Suppl.):80.

Quispel, A. (ed.). 1974. The biology of nitrogen fixation. Elsevier Publishing Co., New York.

Raggio, M., N. Raggio, and J. G. Torrey. 1957. The nodulation of isolated leguminous roots. Amer. J. Bot. 44:325–34.

Raina, J. L., and V. V. Modi. 1975. Genetic transformation in *Rhizobium*. J. Gen. Microbiol. 57:125–30.

Redei, G., G. Acedo, H. Weingarten, and L. D. Kier. 1976. Has DNA corrected genetically thiamineless mutants of *Arabidopsis*? *In* D. Dudits, G. L. Farkas, and P. Maliga (eds.), Proceedings of the international course on cell genetics in higher plants. Pp. 91–94. Académiai Kiadó, Budapest.

Reporter, M. 1976. Synergetic cultures of *Glycine max* root cells and rhizobia separated by membrane filters. Plant Physiol. 57:651–55.

Robertson, J. G., K. J. F. Farnden, M. P. Warburton, and J. M. Banks. 1975. Induction of glutamine synthetase during nodule development in lupin. Aust. J. Plant Physiol. 2:265–72.

Robertson, J. G., M. P. Warburton, and K. J. F. Farnden. 1975. Induction of glutamate synthase during nodule development in lupin. FEBS. Lett. 55:33–37.

Schell, J., M. van Montagu, A. DePicker, D. De Waele, G. Engler, C. Genetello, J. P.

Hernalsteens, M. Holsters, E. Messens, B. Silva, S. Van den Elsacker, N. Van Larebeke, and I. Zaenen. 1976. Crown gall: Bacterial plasmids as oncogenic elements for eukaryotic cells. *In* I. Rubenstein (ed.), Molecular biology of plants. Academic Press, New York (in press).

Schubert, K. R., and H. J. Evans. 1976. Hydrogen evolution: A major factor affecting the efficiency of nitrogen fixation in nodulated symbionts. Proc. Nat. Acad. Sci. USA 73:1207–11.

Scott, D. B., J. G. Robertson, and K. J. F. Farnden. 1976. Ammonia assimilation in lupin nodules. Nature 263:703–5.

Shanmugan, K. T., and R. C. Valentine. 1976. Solar protein. California Agriculture, November, pp. 4–7.

Shannon, M. C., and L. E. Francois. 1977. Influence of seed pretreatments on salt tolerance of cotton during germination. Agron. J. 69:619–22.

Silsbury, J. H. 1977. Energy requirement for symbiotic nitrogen fixation. Nature 267:149–50.

Silver, W. S., and R. W. F. Hardy. 1976. Newer developments in biological dinitrogen fixation of possible relevance to forage production. *In* Biological N fixation in forage-livestock systems. ASA Special Publication No. 28, pp. 1–34.

Simon, E. W. 1974. Phospholipids and plant membrane permeability. New Phytol. 73:377–420.

Soyfer, V. N., N. A. Kartel, N. M. Chekalin, Y. B. Titov, K. K. Cieminis, and N. V. Turbin. 1976. Genetic modification of the waxy character in barley after an injection of wild-type exogenous DNA. Analysis of the second seed generation. Mutat. Res. 36:303–10.

Soyfer, V. N., A. D. Morozkin, V. P. Bogdanov, Y. B. Titov, and N. M. Chekalin. 1977. Starch composition and hordein electrophoretic patterns of grains of genetically transformed barley plants. Envir. Exp. Bot. (in press).

Sprent, J. I. 1969. Prolonged reduction of acetylene by detached root nodules. Planta 88:372–75.

Sprent, J. I. 1971. The effects of water stress on nitrogen-fixing root nodules. I. Effects on the physiology of detached soybean nodules. New Phytol. 70:9–17.

Sprent, J. I. 1972. The effects of water stress on nitrogen-fixing root nodules. II. Effects on the fine structure of detached soybean nodules. New Phytol. 71:443–50.

Streicher, S. L., E. G. Gurney, and R. C. Valentine. 1971. Transduction of the nitrogen-fixation genes in *Klebsiella pneumoniae*. Proc. Nat. Acad. Sci. USA 68:1174–77.

Streicher, S. L., E. G. Gurney, and R. C. Valentine. 1972. The nitrogen fixation genes. Nature 239:495–99.

Streicher, S. L., K. T. Shanmugan, F. Ausubel, C. Morandi, and R. B. Goldberg. 1974. Regulation of nitrogen fixation in *Klebsiella pneumoniae*: Evidence for a role of glutamine synthetase as a regulator of nitrogenase synthesis. J. Bacteriol. 120:815–21.

Thornton, H. G. 1929. The role of the young lucerne plant in determining the infection of the root by nodule-forming bacteria. Proc. Roy. Soc. Lond. 104B:481–92.

Tubb, R. S. 1974. Glutamine synthetase and ammonium regulation of nitrogenase synthesis in *Klebsiella*. Nature 251:481–84.

Turbin, N. V., V. N. Soyfer, N. A. Kartel, N. M. Chekalin, Y. L. Dorohov, Y. B. Titov, and K. K. Cieminis. 1975. Genetic modification of the waxy character in barley under the action of exogenous DNA of the wild variety. Mutat. Res. 27:59–68.

Vincent, J. M. 1965. Environmental factors and the fixation of nitrogen by the legume. *In* W. V. Bartholomew and F. E. Clark (eds.), ASA Monograph No. 10, pp. 385–435.

Wilson, P. W. 1940. The biochemistry of symbiotic nitrogen fixation. University of Wisconsin Press, Madison.

Yao, P. Y., and J. M. Vincent. 1969. Host specificity in the root hair "curling factor" of *Rhizobium* spp. Aust. J. Biol. Sci. 22:413–23.

Zelitch, I. 1975. Pathways of carbon fixation in green plants. Ann. Rev. Biochem. 44:123–45.

Zelitch, I. 1976. Biochemical and genetic control of photorespiration. *In* R. H. Burris, and C. C. Black (eds.), CO_2 metabolism and plant productivity. Pp. 343–58. University Park Press, Baltimore.

Zelitch, I., and P. R. Day. 1973. The effect on net photosynthesis of pedigree selection for low and high rates of photorespiration in tobacco. Plant Physiol. 52:33–37.

DAVID A. EVANS AND ELTON F. PADDOCK

Mitotic Crossing-over in Higher Plants

18

Mitotic crossing-over (MCO), an extensively characterized genetic pheno-menon in fungi (Pontecorvo and Kafer, 1958) and *Drosophila* (Becker, 1976), has not been utilized in higher plants. Pontecorvo (1958) demon-strated the importance of MCO as an integral component in the parasexual life cycle of fungi by utilizing MCO to map genetic loci. As continued research results in a parasexual life cycle of higher plants, as proposed by Carlson et al. (1972), MCO will prove valuable as an aid in somatic cell genetics. In this paper we review the evidence in support of the existence of MCO in higher plants and characterize the phenomenon.

Somatic mosaicism has been observed frequently in plants and animals. These variegations may be the result of chromosome breakage, deletion, nondisjunction, duplication, translocation, MCO, or gene mutation. In higher eukaryotes, though, the trademark of MCO has been the double (Db) spot consisting of two adjacent, equal-sized areas of respectively different mutant phenotypes amidst otherwise wild-type cells. In animals, eye and skin pigmentation and external morphological variations have been observed as Db spots. In plants, on the other hand, most such variegations are the readily detected alterations in chlorophyll or antho-cyanin pigmentation. In lower eukaryotes (fungi), Db spots often consist of biochemical or pigment variations. Double spots are observed in tissue that is heterozygous for the genetic loci involved in spot formation. In most Db spots observed in higher plants (Evans and Paddock, 1976), only one locus

David A. Evans and Elton F. Paddock, Department of Genetics, Ohio State University, Columbus, Ohio 43210 (present address of David A. Evans: Department of Biological Sciences, State University of New York, Binghamton, New York 13901).

Fig. 1.   Double spot in *Xa-2/xa-2* Lycopersicon esculentum.

with incomplete dominance is involved, thereby diminishing the certainty that the Db spot originated by MCO. An example of a Db spot in tomato leaf tissue with background genotype *Xa-2/xa-2* is depicted in figure 1.

Historically, Serebrovsky (1925) first used the term *somatic segregation* to explain variegations on feathers of chickens. Stern (1936) initiated the term *somatic crossing-over*, whereas fungus geneticists (Pontecorvo and Kafer, 1958) used the term mitotic recombination. Since the phenomenon that Stern described as somatic crossing-over may occur in almost all cells undergoing mitosis (Bateman, 1967), including cells in the gonial line (Suzuki, 1965a, 1965b) and cells in tissue culture (Carlson, 1974), we shall refer to the phenomenon as MCO. Mitotic recombination will be reserved for those cases where authors have expressed uncertainty as to the reciprocity of the crossover event.

Painter (1934) first suggested that "Db" spots he induced in *Drosophila melanogaster* using X-rays were the result of mitotic crossing-over. Stern (1936) genetically analysed Db spots in *Drosophila melanogaster* and (1939) showed that mitotic and meiotic crossing-over are genetically identical. Stern examined heterozygotes of both singed bristles (*sn*) and yellow body color (*y*). In the otherwise wild-type female flies of the genotype (sn^+ y/sn y^+), Stern identified single spots that were singed (presumably sn y/sn y^+) or yellow (sn^+ y/sn y). Consistent with his explanation, these same flies also had yellow and singed (sn y/sn y) Db spots. Stern was able to explain the frequencies of spots based on the arrangement of loci on the X-chromosome (fig. 2). Based on chromosome arrangement in a G_2 cell in mitosis, four possible outcomes were conceivable: (1) With no crossovers, two wild-type cells result. (2) With a single crossover between the *y* and *sn* loci, a single yellow spot might arise. (3) With a single crossover between *sn* and the centromere, a Db spot results (this possibility is depicted in fig. 2). (4) Single *sn* spots only resulted from a double crossover; one between the *sn* locus and the centromere and one between the *y* and *sn* loci.

The frequency of observed MCO reflects the distance between a locus and the centromere. As in figure 2, only one-half of the actual crossovers result in genotypically altered cell lines. Kaufmann (1934) reported the first cytological evidence of chiasma-type configurations, suggesting genetic exchange, between homologous chromosomes in *Drosophila* somatic cells. Using deficiency loci in *Drosophila*, Stern (1935, 1936) argued against deletion or elimination as a cause of Db spots. Stern also used X-chromosome genes, and thereby excluded nondisjunction as a possible mechanism, based on the absence of sex phenotype distortions. The first direct cytological evidence for MCO was reported by Huttner and Ruddle (1976) for *Muntiacus muntiac* and mouse A9 cells using the BUdR-Hoechst chromatid-labeling technique (Latt, 1973).

Dahlgren and Ossian (1927), studying apple epidermis, were the first to observe twin spots in plants, including the appearance of a twin spot within

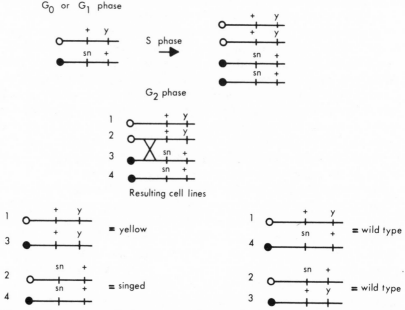

Fig. 2. Arrangement of chromosomes during mitotic crossing-over in *Drosophila melanogaster*.

the dark portion of a twin spot. Jones (1937) proposed that these apple sectors resulted from mitotic segregation, as he cited additional examples of color changes in citrus fruits and emphasized that vegetatively propagated fruits are "notoriously" heterozygous.

Jones (1937, 1938) first attempted to analyze the genetic implications of somatic sectoring in plants. He examined variegations in triploid maize endosperm segregating two at a time for combinations of the *C, C', Pr, Su,* and *Wx* genes. In nearly all genetic combinations, he observed Db spots and computed a combined frequency of 186 Db spots in 4,778 kernels or 0.039 Db spots per kernel. As expected, the frequency of Db spots is dependent upon genetic background and position of the two loci in relation to the centromere. Jones also reported secondary twinning in the dark area of red-white Db spots. He concluded that the Db spots could be due to nondisjunction, translocation, or MCO. Clark and Copeland (1940) made a cytological examination of Jones's material and found that of three types of mosaics studied, (1) simple color losses, (2) high frequency of Db spots, and (3) color losses and color changes, including Db spots, the line (no. 1019) with mostly Db spots had the lowest frequency of abnormal divisions. The frequency of abnormal anaphases in line 1019 was 2.5–4.7 percent. Using very conservative estimates, this frequency of abnormal anaphases may account for those Db spots observed, so that Vig's (1967) conclusion that the number of observed chromosome aberrations was insufficient to account for the number of observed spots, may be questioned. Assuming (1) that only 100 mitoses per seed can result in a visible spot, and (2) that only 1/20 of all abnormalities affect the locus examined (since n = 10 for *Zea mays* and only one-half of all recombinants are visually recovered), and (3) using the lowest abnormality frequency of 2.5 percent, a maximum of 600 spots would be expected among 4,778 kernels, significantly more than the 186 observed.

Hendrychova-Tomkova (1964) observed Db spots in inflorescences of heterozygous *Salvia splendens* (white, red, and violet colors), and suggested that these spots were the result of MCO.

More recently, specific mutants have been used to study Db spots in three higher plant species. Chlorophyll deficiency mutants have been found in *Glycine max* (y_{11}), *Nicotiana tabacum* (*Su*), and *Lycopersicon esculentum* (*Xa-2*), all expressing incomplete dominance. Spots that are either dark green (D, resembling the Y_{11}/Y_{11}, *su/su*, or *xa-2/xa-2* genotypes), aurea (A, resembling the y_{11}/y_{11}, *Su/Su*, or *Xa-2/Xa-2* genotypes), or double (Db, composed of adjacent areas of dark green and aurea) are frequently found on the leaf blades of heterozygotes of all three species.

Based on the heterozygous nature of these mutants, they are most likely regulatory mutants affecting chlorophyll synthesis and not the result of defective enzymes affecting a portion of the pathway of chlorophyll biosynthesis (von Wettstein et al., 1974).

Vig and Paddock (1968) first suggested MCO as the mechanism of origin of Db spots in *Glycine max*. Progeny of self-fertilized heterozygous plants assort in a 1 dark green:2 light green:1 lethal yellow ratio. The y_{11} locus was discovered by Weber and Weiss (1959), and has been used subsequently for teaching basic genetics (Paddock, 1966). Mark (1961) was the first to report Db spots on the surface of Y_{11}/y_{11} leaves, and he increased the frequency of single spots with colchicine. Mark used a small sample size and found that heterozygous seedlings raised following colchicine treatment had a higher frequency of spots at 80° F than at 60° F. Such seedlings raised at each temperature without colchicine had no spots. Such seedlings raised at 65° F with colchicine and red light also had no spots. Vig (1971) confirmed this colchicine effect.

Johnson and Bernard (1963) and Bernard and Weiss (1973) listed eighteen loci governing chlorophyll development in soybean, but only y_{11} has incomplete dominance. Nilan and Vig (1976) suggested that y_9 is allelic to y_{11}, based on the ability of each to form Db spots when heterozygous, but this was not confirmed with traditional genetic tests (Bernard, personal communication).

A number of experimenters have further characterized this locus. Wolf (1963, 1965) analysed pigment content and parameters of photosynthesis in the three genotypes. He found altered carotene content and disruption in the chlorophyll a:chlorophyll b ratio and nearly the same maximum rates of photosynthesis in Y_{11}/Y_{11} and Y_{11}/y_{11} seedlings. Sun (1965) and Crang and Noble (1974) characterized the ultra-structural differences of all three genotypes. In Y_{11}/Y_{11} seedlings the grana are extensively stacked, slightly stacked in Y_{11}/y_{11} seedlings, and unstacked and slightly coiled in y_{11}/y_{11} seedlings. Keck et al. (1970a, 1970b) characterized additional physiological parameters of photosynthesis in these genotypes. Koller and Dilley (1974) examined the rate of CO_2-fixation in all three types of seedlings.

The sulfur (Su) mutant of *Nicotiana tabacum* was reported by Burk and Menser (1964) to segregate in a 1:2:1 ratio for dark green:light green:lethal aurea. They also reported single spots on the surface of heterozygous Su/su plants. Menser et al. (1965) examined the pigment contents of this mutant. Schmid (1967) defined the growth rate of each genotype under different light intensities and concluded that Su/su plants require high light intensity to compete with dark green (su/su) plants. We have cultured leaf explants of all three genotypes and found that prolific shoot formation occurred on all media listed in table 1. The three genotypes, Su/Su, Su/su, and su/su, grow equally well *in vitro* on agar medium with sucrose and auxin (Schaeffer and Menser, 1975) and respond equally well to various concentrations of benzyladenine (Schaeffer, 1977). Burk (1970) regenerated complete haploid plants from cultured anthers of Su/su. Carlson (1974) regenerated complete diploid plants from isolated components of Db spots of Su/su leaves demonstrating the regenerative potential of these genotypes in tobacco, and thus carried out genetic analyses of the spots. He found that 12/13 Db spots resulted from MCO, and 1/13 Db spots resulted from nondisjunction.

Ross and Holm (1959) described an aurea mutant in *Lycopersicon esculentum* whose progeny segregated in a ratio of 1 dark green:2 light green:1 lethal yellow. The locus has subsequently been officially named *Xanthophyllic-2* (Clayberg et al., 1966). Ross and Holm noted Db spots in both colchicine-treated(0.5% in lanolin)and control plants of the light green seedlings. They also noted a predominance of dark green (D) versus yellow (Y) single spots, 442 D to 31 Y single spots in colchicine-treated material and 337 D to 13 Y spots among controls.

Double spots have also been reported in higher plants that are heterozygous at two loci each with complete dominance (as in animals and fungi). For example, Db spots appear on flower petals and in stamen hairs of *Tradescantia hirsuticaulis* heterozygous at the D and E loci. Somatic sectoring on petals of *Tradescantia* (Clone 02) was first reported by Cuany et al. (1959). Mericle and Mericle (1967) first proposed MCO as a mechanism to account for sectoring in *Tradescantia* Clone 02. Sparrow et al. (1974) used Clone 02 for a number of radiation experiments, but Clone 02 is heterozygous at only the *D* locus and therefore has no double spots. Subsequently, a mutant line of *Tradescantia* was isolated that was heterozygous at two loci involved in anthocyanin biosynthesis. The D^+ allele confers the ability to produce a blue pigment, delphinidin, while suppressing production of cyanidin, a red pigment. The E^+ allele results in production of cyanidin, i.e., overcomes suppression by the D^+ allele. Flowers on plants that are double heterozygotes produce both red and blue pigment, resulting in purple petals (Mericle et al., 1974). Christianson (1975) examined the increase in Db spot frequency with gamma-ray dose in *Tradescantia hirsuticaulis* and correlated the radiation results with those found in *Drosophila* for gamma- and X-rays. Based on spot frequency analysis, Christianson (1975) suggested that the *D* and *E* loci are on the same arm of a chromosome.

Barrow et al. (1973) reported MCO in *Gossypium barbadense* (tetraploid cotton). They used the V_1 (isolated by Killough and Horlacker, 1933) and V_7 (isolated by Turcotte and Feaster, 1972) genes, which were shown to be homoeoallelic by Turcotte and Feaster (1973). Crossing double heterozygotes resulted in a 5 green:6 yellow:5 white seedling ratio. Barrow et al. (1973) discovered spots on leaves of tetraploid V_1/V_1, v_7/v_7 and diploid V_1, v_7 plants and proposed that homoeologous MCO would account for Db spots.

Barrow and Dunford (1974) observed multivalents in V_1/V_1, v_7/v_7 cotton that purportedly resulted from reciprocal translocations, i.e., homoeologous MCO.

Dulieu (1974, 1975) observed Db spots on leaves of tetraploid a_1^+/a_1, a_2^+/a_2 and diploid a_1^+, a_2 chlorophyll-deficient tobacco plants.

Deshayes and Dulieu (1974) initiated callus cultures and subsequently produced plants from both sectors of a Db spot of tetraploid *N. tabacum*, but on the results of progeny tests it was suggested that the Db spots originated via reciprocal translocation rather than MCO.

Zaman and Rai (1973) have concluded that MCO is responsible for single albino spots on the leaves of *Collinsia heterophylla* heterozygous for the w_1 gene.

SPONTANEOUS FREQUENCIES OF MITOTIC CROSSING-OVER

To date, Db spots have been observed in a number of organisms and, usually along with additional genetic analyses, have been used as evidence of MCO. A summary of available spontaneous frequencies of Db spots is

TABLE 1

FIVE VARIATIONS OF MEDIUM A ON WHICH LEAF EXPLANTS
FROM ALL THREE GENOTYPES OF SU TOBACCO FORMED SHOOTS PROLIFICALLY

Medium A (Control)	MSBS
	0.64 mg/l kinetin
	20 g/l sucrose
	2 mg/l glycine
	0.5 mg/l pyridoxin
Variation I	A + 4 mg/l IAA
Variation II	A + 2 mg/l IAA; increase sucrose to 30 g/l
Variation III	A + 40 mg/l adenine
Variation IV	A; decrease sucrose to 5 mg/l
Variation V	A; increase sucrose to 50 mg/l

NOTE: Five replicates of each genotype on each variation.

shown in table 2. We have characterized parameters of spontaneous MCO in higher plants (Evans and Paddock, 1976). For Su/su *N. tabacum* and Y_{11}/y_{11} *G. max*, characteristics with documented randomness are summarized in table 3.

In tobacco, randomness is observed for left versus right leaf halves and between different leaves in the shoot, but more Db spots are visible on the distal versus the proximal equal area portion of each Su/su *N. tabacum* leaf (Evans and Paddock, 1976). This inequality, though, reflects the cell number, as the cell density is greater in the distal portion of the leaf. When the distal and proximal halves are reassigned based on equal cell numbers, the spot frequencies are equal (table 4).

In soybean, proximal versus distal, and left versus right leaf halves are random, but simple leaves contain more spots than first compound leaves (especially in aged seed). This reflects the larger number of cells present in the dormant seed in each of the two simple leaf primordia than in the one first compound leaf primordium. Spontaneous total spot frequencies per leaf differ between species from 0.06 (tomato) to 2.30 (tobacco) (Evans and Paddock, 1976).

INDUCED FREQUENCIES OF MITOTIC CROSSING-OVER

To utilize or characterize the phenomenon of MCO, it must be increased in frequency. Several chemical and physical treatments have been used. The two most effective treatments are X-rays and Mitomycin C (MMC).

X-ray–Increased Frequency of Mitotic Crossing-over

Becker (1956) first used X-rays (1,200 R) to increase the frequency of spots in *Drosophila* eyes (table 5) and observed that the third chromosome

TABLE 2

SPONTANEOUS FREQUENCIES OF DOUBLE SPOTS OR MITOTIC CROSSING-OVER

Organism	Loci	Spontaneous Frequency	Reference
Animals			
Drosophila melanogaster	y, sn	0.002–0.06 spots/fly	Stern, 1936
Mus musculus	C^{cn}, C	NA	Bateman, 1967; Grunberg, 1966
Musca domestica	bwb, w	0.00	Nothiger and Dubendorfer, 1971
Plants			
Antirrhinum majus	Inc	0.96×10^{-5}/cell	Harrison and Carpenter, 1977
Aspergillus amstelodami	13 markers	NA	Lewis, 1969
Aspergillus fumigatus	leu, cr, ch, ade, lys	NA	Stromnaes and Garber, 1963
Aspergillus nidulans	many	0.001 spots/conidium	Pontecorvo and Kafer, 1958
Aspergillus niger	many	NA	Pontecorvo et al., 1953
Cochliobolus sativus	An, me, w2, wt, thi, lys, pyro	NA	Tinline, 1962
Coprinus radiatus	many	NA	Prud'Homme, 1970
Dictyostelium discoideum	$MeOH^{r}$	5×10^{-5}/cell	Katz and Kao, 1974; Williams et al., 1974
Glycine max	y_{11}	0.08 spots/leaf	Vig, 1967; Vig and Paddock, 1968
Gossypium barbadense	v_1, v_7	0.03 spots/leaf	Barrow et al., 1973
Lycopersicon esculentum	xa-2	0.01 spots/leaf	Ross and Holm, 1969
Nicotiana tabacum	Su	0.15 spots/leaf	Evans and Paddock, 1976
Neurospora crassa	pan3, pan5, ylo, ad8, typr-2	NA	Pittenger and Coyle, 1963
Penicillium chrysogenum	w,y	1.08×10^{-5}/conidium	Pontecorvo and Sermonti, 1954
Penicillium italicum	wh-4, met-6, wh, ade-l	NA	Stromnaes et al., 1964
Penicillium expansum	b, ws, pyr, br3, nic2, ri, acr	NA	Barron, 1962
Saccharomyces cerevisiae	chx4, ade2	12×10^{-5} survivors	Mayer, 1973; Zimmerman, 1975
Salvia splendens	V, L, Int, P	0.295 spots/plant	Hendrychova-Tomkova, 1964
Tradescantia hirsuticaulis	D, E	0.001 spots/stamen hair	Christianson, 1975
Verticillium albo-atrum	ur-1	0.021 ± 0.005/conidium	Hastie, 1967
Zea mays	C, Pr, Su, Wx	0.039 spots/kernel	Jones, 1937

TABLE 3

RANDOMNESS OF SPONTANEOUS DB SPOT FREQUENCIES ON
SU/SU NICOTIANA TABACUM AND Y_{11}/Y_{11} GLYCINE MAX

	Nicotiana tabacum	*Glycine max*
Leaf surface		
Left vs. Right	Random	Random
Prox. vs. Distal	Nonrandom[1]	Random
(equal area halves)		
Position in the	Random	Nonrandom[2]
leaf sequence		

1 = reflects a difference in cell density; 2 = simple leaves have
more spots than first compound leaves, especially in aged seed. This
inequality reflects a variation in number of cells present in the leaf
primordia of the dormant seed.

TABLE 4

FREQUENCY OF SPOTS WITHIN THE LEAVES OF SU/SU N. TABACUM

	LEFT VERSUS RIGHT HALVES				
	Left	Right	No. of Leaves	X^2-value	P
Equal area	192	181	160	0.324	$0.5 < p < 0.7$
	PROXIMAL VERSUS DISTAL HALVES				
	Proximal	Distal	No. of Leaves	X^2-value	P
Equal area	160	213	160	7.531	< 0.001
Equal cell numbers	181	192	160	0.324	$0.5 < p < 0.7$

is more resistant than the X-chromosome to X-ray–induced MCO. Utilizing inversion heterozygotes, Becker (1969) showed that the greater amount
of heterochromatin present in the X-chromosome was responsible for its
greater sensitivity to X-rays. Becker also demonstrated that first- and
second-instar larvae are more sensitive to X-rays than younger developmental stages. Abbadessa and Burdick (1963) applied X-rays at three
periods of larval development and demonstrated that spot frequency
decreased with increasing time after egg deposition from 48 to 84 to 120 hr
(consistent with Becker), and suggesting that cells in *Drosophila* larvae are
not uniformly sensitive to X-rays over developmental time, unlike those in
tobacco leaf primordia (Evans and Paddock, 1977).

Merriam and Fyffe (1972) found that neither size nor phenotype of spots
was altered by X-ray dosage, implying that X-rays only increased the
frequency of the regularly occurring process of MCO. The effect on cells of
X-rays fitted a dose-squared (nonlinear) curve suggesting an event involving two breaks. They observed a decrease in frequency of MCO using
dosage fractionation and concluded that it requires two breaks to produce

TABLE 5

Effect of X-rays on the Frequency of Double Spots

Species	Genotype	Dosage	Frequency of Mutant Spots for One Locus	Reference
Animals				
Drosophila melanogaster	w/+, female	1,200 R	1.5×10^{-2}	Becker, 1957
Drosophila melanogaster	mwh/+	1,000 R	0.1–1.6×10^{-2}	Garcia-Bellido and Merriam, 1971
Musca domestica	bwb +/+ w	1,000 R	2.7×10^{-2}	Nothiger and Dubendorfer, 1971
Plants				
Saccharomyces cerevisiae	Y_{11}/y_{11}	15,000 R	2.97×10^{-2}	Nakai and Mortier, 1969
Glycine max	Su/su	1,600 R	0.96×10^{-2}	Evans, 1977
Nicotiana tabacum		800 R	0.60×10^{-2}	Evans, unpublished

TABLE 6

Effect of Mitomycin C on the Frequency of Mitotic Crossing-over

Organism	Concentration	Treatment	Duration	Fold-Increase	Frequency/Cell	Reference
Glycine max	0.075 mM	Wetted seeds	12 hr	28.67	0.367×10^{-2}	Vig and Paddock, 1968
			24 hr	110.83	1.419×10^{-2}	
Nicotiana tabacum	0.1 mM	Cell culture	1 hr	4.7	0.216×10^{-2}	Carlson, 1974
Ustilago maydis	1.20 mM	Agar culture	200 min	31.93	0.182×10^{-2}	Holliday, 1964
Saccharomyces cerevisiae	1.20 mM	Agar culture	240 min	8.0	0.05×10^{-2}	Holliday, 1964
				(try') 5.67 (ac')	0.14×10^{-2}	
Drosophila melanogaster	0.30 mM	Injection	–	3.85	NA	Suzuki, 1965a

an MCO and that, by utilizing dose fractionation, the first break is repaired before the second break is induced.

Garcia-Bellido and Merriam (1971) used X-ray–increased MCO at various developmental stages to ascertain the cellular progenitors to such body organs as trichomes, chaetae, and sense organs. Garcia-Bellido (1972) was the first to use X-ray–induced MCO to map a large number of loci in *Drosophila*, as has been done routinely in fungi. He used a trichome marker, two color mutants, and five chaetae morphological mutants. The map of these X-ray–induced markers was comparable to a map produced by spontaneous MCO and to the actual cytogenetic location of loci in salivary chromosomes.

Garcia-Bellido and Meriam (1971) found that the frequency of X-ray–induced spots for *mwh* on the wing increases *exponentially* with the *age of larva* at the time of irradiation. The size of spots (number of cells per clone) *decreases exponentially* with increasing larval age at the time of irradiation, a relationship based on the log growth of cells in the imaginal discs and the decrease in number of divisions per clone prior to morphogenesis. From these observations they were able to estimate the cell cycle time as 8–9 hr in early larval periods (Merriam and Garcia-Bellido, 1972).

Nothiger and Dubendorfer (1971) irradiated larvae of *Musca domesticus* (houseflies) with 500, 1,000, or 2,000 R at 1,210 R/min. No flies survived the 2,000 R treatment. Young larvae (48 to 72 hr after oviposition) were most susceptible to the X-rays as measured by size of spots but not number of spots (probably due to smaller X-ray target size of young larvae).

Stern (1969) first observed *spontaneous intragenic* mitotic recombination in *Drosophila* examining the X-linked *white eye* locus. Kelly (1974) further examined X-ray–induced intragenic recombination in eye pigments using vermilion and apricot eye color pigments as distal markers in relation to mutants of the *lozenge* locus. She showed that MCO induced by X-rays was nonreciprocal within the *lozenge* locus.

X-rays have been used to increase the frequency of MCO in Su/su N. *tabacum* (Evans and Paddock, 1977), Y_{11}/y_{11} G. *max*, and $Xa-2/xa-2$ L. *esculentum* (Evans, 1977). We showed that MCO was increased by X-rays at random over the tobacco leaf surface with respect to cell density. On the other hand, as seven leaf primordia were sensitive at the time of X-irradiation of tobacco shoot apices, all leaf primordia were at different developmental stages (Evans and Paddock, 1977). Although all cells in a leaf primordium apparently are uniformly sensitive to the X-rays, a mutation response curve for X-rays similar to that of other plant species was obtained (e.g., *Tradescantia*, Mericle and Mericle, 1967; *Antirrhinum*, Cuany et al., 1959; *Zea*, Stein and Steffensen, 1959; and *Petunia*, Moore and Haskins, 1935). Between 0 and 300 R, Db spots increase linearly with increasing dosage. Seeds of Y_{11}/y_{11} *G. max* were treated with X-rays on three separate occasions, following the methodology presented by Evans (1977), using dosages from 50 to 8,000 R. It was impossible to count the overabundance of spots on simple leaves from seeds receiving greater than 2,000 R. The frequency of all three types of spots increased with increasing X-ray dosage. For first compound leaves the frequency of Db spots increased linearly (based on regression analysis) using 8 dosages from 0 to 1,600 R. The 1,600 R X-ray treatment resulted in a 282-fold increase in MCO frequency over control (Evans, 1977). For *Xa-2/xa-2 L. esculentum*, when shoot apices were irradiated with 100 and 300 R, an increase in all three types of spots was observed. At least three successive leaf primordia were present in the shoot apex at the time of X-irradiation. The frequency of MCO for tomato was much lower than for soybean or tobacco. This difference in frequencies may reflect three factors. (1) Homoeologous exchanges are not possible in diploid tomato. If homoeologous exchanges can produce Db spots (as suggested by Barrow et al., 1973), then tomato would be expected to have a reduced frequency of Db spots. In support of this hypothesis, it should be noted that 4X *G. barbadense* contains more spots than 2X *G. barbadense* (Barrow et al., 1973). (2) There are differences in genetic background between species, which could alter the frequency of MCO. (3) The Db spot frequency may reflect chromosome distance between the marker and the centromere. In all three species the ratios of types of spots remain identical to control ratios following X-irradiation.

Roper (1952) was the first to observe MCO in induced diploid strains of *Aspergillus nidulans*. Mortimer (1957) has used X-rays to increase the frequency of MCO in *Saccharomyces*. Subsequent mitoses following crossing-over result in haploids (via haploidization) or diploids (used for mitotic recombination analysis). Haploidization has been used to identify the linkage groups of differing markers (Kafer, 1958). In haploids resulting from a diploid that underwent recombination involving identifiable markers, only *ab* and a^+b^+ offspring result if both markers are on the same chromosome, whereas if the markers are in repulsion, only a^+b and ab^+ offspring result. On the other hand, markers on nonhomologous chromosomes produce haploids with all four combinations equally likely (a^+b^+, a^+b, ab^+, and *ab*). In *Aspergillus nidulans* eight linkage groups have been established using this method. Diploid segregants resulting from MCO have been used to document the sequence of markers that are on the same chromosome. There has been some disparity, though, between maps based on meiotic and mitotic crossing-over. This suggests that the two types of crossovers do not occur with equal constancy over the entire chromosome, and has led Holliday (1964) to conclude that the frequency of

MCO, which is less than the frequency of meiotic crossing-over, is dependent upon the extent of mitotic pairing, which is probably not complete. The frequency of coincidence of MCO is rare in different chromosomes, in two arms of one chromosome, in one arm of a chromosome, and in conjunction with nondisjunction (Pontecorvo and Kafer, 1958). Parag and Parag (1975) have isolated mutants affecting the frequency of MCO in *Aspergillus nidulans*. They have isolated *rec* mutants that decrease mitotic recombination as well as mutants that increase mitotic recombination frequency up to 100-fold. They have not yet tested the effect of these recombination mutants on meiosis. Esposito (1968) has shown that gene maps based on X-ray–increased MCO correspond excellently with maps based on meiotic crossing-over.

Mitomycin C–Increased Frequency of
Mitotic Crossing-over

Holliday (1964) used Mitomycin C (MMC) to increase the frequency of MCO in both *Ustilago maydis* and *Saccharomyces cerevisiae*. He used 1.20 mM of MMC in the culture medium and observed the increases summarized in table 6. Therefore, a concentration of MMC that was found not to be fungicidal was an efficient inducer of MCO. Holliday showed that unlike UV, MMC did not induce any mutations in *haploid Ustilago* at the concentrations effectively used to increase MCO. This implies that MMC is not mutagenic in these fungi, although it does increase the frequency of MCO.

Vig and Paddock (1968) observed a preferential increase in Db spots over single dark green and single yellow spots in seedlings of Y_{11}/y_{11} soybeans when seeds were soaked in 0.075 mM MMC for 12–24 hr. Spots were always more prevalent on simple leaves than on first compound leaves. They suggested that a number of the single spots may have been the result of MMC-induced Db spots in which one cell line is inviable.

Vig (1973c), in a continuation of his MMC experiments, observed that the effect of MMC varied with postgermination age of the soybean seeds, with a maximum effect of 28–32 hr postgermination. Unfortunately, when analyzing his data, he did not separate simple and first compound leaves that are present in the dormant seed in different developmental stages. This precluded any conclusions based on cell numbers in different primordia.

Carlson (1974) treated suspension cultures of *Nicotiana tabacum* with 0.1 mM of MMC and observed a 4.7-fold increase over the spontaneous frequency of Db spots. The MMC was added to liquid medium followed by two rinse periods in fresh medium.

Shaw and Cohen (1965) showed that MMC induced chromosome

exchanges in human leukocytes. Chromosome pairs no. 1 and no. 9 were preferentially involved in chromosome interchanges including 14/18 nonhomologous interchanges and 72/110 homologous interchanges. This implies nonrandom association of chromosomes. German and LaRock (1969) advised against the long-term use of MMC but emphasized that for short-term treatments in low concentrations, MMC effectively induced quadiradial configurations, chromosome abnormalities, and mitotic recombination. Cohen and Shaw (1964) suggest that the exchanges induced by MMC are only at heterochromatic regions of the chromosomes. More recently, Craig-Holmes et al. (1975) analyzed variability between 33 individuals in a pedigree. They concluded that somatic mosaicism in C-bands is best explained by MCO and that many of the variations between individuals are the result of unequal MCO in heterochromatic regions. Walen (1964) found in *Drosophila melanogaster* that MCO frequency is positively correlated with the amount of heterochromatin in the chromosomes.

At least partial somatic pairing must occur prior to breakage and exchange of homologous genetic material. As is commonly known, homologous chromosomes are paired at all stages of mitosis in all *Diptera* (Metz, 1916), but somatic pairing has also been observed in plants. For example, Watkins (1935) listed 33 species of plants with somatic pairing. Since heterochromatic regions are preferentially involved in MMC-induced exchanges and since somatic pairing does occur more readily between heterochromatic regions, it has been proposed for this reason that the degree of somatic pairing determines the frequency of MCO (Vogel and Schroeder, 1974).

Besides its mitotic recombinogenic activity, MMC has a wide spectrum of biological effects including inhibition of DNA synthesis (Shiba et al., 1959), degradation of DNA (Reich et al., 1961), increase in sister chromatid exchanges (Latt, 1974), and mutation induction. These effects have been reviewed by Vig (1977). Iijima and Hagiwara (1960) showed that MMC is mutagenic in *E. coli*. Microorganisms vary in sensitivity to MMC, as *Micrococcus radiodurans* is four times more resistant to MMC than *E. coli* (Sweet and Moseley, 1976). Suzuki (1965a) suggested that MMC increased the frequency of MCO in *Drosophila*. This result was complicated, though, when Mukherjee (1965) showed that MMC was also mutagenic in *Drosophila*. Alternately, Holliday (1964) had previously shown, using haploid and diploid *Ustilago maydis*, that MMC increased the frequency of MCO while it was *not* mutagenic. Rao and Natarajan (1967) observed no mutations in *Vicia faba* at concentrations of MMC that were sufficient to cause breakage and exchange between homologous chromosomes.

Iyer and Szybalski (1963) showed that MMC is a bifunctional alkylating agent that produces monoadducts (linkage to a single DNA base) and cross-links with DNA. Several authors have suggested that the chromosome effects of MMC are localized in heterochromatic regions, especially in *Vicia faba* (Arora et al., 1969; Shah et al., 1972) and in tissue-cultured mouse cells (Natarajan and Raposa, 1975). Centromeric heterochromatin is especially sensitive to breakage by MMC (Huttner and Ruddle, 1976; Bochkov, 1972) and stains darkly with the C-banding procedure (Hsu, 1973). Exchanges induced by MMC do preferentially occur at C-band regions in *Vicia faba* (Reiger et al., 1975) and in Chinese hamster cells (Natarajan and Schmid, 1971). An indication of MMC-induced exchange points can be obtained for plants by examining C-banding patterns. Utsumi (1971) has shown that MMC-induced breaks reflect C-band number and intensity as more breaks are observed in *Vicia*, with C-bands, than in *Tradescantia*, lacking C-bands. Son (1977) has shown that nearly all chromsomes contain darkly stained C-bands using *Lilium lancifolium*. Yen and Filion (1976) in their figure 1A depict the central location of C-bands in the chromosomes of *Avena*. Since MMC preferentially causes breaks in the central portion of chromosomes, genetic exchange is maximized. On the other hand, such MMC-induced MCO is of limited usefulness for mapping genes within an arm.

Vig (1967) treated seeds of Y_{11}/y_{11} *G. max* with various concentrations of MMC. We recalculated his data and expressed the results as Db spots per leaf and per μM MMC. The frequency of Db spots increases with increasing concentrations of MMC (fig. 3). This was compared (fig. 4) to the increase in leaf spot frequency after seeds of Y_{11}/y_{11} *G. max* were treated with X-rays (Evans, 1977). In each experiment the same protocol was followed. Seeds were treated for the designated period of time, sown in the same greenhouse, and spots counted when the first compound leaf reached maturity. Based on our regression analyses, the rate of increase of Db spots was linear for the treatments of X-rays and MMC used. The regression coefficients are compared in table 7, where the units of increase for the two kinds of treatment are 1 μM MMC and 1 R of X-rays. (1) The respective regression coefficients for each kind of single spot by X-rays and MMC are nearly equal. (2) Within either the MMC or X-rays, the regression coefficients of the D and A single spots are nearly equal. (3) For Db spots though, the regression coefficient for MMC is about six times the value for either D or A single spots induced by MMC and is nearly 23 times the value for X-ray induced Db spots. These results imply that MMC preferentially increases the frequency of genetic exchange via MCO. This increase by MMC may be the result of (a) preferential breakage and exchange between homologues, (b) localization of MMC breaks in centromeric regions, or (c) preferential effect on anaphase distribution of the four

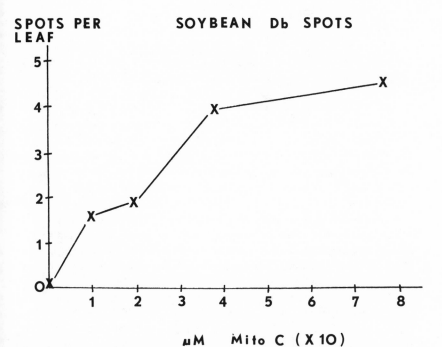

SPOTS PER LEAF

SOYBEAN Db SPOTS

µM　Mito C (X 10)

Fig. 3. Increase in Db spot frequency with increasing concentration of MMC (data from Vig, 1967).

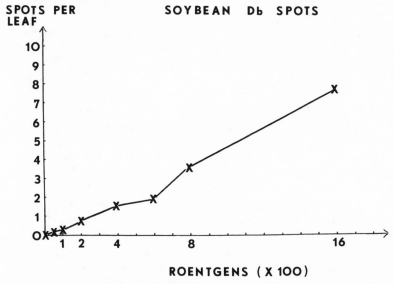

SPOTS PER LEAF

SOYBEAN Db SPOTS

ROENTGENS (X 100)

Fig. 4. Increase in Db spot frequency with increasing dosage of X-rays (data from Evans, 1977).

TABLE 7

REGRESSION COEFFICIENTS FOR THE RATE OF INCREASE OF
EACH TYPE OF SPOT ON LEAVES OF GLYCINE MAX

Spot Type	Mitomycin C	X-rays
D	0.01609 log spots/μM MC	0.01700 log spots/R
A	0.01394 log spots/μM MC	0.01673 log spots/R
Db	0.09287 log spots/μM MC	0.00438 log spots/R

participating chromatids. The effect of preferential distribution alone is insufficient to account for the MMC-increased frequency in Db spots. Reiger et al. (1975) have shown that MMC-induced breaks are more localized than X-ray–induced breaks in *Vicia faba*.

Numerous other chemicals have been used to increase the frequency of MCO in other experimental organisms (table 8), all with only eight species of plants. Vig (1975), in particular, has emphasized the soybean system as ideal for testing chemical agents, and has analyzed numerous agents that increase (table 8) or have no detectable effect (Vig and Mendeville, 1972) on the frequency of MCO. He has found only one chemical, deoxyribose cytidine (d-C), that reduces the frequency of MCO (Vig. 1972). This nucleoside apparently decreases the chances of MCO by disrupting interchromosomal arrangements. In 21-hr treatments, d-C as well as deoxyribose adenine, deoxyribose guanine, and deoxyribose thymine, all significantly decreased the frequency of MCO, whereas in shorter treatments (6 or 16 hr), only d-C decreased MCO.

Inhibitors of DNA synthesis may aid in bringing about homologous pairing of mitotic chromosomes by creating a state of unbalanced growth similar to cells in meiosis (Holliday, 1961). Inhibitors of DNA synthesis have been particularly effective in stimulating MCO. For example, MMC (table 8), caffeine (Vig, 1973a; Holliday, 1961), fluorodeoxyuridine (FUdR) (Esposito and Holliday, 1964; Beccari et al., 1967), and actinomycin D (Vig, 1973b; Suzuki, 1965b). Esposito and Holliday (1964) observed a 14.3-fold increase in MCO by culturing *Ustilago maydis* in medium with 100 μg of FUdR per ml of medium for 120 min while maintaining 100 percent viability. Becarri et al. (1967) used FUdR and fluorouracil to induce MCO in *Aspergillus nidulans*; fluorouracil was especially effective. Parry (1969) has been able to distinguish between intergenic and intragenic crossing-over using ultraviolet (UV) light and ethylmethane sulfonate (EMS). At a low survival level, intergenic crossing-over was increased equally by EMS and UV light, but EMS preferentially increased the frequency of intragenic crossing-over. Brogger (1974) suggested that UV light and MMC both act on DNA in the same manner, implying that the type of breakage and exchange may be identical.

TABLE 8

CHEMICAL AND PHYSICAL AGENTS SHOWN TO INCREASE THE FREQUENCY OF MITOTIC CROSSING-OVER

Agent	Organism	Reference
Chemicals		
Actinomycin D	*Drosophila melanogaster*	Suzuki, 1965a
	Glycine max	Vig, 1973b
Aminotriazole	*Aspergillus nidulans*	Aulicino et al., 1976
Caffeine	*Antirrhinum majus*	Harrison and Carpenter, 1977
	G. max	Vig, 1973a
	Nicotiana tabacum	Carlson, 1974
	Ustilago maydis	Holliday, 1961
Captan	*A. nidulans*	Aulicino et al., 1976
Carofur	*Saccharomyces cerevisiae*	Zimmerman and Vig, 1975
	G. max	Vig and Zimmerman, 1977
Colchicine	*G. max*	Mark, 1961; Vig, 1971
	Lycopersicon esculentum	Ross and Holm, 1959
Dichlorvos	*A. nidulans*	Aulicino et al., 1976
Diepoxybutane	*A. nidulans*	Morpurgo, 1963
	G. max	Vig and Zimmerman, 1977
Diethyl sulfate	*S. cerevisiae*	Zimmerman et al., 1966
Dimethyl nitrosamine	*S. cerevisiae*	Mayer, 1973
Diethyl nitrosamine	*S. cerevisiae*	Mayer, 1973
Ethyleneimine	*S. cerevisiae*	Zimmerman and von Laer, 1967
EMS	*G. max*	Vig, 1975
	S. cerevisiae	Yost et al., 1967
FUdR	*A. nidulans*	Beccari et al., 1967
	U. maydis	Esposito and Holliday, 1964
Fluorouracil	*A. nidulans*	Beccari et al., 1967
Furylfuramide	*S. cerevisiae*	Ong and Shahin, 1974
Hydroxylamine	*S. cerevisiae*	Putrament and Baranowska, 1971
HN-2*	*A. nidulans*	Morpurgo, 1963
MMS	*D. melanogaster*	Mollet and Wurgler, 1974
MMS, MES, MBS†	*G. max*	Nilan and Vig, 1976
Mitomycin C	Listed in table 6	
Naphthylamine	*S. cerevisiae*	Mayer, 1973
Nitrosamides	*S. cerevisiae*	Zimmerman et al., 1966
Nitrous acid	*S. cerevisiae*	Zimmerman et al., 1966
Psoralen	*N. tabacum*	Carlson, 1974
Puromycin	*G. max*	Vig, 1973b
Sodium azide	*G. max*	Vig, 1973a
Sulfanilamide	*A. nidulans*	Bignami et al., 1976
Toluidine	*S. cerevisiae*	Mayer, 1977
Trenimon	*G. max*	Vig and Zimmerman, 1977
Physical agents		
Gamma rays	*Tradescantia hirsuticaulis*	Christianson, 1975
	A. nidulans	Kafer, 1963
	D. melanogaster	Lefevre, 1948
	G. max	Vig, 1974
	N. tabacum	Carlson, 1974, *in vitro*; Dulieu and Dalebroux, 1975
Tritiated water	*G. max*	Vig, 1974; Vig and McFarlene, 1975
Ultraviolet light	*A. fumigatus*	Stromnaes and Garber, 1963
	A. niger	Pontecorvo and Kafer, 1958
	D. melanogaster	Martensen and Green, 1976
	N. tabacum	Carlson, 1974
	S. cerevisiae	Parry, 1969

TABLE 8—*Continued*

Agent	Organism	Reference
X-rays Heat	*U. maydis* *U. violacea* Listed in table 5 *D. melanogaster*	Holliday, 1961 Day and Jones, 1968 Stern and Rentschler, 1936

* HN-2 = methyl-bis (beta-chloroethyl) amine
† MMS = Methyl methanesulfonate; MES = methyl ethanesulfonate; MBS = methyl butane-sulfonate.

More traditional mutagens (table 8) have been effective in inducing MCO (e.g. EMS, nitrosamides, methylmethanesulfonate (MMS), and colchicine) in a variety of organisms.

In plant species, physical agents have been used more often to increase the frequency of MCO than chemical agents. Gamma rays, UV light, and X-rays have all been effective (table 8). Data on increase by gamma rays and UV light, used most often, are summarized in table 9.

Table 10 is a summary of chemical and physical agents known to increase the frequency of MCO in each of at least three species. Blank spaces represent experiments that have not been reported, not negative results. References can be found in tables 5, 6, and 8.

TABLE 9

RATE OF INCREASE IN THE FREQUENCY OF MITOTIC
CROSSING-OVER BY GAMMA-RAYS AND ULTRAVIOLET LIGHT

Organism	Treatment	Fold-Increase
Gamma-rays		
A. nidulans	1,000 R	6% crossovers
G. max	250 R	9.5
N. tabacum (in vitro)	800 R	3.1
	64 R	11.7
T. hirsuticaulis	60 R	6.1
UV-light		
D. melanogaster	52 ergs/mm^2	9.33
A. niger	NA	11.4
N. tabacum (in vitro)	1,000 ergs/mm^2	4.3
S. cerevisiae	1,320 ergs/mm^2	70.0
U. maydis	780 ergs/mm^2	7.0

NOTE: References in table 8.

TISSUE CULTURE EXPERIMENTATION

Two different approaches to tissue culture experimentation may be utilized in the study of MCO. First, experiments may be designed to ascertain if a genetically visible phenomenon is truly the result of MCO. Second, MCO as studied in tissue culture may lead to answers for practical genetic questions.

The Db spots reported in higher plants involve only one locus. Since numerous cytological and genetical hypotheses can account for Db spots,

TABLE 10

CHEMICAL AND PHYSICAL TREATMENTS KNOWN TO INCREASE
MITOTIC CROSSING-OVER IN THREE OR MORE SPECIES

Organism	X-Rays	MC	UV Light	Gamma-Rays	Caffeine	FUdR
Aspergillus sp.	+		+	+		+
Drosophila melanogaster	+	+	+	+		
Glycine max	+	+		+	+	
Lycopersicon esculentum	+	+				
Musca domestica	+					
Nicotiana tabacum	+	+	+	+	+	
Saccharomyces cerevisiae	+	+	+			
Tradescantia hirsuticaulis				+		
Ustilago maydis		+	+		+	+
Antirrhinum majus					+	

Db spots must be shown, in at least one of the reported higher plant loci, to result from MCO. Carlson (1974) has investigated whether Db spots originate via MCO in tobacco. The experimental protocol is summarized in figure 5. He regenerated complete plants from both portions of 12 Db spots. Three of the spots were spontaneous in leaf tissue of the double heterozygous *Su,* + / *su, cl* genotype. Two occurred in photoauxotrophic cell suspensions under high light intensity. The other seven occurred in suspension culture after four of the six treatments that were expected (table 8) to increase the frequency of MCO. Mitomycin C was the most effective of these treatments. Based on their diploid chromosome number, eleven Db spots were the result of MCO, and one (from colchicine treatment) presumably resulted from nondisjunction. In this spot the *Su/Su*-regenerated tissue had 45 chromosomes, and the *su/su*-regenerated tissue had 49 chromosomes. Judging from such regenerated plants, the *cl* locus is either proximal to the *Su* locus or on the opposite arm of the chromosome.

From leaves of *Su/su N. tabacum*, we have progeny-tested four plants regenerated from the D portions of four Db spots. All were D and bred true for the D phenotype. We have also excised and regenerated single spots. The origin of each spot was carefully recorded (fig. 6), the leaf epidermis over the spot removed, the leaf sterilized, and the remaining palisade cells cultured on Murashige and Skoog (1962) shoot medium. We plan to identify and compare the mechanisms of Db and single spot formation. Plantlets obtained from single spots are all morphologically normal, suggesting chromosome number stability in tobacco, but chromosome counts have not been completed. These plants will be progeny-tested.

We have shown that X-rays increase the frequency of MCO in tobacco, tomato, and soybean (Evans, 1977). Many other effects have been ascribed

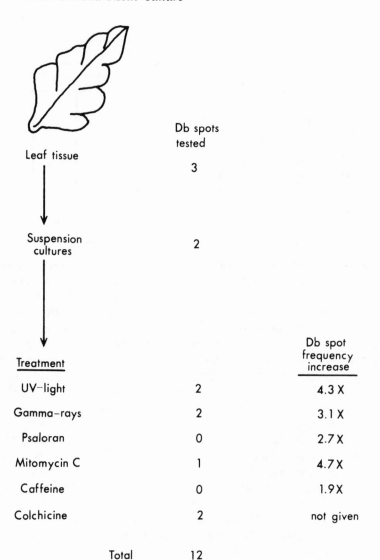

Treatment	Db spots tested	Db spot frequency increase
UV–light	2	4.3 X
Gamma–rays	2	3.1 X
Psaloran	0	2.7 X
Mitomycin C	1	4.7 X
Caffeine	0	1.9 X
Colchicine	2	not given
Total	12	

Fig. 5. Derivation from *Su, +/su, cl N. tabacum* of twelve Db spots regenerated into whole plants and classified for chromosome number by Carlson (1974). One spot from colchicine treatment arose via nondisjunction, the other eleven all via MCO.

to X-rays. For example, X-rays have resulted in cell death (Ichikawa et al., 1969), accelerated germination (Sax, 1963, for review), early flowering (Sax, 1963), and embryogenesis from *Citrus* protoplasts (Vardi et al., 1975).

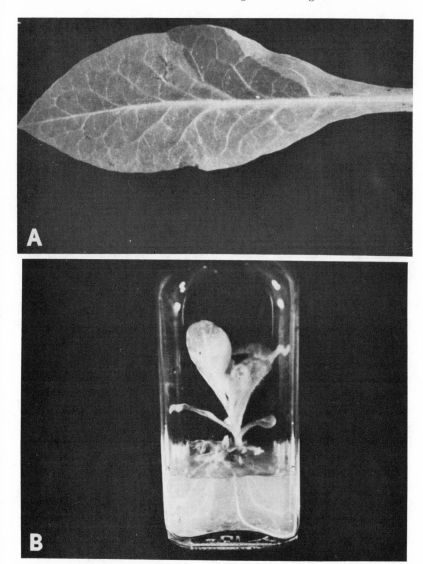

Fig. 6. (A) Dark green single spot in Su/su *N. tabacum*; (B) plantlet regenerated from the spot shown in (A).

We have observed that X-ray treatment of seeds of Y_{11}/y_{11} *G. max* increased the frequency of MCO in the fully expanded simple and first compound leaves. In preliminary experiments low doses of X-rays appeared to increase the formation of roots in tissue-cultured leaf explants from irradiated seeds. We investigated this apparent correlation further. In addition we made growth measurements to monitor more complex X-ray alterations. The experimental protocol (fig. 7) was followed in two experiments. In the first experiment (table 11), 100 seeds at two moisture contents (8 and 55 percent) were X-irradiated with 500, 1,000, 2,000, 4,000, and 8,000 R (Evans, 1977). The seeds were sown in the greenhouse, and spots were counted upon maturation of the first compound leaves. The simple and first compound leaves were then surface sterilized, sectioned, and plated onto soybean high rooting medium (HRM) devised by Evans et al. (1977). Root formation was consistently observed in 90 percent of leaf explants cultured onto HRM. Root number, fresh weight, and dry weight

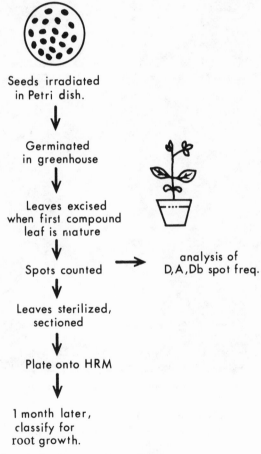

Fig. 7.   Protocol followed after X-ray treatment of seeds.

TABLE 11

ANALYSIS OF SIMPLE LEAVES FROM X-IRRADIATED SEEDS
OF Y_{11}/Y_{11} GLYCINE MAX

DOSAGE	FREQUENCY OF ROOTS/EXPLANT		DB SPOTS/LEAF	
	Moist	Dry	Moist	Dry
0 R	10.63	5.92	0.326	0.138
500 R	16.35	5.21	1.059	0.839
1,000 R	....	....	4.066	3.250
2,000 R	9.84	8.36	TNTC*	TNTC
4,000 R	0.86	14.44	TNTC	TNTC
8,000 R	0.44	4.12	TNTC	TNTC
0 R	3.5		0.0	
50 R	3.8		0.216	
100 R	....		0.250	
200 R	5.5		0.765	
400 R	4.8		1.600	
600 R	....		1.938	
800 R	4.8		3.581	
1,600 R	3.0		7.667	

* TNTC = too numerous to count

were recorded after four wk in culture. Table 11 contains the data from simple leaves. At both moisture contents examined, the fresh weight, dry weight, and root number were all significantly increased over control at low X-ray dosages; but as X-ray dosage was increased, root formation decreased (Evans, 1977). Maximum root formation was observed at a lower X-ray dosage (500 R) for moist seeds than for dry seeds (4,000 R). In a second experiment (table 11), moist seeds (55 percent moisture content) were X-irradiated with 0, 50, 100, 200, 400, 600, 800, 1,600 R to ascertain (1) if the enhancement of root formation was reproducible and (2) a closer approximation of the peak of the X-ray effect. A significant increase in root formation over control was again observed with low dosages of X-rays, and maximum root formation was observed at 200 R.

Alteration also was observed in speed of germination (significantly increased at low dosages), in leaf morphology (distortions observed with greater than 2,000 R), and in final leaf area (decreased with increasing dosage). So, although X-rays altered MCO as measured by the increase in Db spots, X-rays also significantly altered other aspects of soybean growth and development. The root enhancement effect was not correlated with the increase in MCO. Both phenomena increase from 0 to 500 R, but at higher dosages root formation is inhibited whereas MCO is further enhanced (table 11). Also, the increase in MCO is independent of moisture content, whereas the optimum dosage for root formation is dependent on moisture content of the seeds (table 11). As X-rays alter a multitude of cellular phenomena and do *not* preferentially increase the frequency of Db spots (table 7), Mitomycin C may be the preferred agent to increase the frequency of MCO.

Finally, an experiment was designed to ascertain if MCO may be useful in research in plant genetics. As with fungi, MCO will be useful in mapping biochemical mutants in higher plants in tissue culture, but this aspect is dependent upon the isolation of mutants in plant tissue culture. Numerous mutants have been isolated in plant tissue culture (reviewed by Nabors, 1975; Widholm, 1974; Maliga, 1976; Chaleff and Carlson, 1974; Smith, 1974). Few of the mutations isolated have been characterized genetically. The mutations that have been characterized are resistance mutants that are either recessive or semidominant, with semidominant mutants only moderately resistant (Marton and Maliga, 1975; Maliga et al., 1975; Carlson, 1973). In each case, where *N. tabacum* mutants have been characterized, the haploid was used. Haploid cultures must be used for those recessive mutants that cannot be recovered in diploid condition. The pitfalls of using haploid plants to obtain genetically desirable diploid offspring have been elucidated (Walsh, 1974).

We have attempted to design a procedure whereby diploid cells can be mutagenized in tissue culture with recovery of recessive mutants. The experimental protocol is outlined in figure 8, with expected genetic outcomes expressed in table 12. Methylmethane sulfonate (MMS) was found to be a more effective mutagen at sublethal concentrations than either ethylmethanesulfonate or nitrosoguanidine. Early success was achieved with cycloheximide (CHX) as a selective agent. Mutants resistant to CHX have been isolated in yeast (Wilkie and Lee, 1965) and Chinese hamster (Poche et al., 1975). Maliga et al. (1976) mutagenized haploid cells

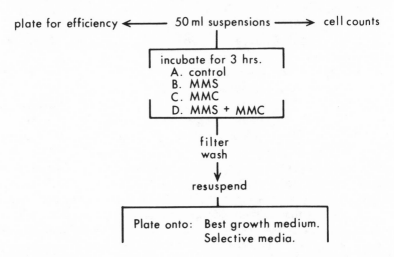

Fig. 8. Mutagenesis protocol for plant tissue cultures using a mutagen (MMS) and a mitotic recombinogen (MMC).

TABLE 12

RESULTS OF MUTAGENESIS PROTOCOL OUTLINED IN FIGURE 8

Incubation	PLATE ONTO: Best Growth Medium*	Selective Media†
Control	Plating efficiency	(a) Spontaneous resistance
		(b) Killing effect of selective agent
MMS	MMS killing effect	MMS-induced variants
MMC	MMC killing effect	(a) MMC-induced variants
		(b) MCO expression of recessives
MMS + MMC	Synergistic killing effect	(a) Synergistic mutant effect, independent of MCO
		(b) Recessives and semidominants induced by MMS, expressed by MMC via MCO

   * Varies with species

   † A number of selective media may be used; we used cycloheximide, hydroxyproline, p-fluorophenylalanine, and fluorouracil.

of *N. tabacum* and isolated variant cell lines resistant to 1 mg/ l CHX, but showed that resistance was unstable and due to differential gene expression. We used 20 mg/ l CHX and isolated variant colonies treated with sublethal concentrations of MMS and MMC (fig. 8). Some small colonies appeared on plates treated with MMS, and larger colonies appeared after treatment with 5 μM MMS and 0.12 mM MMC (suggesting that some variants were semidominant). The number of colonies found on MMS + MMC suggest that some recessive mutants were isolated as well as semidominants. These resistant colonies were grown in the absence of CHX and subsequently formed shoots (fig. 9c) in the presence of 20 mg/ l CHX. Callus from leaf tissue of these plantlets was resistant to 40 mg/ l CHX when retested. The plants are growing in the greenhouse and will eventually be tested genetically for stability of CHX-resistant phenotype.

    Other selective agents have been used with soybean and Paul's Scarlet Rose (PSR) cells with limited success to date, including 194 mg/ l 5-fluorouracil, 195 mg/ l hydroxyproline, and 18.3 mg/ l p-fluorophenylalanine. The PSR cells were extremely sensitive to concentrations of MMC used as a recombinogen in *N. tabacum* and *G. max*. The PSR cells were also naturally resistant to 20 mg/ l CHX and 10 mg/ l streptomycin, i.e., numerous colonies appeared on control plates. Some PSR colonies resistant to hydroxyproline have recently been isolated following the technique in fig. 8. These have not yet been retested.

SUMMARY

    The widespread phenomenon of MCO may become useful as knowledge increases in the somatic cell genetics of higher plants. We have observed MCO at low spontaneous frequencies (10^{-5}) as well as chemically and physically increased frequencies (10^{-2}) that would be more useful in cell

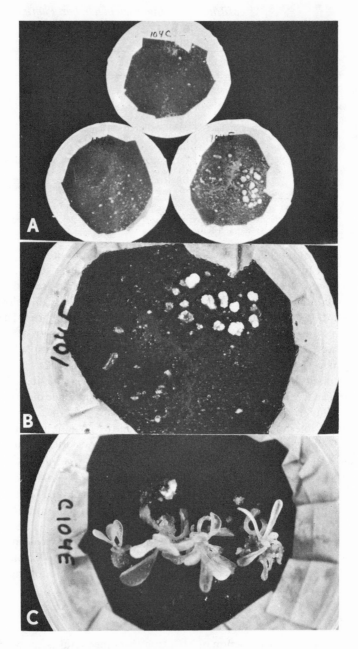

Fig. 9. Selection of cycloheximide-resistant mutants of N. tabacum. (A) Cells on selective medium following treatment with MMC (top), MMS (left), and MMC + MMS (right); (B) close-up of MMC + MMS treatment in (A); (C) plantlets from colonies in (B), on medium containing 20 mg/l CHX.

genetics. The most effective agent appears to be MMC, although MMC preferentially caused breakage and exchange in heterochromatic regions of the chromosome. The phenomenon of spontaneous MCO has been observed in a number of species (table 13), all nearly within one order of magnitude in frequency. When increased in frequency, MCO could be used

TABLE 13

SPONTANEOUS FREQUENCIES OF MITOTIC CROSSING-OVER PER CELL

0.77×10^{-5}	*Nicotiana tabacum* (field-grown)
0.96×10^{-5}	*Antirrhinum majus*
1.08×10^{-5}	*Penicillium chrysogenum*
4.6×10^{-5}	*Nicotiana tabacum* (Carlson, *in vitro*)
5.0×10^{-5}	*Dictyostelium discoideum*
5.68×10^{-5}	*Ustilago maydis*
12.0×10^{-5}	*Saccharomyces cerevisiae*
12.8×10^{-5}	*Glycine max*
13.5×10^{-5}	*Tradescantia hirsuticaulis*

NOTE: Based on double spots and expected genetic outcomes.

to obtain phenotypic expression of much of the genetic variability already present or that may be introduced into the plant genome via mutagens. With more available biochemical markers, MCO will be more easily detected. As a parasexual cycle is developed for higher plants (Carlson et al., 1972), MCO may be essential for mapping genetic loci. The Db spot phenomenon, following further documentation that nearly all Db spots arise via MCO, could be used as a test system for the detection of the effects of chemical and physical agents on plant cells *in vivo* and *in vitro*.

ACKNOWLEDGMENTS

The authors wish to thank Drs. W. R. Sharp and W. C. Myser for helpful assistance in, respectively, tissue culture and X-irradiation experimentation.

REFERENCES

Abbadessa, R., and A. B. Burdick. 1963. The effect of X-irradiation on somatic crossing over in *Drosophila melanogaster*. Genetics 48:1345–56.

Arora, O. P., V. C. Shah, and S. R. V. Rao. 1969. Studies on micronuclei induced by mitomycin C in root tip cells of *Vicia faba*. Exp. Cell Res. 56:443–48.

Aulicino, F., M. Bignami, A. Carere, G. Conti, G. Morpurgo, and A. Velchich. 1976. Mutational studies with some pesticides in *Aspergillus nidulans*. Mutat. Res. 38:138.

Barron, G. L. 1962. The parasexual cycle and linkage relationships in the storage rot fungus, *Penicillium expansum*. Canad. J. Bot. 40:1603–13.

Barrow, J. R., H. Chaudhari, and M. P. Dunford. 1973. Twin spots on leaves of homozygous cotton plants. J. Hered. 64:222–26.

Barrow, J. R., and M. P. Dunford. 1974. Somatic crossing over as a cause of chromosome multivalents in cotton. J. Hered. 65:3–7.

Bateman, A. J. 1967. A probable case of mitotic crossing over in the mouse. Genet. Res. 9:375.

Beccari, E., P. Modigliani, and G. Morpurgo. 1967. Induction of inter and intragenic mitotic recombination by fluorodeoxyuridine and fluorouracil in *Aspergillus nidulans*. Genetics 56:7–12.

Becker, H. J. 1956. On X-ray induced somatic crossing over. Dros. Information Serv. 30:101–2.

Becker, H. J. 1969. The influence of heterochromatin, inversion-heterozygosity, and somatic pairing on X-ray induced mitotic recombination in *Drosophila melanogaster*. Molec. Gen. Genet. 105:203–18.

Becker, H. J. 1976. Mitotic recombination. *In* E. Novitski and M. Ashburner (eds.), Genetics and biology of *Drosophila*. 1C:1019–87. Academic Press, New York.

Bernard, R. L., and M. G. Weiss. 1973. Qualitative genetics. *In* B. E. Caldwell (ed.), Soybeans: Improvement, production, and uses. Pp. 117–86. Amer. Soc. Agron., Madison.

Bignami, M., L. Conti, G. Morpurgo, and A. Velcich. 1976. Comparative analysis of different test systems for somatic recombination with *Aspergillus nidulans*. Mutat. Res. 38:138–39.

Bochkov, N. P. 1972. Distribution of defective chromosomes in human cells after treatment with chemical mutagens *in vitro* and *in vivo*. Sov. Genet. 8:1595–1601.

Brogger, A. 1974. Caffeine-induced enhancement of chromosome damage in human lymphocytes treated with methylmethane sulfonate, mitomycin C, and X-rays. Mutat. Res. 23:353–60.

Burk, L. G. 1970. Green and light-yellow haploid seedlings from anthers of sulfur tobacco. J. Hered. 61:279.

Burk, L. G., and H. P. Menser. 1964. A dominant aurea mutation in tobacco. Tob. Sci. 8:101–4.

Carlson, P. S. 1973. Methionine sulfoximine-resistant mutants of tobacco. Science 180:1366–68.

Carlson, P. S. 1974. Mitotic crossing over in a higher plant. Genet. Res. 24:109–12.

Carlson, P. S., H. H. Smith, and R. D. Dearing. 1972. Parasexual interspecific plant hybridization. Proc. Nat. Acad. Sci. USA 69:2292–94.

Chaleff, R. S., and P. S. Carlson. 1974. Somatic cell genetics of higher plants. Ann. Rev. Genet. 8:267–78.

Christianson, M. L. 1975. Mitotic crossing over as an important mechanism of floral sectoring in *Tradescantia*. Mutat. Res. 28:389–95.

Clark, F. J., and F. C. Copeland. 1940. Chromosome aberrations in the endosperm of maize. Amer. J. Bot. 27:247–51.

Clayberg, C. D., L. Butler, E. A. Kerr, C. M. Rick, and R. M. Robinson. 1966. Third list of known genes in the tomato with revised linkage map and additional rules. J. Hered. 57:189–96.

Cohen, M. M., and M. W. Shaw. 1964. Effects of mitomycin C on human chromosomes. J. Cell Biol. 23:386–95.

Craig-Holmes, A. P., F. B. Moore, and M. W. Shaw. 1975. Polymorphism of human C-band heterochromatin. II. Family studies with suggestive evidence for somatic crossing over. Amer. J. Hum. Genet. 27:178–89.

Crang, R. E., and R. D. Noble. 1974. Ultrastructural and physiological differences in soybean with genetically altered levels of photosynthetic pigments. Amer. J. Bot. 61:903–8.

Cuany, R. L., A. H. Sparrow, and A. H. Jahn. 1959. Spontaneous and radiation-induced somatic mutation rates in *Antirrhinum, Petunia, Tradescantia,* and *Lilium*. Proc. Tenth Intern. Congr. Genet., Montreal 2:62–63. University of Toronto Press, Toronto.

Dahlgren, K., and V. Ossian. 1927. Eine sektorialchimare vom Apfel. Hereditas 9:335–42.

Day, A. W., and J. K. Jones. 1968. The production and characteristics of diploids in *Ustilago violacea*. Genet. Res. 11:63–81.

Deshayes, A., and H. Dulieu. 1974. Study of two genetic chlorophyll mutants of *N. tabacum*. L. *In* Polyploidy and induced mutations in plant breeding. Pp. 85–99. International Atomic Energy Agency, Vienna.

Dulieu, H. L. 1974. Somatic variations on a yellow mutant in *Nicotiana tabacum* L. (a_1^+/a_1, a_2^+/a_2). I. Nonreciprocal genetic events occurring in leaf cells. Mutat. Res. 25:289–304.

Dulieu, H. L. 1975. Somatic variations of a yellow mutant in *Nicotiana tabacum* L. ($a_1^+/a_1,a_2^+/a_2$). II. Reciprocal genetic events occurring in leaf cells. Mutat. Res. 28:69–77.

Dulieu, H. L., and M. A. Dalebroux. 1975. Spontaneous and induced reversion rates in a double heterozygous mutant of *Nicotiana tabacum* var. Xanthi N.C.-dose-response relationship. Mutat. Res. 30:63–70.

Esposito, M. S. 1968. X-ray and meiotic fine structure mapping of the adenine-8 locus in *Saccharomyces cerevisiae*. Genetics 58:507–27.

Esposito, R. E., and R. Holliday. 1964. The effect of 5-fluorodeoxyuridine on genetic replication and somatic recombination in synchronously dividing cultures of *Ustilago maydis*. Genetics 50:1009–17.

Evans, D. A. 1977. Modification of the frequency of mitotic crossing over in *Nicotiana tabacum, Glycine max,* and *Lycopersicon esculentum* using X-rays. Ph.D. dissertation, Ohio State University, Columbus.

Evans, D. A., and E. F. Paddock. 1976. Comparisons of somatic crossing over frequency in *Nicotiana tabacum* and three other crop species. Canad. J. Genet. Cytol. 18:57–65.

Evans, D. A., and E. F. Paddock. 1977. X-ray induced increase of mitotic crossing over frequency in *Nicotiana tabacum*. Environ. Exper. Bot. 17:99–106.

Evans, D. A., W. R. Sharp, and E. F. Paddock. 1976. Variation in callus proliferation and root morphogenesis in leaf tissue cultures of *Glycine max*, Strain T219. Phytomorphology 26:379–84.

Garcia-Bellido, A. 1972. Some parameters of mitotic recombination in *Drosophila melanogaster*. Molec. Gen. Genet. 115:54–72.

Garcia-Bellido, A., and J. R. Merriam. 1971. Clonal parameters of tergite development in *Drosophila*. Devel. Biol. 26:264–76.

German, J., and T. LaRock. 1969. Chromosomal effects of mitomycin, a potential recombinogen in mammalian cell genetics. Texas Rep. Biol. Med. 27:409–18.

Grunberg, H. 1966. The case of somatic crossing over in the mouse. Genet. Res. 7:58–75.

Harrison, B. J., and R. Carpenter. 1977. Somatic crossing over in *Antirrhinum majus*. Heredity 38:169–89.

Hastie, A. C. 1967. Mitotic recombination in conidiophores of *Verticillium albo-atrum*. Nature 214:249–52.

Hendrychova-Tomkova, J. 1964. Local somatic colour changes in *Salvia splendens*. J. Genet. 59:7–13.

Holliday, R. 1961. Induced mitotic crossing over in *Ustilago maydis*. Genet. Res. 2:231–48.

Holliday, R. 1964. The induction of mitotic recombination by mitomycin C in *Ustilago* and *Saccharomyces*. Genetics 50:323–25.

Hsu, T. C. 1973. Longitudinal differentiation of chromosomes. Ann. Rev. Genet. 7:153–76.

Huttner, K. M., and F. H. Ruddle. 1976. Study of mitomycin-C-induced chromosomal exchange. Chromosoma 56:1–13.

Ichikawa, S., A. H. Sparrow, and K. H. Thompson. 1969. Morphologically abnormal cells, somatic mutations and loss of reproductive integrity in irradiated *Tradescantia* stamen hairs. Rad. Bot. 9:195–211.

Iijima, T., and A. Hagiwara. 1960. Mutagenic action of mitomycin C on *Escherichia coli*. Nature 185:395–96.

Iyer, V. N., and W. Szybalski. 1963. A molecular mechanism of mitomycin action: Linking of complementary DNA-strands. Proc. Nat. Acad. Sci. USA 50:355–62.

Johnson, H. W., and R. L. Bernard. 1963. Soybean genetics and breeding. *In* A. G. Norman (ed.), The soybean. Pp. 1–70. Academic Press, New York.

Jones, D. F. 1937. Somatic segregation and its relation to atypical growth. Genetics 22:484–522.

Jones, D. F. 1938. Translocation in relation to mosaic formation in maize. Proc. Nat. Acad. Sci. USA 24:208–11.

Kafer, E. 1958. An 8-chromosome map of *Aspergillus nidulans*. Adv. Genet. 9:105–45.

Kafer, E. 1963. Radiation effects and mitotic recombination in diploids of *Aspergillus nidulans*. Genetics 48:27–45.

Katz, E. R., and V. Kao. 1974. Evidence for mitotic recombination in the cellular slime mold *Dictyostelium discoideum*. Proc. Nat. Acad. Sci. USA 71:4025–26.

Kaufman, B. P. 1934. Somatic mitoses of *Drosophila melanogaster*. J. Morphol. 56:125–55.

Keck, R. W., R. A. Dilley, C. F. Allen, and S. Biggs. 1970a. Chloroplast composition and structure differences in a soybean mutant. Plant Physiol. 46:692–98.

Keck, R. W., R. A. Dilley, and B. Ke. 1970b. Photochemical characteristics in a soybean mutant. Plant Physiol. 46:699–704.

Kelly, P. T. 1974. Non-reciprocal intragenic mitotic recombination in *Drosophila melanogaster*. Genet. Res. 23:1–12.

Killough, D. T., and W. R. Horlacker. 1933. The inheritance of virescent yellow and red plant characters in cotton. Genetics 18:329–34.

Koller, H. R. and R. A. Dilley. 1974. Light intensity during leaf growth affects chlorophyll concentration and CO_2 assimilation of a soybean chlorophyll mutant. Crop Sci. 14:779–82.

Latt, S. A. 1973. Microfluorometric detection of DNA replication in human metaphase chromosomes. Proc. Nat. Acad. Sci. USA 70:3395–99.

Latt, S. A. 1974. Sister chromatid exchanges, indices of human chromosome damage and repair: Detection by fluorescence and induction by mitomycin C. Proc. Nat. Acad. Sci. USA 71:3162–66.

Lefevre, G. 1948. The relative effectiveness of fast neutrons and gamma rays on producing somatic crossing over in *Drosophila*. Genetics 33:113.

Lewis, L. A. 1969. Correlated meiotic and mitotic maps in *Aspergillus amstelogami*. Genet. Res. 14:185–93.

Maliga, P. 1976. Isolation of mutants of cultured plant cells. *In* D. Dudits, G. L. Farkas, and P. Maliga (eds.), Cell genetics of higher plants. Pp. 59–76. Akadémiai Kiadó, Budapest.

Maliga, P., A. Sz-Breznovits, L. Marton, and F. Joo. 1975. Non-mendelian streptomycin-resistant tobacco mutant with altered chloroplasts and mitochondria. Nature 255:401–2.

Maliga, P., G. Lazar, Z. Svab, and F. Nagy. 1976. Transient cycloheximide resistance in a tobacco cell line. Molec. Gen. Genet. 149:267-71.

Mark, C. 1961. Colchicine induced genetic changes in soybean leaf tissue. Proc. S. Dak. Acad. Sci. 40:219-25.

Martensen, D. V., and M. M. Green. 1976. UV-induced mitotic recombination in somatic cells of *Drosophila melanogaster*. Mutat. Res. 36:391-96.

Marton, L., and P. Maliga. 1975. Control of resistance in tobacco cells to 5-bromodeoxyuridine by a simple Mendelian factor. Plant Sci. Lett. 5:77-81.

Mayer, V. W. 1973. Induction of mitotic recombination in *Saccharomyces cerevisiae* by breakdown products of dimethylnitrosamine, diethylnitrosamine, 1-naphthylamine and 2 naphthylamine formed by an *in vitro* hydroxylating system. Genetics 74:433-43.

Mayer, V. W. 1977. Induction of mitotic crossing over in *Saccharomyces* by p-toluidine. Molec. Gen. Genet. 151:1-4.

Menser, H. A., T. Sorokin, and M. E. Engelhaupt. 1965. The pigments, amino acids, alkaloids, growth rate, and response to ozone of a chlorophyll-deficient mutation of tobacco. Tob. Sci. 9:21-25.

Mericle, L. W., and R. P. Mericle. 1967. Mutation induction as influenced by developmental stage and age. Erwin-Baue-Gedachtnisvorlesungen IV. Pp. 65-77. 1966. Abhandl. Deut. Akad. Wissen., Berlin.

Mericle, R. P., M. L. Christianson, and L. W. Mericle. 1974. Prediction of flower color genotype in *Tradescantia* by somatic cell analyses. J. Hered. 65:21-27.

Merriam, J. R., and W. E. Fyffe. 1972. The kinetics of X-ray-induced somatic crossing over in *Drosophila melanogaster* and the effects of dose fractionation. Mutat. Res. 14:309-14.

Merriam, J. R., and A. Garcia-Bellido. 1972. A model for somatic pairing derived from somatic crossing over with third chromosome rearrangements in *Drosophila melanogaster*. Molec. Gen. Genet. 115:302-13.

Metz, C. W. 1916. Chromosome studies in the *Diptera*. II. The paired association of chromosomes in *Diptera*, and its significance. J. Exp. Zool. 21:213-79.

Mollet, P., and W. E. Wurgler. 1974. Detection of somatic recombination and mutation in *Drosophila*—a method for testing genetic activity of chemical compounds. Mutat. Res. 25:421-24.

Moore, C. N., and C. P. Haskins. 1935. X-ray induced modifications of flower color in *Petunia*. J. Hered. 26:349-55.

Morpurgo, G. 1963. Induction of mitotic crossing over in *Aspergillus nidulans* by bifunctional alkylating agents. Genetics 48:1259-63.

Mortimer, R. K. 1957. X-ray-induced homozygosis in diploid yeast. Rad. Res. 7:439-40.

Mukherjee, R. 1965. Mutagenic action of mitomycin C on *Drosophila melanogaster*. Genetics 51:947-51.

Murashige, T., and F. Skoog. 1962. A revised medium for rapid growth and bioassays with tobacco tissue cultures. Physiol. Plantarum 15:473-97.

Nabors, M. W. 1975. Using spontaneous occurring and induced mutations to obtain agriculturally useful plants. Bioscience 26:761-68.

Nakai, S., and R. K. Mortimer. 1969. Studies of the genetic mechanism of radiation-induced mitotic segregation in yeast. Molec. Gen. Genet. 103:329-38.

Natarajan, A. T., and T. Raposa. 1975. Heterochromatin and chromosome aberrations. A comparative study of three mouse cell lines with different karyotype and heterochromatin distribution. Hereditas 80:83–90.

Natarajan, A. T., and W. Schmid. 1971. Differential response of constitutive and facultative heterochromatin in the manifestation of mitomycin induced chromosome aberrations in Chinese hamster cells *in vitro*. Chromosoma 33:48–62.

Nilan, R. A., and B. K. Vig. 1976. Plant test systems for detection of chemical mutagens. *In* A. E. Hollaender (ed.), Chemical mutagens. 4:143–70. Plenum Press, New York.

Nothiger, R., and A. Dubendorfer. 1971. Somatic crossing over in the housefly. Molec. Gen. Genet. 112:9–13.

Ong., T. M., and M. M. Shahin. 1974. Mutagenic and recombinogenic activities of the food additive furylfuramide in eukaryotes. Science 184:1086–87.

Paddock, E. F. 1966. Mendel could have, your students can. Sci. Teacher 33:21–23.

Painter, T. S. 1934. A new method for the study of chromosome aberrations and the plotting of chromosome maps in *Drosophila melanogaster*. Genetics 19:175–88.

Parag, Y., and G. Parag. 1975. Mutation affecting mitotic recombination frequency in haploids and diploids of the filamentous fungus *Aspergillus nidulans*. Molec. Gen. Genet. 137:109–23.

Parry, J. M. 1969. Comparison of effects of UV-light and ethylmethane-sulfonate upon the frequency of mitotic recombination in yeast. Molec Gen. Genet. 106:66–72.

Pittenger, T. H. and M. B. Coyle. 1963. Somatic recombination in pseudo-wild type cultures of *Neurospora crassa*. Proc. Nat. Acad. Sci. USA 49:445–51.

Poche, H., I. Junghahn, E. Geissler, and H. Bielke. 1975. Cycloheximide resistance in Chinese hamster cells. Molec. Gen. Genet. 138:173–77.

Pontecorvo, G. 1958. Trends in genetic analysis. Columbia University Press, New York.

Pontecorvo, G., and E. Kafer. 1958. Genetic analysis based on mitotic recombination. Adv. Genet. 9:71–104.

Pontecorvo, G., J. A. Roper, and E. Forbes. 1953. Genetic recombination without sexual reproduction in *Aspergillus niger*. J. Gen. Microbiol. 8:198–210.

Pontecorvo, G., and G. Sermonti. 1954. Parasexual recombination in *Penicillium chrysogenum*. J. Gen. Microbiol. 11:94–104.

Prud'Homme, N. 1970. Recombinaisons mitotiques chez un basidiomycète: *Coprinus radiatus*. Molec. Gen. Genet. 107:256–71.

Putrament, A., and H. Baranowska. 1971. Induction of intragenic mitotic recombination in yeast by hydroxylamine and its mutagenic specificity. Molec. Gen. Genet. 111:89–96.

Rao, R. N., and A. T. Natarajan. 1967. Somatic association in relation to chemically induced chromosome aberrations in *Vicia faba*. Genetics 57:821–35.

Reich, E., A. J. Shatkin, and E. L. Tatum. 1961. Bacteriocidal action of mitomycin C. Biochim. Biophys. Acta 53:132–49.

Reiger, R., A. Michaelis, I. Schubert, B. Doebel, and H. W. Janak. 1975. Non-random intrachromosomal distribution of chromatid aberrations induced by X-rays, alkylating agents, and ethanol in *Vicia faba*. Mutat. Res. 27:69–79.

Roper, J. A. 1952. Production of heterozygous diploids in filamentous fungi. Experientia 8:14–15.

Ross, J. G., and G. Holm. 1959. Somatic segregation in tomato. Hereditas 46:224–30.

Sax, K. 1963. The stimulation of plant growth by ionizing radiation. Rad. Bot. 3:179–86.

Schaeffer, G. W. 1977. Culture and morphogenetic response of a lethal chlorophyll-deficient mutant of tobacco to hormones, amino acids, and sucrose. In Vitro 13:31–35.

Schaeffer, G. W., and H. A. Menser. 1975. Tissue culture of a tobacco albino mutant. Crop Sci. 14:728–30.

Schmid, G. H. 1967. The influence of different light intensities on the growth of the tobacco aurea mutant *Su/su*. Planta 77:77–94.

Serebrovsky, A. S. 1925. Somatic segregation in domestic fowl. J. Genet. 16:33–42.

Shah, V. C., S. R. Rao, and O. P. Arora. 1972. Effect of mitomycin C on root meristem cells of *Vicia faba* L. II. Induced chromosomal aberration in relation to cell cycle. Indian J. Exp. Biol. 10:431–35.

Shaw, M. W., and M. M. Cohen. 1965. Chromosome exchanges in human leukocytes induced by mitomycin C. Genetics 51:181–90.

Shiba, S., A. Terawaki, T. Taguchi, and J. Kawamata. 1959. Selective inhibition of formation of DNA in *Escherichia coli* by mitomycin C. Nature 183:1056–57.

Smith, H. H. 1974. Model systems for somatic cell genetics. Bioscience 24:269–76.

Son, J. 1977. Karyotype analysis of *Lilium lancifolium* Thunberg by means of C-banding method. Jap. J. Genet. 52:217–22.

Sparrow, A. H., L. A. Schairer, and R. Villalobos-Pietrini. 1974. Comparison of somatic mutation rates induced in *Tradescantia* by chemical and physical mutagens. Mutat. Res. 26:265–76.

Stein. O. L., and D. Steffensen. 1959. Radiation-induced genetic markers in the study of leaf growth in *Zea*. Amer. J. Bot. 46:485–89.

Stern, C. 1935. The effect of yellow-scute gene deficiency on somatic cells of *Drosophila*. Proc. Nat. Acad. Sci. USA 21:374–79.

Stern, C. 1936. Somatic crossing over and segregation in *Drosophila melanogaster*. Genetics 21:625–730.

Stern, C. 1939. Somatic crossing over and somatic translocations. Amer. Nat. 73:95–96.

Stern, C. 1969. Somatic recombination within the white locus of *Drosophila melanogaster*. Genetics 62:573–81.

Stern, C., and V. Rentschler. 1936. The effect of temperature on the frequency of somatic crossing-over in *Drosophila melanogaster*. Proc. Nat. Acad. Sci. USA 22:451–53.

Stromnaes, O., and E. D. Garber. 1963. Heterocaryosis and the parasexual cycle in *Aspergillus fumigatus*. Genetics 48:653–62.

Stromnaes., O., E. D. Garber, and L. Beraha. 1964. Genetics of phytopathogenic fungi. IX. Heterocaryosis and the parasexual cycle in *Penicillium italicum* and *Penicillium digitatum*. Canad. J. Bot. 42:423–27.

Sun, C. N. 1965. The effect of genetic factors on the submicroscopic structure of soybean chloroplasts. Cytologia 28:257–63.

Suzuki, D. T. 1965a. The effects of actinomycin D on crossing over in *Drosophila melanogaster*. Genetics 51:11–21.

Suzuki, D. T. 1965b. Effects of mitomycin C on crossing over in *Drosophila melanogaster*. Genetics 51:635–40.

Sweet, D. M., and B. E. B. Moseley. 1976. The resistance of *Micrococcus radiodurans* to killing and mutations by agents which damage DNA. Mutat. Res. 34:175–86.

Tinline, R. D. 1962. *Cochliobolus sativus*. V. Heterokaryosis and parasexuality. Canad. J. Bot. 40:425–37.

Turcotte, E. L., and C. V. Feaster. 1972. Genetic markers in American Pima cotton. Cott. Gr. Rev. 49:50–56.

Turcotte, E. L., and C. V. Feaster. 1973. The interaction of two genes for yellow foliage in cotton. J. Hered. 64:231–32.

Utsumi,. S. 1971. Localized chromosome breakage induced by mitomycin C in *Tradescantia paludosa* and *Vicia faba*. Jap. J. Genet. 46:125–34.

Vardi, A., P. Spiegel-Roy, and E. Galun. 1975. *Citrus* cell culture: Isolation of protoplasts, plating densities, effect of mutagens, and regeneration of embryos. Plant Sci. Lett. 4:231–36.

Vig, B. K. 1967. Experimental alterations of leaf spot frequencies in *Glycine max* (L.) Merrill, with reference to the mechanism of spot formation. Ph.D. dissertation, Ohio State University, Columbus.

Vig, B. K. 1971. Increase induced by colchicine in the incidence of somatic crossing over in *Glycine max*. Theor. Appl. Genet. 41:145–49.

Vig, B. K. 1972. Suppression of somatic crossing over in *Glycine max* (L.) Merrill by deoxyribose cytidine. Molec. Gen. Genet. 116:158–65.

Vig, B. K. 1973a. Somatic crossing over in *Glycine max* (L.) Merrill: Mutagenicity of sodium azide and lack of synergistic effect with caffeine and mitomycin C. Genetics 75:265–77.

Vig, B. K. 1973b. Somatic crossing over in *Glycine max* (L.) Merrill: Effect of some inhibitors of DNA synthesis on the induction of somatic crossing over and point mutations. Genetics 73:583–96.

Vig, B. K. 1973c. Mitomycin C induced leaf mosaicism in *Glycine max* (L.) Merrill in relation to the post germination age of the seed. Theor. Appl. Genet. 43:27–30.

Vig, B. K. 1974. Somatic crossing over in *Glycine max* (L.) Merrill: Differential response to ^{3}H-emitted beta-particles and ^{60}Co-emitted gamma-rays. Rad. Bot. 14:127–37.

Vig, B. K. 1975. Soybean (*Glycine max*): A new test system for study of genetic parameters as effected by environmental mutagens. Mutat. Res. 31:49–56.

Vig, B. K. 1977. Genetic toxicology of mitomycin C, actinomycins, daunomycin, and adriamycin. Mutat. Res. 39:189–238.

Vig, B. K., and J. C. McFarlene. 1975. Somatic crossing over in *Glycine max* (L.) Merrill: Sensitivity to and saturation of the system at low levels of tritium emitted beta-radiation. Theor. Appl. Genet. 46:331–37.

Vig, B. K., and W. Mendeville, 1972. Ineffectivity of metallic salts in induction of somatic crossing over and mutations in *Glycine max* (L.) Merrill. Mutat. Res. 16:151–55.

Vig, B. K., and E. F. Paddock. 1968. Alteration by mitomycin C of spot frequencies in soybean leaves. J. Hered. 59:225–29.

Vig, B. K., and F. K. Zimmerman. 1977. Somatic crossing over in *Glycine max*: An induction of the phenomenon by carofur, diepoxybutane, and trenimon. Environ. Exper. Bot. 17:113–20.

Vogel, T. M., and F. Schroeder. 1974. The internal order of the interphase nucleus. Humangenetik 25:265–97..

von Wettstein, D., A. Kuhn, O. F. Nielsen, and S. Gough. 1974. Genetic regulation of chlorophyll synthesis analyzed with mutants in barley. Science 184:800–802.

Walen, K. H. 1964. Somatic crossing over in relationship to heterchromatin in *Drosophila melangoaster*. Genetics 49:905–23.

Walsh, E. J. 1974. Efficiency of the haploid method of breeding autogamous diploid species: A computer simulation study. *In* K. J. Kasha (ed.), Haploids in higher plants. Pp. 195–209. University of Guelph Press, Guelph, Ontario, Canada.

Watkins, G. M. 1935. A study of chromosome pairing in *Yucca rupicola*. Bull. Torrey Bot. Club 62:133–50.

Weber, C. R., and M. G. Weiss. 1959. Chlorophyll mutant in soybean provides teaching aid. J. Hered. 50:53–54.

Widholm, J. M. 1974. Selection and characteristics of biochemical mutants of cultured plant cells. *In* H. E. Street (ed.), Tissue culture and plant science. Pp. 287–99. Blackwell, Oxford.

Wilkie, D., and B. K. Lee. 1965. Genetic analysis of actidione resistance in *Sasccharomyces cerevisiae*. Genet. Res. 6:130–38.

Williams, K. L., R. H. Kessin, and P. C. Newell. 1974. Parasexual genetics in *Dictyostelium discoideum*: Mitotic analysis of acriflavin resistance and growth in axenic medium. J. Gen. Microbiol. 84:59–69.

Wolf, F. T. 1963. The chloroplast pigments of certain soybean mutants. Bull. Torrey Bot. Club 90:139–43.

Wolf, F. T. 1965. Photosynthesis of certain soybean mutants. Bull.Torrey Bot. Club 92:99–101.

Yen, S.-T., and W. G. Filion. 1976. Differential giemsa staining in plants. IV. C-banding in *Avena strigosa*. J. Hered. 67:117–18.

Yost, H. T., R. S. Chaleff, and J. P. Finerty. 1967. Induction of mitotic recombination in *Saccharomyces cerevisiae* by ethylmethanesulphonate. Nature 215:660–61.

Zaman, M. A., and K. S. Rai. 1973. Partial floral chimera due to somatic crossing over in *Collinsia heterophylla* Buist, Scrophulariaceae. Bangladesh J. Bot. 2:7–9.

Zimmerman, F. K. 1975. Procedures used in the induction of mitotic recombination and mutation in the yeast *Saccharomyces cerevisiae*. Mutat. Res. 31:71–86.

Zimmerman, F. K., R. Schwaier, and U. von Laer. 1966. Mitotic recombination induced in *Saccharomyces cerevisiae* with nitrous acid, diethylsulfate and carcinogenic alkylating nitrosamides. Z. Vererbungsl. 98:230–46.

Zimmerman, F. K., and B. K. Vig. 1975. Mutagen specificity in the induction of mitotic crossing over in *Saccharomyces cerevisiae*. Mutat. Res. 35:255–68.

Zimmerman, F. K., and U. von Laer. 1967. Induction of mitotic recombination with ethyleneimine in *Saccharomyces cerevisiae*. Mutat. Res. 4:377–79.

E. C. COCKING

Somatic Hybridization by the Fusion of Isolated Protoplasts—An Alternative to Sex

19

As early as 1955 J. B. S. Haldane, in his essay on "Some alternatives to sex," drew attention to the fact that there were at least five methods by which two organisms can give rise to a third, which are so unlike sexual reproduction that he thought that they should be treated separately. In certain fungi, for instance, two hyphae, with genetically different nuclei, may fuse to give rise to a heterokaryon resulting in the production of a mycelium containing nuclei of two kinds. Often, like a heterozygote, such a heterokaryon is more vigorous than either parent. Frequently heterokaryons are unstable and produce sectors that contain the nuclei of only one of the parents. Species formed from heterokaryons contain the nuclei of one or other of the parents, and as a result species formation in a heterokaryon always leads to segregation. Heterokaryosis does not, however, permit the large number of recombinations that are possible when nuclei fuse and the resulting diploid nucleus undergoes meiosis. In certain fungi, following heterokaryosis, various treatments can induce nuclear fusion with the resultant formation of diploids, and from these diploids haploids are sometimes formed. This parasexual process enables recombination to occur even in asexual fungi. No corresponding parasexual process has as yet been discovered in higher plants, but as we shall see later the fusion of plant protoplasts has enabled heterokaryons to be produced both intra- and interspecifically; and in certain instances

E. C. Cocking, Agricultural Research Council Group, Department of Botany, University of Nottingham, Nottingham, England.

hybrid cells and sometimes hybrid plants have been produced following diploidization.

Somatic hybridization, by the fusion of isolated protoplasts, has developed rapidly in recent years, and it will be particularly helpful to see how this has taken place. Particular emphasis will be placed on work carried out by the Agricultural Research Council (ARC) Group at Nottingham. Other work will not be neglected, but firsthand acquaintance with the subject at Nottingham will help greatly in providing a perspective. These studies cannot be divorced from the great amount of previous, and ongoing, work in the general field of plant cell culture and whole plant regeneration; somatic hybridization studies necessitating the use of isolated protoplasts have, however, greatly stimulated this area of investigation, providing an extra meaning and purpose to the many basic cultural studies being carried out.

PROTOPLAST DEVELOPMENT

The protoplast as the basic homeostatic cell unit, whether it be plant or animal, has for long attracted the interests of chemists, physicists, and biologists—so much so that a poem was composed in 1942 (Scarth, 1942) to highlight some of the current problems associated at that time with the protoplasts and their structure. Indeed, plant protoplasts would have remained as cytological curiosities if their properties—particularly whether or not they could regenerate a new cell wall and divide—had not been further studied at Nottingham. Protoplasts isolated mechanically by slicing plasmolyzed plant tissues had been extensively studied cytologically for many years throughout the first half of this century, following the description by Klercker (1892) of the mechanical method for their isolation. Physiological studies mainly centered on plasmolysis phenomena and attempts to detect action of auxins on the plasma membrane (for a detailed account of this earlier work see Cocking, 1967). The introduction of an enzymatic method for the isolation of protoplasts (Cocking, 1960), coupled with ongoing work at Nottingham on their structural and physiological properties, provided extra momentum. It became possible to isolate large numbers of protoplasts from a wide range of plant tissues, including callus and cell suspension cultures, using cell wall degrading enzymes (Cocking and Evans, 1973; Evans and Cocking, 1975).

The first significant step in the development of isolated protoplasts is the regeneration of a new cell wall at the surface of the plasma membrane. Resynthesis of the wall is a key prelude to the further development of the protoplast system as a dividing cell. This resynthesis at the surface of isolated mesophyll protoplasts has been examined ultrastructurally (Willison, 1976); and it seems that there is a fairly rapid process that can begin soon after the transfer of protoplasts to a suitable culture medium. Cellulose microfibrils can be detected being formed at, or in, the plasma membrane (Grout, 1975) by the technique of deep etching. Optically, early stages in the deposition of cellulose microfibrils can be detected using optical brighteners, such as Calcafluor, which bind to the microfibrils and fluoresce when irradiated with blue light (Nagata and Takebe, 1970). Pectins are also deposited at the surface of the plasma membrane, probably by a process of exocytosis. Before the onset of wall regeneration, fine structural studies have demonstrated that very few microfibrils remain on the surface of isolated protoplasts. It should be noted that the fusion of isolated protoplasts with one another is markedly inhibited as a result of the onset of the early stages of cell wall regeneration. As a result of cell wall regeneration, essentially uniform populations of single cells can be obtained. Whether or not these single cells will proceed to divide will depend on numerous factors; it will not be profitable to discuss at length the various factors that influence the division of these single cells, since this has been discussed very comprehensively quite recently (Bhojwani et al., 1977). It is sufficient to emphasize that each species, and also the various varieties of each species, has its own particular cultural peculiarities. As a result, it is difficult to generalize about the possible division capabilities of different species, since there is a strong interplay of genetic, nutritional, and environmental factors. Currently over forty different protoplast systems (including those from several crop species) have been induced to divide and form callus (or embryoids) (Bajaj, 1977), and sometimes plant regeneration has taken place. The general scheme for the culture of isolated protoplasts is shown in figure 1.

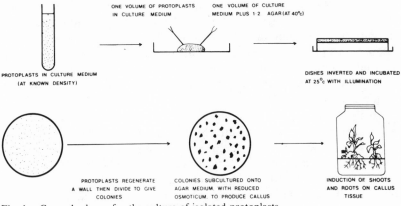

Fig. 1. General scheme for the culture of isolated protoplasts.

INDUCED FUSION OF PROTOPLASTS

There is great interest currently in the use of protoplasts for fusion in interspecies somatic hybrid production. Such an alternative to sexual hybridization will necessitate the use of isolated protoplasts, their fusion, the selection of the often few resultant hybrid cells from the mass of parental cells, and the regeneration of the selected hybrid cells into somatic hybrid plants. This great interest in plant protoplast fusion is clearly largely related to the fascinating prospect of the possibility that wider crosses than are possible by sexual means could be made by protoplast fusions. At the Versailles Symposium on Protoplasts (Cocking, 1973), I emphasized that, although workers were understandably preoccupied with the desire to produce new somatic hybrid plants, they were sometimes not prepared to study in detail the various stages necessarily required for this novel approach to plant breeding to be fully implemented. I also emphasized that the problems are many and arise principally because somatic hybridization is a multistage process. Steward and Krikorian (1971), in discussing the culturing of higher plant cells, its status, problems, and potentialities, argued even more forcefully in rather similar vein. They stressed that, despite some successes, "speculation seems often to be ahead of critically evaluated performance." They cogently argued that further work on protoplast fusion was essential, and that generally applicable selection procedures were essential—but that manual cell isolation was difficult. They also stressed the need for a high frequency of selection. Because of the close parallel between protoplast fusion and the act of sexual fertilization, haploid cell selection was regarded as a prior necessity for fusions or diploidization. But, as we shall see later, haploids are not essential for interspecies somatic hybridization, and the use of diploid protoplasts for fusion may be advantageous. Many workers have tended to view somatic hybridization as being of potential use only in crossing species that are sexually incompatible (Vasil, 1976); but, as we shall also see later, an ability to produce tetraploids directly as a result of diploid somatic fusions will be very useful in certain aspects of plant breeding, even when sexual hybridization is possible.

Freely isolated protoplasts do not normally fuse with one another because they are usually negatively charged at the surface (Grout and Coutts, 1974) and repel one another over short distances. It has been suggested (Melchers, 1976) that if one of the protoplast systems could be altered to have a positive zeta potential at its surface and the other kept negative, it might be possible to create a type of "artificial sexuality." Recently there has been a very comprehensive discussion of the development of the study of the induced fusion of isolated protoplasts (Power,

Evans, and Cocking, 1977). Using protoplasts isolated mechanically, Kuster (1910) described the first fusion of freely isolated protoplasts using deplasmolysis and monitoring protoplasts in close contact for several hours. Although fusion was still a completely random and infrequent event, the basic prerequisites for induced fusion emerged from this early work. Protoplasts had to be freed of their cell walls, brought into contact for a finite period of time, and upon deplasmolysis they might be induced to fuse. As we have discussed in detail (Power, Evans, and Cocking, 1977), further work up to 1970 showed that, although fusion was still basically nonreproducible, the chance observations suggested that there were no barriers to interspecies protoplast fusion, and that, provided an inducer of fusion could be established, interspecies and intergeneric protoplast fusion could be achieved. The availability of enzymatically isolated protoplasts, and in large numbers, created a much broader basis for the examination of fusion methods. Work directed towards more reproducible interspecies fusion was also greatly stimulated by the Bellagio Meeting in 1969, sponsored by the Rockefeller Foundation, which attempted to relate the relevance of such earlier fusion studies to the general problem of the somatic hybridization of plants to be brought about by the fusion of protoplasts (fig. 2) (Cocking, 1971). One of the major perplexing considerations was the apparent lack of any system in which any degree of reproducible fusion was known to take place, which caused Steward et al. (1969) to conclude that naked angiosperm protoplasts do not fuse. In 1969 it seemed to us at Nottingham that the system described by Yoshida, which indicated that subprotoplast fusion within plasmolyzed cells was enhanced by the use of sodium salts, was particularly relevant to the laying of a foundation for the more controlled fusion of freely isolated protoplasts (Power et al., 1970). Using sodium nitrate–induced fusion of highly cytoplasmic, slightly vacuolated root protoplasts, we described in 1970 the first reproducible induced fusion of protoplasts illustrated by heterokaryon formation between oat and maize root protoplasts. In general, the less differentiated the protoplast, the more readily fusion takes place, and fusion can sometimes result in loss of viability. For somatic hybridization, adequate survival and division of heterokaryons are essential. A disadvantage of sodium nitrate-induced fusion was that fusion frequencies, particularly with mesophyll protoplasts, were very low. The presence of chloroplasts, within a thin peripheral layer of cytoplasm surrounding a large vacuole, prevented the complete coalescence of protoplasts (Power and Cocking, 1971).

Calcium ions at high pH (9.5–10.5) (Keller and Melchers, 1973) and polyethylene glycol (PEG) (Kao and Michayluk, 1974; Wallin et al., 1974) have been utilized for the fusion of mesophyll and cultured cell protoplasts.

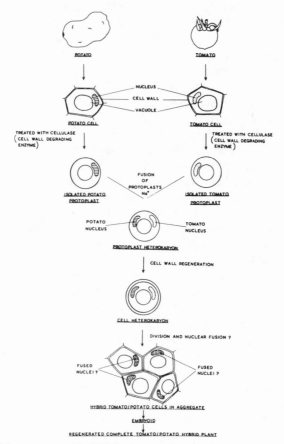

Fig. 2.   Scheme suggested in 1971 for interspecies somatic hybridization.

Increasingly, PEG-induced fusion is being employed for the fusion of plant protoplasts, for the fusion of animal cells (Pontecorvo, 1975), and for the fusion of plant protoplasts with animal cells (Ahkong et al., 1975). For fusion induction PEG solutions of about 20 percent w/v are employed, and there results a severe dehydration and shrinkage of protoplasts. Protoplasts adhere closely, and, probably during the subsequent gradual dilution of the PEG, fusion takes place. The detailed mechanism of PEG action is not understood, but the marked dehydration, with associated decrease in turgidity of vacuoles, may return differentiated cells, such as those of the mesophyll, to a more "structurally meristematic" condition. This may enable cytoplasmic mixing to occur more readily, and deplasmolysis may trigger off membrane interactions. Generally speaking, it is now clear that it is possible to obtain an acceptable level of fusion (1–10 percent) with one or other of these two inducing agents, with any two protoplast systems (Cocking, 1976a). The general laboratory procedure employed is shown in figure 3. The ultimate assessment of a given fusion procedure must lie in the recovery of somatic hybrids following that procedure.

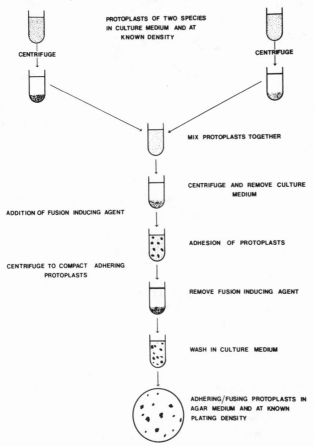

Fig. 3. General laboratory procedure for interspecies-induced fusion of protoplasts.

Fusion of protoplasts will always result in the formation of a cytoplasmic hybrid between the two fusion partners. The species being fused together will greatly influence the stability of hybrid nuclei. If chromosomes of one of the species are all eliminated (or if one of the nuclei dies), then a cybrid will result rather than a somatic hybrid (fig. 4). Moreover, the stages in the mitotic cycle of the nuclei in the heterokaryon will influence the extent to which nuclear fusion takes place. Of great interest is the extent to which division in an interphase nucleus, in a normally nondividing and often highly differentiated protoplast (such as that of cereal leaves), can be

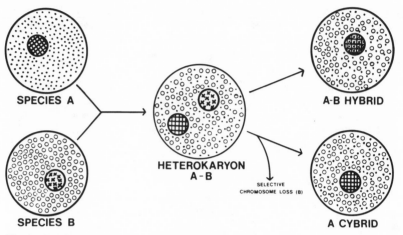

SPECIES A

A-B HYBRID

HETEROKARYON
A - B
SELECTIVE
CHROMOSOME LOSS (B)

SPECIES B

A CYBRID

Fig. 4.   Possible behavior of interspecies heterokaryons.

induced as a consequence of fusion with a mitotic protoplast system, such as that from suspension cultured cells.

PROCEDURES FOR THE SELECTION OF PLANT SOMATIC HYBRIDS

The proven ability of isolated plant protoplasts to fuse and form heterokaryons, and the developmental potential of isolated protoplasts to give rise to entire plants, offer the possibility of the production of hybrid plants by somatic hybridization. The fact that fusion products from such widely divergent species as barley and soybean and barley and tobacco can be obtained and cultured for a limited time (Dudits et al., 1976) suggests that, at least at the initial stages of fusion, there may be no incompatibility barrier. The high levels of heterokaryon formation that are now possible (up to 30 percent or more) might suggest that the recovery of somatic hybrids would be relatively easy, and of high frequency. Experience has shown, however, that this is usually not the case. Heterokaryons may, at an early stage in their existence, be particularly sensitive to the general cultural milieu, and growth at low cell density may be a major problem. Survival of heterokaryons is enhanced if a medium selective for their survival can be employed to enable them to grow and divide while the parental cell lines do not. Such special nutritional selection methods are rare, and usually depend on a knowledge of the sexual hybrid between the two parental species. Following heterokaryon formation, nuclei may fail to fuse during subsequent mitosis; or extensive chromosome loss from one or other, or both, of the fusion partners may occur. It is highly probable, therefore, that problems of the integration of nuclear material will lead to the loss of many of these heterokaryons, such that the final frequency of stable hybrid cells capable of continued development will be low. In order to rcover these somatic hybrids, it will therefore be necessary to select a few colonies from among many thousands (or even sometimes millions) of colonies ensuing from parental protoplasts.

A selection system has been described that involves the selective growth of heterokaryons found as a result of the sodium nitrate-induced fusion of Petunia leaf mesophyll protoplasts and protoplasts isolated from cultured cells of Parthenocissus crown gall (Power et al., 1975). The result here was particularly interesting, since selection resulted in the growth of a cell line that possessed only the nuclei of Parthenocissus, but showed isoenzyme patterns specific for both Parthenocissus and Petunia. These results probably indicate that selective chromosome elimination had taken place with the formation of a cybrid. Differences in mitotic cycle times and other factors in species such as these that are sexually incompatible may result in such chromosome elimination. PEG-induced fusion of mesophyll protoplasts of two sexually compatible tobacco species (probably both at the 2C stage of the mitotic cycle [Banks and Evans, 1976]), coupled with selective growth in a medium already known to be conducive to the growth of the sexual hybrid, resulted in the regeneration of somatic hybrid plants but surprisingly none of the somatic hybrids possessed the amphidiploid number of chromosomes (Smith et al., 1976).

Some form of complementation selection is probably the best way of achieving the desired selection of interspecies plant somatic hybrids. Since we have recently extensively discussed the use of such complementation selection for somatic hybrid selection (Power and Cocking, 1977), it will not be profitable to repeat this again. It is sufficient to remind the reader that genetic complementation in chloroplasts as a consequence of intraspecies protoplast fusion, first reported by Giles (1973), involved the use of mutant lines of *Zea mays*. Fusion of protoplasts (using sodium nitrate) containing green chloroplasts from the striped mutant iojap with protoplasts of the white deficient mutant containing only white undeveloped plastids caused the greening of the white deficient chloroplasts over a period of 72 hours. Selection using drug-resistant mutant complementation (dominant), or auxotrophic mutant complementation (recessive), could readily be employed if such mutants were available. Such complementation is the basis of the Littlefield HAT selection scheme—if one hybridizes a cell that is 8-AZG–resistant and aminopterin-sensitive with one that is BUDR-resistant and aminopterin-sensitive and grows the resulting mixed population on a medium containing hypoxanthine (HAT), both parental cells should die, but the hybrid cells should grow due to complementation of the deficiencies in the hybrid. Half-selection HAT systems can also be employed when the normal cells grow relatively slowly (Power and Cocking, 1977). Unfortunately, because the wild-type plant cells do not grow well in HAT-type media, such a selection is unlikely to be suitable for plant somatic hybrid selection (Cocking, 1976b). The type of auxotrophic (nitrate reductase/deficient) plant mutants described recently (Grafe et al., 1977) should allow extensive genetic analysis, as a result of somatic hybridization and complementation selection of the nitrate reductase–proficient somatic hybrids.

Here in the Agricultural Research Council Somatic Hybridisation Group at Nottingham we have recently developed two such complementation selection procedures and applied them to the production of the fertile interspecies somatic hybrid *Petunia hybrida* × *P. parodii* (table 1). The first of these (Power et al., 1976) involves selection using naturally occurring differences in the sensitivity of these two species to Actinomycin D, coupled with a difference in growth characteristics in a particular culture medium. More recently, this has been modified using a 2,4-dichlorophenoxyacetic acid (2,4-D)–based medium in which selective growth of the somatic hybrid occurs and gives a selection efficienty of somatic hybrids of approximately 6 percent (Power, Berry, Frearson, and Cocking, 1977). The tolerance of relatively high concentrations of 2,4-D in the hybrid is probably due to intergeneric complementation (Izhar and Power, 1977). The other method (Cocking et al., 1977) involved fusing albino suspension culture protoplasts with leaf mesophyll protoplasts of the other species, and selecting green colonies formed as a result of complementation (see Giles, 1973) and selective growth. This selection procedure (including also the complementation to green colonies of two albinos as a consequence of fusion (fig. 5), should be readily applicable to a wide range of interspecies fusions, including those of sexually incompatible species. Work in other laboratories, together with this work from Nottingham, is summarized in table 1. Only work that has produced somatic hybrid plants is included. Somatic hybrids should be distinguishable from most sexual hybrids, since the somatic hybrid may contain in the somatic hybrid plant the cytoplasms of the two parental species. Great interest, therefore, centers on the extent to which biochemical cytoplasmic markers (such as Fraction I) can be employed in the analysis of somatic hybrids (Gatenby and Cocking, 1977).

DESIRED INTERSPECIFIC CROSSES

To what extent genome incompatibility will be a problem in plant interspecies somatic hybridization is not yet clear. Pre-zygotic incompatibilities are likely to be eliminated, but various post-zygotic ones may remain. Some recent work, on barley and wheat sexual crosses, has indicated that directional chromosome elimination in interspecific hybrids

TABLE 1

SOMATIC HYBRID PLANTS FROM FUSION OF PROTOPLASTS

Fusion Inducer	Somatic Hybrid Plant	Selection	Reference
Polyethylene glycol	*P. hybrida* × *P. parodii*	Complementation	Power et al., 1976
Polyethylene glycol	*P. hybrida* (albino) × *P. parodii*	Complementation	Cocking et al., 1977
Polyethylene glycol	*N. glauca* × *N. langsdorffii*	Selective growth	Smith et al., 1976
Ca^{+}/high pH	*N. tabacum* × *N. sylvestris*	Complementation	Melchers, 1976
Polyethylene glycol	*N. tabacum* (male sterile) × *N. tabacum*	Regeneration	Belliard et al., 1977
Ca^{+}/high pH	*N. tabacum$_a$* × *N. tabacum$_b$*	Complementation	Melchers and Labib, 1974
Polyethylene glycol	*N. tabacum$_c$* × *N. tabacum$_d$*	Complementation	Gleba et al., 1975

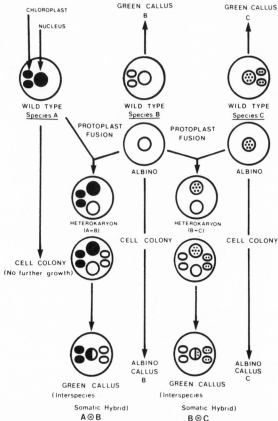

Fig. 5.   Schemes for the albino complementation selection of interspecies somatic hybrids (Cocking et al., 1977; reprinted with permission).

in angiosperms may be more frequent than has been generally realized (Bennett et al., 1976). Although the outcome of nuclear gene interactions in interspecific somatic hybrids is largely unknown, it is even less clear what the extent of cytoplasmic interaction might be following fusion. Cytoplasmically inherited traits are of widespread use in plant breeding, particularly as a source of male sterility; and it is likely that somatic hybridization may be used to transfer cytoplasmic features, such as male sterility, from one species to another. Indeed, intraspecies transfer of male sterility within the tobaccos has recently been reported (Belliard et al., 1977).

Generally speaking, production of intespecific crosses in crop plants, often within the same genus, are desired for two major reasons: the production of new amphidiploids; and the use of intespecific hybridization for gene transfer, for example, resistance to plant diseases or environmental stress (Cocking, 1976c; 1977). A general feature of the production of interspecific crosses is that the usual sterile hybrids become reasonably fertile when their chromosome number is doubled, usually achieved, where it is possible, as a result of colchicine treatment. Such amphidiploids may also become reasonably stable for the hybrid condition and produce progeny that are more or less uniformly like the original hybrid. Consequently, the use of diploid somatic cells for fusion, which subsequently yield amphidiploids directly without, of course, the reduction division of the usual sexual reproduction (necessitating either spontaneous doubling, or the subsequent artificial doubling, of the chromosome complement by colchicine treatment), may be a distinct advantage. Even when the sexual cross is possible, an inability to produce tetraploids from the diploid sexual hybrid by colchicine or other treatments may make somatic hybridization an attractive alternative to sexual hybridization. Somatic hybridization could also be useful for the crossing of vegetatively propagated plants facilitating, for instance, the production of potato tetraploids from potato dihaploids.

FUTURE PROSPECTS

It is now nearly ten years since the Rockefeller Conference at the Villa Serbelloni on Lake Como identified the need for a resurgence of interest in the fusion of plant protoplasts and the extra possibilities for crop improvement that might result from the development of a technology of an alternative to sexual hybridization. Sir Macfarlane Burnet (1971), who sat in on our deliberations, has commented, "What the botanists had in mind was the possibility of an incomparably wide range of hybridisations if a means could be found to fuse somatic cells of almost unrelated species and persuade such unnatural hybrids to develop into complete plants. If cells of man and fowl or man and mouse can produce composite cells which can multiply in tissue culture, why should not a cell of a high-bearing rice hybridise with a desirable strain of sweet potato? And if a tobacco or begonia single cell can produce a complete plant of its proper type, why should not the hybrid cell produce a plant with the virtues of both its parents? So far as I am aware, no such artificial fusion of plant cells has yet produced a complete composite plant, but I know that a variety of possible approaches is being explored."

Clearly, as we have already discussed, there are a number of problems still remaining, many of which relate to our present inability to culture certain species satisfactorily. We also need to find out much more about the consequences of fusion of protoplasts of sexually incompatible species. Nevertheless, what has been accomplished on somatic hybridization since the Rockefeller Conference in 1969 should encourage us to proceed actively with our experimentation, while perhaps heeding the words of A. N. Whitehead (1929): "When we survey the history of thought, and likewise the history of practice, we find that one idea after another is tried, its limits defined, and its core of truth elicited . . . the proper test is not of finality but of progress."

ACKNOWLEDGMENTS

Original research by the ARC Group at Nottingham contained in this paper was supported by a grant from the Agricultural Research Council.

REFERENCES

Ahkong, Q. F., J. I. Howell, J. A. Lucy, F. Safwat, M. R. Davey, and E. C. Cocking. 1975. Fusion of hen erythrocytes with yeast protoplasts induced by polyethylene glycol. Nature 255:66–67.

Bajaj, Y.P.S. 1977. Protoplast isolation, culture, and somatic hybridisation. In J. Reinert and Y.P.S. Bajaj (eds.), Applied and fundamental aspects of plant cell, tissue, and organ culture. Pp. 467–96. Springer-Verlag, Berlin.

Banks, M. S., and P. K. Evans. 1976. The use of heterochromatin spot counts to determine ploidy levels in root tips and callus cultures of *Nicotiana otophora* and its hybrids. Plant Sci. Lett. 7:409–16.

Belliard, G., G. Pelletier, and M. Farault. 1977. Fusion de protoplastes de *Nicotiana tabacum* a cytoplasmes différents: Etude des hybrides cytoplasmiques néo-formé. C. R. Acad. Sci. Paris 284D:749–52.

Bennett, M. D., R. A. Finch, and I. R. Barclay. 1976. The time rate and mechanisms of chromosome elimination in *Hordeum* hybrids. Chromosoma 54:175–200.

Bhojwani, S. S., P. K. Evans, and E. C. Cocking. 1977. Protoplast technology in relation to crop plants: Progress and problems. Euphytica 26:343–60.

Burnet, Sir Macfarlane. 1971. Genes, dreams and realities. Basic Books, New York.

Cocking, E. C. 1960. A method for the isolation of plant protoplasts and vacuoles. Nature 187:927–29.

Cocking, E. C. 1967. Plant protoplasts. In J. D. Carthy and C. L. Duddington (eds.), Viewpoints in biology. 4:170–203. Butterworths, London.

Cocking, E.C. 1971. Fusion of isolated protoplasts: A first step towards the somatic hybridisation of plants. In M. L. Hirth and G. Morel (eds.), Les cultures de tissue de plantes. Pp. 303–17. Editions du C.N.R.S., Paris.

Cocking, E. C. 1973. Plant cell modification: Problems and perspectives. In J. Tempé (ed.), Protoplastes et fusion de cellules somatique végétales. Colloques Intern. C.N.R.S. 212:327–41.

Cocking, E. C. 1976a. Fusion and somatic hybridisation of higher plant protoplasts. In J. F. Peberdy, A. H. Rose, H. J. Rogers, and E. C. Cocking (eds.), Microbial and plant protoplasts. Pp. 189–200. Academic Press, London.

Cocking, E. C. 1976b. A new procedure for the selection of somatic hybrids in plants. *In* D. Dudits, G. L. Farkas, and P. Maliga (eds.), Cell genetics in higher plants. Pp. 141-48. Akadémiai Kiadó, Budapest.

Cocking, E. C. 1976c. Perspectives in the utilisation of protoplasts in genetics and plant breeding. *In* Proceedings of Conference New trends in genetics and plant breeding. Pp. 87-100. Faculté des Sciences Agronomiques de l'Etat et Centre de Recherches Agronomiques, Gembloux, Belgium.

Cocking, E. C. 1977. Plant protoplast fusion: Progress and prospects for agriculture. *In* R. F. Beers and E. G. Bassett (eds.), Recombinant molecules: Impact on science and society. Pp. 195-208. Raven Press, New York.

Cocking, E. C., and P. K. Evans. 1973. The isolation of protoplasts. *In* H. E. Street (ed.), Plant tissue and cell culture. Pp. 100-120. Blackwell Scientific Publications, Oxford.

Cocking, E. C., D. Gerorge, M. J. Price-Jones, and J. B. Power. 1977. Selection procedures for the production of inter-species somatic hybrids of *Petunia hybrida* and *Petunia parodii*. II. Albino complementation selection. Plant Sci. Lett. 10:7-12.

Dudits, D., K. N. Kao, F. Constable, and O. L. Gamborg. 1976. Fusion of carrot and barley protoplasts and division of heterokaryocytes. Canad. J. Genet. Cytol. 18:263-69.

Evans, P. K., and E. C. Cocking. 1975. The techniques of plant cell culture and somatic cell hybridization. *In* R. H. Pain and B. J. Smith (eds.), New techniques in biophysics and cell biology. 2:127-58. John Wiley & Sons, New York.

Gatenby, A. A., and E. C. Cocking. 1977. Polypeptide composition of Fraction I protein subunits in the genus *Petunia*. Plant Sci. Lett. 10:97-101.

Giles, K. L. 1973. Attempts to demonstrate genetic complementation by the technique of protoplast fusion. *In* J. Tempé (ed.), Protoplastes et fusion de cellules somatique vegetales. Colloques Intern. C.N.R.S. 212:485-95.

Gleba, Y. Y., R. G. Butenko, and K. M. Sytnik, 1975. Fusion of protoplasts and parasexual hybridization in *Nicotiana tabacum* L. Dokl. Akad. Nauk. USSR 221:1196-98.

Grafe, R., A. J. Müller, and I. Saalbach 1977. Cell genetic studies on nitrate reductase-deficient mutants. *In* International conference on regulation of developmental processes in plants. P. 128 (abstr. 109). Halle (Saale).

Grout, B. W. W. 1975. Cellulose microfibril deposition at the plasmalemma surface of regenerating tobacco mesophyll protoplasts: A deep etch study. Planta 123(3):275-82.

Grout, B. W. W., and R. H. A. Coutts. 1974. Additives for the enhancement of fusion and endocytosis in higher plant protoplasts: An electrophoretic study. Plant Sci. Lett. 2:397-403.

Haldane, J. B. S. 1955. Some alternatives to sex. *In* M. L. Johnson, M. Abercrombie, and G. E. Fogg (eds.), New biology. 19:7-26. Penguin Books, Harmondsworth, Middlesex.

Izhar, S., and J. B. Power. 1977. Genetic studies with petunia leaf protoplasts. I. Genetic variation to specific growth hormones and possible genetic control on stages of protoplast development in culture. Plant Sci. Lett. 8:375-86.

Kao, K. N., and M. R. Michayluk. 1974. A method for high-frequency intergeneric fusion of plant protoplasts. Planta 155:355-67.

Keller, W. A., and G. Melchers. 1973. The effect of high pH and calcium on tobacco leaf protoplast fusion. Z Naturforsch. 28C:737-41.

Klercker, J. 1892. Eine Methode zur isolierung lebender Protoplasten. Öfvers Vetensk. Akad. Forh. Stokh. 49:463-75.

Kuster, E. 1910. Eine Method zur gewinnung abnorm grosser Protoplasten. Wilhelm Rouxs Arch. Entwicklungsmech. Org. 30:351-55.

Melchers, G. 1976. Microbial techniques in somatic hybridisation by fusion of protoplasts. First International Cell Biology Symposium, Boston.

Melchers, G., and G. Labib. 1974. Somatic hybridization of plants by fusion of protoplasts. I. Selection of light resistant hybrids of "haploid" light sensitive varieties of tobacco. Molec. Gen. Genet. 135:277-94.

Nagata, T., and I. Takebe 1970. Cell wall regeneration and cell division in isolated tobacco mesophyll protoplasts. Planta 92:301-8.

Pontecorvo, G. 1975. Production of mammalian somatic cell hybrids by means of polyethylene glycol treatment. Somatic Cell Genet. 1:397-400.

Power, J. B., S. F. Berry, E. M. Frearson, and E. C. Cocking. 1977. Selection procedures for the production of inter-species somatic hybrids of *Petunia hybrida* and *Petunia parodii*. I. Nutrient media and drug sensitivity complementation selection. Plant Sci. Let. 10:1-6.

Power, J. B., and E. C. Cocking 1971. Fusion of plant protoplasts. Sci. Progress 59:181-98.

Power, J. B., and E. C. Cocking. 1977. Selection systems for somatic hybrids. *In* J. Reinert and Y. P. S. Bajaj (eds.), Applied and fundamental aspects of plant cell, tissue, and organ culture. Pp. 497-505. Springer-Verlag, Berlin.

Power, J. B., S. E. Cummins, and E. C. Cocking. 1970. Fusion of isolated plant protopasts. Nature 225:1016-18.

Power, J. B., P. K. Evans, and E. C. Cocking. 1977. Fusion of plant protoplasts. *In* G. Poste and G. L. Nicholson (eds.), Cell surface reviews. 8:370-85. North-Holland Publishing Co., Amsterdam.

Power, J. B., E. M. Frearson, C. Hayward, and E. C. Cocking. 1975. Some consequences of the fusion and selective culture of *Petunia* and *Parthenocissus* protoplasts. Plant Sci. Lett. 5(3):197-207.

Power, J. B., and E. M. Frearson, C. Hayward, D. George, P. K. Evans, S. F. Berry, and E. C. Cocking. 1976. Somatic hybridisation of *Petunia hybrida* and *Petunia parodii*. Nature 263:500-502.

Scarth, G. W. 1942. Structural differentiation of cytoplasm. *In* W. Seifriz (ed.), A symposium on the strucure of protoplasm. Pp. 99-107. Monograph of the American Society of Plant Physiologists. Iowa State College Press, Ames.

Smith, H. H., K. N. Kao, and N. C. Combatti. 1976. Interspecific hybridization by protoplast fusion in *Nicotiana*: Confirmation and extension. J. Hered. 67:123-28.

Steward, F. C., and A. D. Krikorian. 1971. The culturing of higher plant cells: Its status, problems, and potentialities. *In* H. Y. Mohan Ram, J. J. Shah, and C. K. Shah (eds.), Form, structure, and function in plants. Pp. 144-70. Sarita Prakashan, Meerut, India.

Steward, F. C., M. O. Mapes, and P. V. Ammirato. 1969. Growth and morphogenesis in tissue and free-cell cultures. *In* F. C. Steward (ed.), Plant physiology. 5B:329-76. Academic Press, New York.

Vasil, I. K. 1976. The progress, problems, and prospects of plant protoplast research. *In* N. C. Brady (ed.), Advances in agronomy. 28:119-60. Academic Press, New York.

Wallin, A., K. Glimelius, and T. Eriksson 1974. The induction of aggregation and fusion of *Daucus carota* protoplasts by polyethylene glycol. Z. Pflanzenphysiol. 74:64-80.

Whitehead, A. N. 1929. Process and reality. At the University Press, Cambridge.

Willison, J. H. M. 1976. Synthesis of cell walls by higher plant protoplasts. *In* J. F. Peberdy, A. H. Rose, H. J. Rogers, and E. C. Cocking (eds.), Microbial and plant protoplasts. Pp. 283–97. Academic Press, London.

O. L. GAMBORG, K. OHYAMA, L. E. PELCHER, L. C. FOWKE,
K. KARTHA, F. CONSTABEL, AND K. KAO

Genetic Modification of Plants

20

INTRODUCTION

The production of new hybrids without recourse to sexual reproduction constitutes one of the greatest potentials in plant genetics for crop improvement (Wittwer, 1974; Brown et al., 1975; Nickell and Torrey, 1969).

Current techniques of plant breeding are based on utilizing random variations induced by mutation or using the sexual cycle with the meiotic process to recombine existing variation. If systems could be developed for transferring genetic information between widely different plant species, they would provide the basis for a technology with unlimited potential for construction of new types of crop plants (Nickell and Torrey, 1969; Carlson and Polacco, 1975). The development of hybridization approaches with somatic cells is the major aim of the research to increase genetic diversity and to provide alternate means of genetic recombination. Two possible approaches are being explored. One procedure involves the formation of hybrids by fusion of protoplasts from different plant sources (Gamborg et al., 1974a). A more selective procedure consists of transferring isolated DNA or specified genes from one source into a recipient plant source (Scowcroft, 1978; Holl, 1975a). A considerable technology already is available on genetic manipulation in microorganisms (Raspe, 1972). Plant cells offer a vastly greater scope in somatic genetics and genetic manipula-

O. L. Gamborg, Prairie Regional Laboratory, Saskatoon, Saskatchewan, Canada.

tion than either prokaryotic or animal cells (Ringertz and Savage, 1976) because of their potential for regenerating a complete organism.

The somatic hybridization and genetic transformation methods implicitly depend upon plant tissue culture procedures. These techniques include a broad spectrum of methods for culturing tissues, cells, embryos, shoot tips, or isolated protoplasts (Reinert and Bajaj, 1977). Under appropriate conditions the cells of some species regenerate complete plants. Utilizing genetic manipulation procedures and subsequently growing the cells would provide new approaches to investigate and expand basic knowledge of plant genetics and physiology and the possibility of producing entirely new plants.

PLANT CELL CULTURE

The success in somatic hybridization has come about as a result of major advances in plant cell culture and protoplast technology (Vasil, 1976; Gamborg, 1977b). Cells of most plant species can be grown on chemically defined media containing growth hormones (Gamborg et al., 1976).

The growing of large populations of plant cells in fermentors and plating of single cells have become routine procedures. Morphogenesis has been reported in an increasing number of species (Murashige, 1974; Narayanaswamy, 1977). The differentiation process is induced by cytokinin compounds, and plant regeneration has been achieved in a variety of crop species that include cereals, legumes, oilseeds, coffee, forest trees, and horticultural plants (table 1). In some species the plant regeneration occurs by somatic embryogenesis, in which embryos are produced from single cells in the absence of exogenous growth regulators (Halperin, 1970). Plant regenerations from meristems and shoot tips have become standard procedures for clonal propagation and in plant disease control (Heinz et al., 1977; Mellor and Stace-Smith, 1977; Kartha and Gamborg, 1975).

The culturing of anthers is used in producing haploid cells and plants (Vasil and Nitsch, 1975). The haploid material facilitates selection by producing homozygous lines. By employing populations of haploid cells, it is possible to apply mutagenic treatments followed by chromosome doubling. On subsequent plating of the homozygous cells, it becomes possible to isolate mutants with lesions that are generally recessive and ordinarily masked in heterozygous progeny. In species in which plant regeneration can be realized, it is possible to examine in complete plants the modifications induced in culture.

Along with progress in the use of tissue culture has come the development of a plant protoplast technology (Gamborg, 1976; Vasil, 1976). The elimination of plant cell walls achieved by the use of enzymes results in the

TABLE 1

PLANT REGENERATION FROM SOMATIC CELLS AND PROTOPLASTS

Common Name	Species	Tissue	Reference
Alfalfa	*Medicago sativa*	Callus	Saunders and Bingham, 1972
Almond	*Prunus amygdalus*	Callus	Mehra and Mehra, 1974
Asparagus	*Asparagus officinalis*	Protoplasts	Bui Dang Ha and MacKenzie, 1973
Barley	*Hordeum vulgare*	Calus	Cheng and Smith, 1975
Brussels sprout	*Brassica oleracia*	Callus	Walkey and Woolfit, 1970
Carrot	*Daucus carota*	Protoplasts	Dudits et al., 1976
Citrus	*Citrus sp.*	Protoplasts	Vardi et al., 1975
Coffee	*Coffea arabica*	Callus	Sondahl and Sharp, 1977
Corn	*Zea mays*	Callus	Green and Phillips, 1975
Douglas fir	*Pseudotsuga menziesii*	Callus	Cheng, 1975
Flax	*Linum usitatissimum*	Callus	Gamborg et al., 1976
Forage legume	*Stylosanthes hamata*	Callus	Scowcroft and Adamson, 1976
Millet	*Pennisetum sp.*	Callus	Rangan, 1976
Oat	*Avena sativa*	Callus	Cummings et al., 1976
Onion	*Allium cepa*	Callus	Fridborg, 1971
Pea	*Pisum sativum*	Callus	Gamborg et al., 1974
Potato	*Solanum tuberosum*	Protoplasts	Shepard and Totten, 1977
Rapeseed	*Brassica napus*	Protoplasts	Kartha et al., 1974
Rapeseed	*Brassica campestris*	Microspore	Keller et al., 1975
Rice	*Oryza sativa*	Microspore	Niizeki and Oono, 1968
Rye	*Secale cereale*	Microspore	Thomas et al., 1975
Sorghum	*Sorghum bicolor*	Callus	Gamborg et al., 1977
Sugarcane	*Saccharum officinarum*	Callus	Heinz and Mee, 1969
Tobacco	*Nicotiana tabacum*	Protoplasts	Takebe et al., 1971
Wheat	*Triticum aestivum*	Callus	Shimada et al., 1969

See Pierik, 1975; Murashige, 1974; and Winton and Huhtinen, 1976.

release of protoplasts. Under suitable conditions the cell walls reform, and cell division followed by plant regeneration can occur.

New insight into plant virus infection and replication has been gained (Takebe, 1975), and fruitful investigations on the mode of action of plant pathogenic toxins at the molecular and cellular level were made possible by the aid of protoplasts (Strobel, 1973). However, a major reason for the rapidly expanding interest in protoplasts is their potential use in plant cell genetics, and specifically in cell fusion and transfer of genetic information by DNA uptake and organelle implantations (Cocking, 1972; Gamborg, 1977b).

SOMATIC HYBRIDIZATION

The concept of somatic hybridization was originally formulated and presented as a potential technology to overcome some of the natural limitations inherent in producing wide crosses for plant breeding (Borlaug, 1971; Bates and Deyoe, 1973). Hybrids between widely different plants are sought for the purposes of transferring disease resistance and tolerance to stress conditions and of generally improving product quality and growth characteristics.

Cell fusion and somatic hybridization have become valuable analytical techniques in animal and human genetics (Ringertz and Savage, 1976). A wide variety of animal hybrid cells have been produced, including combinations with cells of human origin. The mammalian cells can be fused directly, but before plant cells can be fused, the pectin and hemicellulose holding cells together, as well as the cellulose cell walls, must be removed. Protoplast fusion and somatic hybridization in plants involve four basic procedures: (1) protoplast isolation; (2) fusion and growth of hybrid cells; (3) hybrid cell selection and cloning; and (4) regeneration of hybrid plants.

Plant Protoplasts

The standard procedures for protoplast isolation consist of treating the tissues with a mixture of wall-degrading enzymes in appropriate solutions with osmotic stabilizers to preserve the structure and viability of the protoplasts. The details of isolation conditions and procedures have been discussed in recent reviews (Gamborg, 1977a, 1977b; Vasil, 1976). Protoplasts have been isolated from a large number of plant tissues. Convenient and suitable materials are leaf mesophyll and cells from liquid suspension cultures. Protoplast yields and viability are profoundly influenced by the growing conditions of plants serving as leaf mesophyll sources. The age of the plant and of the leaf and the prevailing conditions of light, photoperiod, humidity, temperature, nutrition, and watering are contributing factors. Cell suspension cultures may provide a more reliable cell source for obtaining consistent quality protoplasts. It is necessary, however, to establish and maintain the cells at maximum growth rates and utilize the cells at the early log phase. The survival of protoplasts also is affected by the choice of enzymes (Cassells and Barlass, 1976). Some enzymes have deleterious effects. The problems also are related to membrane stability. Leaf protoplasts may have a tendency to be less stable than cell culture protoplasts. The stability and/or survival of protoplasts also vary greatly between species and are influenced by the prevailing growing conditions of the plant source. The protoplasts are isolated under aseptic conditions and carefully freed of enzymes and debris by washing (Constabel, 1975; Larkin, 1976; Shepard and Totten, 1975).

After isolation the protoplasts are suspended in appropriate solutions for maintaining viability and structural integrity. The protoplast media consist of mineral salts, vitamins, carbon sources, and growth hormones, as well as osmotic stabilizers and possibly organic nitrogen sources, coconut milk, and organic acids (Gamborg, 1977b, Reinert and Bajaj, 1977; Vasil, 1976). Protoplasts are usually cultured in liquid media until cell clusters have formed. The cells can then be transferred and plated on agar.

Freshly isolated protoplasts are devoid of cell walls (fig. 1). The absence of microfibrils has been demonstrated by the use of freeze-etching and a platinum/palladium replica technique (Williamson et al., 1977). However, it is conceivable that wall removal may not be complete in all protoplasts of a given population if wide variations exist in cell age and type within the source material.

The viability of protoplasts can be established by detecting cytoplasmic streaming and the use of specific dyes (Larkin, 1976). Unless the protoplasts are capable of wall regeneration and division, they will generally deteriorate within hours or a few days.

Protoplasts begin to deposit cellulose microfibrils immediately after washing and enzyme removal (fig. 2). Williamson et al. (1977) have shown that microfibrils can be detected within 10 min. of culture after washing. However, whereas some plant protoplasts quickly reform cell walls and subsequently enter division, those of other species have consistently been recalcitrant. The protoplasts from leaf mesophyll or root tissue of cereals and grasses are examples.

Division, followed by the formation of callus and cell cultures, has been reported for an increasing number of species. The cells from protoplasts retain the capacity for morphogenesis, and complete plant regeneration has been achieved (table 1).

Fusion

Experimental fusion of plant protoplasts was reported by Kuster (1909) and by Michel (1937). Although fusion products were obtained, they failed to survive because of shortcomings in the understanding of environmental, nutritional, and hormonal requirements of isolated protoplasts.

Fusion was achieved by using a salt solution, and the process was aided by the presence of nitrate. After protoplasts became available in larger quantities with the aid of enzymes, renewed attempts were made at cell fusion (Cocking, 1972; Eriksson, 1970).

Spontaneous fusion can occur during isolation from leaf tissue and particularly from cultured cells (Miller et al., 1971; Withers and Cocking, 1972; Morel et al., 1972). The isolated protoplasts generally fail to fuse. An exception is the protoplasts from microsporocytes of some plants of the Lily family (Ito and Maeda, 1974).

Fusion of protoplasts from different sources requires an inducing agent. Some success was reported using nitrate (Power et al., 1970; Potrykus, 1972), salt mixtures (Binding, 1974), and gelatin (Kameya, 1975).

A procedure involving high calcium levels and alkaline pH was used in the fusion of tobacco protoplasts and subsequent formation of hybrid plants (Keller and Melchers, 1973; Melchers and Labib, 1974). The same conditions also were used in fusion and hybridization of moss (*Physcomitrella patens*) (Grimsley et al., 1977).

Many fusion experiments are performed by a polyethylene glycol (PEG) method (Kao et al., 1974; Wallin et al., 1974; Power et al., 1976). Protoplasts exposed to high concentration of PEG form adhesion aggregates (Fowke et al., 1975a, 1975b; Burgess and Fleming, 1974). Fusion may be initiated at this point, and goes to completion when the PEG concentration is reduced. It is essential that walls are completely

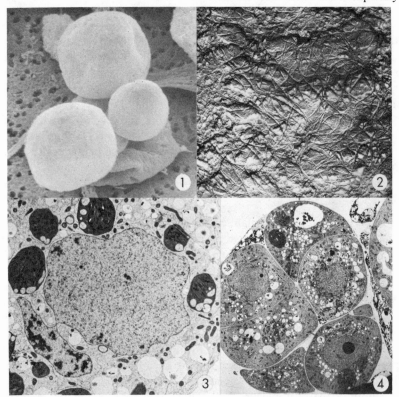

Figs. 1–4. 1. Scanning electron micrograph of soybean protoplasts. 2. Wall regeneration on cultured protoplast. 3. Electron micrograph showing fusion of nuclei of pea (with heterochromatin) and soybean. 4. Hybrid cells of pea and soybean.

removed for fusion to occur (Weber et al., 1976). Evidence from ultra-structural examinations suggests that interaction occurs at several points in the opposing membranes, and cytoplasmic bridges are formed (Fowke et al., 1975). Initially the individual protoplasts can be distinguished, but within a few hours the cytoplasms have merged.

Hybrid Identification and Development

The identification of protoplast fusion products is based on differences between the parental cells with respect to pigmentation, presence of chloroplasts, cytology, and cytoplasmic markers. A system that has been used quite successfully consists of fusing protoplasts of leaf mesophyll that contain chloroplasts with those from cell cultures that lack chloroplasts (Kao, 1975; Kartha et al., 1974b; Constabel et al., 1976). The fusion products contain chloroplasts and can be distinguished from unfused protoplasts. Employing the above approach and a differential staining technique, it has been possible to quantitate fusion frequency and monitor the fusion products (Kao and Michayluk, 1974; Constabel et al., 1975a; Dudits et al., 1976b; Kao, 1977). The fusion frequency is affected by protoplast quality as well as by fusion conditions.

The fusion products (heterokaryons) contain one or more nuclei from each parental protoplast. Constabel et al. (1975a) examined heterokaryons of pea and soybean. The ratio of 1:1 nucleus of each parent occurred most frequently (14 percent), followed by those with a 2:1 ratio. During subsequent culture the nuclei may fuse during interphase through the formation of nuclear membrane bridges (fig. 3) (Fowke et al., 1975a, 1975b; Fowke et al., 1977), or hybridization may occur during mitosis (Kao, 1977; Constabel et al., 1977, Gosch and Reinert, 1976).

The hybrid nature of the cell progeny has been established on the basis of ultrastructural examination (Fowke et al., 1976, 1977), karyology (Kao, 1977; Constabel et al., 1977), and zymograms (Wetter and Kao, 1976; Wetter, 1977). In the initial stages the cells contain chloroplasts originating from the leaf protoplasts and leucoplasts carried over from the cultured cell (fig. 4). The nuclei also contain two types of heterochromatins (fig. 3) (Fowke et al., 1977) and retain chromosomes of both parental species (Kao, 1977; Constabel et al., 1977). The isoenzyme zymograms of different constitutive enzymes further reflect that hybridization has occurred (Wetter, 1977).

The PEG treatment is generally tolerated by plant protoplasts, and a wide variety of hybrid cells has been obtained. These include the intergeneric hybrids of soybean with *Vicia hajastana*, pea, and other genera (Constabel et al., 1975a). The production of hybrid cells of *Nicotiana glauca* and soybean (Kao, 1977), carrot + barley (Dudits et al., 1976b), and soybean + *Brassica napus* (Kartha et al., 1974b) suggest that the incompatibility existing between different plant families is not apparent in somatic cells (Constabel, 1976).

Hybrid Selection and Isolation

In most fusion experiments the division rates are relatively low. At the same time, one or both parental protoplast species also may divide, and very shortly the hybrid cells can no longer be distinguished from parental cells. Some type of selection procedure is essential. The basis for hybrid selection should result in preferential growth, survival of hybrids, and elimination of parental cells.

The first mammalian hybrid cells were discovered and isolated because they proliferated more rapidly than either of the parental cell progeny (Davidson, 1971). Although highly desirable, hybrid vigor of this type has not been observed in plant cell hybrids. In mixtures of hybrid and parental cells, the latter usually predominate, because they adapt more readily to the culture environment and possibly have a shorter generation time.

Several selection methods have been applied. The hybrids of *Nicotiana glauca* + soybean were isolated manually with special plating procedures for single hybrid cells (Kao, 1977). The selection of the hybrids of *Nicotiana glauca* and *N. langsdorffi* was based on auxin autotrophy of the hybrid cells (Carlson et al., 1972). The parental cells required an auxin compound in order to proliferate, whereas the hybrid callus had no such requirement because the cells are auxinautotrophic.

Attempts have also been made to utilize uncommon amino acids as selective agents. Canavanine (Rosenthal, 1977), which occurs in some legumes, inhibits division of soybean and pea cells, but those of sweet clover and alfalfa are unaffected. Heterokaryons obtained by fusion of protoplasts from soybean (sensitive) with those from any one of the resistant plants failed to divide in the presence of canavanine (Constabel, et al., 1975b). It is conceivable that the canavanine sensitivity of soybean may be a dominant characteristic, and thus may be expressed in the hybrids.

Other types of potential chemical selection procedures may involve herbicides, phytotoxins, or antibiotics. Plants possess differences in their capacity to metabolize herbicides. An example is the resistance of rice plants to propanil (3,4-dichloropropionanilide) (Still, 1968). The resistance is based on the ability of rice cells to metabolize propanil. It remains to be demonstrated if propanil or other herbicides can be used effectively for selection (Dudits, 1976).

Phytotoxins isolated from culture filtrates of several pathogens are metabolite analogs and have been shown to be specific for particular cells

(Pelcher et al., 1975). The compounds are active in low concentrations. The toxin of *Helminthosporium maydis* has been used in selection of resistant cell lines of corn (Gengenbach and Green, 1975). The toxins may provide effective systems by utilizing appropriate parental protoplasts and in conjunction with other agents.

Cell strains resistant to antibiotics are relatively easy to obtain, and their usefulness is being explored in hybrid selection (Maliga, 1976). The biochemical basis for antibiotic resistance has not been established. A highly bromodeoxyuridine-resistant soybean cell line has been carefully examined by Ohyama (1976). The results failed to fully explain the resistance, but the mechanism is unlike any of those observed in mammalian cells (Chu and Powell, 1976).

The drug actinomycin D was used by Power et al. (1976) in the selection of somatic hybrids of two *Petunia* species. Cells from fusion products of protoplasts from *P. parodii* and *P. hybrida* were grown into complete plants. The cells of *P. hybrida* failed to grow in the presence of actinomycin D. Adjustments in medium resulted in preferential growth of the hybrid cells and subsequent plant regeneration, whereas the *P. parodii* failed to regenerate plants.

Systems based on genetic complementation are perhaps the most reliable and effective for the recovery of somatic hybrids. Melchers and Labib (1974) fused protoplasts of two chlorophyll deficient, light-sensitive mutants of *Nicotiana tabacum*. The hybrid plants regenerated from the fusion products had normal leaf color and had normal reaction to light. Gleba et al. (1975) recovered green (wild type) tobacco hybrids after fusing protoplasts from a chlorophyll deficient genome mutant with those from a plastome mutant.

Nutritional (auxotrophic) mutants would be the most attractive material, because hybrids could be selected at the cellular level and plant regeneration would not be an essential part of the selection procedure. Nutritional mutants were used in experiments with liverwort (*Sphaerocarpus donnellii*) (Schieder, 1974). Hybrids obtained by fusion of protoplasts from nicotinic acid- and glucose-requiring mutants were selected on minimal media. The regenerated hybrid plants were identified on the basis of morphology and karyotype (Schieder, 1974). Nutritional mutants also have been used in somatic hybridization with moss (*Physcomitrella patens*) (Grimsley et al., 1977).

Analogous metabolic mutants have not been available in higher plants. Recently a report described a series of nitrate reductase–deficient mutants (Muller et al., 1976). The mutants were selected from mutagenized haploid cells of *Nicotiana tabacum* cultured on media containing chlorate and with amino acids as the nitrogen source. Cells with active nitrate reductase allegedly convert chlorate to chlorite, which is cytotoxic (Cove, 1976). The isolated mutants were unable to grow on nitrate and lacked nitrate reductase and other molybdenum-protein-containing enzymes. Such mutants might be suitable for hybrid cell selection based on genetic complementation. Adequate methods are lacking for obtaining auxotrophic cell mutants of seed-bearing plants. Metabolic mutants of both liverwort (Schieder, 1976a) and moss (Ashton and Cove, 1977) have been reported, indicating the feasibility of obtaining similar mutants in higher plants. Schieder (1976b) has obtained chlorophyll-deficient mutant plants regenerated from haploid cells of *Datura innoxia* after radiation treatment.

A variety of plant mutants of *Arabidopsis* has been isolated and characterized (Redei and Acedo, 1976), and a proline-requiring mutant of corn has been reported (Gavazzi et al., 1975). A bromodeoxyuridine and light treatment procedure for elimination of wild-type cells has been used to obtain metabolic mutants of mammalian cells (Chu et al., 1972). The method was employed in experiments with diploid tobacco cells, but the mutants obtained were leaky auxotrophs (Carlson, 1973). No further reports have appeared on the use of this procedure with plant cells, although it may have potential if haploid cell material is used. Using mutagenic treatment on haploid cells followed by chromosome doubling yields homozygous materials, and mutations that are generally recessive can be selected directly. Chu and Powell (1976) have suggested other potential selection systems for eukaryotic cell mutants.

Plant Regeneration

The ultimate objective in somatic hybridization is the reconstruction of plants from the hybrid cells. In the first reports on production of plants from fused protoplasts the investigators used *N. glauca* and *N. langsdorffi* (Carlson et al., 1972; Kao et al., 1974). These species can be crossed sexually and also readily manifest plant regeneration. The hybrid cells from the fused protoplasts did not require hormones for growth because they are auxinautotrophic (oncogenic), and the hybrid plants could be checked against natural sexual hybrids, which form genetic tumors.

A study of karyotypes of the somatic hybrid plants from these species of tobacco showed that most plants had a chromosome number of 60 ± 4, which corresponded to the combination of one *N. glauca* and two *N. langsdorffi* or vice versa (Smith et al., 1976). The plants may therefore have arisen from cell progeny of triple-fusion products. Alternatively, a nuclear division may have occurred prior to cell division (Nandi and Eriksson, 1977).

A large number of normal green plants have also been obtained from cell progeny of fusion products from mesophyll protoplasts of chlorophyll-deficient tobacco mutants (Melchers and Labib, 1974). Potentially the chlorophyll-deficient and particularly albino mutants can be used with any combination of species when plant regeneration from the fusion products is possible. The correction of lesions in chlorophyll production can be detected reliably only in shoots or plants.

Somatic hybrid plants have also been obtained in sexually compatible species of *Petunia*. Fusion products from protoplasts of *P. hybrida* and *P. parodii* formed callus. Plants regenerated from the callus corresponded in flower shape and color with those of sexual hybrids (Power et al., 1976).

There are several intergeneric crosses that have been produced by sexual means with the aid of *in vitro* embryo culture or chromosome doubling with colchicine. The best known crosses are Triticale, which is the wheat-rye cross (Hulse and Spurgeon, 1974), and *Raphanobrassica* (Tokumaso, 1976). The latter is a cross between *Brassica japonica* Sieb and *Raphanus sativus* L. Other crosses include wheat × oat (Kruse, 1969)., parsley × celery (Madjarova et al., 1971), and hybrids between other cereals and grasses (Tsitsina, 1976). These crosses attest to the potential feasibility of producing plants from intergeneric crosses.

The basic requirement initially for reconstructing plants from somatic hybrid cells is a capacity of the cell material to manifest morphogenesis. In species of *Solanaceae (Nicotiana, Solanum, Lycopersicon, Datura, Petunia)*, of *Umbelliferae (Daucus carota* [carrot], *Pimpinella anisum* [anise], *Carum carvi* [caraway], *Apium graveolens* [celery]), and of *Brassica* the process of plant regeneration can be induced fairly readily.

The use of materials with the capacity for somatic cell embryogenesis (such as the carrot) offers an apparent advantage. It has been shown in several species that the embryogenic process and plant regeneration occur in protoplasts directly from the single cell. The process similarly should be expected to occur in the developing fusion products unless the general capacity for morphogenesis is lost. Achieving morphogenesis at the very early stage in the developing and dividing hybrid cell fusion product may ensure plant regeneration in the hybrid before possible chromosome elimination occurs or becomes extensive. The phenomenon of chromosome elimination is common in mammalian somatic hybrids (Ringertz and Savage, 1976) and may be anticipated also in plant hybrid cells (Kao, 1977).

Achieving plant regeneration from single cells and protoplasts of cereals, grasses, grain legumes, and many other major forage crops is plagued by difficulties. However, considerable progress has been made in recent years (table 1), attesting to the continued advancement in the knowledge and technology of growing plants from somatic cells.

GENE TRANSFER METHODS

Genetic modification in plants is also being considered through uptake of DNA and organelle implantation. The uptake of organelles such as chloroplasts, nuclei, or chromosomes into protoplasts may provide new and effective approaches to study nuclear-organelle and nuclear-cytoplasmic interactions that are involved in differentiation and regulatory processes in plants.

Genetic transformation by DNA uptake implies that DNA from one source be taken up and incorporated into the host genome and that genetic information encoded in the DNA be expressed as a new, stable characteristic in the recipient organism (Merril and Stanbro, 1974). The process has been very successful in certain bacterial systems and has become a routine procedure in some microorganisms used for gene cloning (Tomasz, 1975; Cohen, 1975).

Several reports have also appeared on genetic transformation in higher eukaryotes (Merril and Stanbro, 1974). There is evidence for stable phenotypic changes in *Drosophila* achieved by injection of DNA (Fox et al., 1975). Several reports have appeared on experiments with higher plants in which phenotypic changes had occurred in recipient plants after feeding DNA (Ledoux, 1975; Hess, 1977).

There are two complementary approaches that have been employed in plant experiments. One approach focuses on using DNA labeled with radioisotopes to follow uptake and the fate of DNA by biochemical techniques. The other is based on detection of changes in phenotype in recipient plant materials (Ledoux, 1975; Hess, 1977).

DNA Uptake

The uptake of DNA has been investigated with a range of plant materials including seeds (Ledoux et al., 1971; Turbin et al., 1975), pollen (Hess, 1977), cultured cells (Lurquin and Hotta, 1975), protoplasts (Ohyama et al., 1972; Uchimiya and Murashige, 1977; Suzuki and Takebe, 1976), and nuclei (Ohyama et al., 1977; Ohyama, 1978).

The anticipated steps in the uptake process into cells or protoplasts include binding to specific sites on the plasma membrane and passage through the membrane and cytoplasm (Merril and Stanbro, 1974; Holl et al., 1974). This is followed by a binding to specific binding sites, passage through the nuclear membrane into the nucleus, and ultimate integration into the host cell genome (fig. 5) (Ohyama et al. 1977).

The evidence for uptake of labeled, intact donor DNA is based on radioactivity in DNA isolated from the recipient cells after a period of

EVENTS IN DNA UPTAKE BY PROTOPLASTS

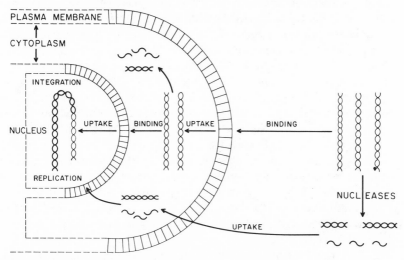

Fig. 5. Diagram of events in DNA uptake by plant protoplasts.

incubation. When DNA is fed to cells, the cell walls may constitute a barrier. Using autoradiographic methods, Kool and Pelcher (unpublished) observed that [^{14}C-thymidine]-labeled *Salmonella typhimurium* DNA was bound to the walls of cultured cells of soybean and tobacco. More of the DNA was bound on younger than on older cells. The binding was stimulated by calcium ions, and chelating agents were inhibitory. The binding reaction exhibited pH optima near 5.5 and 9. The bound DNA could not be removed by washing, but was degraded after prolonged treatment with DNAase. It has not been established if wall binding represents a step in DNA uptake, but the existence of the phenomenon suggests that caution should be exercised when attempts are made to interpret data on DNA uptake based on biochemical analyses. Some of the obstacles inherent in using cell materials may be avoided by feeding DNA to protoplasts. Ohyama et al. (1972) showed that protoplasts of soybean, carrot, and *Ammi visnaga* took up labeled *E. coli* DNA, and radioactivity was detected in the high molecular weight DNA fraction. In a recent report Uchimiya and Murashige (1977) recorded the uptake of labeled DNA by protoplasts of *N. glutinosa* and found that up to 10 percent of the DNA entered the protoplasts. A substantial percentage of the absorbed DNA was associated with the nuclear and a mitochondrial fraction. Only 15 percent of the recorded radioactivity was associated with DNA. Extensive degradation and reutilization of the molecular products into other cell constituents had occurred.

Suzuki and Takebe (1976) have reported on the uptake of single-stranded bacteriophage fd-DNA by leaf protoplasts of tobacco (*N. tabacum* cv. Xanthi). Up to 30 percent of the supplied DNA was taken up, and approximately one-third retained the size of intact fd-DNA. There was no evidence of fd-DNA replication in the cells.

The recent investigations by Ohyama et al. (1977) have provided the first information on DNA uptake by isolated nuclei. The results indicate the involvement of separate events of binding and passage into the nuclei. The binding and uptake of homologous DNA by nuclei from soybean protoplasts required Ca^{++} or Mg^{++}. The uptake was enhanced by poly-L-lysine and inclusion of an ATP-generating system. Maximum uptake occurred between pH 6 and 7. The results suggest that uptake into nuclei is an active metabolic process and may involve binding proteins at specific binding sites (Tomasz, 1975). In competition experiments with several types of DNA, that of *Salmonella typhimurium* DNA showed a high degree of competition with soybean DNA for uptake, whereas calf thymus DNA exhibited no competition.

In other investigations it was shown that binding and uptake of bacteriophage fd-DNA as well as double-stranded *S. typhimurium* DNA required Ca^{++}, and poly-L-lysine was stimulatory. The uptake of fd-DNA was not influenced by temperature or the presence of an ATP-generating system. On the basis of competition experiments, Ohyama (1978) proposed that the binding sites were preferentially occupied by single-stranded DNA.

In DNA uptake experiments with protoplasts of nuclei, there is extensive and rapid degradation by nucleases (Holl, 1973; Uchimiya and Murashige, 1977; Ohyama et al., 1977).

The Use of Vectors for Gene Transfer

A major obstacle in feeding isolated DNA to plant materials is the presence of enzymes that degrade DNA. Up to 45 percent of free DNA may be destroyed within 20 minutes (Ohyama et al., 1977). Enzyme preparations used in isolation of protoplasts contain very high nuclease activity, and also the degradation takes place after uptake. It is clear that means must be found to circumvent the problems to ensure protection of the donor DNA and facilitate its stabilization in recipient cells. The implementation of gene vectors may reduce or possibly alleviate some of the problems.

Two types of vectors are being considered, a plant virus and bacterial plasmids. The plasmids are double-stranded, closed-circular extrachromosomal DNA found in bacteria (Primrose, 1977; Cohen, 1975). The plasmids replicate independently, but the plasmid genes can integrate into the bacterial chromosome.

The research with plasmids has developed rapidly in recent years and has resulted in new and revolutionary techniques in genetics (Cohen, 1975; Hollaender, 1977; Mach, 1977).

A series of endonucleases from bacteria degrade DNA at points with specified nucleotide sequences (Primrose, 1977). These restriction enzymes cleave the plasmid DNA to linear, double-stranded sections with overlapping and complementary nucleotide sequences. By appropriate choice of chemical methods and enzymes, the sections can be rejoined and circularized with reconstitution of the plasmid. Alternatively, the linear DNA sections of the plasmid can be mixed with endonuclease-cleaved sections of DNA from other sources. In the process of reconstituting the plasmid, the foreign DNA is inserted into the plasmid (Cohen and Chang, 1974; Lomax and Helling, 1977). After the plasmid is taken up by the bacterial cell it is replicated, and the genetic information that is encoded in the foreign DNA can be transcribed and eventually expressed in the bacteria. Such gene-cloning procedures are being explored for use in gene analysis of the yeast genome (Beckmann et al., 1977) and of genetic materials of higher eukaryotes (Cohen and Chang, 1974; Lomax and Helling, 1977).

The plasmids that are most commonly employed are obtained from *E. coli*. There is relatively little theoretical basis for assuming that *E. coli* plasmids would replicate in plant cells (Goebel and Scheiss, 1975). The important discovery made recently has established that the bacterium causing crown gall tumor formation in plants (*Agrobacterium tumefaciens*) possesses plasmids (Schell and Van Montagu, 1977). The Ti-plasmids contain the genes that code for tumor formation in the plant. They also possess genes that code for the production of compounds specifically produced by plants infected with the pathogen (Bomhoff et al., 1976). There are indications that plasmids replicate within the plant cells, and the genetic information is expressed by utilizing the biochemical facilities of the plant.

The A. tumefaciens Plasmids

The plasmids of *A. tumefaciens* possess the genes that determine oncogenicity, and also for new and specific pathways that are expressed only in transformed plants (Schell and Van Montagu, 1977; Bomhoff et al., 1976). Agrobacteria with one of three types of Ti-plasmids have been identified, those coding for bacterial octopine metabolism, those coding for nopaline metabolism, and those that are able to metabolize either compound. The Ti-plasmids thus possess genes that determine oncogenicity, synthesis of specific opines in transformed plants, and also the catabolism of the same opine in the *A. tumefaciens* strain. These properties provide potential biochemical markers.

There is supportive evidence that hybridization of portions of the Ti-plasmid occurs with plant DNA (Schilperoort et al., 1967; Schell and Van Montagu, 1977). Schell and Van Montagu (1977) have proposed a two-step mechanism for transfer of Ti-plasmid DNA to plant cells. The initial step involves the intake of the Ti-plasmid DNA into the cell. The integration that follows would occur by transposition of one or more segments of the Ti-plasmid DNA to the plant genome. There is evidence that Ti-plasmids contain transposons, which are DNA segments involved in translocating nonhomologous DNA molecules (Cohen, 1976). The transposons have repeated DNA sequences known as insertion elements flanked by a transposable set of genes in opposite orientation forming inverted repeat structures. The Ti-plasmids have been hybridized with R-type conjugative plasmid RP4.

The Ti:RP4 plasmid cointegrates consist of RP4 and a Ti-plasmid and have a DNA sequence in common (Van Larebeke et al., 1977; Schell and Van Montagu, 1977). There is evidence that the same DNA sequence into which RP4 integrates in the Ti-plasmid also is involved in the plant transformation mechanism. Alternatively the Ti-plasmids may have several types of insertion-like DNA sequences, some of which have specific properties for cointegration with plant DNA.

This mechanism has been invoked to explain transformation in plant cells infected with *A. tumefaciens* and now has been presented as a working model for genetic transformation in plants (fig. 6).

Plant DNA Virus

An alternate potential vector that is being considered is the plant virus. Most plant viruses are RNA, but a small group consists of DNA (Shepherd, 1976). An example is the cauliflower mosaic virus (CaMV), which may be a potential replicon for achieving gene transfer. Information on molecular structure with respect to replication sites and mode of replication is very limited (Shepherd, 1976).

Szeto et al. (1977) treated CaMV-DNA with restriction enzymes and isolated two segments cloned and replicated by *E. coli* plasmid procedures. The segments reisolated from the plasmid failed to infect leaves of *Brassica campestris*, which had unambiguous symptoms when infected with intact CaMV.

There is insufficient information available to predict the usefulness of the

PLASMIDS FOR GENE TRANSFER IN PLANTS

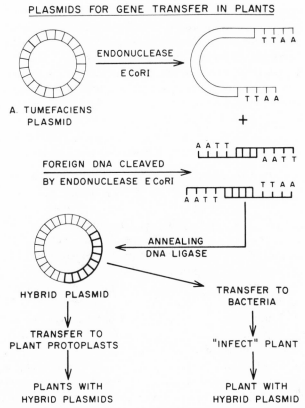

Fig. 6. Diagram projecting the use of plasmids (*A. tumefaciens*) as vectors in gene transfer in plants.

CaMV or other plant DNA viruses for gene transfer experiments. After mapping of the DNA virus has been accomplished, it may be possible to devise methods to replace nonessential regions with foreign DNA and effect coupling to the initiation site and replicase gene of the virus. The location of DNA replication for some of the DNA virus is believed to be the nucleus, but there is also evidence for the presence of inclusion bodies in the cytoplasm in which DNA virus replication occurs (Shepherd, 1976). It clearly is not possible to predict if DNA virus would provide a suitable vector for plant genetic transformation.

Langridge (1977) has proposed the construction of a hybrid consisting of a bacterial plasmid + plant DNA-virus. The plasmid carries the bacterial genes for Kanamycin resistance. Since the plant cells originally are sensitive to Kanamycin, the plasmid-viral DNA transfer and subsequent gene expression is anticipated to confer Kanamycin resistance to the plant cell (Langridge, 1977). The proposal is ambitious, but entails an extremely interesting experimental approach. The uptake and replication of RNA virus have been investigated with plant protoplasts and have become routine procedures for some virus species (Takebe, 1975). Protoplasts of *Brassica* (Kartha et al., 1974a, 1974c) or other appropriate species similarly would be attractive materials to elucidate details of DNA virus replication and to establish an experimental system for assaying genetically modified viral DNA.

DNA Feeding and Phenotypic Expression

The genetic transformation approach is handicapped by the lack of clear-cut biological markers in cultured cells. As an alternative, various plant mutants have been used in DNA feeding experiments. Ledoux et al. (1971) reported on the correction of thiamine requiring mutants of *Arabidopsis* after allowing mature seeds to imbibe bacterial DNA. Hess (1977) and colleagues fed DNA from pure lines of red-flowering plants to seedlings of white-flowering mutants of *Petunia*. The treatment produced red-flowering plants. Using a non-nodulating line of field pea (*Pisum sativum*), Holl (1975b) has observed corrections resulting in *Rhizobium* infection in plants after feeding DNA from normal nodulating and nitrogen-fixing plants to seeds of the mutant. The capacity for nodulation is determined by a single gene (Holl, 1975b).

Soyfer et al. (1976) and Turbin et al. (1975) have reported on correction of biochemical lesions in waxy mutants of barley (*Hordeum vulgare*). High molecular weight DNA from normal barley endosperm was injected into immature seeds of waxy mutant plants. The pollen of normal barley contains amylose and amylopectin, whereas the mutant pollen contains only amylose. The two pollen types can be detected by iodine staining procedures. A relatively high frequency of corrections was observed in the initial experiments. Analyses of plants in the F_1 indicate that the changes persist, but that reversion is prevalent. The treatment also resulted in a change from two-rowed to six-rowed barley, a feature that similarly reverted at high frequency in the F_1. Other examples of phenotypic changes include the alteration in fruiting characters of red pepper plants (*Capsicum annum* L.) (Nawa et al., 1975) and flower induction in shoots of *Nicotiana tabacum* L. cv. Wisconsin 38 after feeding DNA extracted from stems in the floral state (Wardell, 1976).

These reports suggest that changes are observed in recipient plants after DNA treatment. The changes may be brought about by quantities of DNA that are too small to detect by existing biochemical procedures. The DNA may be inadequately stabilized and replicated and thus may account for the instability of the changes. Present information on translocation of macro-molecules and on gene/chromosome structure with respect to insertion mechanisms in plants is extremely limited. It is therefore not surprising that the frequency of the correction phenomenon often is unpredictable and that explanation of the observation in molecular genetic terms is at best inadequate.

Uptake of Single Cells and Organelles by Protoplasts

The availability of plant protoplasts has made it possible to consider uptake of unicellular organisms, implantation of nuclei, chloroplasts, chromosomes, or mitochondria. These subjects have been reviewed recently (Davey, 1977; Giles, 1977). Bacterial cell uptake by plant protoplasts has been investigated with species of *Rhizobium* (Davey and Cocking, 1972) and *Spirillium* (see Vasil et al., 1977). There are reports based on ultrastructural examinations that bacteria enter the cells by endocytosis and may become embedded in vesicles in the cytoplasm of protoplasts. Similar uptake studies were performed with yeast (*Saccharomyces cerevisiae*) and blue-green algal cells (Davey and Power, 1975; Burgoon and Bottino, 1976). There has been no reported evidence of survival or development of any of the organisms within the protoplasts, and there is perhaps very little if any reason to assume that the particular yeast or bacterial cell should function within cells of a higher plant. It is well known that plant cell cultures contaminated with bacteria or yeast quickly deteriorate, and there is no evidence that any portion of the microbial genome can replicate and function within plant cells. The recently discovered Ti-plasmids of *A. tumefaciens* may provide the needed material to ascertain if foreign genes can be inserted into plant DNA. The Ti-plasmids are replicated in crown gall cells by the plant biochemical system. Some of the plasmids of *Rhizobia* similarly may be replicated within plant cells, although there is at present no evidence that *Rhizobia* plasmids are replicated outside the vesicles containing the bacteriods. Several reports have described uptake of chloroplasts and nuclei by protoplasts (Potrykus, 1975; Giles, 1977). The uptake process is enhanced by polyethylene glycol (PEG) (Bonnett and Eriksson, 1974; Davey et al., 1976). The most complete data on chloroplast uptake were provided by Bonnett and Eriksson (1974). Chloroplasts isolated from the alga (*Vaucheria dichotoma* L. Ag) were implanted into carrot cell culture protoplasts. Under optimum conditions in the presence of PEG, up to 16 percent of the protoplasts contained one or more chloroplasts and up to 80 percent of the protoplasts remained viable after the experiments. The chloroplasts may enter the cytoplasm enclosed in membrane-bound vesicles (Davey et al., 1976), although the enclosing membrane in some cases is absent (Bonnett, 1976). Plant regeneration from albino leaf protoplasts after feeding chloroplasts has been reported. Biological evidence of chloroplast gene expression was presented, but the experiments have not been confirmed (Kung et al., 1975).

The viability and ability of chloroplasts to survive and multiply in recipient protoplasts have not been unequivocally demonstrated, although limited replication has been reported (Giles, 1977). Potentially the chloroplast uptake procedure offers an excellent approach to study chloroplast /cytoplasm and nuclear interrelationships, genetic and physiological autonomy, and specificity of functions of the organelles (Birky et al., 1975).

The uptake of isolated metaphase chromosomes has proven successful with mammalian cells (Burch and McBride, 1975; Willecke and Ruddle, 1975). Similar procedures should also be feasible in plants through the use of protoplasts and should provide a valuable method for genetic information transfer and gene analysis.

LITERATURE CITED

Ashton, N. W., and D. J. Cove. 1977. The isolation and preliminary characterization of auxotrophic and analog resistant mutants of moss *Physcomitrella patens*. Molec. Gen. Genet. 154:87–96.

Bates, L. S., and C. W. Deyoe. 1973. Wide hybridization and cereal improvement. Econ. Bot. 27:401–12.

Beckmann, J. S., P. F. Johnson, and J. Abelson. 1977. Cloning of yeast transfer RNA genes in *Escherichia coli*. Science 196:205–8.

Binding, H. 1974. Fusion experiments with isolated protoplasts of *Petunia hybrida* L. Z. Pflanzenphysiol. 72:422–26.

Birky, C. W., P. S. Perlman, and T. J. Byers. (eds.). 1975. Genetics and biogenesis of mitochondria and chloroplasts. Ohio State University Press, Columbus.

Bomhoff, G., P. M. Klapwijk, H. C. M. Kester, and R. A. Schilperoort. 1976. Octopine and nopaline synthesis and breakdown genetically controlled by a plasmid of *Agrobacterium tumefaciens*. Molec. Gen. Genet. 145:177–81.

Bonnett, H. T. 1976. Mechanism of uptake of *Vaucheria* chloroplasts by carrot protoplasts treated with polyethylene-glycol. Planta 131:229–33.

Bonnett, H. T., and T. Eriksson. 1974. Transfer of algal chloroplasts into protoplasts of higher plants. Planta 120:71–79.

Borlaug, N. E. 1971. The Green Revolution, peace and humanity. Cereal Sci. Today 16:401–11.

Brown, A. W. A., T. C. Byerly, M. Gibbs, and A. San Pietro. 1975. Crop productivity: Research imperatives. Charles F. Kettering Foundation, Yellow Springs, Ohio.

Bui Dang Ha, D., and I. A. MacKenzie. 1973. Division of protoplasts from *Asparagus officinalis* L. and their growth and differentiation. Protoplasma 78:215–21.

Burch, J., and O. W. McBride. 1975. Human gene expression in rodent cells after uptake of isolated metaphase chromosomes. Proc. Nat. Acad. Sci. USA 72:1797–1801.

Burgess, J., and E. N. Fleming. 1974. Ultrastructural studies of the aggregation and fusion of plant protoplasts. Planta 118:183–93.

Burgoon, A. C., and P. J. Bottino. 1976. Uptake of the nitrogen fixing blue-green algae *Gloeocapsa* into protoplasts of tobacco and maize. J. Hered. 67:223–26.

Carlson, P. S. 1973. Somatic cell genetics of higher plants. *In* F. H. Ruddle (ed.), Genetic mechanisms of development. Pp. 329–53. Academic Press, New York.

Carlson, P. S., and J. C. Polacco. 1975. Plant cell cultures: Genetic aspects of crop improvement. Science 188:622–25.

Carlson, P. S., H. H. Smith, and R. D. Dearing. 1972. Parasexual interspecific plant hybridization. Proc. Nat. Acad. Sci. USA 69:2292–94.

Cassells, A. C., and M. Barlass. 1976. Environmentally induced changes in the cell walls of tomato leaves in relation to cell and protoplast release. Physiol. Plantarum 37:239–46.

Cheng, T. 1975. Adventitious bud formation in culture of Douglas fir. Plant Sci. Lett. 5:97–102.

Cheng, T. Y., and H. H. Smith. 1975. Organogenesis from callus culture of *Hordeum vulgare*. Planta 123:307–10.

Chu, E. H. Y., and S. S. Powell. 1976. Selective systems in somatic cell genetics. Adv. Human Genet. 7:189–258.

Chu, E. H. Y., N. C. Sun, and C. C. Chang. 1972. Induction of auxotrophic mutations by treatment of Chinese hamster cells with 5-bromodeoxyuridine and black light. Proc. Nat. Acad. Sci. USA 69:3459–63.

Cocking, E. C. 1972. Plant cell protoplasts—isolation and development. Ann. Rev. Plant Physiol. 23:29–50.

Cohen, S. N. 1975. The manipulation of genes. Scientific Amer. 233:24–33.

Cohen, S. N. 1976. Transposable genetic elements and plasmid evolution. Nature 263:731–38.

Cohen, S. N., and C. Y. Chang. 1974. A method for selective cloning of eukaryotic DNA fragments in *E. coli* by repeated transformation. Molec. Gen. Genet. 134:133–41.

Constabel, F. 1975. Isolation and culture of plant protoplasts. *In* O. L. Gamborg and L. R. Wetter (eds.), Plant tissue culture methods. Pp. 11–21. National Research Council of Canada,Ottawa.

Constabel, F. 1976. Somatic hybridization in higher plants. In Vitro 12:743–48.

Constabel, F., D. Dudits, O. L. Gamborg, and K. N. Kao. 1975a. Nuclear fusion in intergeneric heterokaryons. Canad. J. Bot. 53:2092–95.

Constabel, F., J. W. Kirkpatrick, K. N. Kao, and K. K. Kartha. 1975b. The effect of canavanine on the growth of cells from suspension cultures and on intergeneric heteroka-ryocytes of canavanine sensitive and tolerant plants. Biochem. Physiol. Pflanzen, 168:319–25.

Constabel, F., G. Weber, and J. W. Kirkpatrick. 1977. Chromosome compatibility in intergeneric cell hybrids, *Glycine max + Vicia hajastana*. C. R. Acad. Sci. Paris 285:319–22.

Constabel, F., G. Weber, J. W. Kirkpatrick, and K. Pahl. 1976. Cell division of intergeneric protoplast fusion products. Z. Pflanzenphysiol. 79:1–7.

Cove, D. J. 1976. Chlorate toxicity in *Aspergillus nidulans*. Molec. Gen. Genet. 146:147–59.

Cummings, P. D., C. E. Green, and D. D. Stuthman. 1976. Callus induction and plant regeneration in oats. Crop Sci. 16:465–70.

Davey, M. R. 1977. Bacterial uptake and nitrogen fixation. *In* J. Reinert and Y. P. S. Bajaj (eds.), Applied and fundamental aspects of plant cell, tissue, and organ culture. Pp. 551–62. Springer-Verlag, Berlin.

Davey, M. R., and E. C. Coking. 1972. Uptake of bacteria by isolated higher plant protoplasts. Nature 239:455–56.

Davey, M. R., E. M. Frearson, and J. B. Power. 1976. Polyethylene glycol-induced transplantation of chloroplasts into protoplasts: An ultrastructural assessment. Plant Sci. Lett. 7:7–16.

Davey, M. R., and J. B. Power. 1975. Polyethylene glycol-induced uptake of micro-organisms into higher plant protoplasts: An ultrastructural study. Plant Sci. Lett. 5:269–74.

Davidson, R. L. 1971. Regulation of gene expression in somatic cell hybrids: A review. In Vitro 6:411–26.

Dudits, D. 1976. The effect of selective conditions on the products of plant protoplast fusion. *In* D. Dudits, G. L. Farkas, and P. Maliga (eds.), Cell genetics in higher plants. Pp. 153–62. Akadémiai Kiadó, Budapest.

Dudits, D., K. N. Kao, F. Constabel, and O. L. Gamborg. 1976a. Embryogenesis and formation of tetraploid and hexaploid plants from carrot protoplasts. Canad. J. Bot. 54:1063–64.

Dudits, D., K. N. Kao, F. Constabel, and O. L. Gamborg. 1976b. Fusion of carrot and barley protoplasts and division of heterokaryocytes. Canad. J. Genet. Cytol. 18:263–69.

Eriksson, T. 1970. Isolation and fusion of plant protoplasts. Colloques Intern. C.N.R.S. 193:297–302.

Fowke, L. C., C. W. Bech-Hansen, O. L. Gamborg, and F. Constabel. 1975a. Electron microscopic observations of mitosis and cytokinesis in multinucleate protoplasts of soybean. J. Cell Sci. 18:491–507.

Fowke, L. C., F. Constabel, and O. L. Gamborg. 1977. Fine structure of fusion products from soybean cell culture and pea leaf protoplasts. Planta 135:257–66.

Fowke, L. C., P. J. Rennie, J. W. Kirkpatrick, and F. Constabel. 1975b. Ultrastructural characteristics of intergeneric protoplast fusion. Canad. J. Bot. 53:272–78.

Fowke, L. C., P. J. Rennie, J. W. Kirkpatrick, and F. Constabel. 1976. Ultrastructural characterization of soybean × sweet clover fusion products. Planta 130:39–45.

Fox, A. S., S. Parzen, H. Salverson, and S. B. Yoon. 1975. Gene transfer in *Drosophila melanogaster:* Genetic transformations induced by the DNA of transformed stocks. Genet. Res. 26:137–47.

Fridborg, G. 1971. Growth and organogenesis in tissue cultures of *Allium cepa* var. *proliferum*. Physiol. Plantarum 25(3):436–40.

Gamborg, O. L. 1976. Plant protoplast isolation, culture, and fusion. *In* D. Dudits, G. L. Farkas, and P. Maliga (eds.), Cell genetics in higher plants. Pp. 107–27. Akadémiai Kiadó, Budapest.

Gamborg, O. L. 1977a. Culture media for plant protoplasts. *In* M. Recheigl (ed.), CRC Handbook in Nutrition and Food. Pp. 415–22. CRC Press, Cleveland.

Gamborg, O. L. 1977b. Somatic cell hybridization by protoplast fusion and morphogenesis. *In* W. Barz, E. Reinhard, and M. H. Zonk (eds.) Plant tissue culture and its biotechnological application. Pp. 287–301. Springer-Verlag, Berlin.

Gamborg, O. L., F. Constabel, L. C. Fowke, K. N. Kao, K. Ohyama, K. K. Kartha, and L. E. Pelcher. 1974a. Protoplast and cell culture methods in somatic hybridization in higher plants. Canad. J. Genet. Cytol. 16:737–50.

Gamborg, O. L., F. Constabel, and J. P. Shyluk. 1974b. Organogenesis in callus from shoot apices of *Pisum sativum*. Physiol. Plantarum 30:125–28.

Gamborg, O. L., T. Murashige, T. A. Thorpe, and I. K. Vasil. 1976. Plant tissue culture media. In Vitro 12:473–78.

Gamborg, O. L., and J. P. Shyluk. 1976. Tissue culture, protoplasts, and morphogenesis in flax. Bot. Gaz. 137:301–6.

Gamborg, O. L., J. P. Shyluk, D. S. Brar, and F. Constabel. 1977. Morphogenesis and plant regeneration from callus of immature embryos of sorghum. Plant Sci. Lett. 10:67–74.

Gavazzi, G., M. Nava-Racchi, and C. Tonelli. 1975. A mutation causing proline requirement in *Zea mays*. Theor. Appl. Genet. 46:339–45.

Gengenbach, B. G., and C. E. Green. 1975. Selection of T-cytoplasm maize callus cultures resistant to *Helminthosporium maydis* race-T pathotoxin. Crop Sci. 15:645–49.

Giles, K. L. 1977. Chloroplast uptake and genetic complementation. In J. Reinert and Y. P. S. Bajaj (eds.), Applied and fundamental aspects of plant cell, tissue, and organ culture. Pp. 536–50. Springer-Verlag, Berlin.

Gleba, Y. Y., R. G. Butenko, and K. M. Sytnik. 1975. Fusion of protoplasts and parasexual hybridization in *Nicotiana tabacum*. Dokl. Akad. Nauk USSR 221:1196–98.

Goebel, W., and W. Schiess. 1975. The fate of a bacterial plasmid in mammalian cells. Molec. Gen. Genet. 138:213–23.

Gosch, G., and J. Reinert. 1976. Nuclear fusion in intergeneric heterokaryocyte and subsequent mitosis of hybrid nuclei. Naturwiss. 63(11):534.

Green, C. E., and R. L. Phillips. 1975. Plant regeneration from tissue cultures of maize. Crop. Sci. 15:417–21.

Grimsley, N. H., N. W. Ashton, and D. J. Cove. 1977. The production of somatic hybrids by protoplast fusion in the moss *Physcomitrella patens*. Molec. Gen. Genet. 154:97–100.

Halperin, W. 1970. Embryos from somatic plant cells. Symp. Intern. Soc. Cell Biol. 9:169–91.

Heinz, D. J., M. Kirshnamurthi, L. G. Nickell, and A. Maretzki. 1977. Cell, tissue, and organ culture in sugarcane (*Saccharum*) improvement. In J. Reinert and Y. P. S. Bajaj (eds.), Applied and fundamental aspects of plant cell, tissue, and organ culture. Pp. 3–17. Springer-Verlag, Berlin.

Heinz, D. J., and G. W. P. Mee. 1969. Plant differentiation from callus tissue of *Saccharum* species. Crop Sci. 9:346–48.

Hess, D. 1977. Cell modification by DNA uptake. In J. Reinert and Y. P. S. Bajaj (eds.), Applied and fundamental aspects of plant cell, tissue, and organ culture. Pp. 506–35. Springer-Verlag, Berlin.

Holl, F. B. 1973. Cellular environment and the transfer of genetic information. Colloques Intern. C.N.R.S. 212:509–16.

Holl, F. B. 1975a. Innovative approaches to genetics in agriculture. Canad. J. Genet. Cytol. 17:517–24.

Holl, F. B. 1975b. Host plant control of the inheritance of dinitrogen fixation in the *Pisum-Rhizobium* symbiosis. Euphytica 24:767–70.

Holl, F. B., O. L. Gamborg, K. Ohyama, and L. E. Pelcher. 1974. Genetic transformation in plants. In H. E. Street (ed.), Tissue culture and plant science. Pp. 301–27. Academic Press, New York.

Hollaender, A. (ed.). 1977. Genetic engineering for nitrogen fixation. Plenum Press, New York.

Hulse, J. H., and D. Spurgeon. 1974. Triticale. Scientific Amer. pp. 72–80.

Ito, M., and M. Maeda. 1974. Meiotic division and fusion of nuclei in multinucleate cells from induced fusion of meiotic protoplasts of Liliaceous plants. Bot. Mag. 87:219–28.

Kameya, T. 1975. Induction of hybrids through somatic cell fusion with dextran sulfate and gelatin. Jap. J. Genet. 50:235–46.

Kao, K. N. 1975. A method for fusion of plant protoplasts with polyethylene glycol. *In* O. L. Gamborg and L. R. Wetter (eds.), Plant tissue culture methods. Pp. 22–27. National Research Council of Canada, Prairie Regional Laboratory, Saskatoon.

Kao, K. N. 1977. Chromosomal behavior in somatic hybrids of soybean-*Nicotiana glauca*. Molec. Gen. Genet, 50:225–30.

Kao, K. N., F. Constabel, M. R. Michayluk, and O. L. Gamborg. 1974. Plant protoplast fusion and growth of intergeneric hybrid cells. Planta 120:215–27.

Kao, K. N., and M. R. Michayluk. 1974. A method for high frequency intergeneric fusion of plant protoplasts. Planta 115:355–67.

Kartha, K. K., and O. L. Gamborg. 1975. Elimination of cassava mosaic disease by meristem culture. Phytopathology 65:826–28.

Kartha, K. K., O. L. Gamborg, and F. Constabel. 1974a. *In vitro* plant formation from stem explants of rape (*Brassica napus* cv. Zephyr). Physiol. Plantarum 31:217–20.

Kartha, K. K., O. L. Gamborg, F. Constabel, and K. N. Kao. 1974b. Fusion of rapeseed and soybean protoplasts and subsequent division of heterokaryocytes. Canad. J. Bot. 52:2435–36.

Kartha, K. K., M. R. Michayluk, K. N. Kao, and O. L. Gamborg. 1974c, Callus formation and plant regeneration from mesophyll protoplasts of rape plants (*Brassica napus* L. cv. Zephyr). Plant Sci. Lett. 3:265–71.

Keller, W. A. and G. Melchers. 1973. Effect of high pH and calcium on tobacco leaf protoplast fusion. Z. Naturforsch. 28:737–41.

Keller, W. A., T. Rajhathy, and J. Lacapra. 1975. In vitro production of plants from pollen in *Brassica campestris*. Canad. J. Genet. Cytol. 17:655–66.

Kool, A., and L. E. Pelcher. Unpublished.

Kruse, A. 1969. Intergeneric hybrids between *Triticum aestivum* L. (v. Koga II, 2n = 42) and *Avena sativa* L. (v. Stal, 2n = 42) with pseudogamous seed formation. *In* Yearbook, Royal Veterinary and Agricultural University, Copenhagen, Denmark, pp. 118–200.

Kung, S. D., J. C. Gray, S. G. Wildman, and P. S. Carlson. 1975. Polypeptide composition of fraction protein from parasexual hybrid plants in the genus *Nicotiana*. Science 187:353–54.

Kuster, E. 1909. Uber die Verschmelzung nachter protoplasten. Ber. Dtsch. Bot. Ges. 27:589–98.

Langridge, J. 1977. Genetic engineering in plants. Search 8:13–15.

Larkin, P. J. 1976. Purification and viability determinations of plant protoplasts. Planta 128:213–16.

Ledoux, L. 1975. Genetic manipulations with plant material. NATO Advanced Study Institutes, vol. 3. Plenum Press, New York.

Ledoux, L., R. Huart, and M. Jacobs. 1971. Fate of exogenous DNA in *Arabidopsis thaliana*: Translocation and integration. Eur. J. Biochem. 23:96–108.

Lomax, M. I., and R. B. Helling. 1977. Cloned ribosomal RNA genes from chloroplasts of *Euglena gracilis*. Science 196:202–5.

Lurquin, P. F., and Y. Hotta. 1975. Reutilization of bacterial DNA by *Arabidopsis thaliana* cells in tissue culture. Plant Sci. Lett. 5:103–12.

Mach, B. 1977. Genetic engineering and plasmids. Experientia 33:105–9.

Madjarova, D., G. Georgiev, E. Benbassat, M. Boubarova, and I. Chavdarov. 1971. New plant forms, obtained by parsley (*Petroselinum hortense* Hoffm.) and celery (*Apium graveolens* L.) hybridization. C. R. Acad. Sci. Agric. Bulg., Vol. 4. No. 3.

Maliga, P. 1976. Isolation of mutants from cultured plant cells. *In* D. Dudits, G. L. Farkas, and P. Maliga (eds.), Cell genetics in higher plants. Pp. 59–76. Akadémiai Kiadó, Budapest.

Mehra, A., and P. N. Mehra. 1974. Organogenesis and plantlet formation in vitro in almond. Bot. Gaz. 135(1):61–73.

Melchers, G., and G. Labib. 1974. Somatic hybridization of plants by fusion of protoplasts. Molec. Gen. Genet. 135:277–94.

Mellor, F. C., and R. Stace-Smith. 1977. Virus-free potatoes by tissue culture. *In* J. Reinert and Y. P. S. Bajaj (eds.), Applied and fundamental aspects of plant cell, tissue, and organ culture. Pp. 616–35. Springer-Verlag, Berlin.

Merril, C. R. and H. Stanbro. 1974. Intercellular gene transfer. Z. Pflanzenphysiol. 73:371–88.

Michel, W. 1937. Uder die experimentelle fusion pflanzlicher protoplasen. Arch. Exp. Zellf. 20:230–52.

Miller, R. A., O. L. Gamborg, W. A. Keller, and K. N. Kao. 1971. Fusion and division of nuclei in multinucleated soybean protoplasts. Canad. J. Genet. Cytol. 13:347–53.

Morel, G., J. P. Bourgin, and Y. Chupeau. 1972. The use of protoplasts in plant biology. *In* J. R. Villaneuva et al. (eds.), Yeast, mould, and plant protoplasts. Pp. 333–44. Academic Press, New York.

Muller, A. J., R. Grafe, R. R. Mendel, and I. Saalbach. 1976. Nitrate reductase-deficient mutants isolated from cultured *Nicotiana tabacum* cells. *In* Proc. Symposium on Experimental Mutagenesis in Plants, October 1976, Varna, Bulgaria.

Murashige, T. 1974. Plant propagation through tissue cultures. Ann. Rev. Plant Physiol. 25:135–66.

Nandi, S., and T. Eriksson. 1977. Nuclear behaviour of pea leaf protoplasts. Hereditas 85:49–55.

Narayanaswamy, S. 1977. Regeneration of plants from tissue cultures. *In* J. Reinert and Y. P. S. Bajaj (eds.), Applied and fundamental aspects of plant cell, tissue, and organ culture. Pp. 179–248. Springer-Verlag, Berlin.

Nawa, S., M. Yamada, and Y. Ohta. 1975. Hereditary changes in *Capsicum annuum* L. induced by DNA treatment. Jap. J. Genet. 50:341–44.

Nickell, L. G., and J. G. Torrey. 1969. Crop improvement through plant cell and tissue culture. Science 166:1068–70.

Niizeki, H., and K. Oono. 1968. Induction of haploid rice plants from anther culture. Proc. Jap. Acad. 44:554–52.

Ohyama, K. 1976. Basis for bromodeoxyuridine resistance in plant cells. Envir. Exper. Bot. 16:209–16.

Ohyama, K. 1978. DNA binding and uptake by nuclei isolated from plant protoplasts: Fate of single-stranded bacteriophage fd / DNA. Plant Physiol. 61:515–20.

Ohyama, K., O. L. Gamborg, and R. A. Miller. 1972. Uptake of exogenous DNA by plant protoplasts. Canad. J. Bot. 50:2077–80.

Ohyama, K., L. E. Pelcher, and D. Horn. 1977. DNA binding and uptake by nuclei isolated from plant protoplasts: Factors affecting DNA binding and uptake. Plant Physiol. 60:179–81.

Pelcher, L. E., K. N. Kao, O. L. Gamborg, O. C. Yoder, and V. E. Gracen. 1975. Effects of

Helminthosporium maydis race T toxin on protoplasts of resistant and susceptible corn (*Zea mays* L.) Canad. J. Bot. 73:427–31.

Pierik, R. L. M. 1975. Vegetative propagation of horticultural crops *in vitro* with special attention to shrubs and trees. Acta Horticulturae 54:71–82.

Potrykus, I. 1972. Fusion of differentiated protoplasts. Phytomorphology 22:91–96.

Potrykus, I. 1975. Uptake of cell organelles into isolated protoplasts. *In* R. Markham et al. (eds.), Modification of the information content of plant cells. Pp. 169–79. North-Holland Publishing Co., Amsterdam.

Power, J, B., S. E. Cummins, and E. C. Cocking. 1970. Fusion of isolated plant protoplasts. Nature 225:1016–18.

Power, J. B., E. M. Frearson, C. Hayward, D. George, P. K. Evans, S. F. Berry, and E. C. Cocking. 1976. Somatic hybridization of *Petunia hybrida* and *P. parodii*. Nature· 263:500–502.

Primrose, S. B. 1977. Genetic engineering. Sci. Prog. 64:293–321.

Rangan, T. S. 1976. Growth and plantlet regeneration in tissue cultures of some Indian millets: *Paspalum scrobiculatum* L., *Eleusine coracana* Gaertn., and *Pennisetum typoideum* Pers. Z. Pflanzenphysiol. 78:208–16.

Raspe, G. 1972. Workshop on mechanisms and prospects of genetic exchange. Adv. in Biosciences. I. Pergamon Press, New York.

Redei, G. P., and G. Acedo. 1976. Biochemical mutants in higher plants. *In* D. Dudits, G. L. Farkas, and P. Maliga (eds.), Cell genetics in higher plants. Pp. 39–58. Akadémiai Kiadó, Budapest.

Reinert, J., and Y. P. S. Bajaj. 1977. Applied and fundamental aspects of plant cell, tissue, and organ culture. P. 803. Springer-Verlag, Berlin.

Ringertz, N. R., and R. E. Savage. 1976. Cell hybrids. Academic Press, New York.

Rosenthal, G. A. 1977. The biological effects and mode of action of L-canavanine, a structural analogue of L-arginine. Quart. Rev. Biol. 52:155–78.

Saunders, J. W., and E. T. Bingham. 1972. Production of alfalfa plants from callus tissue. Crop Sci. 12:804–8.

Schell, J., and M. Van Montagu. 1977. The Ti-plasmid of *Agrobacterium tumefaciens*, a natural vector for the introduction of *Nif* genes in plants. *In* A. Hollaender (ed.), Genetic engineering for nitrogen fixation. Pp. 159–80. Plenum Press, New York.

Schieder, O. 1974. Selection of a somatic hybrid between auxotrophic mutants of *Sphaerocarpos donnellii* Aust. using the method of protoplast fusion. Z. Pflanzenphysiol. 74:357–65.

Schieder, O. 1976a. The spectrum of auxotrophic mutants for the liverwort *Sphaerocarpos donnellii* Aust. Molec. Gen. Genet. 144:63–66.

Schieder, O. 1976b. Isolation of mutants with altered pigments after irradiating haploid protoplasts from *Datura innoxia* Mill. with X-rays. Molec. Gen. Genet. 149:251–54.

Schilperoort, R. A., H. Veldstra, S. O. Warnaar, G. Mulder, and J. A. Cohen. 1967. Formation of complexes between DNA isolated from tobacco crown gall tumors and RNA complementary to *Agrobacterium tumefaciens* DNA. Biochim. Biophys. Acta 145:523–25.

Scowcroft, W. F. 1978. Somatic cell genetics and plant improvement. Adv. Agron. 29:39–81.

Scowcroft, W. F., and J. A. Adamson. 1976. Organogenesis from callus cultures of the legume *Stylosanthes Hamata*. Plant Sci. Lett. 7:39–42.

Shepard, J. F., and R. E. Totten. 1975. Isolation and regeneration of tobacco mesophyll cell protoplasts under low osmotic conditions. Plant Physiol. 55:689-94.

Shepard, J. F., and R. E. Totten. 1977. Mesophyll cell protoplasts of potato: Isolation, proliferation, and plant regeneration. Plant Physiol. 60:313-16.

Shepherd, R. J. 1976. DNA viruses of higher plants. Adv. Virus Res. 20:305-39.

Shimada, T., T. Sasakuma, and K. Tsuenwaki. 1969. In vitro culture of wheat tissues. I. Callus formation, organ redifferentiation, and single cell culture. Canad. J. Gen. Cytol. 11:294-304.

Smith, H. H., K. N. Kao, and N. C. Combatti. 1976. Interspecific hybridization by protoplast fusion in *Nicotiana*. J. Hered. 67:123-28.

Sondahl, M. R., and W. R. Sharp. 1977. High frequency induction of somatic embryos in cultured leaf explants of *Coffea arabica* L. Z. Pflanzenphysiol. 81:395-408.

Soyfer, V. N., N. A. Kartel, N. M. Chekalin, Y. Titov, K. K. Cieminis, and N. V. Turbin. 1976. Genetic modification of the waxy character in barley after an injection of wildtype exogenous DNA. Analysis of the second seed generation. Mutat. Res. 36:303-10.

Still, G. G. 1968. Metabolic fate of 3,4-dichloropropionanilide in plants. The metabolism of the propionic acid moiety. Plant Physiol. 43:543-46.

Strobel, G. A. 1973. Biochemical basis of the resistance of sugarcane to eyespot disease. Proc. Nat. Acad. Sci. USA 70:1693-96.

Suzuki, M., and I. Takebe. 1976. Uptake of single-stranded bacteriophage DNA by isolated tobacco protoplasts. Z. Pflanzenphysiol. 78:421-33.

Szeto, W. W., D. H. Hamer, P. S. Carlson, and C. A. Thomas. 1977. Cloning of cauliflower mosaic virus (CaMV) DNA in *Escherichia coli*. Science 196:210-12.

Takebe, I. 1975. The use of protoplasts in plant virology. Ann. Rev. Phytopath. 13:105-25.

Takebe, I., G. Labib, and G. Melchers. 1971. Regeneration of whole plants from isolated mesophyll protoplasts of tobacco. Naturwiss. 58:318-20.

Thomas, E., F. Hoffman, and G. Wenzel. 1975. Haploid plantlets from microspores of rye. Z. Pflanzenzucht. 75:106-13.

Tokumasu, S. 1976. The increase of seed fertility of *Brassicoraphanus* through cytological irregularity. Euphytica 25:463-70.

Tomasz, A. 1975. The mechanism of competence for DNA uptake and transformation in pneumococci. *In* L. Ledoux (ed.), Genetic manipulations with plant material. P. 27. Plenum Press, New York.

Tsitsine, N. V. 1976. Role of remote hybridization in the evolution of plants, pp. 87-98. Vestuik Akad. Nauk, USSR.

Turbin, N. V., V. N. Soyfer, N. A. Kartel, N. M. Chekalin, Y. L. Dorohov, Y. B. Titov, and K. K. Cieminis. 1975. Genetic modification of the *waxy* character in barley under the action of exogenous DNA of the wild variety. Mutat. Res. 27:59-68.

Uchimiya, H., and T. Murashige. 1977. Quantitative analysis of the fate of exogenous DNA in *Nicotiana* protoplasts. Plant Physiol. 59:301-8.

van Larebeke, N., C. Genetello, J. P. Hernalsteens, A. DePicker, I. Zaenen, E. Messens, M. Van Montague, and J. Schell. 1977. Transfer of Ti-plasmids between *Agrobacterium* strains by mobilization with conjugative Plasmid RP4. Molec. Gen. Genet. 152:119-24.

Vardi, A., P. Spiegel-Roy, and E. Galun. 1975. Citrus cell culture: Isolation of protoplasts, plating densities, effect of mutagens, and regeneration of embryos. Plant Sci. Lett. 4:231-36.

Vasil, I. K. 1976. The progress, problems, and prospects of plant protoplast research. Adv. Agron. 28:119–60.

Vasil, I. K., and C. Nitsch. 1975. Experimental production of pollen haploids and their uses. Z. Pflanzenphysiol. 76:191–212.

Vasil, I. K., V. Vasil, and D. H. Hubbell. 1977. Engineered plant cell or fungal association with bacteria that fix nitrogen. *In* A. Hollander (ed.), Genetic engineering for nitrogen fixation. Pp. 197–212. Plenum Press, New York.

Walkey, D. G. A., and J. M. G. Woolfitt. 1970. Rapid clonal multiplication of cauliflower by shake culture. J. Hort. Sci. 54:205–6.

Wallin, A., K. G. Glimelius, and T. Erksisson. 1974. The induction of aggregation and fusion of *Daucus carota* protoplasts by polyethylene glycol. Z. Pflanzenphysiol. 74:64–80.

Wardell, W. L. 1976. Floral activity in solutions of deoxyribonucleic acid extracted from tobacco stems. Plant Physiol. 57:855–61.

Weber, G., F. Constabel, F. Williamson, L. Fowke, and O. L. Gamborg. 1976. Effect of preincubation of protoplasts on PEG-induced fusion of plant cells. Z. Pflanzenphysiol. 79:459–64.

Wetter, L. R. 1977. Isoenzyme patterns in soybean-*Nicotiana* somatic hybrid cell lines. Molec. Gen. Genet. 150:231–35.

Wetter, L. R., and K. N. Kao. 1976. The use of isoenzymes in distinguishing the sexual and somatic hybrids in callus cultures derived from *Nicotiana*. Z. Pflanzenphysiol. 80:455–62.

Willecke, K., and F. H. Ruddle. 1975. Transfer of the human gene for hypoxanthine-guanine phosphoribosyltransferase via isolated human metaphase chromosomes into mouse L-cells. Proc. Nat. Acad. Sci. USA 72:1792–96.

Williamson, F. A., L. C. Fowke, G. Weber, F. Constabel, and O. Gamborg. 1977. Microfibril deposition on cultured protoplasts of *Vicia hajastana*. Protoplasma 91:213–19.

Winton, L., and O. Huhtinen. 1976. *In* J. P. Miksche (ed.), Modern methods in forest genetics. Pp. 243–64. Springer-Verlag, Berlin.

Withers, L. A., and E. C. Cocking. 1972. Fine structural studies on spontaneous and induced fusion of higher plant protoplasts. J. Cell Sci. 11:59–75.

Wittwer, S. H. 1974. Maximum production capacity of food crops. Bioscience 24:216–24.

Part 4

Agricultural Applications

L. R. KRUSBERG AND D. E. BABINEAU

Application of Plant Tissue Culture
to Plant Nematology

21

INTRODUCTION

Most of the existing information about plant parasitic nematodes and their relations with plant tissues has been gained from studies on nematodes feeding and reproducing on plants growing in soil. However, early workers in nematology recognized that certain kinds of information would be difficult or impossible to obtain with nematodes on or in plants in soil. By the early 1900s sterile techniques were well established for culturing many bacteria and fungi, and certain of these techniques were adapted to study nematode parasites of plants.

Byars in 1914 had succeeded in carrying a root knot nematode through a complete life cycle, from egg to egg, on sterile tomato seedlings growing in test tubes on nutrient agar medium. In doing so, he demonstrated that the nematode was the cause of the root galls and not associated microorganisms. Some years later Tyler (1933) demonstrated that root knot nematode is parthenogenetic by serial culture of single infective larvae for twelve generations using excised tomato roots growing on nutrient agar medium.

Another twenty years elapsed before Mountain (1954, 1955) developed a technique for serial propagation of a lesion nematode using root cultures of corn and tobacco. Then Darling et al. (1957) propagated the potato rot nematode on callus tissues of several plants. This latter report initiated the

L. R. Krusberg and D. E. Babineau, Department of Botany, University of Maryland, College Park, Maryland 20742.

current period in which many species of plant parasitic nematodes have come to be propagated on a number of different plant callus tissues. Recent reviews on this topic are those of Zuckerman (1971) and Ingram (1973).

PLANT TISSUE CULTURES AS A TOOL TO
PROPAGATE PLANT NEMATODES

Most root-feeding parasitic nematodes can be propagated in a greenhouse on host plants growing in soil. This method of propagation proved to be less than ideal because of problems such as demands on greenhouse space, risk of contamination with unwanted nematode species and/or other organism, difficulty in recovering nematodes from soil and/or roots, and erratic and often undependable buildup of nematode populations. However, many nematode species must still be propagated by such means becuause better methods have not been developed. In fact, certain nematode species are still difficult or impossible to propagate in the greenhouse.

Root Cultures

Interest in propagating nematodes on plant tissues in culture was stimulated by the success of Mountain (1954) in obtaining reproduction of the lesion nematode, *Pratylenchus minyus*, on root cultures of corn and tobacco. The main objective of this study was to demonstrate that *P. minyus* was a primary pathogen and caused root lesions in the absence of other microorganisms. A by-product of this study was the elucidation of the complete life cycle of that nematode. This initial success prompted further studies (Mountain, 1955) using red clover root cultures as well as root cultures of corn and tobacco with *P. minyus*. Nematodes were continuously cultured for four months by subculturing every two weeks. Tiner (1960, 1961a) refined the techniques for indefinite propagation of lesion nematodes on corn roots in culture using *P. penetrans*. He obtained a maximum of 64,000 nematodes/g fresh weight of roots from three-month-old cultures incubated at 25°C.

The burrowing nematode, *Radopholus similis*, completed its life cycle in roots of citrus seedlings growing on water agar according to Feder and Feldmesser (1955). Dasgupta et al. (1970) utilized root cultures of *Sorghum vulgare* to propagate *Hoplolaimus indicus* and to define its embryology and life cycle.

Nematodes of the family Heteroderidae seem to require vascular development in plant tissues in order to complete their life cycle. Sayre (1958) found that *Meloidogyne incognita*, would not grow on purely undifferentiated tissues of tomato, but would grow and complete its life cycle in tissues having some degrees of vascular development. Miller (1963) obtained all stages of development of *M. hapla* in tomato callus, but based on his techniques vascular tissue could have been present in the callus. Reproduction of both *M. hapla* and *M. incognita acrita* were readily obtained on root cultures of tomato and cucumber, respectively.

The potato cyst nematode, *Heterodera rostochiensis*, developed to mature females in tomato roots in culture (Widdowson et at., 1958). Apparently, males required to fertilize the females were lacking in these cultures so the females produced no eggs. This economically important nematode has still not been carried through a complete life cycle in a plant tissue culture. Moriarty (1964) obtained development of the beet cyst nematode, *H. schachtii*, on excised sugar beet roots in culture, but no eggs were produced by the females because the males all settled to the bottom of the culture dishes before fertilizing the females. Johnson and Viglierchio (1969b) obtained embryonated eggs produced by females of this species growing on excised beet roots, but did not carry the nematodes through a second or succeeding generations on the roots. Brown (1974) found embryonated eggs in females of the oat cyst nematode, *H. avenae*, growing in the roots of wheat seedlings in test tube cultures on nutrient agar medium. Reversat (1975) also obtained egg production in females of the rice cyst nematode, *H. oryzae*, growing on intact rice plants in test tubes. Many larvae penetrated roots, and nearly all developed to adults. However, on excised rice roots few *H. oryzae* larvae developed to the adult stage.

No reports have been published of serial subculturing of either *Meloidogyne* sp. or *Heterodera* sp., although at least one investigator maintains *M. incognita* on tomato root cultures (Sayre, personal communication). On medium supplemented with 2,4-D (2,4-dichlorophenoxyacetic acid), or even when additionally supplemented with IAA (indoleacetic acid), *H. rostochiensis* failed to reproduce in undifferentiated potato callus or cause syncytia to form in the callus (Webster and Lowe, 1966). Obviously, some kind of barrier, biochemical and/or physical, exists that prevents these nematodes from establishing a successful developmental relationship with plant tissues that are undifferentiated.

Callus Tissue

The term *callus tissue* as used in reference to plant nematology loosely means a friable mass of plant cells arising from excised plant tissues or whole seedlings that have been plated on agar medium containing plant growth promoters, usually 2,4-D. These tissues contain many differentiated as well as undifferentiated plant cells. Furthermore, new seedlings are usually used to initiate new plant tissue cultures for subculturing of the nematodes. Thus the term *callus tissue* as used here is considerably different from that defined by Street (1973), where the plant tissue is the primary object of concern. The callus tissue in this case is used by the nematodes as their source of food.

Mountain (1955) probably recognized that the rapid growth of roots of many species of plants was a major drawback of excised root cultures as a medium for propagating nematodes. He found that his cultures had to be subcultured about every two weeks because the plant roots, especially corn, quickly overwhelmed their containers. Darling et al. (1957) reported that the potato rot nematode, *Ditylenchus destructor*, reproduced well on cultures of undifferentiated tissues of potato, carrot, clover, and tobacco as well as on 37 species representing 15 genera of fungi. Large populations of nematodes developed on the plant tissues or fungi within three to four months. The report by Krusberg (1961) that alfalfa callus tissue supported the rapid reproduction of several different species of plant parasitic nematodes caught the attention of a number of nematologists. This technique offered a much more manageable type of plant tissue culture for propagating nematodes for those who needed such a tool.

Tissues from Different Plants as Substrates for Nematode Propagation

A variety of plants has been tested in searching for those that are easy to handle and provide good nematode reproduction. Faulkner and Darling (1961) reported that *Ditylenchus destructor* reproduced well on undifferentiated callus of potato, carrot, clover, and tobacco, reaching populations of up to 200,000/culture in 4 months at 25°C. They did not indicate any differences among plants as to supporting greater or lesser nematode reproduction. Krusberg (1961) experimented with alfalfa, carrot, sweet potato, tobacco, tomato, and soybean, and found that alfalfa provided the best reproduction of *Ditylenchus dipsaci* and *Pratylenchus zeae*. *Aphelenchoides ritzemabosi* was reported to reproduce rapidly on undifferentiated callus cultures of tobacco, carrot, periwinkle, marigold, tomato, grape, sunflower, and Paris daisy (Dolliver et al., 1962). Okra root callus proved to be a good substrate for *Radopholus similis*. In fact, 5 test tube cultures of okra callus provided as many nematodes as 100 citrus seedlings growing in 8-inch pots in the greenhouse (Feder et al., 1962), but later studies showed that alfalfa callus was even superior to okra callus (Myers et al., 1965). Many clones of several varieties of clover and alfalfa proved to be good substrates for *D. dipsaci* (Bingefors and Eriksson, 1963).

Webster and Lowe (1966) reported that *A. ritzemabosi* reproduced rapidly on alfalfa callus produced from both stems and roots of seedlings. *Tylenchus agricola* and *Tylenchorhynchus claytoni* reproduced well on alfalfa callus, but not on callus of tomato, broccoli, carrot, cabbage, rye, or corn (Khera and Zuckerman, 1962). *Dolichodorus heterocephalus* reproduced on corn callus developed from root tips, but not on alfalfa callus developed from seedlings (Paracer and Zuckerman, 1967). *Paratylenchus projectus* reproduced well on white clover callus, but not on alfalfa callus (Townshend, 1974). A list of plant-feeding nematodes and the plant tissues on which they have been propagated is presented in table 1.

So it is seen that though alfalfa callus tissue is a good substrate for propagating many diverse nematodes, it is not the "universal substrate plant." Unsuccessful attempts have been made to propagate many other nematodes on plant tissues in culture. These failures may have been due to incompatible plants in tissue culture or perhaps to other factors in the culture environment.

Seedlings or Root Tissues versus Callus

A number of studies have compared the reproductive rates of various nematodes on seedlings or roots in culture with callus cultures of the same plant. Krusberg (1961) found that reproduction of *A. ritzemabosi* and *D. dipsaci* was better on alfalfa callus than on seedlings by factors of 4 and 1.5, respectively. *Pratylenchus zeae* reproduced well on alfalfa callus, but not at all on intact seedings. *P. penetrans* reproduced better on callus than root cultures of eleven different plants according to Schroeder and Jenkins (1963); only in root cultures of snap bean did this nematode reproduce better than in bean callus cultures. *D. destructor* reproduced well on undifferentiated callus cultures of clover, but failed to reproduce on root cultures of clover or tomato (Faulkner and Darling, 1961). Feder et al. (1962) obtained about the same reproductive rate for *Radopholus similis* on root or callus cultures of okra. Thus, as a general rule, with the exception of members of the family Heteroderidae, nematodes usually reproduce more rapidly on callus tissues of a plant than on excised root cultures or intact seedlings. Furthermore, callus tissues of plants that are resistant to, or are nonhosts of, a nematode in nature will frequently support good reproduction of that nematode (Webster and Lowe, 1966; Krusberg, 1961; Bingefors and Bingefors, 1976; Viglierchio et al., 1973; Krusberg and Blickenstaff, 1964). The nature of the plant growth promoter-induced alteration in these cells making incompatible cells compatible to nematodes is unknown.

TABLE I

Nematodes Matured or Propagated on Plant Tissues in Culture

Nematode	Plant Tissue	Reference
Aphelenchoides bicaudatus	Alfalfa callus	Wood, 1973
A. fragariae	Alfalfa, red clover, and tobacco callus	Bingefors and Bingefors, 1976
A. ritzemabosi	Alfalfa callus	Krusberg, 1960, 1961; Webster, 1967; Corbett, 1970; Webb, 1971
	Carrot, grape, marigold, Paris-daisy, periwinkle, sunflower, tobacco, and tomato callus	Dolliver et al., 1962
	Oat callus	Webster, 1966
	Alfalfa, apple, potato, red clover, and rose callus	Webster and Lowe, 1966
	Alfalfa, red clover, and tobacco callus	Bingefors and Bingefors, 1976
A. sacchari	Alfalfa callus	Myers, 1967
Aphelenchus avenae	Tobacco callus	Barker, 1963
	Carrot, periwinkle, tobacco, and tomato callus	Barker and Darling, 1965
	Alfalfa callus	Wood, 1973
Bursaphelenchus lignicolus	Alfalfa callus	Tamura and Mamiya, 1975
Ditylenchus destructor	Excised clover and tomato roots	Darling et al., 1957
	Carrot, clover, potato, and tobacco callus	Darling et al., 1957; Faulkner and Darling, 1961
	Alfalfa, red clover, and tobacco callus	Bingefors and Bingefors, 1976
D. dipsaci	Alfalfa callus	Krusberg, 1960, 1961; Krusberg and Blickenstaff, 1964; Eriksson, 1972; Faulkner et al., 1974
	Alfalfa and red clover callus	Bingefors and Eriksson, 1963; Eriksson, 1965; Webster and Lowe, 1966; Bingefors and Eriksson, 1968
	Onion callus	Mai and Thistlethwayte, 1967; Riedel and Foster, 1970; Riedel et al., 1973
	Alfalfa, red clover, and white clover callus	Croll, 1968
	Alfalfa, broad bean, and clover callus	Viglierchio et al., 1973
	Alfalfa, red clover, and tobacco callus	Bingefors and Bingefors, 1976
D. myceliophagus	Alfalfa callus	Khera et al., 1968
Dolichodorus heterocephalus	Corn callus	Paracer and Zuckerman, 1967

TABLE 1—*Continued*

Nematode	Plant Tissue	Reference
Heterodera avenae	Wheat seedling culture	Brown, 1974
H. marioni (= *Meloidogyne* sp.)	Excised tomato roots	Tyler, 1933; Ferguson, 1948
H. oryzae	Excised rice roots	Reversat, 1975
H. radicicola (=*Meloidogyne* sp.)	Cowpea and tomato seedling culture	Byars, 1914
H. rostochiensis	Excised tomato roots	Widdowson et al., 1958
H. schachtii	Excised sugar beet roots	Moriarty, 1964; Johnson and Viglierchio, 1969a, 1969b
Hoplolaimus coronatus	Alfalfa callus	Krusberg, 1961
H. indicus	Excised *Sorghum vulgare* roots	Dasgupta et al., 1970
Meloidogyne arenaria	Excised tomato roots	Dropkin and Boone, 1966
M. hapla	Excised tomato roots	Wieser, 1955; Schuster and Sullivan, 1960; Dropkin and Boone, 1966
	Cucumber and tomato callus	Miller, 1963
	Tomato seedling culture	Russell and Morrison, 1975
	Potato and tobacco pith, excised tomato roots, and tomato callus with roots	Sayre, 1958
M. incognita	Excised tomato roots	Peacock, 1959; Schuster and Sullivan, 1960; Dropkin and Boone, 1966
	Carrot discs	Sandstedt and Schuster, 1963, 1965
	Excised cucumber roots	McClure and Viglierchio, 1966a, 1966b
	Excised tomato roots and excised tobacco pith discs	Sandstedt and Schuster, 1966a
	Excised tobacco stem segments	Sandstedt and Schuster, 1966b
M. incognita acrita	Cucumber and tomato callus	Miller, 1963
	Excised tomato roots	Dropkin, 1966; Dropkin and Boone, 1966
M. javanica	Excised tomato roots	Dropkin and Boone, 1966
Nacobbus batatiformis	Excised tomato roots	Schuster and Sullivan, 1960
N. serendipiticus	Excised tomato roots	Prasad and Webster, 1967

TABLE 1—*Continued*

Nematode	Plant Tissue	Reference
Paratylenchus projectus	White clover callus	Townshend, 1974
Pratylenchus brachyurus	Excised corn roots	Boswell, 1963
	Carrot discs	O'Bannon and Taylor, 1968
P. coffeae	Citrus leaf callus	Inserra and O'Bannon, 1975
P. fallax	Alfalfa callus	Corbett, 1970
P. minyus	Excised corn, red clover, and tobacco roots	Mountain, 1954, 1955
P. penetrans	Excised corn roots	Tiner, 1960, 1961a, 1961b; Schroeder, 1963
	White clover seedling culture	Chen et al., 1961
	Alfalfa callus	Krusberg, 1961; Krusberg and Blickenstaff, 1964; Mai and Thistlethwayte, 1967; Högger, 1969; Riedel and Foster, 1970; Andersen, 1972; Dunn, 1973; Riedel et al., 1973
	Alfalfa, cabbage, celery, cucumber, lettuce, onion, pea, pepper, rye, snap bean, and soybean callus; excised roots of above except cabbage	Schroeder and Jenkins, 1963
P. thornei	Alfalfa callus	Andersen, 1972
P. vulnus	Alfalfa callus	Lownsbery et al., 1967
P. zeae	Alfalfa callus	Krusberg, 1961; Krusberg and Blickenstaff, 1964
Radopholus similis	Citrus seedling culture	Feder and Feldmesser, 1955
	Excised okra roots	Feder, 1958
	Okra callus	Feder et al., 1962
	Alfalfa, grapefruit, and okra callus	Myers et al., 1965
	Carrot discs	O'Bannon and Taylor, 1968
	Citrus leaf callus	Inserra and O'Bannon, 1975
Telotylenchus indicus	Alfalfa callus	Khera et al., 1969
Tylenchorhynchus capitatus	Alfalfa callus	Krusberg, 1961
T. claytoni	Alfalfa callus	Khera and Zuckerman, 1962
Tylenchus agricola	Alfalfa callus	Khera and Zuckerman, 1962
T. hexalineatus	Excised alfalfa, moosewood, plantago, poplar, and sugar maple roots	Savage and Fisher, 1966
T. neozelandicus	Alfalfa callus	Wood, 1973

Growth Promoters

The initial plant growth regulator used to stimulate callus formation by plant tissues was 2,4-D, and it is still in common use. Krusberg (1961) in early studies employed the nutrient agar medium of Hildebrandt et al. (1946) that had been developed for culturing tobacco tissues, and alfalfa was selected as the substrate plant because it supported the greatest reproduction of several nematodes in initial tests. When Dolliver et al. (1962) omitted 2,4-D from the medium, reproduction of *aphelenchoides ritzemabosi* decreased on tissues of eight different plants. Similar results were obtained by Krusberg (1961) with *D. dipsaci* and *A. ritzemabosi* on alfalfa; Webster (1967) confirmed this finding with *A. ritzemabosi* on alfalfa. The concentration of 2,4-D used in these studies was 1–2 μg/ml. Webster and Lowe (1966) found that alfalfa callus grew best and *A. ritzemabosi* reproduced to the highest numbers in eight weeks with 2,4-D at 0.125 μg/ml in the medium. Faulkner et al. (1974) used 0.06 μg/ml of 2,4-D in their medium for the mass rearing of *D. dipsaci*. These reports suggest the need for additional studies on the influence of 2,4-D concentration on reproduction of various nematodes propagated monoxenically on plant tissue cultures.

Few studies have dealt with the effects of other growth promoters on nematode reproduction in plant tissue cultures. Krusberg and Blickenstaff (1964) reported that kinetin at 0.5 μg/ml in medium also containing 2.0 μg/ml of 2,4-D inhibited reproduction of *Pratylenchus penetrans* and *P. zeae* on alfalfa tissues, but reproduction of *D. dipsaci* was enhanced. This latter report also determined that NAA (naphthaleneacetic acid) could be omitted from the medium with no adverse effect on nematode reproductive rates. However, kinetin in the medium at the same concentration did not affect reproduction of *P. vulnus* on alfalfa tissue according to Lownsbery et al. (1967). Webster (1967) performed a series of growth regulator studies using *A. ritzemabosi* and alfalfa as the host plant. In medium lacking 2,4-D nematode reproduction increased when kinetin, GA (gibberellic acid), IAA, or tryptophan were added to the medium. Furthermore, the combinations GA + IAA and GA + tryptophan were better than GA alone. Barker and Darling (1965) reported that *Aphelenchus avenae* reproduced poorly on tissue cultures of several plants growing on synthetic medium supplemented with kinetin and IAA. Faulkner et al. (1974) included 1 μg/ml of NAA in their medium for mass propagation of *D. dipsaci* on alfalfa tissues. Only Webster (1967) has investigated the influence of plant growth inhibitors on nematodes in plant tissue cultures. Reproduction of *A. ritzemabosi* on alfalfa tissue was decreased when CCC (2-chloroethyl-trimethylammonium chloride), 7-azoindole, or 2-hydroxy-

5-nitrobenzyl bromide were in the medium especially when GA was also present. Webster (1966) reported that callus formed from oat seedlings only in the presence of a combination of IAA, NAA, and 2,4-D. Obviously, knowledge is scarce dealing with the effects of most plant growth regulators on the reproduction of nematodes in plant tissue cultures.

Coconut milk (CM) is added by some and not by others to media for the propagation of nematodes on plant tissues in culture. Krusberg (1961) routinely used CM, but later (Krusberg and Blickenstaff, 1964) determined that it was not an essential component of the medium for good reproduction of D. dipsaci, P. penetrans, or P. zeae on alfalfa tissues. Riedel and Foster (1970) similarly found that CM in the medium did not provide statistically greater populations of D. dipsaci on onion tissues or P. penetrans on alfalfa tissues than when it was omitted. A. avenae reproduced best on tissues of four different plants growing on medium supplemented with CM (Barker and Darling, 1965). CM also enhanced the reproduction of A. ritzemabosi on tissues of eight different plants (Dolliver et al., 1962), and was essential in two media, but not in a third medium, for the reproduction of Bursaphelenchus lignicolus on alfalfa tissues (Tamura and Mamiya, 1975). Faulkner et al. (1974) list CM as a constituent of their medium for mass propagation of D. dipsaci on alfalfa tissues, as do Bingefors and Bingefors (1976). The latter investigators indicate that there is no difference in milk from green coconuts as compared with mature coconuts. It appears then that CM probably improves the reproductive rate of most nematodes on plant tissues in culture, but is not essential and its addition may not be justified based on the slightly enhanced nematode reproductive rate.

Some reports state that additives to the medium that stimulate the most rapid and greatest plant tissue growth also support the most rapid and greatest nematode reproduction on those tissues (Webster and Lowe, 1966; Viglierchio et al., 1973; Dolliver et al., 1962); but others have found that this is not always true (Krusberg and Blickenstaff, 1964; Webster, 1967).

Related to the rate of plant tissue growth is the time of inoculation of plant tissue cultures with nematodes. According to Bingefors and Eriksson (1968), the callus needs to be in the log phase of growth when the nematodes are added. Callus that is too old (three weeks to two months) provides low, or no, nematode reproduction. Alfalfa tissue cultures are often inoculated with nematodes before they are two weeks old (Krusberg and Blickenstaff, 1964; Riedel and Foster, 1970). Barker and Darling (1965) found that A. avenae reproduced most rapidly on clones of tobacco callus that, when inoculated with nematodes, were only two to four days old. On seven- to ten-day-old clones, reproduction was moderate; and on two- to three-week-old clones, the nematodes rarely reproduced.

Viglierchio et al. (1973), using *D. dipsaci*, reported that age of the tissue culture at the time of inoculation with nematodes profoundly alters the ratios of nematodes in the various developmental stages. Thus the age of plant tissue cultures when inoculated with nematodes is important to obtain the maximum buildup of nematodes.

One criticism of using plant tissue cultures for propagating plant parasitic nematodes has been the complexity of the tissue culture medium. Riedel et al. (1973) developed a medium containing only five components—sucrose, yeast extract, 2,4-D, agar, and water—that provided nearly as good reproduction of *D. dipsaci* and *P. penetrans* on onion and alfalfa tissue cultures as the more complex media. Tamura and Mamiya (1975) obtained good reproduction of *Bursaphelenchus lignicolus* on alfalfa tissue cultures growing on this medium.

The above discussion points out how primitive our knowledge is concerning the propagation of nematodes monoxenically on plant tissues in culture. Many detailed systematic studies are needed to identify precisely the medium composition, the plant tissues, and the conditions optimum for the propagation of diverse nematodes in such cultures. Obviously there has been no compelling need to propagate nematodes by such means based on the low level of effort expended. However, if we could propagate specific nematodes easily, knowledge in many areas of nematology might very well advance at a far more rapid rate than it is advancing at present.

INFLUENCE OF NEMATODES ON PLANT TISSUES IN CULTURE

Nematode feeding leads to a rapid decline of plant callus cultures. The tissues rapidly take on a light tan coloration, become water-soaked in appearance, the cells stop dividing, and the callus deteriorates. The lesion nematode *Pratylenchus penetrans* seems to have a particularly adverse effect on alfalfa callus tissue, causing it to turn black (Anderson, 1972; Krusberg, personal observation), that has not been observed with any other species of *Pratylenchus* or nematodes of other genera reared on that tissue.

Several investigators have used plant tissue cultures to study various aspects of host-parasite relations of root knot nematodes with plant tissues. Infective larvae were found to induce galls on excised tomato roots in culture (Schuster and Sullivan, 1960) and on roots of sterile intact tomato seedlings (Russell and Morrison, 1975) through surface feeding as well as when they penetrated roots. Schuster and Sullivan (1960) further observed that no root hairs formed on these galls when caused by *Meloidogyne incognita acrita*, but the galls caused by *M. hapla* were covered with root hairs. Visible galling was seen in as little as 21 hr following inoculation with nematodes.

Dropkin and Boone (1966) performed a series of experiments with several species of root knot nematodes on excised tomato roots in culture. Some of their more pertinent findings were: (1) out of 250 single larva inoculations the first root gall appeared in 24 hr, most appeared on the third day, and the last appeared on the ninth day; (2) more larvae penetrated and caused galls in freshly excised roots than in seven-day-old excised roots; and (3) a tomato cultivar resistant to four root knot nematode species and susceptible to a fifth species when grown in soil retained the same reaction pattern as excised roots in culture.

McClure and Viglierchio (1966a) using excised cucumber roots found in general that concentrations of sucrose, macronutrient salts, iron chelate and vitamins in the medium that yielded best root growth, also, promoted the greatest percentage penetration by root knot nematode larvae and the largest number of galls formed. Adding NAA or 2,4-D to the medium at several concentrations of each did not affect the percent of larvae penetrating the roots. Only one larva in all the NAA treatments caused syncytium formation and none in the 2,4-D treatments six days after inoculation with nematodes. McClure and Viglierchio (1966b) then related some of these host nutritional factors to rate of development of the root knot nematode in the excised cucumber roots. Low levels of sucrose or iron chelate in the medium slowed the rate of nematode development, whereas low vitamins or macronutrients increased the rate. Low sucrose was the only component to affect sex ratio, causing all the developing nematodes to become males. Rate of nematode development was also inversely proportional to the number of nematodes in the inoculum and the number in each gall. Sandstedt and Schuster (1963, 1965) inoculated carrot sections growing on nutrient agar medium with root knot nematodes. In the early experiments (1963) sterile egg masses were placed on top of carrot discs. Larvae, as they hatched, penetrated the carrot, inducing lumps of callus tissue containing nematodes and syncytia; these nematodes produced eggs in about six weeks. After four weeks nematode-infected discs weighed 150 percent as much as control discs, and they suggested that the nematodes provided plant growth promoters to account for the increased plant tissue weight. In the later experiments (1965) they found using longitudinal sections of the carrot tissue that the nematodes induced callus sooner and matured earlier in the radicle ends than in the foliar ends. Earlier callusing and nematode maturation also occurred in discs when the radicle end was up rather than down on the agar. These findings suggested that accumulation of downward-transported plant growth regulators in radicle ends of carrot sections was responsible for this more rapid nematode development.

Sandstedt and Schuster (1966a) also used tobacco pith tissue cultures to attempt to assay for nematode-produced growth substances. They were unable to detect auxin or cytokinin secretion by *Meloidogyne incognita*, but the nematode in the presence of added IAA or kinetin promoted tissue growth that was more friable and irregular than that induced by only the plant growth promoters. Nematodes failed to develop past the second stage larva or form syncytia in the absence of added growth promoters. This suggested that nematodes secrete some unknown substance that acts as a synergist in aiding gall and syncytium formation. In addition they (Sandstedt and Schuster, 1966b) determined that *M. incognita* did not cause effects similar to those caused by IAA on peeled tobacco stem sections. Nematodes affected the tissues more like 2,3,5-triiodobenzoic acid, an inhibitor of IAA transport, in the callus induced. They concluded that nematodes probably did not inject or release "bound" auxins in the plant tissues, but rather that nematode secretions of unknown composition inhibit diffusion or transport of such substances and thereby induce cell proliferation around the nematodes.

INFLUENCE OF PLANT TISSUES IN CULTURE ON NEMATODES

Concern has been expressed at times that nematodes propagated under such artificial conditions on plant tissue cultures might change in their virulence toward host plants when tested under more natural conditions. Evidence collected by several investigators indicates that no change in virulence occurs in nematodes propagated on plant tissue cultures. This has been found to be so for *Ditylenchus destructor* propagated for four years on undifferentiated callus tissues of four host plants (Faulkner and Darling, 1961), for *Radopholus similis* propagated on okra root callus (Feder et al., 1962), for *Pratylenchus penetrans* propagated on alfalfa callus tissues (Mai and Thistlethwayte, 1967; Högger, 1969, and for *D. dipsaci* propagated on alfalfa callus tissues (Mai and Thistlethwayte, 1967), even after 14 years' continuous subculturing (Bingefors and Bingefors, 1976). Bingefors and Eriksson (1968) also reported that after five years of subculturing on alfalfa callus the red clover and alfalfa races of *D. dipsaci* were still specifically pathogenic toward red clover and alfalfa, respectively.

UTILIZATION OF NEMATODES PROPAGATED ON
PLANT TISSUE CULTURES

The greatest use made of plant tissue cultures in nematology is to propagate nematodes for various purposes. Probably the largest continuous operation for mass rearing of nematodes is the one in Uppsala, Sweden, for propagating races of *D. dipsaci*, as reported by Bingefors and Bingefors (1976). Primary emphasis has been to propagate the alfalfa and red clover races of this nematode to use in screening plants of these crops to find nematode resistance. These nematodes have been primarily used in breeding programs in Sweden and Denmark, but some have been used in five other European countries. Faulkner et al. (1974) also recently published on techniques for mass rearing of the alfalfa race of *D. dipsaci* in the United States for their alfalfa breeding program.

On a smaller scale various species of nematodes propagated on plant tissue cultures have been used for diverse purposes too numerous to attempt to document here. They have included studies of nematode taxonomy, anatomy and morphology, biological control, resistance in plants, chemical control, biochemistry, and various aspects of biology, to name some.

Plant tissue cultures have been little used to study nematode biology except as they concern propagation of nematodes. As was mentioned earlier, the life cycle of some nematodes has been elucidated using plant tissue cultures. Callus cultures of alfalfa and red clover were used by Eriksson (1965, 1972) to make reciprocal crosses of the alfalfa and red clover races of *D. dipsaci*. Females of the alfalfa race and males of the red clover race produced fertile offspring that attacked both alfalfa and red clover, but alfalfa was not attacked as vigorously as by the parent alfalfa race. Some hybrid specimens possessed abnormal tails and body thickening. The reciprocal cross, red clover females and alfalfa males, failed to produce fertile progeny, apparently due to some sexlinked incompatibility factor. Studies by other investigators using host plants in the greenhouse on which to cross populations of *D. dipsaci* have also found that these races are quite stable genetically and pathogenically.

CURRENT STATUS, FUTURE, AND CONCLUSIONS

A small survey by letter in December 1976 was conducted to obtain some measure of the extent of utilization of plant tissue cultures in nematology and some consensus of the potential use of such cultures for solving nematological problems. From the 68 letters sent out, 45 replies were received, and 22 respondents indicated that they currently used some form of plant tissue culture for propagating various nematodes. Over 70 percent of the respondents thought that more effort should be expended on propagating nematodes on plant tissue cultures.

Some areas in which several investigators thought advances were needed or would be useful included:

1. Practical systems for propagating root knot (*Meloidogyne* spp.) and cyst (*Heterodera* spp.) nematodes on plant tissue cultures.
2. Practical systems for propagating on plant tissue cultures several

other nematodes, especially species on the genera *Xiphinema, Trichodorus, Belonolaimus,* and *Rotylenchulus.*

3. More-simplified and less-time-consuming methods for propagating nematodes on these cultures. Several respondents complained that they simply did not have the manpower resources required for current culturing techniques. Riedel et al. (1973) devised a much-simplified agar medium for culturing of plant tissues that has helped considerably. Fungal and bacterial contamination of cultures remains a serious problem.

4. Methods for long-term preservation of cultures is needed to reduce the time consumed in short term subculturing (e.g., every two weeks to three months). Storage at low temperatures (5-15°C) should be investigated for prolonging culture life (Riedel, personal communication).

Most respondents thought that a large number of nematological questions were amenable to study using plant tissue and/or cell cultures; in fact they were so numerous as to seem limited only by the imagination and ingenuity of the individual investigator. Fassuliotis (1975) found that roots of eggplant regenerated from isolated stem parenchyma cells retained their resistance to *M. incognita.* He is currently attempting to use protoplast fusion to incorporate nematode resistance into certain plants (Fassuliotis, personal communication).

As measured by the number of publications being issued, we seem to be in a lull period of nematology research with plant tissue cultures. Several people who were working in this area seem to have changed their research, or their priorities have been changed, toward other directions. Probably another widely applicable advance will be required to stimulate interest and activity in applying plant tissue cultures to nematology.

In conclusion, plant tissue cultures are being used in plant nematology research on a relatively small scale. Development of methods for propagating two of the most economically important groups of nematodes, root knot and cyst, on plant tissue cultures and further simplification of current techniques for propagating nematodes on these cultures would probably stimulate a resurgence of interest in this area of research.

LITERATURE CITED

Andersen, H. J. 1972. Culturing of *Pratylenchus penetrans* and *Pratylenchus thornei* on alfalfa callus-tissue. Tidssk. Plant. 76:375–77.

Barker, K. R. 1963. Parasitism of tobacco callus and Kentucky bluegrass by *Aphelenchus avenae.* Phytopathology 53:870 (abstr.).

Barker, L. R., and H. M. Darling. 1965 Reproduction of *Aphelenchus avenae* on plant tissue in culture. Nematologica 11:162–66.

Bingefors, S., and S. Bingefors. 1976. Rearing stem nematode inoculum for plant breeding purposes. Swedish J. Agric. Res. 6:13–17.

Bingefors, S., and K. B. Erikkson. 1963. Rearing stem nematode inoculum on tissue culture. Preliminary report. Lantbrugshögsk. Ann. 29:107–18.

Bingefors, S., and K. B. Eriksson. 1968. Some problems connected with resistance breeding against stem nematodes in Sweden. Z. Pflanzenzüchtg. 59:359–75.

Boswell, T. E. 1963. A method for separation and sterilization of *Praylenchus brachyurus* from mixed population of nematodes. Phytopathology 53:622 (abstr.).

Brown, J. A. A. 1974. Test tube reproduction of *Heterodera avenae* on resistant and susceptible wheats. Nematologica 20:192–203.

Byars, L. P. 1914. Preliminary notes on the cultivation of the plant parasitic nematode *Heterodera radicicola*. Phytopathology 4:323–26.

Chen, T., R. A. Kilpatrick, and A. E. Rich. 1961. Sterile culture techniques as tools in plant nematology research. Phytopathology 51:799–800.

Corbett, D. C. M. 1970. Maintaining nematode cultures under mineral oil. Nematologica 16:156.

Croll, N. A. 1968. The use of callus tissue culture to support speculations on the mechanism of plant resistance to nematodes. Proc. N. W. Nematol. Workshop. Vancouver, B.C., pp. 19–20.

Darling, H. M., L. R. Faulkner, and P. Wallendal. 1957. Culturing the potato rot nematode. Phytopathology 47:7 (abstr.).

Dasgupta, D. R. R., S. Nand, and A. R. Seshadri. 1970. Culturing, embryology, and life history studies on the lance nematode, *Hoplolaimus indicus*. Nematologica 16:235–48.

Dolliver, J. S., A. C. Hildebrandt, and A. J. Riker. 1962. Studies of reproduction of *Aphelenchoides ritzemabosi* (Schwartz) on plant tissues in culture. Nematologica 7:294–300.

Dropkin, V. H. 1966. A culture unit for the analysis of host-parasite relationships of root-knot nematodes (*Meloidogyne* sp.). Nematologica 12:89 (abstr.).

Dropkin, V. H., and W. R. Boone. 1966. Analysis of host-parasite relationships of root-knot nematodes by single-larva inoculations of excise tomato roots. Nematologica 12:225–36.

Dunn, R. A. 1973. Extraction of eggs of *Pratylenchus penetrans* from alfalfa callus and relationship between age of culture and yield of eggs. J. Nematol. 5:73.

Eriksson, J. B. 1965. Crossing experiments with races of *Ditylenchus dipsaci* on callus tissue cultures. Nematologica 11:244–48.

Erikssson, K. B. 1972. Studies on *Ditylenchus dipsaci* (Kühn) with reference to plant resistance. Ph.D. thesis, Agricultural College of Sweden, Uppsala. 108pp.

Fassuliotis, G. 1975. Regeneration of whole plants from isolated stem parenchyma cells of *Solanum sisymbriifolium*. J. Amer. Soc. Hort. Sci. 100:636–38.

Faulkner, L. R., D. B. Bower, D. W. Evans, and J. H. Elgin, Jr. 1974. Mass culturing of *Ditylenchus dipsaci* to yield large quantities of inoculum. J. Nematol. 6:126–29.

Faulkner, L. R., and H. M. Darling. 1961. Pathological histology, hosts, and culture of the potato rot nematode. Phytopathology 51:778–86.

Feder, W. A. 1958. Aseptic culture of the burrowing nematode, *Radopholus similis* (Cobb) Thorne on excised okra root tissues. Phytopathology 48:392–93.

Feder, W. A., and J. Feldmesser. 1955. Progress report on studies on the reproduction of the burrowing nematode, *Radopholus similis* (Cobb) Thorne, on citrus seedlings growing in petri dishes. Plant Dis. Report. 39:395-96.

Feder, W. A., P. A. Hutchins, and R. Whidden. 1962. Aseptic growth of *Radopholus similis* (Cobb) Thorne on okra root callus tissue. Proc. Fla. State Hort. Soc. 75:74-76.

Ferguson, M. S. 1948. Culture experiments with *Heterodera marioni*. J. Parasitol. Suppl. 34:32-33 (abstr.).

Hildebrandt, A. C., A. J. Riker, and B. M. Duggar. 1946. The influence of the composition of the medium on growth in vitro of excised tobacco and sunflower tissue cultures. Amer. J. Bot. 33:591-97.

Högger, C. H. 1969. Comparison of penetration of potato roots by *Pratylenchus penetrans* grown in tissue cultures and from field populations. J. Nematol. 1:10 (abstr.).

Ingram, D. S. 1973. Growth of plant parasites in tissue culture. *In* H. E. Street (ed.), Plant tissue and cell culture. Pp. 392-421. University of California Press, Berkeley.

Inserra, R. N., and J. H. O'Bannon. 1975. Rearing migratory endoparasitic nematodes in cirtus callus and roots produced from citrus leaves. J. Nematol. 7:261-63.

Johnson, R. N., and D. R. Viglierchio. 1969a. Sugar beet nematode (*Heterodera schachtii*) reared on axenic *Beta vulgaris* root explants. I. Selected environmental factors affecting penetration. Nematologica 15:129-43.

Johnson, R. N., and D. R. Viglierchio. 1969b. Sugar beet nematode (*Heterodera schachtii*) reared on axenic *Beta vulgaris* root explants. II. Selected environmental and nutritional factors affecting development and sex-ratio. Nematologica 15:144-52.

Khera, S., G. C. Bhatnagar, N. Kumar, and M. G. Tikyani. 1968. Studies on the culturing of *Ditylenchus myceliophagus* Goodey, 1958. Indian Phytopath. 21:103-6.

Khera, S., G. C. Bhatnagar, M. G. Tikyani, and C. Nandalkumar. 1969. Culturing of *Telotylenchus indicus* Siddiqi, 1960 on alfalfa callus tissue. Labdev J. Sci. Technol. 7B:330-31.

Khera, S., and B. M. Zuckerman. 1962. Studies on the culturing of certain ectoparasitic nematodes on plant callus tissue. Nematologica 8:272-74.

Krusberg, L. R. 1960. Culturing, histopathology, and biochemistry of *Ditylenchus dipsaci* and *Aphelenchoides ritzemabosi* on alfalfa tissues. Phytopathology 50:642 (abstr.).

Krusberg, L. R. 1961. Studies on the culturing and parasitism of plant-parasitic nematodes, in particular *Ditylenchus dipsaci* and *Aphelenchoides ritzenabosi* on alfalfa tissues. Nematologica 6:181-200.

Krusberg, L. R., and M. L. Blickenstaff. 1964. Influence of plant growth regulating substances on reproduction of *Ditylenchus dipsaci, Pratylenchus penetrans,* and *Pratylenchus zeae* on alfalfa tissue cultures. Nematologica 10:145-50.

Lownsbery, B. F., C. S. Huang, and R. N. Johnson. 1967. Tissue culture and maintenance of the root-lesion nematode, *Pratylenchus vulnus.* Nematologica 13:390-94.

McClure, M. A., and D. R. Viglierchio. 1966a. Penetration of *Meliodogyne incognita* in relation to growth and nutrition of sterile, excised cucumber roots. Nematologica 12:237-47.

McClure, M. A., and D. R. Viglierchio. 1966b. The influence of host nutrition and intensity of infection on the sex ratio and development of *Meliodognye incognita* in sterile agar cultures of excised cucumber roots. Nematologica 12:248-58.

Mai, W. F., and B. Thistlethwayte. 1967. The feasibility of using plant pathogenic nematodes produced in monoxenic culture for nematocide screening. Phytopathology 57:820 (abstr.).

Miller, C. E. 1963. A method of obtaining *Meloidognye* sp. in aseptic cultures. Phytopathology 53:350–51 (abstr.).

Moriarty, F. 1964. The monoxenic culture of beet eelworm (*Heterodera schachtii* Schm.) on excised roots of sugar beet (*Beta vulgaris* L.). Parasitology 54:289–93.

Mountain, W. B. 1954. Studies of nematodes in relation to brown root rot of tobacco in Ontario. Canad. J. bot. 32:737–59.

Mountain, W. B. 1955. A method of culturing plant parasitic nematodes under sterile conditions. Proc. Helm. Soc. Wash. 22:49–52.

Myers, R. F. 1967. Axenic cultivation of *Aphelenchoides sacchari* Hooper. Proc. Helm. Soc. Wash. 34:251–55.

Myers, R. F., W. A. Feder, and P. C. Hutchins. 1965. The rearing of *Radopholus similis* (Cobb) Thorne on grapefruit, okra, and alfalfa root callus tissues. Proc. Helm. Soc. Wash. 32:94–95.

O'Bannon, J. H., and A. L. Taylor. 1968. Mirgratory endoparasitic nematodes reared on carrot discs. Phytopathology 58:385.

Paracer, S. M., and B. M. Zuckerman, 1967. Monoxenic culturing of *Dolichodorus hetereocephalus* on corn root callus. Nematologica 13:478.

Peacock, F. C. 1959. The development of a technique for studying the host-parasite relationship of the root-knot nematode *Meloidogyne incognita* under controlled conditions. Nematologica 4:43–55.

Prasad, S. K., and J. M. Webster. 1967. Effect of temperature on the rate of development of *Nacobbus serendipiticus* in excised tomato roots. Nematologica 13:85–90.

Reversat, G. 1975. Monoxenic culture of *Heterodera oryzae* on rice. Ann. Zool-Ecol. Anim. 7:81–89.

Riedel, R. M., and J. G. Foster. 1970. Monoxenic culture of *Ditylenchus dipsaci* and *Pratylenchus penetrans* with modified Krusberg's and White's media. Plant Dis. Report. 54:251–54.

Riedel, R. M., J. G. Foster, and W.F. Mai. 1973. A simplified medium for monoxenic culture of *Pratylenchus penetrans* and *Ditylenchus dipsaci*. J. Nematol. 5:71–72.

Russell, C. C., and L. S. Morrison. 1975. Induced ectoparasitic feeding of *Medoidognye hapla* on tomato. J. Nematol. 7:392 (abstr.).

Sandstedt, R., and M. L. Schuster. 1963. Nematode-induced callus on carrot discs grown in vitro. Phytopathology 53:1309–12.

Sandstedt, R., and M. L. Schuster. 1965. Host-parasite interaction in root-knot nematode-infected carrot tissue. Phytopathology 55:393–95.

Sandstedt, R., and M. L. Schuster. 1966a. Excised tobacco pith bioassays for root-knot nematode-produced plant growth substances. Physiol. Plantarum 19:99–104.

Sandstedt, R., and M. L. Schuster. 1966b. The role of auxins in root-knot nematode-induced growth on excised tobacco stem segments. Physiol. Plantarum 19:960–67.

Savage, H. E., and K. D. Fisher. 1966. Host-parasite relations of *Tylenchus hexalineatus*. Phytopathology 56:898 (abstr.).

Sayre, R. M. 1958. Plant tissue culture as a tool in the study of the physiology of the root-knot nematode, *Meloidogyne incognita* Chit. Ph.D. thesis, University of Nebraska. 56pp.

Schroeder, P. H. 1963. Reproduction of *Pratylenchus penetrans* on corn roots cultured on various nutrient media. Phytopathology 53:888–89 (abstr.).

Schroeder, P. H., and W. R. Jenkins. 1963. Reproduction of *Pratylenchus penetrans* on root tissues grown on three media. Nematologica 9:327–31.

Schuster, M. L., and T. Sullivan. 1960. Species differentiation of nematodes through host reaction in tissue culture. I. Comparisons of *Meloidogyne hapla, Meloidogyne incognita incognita,* and *Nacobbus batatiformis.* Phytopathology 50:874–76.

Tamura, H., and Y. Mamiya. 1975. Reproduction of *Bursaphelenchus lignicolus* on alfalfa callus tissue. Nematologica 21:449–54.

Tiner, J. D. 1960. Cultures of the plant parasitic nematode genus *Pratylenchus* on sterile excised roots. I. Their establishment and maintenance. Exp. Parasitol. 9:121–26.

Tiner, J. D. 1961a. Laboratory cultures of the plant parasitic nematode genus *Pratylenchus*: A new instrument and a new nematocide test. J. Parasitol. 47:25 (abstr.)

Tiner, J. D. 1961b. Cultures of the plant-parasitic nematode genus *Pratylenchus* on sterile excised roots. II. A trap for collection of axenic nematodes and quantitative initiation of experiments. Exp. Parasitol. 11:231–40.

Townshend, J. L. 1974. Monoxenic culture of *Paratylenchus projectus.* Nematologica 20:264–66.

Tyler, J. 1933. Reproduction without males in aseptic root cultures of the root-knot nematode. Hilgardia 7:373–88.

Viglierchio, D. R., I. A. Siddiqui, and N. A. Croll. 1973. Culturing and population studies of *Ditylenchus dipsaci* under monoxenic conditions. Hilgardia 42:177–214.

Webb, R. M. 1971. Extraction of nematodes from sterile culture. Nematologica 17:173–74.

Webster, J. M. 1966. Production of oat callus and its susceptibility to a plant parasitic nematode. Nature 212:1472.

Webster, J. M. 1967. The influence of plant-growth substances and their inhibitors on the host-parasite relationships of *Aphelenchoides ritzemabosi* in culture. Nematologica 13:256–62.

Webster, J. M., and D. Lowe. 1966. The effect of the synthetic plant-growth substance, 2,4-dichlorophenoxyacetic acid, on the host-parasite relationships of some plant-parasitic nematodes in monoxenic callus culture. Parasitology 56:313–22.

Widdowson, E., C. C. Doncaster, and D. W. Fenwick. 1958. Observations on the development of *Heterodera rostochiensis* Woll. in sterile root cultures. Nematologica 3:308–14.

Wieser, W. 1955. The attractiveness of plants to larvae of root-knot nematodes. I. The effect of tomato seedlings and excised roots on *Meloidogyne hapla* Chitwood. Proc. Helm. Soc. Wash. 22:106–12.

Wood, F. H. 1973. Nematode feeding relationships. Feeding relationships of soil-dwelling nematodes. Soil Biol. Biochem. 5:593–601.

Zuckerman, B. M. 1971. Gnotobiology. *In* B. M. Zuckerman, W. F. Mai, and R. A. Rohde (eds.), Plant parasitic nematodes. 2:159–84. Academic Press, New York.

G. BRUENING, S.-L. LEE, AND H. BEIER

Immunity to Plant Virus Infection

22

INTRODUCTION

Plant viruses can reduce yields of various crops. They can make the cultivation of a particular crop impractical in a given area. However, there is a widely applied approach to reducing the adverse effects of a plant virus. It is economical. It can produce results of long duration and with minimal adverse effect on the environment or quality of the cultivar. The approach is to discover one or more "sources of resistance" among plants that can be genetically crossed to the cultivar in question. Not only resistance (the ability on the part of the plant to restrict the virus increase to a low level) but also tolerance (the plant supports virus increase to a level comparable to the level attained in other susceptible plants but the crop yield is not greatly reduced; reviewed by Schafer, 1971) and immunity (no detectable increase in virus concentration following inoculation) can be so introduced into the cultivar, it is hoped with salutary effect. Although the genetic introduction of immunity to a plant virus into a cultivar is a sound and proven practice, no biochemical mechanism of immunity to a plant virus has yet been elucidated. Knowing such mechanisms might be of value, eventually, in inducing immunity where a natural source is not available or in retarding the ability of viruses to overcome existing immunity or resistance (reviewed by Gibbs and Harrison, 1976, pp. 225–27).

Immunity, rather than resistance or tolerance, would seem a priori to be a reasonable phenomenon to study biochemically because of the contrast it represents to full susceptibility. The concept of host range is closely related

G. Bruening, Department of Biochemistry and Biophysics, University of California, Davis, California 95616.

to immunity. The most important difference between the two concepts is that an organism that fails to support the increase of a particular virus, but is killed by it, is outside the host range of the virus but is not immune to it. The concept of host range is useful because some causes of restricted host range and some causes of immunity could be similar, at the molecular level. Possible mechanisms of immunity can be postulated most easily for viruses for which the replication cycle is most completely understood. The details of the replication cycle of bacterial and animal virus systems are better understood than those of plant virus systems at the present time. Two examples with potential parallels to plant virus systems are described below.

The bacteriophage T7 certainly stands among the viruses that have been most intensively studied. Chamberlin (1974) isolated mutants of *Escherichia coli* B that could be killed by T7 but did not support formation of mature virus. (That is, cells of the mutant strains were outside the host range of T7 but were not immune to it.) For one class of such mutants, named tsnC, the synthesis of T7 DNA did not occur following inoculation. Extracts of T7-inoculated tsnC cells provided an assay system for a putative "tsnC protein" in extracts of wild-type cells. A protein was found in uninfected, wild-type cells that stimulated DNA synthesis in the extracts of T7-inoculated tsnC cells (Modrich and Richardson, 1975a). Mixtures of the tsnC protein from wild-type cells and a protein that accumulated in T7-inoculated tsnC cells (and that proved to be the product of T7 gene 5) were active in the synthesis of T7 DNA (Modrich and Richardson, 1975b). Neither the gene 5 nor the tsnC protein alone showed T7 DNA polymerase activity; they formed a 1:1 molar complex that had activity.

The tsnC protein has been shown to be thioredoxin (Mark and Richardson, 1976), a participant in the ribonucleoside diphosphate reductase reaction by which deoxyribonucleotides are formed. At least some of the tsnC mutants lack thioredoxin, as judged by both activity and immunological assays. Apparently an alternative hydrogen donor system, dependent upon glutathione, acts in the essential ribonucleoside diphosphate reductase system and allows the tsnC mutants to survive (Holmgren, 1976; Fuchs, 1977). Conceivably, a multicellular organism with a mutation analogous to the tsnC mutation would be operationally immune to a virus that required a "tsnC protein." That is, at levels of virus inoculum that were not too high, the organism would perhaps lose only the few cells actually entered by the virus. Other host mutations have been studied that prevent or reduce T7 replication (see review by Hausmann, 1976).

Moehring and Moehring (1976) have studied a strain of the KB line of human epidermoid carcinoma cells that is resistant to diphtheria toxin. These cells supported the replication of Sindbis virus to less than 3 percent of the level attained by that virus in the original KB cells (sensitive to diphtheria toxin). The diphtheria toxin-resistant cells showed varying degrees of resistance to other viruses but were as susceptible as the wild-type KB cells to some viruses. The mechanism of the resistance is not known. However, the viruses absorb equally well to the resistant and the susceptible KB cells; the block to efficient replication presumably lies at a step after adsorption. These results imply that it may be possible to use agents other than the virus itself to efficiently select variants of plants that would be resistant or immune to a plant virus.

A few systems that exhibit resistance or immunity to a plant virus have been investigated. Levy et al. (1974) and Nachman et al. (1971) studied the nearly isogenic cucumber cultivars Elem and Bet Alpha. Virus accumulated in Elem to less than one-tenth the level achieved in Bet Alpha after inoculation with cucumber mosiac virus (CMV). The virus accumulation in Elem was increased 3- to 5-fold if in the period 1 to 3 days after inoculation with CMV the leaves were irradiated with ultraviolet light ($\sim$ 254 nm). Irradiation of Elem at later times had little effect. CMV accumulation in Bet Alpha was not significantly influenced by similar irradiations. Approximately parallel results were obtained by treating leaves with actinomycin D rather than ultraviolet irradiation. Perhaps the resistance of Elem to CMV is dependent upon the transcription from the host DNA of new RNA(s).

Two disease-resistant cultivars of beans are Robust and Corbett Refugee (reviewed by Zaumeyer and Meiners, 1975), both immune to mechanically inoculated bean common mosaic virus (BCMV). The Corbett Refugee trait is inherited as a single dominant gene, and Corbett Refugee plants can be infected with BCMV if the virus is introduced by grafting Corbett Refugee to a susceptible, infected bean plant. In contrast, the Robust trait is inherited as a single recessive gene, and it confers immunity to graft-inoculated as well as to mechanically inoculated BCMV. A number of explanations for the immunity of Robust are conceivable. One that is consistent with the control by a recessive gene and the graft immunity, and that is amenable to tests, is that the immune Robust plants lack a "receptor site" for the virus that susceptible plants have. BCMV introduced into the vascular system (e.g., by grafting) cannot enter the cells because they lack the receptor site. In the heterozygous state (Robust crossed to a susceptible bean) receptor sites would be present.

IMMUNITY OF COWPEA SEEDLINGS TO CPMV CONTRASTED
WITH SUSCEPTIBILITY OF PROTOPLASTS

Cowpea mosaic virus (CPMV) is one of several plant viruses that produce, in infected tissue, two kinds of ribonucleoprotein particles, each with a distinct strand of RNA (see review, Bruening, 1977). Both kinds of particles are necessary in order to induce an infection. Most strains of CPMV can be classified as belonging either to a "yellow" or a "severe" subgroup on the basis of symptoms induced on cowpeas (*Vigna sinensis*) and/or serological relationships. The SB strain (CPMV-SB) is considered to be the type member of the comovirus group (Harrison et al., 1971) of plant viruses; it is a member of the yellow subgroup of CPMV. CPMVs of the yellow subgroup generally are of Old World origin (J. P. Fulton, University of Arkansas, personal communication), whereas CPMVs of the severe subgroup are of New World origin. The DG strain (CPMV-DG) was obtained by multiple local lesion transfer from infected tissue donated by N. Vakili (Mayaguez, Puerto Rico). It shares many chemical characteristics with CPMV-SB, including similar sizes of capsid proteins and RNAs. CPMV-DG shows serological cross-reaction with CPMV-SB, but is more closely related to other severe CPMVs than to CPMV-SB.

After inoculation, the Blackeye 5 line of cowpeas becomes systemically infected by CPMV-SB. A chance observation that seedlings of the Black cowpea line did not develop symptoms after inoculation with CPMV-SB led to a survey of other cowpea lines. Blackeye 5 seedlings consistently become infected following inoculation with CPMV-SB at $1 \mu g/ml$. We (Beier et al., 1977) defined a criterion of operational immunity to CPMV-SB as follows: a line is classified as "immune" if at 7 to 12 days following inoculation with CPMV-SB at 250 $\mu g/ml$ no virus can be detected by transfer to Blackeye 5 seedlings or by immunodiffusion against antiserum to CPMV-SB. (Preliminary tests established that extracts of seedlings from several cowpea lines did not contain any detectable inhibitor of transmission of CPMV-SB to Blackeye 5 seedlings. If such an inhibitor had been present, transfer to Blackeye 5 might have been invalidated as a test for the presence of virus in the tissues of the other cowpeas.) The results of a survey of 1,031 cowpea lines are summarized in table 1. Sixty-five lines (6.3 percent of the total) were immune by our criteria. As expected, most of the lines supported the replication of CPMV-SB either locally or systematically.

Perhaps less expected were the results of a survey of susceptibility of protoplasts from the immune lines. We (Beier and Bruening, 1975) have developed a simplified procedure for isolating cowpea primary leaf protoplasts that omits the common and tedious step of epidermal stripping. Seed of sufficient quantity and quality was available from 55 of the 65 immune lines, and viable protoplasts were recovered from seedlings of all 55 lines. Protoplasts were inoculated with CPMV-SB at ~ 10 $\mu g/ml$

TABLE 1

SURVEY OF SUSCEPTIBILITY OF COWPEA SEEDLINGS AND LEAF PROTOPLASTS TO CPMV-SB

Class of Symptoms	Number of Lines	Percentage of Lines	Number Tested as Protoplasts	Recovered Infectivity, Local Lesions/Leaf
Sensitive: local lesions on inoculated primary leaves; virus was recovered from secondary leaves	846	82	1	840
Resistant: local lesions on primary leaves; virus was not recovered from secondary leaves	120	12	1	800
Immune: at 7 to 12 days after inoculation no symptoms were detected, no virus could be recovered	65	6	55 ⎰54 ⎱1	210 to 830 / 12

One thousand thirty-one lines of cowpeas were tested by inoculating seedlings, at the two-leaf stage, with CPMV-SB at 2.5 μg/ml in 50 mM potassium phosphate buffer, pH 7, using 320 mesh carborundum and cotton swabs to abrade the leaf upper epidermis. Seedlings from lines that appeared to be immune (no symptoms) to CPMV-SB at 2.5 μg/ml were tested at 250 μg/ml. Extracts of inoculated and secondary tissues were tested by transfer to Blackeye 5 cowpeas and by immunodiffusion 7 to 12 days after inoculation. Sixty-five lines were considered to be operationally immune to CPMV-SB because no virus could be detected under these circumstances. Protoplasts were isolated from primary leaves of seedlings of 55 of the immune lines and were inoculated with CPMV-SB. The protoplasts were incubated for 24 hr and were washed, disrupted, and assayed for virus by local lesion assay as described (Beier and Bruening, 1975). Data from Beier et al., 1977.

and the yield of virus was estimated by local lesion assay of protoplasts extracts after 24 hours of incubation. Protoplasts from 54 of the immune lines yielded an amount of virus that was within a factor of four of the amount recovered from Blackeye 5 protoplasts (840 local lesions per leaf, corrected for dilution of the extract; table 1). Protoplasts from a local lesion host, Chinese red × Iron cowpea, supported virus replication as well (800 local lesions per leaf) as Blackeye 5 protoplasts. However, protoplasts from one of the immune lines, Arlington, supported CPMV-SB increase, or at least retention, to less than 2 percent of the level for Blackeye 5 protoplasts (also incubated 24 hours). Thus, among the lines for which the seedlings are immune, susceptibility, rather than immunity or resistance, of the protoplasts to CPMV-SB was the common condition.

Although our survey of protoplasts from cowpea lines probably represents the most extensive such survey of lines from a single species, the rarity of resistance to plant virus infection by protoplasts could have been anticipated from the existence of other systems in which protoplasts from plants that are poor or restrictive hosts for a particular virus have proved to be fully susceptible to it (Takebe, 1975). That is, the host range of plant viruses seems to be wider among protoplasts than among the corresponding intact plants. For example, although CPMV apparently multiplies in Samsun tobacco leaves only to a very limited extent after inoculation at a very high concentration (1 mg/ml), more than 80 percent of Samsun tobacco mesophyll protoplasts can be infected with CPMV. The increase of CPMV in the tobacco protoplasts was delayed by comparison with the increase of tobacco mosaic virus (TMV) in tobacco protoplasts (or the increase of CPMV in cowpea protoplasts), requiring about twice as long as TMV to complete the phase of rapid increase. The yields of CPMV from infected cowpea and tobacco protoplasts were comparable (Huber et al., 1977). Similarly, protoplasts from Samsun NN tobacco were as susceptible as, and supported TMV increase with similar kinetics as, Samson tobacco (Otsuki et al., 1972). Intact Samsun tobacco is a systemic host for TMV, whereas Samsun NN tobacco gives a hypersensitive reaction, forming necrotic local lesions. That is, the extent of replication and movement of TMV in intact Samsun NN is severely restricted. TMV infection did not cause cell death in either the Samsun or the Samsun NN protoplasts.

CHARACTERISTICS OF IMMUNE LINES THAT ARE SOURCES
OF SUSCEPTIBLE PROTOPLASTS

Here we briefly review previously reported experiments (Beier et al., 1977, 1978) in which several lines of cowpeas were compared as seedlings and as protoplasts (after inoculation by several routes) for their response to CPMV. The lines (and their United States Department of Agriculture Plant Introduction numbers) include Brabham (145198), Groit (293514), Guarentana (142779) and Black (no P.I. number), all of which are immune as seedlings and fully susceptible as protoplasts, plus Chinese red × Iron (194207), a hypersensitive local lesion host, and the systemic host Blackeye 5 (293458).

The immunity of Brabham, Groit, Guarentana, and Black seedlings to high concentrations of mechanically inoculated CPMV-SB was detected at 27C. In several other virus-host systems, incubation at a constant elevated temperature, brief exposure to high temperture (heat shock), or treatment with actinomycin D has been shown to influence virus accumulation and spread and symptom development. In the various systems, local infections have been caused to become systemic, local lesions have been induced where either no infection or systemic infection would otherwise have been observed, or local lesion formation has been inhibited (Foster and Ross, 1968; papers reviewed by Loebenstein, 1972). Although virus could not be detected in the secondary tissue of Chinese red × Iron cowpeas that had been inoculated on the primary leaves with CPMV-SB and incubated at 27C, when they were incubated at 34°C (with or without injection of actinomycin D into the midveins of the inoculated primary leaves) or when the leaves and upper stem were dipped in water at 50C for 50 seconds at 6 hours after inoculation, virus was recovered from the secondary tissue at later times. No virus was detected in either the primary or secondary tissue of Brabham, Groit, Guarentana or Black seedlings that were similarly treated, nor were any virus-associated symptoms observed. These results do not provide support for the hypothesis that the immunity exhibited by seedlings of the latter four lines might be due to "invisible hypersensitivity" (A. Siegel, Wayne State University, personal communication), similar to the visible hypersensitivity of Chinese red × Iron. That is, the immunity does not seem to be an exaggerated hypersensitive reaction limiting the "local lesions" to undetectable proportions. All of these lines became systematically infected at 27C following mechanical inoculation of seedlings with CPMV-DG. J. P. Fulton (personal communication) has observed that CPMVs of the severe symptom subgroup generally infect most lines of cowpeas.

Since, in some instances, plants that are immune to mechanical inoculation can be infected by grafting them to infected plants (e.g., Corbett Refugee beans, discussed above), seedlings of the various lines were grafted to Blackeye 5 seedlings that subsequently were inoculated. CPMV-DG was readily transmitted across such grafts, and CPMV-SB was readily transmitted from Blackeye 5 to Blackeye 5. Transmission from Blackeye 5 to Chinese red × Iron was inconsistently observed; when it was observed, virus accumulated in the Chinese red × Iron secondary tissue (local lesion assay of extracts) to a level that was less than 1 percent of the amount that accumulated in graft-inoculated Blackeye 5 tissue. CPMV-SB was sporadically and weakly detected (varying from no cases with Brabham to a few local lesions in half the cases with Groit) after graft inoculation to the four immune lines. Slices from the secondary leaves of the graft-inoculated plants were incubated with [^{32}P]-inorganic phosphate in a test for possible replication of CPMV-SB in that tissue. The tissue was homogenized with unlabeled carrier CPMV-SB and virus was recovered. Radioactivity was at the control level with the possible exception of one out of three experiments with Black and two out of three experiments with Groit.

Inoculation of protoplasts was attempted by both *in vivo* and *in vitro* modes. In the *in vivo* mode, intact primary leaves were inoculated with CPMV-SB and at various times protoplasts were isolated. The protoplast suspension was split into two portions. From one portion extract was prepared and assayed immediately on Chinese red × Iron. The other portion was incubated for 40 hours before local lesion assay. A steady increase in protoplast-associated virus from Blackeye 5 was observed over the period 10 to 45 hours (between inoculation of the leaves and isolation of the protoplasts). As expected, the increase in protoplast-associated virus upon incubation of the isolated protoplasts was most marked at the early times. The ultimate yield of virus per protoplast (protoplasts isolated about 90 hours after inoculation and incubated for an additional 40 hours) was comparable to that observed from *in vitro* inoculated protoplasts. Similar results were obtained with Chinese red × Iron, except that the amount of recovered virus declined after about 50 hours after inoculation of the leaves, perhaps reflecting a hypersensitive reaction in the leaf. The yield from Black and Groit was less than 1 percent of the yield from Blackeye 5. No virus was recovered from Brabham or Guarentana protoplasts from inoculated leaves.

Growth curves for CPMV-SB in *in vitro* inoculated protoplasts are presented in figure 1. The curves for Blackeye 5 and Chinese red × Iron (not shown) protoplasts were practically superimposed. This is in agreement with the previously discussed results of Otsuki et al. (1972) for TMV in protoplasts of Samsun and Samsun NN tobacco. That is, the hypersensitive reaction was not expressed in the *in vitro* inoculated protoplasts. The curves (fig. 1) for Brabham, Groit, and Blackeye 5 were similar. However, the increases of CPMV-SB in Guarentana and Black were reproducibly and significantly delayed compared to the phase of rapid increase in Blackeye 5. The delay cannot be essential to immunity, since it is not exhibited by all the immune lines. However, the curves for CPMV-DG in Black and Blackeye 5 protoplasts were comparable, implying that the delay may contribute to the immunity of Black to CPMV-SB.

IMMUNITY OF ARLINGTON COWPEAS TO CPMV-SB

In our initial survey (Beier et al., 1977) of cowpea lines, Arlington proved to be unique. Among the 55 lines that were tested for susceptibility to CPMV-SB as protoplasts, only Arlington protoplasts showed significant resistance to CPMV-SB when assayed 24 hours after inoculation (last line of table 1). Whether the small amount of observed infectivity associated with the protoplasts was residual inoculum or whether there was a contribution from virus that had replicated in the protoplasts cannot be decided with assurance. However, in the period from 15 to 45 hours after inoculation the infectious titer increased (fig. 2A), though to a level that was less than 1 percent of the level achieved in Blackeye 5 protoplasts incubated for the same time (Beier et al., 1978). It may be argued that the small increase was the result of some sort of activation process increasing the infectivity of virions from the original inoculum. However, the protoplast-associated *virions*, as opposed to infectious *virus*, also were assayed by serologically-specific electron microscopy. The growth curves determined by the two assays were similar in shape. These results support

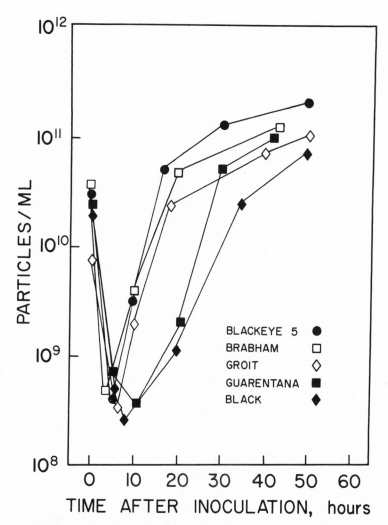

Fig. 1. Virus associated with cowpea leaf protoplasts inoculated with CPMV-SB. Freshly isolated protoplasts were inoculated with virus at 10 μg/ml in pH 5.6 potassium citrate buffer as described (Beier et al., 1977). Aliquots of the protoplast suspension were taken for preparation of extracts and local lesion assay at the times indicated. Local lesion counts were translated into virion particle counts by comparison with local lesion counts induced by inoculated serial dilutions of a purified CPMV-SB solution. The ordinate corresponds to $\sim 2 \times 10^6$ protoplasts per ml. The growth curve for CPMV-SB in Chinese red $\times$ Iron cowpea protoplasts was superimposable with the curve for Blackeye 5 protoplasts in this figure. Data from Beier et al., 1978.

the notion that the synthesis of new virions had occured and that activation alone does not account for the observed increase in infectivity. In the presence of cycloheximide the titer of protoplast-associated infectivity declined steadily after the inoculation of Blackeye 5 protoplasts with CPMV-SB, with no indication of activation at late times (fig. 2B). Cycloheximide, an inhibitor of protein synthesis on cytoplasmic 80S ribosomes, inhibits the synthesis of CPMV-SB capsid protein in infected leaf slices at concentrations that fail to inhibit protein synthesis on chloroplast ribosomes (Owens and Bruening, 1975). It has been shown to inhibit the replication of CPMV in cowpea protopasts (Hibi et al., 1975).

The assays for both infectious virus (on Chinese red × Iron cowpeas) and serologically recognized virions were related, on a per particle basis, to the same preparation of purified CPMV-SB (fig. 2A). It is possible that the higher particle concentration determined by electron microscopy is an indication that virions of reduced specific activity were formed in the Arlington protoplasts. Conceivably, the progeny virus from Arlington protoplasts could have been variants of CPMV-SB selected for the ability to replicate in Arlington. To test for such a variant, Arlington seedlings were inoculated with extracts of the inoculated Arlington protoplasts. No virus was detected, even after serial transfer. Therefore, we have no evidence of such a selection process.

Results in figures 2C and 2D tend to rule out two possible explanations for the low recovery of CPMV-SB from inoculated Arlington protoplasts. A partial growth curve for CPMV-DG in Arlington protoplasts was parallel to a curve for CPMV-DG in Blackeye 5 protoplasts. Therefore, a physiological infirmity, or other general obstruction to virus replication, cannot be the explanation for the poor replication of CPMV-SB in Arlington protoplasts. A postulated defect in whatever process leads to the exposure of virus RNA also fails to account for the resistance of Arlington protoplasts. Isolated CPMV-SB RNA readily infected Blackeye 5 protoplasts, but was no more effective than CPMV-SB virions in infecting Arlington protoplasts (fig. 2D).

Alternative inoculation and assay routines were tested on Arlington seedlings (Beier et al., 1978). We were not able to recover virus from protoplasts (assayed directly or after a 40-hour incubation) derived from Arlington primary leaves that had been inoculated with 100 μg/ml CPMV-SB. In grafting experiments, virus was detected (as just four local lesions upon transfer to Chinese red × Iron) in the Arlington secondary leaves of only one out of eight plants that had been grafted to Blackeye 5 seedlings (that subsequently were inoculated with CPMV-SB). Radiophosphorus incorporation into virus in leaf slices from the grafted Arlington plants was at the background level. Therefore, Arlington seedlings seem to be highly

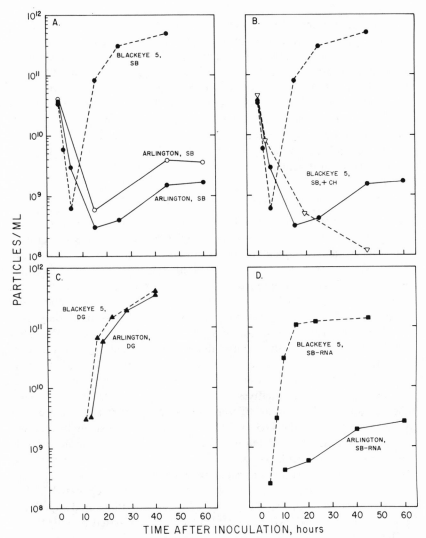

Fig. 2. Virus increase in Arlington and Blackeye 5 cowpea protoplasts inoculated with CPMV or CPMV RNA. Viruses were inoculated in pH 5.6 potassium citrate buffer and viral RNAs in pH 6.0 potassium phosphate buffer. Particle counts were obtained by serologically specific electron microscopy (Beier and Shepherd, 1978) in one case (0) but otherwise by means of local lesions assays. Data points from experiments with Arlington protoplasts are connected by solid lines, whereas dashed lines connect points from experiments with Blackeye 5 protoplasts. In one experiment (panel B, ▽), cycloheximide was included in the protoplast incubation medium at a concentration of 10 μg/ml. The curves for Blackeye 5 and Arlington protoplasts (data from local lesion assays) are reproduced in panel B from panel A. (From Beier et al., 1978; reprinted with permission.)

resistant or immune to graft inoculation with CPMV-SB. (CPMV-DG was readily graft-transmitted from Blackeye 5 to Arlington.) Inoculated Arlington seedlings were incubated at 34C with and without actinomycin D treatment; others were dipped in water at 50C for 50 seconds at 6 hours or 24 hours after inoculation. In none of these experiments was virus detected by local lesion assay of extracts. Therefore, we obtained no evidence for temperature sensitivity of Arlington immunity to CPMV-SB. Preliminary data from genetic crosses indicate that the immunity of Arlington and the immunity of some other lines are controlled by genes that reside on different chromosomes (Beier et al., 1978).

INTERACTION BETWEEN CPMV-SB AND CPMV-DG IN ARLINGTON

The immunity to infection by CPMV-SB that Arlington cowpeas exhibit has proved to be especially general: mechanical (with and without heat treatment) and graft inoculation of seedlings failed to overcome it, and the protoplasts of Arlington proved to be highly resistant to infection by either virions or RNA. An important question about Arlington immunity concerns the preparedness of Arlington cells. Do they stand able, constantly, to prevent extensive replication of CPMV-SB or do they react to incoming virus particles, or to a product associated with CPMV-SB replication, by triggering some defense mechanism? This question probably cannot be answered with certainty until information about molecular level events following inoculation have been obtained. Nevertheless, the alternatives of preexisting and triggered defense mechanisms could lead to different responses to mixed inoculations of CPMV-SB with CPMV-DG. If CPMV-SB replication is strongly inhibited by a preexisting, static defense, CPMV-DG replication should be little influenced by co-inoculation with CPMV-SB (because CPMV-DG would encounter the preexisting, static defense when it is inoculated alone, under conditions already known to be suitable for its replication.) The (extremely) contrasting situation would be a defense that could impede the replication of CPMV-SB and CPMV-DG but is initiated by CPMV-SB only.

Local lesions developed after primary leaves of Arlington seedlings were inoculated with CPMV-DG. They also formed when mixtures of CPMV-DG and CPMV-SB were inoculated. However, the number of local lesions declined monotonically with increasing concentration of CPMV-SB in the inoculum, at constant CPMV-DG concentration (table 2). As a control, CPMV-DG and the top component of CPMV-SB were co-inoculated. The top component of CPMV-SB seems to be an empty capsid: it has the same composition of polypeptides as the ribonucleoprotein components but

TABLE 2

MIXED INOCULATION OF ARLINGTON COWPEA SEEDLINGS

Inoculum		Local Lesions/Leaf
CPMV-DG	CPMV-SB	
7.5 μg/ml	. . . . . .	> 200
7.5 μg/ml	0.75 μg/ml	> 200
7.5 μg/ml	7.5 μg/ml	68
7.5 μg/ml	12.5 μg/ml	36
7.5 μg/ml	47 μg/ml	16
7.5 μg/ml	75 μg/ml	4
7.5 μg/ml	0.5 μg/ml (top)	> 200
7.5 μg/ml	10 μg/ml (top)	> 200
7.5 μg/ml	100 μg/ml (top)	> 200
7.5 μg/ml	50 μg/ml (RNA)	101
. . . . . .	75 μg/ml	0
. . . . . .	50 μg/ml (RNA)	0

Inocula were in 100 mM potassium phosphate, pH 7. Local lesions were counted four days after inoculation. "Top" refers to the top component (RNA-free capsids) of CPMV-SB.

lacks RNA (as judged by lack of labeling by ^{32}P). If the interference of CPMV-SB with local lesion induction by CPMV-DG were due to competition for putative adsorption sites on the leaf, top component ought also to interfere with CPMV-DG. It did not (table 2). CPMV-SB RNA did interfere. As expected, neither CPMV-SB nor its RNA induced local lesions on Arlington seedlings in these experiments.

The results described in the previous paragraph, when considered together with the evidence for a low level of replication of CPMV-SB in Arlington protoplasts, are consistent with a defense against extensive CPMV replication being activated by a low (not detectable in intact leaves) level of replication of CPMV-SB in Arlington cells. Interference (also called "cross protection") between different plant viruses (that both replicate well in a particular host) is a common, but not universally observed, reaction to co-inoculation of plant viruses. Interference phenomena have been observed between both similar and (less commonly) dissimilar viruses and seem to require infectious virus; inactivated virions will not substitute (reviewed by Gibbs and Harrison, 1976, pp. 153–56). In the case of tobacco mosaic virus (TMV), even some defective strains incapable of inducing the synthesis of active capsid protein can reduce the yield of common TMV inoculated several days after the defective strain (Zaitlin, 1976). Therefore, it is conceivable that the obstruction of CPMV-DG replication by co-inoculation with (and subsequent limited replication of) CPMV-SB is a cross-protection phenomenon not related to the immunity of Arlington to CPMV-SB. We have not been able to eliminate this latter (not necessarily mutually exclusive) alternative. However, CPMV-DG and CPMV-SB failed to show interference when co-inoculated on Chinese red × Iron cowpeas. The two viruses form very distinct lesions on this host (pinpoint necrotic local lesions by CPMV-SB and chlorotic local lesions by CPMV-DG). The numbers of local lesions induced by CPMV-SB and CPMV-DG inoculated separately and then together on Chinese red × Iron were additive (data not presented). Therefore, the activation of a defense mechanism in Arlington by CPMV-SB remains a favored hypothesis.

A triggered defense mechanism might be expected to require both viruses to be present in the same cell for inhibition of CPMV-DG replication to occur. Some Arlington seedlings were inoculated with 2.5 μg/ml CPMV-DG and 25 μg/ml CPMV-SB in the same solution. Others were inoculated with CPMV-SB (25 μg/ml) and at a later time with CPMV-DG (2.5 μg/ml). As expected, only an average of 2 lesions per leaf were observed after the simultaneous inoculation. An average of 20 lesions per leaf were observed for an interval of 5 to 10 minutes between inoculations and an average of about 100 lesions per leaf for inoculation intervals of 4, 8, and 24 hours or for inoculation of CPMV-DG without previous inoculation of CPMV-SB. Since the majority of susceptible sites presumably induced on a leaf by the process of mechanical inoculation become refractory with an interval ranging from seconds to several minutes, depending on the host (Jedlinski, 1956; Furumoto and Wildman, 1963) the probability of two inocula being introduced into the same cell in consecutive inoculations ought to decline with increasing time intervals between inoculation. The absence of an interference at long intervals between inoculations of the two viruses may reflect a requirement for both to be in the same cell. Cross protection, in contrast, can be observed in experiments with inoculation intervals of days, even with a defective virus as the interfering agent (Zaitlin, 1976).

Introducing both viruses into the same cell would be expected to be the norm during the inoculation of protoplasts, *in vitro*, with a mixture of CPMV-SB and CPMV-DG. Virtually no CPMV-SB (as detected by formation of necrotic local lesions on Chinese red × Iron seedlings inoculated with extracts of the protoplasts) was recovered from the Arlington protoplasts used in these particular experiments (table 3). Nevertheless, a significant reduction in the yield of CPMV-DG was observed when CPMV-SB was present in the inoculum. Therefore, the interference between the CPMVs has been found to hold in protoplasts of Arlington as well as intact leaves. Tobacco mesophyll protoplasts have been co-infected with several pairs of dissimilar viruses (Otsuki and Takebe, 1976; Barker and Harrison, 1977). Neither antagonism nor synergism of infection was observed in these systems. However, two strains of TMV did show some interference in tobacco protoplasts. The strain-

TABLE 3

MIXED INOCULATION OF ARLINGTON COWPEA PROTOPLASTS

Inoculum	Virus Recovered, Local Lesions on Arlington	Virus Recovered, Necrotic Local Lesions on Chinese Red × Iron
5.6 μg/ml DG	138	0
11.2 μg/ml SB	0	1
5.6 μg/ml DG + 11.2 μg/ml SB	27	0
5.6 μg/ml DG + 28 μg/ml SB	21	0

Arlington protoplasts from 7-day-old seedlings were incubated for 48 hr after inoculation of the virus concentrations indicated, according to the procedure of Beier and Bruening, 1976. Arlington and Chinese red × Iron cowpea were inoculated with diluted extracts of the protoplasts, and local lesions were counted 5 days later.

inoculated second (or co-inoculated at a lower concentration) showed reduced yield (Takebe, 1975).

The two CPMVs and Arlington cowpeas form a system that is closely analogous to a system consisting of TMV strains U1, VM, and pinto bean. TMV-U1 (the common strain of TMV) induces necrotic local lesions on pinto bean, whereas VM fails to cause any observable symptoms, even after heat shock treatment (Wu, 1963) of the inoculated leaves. However, VM decreased local lesion formation by U1 (Wu et al., 1962), reducing the local lesion counts in a manner parallel to that seen in table 2 (for corresponding TMV-VM/TMV-U1 and CPMV-SB/CPMV-DG ratios). A related system is TMV-U2 and TMV-U1 on pinto bean. TMV-U2 fails to induce visible local lesions on pinto bean under most ordinary conditions for culturing pinto bean (Rappaport and Wu, 1963; Helms and McIntyre, 1972), but local lesions can be induced by heat shock (Wu, 1963). TMV-U2 interferes with TMV-U1 local lesion formation on pinto bean (no heat shock) in a fashion parallel to the action of TMV-VM (Wu et al., 1962). TMV-U2 partially inactivated by ultraviolet irradiation also reduced the number of TMV-U1 local lesions formed after mixed inoculation. However, the irradiated TMV-U2 was effective only to the extent that its infectivity (local lesion assay on *Nicotiana glutinosa*) had survived (Wu and Rappaport, 1961). TMV-U2 capsid protein was ineffective in reducing TMV-U1 local lesion counts. Apparently some degree of TMV-U2 (and, presumably, TMV-VM) replication is necessary before an intervention in the replication of TMV-U1 is observed. The analogy between the CPMV and TMV systems is incomplete in the following respect: TMV-U2, TMV-VM, and TMV-U1 interfere with each other in hosts on which both members of the pair being tested produce symptoms (Siegel, 1959; Wu et al., 1962); CPMV-SB and CPMV-DG apparently replicated independently, without interference, in Chinese red × Iron cowpeas, on which both viruses induce local lesions.

As was previously discussed, the host range of plant viruses seems to be wider among protoplasts than among the corresponding intact plants. The susceptibility of pinto bean protoplasts to the TMV strains discussed in the previous paragraph is unknown. However, some tomato lines and TMV strains comprise a system analogous to cowpeas, CPMV-SB, and CPMV-DG in that both fully susceptible and highly resistant or immune plants are available and the strong resistance is expressed, in one instance, by protoplasts (Motoyoshi and Oshima, 1977). Tomato lines homozygous for gene Tm-1 have been reported to be highly resistant to a tomato strain of TMV, TMV-L, allowing only low levels of virus to accumulate, Motoyoshi and Oshima (1977) detected no virus in extracts in TMV-L inoculated leaf discs by a local lesion assay that gave $\sim$ 1,300 local lesions from leaf discs inoculated with TMV-CH2 and similarly incubated. TMV-CH2 is a tomato TMV that was isolated from a tomato plant bearing the Tm-1 gene. Protoplasts from tomatoes homozygous for Tm-1 were inoculated with TMV-L and, in separate experiments, with TMV-CH2. The amounts of recovered virus (local lesion assay) declined continuously in the TMV-L inoculated protoplasts, with no virus recovered at 66 hours after inoculation. There was no indication of a delayed, small increase as is seen in CPMV-SB inoculated Arlington protoplasts. In contrast, virus from TMV-CH2 inoculated protoplasts increased continuously over the same period. Both viruses increased in protoplasts and leaf discs from susceptible tomatoes. Thus, although resistance of the protoplasts to a plant virus seems to be a rare occurrence among lines from plant species generally considered to be natural hosts for the virus, at least two instances have now been reported.

SUMMARY AND CONCLUSIONS

An operational definition of immunity (no virus recovered after inoculation of seedlings with a concentration of virus greater than 100 times that which uniformly infected susceptible seedlings) has been applied to lines of cowpeas and CPMV-SB. Immunity to CPMV-SB was found, being exhibited by about 6 percent of the lines. Among 55 immune (as seedlings) lines, 54 were sources of fully susceptible protoplasts.

In further tests, the immunity of lines Brabham, Groit, Guarentana, and Black (all of which are sources of susceptible protoplasts) was not significantly overcome by stress due to elevated temperature or by graft inoculation. The CPMV-SB inoculated on primary leaves of these lines apparently did not reach the cytoplasm of mesophyll cells in a sufficient concentration to initiate infections; very little or no virus was recovered from protoplasts isolated from inoculated leaves. Incubation of these protoplasts, under conditions that would have resulted in virus increase had the protoplasts been inoculated *in vitro*, did not result in the accumulation of virus to more than a few percent of the levels achieved in *in vitro* inoculated protoplasts. In these tests Brabham and Guarentana seemed to be more resistant than Groit and Black, although the results were not conclusive. However, intact seedlings of all four lines proved to be very refractory to infection by CPMV-SB under a variety of conditions. When protoplasts of the four lines were inoculated *in vitro*, the phase of rapid virus increase was delayed in Guarentana and Black, compared to Brabham and Groit. The delay phenomenon may contribute to immunity, but it apparently is not essential for immunity of the intact seedling. One hypothesis that would explain the results summarized above, but for which there is no direct evidence, is that there exists in the intercellular spaces or on the surfaces of cells, of lines that are immune as seedlings but susceptible as protoplasts, a substance that adsorbs and/or inactivates CPMV-SB but not CPMV-DG.

The protoplasts of one line, Arlington, proved to be highly resistant to CPMV-SB, but not immune. The virus increased in Arlington protoplasts to only about 1 percent of the level achieved in protoplasts from the susceptible line Blackeye 5, and only after a long delay. The resistance of Arlington protoplasts was demonstrated *not* to be due to a physiological incapability of the protoplasts to support the replication of any virus. Since RNA of CPMV-SB failed to overcome the resistance of Arlington protoplasts, the site of the resistance does not appear to be in an uncoating step. Although CPMV-SB did not detectably replicate in Arlington seedlings, under a variety of inoculation conditions, and replicated only poorly in Arlington protoplasts, it did interfere in the replication of CPMV-DG when the two viruses were co-inoculated on either seedlings or protoplasts. The interference was observed when CPMV-SB RNA (but not when empty capsids) was substituted for the virions. Therefore, it is likely that some degree of virus replication is essential to the interference phenomenon and that CPMV-SB replicates to an otherwise undetected level in the inoculated leaves of Arlington seedlings. Among the possible explanations for the interference are that: (1) the interference is analogous to the cross-protection phenomena frequently observed in systems in which two similar viruses interfere in a plant that is a good host for either virus alone, and thus the interference of CPMV-SB with CPMV-DG replication is not directly related to the immunity of Arlington; (2) a low level of CPMV-SB replication in Arlington activates a mechanism that restricts further replication of either CPMV-SB or CPMV-DG; CPMV-DG alone does not activate the mechanism. For neither explanation is there direct support. The second explanation is favored by the apparently independent (i.e., no cross protection) replication of CPMV-SB and CPMV-DG in Chinese red × Iron seedlings.

ACKNOWLEDGMENTS

The research work from this laboratory that is described in this chapter was supported by the Agricultural Experiment Station of the University of California, National Science Foundation Grant No. BMS73-06783, and National Institutes of Health Grant No. AI-13708. We thank Milton Zaitlin and Albert Siegel for helpful discussion and comments.

LITERATURE CITED

Barker, H., and B. D. Harrison. 1977. Infection of tobacco mesophyll protoplasts with raspberry ringspot virus alone and together with tobacco rattle virus. J. Gen. Virol. 35:125–33.

Beier, H., and G. Bruening. 1975. The use of an abrasive in the isolation of cowpea leaf protoplasts which support the multiplication of cowpea mosaic virus. Virology 64:272–76.

Beier, H., and G. Bruening. 1976. Factors influencing the infection of cowpea protoplasts by cowpea mosaic virus RNA. Virology 72:363–69.

Beier, H., G. Bruening, M. L. Russell, and C. L.Tucker. 1978. Replication of cowpea mosaic virus in protoplasts from immune plants. Virology (submitted).

Beier, H., and R. J. Shepherd. 1978. Serologically specific electron microscopy in the quantitative measurement of two isometric viruses. Phytopathology 68:533–38.

Beier, H., D. J. Siler, M. L. Russell, and G. Bruening. 1977. Survey of susceptibility to cowpea mosaic virus among protoplasts and intact plants from *Vigna sinensis* lines. Phytopathology 67:917–21.

Bruening, G. 1977. Plant covirus systems. Two component systems. *In* H. Fraenkel-Conrat and R. R. Wagner (eds.), Comprehensive virology. Pp. 55–141. Plenum Press, New York.

Chamberlin, M. 1974. Isolation and characterization of prototrophic mutants of *Escherichia coli* unable to support the intracellular growth of T7. J. Virol. 14:509–16.

Foster, J. A., and A. F. Ross. 1968. Local lesion development in Turkish tobacco leaves heated after inoculation with tobacco mosaic virus. Phytopathology 68:1050.

Fuchs, J. 1977. Isolation of an *Escherichia coli* mutant deficient in thioredoxin reductase. J. Bacteriol. 129:967–72.

Furumoto, W.A., and S. G. Wildman. 1963. Studies on the mode of attachment of tobacco mosaic virus. Virology 20:45–52.

Gibbs, A., and B. D. Harrison. 1976. Plant virology, the principles. John Wiley & Sons, New York.

Harrison, B.D., J. T. Finch, A. J. Gibbs, M. Hollings, R. J. Shepherd, V. Valenta, and C. Wetter. 1971. Sixteen groups of plant viruses. Virology 45:356–63..

Hausmann, R. 1976. Bacteriophage T7 genetics. Curr. Topics Microbiol. Immunol. 75:77–110.

Helms., K., and G. McIntyre. 1962. Studies on size of lesions of tobacco mosaic virus on pinto bean. Virology 18:535-45.

Hibi, T., G. Rezelman, and A. van Kammen. 1975. Infection of cowpea mesophyll protoplasts with cowpea mosaic virus. Virology 64:308-18.

Holmgren, A. 1976. Hydogen donor system for *Escherichia coli* ribonucleoside diphosphate reductase dependent upon glutathione. Proc. Nat. Acad. Sci. USA 73:2275-79.

Huber. R., G. Rezelman, T. Hibi, and A. van Kammen. 1977. Cowpea mosaic virus infection of protoplasts from Samsun tobacco leaves. J. Gen. Virol. 34:315-23.

Jedlinski, H. 1956. Plant virus infection in relation to the interval between wounding and inoculation. Phytopathology 46:673-76.

Levy, A., G. Loebenstein, M. Smookler, and T. Drori. 1974. Partial suppression by UV irradiation of the mechanism of resistance to cucumber mosaic virus in a resistant cucumber cultivar. Virology 60:37-44.

Loebenstein, G. 1972. Localization and induced resistance in virus-infected plants. Ann. Rev. Phytopath. 10:177-206.

Mark, D. F., and C. D. Richardson. 1976. *Escherichia coli* thioredoxin: A subunit of bacteriophage T7 DNA polymerase. Proc. Nat. Acad Sci. USA 73:780-84.

Modrich, P., and C. C. Richardson. 1975a. Bacteriophage T7 deoxyribonucleic acid replication *in vitro*. A protein of *Esherichia coli* required for bacteriophage T7 DNA polymerase activity. J. Biol. Chem. 250:5508-14.

Modrich, P., and C. C. Richardson. 1975b. Bacteriophage T7 deoxyribonucleic acid replication *in vitro*. Bacteriophage T7 DNA polymerase: An enzyme composed of phage- and host-specified subunits. J. Biol. Chem. 250:5515-22.

Moehring, J. M., and T. J. Moehring. 1976. The spectrum of virus resistance of a KB cell strain resistant to diphtheria toxin. Virology 69:786-88.

Motoyoshi, F., and N. Oshima. 1977. Expression of genetically controlled resistance to tobacco mosaic virus infection in isolated tomato leaf mesophyll protoplasts. J. Gen. Virol. 34:499-506.

Nachman, I., G. Loebenstein, T. van Praagh, and A. Zelcer. 1971. Increased multiplication of cucumber mosaic virus in a resistant cucumber caused by actinomycin D. Physiol. Plant Pathol. 1:67-72.

Otsuki, Y., T. Shimomura, and I. Takebe. 1972. Tobacco mosaic virus multiplication and expression of the N gene in necrotic responding tobacco varieties. Virology 50:45-50.

Otsuki, Y., and I. Takebe. 1976. Double infection of isolated tobacco mesophyll protoplasts by unrelated plant viruses. J. Gen. Virol. 30:309-16.

Owens, R. A., and G. Bruening. 1975. The pattern of amino acid incorporation into two cowpea mosaic virus proteins in the presence of ribosome-specific protein synthesis inhibitors. Virology 64:520-30.

Rappaport, I., and J.-H. Wu. 1963. Activation of latent virus infection by heat. Virology 20:472-76.

Schafer, J. F. 1971. Tolerance to plant disease. Ann. Rev. Phytopath. 9:235-52.

Siegel, A. 1959. Mutual exclusion of strains of tobacco mosaic virus. Virology 8:470-77.

Takebe, I. 1975. The use of protoplasts in plant virology. Ann. Rev. Phytopath. 13:105-25.

Wu, J.-H. 1963. Extension of the host range of tobacco mosaic virus by heat activation of latent infections. Nature 200:610-11.

Wu, J.-H., W. Hudson, and S. G. Wildman. 1962. A quantitative analysis of interference produced by related strains of TMV compared with non-infectious components. Phytopathology 52:1264–66.

Wu, J.-H., and I. Rappaport. 1961. An anlysis of the interference between two strains of tobacco mosaic virus on *Phaseolus vulgaris*. Virology 14:259–63.

Zaitlin, M. 1976. Viral cross protection: More understanding is needed. Phytopathology 66:382–83.

Zaumeyer, W. J., and J. P. Meiners. 1975. Disease resistance in beans. Ann. Rev. Phytopath. 13:313–34.

JEANNE BARNHILL JONES

Commercial Use of Tissue Culture for the Production of Disease-free Plants

23

University demonstration of plant propagation through tissue culture of many floricultural and horticultural crops (Murashige, 1974; Sagawa, 1976) has resulted in a 1970s boom of privately owned plant tissue culture laboratories. The California Association of Nurserymen lists twenty-three facilities involved in the rapid multiplication of plants and/or the recovery of plants free from known pathogens. Additional laboratories in Florida as well as other plant-producing states are being completed each year. Tissue culture propagation methods have become standard nursery operational procedures.

Those facilities using plant tissue culture for rapid multiplication produce plant material *in vitro* by initiation of adventitious shoots, bulbs, corms. tubers, asexual embryos, or by enhancing axillary shoot growth. During the cultural processes, one can easily detect most bacterial and fungal organisms, and those infested cultures are discarded without proving the pathogenicity of the contamination. Among the more-persistent contaminating organisms have been *Erwinia, Pseudomonas,* and *Bacillus* sp. (Knauss, 1977). A selection of plants being commercially propagated through tissue culture is shown in table 1.

Unfortunately, early detection of fungal and bacterial contamination in commercial tissue culture systems has not been always possible. Some cultures apparently clean in the beginning develop symptoms after one or more reculture periods. Symptoms can be a milky-white haze in the agar

Jeanne Barnhill Jones, Paul Ecke Poinsettias, P.O. Box 483, Encinitas, California 92024.

TABLE 1

COMMERCIALLY TISSUE-CULTURED PLANT GENERA

Araceae	Liliaceae
Anthurium andreanum	*Cordyline terminalis*
Dieffenbachia amoena 'Snow'	*Dracaena deremensis*
D. picta 'Perfection'	*D. godseffiana*
Monstera deliciosa	*D. goldieana*
Philodendron oxycardium	*D. marginata*
Scindapsus aureus	*Hemerocallis* sp.
Syngonium podophyllum	*Lilium* sp.
Araliaceae	Marantaceae
Tupidanthus calyptratus	*Maranta leuconeura*
Asteraceae	Moraceae
Chrysanthemum morifolium	*Ficus benjamina*
Gerbera jamesonii	*F. elastica* 'Decora'
Begoniaceae	Polypodiaceae
Begonia sp.	*Adiantum raddianum*
Bromeliaceae	*Asplenium nidus*
Aechmea fasciata	*Nephrolepis exaltata* 'Bostoniensis'
Cryptanthus sp.	*N. exaltata* 'Fluffy Ruffles'
Neoregelia sp.	*N. exaltata* 'Trevillian'
Caryophyllaceae	*N. exaltata* 'Whitmanii'
Dianthus caryophyllus	*N. exaltata* 'Rooseveltii'
Gypsophila sp.	*Platycerium* sp.
Dicksoniaceae	*Pteris argyrea*
Alsophila australis	*Woodwardia fimbriata*
Euphorbiaceae	Orchidaceae
Euphorbia pulcherrima	*Cattleya* sp.
Geraniaceae	*Cymbidium* sp.
Pelargonium × *hortorum*	*Phalaenopsis* sp.
Pelargonium peltatum	Rosaceae
Gesneriaceae	*Fragaria* sp.
Saintpaulia ionantha	Saxifragaceae
S. sp.	*Hydrangea macrophylla*
	Saxifraga sarmentosa 'Tricolor'
	Taxodiaceae
	Sequoia sempervirens

surrounding the explant, cloudy liquid medium that previously had been clear, and rapid decline in the vigor of the plant in culture. *Dieffenbachia picta* 'Perfection' is an example where methods have been developed to detect fungal and bacterial problems *in vitro* (Knauss, 1977). Growers that already have tissue culture facilities may find the method of maintaining mother block stock *in vitro* of ornamental tropical foliage plants, found to be free of known pathogens, advantageous when compared with currently employed methods. Many of the plants listed have a significant increase in commerical value because of a more commercially desirable form (increased vigor, more prolific branching, more shoots terminating in flowers, earlier flowering, denser foliage). These improved qualities may be due to the elimination of bacterial and fungal pathogens.

In contrast to facilities involved in the mass multiplication aspects of commercial plant tissue culture are those concerned with the recovery of plants free from known pathogens. Although several reviews (Hollings, 1965; Murashige and Jones, 1974) list many plants where tissue culture attempts have been successful in obtaining virus-free clones, the commercial nursery industry has seen only some usage and value from such research. Limitations may be due to some of the following:

1. University limitations on performing service work.

2. Lack of academic plant pathology involvement with plant introduction systems.

3. Inadequate training, education, and appreciation of nurserymen for the complexities of pathogen controls.

4. Industry disagreement on the value of achieving and sustaining the use of pathogen-free mother block materials.

5. Lack of industry support in funding necessary research.

6. Previous lack of private nursery–related laboratories staffed by trained personnel in plant tissue culture and plant pathological practices, capable of utilizing university developed techniques.

To commercially involve tissue culture for the production of plants free from known virus, viroid, and mycoplasma pathogens, a nursery must utilize tissue culture techniques as an integral part of its plant production. A company must recognize the need for, and have the desire to produce, a better quality of plant material than is presently available. Commercial examples are firms producing chrysanthemums (*Chrysanthemum morifolium*), carnations (*Dianthus caryophyllus*), orchids (*Phalaenopsis, Cymbidium, Cattleya* sp.), geraniums (*Pelargonium* × *hortorum*), lilies (*Lilium* sp.), hydrangeas (*Hydrangea macrophylla*), and strawberries (*Fragaria* sp.). Spreading of pathogens has occurred easily in these and other plants were conventional propagation techniques allow mechanical transmission in the taking of cuttings or the transmission by insect vetors. Brief descriptions of procedures used by our industry for carnation, chrysanthemum, and strawberry are as follows:

One tissue culture procedure being employed in the recovery of carnation plants free from some viruses involves the use of apical meristem and shoot apex (shoot apical meristem + 1–3 leaf primordia) explants. Shoot tips from infested plants (no heat treatment) are removed, surface disinfested with dilute bleach solution and rinsed in sterile water. After the aseptic removal of additional leaves and stem, the apical meristem (or shoot apex) is placed on a simple nutrient agar solution. Four to five weeks later, the explants are recultured one-for-one onto a multiplication medium for the production of, on the average, four good shoots that can be separated into individual shoots and again recultured. After small increases (up to twenty shoots) from one initial explant are achieved, individual shoots are placed on an agar rooting medium for 10–14 days and then transplanted into 3-inch pots containing a peat/perlite mix. Plants are placed in a polyethylene tent and kept misted for one week. After that time the tent is removed and plants are allowed to acclimate another week. During this time low levels of fertilizer are given. After 1–1 1/2 months in the greenhouse, plants are ready for replanting into larger 7-inch pots. When 6–8 inches tall, plants are checked for viruses.

Carnation virus problems are mottle, ringspot, latent, vein-mottle, and mosaic. Streak and etched-ring types are not checked. The indicator plant used in detecting mottle, vein-mottle, and ringspot is *Chenopodium amaranticolor*; for ringspot and latent, *C. quinoa*; and for mottle and mosaic, *Gomphrena globosa*. Streak and etched-ring types are not checked. Any of the nucleus block of plants derived from tissue cultures showing positive symptoms on indicator plants are discarded. A level of 75 percent and higher has been achieved in recovering carnations free from indexed pathogens using this procedure. Nucleus block plants are annually replaced with new tissue culture-derived plants, with those having positive indicator plant symptoms being discarded. Cuttings from the nucleus block are used for increasing numbers of plants that serve as the mother block. Constant roguing of infested plants in the nucleus block allows for the production of high-quality cuttings known to be free of certain viruses. Plants within the increase block are also checked for systemic fungi and bacteria. *Verticillium, Fusarium oxysporum, Phialophora* wilt, and bacterial wilt (*Pseudomonas* the principal causal agent) as well as *Rhizoctonia* and *Fusarium roseum* are tested for. Methods developed by Colorado State University (Nelson et al., 1960) are employed. Simply a plus or minus evaluation is given, and plants identified as pathogen-infected are eliminated from production.

The chrysanthemum industry uses tissue culture techniques to recover plants free from tomato aspermy, mosaic complexes, and stunt, chlorotic mottle, and a latent strain. Small shoot tips 0.1 mm or less in size are aseptically isolated from plants not subjected to any heat treatment. *Erwinia* and *Verticillium* detection along with the virus and viroid checks are made. Nonselective media are used for systemic fungal and bacterial problems, and indicator varieties are subjected to grafting techniques patterned after Dimock's Cornell program. Plants showing positive results on indicator plants or media tests are eliminated from foundation block plantings.

In contrast to using tissue culture in floricultural crops is the work being done commercially with strawberries. Shoot apical meristems 0.3mm or less in size are taken from plants subjected to 4–6 weeks of heat treatment 100–105°F). One plant is recovered from each surviving meristem. All derived plants are checked for viruses by specially developed indicators of *Fragaria virginiana* and *F. vesca*. Viruses being checked for are mild yellow edge, mottle, crinkle, necrotic shock, palladosis, latent A and C, fan roll, and witches broom. The mycoplasmic aster yellows are also a problem. Plants that have no symptoms on the indicator plants then undergo conventional propagation by stolon production. Every other winter a new set of meristem-derived plants is developed for the foundation mursery. Characteristics of virus-free plants interestingly make it more costly to the eventual berry producer to utilize virus-free material. Clean plants have more foliage, which makes it harder and more time consuming to pick the berries. However, since there is a synergistic effect with strawberry viruses, the likelihood of rapid plant decline from pathogens makes it necessary to use as clean a stock as possible for plant and berry production.

One floricultural crop for which tissue culture procedures are presently being refined is the florist hydrangea. *Hydrangea macrophylla* Thunb. as described by Shanks (1972) was one of the first flowering potted plants sold in chain stores on the spring market. Poor understanding on the part of the producer, distributor, and sales outlets, plus attempts to produce flowering plants more cheaply, have resulted in the presence on the retail market of plants of poor quality. The hydrangea is in need of being upgraded in the eyes of the retailer and the consumer as a high-quality, long-lasting potted plant. Efforts by Paul Ecke Poinsettias have centered on the production of uniform quality plants in the greenhouse, the introduction of new European varieties better adapted to the home, the recovery of known pathogen-free plants through the use of tissue culture, the multiplication of plants through tissue culture, and research procedures for extending the period of blooming plant availability to other than spring months.

Hydrangea macrophylla is susceptible to infection by several viruses: hydrangea ringspot virus (HRSV), tomato ringspot virus (TomRSV), tobacco ringspot virus (TobRSV), and cucumber mosaic virus (CMV) (Zeyen and Stienstra, 1973). Foliar ringspot symptoms appear as yellowish blotches on the lower leaves just before flowering or occasionally as reddish-brown or green rings on yellowish leaves. Recent research (Hearon et al., 1976) shows the association of a mycoplasmic organism (MLO) associated with hydrangea virescence. Bright green cymes and proliferation of leafy structures from dwarfed florets are typical disease symptoms.

Research efforts began three years ago on the use of tissue culture for the recovery of pathogen-free plants. Viruses were thought to be present in some stock plants based on visual observations, but no symptoms of hydrangea virescence were seen. Earlier research (Brierly and Lorentz, 1957) had indicated that hydrangeas are unfavorably affected by heat treatment. Terminal shoots frequently die back, so that no vigorous tip cuttings are available. Therefore, thoughts of heat treatment prior to tissue culture were discarded. The development of the tissue culture program is now described.

THE STARTING TISSUE

Shoot tips of 10–15 mm are removed from vegetatively growing stock plants. All but the last 4–5 primordial leaves surrounding the apical meristem are removed, and the remaining tip is placed in a Petri dish with moistened Whatman No. 50 filter paper. After a sufficient number of shoot tips has been obtained, they are wrapped in small squares, about 4 in, of cheesecloth and transferred to 25 × 150 mm test tubes for surface disinfection. A generous quantity of dilute bleach (Purex Pool Chlorine) diluted 10-fold with deionized water and containing 1–2 drops of Tween-20 emulsifier per 100 ml of disinfectant solution is added to the tube, covering all contents within the tube. The tubes are capped with polypropylene closures (Bellco kaputs). After 10 min the bleach solution is decanted and the contents of the tube are rinsed 3 times with autoclaved deionized water.

Shoot tips are transferred individually into a sterile Petri dish with moistened Whatman No. 50 filter paper for further dissection. Aseptic procedures are followed, and the surgical and transfer steps are performed in a laminar-flow clean-air hood to minimize contamination by airborne microorganisms. The surgical instruments include a pair of fine-tipped long (about 25 cm) forceps, a pair of extra-fine short (about 10 cm) forceps, a scalpel with No. 11 blade, and a Beaver chuck handle scalpel adapted to hold razor blade slivers. The surgical instruments are sterilized by immersion in 99 percent alcohol, followed by flaming. Final dissection using a microscope includes the removal of additional primordial leaves, leaving the shoot apical meristem with 1–3 leaf primordia. The explant is then transferred to the nutrient tube (fig. 1A).

NUTRIENT MEDIA

Three different nutrient formulations are recommended for the shoot apex culture and propagation of hydrangeas in tissue culture, one for each of the three major steps *in vitro*: (1) the establishment of initial culture from freshly excised shoot apices, (2) the subsequent step of multiplication, and (3) the final step of preparing the multiplied shoots for their transfer to soil.

Fig. 1. (A) Initial shoot apex explant (left), enlarged shoot apex (center), plant developed from shoot apex culture (right). (B) Pretransplant step in Mason jar. (C) Established plants in greenhouse. (D) Pinched plants from tissue culture program.

The three formulations contain the Murashige and Skoog salt mixture; the composition of this mixture is reproduced in table 2. Table 3 lists other constituents of the medium of the first step. In table 4 are found the addenda that are employed for the second, or shoot multiplication step; and in table 5, the substances that are added to the nutrient formulation of the third, or pretransplant step.

TABLE 2

THE MURASHIGE-SKOOG INORGANIC SALT FORMULATION EMPLOYED IN
TISSUE CULTURE OF HYDRANGEA MACROPHYLLA THUNB.

Compound	Mg/1	Compound	Mg/1
NH_4NO_3	1,650.0	H_3BO_3	6.2
KNO_3	1,900.0	$MnSO_4.H_2O$	16.9
$CaCl_2.H_2O$	440.0	$ZnSO_4.7H_2O$	8.6
$MgSO_4.7H_2O$	370.0	KI	0.8
KH_2PO_4	170.0	$Na_2MoO_4.2H_2O$	0.25
$Na_2.EDTA$	37.3	$CuSO_4.5H_2O$	0.025
$FeSO_4.7H_2O$	27.8	$CoCL_2.6H_2O$	0.025

TABLE 3

NUTRIENT ADDENDA EMPLOYED IN INITIATING CULTURES FROM
FRESHLY EXCISED HYDRANGEA SHOOT APICES (STEP 1)

Addendum	Mg/1	Addendum	Mg/1
Sucrose	30,000.0	IAA (Indole-3-acetic acid)	0.1
Thiamine HCl	0.4	Kinetin	1.0
i-inositol	100.0	Difco Bacto agar	6,000.0

TABLE 4

NUTRIENT ADDENDA EMPLOYED IN MULTIPLICATION OF
HYDRANGEA SHOOTS (STEP 2)

Addendum	Mg/1	Addendum	Mg/1
Sucrose	30,000.0	IAA (Indole-3-acetic acid)	0.1
Thiamine HCl	0.4	Kinetin	10.0
i-inositol	100.0	Difco Bacto agar	8,000.0
Adenine sulfate.$2H_2O$	10.0		

TABLE 5

NUTRIENT ADDENDA EMPLOYED IN PREPARING HYDRANGEA SHOOTS
FOR TRANSFER TO SOIL (STEP 3)

Addendum	Mg/1	Addendum	Mg/1
Sucrose	30,000.0	IAA (Indole-3-acetic acid)	1.0
Thiamine HCl	0.4	Difco Bacto Agar	6,000.0
i-inositol	100.0		

Agar formulations are advisable for tissue culture of *Hydrangea*, with an initial pH of 5.7 accomplished by adding a few drops of 1N NaOH or HCl before the agar addition. The media are sterilized by autoclaving 15 min at 121°C (15 lb pressure). The nutrient solution of steps 1 and 2 is contained in 25 × 150 mm culture tubes, in 25 ml aliquots. The tubes are capped with polypropylene closures (Bellco K-25 kaputs). During the final pretransplant step *in vitro*, the nutrient medium is placed in narrow mouth 1 qt Mason jars, each jar provided with 75 ml of medium. Jars are capped with aluminum foil, and after explants are transplanted, a rubber band holds the foil.

THE CULTURE ENVIRONMENT

All three steps of the tissue culture procedure are accomplished while providing a constant temperature of 27°C (about 80°F). The cultures are illuminated 16 hr daily with 1,000 lux (100 ft-c) light from Gro Lux lamps. The final pretransplant step is accomplished under higher light intensities of 6,000–10,000 lux provided by Cool White fluorescent lamps.

THE MULTIPLICATION PROCESS

After 6–12 weeks, the shoot apex explants should have enlarged considerably and be ready for transfer to the shoot multiplication medium (fig. 1A). Those that show only little growth, but are still alive, may require further reculture in a freshly prepared step 1 nutrient solution. Transfer of the enlarged shoot apex to the shoot multiplication medium results in the production of, on the average, 2.5 new shoots every 6 weeks. After 3 reculture periods, increase multiplication rates to 4× occur with the individual reculture of large (5–10 mm) single shoots on the multiplication medium. After multiplication to desired shoot numbers, the tissue cultures are ready for step 3 and are prepared for their transfer to soil.

THE PRETRANSPLANT STEP

The pretransplant step is accomplished by simply subdividing the shoots of the multiplication step and reculturing individual shoots in the step 3 medium (fig. 1B). A period of 10–14 days in this medium has been about optimum for hydrangeas. The recultures are placed under the higher light intensities at this time. Shoots elongate substantially and have well-developed leaves, and most will have roots.

ESTABLISHMENT OF PLANT IN SOIL

The individual plantlets are set into a mixture of peat/perlite/vermiculite (table 6) contained in California Rooting Trays. Trays of plantlets are placed under intermittant mist tents in the greenhouse for two weeks. Then the mist is turned off and the plants are allowed to acclimate further for 2 weeks (fig. 1C). Plants are then ready for transplanting into 6-inch pots. At this time, or in the next week depending on shoot size, plants are soft-pinched to yield four to six shoots (fig. 1D). After 2–4 weeks in the greenhouse, plants are ready for moving to seran-protected outdoor area for hardening off before going into full outdoor sun conditions. Plants produced from tissue culture and transplanted during spring and summer months set flower buds in early fall. During November plants are placed in a cooler at 4°C (40°F) for 6 weeks of chilling, then forced commercially for spring bloom.

INDEXING PROGRAM

Preliminary attempts to produce ringspot symptoms on either *Chenopodium amaranticolor* or *C. quinoa* met with failure. *Gomphrena globosa*, however, showed very good ringspot symptoms and therefore was used for routine indexing. Future plans call for the use of *Nicotiana glutinosa* in TobRSV detection.

Seeds of indicator plants are sown in peat/perlite mix and grown until the four-leaf stage, at which time innoculation occurs. Leaves of numbered *Hydrangea* plants are removed and ground in a clean pestle and mortar with phosphate buffer of pH 7.2. Approximately 1 g of leaf material and 3 ml of buffer are used. Macerated tissue is then transferred with a pipe cleaner to the indicator plant. Leaves have first been dusted with fine carborundum powder. For each indexing period, a control leaf is inoculated with phosphate buffer as well as a leaf with known ringspot symptom-producing sap. Individual leaves are numbered during inoculation procedures so that individual plants can be indexed for viruses. Ringspot symptoms appear on indicator plants within 7 days after

TABLE 6

PEAT/PERLITE/VERMICULITE MIXTURE EMPLOYED IN
ESTABLISHING PLANTS IN THE GREENHOUSE

Addendum	Cu m	Addendum	G
Sphagnum peat moss	0.4* (1.5 cu ft)	Dolomite lime	127.4
No. 5 Perlite	0.4 (1.5 cu ft)	Calcium carbonate	127.4
No. 3 Vermiculite	0.3 (1.0 cu ft)	NH_4NO_3	122.5
		KNO_3	302.8
		Osmocote 19-6-12	339.6

* Amounts given are for .84 cu m (3 cu ft).

inoculation. Presently plants used for obtaining tissue culture shoot apices are being checked for HRSV, TomRSV, and CMV and are causing positive results on indicator plants. Plants established from tissue cultures in 1977 will be checked every 6 weeks after pinching until full bloom to measure the recovery rate of virus-free plants.

The production of plants from tissue culture has been possible with *Hydrangea macrophylla* 'Bodensee,' 'Rosa Rita,' 'Buffie,' 'Sidylia,' and 'Blue Donau.' Cultivars of 'Dr. Bernhard Steiniger,' 'Hildegard,' 'Taspo,' 'Sara,' and 'Brugg' have met with little success using the techniques and media described previously. Present visual observations of tissue culture–derived plants indicate significantly improved commercial qualities. These include increased axillary bud development into potential flowering shoots from a single pinch, improved foliage quality, and over-all improved vigor. These commercially desirable characteristics greatly aid the upgrading of the florist hydrangea.

Indexing is now under way on plants produced from tissue culture during the spring and summer of 1977. Plants yielding negative results (no virus present) will be kept as stock plants and utilized for the 1978 shoot tip material. Establishment of known pathogen-free stock plants will allow the expansion of tissue culture techniques for the rapid multiplication of new varieties. It is anticipated that a large part of the blooming plants' requirements can be met through tissue culture production. Thus, not only can tissue culture be utilized in the potential recovery of known pathogen-free plants, it can also be utilized as a means of rapid multiplication, replacing the conventional procedure of taking cuttings from shoots not forming flowers (blind wood) during spring forcing. Plants from tissue culture can be transplanted to greenhouse conditions throughout the year, thereby extending the availability period of plant material.

ACKNOWLEDGMENTS

The author wishes to thank the various members of the California Association of Nurserymen's Plant Tissue Culture Committee for their contributions of information for this paper. Also, special thanks to those persons involved with the *Hydrangea* tissue culture program at Paul Ecke Poinsettias during various times: C. Simpson, C. Morishita, R. Sears, P. Nolan, and R. Castro.

LITERATURE CITED

Brierley, P., and P. Lorentz. 1957. Hydrangea ringspot virus: The probable cause of "running-out" of the florists' Hydrangea. Phytopathology 47:39–43.

Hearon, S. S., R. H. Lawson, F. F. Smith, J. T. McKenzie, and J. Rosen. 1976. Morphology of filamentous forms of a mycoplasmalike organism associated with Hydrangea virescence. Phytopathology 66:608–16.

Hollings, M. 1965. Disease control through virus-free stock. Ann. Rev. Phytopath. 3:337–96.

Knauss, J. F. 1976. A tissue culture method for producing *Dieffenbachia picta* cv. 'Perfection' free of fungi and bacteria. Proc. Fla. State Hort. Soc. 89:293–96.

Knauss, J. F., and J. W. Miller. 1978. *Erwinia caratovora*, a contaminant affecting commercial plant tissue cultures. In Vitro (in press).

Murashige, T. 1974. Plant propagation through tissue cultures. Ann. Rev. Plant Physiol. 25:135–66.

Murashige, T., and J. B. Jones. 1974. Cell and organ culture methods in virus disease therapy. *In* R. H. Lawson and M. K. Corbet (eds.), Proc. 3rd Int. Conf. Ornamental Plant Viruses. Pp. 207–21. ISHS, The Hague.

Nelson, P., J. Tammer, and R. Baker. 1960. Control of vascular wilt diseases of carnations. Phytopathology 50:356–59.

Sagawa, Y. 1976. Potential of *in vitro* culture techniques for improvement of floricultural crops. Acta. Hort. 63:61–66.

Shanks, J. B. 1972. Hydrangeas. *In* V. Ball (ed.), The Ball red book. Pp. 352–68. George Ball, Chicago.

Zeyen, R. J., and W. C. Stienstra. 1973. Phyllody of florists' Hydrangea caused by Hydrangea ringspot virus. Plant Dis. Report. 57:300–304.

S. H. SMITH AND W. A. OGLEVEE-O'DONOVAN

Meristem-tip Culture from Virus-infected Plant Material and Commercial Implications

24

As plant tissue culturists are making serious attempts at commercialization, we are faced with the need to separate scientific hopes from commercial realities. Springing from scientific hopes to a naïve concept, which scientists usually hedge or qualify with the proper disclaimers, that have given false impressions concerning the ease of commercialization of plant tissue culture systems. Unfortunately, most individuals involved in commercial floriculture do not listen to the quietly stated scientific disclaimers and are left with the naïve impression that with any system of plant tissue culture thousands of uniformly high-quality commercially saleable, pathogen-free plants can be obtained in a relatively short period of time regardless of the genetic stability of the crop or its disease problems. Furthermore, it is often assumed that any commercialization of a tissue culture system, no matter how specialized in its intent, will achieve all of these goals, which unfortunately is not correct.

This concept will be examined in relation to meristem-tip culture from virus-infected plant material and its commercial implications. Although the authors have participated in similar programs with nonfloricultural crops aimed at producing plants freed from recognized viruses (Smith et al., 1970; Mullin et al., 1974; Smith and Stouffer, 1975), our comments will

S. H. Smith, Department of Plant Pathology, Pennsylvania State University, University Park, Pennsylvania 16802; W. A. Oglevee-O'Donovan, Campbell Institute for Agricultural Research, 2611 Branch Pike, Cinnaminson, New Jersey 08077 (presently employed by Phytolab, Inc., 306 South Pittsburgh Street, P. O. Box 178, Connellsville, Pennsylvania 15425).

be limited primarily to floricultural crops while recognizing the possibilities and limitations of extrapolation to other commercially used plant species. After examination of the concept, which we have labeled as naïve, we will describe what we hope to be a realistic system for the commercial utilization of meristem-tip cultures derived from virus-infected plants.

The first major portion of the naïve concept that we will examine is the idea of uniformity. It is seldom stated without qualifying comments, but the impression is given that thousands of identical plants can be produced from any tissue culture system. Although the goal of meristem-tip culture from virus-infected plants is usually only a few plants of each variety without the virus in question, the whole idea of uniformity requires comment. It is well recognized that commercial vegetative propagation will produce plants that vary in their growth and morphological characteristics. One has only to look at a greenhouse containing a few thousand plants of any vegetatively propagated floricultural crop to recognize this variability. This variability should not be confused with mutability, which creates the easily recognized aberrant plant. Major mutations, which normally arise at relatively low rates with most plants under commercial vegetative propagating conditions, probably do not occur at a significantly higher rate with meristem-tip cultures, which are in reality only minute commercial cuttings.

Selection is a well-recognized method of taking full advantage of the normal variability within vegetatively propagated crops to obtain a more desirable product. Thus, it is not surprising that propagating only a very limited number of meristem-tips of any particular variety selects for distinct variants.

In figure 1 are two geraniums (*Pelargonium* × *hortorum* Bailey) of the same variety with the one on the left obviously growing more vigorously and flowering better than the one on the right. This comparison, of course, looks like the standard one used to convince commercial firms of the value of meristem-tip culture. The one on the left is usually used to demonstrate the advantages of meristem-tip culture with the subsequent removal of recognized viruses. However, both of these plants originated from meristem-tips taken from the same plant and both have been found to be free of recognized viruses. This degree of variation has been found to be quite common in meristem-tip-culture–derived lines from over twenty-five geranium varieties compared in commercial flowering trials. Whether this variation, which is consistent in cuttings taken from these plants, is due only to selection from within the normal range seen in vegetatively propagated geraniums or due to the fact that tissue-cultured plant cells are characterized by instability of chromosome number and structure (Bayliss,

Fig. 1.   Meristem-tip-culture-derived geraniums of the cultivar 'Snowmass.'

1973) is still unknown. It should be pointed out that tissue culture has actually been used to induce variation within clones of scented geraniums (Skirvin and Janick, 1976).

Uniformity of the plants produced by any means of tissue-culturing must also be considered for commercial and consumer qualities that may be subtle and unrecognizable to individuals specializing in tissue culture who do not have an understanding of the commercial aspects of the crop in question. The plants derived from tissue cultures may look as good as, if not better than, the original; but unless such plants perform well in commercial propagating and growing systems, their value is indeed questionable.

In our original statement was included the term *pathogen-free*, which in addition to being a most illogical term is also one of questionable legality. Pathogen-free implies that the plant material is free of all recognized and nonrecognized pathogens, when indeed the only statement that can actually be made is that the material probably is free from only those pathogens for which it has actually been indexed or tested. Any system, tissue culture or other, aimed at reducing or removing recognized pathogens is only as good as the sensitivity and reliability of the available indexing or testing systems. The use of pathogen-free as a commercial statement of condition in most instances probably exposes the individual selling these plants to legal recourse, as nonrecognized pathogens may in fact be present. As a standard practice, we warn growers that any company that advertizes their material as pathogen-free is publicly stating their lack of knowledge of plant pathology.

The concept expressed in our earlier statement, that tissue culture systems will produce a saleable product within a short period of time, is quite misleading, particularly when dealing with meristem-tip culture from virus-infected plants. The goals of such meristem-tip–culturing programs must not be confused with the claims associated with "rapid propagation" systems. Any commercial firm interested in producing quality plants through meristem-tip culture from virus-infected stock must be prepared to invest approximately two to five years before selling their first plants.

The type of a program we are currently using to produce geraniums that index negatively for the recognized viruses infecting geraniums hopefully avoids many of the concepts stated earlier as the naïve impression of commercial tissue culturing.

MERISTEM-TIP CULTURE OF GERANIUMS

As with other vegetatively propagated crops that have not been exposed to a virus indexing system, over 95 percent of the commercial cultivars were found to be virus-infected. Widespread viruses within geraniums such as tobacco ringspot virus (Oglevee et al., 1975) caused very mild or no visible symptoms in commercial lines (Walsh et al., 1974). Those viruses that induced dramatic foliar symptoms were undoubtedly rogued out of commercial propagation. Although the viruses infecting geraniums did not produce diagnostic foliar symptoms, they did severely affect flowering. To ascertain if viruses such as tomato and tobacco ringspot viruses, which are quite common in geraniums, were actually worth the effort of removing by meristem-tip culture, both viruses were inoculated into geraniums and their horticultural (Scarborough and Smith, 1977) as well as their anatomical (Murdock et al., 1976) effects studied. It was apparent from these studies that a meristem-tip culture system for the removal of those viruses commonly found in commercially propagated geraniums would indeed be worthwhile.

The term *meristem-tip* is used to describe the amount and type of tissue we use in our program. A meristem-tip as defined by Hollings (1965) is "the meristem dome plus the first pair of leaf primordia; this ranges from about 0.1 to 0.5 mm in length and of similar width."

Culture-indexed plants were obtained by the methods described by Nelson et al. (1971) and used to avoid the difficulties of working with plants containing the vascular wilt pathogens and in particular *Xanthomonas pelargonii.* Unless the vascular wilt pathogens are first removed, the plants often will die during heat treatment and microbial contamination of cultures is a problem.

When initiating a meristem-tip culture program, a major decision must be made concerning the use of rapid propagation by tissue cultures. This decision must be based on the time and cost necessary to produce saleable plants. Having demonstrated marked variation due to meristem-tip culturing with geraniums, it was decided that rapid-propagation tissue culture methods contained too high a risk of variability within the plants that would be produced. Also meristem-tip–cultured plants when indexed and selected were able to be increased rapidly by cutting production, enabling the rapid build-up in plants without sacrificing quality.

The first phase of any commercial meristem-tip culture of virus infected plants is the easiest; that is, the production of young plants. The methods we used for this phase are those that are quite standard for production of young plantlets from meristem-tips derived from previously heat-treated plants to increase the probability of removing viruses (Horst et al., 1976). To introduce these plants into a commercial propagation system that will maintain quality while avoiding reinfection is most difficult and time-consuming (fig. 2).

The young plantlets must be indexed for the presence of known viruses and maintained in isolation until the results of the indexing are available. Those plants that index negatively must then undergo extensive selection for plant quality, with only the highest quality plants being retained for immediate use.

Selection for horticultural and commercial qualities in a meristem-tip culture program should be considered as an absolute necessity. The variation between plants derived by such a program can be a very positive factor, as it is possible in most cases to improve the quality of a variety as a cultivar by rigorous selection.

In order to index and select with only minimum risk of contamination, cuttings are taken from the meristem-tip-culture–derived plants. If the cutting indexes positive for virus, both the cutting and the plant it originated from should be destroyed or reintroduced into the system by again heat-treating and meristem-tip culturing to attempt elimination of the viruses in question. Once the cutting has indexed negatively, the plant from which it was derived is termed an "Elite Mother Plant (EMP)." This can then be used for cuttings to be entered in flowering trials where only the best EMP are selected. If the EMP is only of marginal quality, it may be retained as a source of meristem-tips in hopes that through their variability a higher-quality plant will be obtained.

The importance of the selection of EMPs cannot be overemphasized.

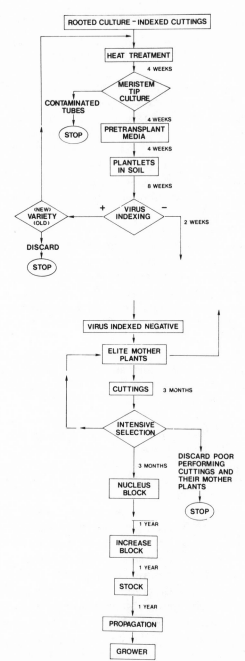

Fig. 2. Schematic diagram of the system used to incorporate heat-treated, meristem-tip–cultured, virus-indexed, and horticulturally selected geraniums into commercial propagation.

This selection should be done under commercial growing conditions. Unfortunately, this selection is not just a matter of picking the biggest and most vigorous plants. In many cases the variety may need to be improved by selecting for such qualities as compactness, cutting production, rooting, or other commercial qualities. For a single variety to be heat-treated, meristem-tip–cultured, virus-indexed, and selected, often at least one year is required even with an established program.

The EMPs that are selected as superior are individually maintained in screen cages and are the source of cuttings that form the Nucleus Block. The Nucleus Block, like the EMPs, require a great deal of care to maintain in isolation to prevent infection by viruses. Although the Nucleus Block is not maintained in individual screened cages, as is the case with the EMPs, it must be maintained in an isolated greenhouse with periodic virus indexing.

For the final phase, cuttings are taken from the Nucleus Block to form an Increase Block, which is shown to be composed of only plants that have been derived from the Nucleus Block. Later these plants act as a source of cuttings for Stock Plants that will eventually provide cuttings that are propagated and sold to the commercial grower.

If this system is to be fully exploited, new EMPs must be produced each year. Those that index negatively for viruses and are intensively selected each year should allow for a continuing increase in the quality of the plants that reach the grower.

Many researchers have produced meristem-tip–cultured and virus-indexed negative plants. Unless these plants can be placed in the hands of a commercial firm to demonstrate their qualities while preventing their reinfection by viruses, their production has been largely an academic exercise. The commercial firm must be one that understands the entire system and is willing to undertake the additional costs without taking any shortcuts. Any system of producing saleable meristem-tip–cultured, virus-indexed plants undoubtedly can be improved, but improvements should be made on the basis of experimental evidence and not on arbitrary management decisions.

REFERENCES

Bayliss, M. W. 1973. Origin of chromosome number variation in cultured plant cells. Nature 246:529–30.

Hollings, M. 1965. Disease control through virus-free stock. Ann. Rev. Phytopath. 3:367–96.

Horst, R. K., S. H. Smith, H. T. Horst, and W. A. Oglevee. 1976. *In vitro* regeneration of shoot and root growth from meristematic tips of *Pelargonium* × *hortorum* Bailey. Acta Hort. 59:131–42.

Mullin, R. H., S. H. Smith, N. W. Frazier, D. E. Schlegel, and S. R. McCall. 1974. Meristem culture frees strawberries of mild yellow edge, pallidosis, and mottle diseases. Phytopathology 64:1425–29.

Murdock, Deborah J., P. E. Nelson, and S. H. Smith. 1976. A histopathological examination of Pelargonium infected with tomato ringspot virus. Phytopathology 66:844–50.

Nelson, P. E., L. P. Nichols, and J. Tammen. 1971. Vascular wilts culture-indexing. In J. W. Mastalerz (ed.), Geraniums: A manual on the culture, diseases, insects, economics, taxonomy, and breeding of geraniums. Pp. 241–48. Pennsylvania Flower Growers, University Park, Pa.

Oglevee, W. A., S. H. Smith, and R. K. Horst. 1975. Isolation of tobacco ringspot virus from symptomless geraniums of the cv. "Diddens Improved Picardy." Proc. Amer. Phytopath. Soc. 2:103 (abstr.).

Scarborough, B. A., and S. H. Smith. 1977. Effects of tobacco- and tomato-ringspot viruses on the reproductive tissue of *Pelargonium* X *hortorum*. Phytopathology 67:292–97.

Skirvin, R. M., and Jules Janick. 1976. Tissue culture-induced variation in scented *Pelargonium* spp. J. Amer. Soc. Hort. Sci. 101:281–90.

Smith, S. H., R. E. Hilton, and N. W. Frazier. 1970. Meristem culture for elimination of strawberry viruses. California Agriculture 24:8–9.

Smith, S. H., and R. F. Stouffer. 1975. Prunus stem pitting. In N. F. Childers (ed.), The peach. Pp. 387–96. Rutgers, The State University, New Brunswick, New Jersey.

Walsh, D. M., R. K. Horst, and S. H. Smith. 1974. Factors influencing symptom expression and recovery of tobacco ringspot virus from geranium. Phytopathology 64:588 (abstr.).

HARRY E. SOMMER AND CLAUD L. BROWN

Application of Tissue Culture to Forest Tree Improvement

25

INTRODUCTION

The application of tissue culture to tree improvement offers forest biologists unprecedented challenges for increasing forest productivity. The early practice of European foresters in selecting the best trees available in natural stands, propagating these by grafting to establish clonal seed orchards, followed by breeding and early progeny testing, has led to the production of improved seed for afforestation.

Traditionally, similar procedures were adopted in America and elsewhere following exploitation of virgin forests and subsequent realization of the need to conserve valuable germplasm for future generations. Today, federal and state forestry agencies and a large number of private forest industries are engaged in some form of tree improvement activities. Almost all these programs have initially relied upon mass selection of phenotypically superior trees followed by the establishment of considerable acreages in clonal seed orchards. Once such orchards are established by vegetative propagation, usually by grafting because of the difficulty in rooting mature forest trees, the general combining ability of each clone is assessed by the performance of half- or full-sib progeny tests (Wright, 1976).

Although conventional methods of selection, breeding, and progeny testing of forest trees have proved to be economically feasible, the time involved to produce succesive generations of improved seed is still appreciable because of the long rotation ages of most forest crops. Even in the southeastern United States, where growing seasons are long and seasonal

Harry E. Sommer and Claud L. Brown, School of Forest Resources, University of Georgia, Athens, Georgia 30602.

rainfall adequate, pine plantations for pulp and paper are commonly managed on rotations of 25 to 30 years, and the rotation ages for sawtimber and veneer may exceed 50 years. Consequently, the age seedling progeny must reach before one is able to predict with reliability their performance at the end of the rotation is still a matter of contention among forest geneticists.

Irrespective of these and many other technical problems associated with present-day tree improvement activities, the time is rapidly approaching when the bulk of cellulose used throughout the world for pulp and paper, reconstituted wood products, plastic feed stocks, or fuel, will be produced from intensively managed, highly specialized plantations under rotation ages of 15 years or less. To see that this trend is evident, one has only to view the tremendous productive capacity of the forested warm temperate, subtropical, and tropical areas of the world where total biomass production exceeds that of northern tempeate zones manyfold. Already the feasibility of producing 8 to 12 metric tons of dry above-ground biomass /hectare/year was demonstrated more than ten years ago in the southeastern United States by harvesting of hardwood stump sprouts at 1- to 5-year intervals (McAlpine et al., 1967). Currently numerous projects are under way in this country and elsewhere to exploit the concept of short-rotation forestry where hardwoods capable of repeated sprouting are planted at row-crop spacings of 4 × 4 feet, fertilized periodically, and machine-harvested at 3 to 5 year intervals following resprouting of the original stumps (Steinbeck and Brown, 1976).

The predicted demands for pulp and paper alone in the United States will more than double by the year 2000 (Keays, 1975). Furthermore, the expanded uses of cellulose for plastic feed stocks, animal feed, or conversion to fuels of one sort or another will undoubtedly increase the demands for cellulose severalfold during the next few decades (Glesinger, 1949; Keays, 1975; Brown, 1976; Goldstein, 1977). Fortunately, cellulose is the most abundant organic material in the world; and because it is a renewable resource, our forests are capable of supplying it in vast quantities, along with other natural products, under intensive systems of short-rotation forest management. Perhaps we are over optimistic about the future use of cellulose in an energy-scarce society because we realize the many potentialities offered by tissue culture research for increasing the biological productivity of the world severalfold. A brief discussion of these potentialities is now appropriate.

VEGETATIVE PROPAGATION

Cloning Selected Phenotypes

The practice of fixing the genotype of perennial plants possessing one or more desirable traits by vegetative means has been used by early man and practicing horticulturists for several hundred years. Similar techniques have rarely been used in forest production because the final product sought was yield of wood rather than some edible fruit or seed. Hence, the millions of plants needed in reforestation programs were more easily produced in nurseries from seed; and, as previously mentioned, most forest trees are difficult to propagate vegetatively by rooted cuttings. The latter reason alone accounts in large measure for the establishment of clonal seed orchards by grafting, not only for seed production orchards but for breeding orchards and for conserving valuable germplasm, i.e., gene storage.

More recently, Libby (1974) has pointed out many advantages of establishing clones from twig cuttings of trees in juvenile and later stages of development. Such clones are useful in (1) exposing the same genotype to different environmental influences to increase the precision of various physiological studies, (2) to study competition among selected genotypes at given spacings, (3) correlation of phenotypic expression of juvenile traits with later stages of phasic development, and (4) possible use in production forestry. With the recent progress made by Sommer et al. (1975), Cheng (1975), and others, in cloning several coniferous species from cotyledons and hypocotyls of mature embryos, and of cloning *Populus* and *Ulmus* from cell suspension cultures (Riou et al., 1975; Durzan and Lopushanski, 1975), it appears only a matter of time until these procedures will be adapted to mass propagation of many forest species.

It takes little imagination to foresee the impact that cloning techniques could have upon forest productivity within the next two decades if the forest industries could immediately take advantage of the most vigorous genotypes nature has to offer. The potentialities are enhanced even more under short-rotations of 5 to 15 years where high selection indices are initially used to select the best phenotypes available from among seedlings grown in plantations at close spacings on a variety of sites. By cloning several hundred of the most vigorous selections, followed by performance tests on different sites, one could quickly assay those clones best adapted to biomass production under short-rotation regimes. Hence, rapid genetic gains could be made by fixing these genotypes through cloning to establish successive generations of plantations. Random mixing of numereous clones in production plantations would minimize the effect of monoculture, and older clones would be replaced with newer, more improved selections in successive generations.

Other methods of tree improvement such as hybridization and introduction of exotics, although highly successful in other countries, have not been used to any appreciable extent in the United States. For example, in Korea the F_1 hybrid of *Pinus rigida* × *P. taeda*, both native to America, has been mass-produced for several years by controlled pollinations in breeding orchards to reforest much of South Korea (Hyun and Kun-Yong, 1959). The production of interspecific hybrids, followed by selection and cloning of the best offspring from among each cross, could likewise result in rapid genetic gain.

In view of three decades of rather intensive tree improvement in this country it is somewhat surprising that the F_1 hybrid between longleaf pine (*Pinus palustris*) and slash pine (*Pinus elliottii*) has not been produced for commercial planting by the forest industries throughout the Southeast. Mass pollinations could be carried out among selected half-sib families in separate seedling seed orchards established for this purpose. The F_1 hybrid, unlike its longleaf parent which possesses a 5- to 6-year juvenile "grass-stage" period of practically no height growth, makes initial height growth the first season, possesses good crown and bole form, and shows some resistance to fusiform rust (*Cronartium fusiforme*), a stem canker to which the slash pine parent is highly susceptible (Brown, 1964). Because there is considerable variation within the growth rates of individual F_1 progenies among specific crosses, one could quickly take advantage of "hybrid vigor" in field selections by cloning selected hybrids at age 5 or less for testing on various sites, followed by cloning the most promising hybrid genotypes for production forestry. Planting the best F_1 hybrids would also likely circumvent many of the problems currently encountered in disease susceptibility and reduced growth of slash pines when planted to the drier, deeper sands of former longleaf pine sites. Numerous other examples could be cited where the genotypes of interspecific hybrid progeny of forest trees could be fixed by cloning to increase yields, resistance to disease, or adaptation to environmental stresses such as cold or drought hardiness.

Protoplast Fusion and Production of Parasexual Hybrids

Some of the most challenging and potentially productive areas of tissue culture research go beyond the immediate practical uses of cloning just discussed. Instead of narrowing the genetic base by intensive selection of plus phenotypes followed by cloning, it is possible to extend the present range of genetic variation in cultured cells by selecting for mutations and by protoplast fusion or hybridization of cells in the laboratory. The latter process was first achieved some five years ago with tobacco cells (Carlson et al., 1972) and similar approaches are currently being tried with diploid cells of conifers, although no tetraploid cells have yet been recovered (Winton et al., 1975).

The induction of ploidy in forest trees may or may not result in improved quality or yields. Where natural polyploidy has been observed or induced in conifers, notably in pine species, the results have been disappointing because of a reduction in overall growth and the formation of morphologically abberrant seedlings (Mergen, 1958; 1959).

In contrast to most diploid gymnosperms polyploidy has been a major evolutionary process giving rise to entire families and genera of dicotyledonous trees (Stebbins, 1950). Even in angiosperms, however, the artificial induction of polyploidy often results in reduced growth rates similar to those observed in conifers (Wright, 1976). In angiosperms, triploids have proved of much greater immediate use in forest practice than tetraploids. Several European and American triploid clones of aspen have produced exceptionally fast growth rates. Hence, the fusion of haploid and diploid protoplasts in cell suspension cultures followed by induced organogenesis could result in the recovery of triploid plantlets that may possess highly desirable, sought-after traits. Similarly, protoplast fusion of haploid cells to form intespecific hybrids followed by chromosome doubling to form allopolyploids could hold much future promise in forest productivity. In most instances, the 4N allopolyploids will be fertile; hence, those hybrids with desirable traits can then be reproduced by seed. Although many technical problems still remain to be overcome, e.g., in the screening and recovery of fused protoplasts followed by the formation of intact plants, or vice versa, tissue culture now makes these possibilities a reality.

Haploid Breeding, Haploid Plantlets, and Homozygotes

The use of tissue culture in establishing and maintaining haploid cell cultures of higher plants, and especially trees, presents an array of breeding options and techniques heretofore unavailable to forest geneticists. Inbreeding in long-lived tree species to obtain homozygous lines for controlled crosses has not been practical in programs of tree improvement because of the length of time required to produce successive inbred generations. With the culture of haploid pollen grains from both gymnosperms and angiosperms or the culture of haploid female gametophytes of gymnosperms, new and different approaches to obtaining haploid plantlets are readily available. Obviously, in the case of microspore cultures millions of different genomes are present as a result of meiotic segregation; hence, the possibility exists for immediately screening myriads of genotypes for vegetative vigor or perhaps such physiological traits as lower nutritional requirements, cold hardiness, or synthesis of specific natural products. Likewise, if any favorable mutants were to arise spontaneously, or be induced by mutagenic agents, these would be immediately expressed and possibly recovered by plating techniques. The more vigorous haploid lines might be fused to produce new diploid combinations some of which may show outstanding heterosis following induced organogenesis. Of even more immediate value in tree improvement is the possibility of inducing the differentiation of haploid plantlets followed by chromosome doubling (or vice versa) to produce viable homozygous diploids, some of which would likely possess desirable traits of growth and form when tested under field conditions.

The uniform haploid genome of the megagametophyte surrounding the embryo in conifers and other gymnosperms offers an immediate practical approach to the forest geneticist in screening for possible superior genotypes prior to hybridizing haploid cell lines in suspension culture. For example, after superior half- or full-sib families are identified for general combining ability in replicated progeny tests, seed from the most outstanding individual trees within each family would be obtained and the genetic value of each embryo determined by cloning and testing of the ramets under diffrent environmental regimes of nutrition, photoperiod, and so on, in short-term greenhouse or growth chamber studies. Concomitantly, upon excision and cloning of the embryo, the surrounding female gametophyte tissue would be placed in cell culture as a routine method of gene storage. Upon early evaluation of the growth response of each clone under test, one would be able to assess the female gametic contribution of that particular clone and then, by judicious hybridization of the best haploid genomes, perhaps create new intraspecific combinations with outstanding hybrid vigor, or conversely attempt to produce homozygotes from the best haploid genomes under test. These are only some of many possibilities now capable of being researched and exploited in tree improvement programs.

Embryo Cultures

Many of the early attempts to grow embryos to maturity resulted from the needs of geneticists to circumvent the problem of incompatibility between maternal tissues and the newly formed embryo following inerspecific hybridization. Various hybrid embryos in *Prunus, Sorbus, Malus,* and *Citrus* have been excised and cultured *in vitro* when they would have otherwise been lost (Maheshwari and Rangaswamy, 1965).Although numerous embryos of forest trees have been cultured *in vitro*, most have been grown for specific physiological or morphogenetic studies rather than to circumvent genetic incompatability. In at least one study, the second-year female strobili of two pine species (*P. strobus* and *P. peuce*), along with several hybrid cones, were grown *in vitro* to maturity (Kriebel and Shafer, 1971). The embryos and female gametophyte developed from the early stages following fertilization to maturity within the intact cultured cones.

HISTORICAL NOTES

Some people seem to think the growth of woody plants in tissue culture is a very new development, when in fact trees were among the first plants cultured. Gautheret (1934) cultured cambial tissue of *Pinus maritima, Abies pectinata, Acer pseudoplatanus, Populus nigra, Salix caproea, Alnus glutinosa, Quercus robur, Fagus silvatica, Ulmus campestris,* and *Fraxinus exelsior.* Tissues from *Populus, Acer, Ulmus,* and *Salix* developed best. In 1936, LaRue (1936) grew embryos of *Pinus resinosa, Thuja occidentalis, Picea canadensis, Tsuga canadensis,* and *Pseudotsuga menziesii* to normal-appearing seedlings in culture. These embryos had been removed from seeds of immature cones.

In 1940, Gautheret (1940a, 1940b) reported on the growth of buds obtained from cambial tissues of *Ulmus campestris* cultured *in vitro.* In 1949, additional work in this area was done by Jacquiot (1949). Morel (1948), reported the culturing of tissues from *Crataegus monogyna, Pyrus communis,* and *Syringa vulgaris* and their use in plant pathology studies. Later Morel also grew *Pinus strobus* callus in Wetmore's laboratory.

In 1950, Ernest Ball reported the differentiation of buds in somatic tissue cultures of *Sequoia* (Ball, 1950). After the third or fourth subculture, he studied the differentiation of buds in the parenchymatous callus tissue. The differentiation of buds started with the division of several parenchymatous cells in a group. Tannin cells, which were abundant in these tissues, did not contribute to the initiation of the meristem, but were often found on the periphery of the latter. Later stages showed the development of leaf primordia with a flat apical meristem, and later the formation of typical conical apical meristems. In later discussions, Ball indicated that the buds did not grow into shoots (Skoog, 1954). In addition, some callus did not form buds, and in those that did the number of buds formed decreased rapidly after the fourth subculture (Gautheret, 1957).

Wetmore (1954) reported on the growth of *Syringa vulgaris* apices in culture. Using a Knop's solution with sucrose, the apices could be grown only if planted upon a nurse culture. When 15 percent autoclaved coconut milk and 0.4 mg/1 casein hydrolysate were added to the nutrient, the apices grew into shoots and rooted.

Konar and Oberio (1965) reported they had found on the cotyledons of *Biota orientalis* embryos in culture, structures they call "embryoids." When removed from the cotyledons the structures did form shoots, but not roots. This was the first indication that buds obtained in gymnosperm tissue cultures could form shoots.

The work of these men represent the start of tree tissue cultures as well as indicating the potential of these techniques. It appeared that with some additional effort it should be possible to obtain diploid plants from the cultures of various woody plants. In the final analysis, however, these studies did not provide either viable propagules that could be grown into full-size plants or information on how to obtain or control organogensis in these cultures with predictability.

We might also add that the development of tree tissue culture has been aided by the development of better media. Reinert and White (1956) developed a medium for *Picea glauca* that consisted of White's salts, sucrose, 11 vitamins, 16 amino acids, an auxin, and agar. Later, Risser and White (1964) refined this medium to a modified White's salts with 4 vitamins, glutamine, an auxin, sucrose, and agar. Even then, not many long-term cultures of gymnosperms were conducted. Soon thereafter, Brown and Lawrence (1968) developed a medium based on Murashige and Skoog (MS) salts with 4 vitamins, asparagine, 2,4-D, surcrose, and agar that was suitable for subculturing slash pine (*P. elliottii*) callus for a period of 5 years. This medium when modified in 2,4-D content was also suitable for pine cell suspension growth. Several other genera of conifers could be grown as callus on this media, and today many of the cultures of trees are maintained on a high-salt medium with a reduced nitrogen source. Other developments such as the discovery of the cytokinins and their use in the tissue culture of trees had their impact, particularly on vegetative propagation via tissue culture techniques.

VEGETATIVE PROPAGATION BY TISSUE CULTURE

Propagation of Gymnosperms

We now turn our attention to the propagation of gymosperms via tissue culture. This is one of the fastest-developing fields in forest tree improvement, and because Dr. Cheng will be covering this subject in this volume, we intend only to outline the several approaches that have been taken to achieve regeneration in gymnosperms. We have mentioned already that Konar and Oberoi (1965) used modified embryo cultures that produced "embryoids" on the cotyledons of *Biota orientalis*. This species has been reinvestigated recently (Thomas et al., 1977), and the adventitious organs found were not embryoids but rather buds, i.e., no structure resembling a radicle was found. We have used a modified embryo culture procedure and obtained adventitious buds for several species (Sommer et al., 1975; Sommer, 1975; Brown and Sommer, 1977). , including 10 species of pines, one of *Taxus* and *Pseudotsuga menziesii*. Plantlets were obtained in 7 pine species and *Pseudotsuga menziesii*. The pine species produced adventitious buds, even if in low percentage, on modified Gresshoff and Doy's medium. More buds were obtained with *Pseudotsuga menziesii* with lower levels of

BA and no, or less than 0.01 ppm, NAA. Cheng (1975) developed a three-step embryo culture procedure. The embryos were first cultured on Cheng's modification of MS medium with 5μM each IAA, IBA, BA, and 2iP. After about 4 weeks, the cultures were flooded with 0.5–1mM BA, and soon the primordia present started to develop into buds. After one week, the cultures were transferred to a one-half strength basal medium for bud and shoot growth. Plantlets were then produced by rooting the shoots.

A two-step method has been used to produce adventitious buds on embryos of *Pinus radiata* (Reilly and Brown, 1976). The embryo is first pretreated in an agitated liquid medium containing 5 ppm BA for up to 4 weeks, then transferred to a one-half strength basal medium for adventitious bud development. This method also works for slash and longleaf pine embryos (Brown and Sommer, unpublished).

Some variations on the direct use of isolated embryos for propagation have been used. For instance, Chalupa (1975) exicsed the embryo from germinating Norway spruce seeds when the cotyledons started to emerge, cut off and placed the upper part of the hypocotyl on Wolter's or Gamborg and Eveleigh's medium with 1–5 ppm BA or kinetin. Adventitious buds were formed. When buds were transferrred to a one-half strength Linsmaier and Skoog's basal medium, short shoots developed. Some shoots formed roots on modified Wolter's medium with IBA. Isikawa (1974) used the hypocotyl sections from 2-week-old seedlings of *Cryptomeria*. Using Wolter's basal medium, adventitious buds were formed on 12 percent of the hypocotyls; with 0.02 ppm NAA buds were formed and grew into shoots on 17 percent of the hypocotyls. Root-like appendages were formed with 1 ppm NAA plus 0.1 ppm BA. With Grinblat's basal medium plus 500 ppm malt extract and 1 ppm ABA, 33 percent of the cultures formed buds, whereas, without ABA only 6 percent of the cultures formed buds. At 10 ppm BA the percentage of cultures forming buds was not significantly different from the percent using ABA. Apparently *Cryptomeria* very readily forms buds, even under some unusual circumstances that need further investigation. Campbell and Durzan (1975, 1976a, 1976b) have made extensive use of hypocotyls in development of a system to obtain plantlets of *Picea glauca*. If the hypocotyl is placed on nutrient with the apical end exposed and supplied with BA at 10^{-5} M and NAA at 0, 10^{-7}, or 10^{-5} M, adventitious buds are produced. On a basal medium the buds grew into small shoots. If the hypocotyl was placed in the nutrient, apical end down, and this time NAA supplied at 10^{-5} M and BAP at 10^{-5}, 10^{-7}, or withheld completely, then roots could be obtained on the basal end of the hypocotyl. Some of the shoots obtained rooted spontaneously on basal media and formed plantlets.

David and David (1977) used a slight modification of hypocotyl culture with *Pinus pinaster* in that they left the apex intact and sometimes also the cotyledons. On Campbell and Durzan's basal medium with 10^{-5} M BA, adventitious buds were formed on the hypocotyl or near the cotyledons. In a series of experiments by Thomas et al. (1977), the shoot, including the hypocotyl of 2–4-week-old seedlings were used, these were placed in Murashige and Skoog's, Heller's, or Lin and Staba's basal media with Nitsch and Nitsch's vitamins and various combinations of NAA, IAA, 2,4-D, zeatin, kinetin, and BA. Only BA was effective in inducing buds. Depending on seedling development, adventitious buds were formed on the cotyledons, primary needles, or hypocotyls. Using different concentrations of 6-BA, similar explants of *Thuja occidentalis, Cupressus sempervirens, Cupressus macrocarpa*, and *Cupressus arizonica* differentiated adventitious buds.

It would be desirable to have propagules of the parent tree and tissue culture–derived trees both available for comparisons of their development. If a mature tree were the source of the explant, this would be easier; however, somatic tissue from mature gymnosperms have not yet been made to regenerate in tissue culture. One approach was taken by Momot (1976) by using leaf explants of *Larix dahurica* which grew callus that regenerated roots. When the seedling apex was used, the callus formed roots and shoots. Similarly, Isikawa (1972) was able to root apex cultures of *Cryptomeria*, Po-Jen Wong went one step further: when the axillary buds of the decapitated *Cryptomeria* seedling grew out, he rooted these shoots. Chalupa (1975) has grown embryo callus of *Picea abies* and had adventitious buds differentiate. Short shoots could be obtained from the buds transferred to one-half strength Linsmaier and Skoog basal medium. When shoots were transferred to Wolter's medium with low auxin, some shoots formed short roots. If this can be extended to other than embryo callus, a system will be available for comparing parent and plantlet.

Two routes to genome multiplication that depend on the presence of preformed buds have been attempted in tissue culture. The Davids have taken needle fascicles of *Pinus pinaster* from 2–3-year-old trees and placed them in culture on the basal medium of Campbell and Durzan with BA at 10^{-5} M (David and David, 1977). After the dormant bud at the base of the needle fascicle started to grow, the developing bud was transferred to a rooting medium based on Heller's salts with asparagine, glutamine, and gibberellin. Plantlets were obtained. The other approach that has been tried is to root isolated buds. This has been done with *Picea abies* (Chalupa, 1975) using modified Wolter's medium.

Returning to some of the methods that worked for the induction of buds on embryos, some of these can be extended for use on isolated cotyledons, needles, or other expendable parts of the plant. Cheng's (1975) method can be used directly or as modified (Cheng, 1977) to produce adventitious buds on proliferating cotyledons or primary needle pieces of *Pseudotsuga menziesii. Tsuga heterophylla* will produce buds. In Cheng's system these can be rooted (Cheng, 1976). The pretreatment method of Reilly and Brown (1976) may be used directly also for *Pinus radiata* cotyledons, primary and secondary needles, and with modification for *Pseudotsuga menziesii* needles. Mott and coworkers (Mott et al., 1977), have taken the cotyledons directly off the embryos of *Pinus taeda* and, by placing them on agar with a low level of BA, obtained adventitious buds. These could be grown and rooted to form plantlets.

Instead of going through several steps to obtain plantlets, it would be convenient to obtain them directly from embryoids in culture. The only report to date presenting acceptable evidence that embryoids have been obtained in culture is by Bonga (1977). He took the dormant shoot from the buds of *Abies balsamea* and soaked them in water for 24 hours, prior to transfer to a modified Romberger's nutrient with 1ppm IBA added. In culture the unsoaked shoot elongated. The lower needles of the soaked shoot often developed yellow-green, opaque structures that were "embryo-like." When sectioned and observed microscopically they were composed of small dense cells, and occasionally differentiated tissues resembling both cotyledon and radicle primordia were present at opposite ends of the structure.

Propagation of Angiosperms

Although horticulturists have made considerable progress in propagating palms, *Citrus, Prunus*, almond, and apple via tissue culture, the production of plantlets from dicotyledonous forest trees has lagged behind. One group that has received considerable attention in *Eucalyptus*; Dr. Caldas will cover them in another part of this volume.

Considerable effort has been devoted to organogenesis in various species of *Populus*. Mathes (1964) was probably the first to achieve the growth of both roots and shoots on callus of *Populus tremuloides*. Callus was grown on Mathes basal medium with 10 percent coconut milk; whereas, 0.8–1.0 ppm kinetin gave 17 percent shoots on the callus after 8 weeks, and with 1 ppm IAA shoot initiation occurred on 13 percent of the callus pieces. The shoots were small. If, however, the callus was maintained on basal medium containing 0.5 ppm NAA, shoot growth could not be stimulated. Studies on root initiation indicated that 6–9 percent green tomato juice, or IAA (1–12 ppm), or IBA (0.5–1.0 ppm) stimulated root initiation. Citric acid also stimulated root initiation. Conversely, 2,3,5-TIBA, 2,4-D, NAA, or kinetin inhibited root initiation. Some cultures initiated both roots and shoots; however, no indication is given as to whether the roots and shoots were connected. Later, Wolter (1968) reported on the formation of shoots and roots from *Populus tremuloides* callus. The callus had been maintained for 7 years on Wolter's medium, and shoots could be initiated by adding less that 0.5 ppm BA in the absence of auxin. Roots could be initiated on callus if NAA was substituted for 2,4-D in Wolter's medium; also, IAA, IBA, and 2,3,6-trichlorobenzoic acid could be substituted for NAA. Kinetin, BA, and 2,4-D (0.02–2 ppm) all inhibited root formation. Shoots could be rooted on Wolter's medium without a cytokinin or auxin.

Winton (1968) has cultured triploid *Populus tremuloides* (and other members of *Populus*) and obtained plantlets. Shoots were often produced on callus with 0.05 ppm BA in Wolter's basal medium. In contrast to other systems roots were obtained with 0.04 ppm 2,4-D and kinetin at 1.0 ppm.

Berbee and coworkers (1972) started shoot tip callus from two hybrid poplar clones on Hildebrandt's D medium with NAA, 2,4-D, and coconut milk. These were transferred to MS medium and finally to MS with BA (0.05 ppm) replacing kinetin. Shoots were formed and were excised and placed on 10 percent Hoaglands solution with 0.3 percent sucrose and 0.6 percent agar. The plantlets obtained have been planted in mother stock blocks and are being evaluated for growth, chromosomal abnormalities, and disease.

More extensive studies of adventitious root and shoot formation in callus of *Populus nigra* L. 'Italica' were made by Venerloo (1973) and Brand and Venerloo (1973). The callus was grown on a modified MS medium with 2,4-D as the auxin. Root formation did not occur at the concentration of 2,4-D (0.5 ppm) used for callus growth, but did occur at a tenfold lower concentration or when NAA was substituted for 2,4-D or on a medium without auxin. At 10 μM the cytokinins inhibited root formation and growth from the callus, but cytokinins were obligatory for bud initiation. The various cytokinins showed differences in their ability to induce leaf, shoot, and buds. Zeatin was effective at 10^{-8} M, kinetin was only marginally effective at 10^{-7} M, and zeatin and BA at 10^{-5} M inhibited the differentiation of normal shoots. Kinetin and 2iP did not inhibit shoot differentiation at 10^{-5}M. Some of the buds differentiated from near root apices formed in the callus (Brand and Venverloo 1973), possibly mimicking the sprouting of poplars in nature.

Venverloo (1973) also studied the carry-over effects of 2,4-D from callus media to differentiation media, with 5 strains of callus of different ages. Callus grown on 0.5 ppm 2,4-D was slow to differentiate roots or shoots when transferred; and if grown on lower 2,4-D concentrations, the differentiation lag was less. Five callus strains obtained 1–4 years earlier were tested for ability to differentiate shoots by transfer to a media with 0.2 ppm BA. Substrains of the oldest strain grown with 0.5 and 0.1 ppm 2,4-D did not form true elongated shoots on the differentiation medium. However, the substrains taken from 0.05 ppm 2,4-D or 1.0 ppm NAA did form true shoots on the differentiation medium. When these results are compared with earlier work, they point out the differences in species sensitivity to NAA and the care that must be taken in adjusting the growth medium so it does not inhibit differentiation, as well as choosing the differentiation medium.

Whitehead and Giles (1976) considered the propagation of poplar via callus cultures slow and not capable of producing enough propagules. They cultured the axillary buds of *Populus nigra 'Italica,' P. deltoides × P. nigra* ('Flevo'), and *P. yunnanensis* on a modified MS medium with 0.2 ppm BA. Within 4 weeks bud break had occurred and initial shoots had lengthened enough to be cut into 5 mm sections and placed again on the same medium. Adventitious buds started to form, and the culture was transferred to a modified MS medium with 0.1 ppm BA and 0.02 ppm NAA. After 6 to 8 weeks, 120–220 shoots had been formed for each bud explanted. These shoots could either be rooted or used for the production of more shoots. Three months after rooting, the trees were 1–1 1/2 meters tall. They estimated that one million plantlets per bud per year could be produced using this approach.

Thompson and Gordon (1977) devised a different system. They emphasized the possibility of enodpolyoidy in callus cultures and the production of plantlets inferior to the selected parent and genetically different. They removed 1–2 mm-tall shoot tips from buds of *Populus tristis × P. balsamifera* cv. Tristis no. 1 and cultured them on a modified MS medium. During the third to fourth week in culture, the first sign of shoot initiation appeared as small green areas on the callus. Adventitious shoots differentiated from these green areas, and 8–12 adventitious shoots were formed on the shoot callus when supplied with 1.2×10^{-6} M BA and 5.7×10^{-7} M IAA. The shoots were stunted, remaining as whorls of leaves without stem growth; however, when placed on a medium containing either 5×10^{-7} M or 5.7×10^{-6} M IAA, internode elongation took place. Rooting could be accomplished by adding 4.9×10^{-7} M IBA to the medium.

Chalupa (1947a, 1975) has regenerated plantlets from the callus of *Populus euroamericana* cv. *robusta, Populus nigra* var. *typica, Populus tremula,* and *Populus canescens.* Both modified Wolter's and modified Linsmaier and Skoog's media were used for growth of the calluses. The data given (Chalupa, 1974a) was for *P. euroamericana* cv. *robusta*. Root formation from callus was optimum with 0.4 ppm NAA on Wolter's medium, whereas shoot formation was optimum with 0.3 ppm BA on a modified Linsmaier and Skoog medium. The shoots could be rooted on modified Wolter's medium with 0.4 ppm NAA or placed in Perlite and sand with 1/2 strength modified Wolter's medium with 0.2 ppm NAA, and 0.2 percent sucrose. The second method produced the longer, strongest roots.

Progress toward mass production of *Populus* using suspension culture has been reported by Riou et al. (1975). A friable callus was obtained on modified MS salts, Nitsch and Nitsch's vitamins, glutamine 200 ppm, casein hydrolysate 5 g/l, sucrose 2 percent, and 2,4-D at 1.0 ppm. The friable callus was suspended to form cell clumps and single cells, sieved, suspended in nutrient agar at 40°C, and plated out in a Petri dish. Colonies of callus grew from the cell clumps in the dark. The medium for the growth of the colonies was as before except sucrose was increased to 30 g/l; K_2HPO_4, 500 mg/l; aspartic acid, 6 g/l; methionine, 492 mg/l; cysteine, 560 mg/l; alanine, 294 mg/l; glutathione, 1.1 g/l; tyrosine, 300 mg/l; serine, 173 mg/l; arginine, 287 mg/l; 2,4-D, 1 ppm; and BA, 1 ppm. After 35 days, small colonies had formed. Next, shoots were induced on these colonies after transferring to media of the original composition but with the homones changed to 0.1 ppm NAA and 0.1 ppm BA. The shoots were excised and transferred to another medium different from the preceeding one in that NAA was present at 1 ppm, the sucrose at 10 g/l, and 0.4 percent activated carbon was added. Plantlets were obtained.

It would seem that *Populus* propagation via tissue culture should be operational by now. The callus can be readily started. Plantlets can be produced in high yields, and even from callus after many subcultures. The plantlets do survive in the field; however, no one has yet demonstrated that the plantlets are true to type both from the initial culture and its subculture, i.e., that they have retained the desirable characteristics of the original selection and maintained the same chromosome number and genetic traits. In addition, for many clones of *Populus*, tissue culture is expected to clear up poplar decline disease problems. That the plantlet route through tissue culture can do this has yet to be demonstrated. What is needed are more field studies with the plantlets, along with laboratory studies on chromosome composition and retention or loss of disease organisms during culture.

Plantlets have been obtained from other angiospermous trees in culture, e.g., *Ulmus campestris* (Chalupa, 1975), *Ulmus americana* (Durzan and Lopuchanski, 1975), and *Acacia koa* (Skolmen and Mapes, 1976), but one of the most exciting is the production of birch plantlets by Huhtinen and

Yahyaoglu (1974; Huhtinen, 1976a). For these experiments they used an early-flowering strain of *Betula pendula*. These flower within their first year and bear only reproductive buds. A cambial callus was initiated on MS medium plus 25 ppm IAA and 0.5 ppm kinetin. Both roots and shoots were initiated independently on this callus. Shoots were excised and rooted on MS medium with 0.1 ppm 2,4-D. The plantlets were potted, grew rapidly, and initiated many lateral shoots. In five months male flowers started to develop and the pollen appeared normal; however, there were noticeable differences from the parent plants. Most outstanding was the observation that fast-growing plantlets produced some vegetative buds and could continue vegetative growth after flowering. The parents had produced only reproductive buds. By virture of the early-flowering characteristic, which was maintained by all the plantlets, it should be possible to do genetic studies and to obtain a better understanding of genetic changes that occur when a tree is propagated via callus culture.

Morselli et al. (1974) found GA_3 would release the dormancy of isolated apical shoots of *Acer saccharum*. In two weeks, four leaves had grown; four weeks later root primordia appeared. A defined medium supplemented with IAA or kinetin did not stimulate dormancy release.

We have taken a slightly different approach to bud culture. Buds of tulip poplar (*Liriodendron tulipifera*) and sweetgum (*Liquidambar styraciflua*) were taken in mid-February, when winter dormancy had been completed. They were surface-sterilized, the bud scales removed, and the exposed shoot tip placed on modified Risser and White's or modified MS media. Bud opening and growth occurred with little delay. After transfer to fresh media for bud growth, the buds were transferred to modified Morel's medium (Start and Cummings, 1976) with IBA for root initiation. After two weeks the cultures were transferred to Risser and White's medium without auxin. A few buds from each species rooted readily. Yellow poplar buds started extension growth and have been potted in soil and are in the growth room. Sweetgum buds have not yet started extension growth.

HAPLOID CULTURES IN WOODY SPECIES

The production and use of haploids has not been exploited as a tool in forest tree improvement (Winton and Stettler, 1974; Stettler, 1976). Since haploids have proved useful for the production of homozygous lines of corn (Chase, 1974), in the breeding of improved varieties of barley (Kasha, 1974b), and in cutting the production time for new varieties of tobacco by one-half (Nakamura et al., 1974), haploids might prove useful in similar ways in tree improvement.

Haploids have been found in *Picea abies* (Illes, 1964), *Prunus persica*

(Toyama, 1974), *Theobroma cacao (Dublin, 1974)*, and *Thuja plicata* (Simak et al., 1974). They occur in these and other species at low frequency. In *Prunus persica*, 24 haploids were found among 20,053 seedlings. Higher rates of haploid progeny sometimes can be obtained by artificial means. For example, Illes (1974a, 1974b) has obtained haploids of *Populus* through parthenogenesis by inactivating the male gametes with Toluidine-blue-O after pollen tube growth has started, but before fertilization has taken place.

Even greater rates of haploid production should be obtainable through the use of anther, microspore, or megaspore cultures. We need only to note the enormous numbers of haploid individuals that can be obtained from the anthers of a single plant of tobacco. However, we should note that anther culture has been most successful in the production of haploid plants from inbred lines that presumably are lethal free. Such lines are not available for forest trees. For instance, Sorensen (1971) calculated the minimum number of lethal genes required to account for seed loss resulting from selfing of several Douglas fir trees. The results ranged from 3 to 27 lethal genes. Even with 3 lethal genes present, only about 12 percent of the products of meiosis would be lethal-free, thus leading to high losses based on the numbers that might be expected under more favorable circumstances as in the case of tobacco. But this limitation is true regardless of the method used to obtain haploids from trees. To look at the situation in a more positive light, we need only to think of the possible impact a few homozygous trees could have on the understanding of inheritance in a tree-improvement program (Kasha, 1974a).

C. D. LaRue (1954) was probably the first to attempt to obtain the continued growth of pollen beyond the mature grain stage. During his studies on the microgametophytes of *Zamia floridana*, he reported the development of up to 8 similar cells within the pollen exine. To LaRue, these groups of cells resembled the primordia of adventive embryos. He conducted experimens with pollen from 18 species of conifers, but was unable to obtain continued growth of the pollen cells. These experiments were performed on White's medium both with and without IAA. At the same time, one of LaRue's students, Walter Tulecke (1953), was studying the growth of *Ginkgo* pollen *in vitro*. Among the abnormalities he found was the development of a tissue from the mass of germinating pollen grains. The tissue was reported to have been subcultured 13 times. Its original chromosome complement was 12, the haploid number for *Ginkgo*. About 4 percent of the 634 cultures tried produced a tissue. As before, White's medium was used, but the best growth was obtained by adding 0.25 percent yeast extract and 1 ppm IAA. Later Tulecke (1957) reported additional information on the formation of tissue masses on different media tested. In the case of White's medium plus yeast extract and 1 ppm IAA, he estimated one tissue mass was formed per 1,450,000 pollen grains in culture. Using White's medium with 20 percent autoclaved coconut milk, the ratio was 1:9000. Tissue formation could also be obtained using synthetic White's medium with 1 ppm calcium pantothenate, and 0.1 ppm NAA, with a ratio of 1:380,000. Serial subculture of the tissue required yeast extract or coconut milk.

In contrast to more recent work with angiosperms, Tulecke believed the tissue masses were derived by divisions in, and septation of, the tube cell. Tulecke (1960) attempted to devise a synthetic media to replace the yeast extract or coconut milk without success, so he next attempted to develop new strains of callus from pollen on synthetic medium. These experiments produced a *Ginkgo* callus on White's basal medium supplemented with 0.1 ppm NAA and arginine. Ammonium sulfate added at 0.5 mM during three transfers gave growth rates only slightly lower than those with 0.5 mM L-arginine as a source of reduced nitrogen. Nitrate was not utilized. Ammonium sulfate when added to the medium with proline or arginine greatly stimulated the growth rate of the callus. Though not an explanation offered by Tulecke, it seems that a reduced nitrogen source may have been one of the factors limiting attempts at obtaining a tissue from pollen.

Tulecke (1959) reported on a tissue obtained from *Taxus* pollen by LaRue. The synthetic medium developed for *Ginkgo* supported growth of the *Taxus* callus, and a green stain of callus was selected from this culture. All previous tissue obtained from microspores had not been green. Tulecke and Sehgal (1963) reported obtaining a tissue from *Torreya nucifera* pollen. It grew well not only on the synthetic *Ginkgo* medium but also on a modified Wood and Braun's *Vinca* medium. The pioneering work of Tulecke and LaRue demonstrated that tissues could be obtained from microspores and grown on a synthetic medium.

There was not extensive interest in the growth of tissues from gymnosperm pollen until after the discovery that haploid tobacco plants could be obtained via anther culture. Bonga and Fowler (1970) started some studies on the culture of microsporangia, microsporophylls, and mature pollen of *Pinus resinosa*. This species was chosen because it is believed to carry fewer deleterious genes than other pines. Collections were made from early May until pollen shed in June. Six media were used (White's medium with 10 percent coconut milk, 250 ppm casein hydrolysate, and 1 ppm IAA; Harvy's; Brown and Lawrence's; Tulecke's *Ginkgo*; Tulecke and Seghals'; and Tulecke's female gametophyte). The microsporangiate strobili produced large amounts of callus. Most of the callus came from somatic tissue, but some of it came from the microspores as confirmed by chromosome counts. The callus originated from many microspores. Field observations showed meiosis had taken place on 21 May, and haploid callus was found only in cultures of material collected from 16 May to 5 June. In some cases meiosis must have taken place in culture. The haploid callus was randomly distributed and varied in size and shape. Mature pollen did not form a callus. Further studies (Bonga, 1974) were done on microsporophyll cultures using Brown and Lawrence's medium. The strobili were stored in the cold for 1–4 weeks. In these studies no callus was formed from microspores collected before or during meiosis; however, in the case of those collected 2–9 days after meiosis, up to 80 percent of the microsporangia in 3–6 weeks split open and a yellow callus grew out. The callus contained both haploid and diploid cells. The cultures from the ninth day after meiosis had the highest proportion of haploid to diploid (536 haploid to 119 diploid) cells. Cultured material collected 7 days after meiosis produced more diploid than haploid metaphases, while those collected 11 and 14 days after meiosis did not produce a callus. The calluses could be subcultured. Later, Bonga and McInnis (1975) determined that the number of sporophylls forming callus could be increased if the sporophylls were first centrifuged at 2,000 rpm for 30 minutes before being placed in culture on Brown and Lawrence's medium with 1 ppm 2,4-D.

The only instance where organogenesis may have been obtained from sporophyll callus is the formation of roots from *Cryptomeria japonica* (Isikawa, 1972); however, the ploidy of the roots was not determined.

Duhoux and Norrell (1974) isolated the pollen grains of *Juniperus chinensis* L., *Juniperus communis* L., and *Cupressus arizonica* on several media. Some haploid cultures were obtained. Growth induction was best when somatic tissue of the sporophyll was present. There appeared to be synergistic interaction betwen the amino acids and growth substances.

We have attempted sporophyll cultures of loblolly, longleaf, and Virginia pines, *Cunninghamia, Cedrus deodara,* and *Cedrus atlantica.* Some callus has been observed with all species tried on at least one of twelve media tested. So far, the only verified haploid callus obtained has been from Virginia pine. The strobili were collected shortly after they had burst the bud scales. Meiosis was in progress at the time. Callus was produced on both Brown and Lawrence's (1968) medium and on Gresshoff and Doy's (1972) number 2 synthetic tomato medium. In both instances the callus contained mitotic figures with the haploid chromosome number.

The proposition that conifer pollen can be induced to continue to divide and form a tissue has been established, and in addition, in one case the tissue was green. As has more recently been shown by Rohr (1973), plastids are present in *Taxus* pollen and the ability to synthesize chlorophyll is only inhibited, not lost. Some difficulties in using sporophyll cultures for obtaining haploids is the apparent lack of confirmed organogenesis by gymnosperm microspore tissues. In all calluses obtained to date there has been a mixture of cells of differing origin and ploidy number. Thus the callus may contain cells not only of somatic origin but also all the possible segregation products of meiosis. Thus, if a plantlet were obtained, it appears highly possible it could be chimeric or not of microspore origin. So until the culture of microspores of conifers has been refined to the point where individual microspores can be cultured and undergo embryogenesis or organogenesis, its usefulness as a system to obtain haploid plants is limited.

Far less work has been done on the microspore culture of angiospermous trees. LaRue was not able to obtain any continued growth from the pollen of *Catalpa, Hibiscus,* or *Virburnum* (Tulecke, 1959). More recently, several reports have been made on the culture of anthers from trees. Jordan (1974) has worked with *Prunus avium*, and the pollen produced multicellular pollen grains that after 4 weeks burst the exine-producing small clumps of cells. These anthers were cultured on chromatography paper over a liquid medium after Nitsch with 1 ppm NAA and 1 ppm BA. Tissue formation was not obtained with an agar medium. Additional studies were done with *Prunus persica* and *Prunus amygdalius* by Michellon and coworkers (1974). Only *Prunus persica* var. Nectared IV produced a haploid callus. Miller's salts proved to give more callus than Nitsch's, and supplementing this with 0.2 ppm kinetin, 1.1 ppm 2,4-D, and 0.1 ppm IAA gave the greatest numbers of calluses for all species and varieties used. Most of the calluses produced were formed on the filament rather than originating from the microspores. Zenkteller and coworkers (1975) later obtained a multicellular pollen from *Prunus avium* on MS medium with 12 percent sucrose, 1 ppm IAA, and 1 ppm BA. Cell division was sufficient to burst the exine. During the first 2 weeks in culture, all except 0.1 percent of the pollen degenerated leaving a small sample of the total for multicellular pollen formation.

Sato (1974) has obtained root formation from the callus of anther cultures of *Alnus tinctoria, Prunus apetala, P. edoensis,* and *Prunus lannesiana,* and complete plantlets from anther cultures of three hybrid poplars. However, the chromosome counts from the poplar root tips were diploid.

The Tree Improvement Laboratory, Heilungkiang Institute of Forestry (1975), has worked with anther cultures of *Populus ussuriensis* and hybrid *Populus simonii* × *P. nigra.* Using Bourgin and Nitsch's medium with 1 percent agar, callus induction from the hybrid was 34.9 percent when 2 ppm of 2,4-D and 1 ppm kinetin was added. In a study of 142 calluses, 33.8 percent were from pollen, 52.8 percent from filaments, and 13.4 percent from walls and other origins. Under similar conditions only 15 of the *P.*

ussuriensis anthers formed calluses; however, no callus was formed from the filaments. It was found that if the hormones were changed to 3 ppm kinetin and 0.5 ppm IAA or 2 ppm kinetin and 0.3 ppm IAA, 80 percent of the calluses would differentiate shoots. The shoots could be rooted in a medium with 2 ppm IAA and 0.5 ppm kinetin. Chromosome counts of the root tips gave the compliment as 19, confirming that the roots were haploid.

Several people have asked about a report of Winton and Huhtinen (1976) that we have cultured elm anthers. These anthers were cultured either on Staritsky's coffee medium (1970) or Sharp's medium (1972). Those cultured on Staritsky's medium formed a callus from the filament, but this did not occur on Sharp's medium. The callus produced on Staritsky's medium was hard and white to light brown in color, whereas the callus produced on Sharp's medium was friable and dark brown. Both lines were periodically subcultured for three months both on their initiation medium and the other medium. Each retained its original texture and color on both media. No attempt was made to obtain chromosome counts. This experiment should be repeated before any conclusion is drawn on the origin of the friable callus.

Looking at the possibilities of using anther culture to obtain haploids of angiospermous trees, many of the problems found with gymnosperms still exist. However, the reported differentiation of plantlets from two members of *Populus* should encourage further work, and does indicate that success might be possible with other species of angiosperms.

Gymnosperms offer a unique opportunity for the establishment of haploid cultures. The megagametophyte is present in the seed as a multicellar structure, usually occupying a greater volume of the seed than the embryo. Not only is the megagametophyte haploid but it is derived from only one megaspore. Thus all cells are of one genotype and not of numerous genotypes as with the pollen in the anther.

Again, the first studies done on the culture of the female gametophytes were done by LaRue and his students Tulecke and Norstog. In 1948 LaRue reported the regeneration of roots and shoots from the megagametophyte of *Zamia floridana*. Approximately one-half of the gametophytes formed a layer of cork on the outside, a few showed some greening, and less that 1 percent showed regeneration of organs. If the megagametophyte was taken at about the time of fertilization, the differentiation of organs took about one year. If it was taken 2 months before fertilization, regeneration of organs took at least 3 months; and if taken 3 months before fertilization, regeneration took at least 5 months. Younger megagametophytes could not be cultured. So there were apparently subtle differences between gametophytes and stage of development that determined their ability to differentiate organs. No plants that continued to grow were recovered. The work was successful in that it did demonstrate that at least apogamy was possible in the case of gymnosperm megagametophytes.

Later in 1954, LaRue summarized his work on the culture of the megagametophytes and microgametophytes of gymnosperms. He had continued the work with *Zamia* and found another type of regeneration in the form of little balls of tissue he called "pseudo-bulbils." When detached from the gametophyte and placed in culture, the pseudobulbils produced more pseudobulbils. After 2 years some of these pseudobulbils elongated and formed one- or two-pointed structures at one end. In time the single points formed roots, and the double points formed leaves. In a few instances where both types of points were present on the same structure, small plants were formed but did not grow further. This work was extended to *Cycas revoluta* megagametophytes, with a similar regeneration of buds and roots after being cultured on a simple nutrient, using sand as a support. In this instance regeneration required 2 years. In 11 other genera of gymnosperms tried, many of them conifers, no regeneration of organs by the megagametophyte occured.

In 1965 a former student of LaRue's, Norstog (1965), again studied the problem of apogamy in the megagametophyte of *Zamia*. Several media, ranging from water and agar alone to White's salts with sucrose and various hormones and amino acids, were tried. The culture period was generally 2 months (1 month shorter than LaRue), although some ran for six months; none extended to years. During the culture period roots were differentiated only on media containing an auxin and a cytokinin. If glutamine and asparagine were added, root production increased and leaves also were differentiated. The percentage of cultures with roots was increased when, in addition to the above, adenine and alanine were added to the media. In addition in this study, the roots were shown to be differentiated from meristematic areas on the periphery of the megagametophyte. Chromosome counts showed the roots were haploid ($n = 8$). With the best treatments 82 percent of the cultures formed roots, leaves, or both. This represents quite an improvement over LaRue's results, which were less that 1 percent. This study demonstrated for the first time the ability to control organogenesis in a predictable manner in a gymnosperm tissue culture.

Further studies by Norstog and Rhamstine (1967) resulted in a system that produced callus, pseudobulbils, and plantlets with both haploid and diploid tissues of *Zamia*. Callus was initiated on medium 59 (MS salts, glutamine, alanine, 2,4-D, and kinetin). When the callus was transferred to

media 21, a medium similar to 59 but without 2,4-D or kinetin, pseudobulbils were formed. For growth of the pseudobulbils to plantlets, they were
transferred to medium 71, which contained only the mineral salts. However, the plantlets did not develop to more than a few centimeters in size. This
demonstrated it was possible to form plantlets from a gymnosperm after
the production of callus, but left as a major concern whether the plantlets
would form full-size plants.

Most of the work we have spoken of up to this time has been done with
Zamia, a gymnosperm, but quite different in many respects from members
of the coniferales.

In 1964 Tulecke reported the successful culture of the female gametophyte of *Ginkgo biloba*. Callus could be started on either a White's medium
with 6 ppm 2,4-D, and 18 percent coconut water or on a modified Wood
and Braun *Vinca* medium. The synthetic medium was best for the
continued growth of the callus. The chromosome count was 12, confirming
the haploid nature of the callus, and the callus was green. The best stage of
development of the female gametophyte for starting the cultures was 11–13
weeks after pollination. Fertilization does not take place until 4–8 weeks
later. In 1965 Tulecke reported that 15 roots had been observed from the
cultures of the female gametophytes cultured on several media and that one
culture had produced 5 abortive shoots.

Borchert (1968) cultured megagametophytes of *Pinus lambertiana* and
studied the relative amounts of DNA in the nucleus and the chromosome
number of the callus. Callus was initiated from mature female gametophytes on Brochert's high salts medium with 1 ppm 2,4-D, 0.2 ppm kinetin,
and 10 ppm adenine. The callus was maintained in Heller's minerals with 2
percent sucrose, 15 percent coconut milk, 1 ppm 2,4-D, and 0.2 ppm
kinetin. Upon subculture the relative DNA content and chromosome
numbers of the cells were determined. The cells had converted to diploid
and even, rarely, to tetraploid.

Zenkteler and Guzowska (1970) cultured *Taxus baccata* female gametophytes– derived callus on modified White's medium with 2 percent sucrose,
0.8 percent agar, 500 ppm casein hydrolysate, and 5 ppm 2,4-D. Only 2–5
percent of the female gametophytes produced a callus. The callus was
subcultured for 3 years. Female gametophyte cells in this case were
multinucleate, whereas callus cells were uninucleate. Younger cells at the
periphery of the callus contained chloroplasts. The chromosome number
of the cells in the fourteenth and fifteenth passage varied from haploid to
octoploid, and haploid and diploid frequencies were low.

During the last few years Huhtinen (1972, 1976b; see also Thomas and
Davey, 1975) has done extensive work on the culture of the female
gametophyte of *Picea abies*. He examined the stages of female gameto-

phyte development best suited for callus production in culture. Cultures could be established from the mature female gametophyte; however, better rates of success (0.1–1.0 percent) were achieved with female gametophytes that had just ceased mitosis but had not changed from a "liquid" to a "solid" consistency. The fertilized archegonia and embryos were hard to detect and remove at this stage. Some female gametophytes at this stage placed on Risser and White's medium with 0.5 ppm 2,4-D, and 1 ppm kinetin grew slowly; and when transferred to Risser and White's medium with 0.5 pm 2,4-D and glutamine as an organic nitrogen source, the cultures became brown and root-like structures were formed. Some tissues formed a light, yellow-green pigment that turned brown in time. To get around the difficulty of removing embryos and fertilized archegonia, Huhtinen vernalized the seeds in the cones at high humidity under refrigeration. Under these conditions the embryos continued to grow, making their later removal easier. The female gametophyte remained soft during the period of vernalization. Successful establishment of cultures ranged from 0.1–5 percent on a modified MS medium with 500 ppm glutamine, 0.1–1 ppm kinetin, and 1.0–10 ppm 2,4-D. Ninety percent of the original calluses were haploid, the remainder diploid or polyploid. Some of the lines remained haploid upon subculturing. Some cultures when transferred to a medium with 1 or 5 ppm kinetin formed small nodules, and these developed into embryo-like shoots bearing cotyledons, although none of these shoots rooted.

Bonga and Fowler (1970) have cultured the immature female gametophyte of *Pinus resinosa*. Callusing was not common, but development of the gametophyte did continue on White's medium supplemented with either "Medium 199 with 1-glutamine" (an animal tissue culture medium), coconut milk, casein hydrolysate, and 2,4-D or IAA. Later, Bonga (1974) worked with Austrian and Mugo pine female gametophytes collected one month after fertilization. The cultures were on Brown and Lawrence's medium. About 10 percent of the Austrian and 15 percent of the Mugo pine cultures produced a bright green callus. Among the calluses examined nearly all showed only haploid metaphases. The fastest-growing haploid lines were subcultured and examined for cellular differentiation. Only Mugo pine lines showed circular growth centers that sometimes contained tracheids.

Later, Bonga (1977) published a photo of a plantlet arising from a female gametophyte callus of *Picea abies* cultured on Brown and Lawrence's medium with 0.2 ppm IAA in place of 2,4-D.

We have cultured the female gametophytes of longleaf pine, pinyon pine (*Pinus edulis*), and white pine (*Pinus strobus*). The majority of our work has been with longleaf pine. We were able to obtain at least sporadic callusing of the female gametophyte with several media including Sharp's (Sharp et al., 1972), Brown and Lawrence's (1968), and Gresshoff and Doy's TMS2 (GDII) (1972). The GDII medium gave the best results, producing a callus within one week to one month's time. The calluses were small, and they usually did not grow appreciably on transfer to fresh medium. Two of our cultures have developed a green callus. If kept on agar, the callus was lost by the fourth subculture; however, if the proliferating female gametophyte was transferred to liquid GDII on a shaker, often the callus would continue to grow. Growth was slow initially, requiring two months to build up a sufficient population of cells in the medium to require a change to fresh medium. The cells were predominately haploid at that time. After the suspension cultures were established, they showed a 3- to 6-fold increase in growth every eight weeks. These cultures have now been maintained for more than three years, are no longer haploid, and are still bright green. They now grow on agar and require transfer every three weeks, just as our somatically derived pine cultures do.

As we view what has been done with the female gametophytes of gymnosperms, the situation seems very encouraging. Organogenesis of varying degrees has been obtained. Some lines show stability in maintaining the haploid chromosome number, but in other cases there appears to be high variability in chromosome complement, particularly to the extent that the female gametophyte cells may be multinucleate. With sufficient effort expended we expect that haploid conifer plantlets will be produced from female gametophytes in the near future.

FUTURE ROLE OF TISSUE CULTURE IN TREE IMPROVEMENT

There are relatively few publications (Chalupa, 1974b; Rona and Grigmon, 1972; Saito, 1976; Wakasa, 1973; Winton et al., 1975) concerning the production of protoplasts from tree tissue, and none that deal with the production of protoplasts from haploid tissue of trees. However, in the future we foresee a substantial role for fusing haploid protoplasts in the breeding of economically important trees. Current methods for the hybridization of pines are costly and relatively inefficient except under special circumstances where, for example, in South Korea, reforestion is done with the *Pinus rigida* F_1 hybrid from *P. taeda* crosses that, ideally, retain the cold resistance of *P. rigida* and the better growth characteristics of *P. taeda*. Low labor costs allowed enough controlled crosses to be made to produce enough seed for mass planting of the hybrid.

In the southern pines there are also other desirable crosses known, viz., shortleaf × loblolly pines and longleaf × slash pine, for improved resistance to fusiform rust. In some species such as Douglas fir there are problems with the storage of pollen from one breeding season to another. Parasexual hybridization would circumvent having to wait until the next breeding season to make the crosses, the expense of making controlled crosses, and the risk of unfavorable conditions affecting pollen and seed formation. In angiosperms some triploids have shown exceptional growth rates; however, natural triploids are rare. Again, if the parasexual route were taken, the fusion of a haploid and diploid cell or of 3 haploid cells would give rise to triploids.

Before parasexual hybridization becomes a reality, we need to learn more about the behavior of tree protoplasts, their controlled fusion, and the regeneration of plantlets from the fusion products. Of course, the first step is try it. There is still another method of producing polyploid angiosperms, i.e., culture of the endosperm and regeneration of plantlets (Zenkteler et al., 1970; Bhojwani and Johri, 1970; Johri and Srivastava, 1970).

Much speculation has taken place concerning the use of protoplast fusion or transgenosis to induce pines and other trees to fix nitrogen. Giles addresses this topic later in this volume, but I feel we should mention here that Giles and Whitehead (1975) have induced the protoplasts of ectomy-corrhizal fungus to take up *Azotobacter* cells and have demonstrated acetylene reduction by the fungus. Perhaps since the conifers, and many dicotyledenous trees, have ectotrophic mycorrhizae, one of the best ways to add nitrogen fixation ability is to the mycorrhizal fungus rather than to the tree itself.

There are many areas of tissue culture we have not covered that could play a significant role in tree improvement, currently or in the future, e.g.: *in vitro* studies of flowering and fertilization; meristeming; comparative studies of growth *in vitro* vs. *in vivo* growth in the field; screening for disease resistance, natural products production, isolation of disease-free or disease-resistant strains, production of aneuploids, polyploids, and other abnormal forms for genetics research; and studying physiological problems such as phasic development, processes of aging, and mycorrhizal symbiosis, to mention only a few.

But at present our most-pressing problems are the regeneration of plantlets from somatic and gametic tissue, followed by field testing, so that tissue culture may take its proper place among the techniques available for forest tree improvement. From what we have said previously, you may feel that the major problems in regeneration have been solved; however, regeneration for tree improvement implies more than just producing one or more plantlets sporadically. It implies not only obtaining plantlets from juvenile material but also from "mature" sources; being able to regenerate the species from many selections, families or provenances; to regenerate plantlets that duplicate parent phenotypes and genotypes; to produce enough plantlets for genetic testing; and to produce plantlets that will survive in the greenhouse and in the field.

In conclusion, let us summarize by saying that today programs for forest tree improvement are needed more urgently than ever before, and that they must employ every tool available to the tree-breeder, including tissue culture. But since we are dealing with a crop, worldwide, that consists of hundreds of species composed of highly heterozygous populations, the job will not be as easy as producing a new variety of a single herbaceous plant such as tobacco. It can be done, and the possibilities are exciting; but the road will be long, and it will demand a long-term commitment.

ACKNOWLEDGMENTS

Much of the authors' tissue culture research has been supported over the past five years by the Georgia Forest Research Council, Macon, Georgia.

REFERENCES

Ball, E. 1950, Differentiation in a callus culture of *Sequoia sempervirens*. Growth 14: 295–325.

Berbee, F. M., J. G. Berbee, and A. C. Hildebrandt. 1972. Introduction of callus and trees from stem tip cultures of a hybrid poplar. In Vitro 7:269.

Bhojwani, S. S., and B. M. Johri. 1970. Cytokinin-induced shoot bud differentiation in mature endosperm of *Scurrula pulverulenta*. Z. Pflanzenphysiol. 63:269–75.

Bonga, J. M. 1974. In vitro culture of microsporophylls and megagametophyte tissue in *Pinus*. In Vitro 9:270–77.

Bonga, J. M. 1977. Applications of tissue culture in forestry. *In* J. Reinert and Y. P. S. Bajaj (eds.), Applied and fundamental aspects of plant cell, tissue, and organ culture. Pp. 93–108. Springer-Verlag, Berlin.

Bonga, J. M., and D. P. Fowler. 1970. Growth and differentiation in gametophytes of *Pinus resinosa* cultured in vitro. Canad. J. Bot. 48:2205–7.

Bonga, J. M., and A. H. McInnis. 1975. Stimulation of callus development from immature pollen of *Pinus resinosa* by centrifugation. Plant Sci. Lett. 4:199–203.

Borchert, R. 1968. Spontane Diploidisierung in Gewebekulturen des Megogametophyten von *Pinus Lambertiana*. Z. Pflanzenphysiol. 59:389–92.

Brand, R., and C. J. Venverloo. 1973. The formation of adventitious organs. II. The origin of buds formed on young adventitious roots of *Populus nigra* L. 'Italica.' Acta Bot. Neerl. 22:399–406.

Brown, C. L. 1964. The place of longleaf pine in southern forest management. Forest Farmer 24(3):6–9, 18–19.

Brown, C. L. 1976. Forests as energy sources in the year 2000: What man can imagine, man can do. J. For. 74(1):1–6.

Brown, C. L., and R. H. Lawrence. 1968. Culture of pine callus on a defined medium. For Sci. 14:62–64.

Brown, C. L., and H. E. Sommer. 1977. Bud and root differentiation in conifer cultures. TAPPI 60:72-73.

Campbell, R. A., and D. J. Durzan. 1975. Induction of multiple buds and needles in tissue cultures of *Picea glauca*. Canad. J. Bot. 53:1652-57.

Campbell, R. A., and D. J. Durzan. 1976a. Vegetative propagation of *Picea glauca* by tissue culture. Canad. J. For. Res. 6:240-43.

Campbell, R. A., and D. J. Durzan. 1976b. The potential for cloning white spruce via tissue culture. *In* USDA Forest Service Gen. Tech. Rept. NC-26. Proc. Twelfth Lake States Forest Tree Improvement Conference. Pp. 158-66. Forest Service, St. Paul, Minnesota.

Carlson, P. S., H. H. Smith, and R. D. Dearing. 1972. Parasexual interspecific plant hybridization. Proc. Nat. Acad. Sci. USA 69:2292-94.

Chalupa, V. 1974a. Control of root and shoot formation and production of trees from poplar callus. Biol. Plantarum 16:316-20.

Chalupa, V. K. 1974b. Isolation and division of protoplasts of some forest tree species. Abstract No. 174. 3rd Intern. Cong. of Plant Tissue and Cell Culture, Leicester.

Chalupa, V. 1975. Induction of organogenesis in forest tree cultures. Commun. Inst. For., Czechosloveniae, 9:39-50.

Chase, S. S. 1974. Utilization of haploids in plant breeding: Breeding diploid species. *In* K. J. Kasha (ed.), Haploids in higher plants. Pp. 211-30. University of Guelph, Guelph, Ontario, Canada.

Cheng, T. 1975. Adventitious bud formation in culture of Douglas fir (*Pseudotsuga menziesii* (Mirb.) Franco). Plant Sci. Lett. 5:97-102.

Cheng, T. 1976. Vegetative propagation of western hemlocks (*Tsuga heterophylla*) through tissue culture. Plant and Cell Physiol. 17:1347-50.

Cheng, T. 1977. Factors effecting adventitious bud formation of cotyledon culture of Douglas fir. Plant Sci. Lett. 7:179-87.

David, A., and H. David. 1977. Manifestations de diverses potentialités organogènes d'organes ou de fragments d'organes de Pin maritime (Pinus pinaster Sol.) en cultivé in vitro. C. R. Acad. Sci. Paris 284D:627-30.

Dublin, P. 1974. Les haploides de *Theobroma Cacao* L. diploidisation et obtention d'individus homozygotes. Café Caco Thé 18:82-96.

Duhoux, E., and B. Norreel. 1974. Sur l'isolement de colonies tissulaires d'origine pollinique à partir de cônes mâles du *Juniperous chinensis* L., du *Juniperus communis* L. et du *Cupressus arizonica* G., cultivés in vitro. C. R. Acad. Sci. Paris 279D:651-54.

Durzan, D. J., and S. M. Lopushanski. 1975. Propagation of American elm via cell suspension cultures. Canad. J. For. Res. 5:273-77.

Gautheret, R. J. 1934. Culture du tissu cambial. C. R. Acad. Sci Paris 198:2195-96.

Gautheret, R. 1940a. Recherches sur le bourgeonnement du tissu cambial d'*Ulmus campestris*, cultivé in vitro. C. R. Acad. Sci. Paris 210:632-34.

Gautheret, R. 1940b. Nouvelles recherches sur le bouregonnement du tissu cambial d'*Ulmus campestris* cultivé in vitro. C. R. Acad. Sci. Paris 210:744-46.

Gautheret, R. J. 1957. Histogenesis in plant tissue cultures. J. Nat. Cancer Inst. 19:555-90.

Giles, K. L., and H. C. M. Whitehead. 1975. The transfer of nitrogen fixing ability to a eukaryote cell. Cytobios. 14:49-61.

Glesinger, E. 1949. The coming of age of wood. Simon and Schuster, New York. 279 pp.

Goldstein, I. S. 1977. The place of cellulose under energy scarcity. *In* J. C. Arthur, Jr. (ed.),

Cellulose chemistry and technology. Symposium on International Developments in Cellulose, Paper and Textiles. Pp. 382–87. ACS Symp. Series 48. Washington, D. C.

Gresshoff, P. M., and C. H. Doy. 1972. Development and differentiation of haploid *Lycopersicon esculentum* (Tomato). Planta 107:161–70.

Heilung Institute of Forestry, Tree Improvement Laboratory. 1975. Induction of haploid poplar plants from anther culture *in vitro*. Scientia Sinica 18:769–77.

Huhtinen, O. 1972. Production and use of haploids in breeding conifers. IUFRO Genetics. SABRAO Joint Symposia, Tokyo. D-3 (I):1–8.

Huhtinen, O. 1976a. Early flowering of birch and its maintenance in plants regenerated through tissue cultures. Acta Hort. 56:243–49.

Huhtinen, O. 1976b. *In vitro* of haploid tissue of trees. *In* XVI IUFRO World Congress Proceedings Division II. Pp. 28–30. Norwegian IUFRO Congress Committee os-NLH, Norway.

Huhtinen, O., and Z. Yahyaoglu. 1974. Das frühe Blühen von aus Kalluskulturen herangezogenen Pflanzchen bei der Birke (*Betula pendula* Roth). Silvae Genetica 23:32–34.

Hyun, S. K., and A. Kun-Yong. 1959. Mass production of pitch-loblolly hybrid pine (X *Pinus rigitaeda*) seed. Inst. Forest Genet Res. Rept. 1:11–24. Suwon, Korea.

Illes, Z. M. 1964. Auftreten haploider Keimlinge bei *Picea abies*. Naurwis. 5:422.

Illes, Z. M. 1974a. Induction of haploid parthenogenesis in aspen by post pollination treatment with Toluidine-blue. Silvae Genet 23:221–26.

Illes, Z. M. 1974b. Experimentally induced haploid parthenogenesis in the *Populus* section *Leuce* after late inactivation of the male gamete with toluidin-blue-O. *In* H. F. Linskens (ed.), Fertilization in higher plants. Pp. 335–40. North-Holland Publishing Co., Amsterdam.

Isikawa, H. 1972. Culture of cells and tissues and differentiation of organs in forest trees. IUFRO Genetics SABRAO Joint Symposia, Tokyo D-2(I):1–13.

Isikawa, H. 1974. In vitro formation of adventitious buds and roots on the hypocotyl of *Cryptomeria japonica*. Bot. Mag. Tokyo 87:73–77.

Jacquiot, C. 1949. Observations sur la néoformation de bourgeons chez le tissu cambial d'*Ulmus campestris* cultivé in vitro. C. R. Acad. Sci. Paris 229:529–30.

Jacquiot, C. 1951. Action du mésoinositol et de l'adénine sur la formation de bourgeons par le tissu cambial d'*Ulmus campestris* cultivé *in vitro*. C. R. Acad. Sci. Paris 233:815–17.

Johri, B. M., and P. S. Srivastava. 1973. Morphogenesis in endosperm cultures. Z. Pflanzenphysiol. 70:285–304.

Jordan, M. 1974. Multizelludäre Pollen bei *Prunus avium* nach in-vitro-Kultur. Z. Pflanzenzüchtg. 71:358–63.

Kasha, K. J. 1974a. Haploids in higher plants. University of Guelph, Guelph, Ontario, Canada. 421 pp.

Kasha, K. J. 1974b. Haploids from somatic cells. *In* K. J. Kasha (ed.), Haploids in higher plants. Pp. 67–87. University of Guelph, Guelph, Ontario, Canada.

Keays, J. L. 1975. Projection of world demand and supply for wood fiber to the year 2000. *In* T. E. Timell (ed.), Proc. Eighth Cellulose Conf. I. Wood chemicals—a future challenge. Applied Polymer Symposia No. 28. Pp. 29–45. John Wiley & Sons, New York.

Konar, R. N., and Y. P. Oberoi. 1965. In vitro development of embryoids on the cotyledons of *Biota orientalis*. Phytomorphology 15:137–40.

Kriebel, H. B., and T. H. Shafer, 1971. *In vitro* culture of second year cones of white pines—preliminary results. *In* Proc. on reproduction in forest trees, XV IUFRO Cong. Gainesville, Florida.

LaRue, C. D. 1936. The growth of plant embryos in culture. Bull. Torrey Bot. Club 63:365–82.

LaRue, C. D. 1948. Regeneration in the megagametophyte of *Zamia floridana*. Bull. Torrey. Bot. Club 75:597–603.

LaRue, C. D. 1954. Studies on growth and regeneration in gametophytes and sporophytes of gymnosperms. *In* Abnormal and pathological plant growth. Brookhaven Symposia in Biology No. 5. Pp. 187–208. Brookhaven National Laboratory, Upton, N.Y.

Libby, W. J. 1974. The use of vegetative propagules in forest genetics and tree improvement. N. Z. For. Sci. 4(2):440–53.

McAlpine, R. L., C. L. Brown, A. M. Herrick, and H. E. Ruark. 1967. "Silage" Sycamore. Forest Farmer 26(1):6–7, 16.

Maheshwari, P., and N. S. Rangaswamy. 1965. Embryology in relation to physiology and genetics. *In* R. D Preston (ed.), Advances in botanical research. 2:219–321. Academic Press, New York.

Mathes, M. C. 1964. The in vitro formation of plantlets from isolated aspen tissue. YTON 21:137–41.

Mergen, F. 1958. Natural polyploidy in slash pine. For. Sci. 4:283–93.

Mergen, F. 1959. Colchicine-induced polypolidy in pines. J. For. 57:180–90.

Michellon, R., J. Hugard, and R. Jonard. 1974. Sur l'isolement de colonies tissulaires de Pêrcher (*Prunus persica* Batsch, cultivare Dixired et Nectared IV) et d'Amandier (*Prunus amygdalus* Stokes, Cultivar Ai) á partir d'anthères cultivées in vitro. C. R. Acad. Sci. Paris 278D:1719–22.

Momot, T. S. 1976. Organogenesis of leaf tissues of *Larix dahurica* Turez, cultured in vitro. Lesnop Zhurnal No. 5, pp. 27–29.

Morel, G. 1948. Recherches sur la culture associée de parasites obligatoires et de tissus végétaux. Annales des Épiphyties XIV (N.S.) Serie Pathologie Végétale-Ménoire. No. 5, p. 112.

Morselli, M., K. J. Katagiri, and A. D. Ehrlick. 1974. Gnotoculture of apical meristems of sugar maple (*Acer saccharum*, Marsh). Abstract No. 159. 3rd Intern. Cong. of Plant Tissue and Cell Culture, Leicester.

Mott, R. L., R. H. Smeltzer, A. Mehra-Palta, and B. J. Zobel. 1977. Production of forest trees by tissue culture. TAPPI 60:62–64.

Nakamura, A., T. Yamada, N. Kadontani, and R. Itagaki. 1974. Improvement of fluecured tobacco variety M. C. 1610 by means of haploid breeding methods and some problems of this method. *In* K. J. Kasha (ed.), Haploids in higher plants. Pp. 277–78. University of Guelph, Guelph, Ontario, Canada.

Norstog, K. 1965. Induction of apogamy in megagametophytes of *Zamia integrifolia*. Amer. J. Bot. 52:993–99.

Norstog, K., and E. Ramstine. 1967. Isolation and culture of haploid and diploid cycad tissues. Phytomorphology 17:374–81.

Reilly, K., and C. L. Brown. 1976. *In vitro* studies of bud and shoot formation in *Pinus radiata* and *Pseudotsuga menziesii*. Georgia For. Res. Pap. 86:1–9.

Reinert, J., and P. R. White. 1956. The cultivation *in vitro* of tumor tissues and normal tissues of *Picea glauca*. Physiol. Plantarum 9:177–89.

Riou, A., H. Harada, and B. Taris. 1975. Production des plantes entière à partir de cellules séparées de cals de *Populus*. C. R. Acad. Sci. Paris 280D:2657-59.

Risser, P. G., and P. R. White. 1964. Nutritional requirements of spruce tumor cells in vitro. Physiol. Plantarum 17:620-35.

Rohr, R. 1973. Production de cals par les gametophytes males de *Taxus baccata* L. Cultivés sur un milieu artificiel. Etude en microscopie photonique et éléctronique. Caryologia (Suppl.) 25:177-89.

Rona, J. P., and C. Grignon. 1972. Obtention de protoplastes à partir de suspensions de cellules d'*Acer pseudoplatanus* L. C. R. Acad. Sci. Paris 274D:2976-79.

Saito, A. 1976. Isolation of protoplasts from mesophyll cells of *Paulownia Fortunei* Hemsl. and *Populus euramericana* Cv. 1-45/51. J. Jap. For. Soc. 58:301-5.

Satô, T. 1974. Callus induction and organ differentiation in anther culture of poplars. J. Jap. For. Soc. 56:55-62.

Sharp, W. R., R. S. Raskin, and H. E. Sommer. 1972. The use of nurse culture in the development of haploid clones in tomato. Planta 104:357-61.

Simak, M., A. Gustafsson, and W. Rautenberg. 1974. Meiosis and pollen formation in haploid *Thuja plicata gracilis* Oud. Hereditas 76:227-38.

Skolmen, R. G., and M. O. Mapes. 1976. *Acacia koa* Gray plantlets from somatic callus tissue. J. Hered. 67:114-15.

Skoog, F. 1954. Substances involved in normal growth and differentiation of plants. *In* Abnormal and pathological plant growth. Brookhaven Symposia in Biology No. 6. Pp. 1-21. Brookhaven National Laboratory, Upton, N.Y.

Sommer, H. E. 1975. Differentiation of adventitious buds on Douglas-fir embryos in vitro. Proc. Intern. Plant Prop. Soc. 25:125-27.

Sommer, H. E., C. L. Brown, and P. P. Kormanik. 1975. Differentiation of plantlets in longleaf pine (*Pinus palustris* Mill.) tissue cultured *in vitro*. Bot. Gaz. 136:196-200.

Sorensen, F. 1971. Estimate of self-fertility in coastal Douglas-fir from inbreeding studies. Silv. Genet. 20:115-20.

Staritsky, G. 1970. Embryoid formation in callus tissues of coffee. Acta Bot. Neerl. 19:509-514.

Start, N. D., and B. G. Cummings. 1976. *In vitro* propagation of *Saintpaulia ionantha* Wendl. Hort. Sci. 11:204-6.

Stebbins, G. L. 1950. Variation and evolution in plants. Columbia University Press, New York. 643 pp.

Steinbeck, K., and C. L. Brown. 1976. Yield and utilization of hardwood fiber grown on short rotations. Applied Polymer Symposium No. 28. Pp. 394-401. John Wiley & Sons, New York.

Stettler, R. F. 1976. Haploidy and forest-tree breeding. *In* XVI IUFRO World Congress Proceedings Division II. Pp. 260-66. Norwegian IUFRO Congress Committee, A°s-NLH, Norway.

Thomas, E., and M. R. Davey. 1975. From single cells to plants. Pp. 119-20. Wrjkeham Publications, London.

Thomas, M. J., E. Duhoux, and J. Vazart. 1977. *In vitro* organ initiation in tissue culture of *Biota orientalis* and other species of the Cupressaceae. Plant Sci. Lett. 8:395-400.

Thompson, D. G., and J. C. Gordon. 1977. Propagation of poplars by shoot apex culture and nutrient film technique. *In* TAPPI conference papers-forest biology wood chemistry conference 1977. Pp. 77–82.

Toyama, T. K. 1974. Haploidy in peach. Hort. Sci. 9:187–88.

Tulecke, W. 1953. A tissue derived from the pollen of *Ginkgo biloba*. Science 117:599–600.

Tulecke, W. 1957. The pollen of *Ginkgo biloba*: In vitro culture and tissue formation. Amer. J. Bot. 44:602–8.

Tulecke, W. 1959. The pollen cultures of C. D. LaRue: A tissue from the pollen of *Taxus*. Bull. Torrey Bot. Club 86:283–89.

Tulecke, W. 1960. Arginine-requiring strains of tissue obtained from *Ginkgo* pollen. Plant Physiol. 35:19–24.

Tulecke, W. 1964. A haploid tissue culture from the female gametophyte of *Ginkgo biloba* L. Nature 203:94–95.

Tulecke, W. 1965. Haploidy vs. diploidy in the reproductive cell type. *In* Reproduction: Molecular, subcellular, cellular. Pp. 217–41. Symp. Soc. Develop. Biol. 24. Academic Press, New York.

Tulecke, W., and N. Sehgal. 1963. Cell proliferation from the pollen of *Torreya nucifera*. Contr. Boyce Thompson Inst. 21:153–63.

Venverloo, C. J. 1973. The formation of adventitious organs. I. Cytokinin-induced formation of leaves and shoots in callus cultures of *Populus nigra* L. 'Italica.' Acta Bot. Neerl. 22:390–98.

Wakasa, K. 1973. Isolation of protoplasts from various plant organs. Jap. J. Genet. 48:279–89.

Wetmore, R. H. 1954. The use of "in vitro" cultures in the investigation of growth and differentiation in vascular plants. *In* Abnormal and pathological plant growth. Brookhaven Symposia in Biology No. 6. Pp. 22–40. Brookhaven National Laboratory, Upton, N.Y.

Whitehead, H. C. M., and K. L. Giles. 1976. Rapid propagation of poplars by tissue culture methods. Proc. Intern. Plant Prop. Soc. 26:340–43.

Winton, L. L. 1968. Plantlets from aspen tissue cultures. Science 160:1234–35.

Winton, L., and O. Huhtinen. 1976. Tissue culture of trees. *In* J. P. Miksche (ed.), Modern methods in forest genetics. Pp. 243–64. Springer-Verlag, Berlin.

Winton, L. L., R. A. Parham, and H. M. Kaustinen. 1975. Isolation of conifer protoplasts. Genetics and Physiology Notes No. 20. Inst. of Paper Chem., Appleton, Wis. 9 pp.

Winton, L. L., and R. F. Stettler. 1974. Utilization of haploidy in tree breeding. *In* K. J. Kasha (ed.), Haploids in higher plants. Pp. 259–73. University of Guelph, Guelph, Ontario, Canada.

Wolter, K. E. 1968. Root and shoot initiation in aspen callus culture. Nature 219:509–10.

Wright, J. W. 1976. Introduction to forest genetics. Academic Press, New York. 463 pp.

Zenkteler, M. A., and I. Guzowska. 1970. Cytological studies on the regenerating mature female gametophyte of *Taxus baccata* L. and mature endosperm of *Tilia platyphyllos* Scop. in *in vitro* culture. Acta Soc. Bot. Pol. 39:161–73.

Zenkteler, M., E. Misiura, and A. Ponitka. 1975. Induction of androgenetic embryoids in the *in vitro* cultured anthers of several species. Experientia 31:289–91.

Recent Advances in Development of
In Vitro Techniques for Douglas Fir

26

INTRODUCTION

The potential use of *in vitro* techniques in forest tree improvement programs has been described (Bonga, 1974; Brown, 1976; Cheng, 1976a; Duzan and Campbell, 1974; Konar and Magmani, 1974; Rediske, 1974). The immediate application of tissue culture methods is for mass propagation of genetically superior populations and clones. Morphogenesis in culture of coniferous tree species has been reported for *Sequoia sempervirens* (Ball, 1950), *Biota orientalis* (Konar and Oberoi, 1965; Thomas et al., 1977), *Pinus gerardiana* (Konar, 1972), *Cryptomeria japonica* (Isikawa, 1974), *Picea glauca* (Campbell and Durzan, 1975), *Pinus palustris* (Sommer et al., 1975), *Pseudotsuga menziesii* (Mirb.) Franco (Cheng, 1975, 1977; Sommer, 1975; Winton and Verhagen, 1977), and *Tsuga heterophylla* (Cheng, 1976b).

In this laboratory, *in vitro* initiation of adventitious bud development of gymnosperms has been accomplished using various plant materials as shown in table 1. The main effort of this research has been focused on development of *in vitro* techniques for Douglas fir. Some culture samples showing the production of adventitious buds derived from various Douglas fir plant materials, such as excised mature embryo, cotyledon, needle, and stem, are presented in figure 1. The extent of the applicability of *in vitro*

Tsai-Ying Cheng, Department of Chemistry and Biochemical Sciences, Oregon Graduate Center, Beaverton, Oregon 97005.

TABLE 1

CONIFEROUS TREE SPECIES FOR WHICH PRODUCTION OF
ADVENTITIOUS BUDS IN CULTURE HAS BEEN ACCOMPLISHED

Plant	Explant Source	Ages of Tissues Used
Pseudotsuga menziesii	Excised embryo	. . . .
	Cotyledon	Up to 3 months
	Hypocotyl	Up to 3 months
	Stem	Up to 4 years
		Juvenile
		Adult
	Needle	Up to 2 years
		Juvenile
		Adult
Tsuga heterophylla	Cotyledon	Up to 3 months
	Stem	Up to 2 years
	Needle	Up to 2 years
Pinus taeda	Excised embryo	. . . .
	Cotyledon	Up to 3 months
	Stem	Up to 3 months
Pinus Ponderosa	Excised embryo	. . . .
Pinus caribaea	Excised embryo	. . . .
	Cotyledon	Up to 3 months
Cryptomeria japonica	Cotyledon	Up to 3 months
	Stem	Up to 3 months
Cupressus arizonica	Stem	Up to 3 months

techniques towards cloning and mass propagation of superior genotypes of forest tree species is dependent entirely upon a reliable culture system capable of production of plants at a high frequency. Thus, cotyledon explants of Douglas fir were chosen as a model system for investigating various factors influencing morphogenetic processes. The biochemical mechanisms involved in morphogenesis are also under study.

IN VITRO CLONAL PROPAGATION

Tissue Culture System

Cotyledons obtained from 2- to 4-week-old Douglas fir (*Pseudotsuga menziesii* [Mirb.] Franco) seedlings of open pollinated seed (supplied by the Forestry Research Center, Weyerhaeuser Co., Centralia, Wash.) were used as experimental material. The conditions for growth of seedlings, the composition of the defined basal medium, methods of culturing cotyledons, and physical environment for culturing cotyledons have been established and described in detail (Cheng, 1975, 1977). The basal medium contains, per liter, (1) inorganic compounds: 825 mg NH_4NO_3, 950 mg KNO_3, 220 mg $CaCl_2$ $2H_2O$, 185 mg $MgSO_4$ $7H_2O$, 85 mg KH_2PO_4, 6 mg $FeSO_4$ $7H_2O$, 7.2 mg Na_2EDTA, 3.1 mg H_3BO_3, 11.15 mg $MnSO_4$ $4H_2O$,

Fig. 1. Regeneration of adventitious shoots *in vitro* from Douglas fir. Initial inoculum was derived from (a) excised embryo, (b) cotyledon from 2-week-old seedling, (c) needle from 6-month-old seedling, and (d) stem from 2-year-old tree.

5.25 mg ZnSo$_4$ 7H$_2$O, 0.4 mg KI, 0.15 mg NaMoO$_4$ 2H$_2$O, 0.013 mg CuSO$_4$ 5H$_2$O, 0.013 mg CoCl$_2$ 6H$_2$O; and (2) organic substances; 250 mg myo-inositol, 2.5 mg thiamine HCl, and 30 g sucrose. Plant growth regulators added to the culture media consisted of auxins, indole-3-acetic acid (IAA), indole-3-butyric acid (IBA), naphthaleneacetic acid (NAA), and 2,4-dichlorophenoxyacetic acid (2,4-D), and cytokinins, N$_6$-benzylaminopurine (BAP), N$_6$-(2-isopentenyl) aminopurine (2iP), and kinetin. The final pH of the culture medium was adjusted to 5.5.

Before cotyledons were established in culture, a preconditioning treatment was applied to sodium hypochlorite sterilized seedlings (excised at the upper region of the hypocotyl) by placing them for 3–6 days on an agar-solidified nutrient medium. This preconditioning step ensured the selection of vigorously growing cotyledons for use in the establishment of cultures and allowed the elimination of contaminated and injured tissues.

Cotyledon tissues were cultured either on an agar-solidified nutrient medium (Cheng, 1977) or on a fabric tissue support (100 percent polyester fleece of 3-mm thickness, Pellon Corp., Lowell, Mass.). This tissue support was placed inside a plastic petri dish (either 60 × 15 mm, or 100 × 20 mm) filled with liquid nutrient medium to such a level that the fabric tissue support was well moistened and served as a bridge between tissue explants and nutrient (Cheng and Voqui, 1977).

This tissue culture system, using the fabric tissue support, incorporates important advantages of both solid- and liquid-media techniques. Replacing the solidified agar as a supporting material for tissue explants with a fabric tissue support allows complete flexibility for nutrient flow in a liquid state, and thus facilitates the processes involved in supplying tissue explants with the proper nourishment required for each developmental stage without involving transfer of cultured tissues. Because of the flexibility and simplicity of this culture method, it has potential for use in industrial implementation of mass clonal propagation.

Factors Influencing Adventitious Bud Formation

The effects of (1) the genetic diversity of plant materials used and (2) the use of various combinations of plant growth regulators (auxins and cytokinins) on development of adventitious buds have been described (Cheng, 1975, 1977). The influence of variations in physiological states and genomic composition of Douglas fir cotyledons on morphogenetic responses *in vitro* was more dramatically expressed when natural auxins, IAA and IBA, were used. For example, different regions of whole cotyledons subjected to the same culture treatment showed a wide variation in morphogenetic response; cotyledons from near the stem axis (basal region) were most responsive in producing adventitious buds (81 percent). This potential decreased to 69 percent and 52 percent, respectively, for cotyledons derived from middle and distal regions. Considerable variation among Douglas fir clones was also observed; the frequency of morphogenesis varied from 13 to 91 percent. The most likely interpretation of these observed differences was the existence, in the heterogeneous plant population used, of different amounts of auxin-degrading enzymes. Thus, even though the same natural auxin level was applied to all cotyledon explants, depending on the extent of auxin-degrading activity, the actual functional auxin concentration among individual explants differed. Synthetic auxins are much less susceptable to enzymatic degradation in these plant materials. Therefore, application of either NAA or 2,4-D at much lower concentrations than is necessary when the natural auxins IAA or IBA are used, produced uniformly good responses.

The degree of morphogenetic expression was greatly influenced by the concentration of exogenously added, synthetic auxins. A low frequency of morphogenetic response was observed with the cytokinin (BAP) in the absence of auxin. The hormonal behavior of NAA and 2,4-D are comparable in that, at the nanamolar level, both auxins stimulated maximum adventitious bud production, but at the micromolar level, organ differentiation was completely suppressed but rapid cellular growth occurred. Thus, rapid cellular growth, stimulated by high concentration of NAA or 2,4-D, is not a prerequisite for induction of differentiation *in vitro*. Rather, it is essential to provide optimal culture conditions for production of competent cells, capable of responding to *in vitro* stimuli that trigger genes involved in differentiation processes.

Among the commonly used cytokinins (BAP, kinetin, and 2iP), BAP always exhibited the highest percentage of adventitious bud formation and 2iP was the least effective. The relative ineffectiveness of 2iP in this Douglas fir system contrasts sharply with results obtained in the tobacco system (Skoog et al., 1967). Suitable concentrations of plant growth regulators required for stimulation of adventitious bud formation were 5 μM BAP plus 0.5–5.0 ηM NAA, or 5 μM BAP plus 0.25–5.0 μM each of IAA plus IBA.

For stimulating bud primordia formation, BAP was more effective than kinetin or 2iP. Kinetin, however, stimulated significantly more rapid formation of needle primordia. These results suggest that combined use of BAP and kinetin may be advantageous.

Factors Influencing Adventitious Root Formation

During the course of establishing optimal plant growth regulator requirements for stimulation of adventitious bud formation of Douglas fir cotyledons, it was discovered that an appropriate combination of 2iP and NAA stimulated adventitious root formation. Thus, the effectiveness of different types of cytokinins in combination with NAA for adventitious root formation was investigated (table 2). The experimental results showed that only 2iP exhibited a stimulatory effect. In control cultures (0.25 μM NAA without 2iP) 19 percent of explants produced roots. A 5.0 μM concentration of 2iP acted synergistically with either 0.5 μM or 0.25 μM NAA in stimulating adventitious root formation (72 percent and 78 percent respectively).

TABLE 2

STIMULATION OF ADVENTITIOUS ROOT FORMATION IN CULTURE
OF COTYLEDON EXPLANTS BY VARIOUS CYTOKININS: EFFECT
OF VARYING CYTOKININ CONCENTRATION

Cytokinin	Concentration (μM)	Root Formation (% of Explants)	
		a	b
BAP	5.0	0	0
	25.0	0	0
	50.0	0	0
2iP	5.0	72	78
	25.0	0	0
	50.0	0	0
Kinetin	5.0	0	0
	25.0	0	0
	50.0	0	0
Control		. . .	19

Various concentrations of BAP, 2iP, and kinetin were added to defined basal medium supplemented with 5 μM IBA plus two concentrations of NAA: (a) 0.5 μM and (b) 0.25 μM. Each treatment consisted of 18 tissue pieces; 27 were used for each control culture.

The failure of BAP and kinetin to stimulate root formation (table 2) suggested the need for testing wide ranges of NAA concentrations. Regardless of NAA concentrations used (0.5 μM to 30 μM) BAP and kinetin still failed to stimulate root formation (table 3). A stimulatory effect of 2iP was again observed; however, no stimulatory effect was observed if the NAA concentration was below 0.25 μM or higher than 0.5 μM. A combination of 2iP and NAA at 0.5–5.0 μM and 0.5 μM, respectively, exhibited the maximum stimulatory effect (table 4) and the numerous elongated roots produced showed a geotropical growth behavior (fig. 2a). When the NAA concentration was increased to 5.0 μM, root primordia were initiated, but root development was retarded, resulting in a root nodulelike structure (fig. 2b). NAA at a concentration of 0.25 μM, alone without 2iP, resulted in formation of few roots that did not undergo rapid elongation (fig. 2c).

TABLE 3

STIMULATION OF ADVENTITIOUS ROOT FORMATION IN CULTURE
OF COTYLEDON EXPLANTS BY VARIOUS CYTOKININS: EFFECT
OF VARYING NAA CONCENTRATION

NAA Concentration	Root Formation (% of Explants)		
	BAP	2iP	Kinetin
μM			
30.0	0	0	0
15.0	0	1	0
5.0	0	26	0
0.5	0	82	0
ηM			
50.0	0	5	0
5.0	0	0	0
0.5	0	0	0
Control	0	0	0

Various concentrations of NAA were added to defined basal medium containing 5 μM of BAP, 2iP, or kinetin, respectively. No NAA was added to the control cultures. Each treatment consisted of 27 tissue pieces.

TABLE 4

EFFECTS OF VARIOUS CONCENTRATIONS OF 2iP AND NAA
ON ADVENTITIOUS ROOT FORMATION IN CULTURE OF
COTYLEDON EXPLANTS

Concentration (μM)		Root Formation (% of Explants)
2iP	NAA	
5.0	0.5	75
	0.05	0
	0.005	0
2.5	0.5	71
	0.05	1
	0.005	0
0.5	0.5	82
	0.05	0
	0.005	0

Concentrations of 2iP and NAA added to defined basal medium were as listed. Each treatment consisted of 27 tissue pieces.

Regeneration of Plantlets

Regeneration of Douglas fir plantlets under defined conditions has been accomplished (Cheng and Voqui, 1977). For production of Douglas fir plantlets, each excised adventitious shoot was placed in an agar-solidified rooting medium. In addition to an appropriate concentration of auxin (NAA), a sucrose concentration 6-fold lower than that used for bud initiation favored root formation. Results obtained by varying concentrations of auxin (NAA) and sucrose in the culture medium indicated that the optimal concentrations of these two compounds for stimulation of root

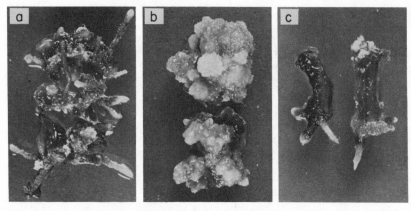

Fig. 2. Development of adventitious roots *in vitro* from Douglas fir cotyledons. Cotyledon explants were cultured on a defined nutrient medium supplemented with plant growth regulators at concentrations of (a) 0.5 μM NAA plus 5 μM 2iP, (b) 5 μM NAA plus 5 μM 2iP, and (c) 0.25 μM NAA without 2iP. The culture environment for root formation was 24°C with 16-hr photoperiod (100 ft-c) and 8-hr dark period.

formation were 0.5 percent sucrose and 0.25 μM NAA. Using this medium, regeneration of plantlets (fig. 3) as high as 80 percent was obtained. Failure to produce plantlets occurred when a concentration of either NAA or sucrose higher than optimal was used; high levels of NAA caused prolific callus growth at the basal end of stem, whereas excess sucrose reduced the vitality of the shoots.

The incubation temperature also had a profound influence on the frequency of plantlet production and on the morphological appearance of the plantlets that were formed. Experiments were performed in which

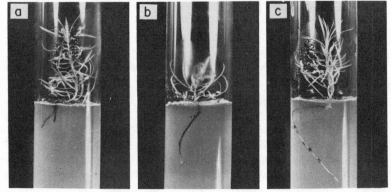

Fig. 3. Regeneration of plantlets *in vitro* from cotyledons of Douglas fir. Regeneration of plantlets was accomplished by rooting adventitious shoots on a 2-fold diluted basal medium containing 0.5% sucrose plus 0.25 μM NAA. The incubation environment was maintained at 19°C with a 16-hr photoperiod (200 ft-c) and 8-hr dark period.

excised shoots were incubated at two different temperatures (24°C or 19°C). At 24°C , relatively few plantlets were produced and, more seriously, these plantlets were abnormal, exhibiting a discontinuity in their anatomical structure caused by a proliferation of friable callus at the transition region between stem and root. In contrast, at the lower incubation temperature of 19°C, a high frequency of plantlet formation was observed and all of the plantlets seemed to possess a normal morphological appearance. Most importantly, establishment of plantlets generated at 19°C in soil was successful, exhibiting more than 90 percent survival, whereas plantlets produced at 24°C all died.

These experimental results demonstrate that the mode of action of auxin (in this case, NAA) is influenced by changes in incubation temperature. Since cells cultured at the higher temperature (24°C) are expected to be metabolically more active than those cultured at the lower temperature (19°C), we conclude that auxin affects two different cell types. Thus, the hormonal effect of auxin (NAA) on cultured cells is expressed either as (1) stimulation of adventitious root formation at the lower temperature, or (2) stimulation of unorganized cell proliferation at the elevated temperature.

Previous reports of plantlet regeneration involved only two conifer species, longleaf pine (*Pinus palustris* Mill.), for which no quantitative data were presented (Sommer et al., 1975), and western hemlock (*Tsuga heterophylla*), for which an undefined rooting medium (soil) was used (Cheng, 1976b). In the present work, using chemically defined media, high-frequency regeneration of plantlets from somatic cells of Douglas fir (*Pseudotsuga menziesii* [Mirb.] Franco) was achieved.

HISTOLOGICAL STUDIES OF ADVENTITIOUS BUD DEVELOPMENT

Development of adventitious buds of Douglas fir cotyledons *in vitro* required an appropriate combination of auxin and cytokinin, and any deviation of concentrations from that optimal for morphogenesis resulted in the production of various morphologically distinct cultures (Cheng, 1977). Four different growth regulator–induced culture types (table 5) have been used for two histological studies: (1) establishment of sequential events in adventitious bud development, and (2) hormonal effects on cellular growth behavior (Cheah and Cheng, 1978).

Systematic analysis of longitudinal microscopic sections of bud culture, from culture day zero to day 21, showed that adventitious buds are derived from hypodermal cells of Douglas fir cotyledon. The developmental process could be divided into 4 anatomically distinguishable stages: (1) meristemoid stage, (2) bud primordium stage, (3) early, and (4) late stage of adventitious bud development. For bud culture, active cell proliferation

TABLE 5

DEVELOPMENT OF FIVE MORPHOLOGICALLY DISTINCT DOUGLAS
FIR COTYLEDON CULTURE TYPES CONTROLLED BY EXOGENOUSLY
ADDED PLANT GROWTH REGULATORS

Culture	EXOGENOUS GROWTH REGULATOR REQUIREMENT		
Type	Cytokinin (BAP)	Auxin (NAA)	Growth Responses
Bud	5 μM	5 ηM	Adventitious bud formation
Bud-Callus	5 μM	5 μM	Reduced adventitious bud formation and callus
Callus	0	5 μM	Callus formation
Callus	5 μM	30 μM	Callus formation
Control	0	0	Minimal growth

Concentrations of BAP and NAA added to a defined medium were as listed. All cotyledon explants were placed on the surface of a tissue support saturated with liquid medium and were incubated at a constant temperature of 24°C with a 24-hr cycling period consisting of 18 hr of light at approximately 200 ft-candles and 6 hr of darkness. The respective culture types were fully developed after approximately 6 weeks in culture.

started to occur in the hypodermal layer as early as day 4. By day 12, continued proliferation of these cells in an organized fashion led to the formation of a meristemoid-like structure, the precursor of an adventitious bud. Well-developed bud primordia became visible in 21-day-old cultures. As the morphogenetic processes advanced further, cells located in the outer layer of the peripheral zone adjacent to the apical dome started to divide vigorously in a periclinal fashion, and eventually gave rise to leaf primordia on both sides of the shoot apex. In this way bud primordia progressed into the early stage of adventitious bud development. The stem axis of the adventitious bud continued to elongate, needle primordia progressed towards maturation, and concurrently, newly formed needle primordia continued to emerge from near the shoot apex. Adventitious buds in the late developmental stage were indistinguishable from those derived from the intact plant (Allen and Owens, 1972). Beneath the shoot apex, the stem consisted of a well-differentiated pith and vascular bundles and this vascular system progressed further from the stem axis acropetally into the needles.

The mode of action of the specific auxin-cytokinin combinations capable of stimulating high-frequency differentiation of cotyledons was elucidated by comparing histogenesis of bud culture prior to the meristemoid stage with that of morphologically different culture types induced by different concentrations of auxin and cytokinin. Systematic comparison of microscopic sections of bud culture, bud-callus culture, and callus culture showed characteristic differences in the modes of cell activation as detected at the time of first cell division (approximately day 4) and in the patterns of subsequent cellular growth. Activation of cell division was observed for all 4-day-old cotyledon cultures, regardless of the concentrations of exogenously added plant growth regulators. However, the distribution of sites of cell division among the tissue layers differed.

For bud culture, cell division occurred mainly at the hypodermal layer and to an almost negligible extent for mesophyll cells. Endodermis and transfusion tissue showed a relatively low activity in cell division, which was similar to that observed for cotyledons cultured without any growth regulators. These results suggest that concentrations of endogenous phytohormones were sufficient to maintain a low level of cellular activity for these two tissue layers and that low levels of exogenous phytohormones are essential for activation of cells of the epidermis.

For bud-callus and callus cultures, cell division occurred throughout the epidermis, mesophyll, endodermis, and transfusion tissue, but not in the vascular bundle. As the cultures progressed, the characteristic differences among these cultures became increasingly pronounced. Cell division for bud culture remained restricted to the epidermis and these localized proliferating cell masses gave rise to meristemoid-like structures. Cell division of other tissue layers was noticeably slow and orderly, thus the original cellular organization of the cotyledon structure was retained. In bud-callus culture, rapid proliferation of mesophyll cells completely occupied intercellular space, and eventually these cells forced through the weaker region of the epidermis and grew continuously as a friable callus mass. Cells of the epidermis, endodermis, and transfusion tissue proliferated slowly, forming compact cells with an orderly arrangement. Only a few meristemoid-like structures were observed at the hypodermal layer.

Although the pattern of cellular growth for callus culture initially was quite similar to that of bud-callus culture, in the callus culture, cell division gradually ceased. The appearance of ghost cells lacking nuclei and cytoplasm and the accumulation of tannin-like compounds indicated degeneration of mesophyll cells. The production of tannin-like compounds *in vitro* appears to be a common phenomenon among coniferous species (Ball, 1950; Constabel, 1963; Jorgensen and Balsillie, 1969; Baur and Walkinshaw, 1974).

This histogenetic analysis of the growth regulator-induced, morphologically distinct culture types clearly demonstrates hormonal control of cell activation and growth behavior. In the callus system, cotyledon explants were cultured on a medium supplemented with auxin (NAA) but completely lacking cytokinin (BAP). Thus, continued cell proliferation depended on the ability of the cultured cells to synthesize needed cytokinin. Since all cell division ceased after 2–3 weeks, it seems likely that the initial cell division was stimulated by endogenous cytokinin (present in the initial cotyledon explants) that was not augmented by biosynthesis. Presence of 5 μM BAP in the bud-callus culture medium supported continual cell growth. For bud culture, a combination of cytokinin-auxin at 5 μM BAP plus 5 η M NAA activates selective cell division. Since each tissue layer is composed of different types of highly differentiated cells, and cytokinin at 5 μM BAP is sufficient for cell growth, it is reasonable to assume that 5 η M NAA is just sufficient to activate hypodermal cells, but not other types of cells that require a higher concentration of auxin for proliferation. Therefore, morphogenetic processes *in vitro* appear to be under stringent regulation by plant growth regulators. Concentrations of cytokinin and auxin that deviate from the optimum cause cells to lose their morphogenetic potential. This work also clearly demonstrates that adventitious buds arise specifically from the hypodermis of the cotyledon. Similar observations have been made for other coniferous species (Isikawa, 1974; Sommer et al., 1975; Campbell and Durzan, 1975).

IDENTIFICATION OF GENE PRODUCTS ASSOCIATED WITH
ADVENTITIOUS BUD FORMATION

Initiation of adventitious bud formation *in vitro* is a *de novo* phenomenon for cotyledon tissue, and activation of genes involved in morphogenetic processes is required to achieve the cellular metamorphosis. Specific gene products (i.e., newly synthesized proteins) derived from translation of messages transcribed from these activated genes have been detected (Hasegawa et al., 1978). High-resolution techniques for fractionating double-labeled samples of newly synthesized cytoplasmic soluble proteins involving electrophoresis through SDS polyacylamide gels allowed us to differentiate the synthesis of various proteins. This high-sensitivity technique has been used successfully to analyze unfractionated protein samples in other studies (Dhindsa and Cleland, 1975; Eppig and Eckhardt, 1976; Snashoua, 1976; Terman, 1970). Cytoplasmic soluble proteins obtained from Douglas fir cotyledon fractionated by SDS polyacrylamide gel could be roughly divided into 5 regions based on molecular weight: region 1: 76,000–84,000 daltons; region 2: 52,000–58,000 daltons; region 3: 37,000–42,000 daltons; region 4: 24,000–27,000 daltons, and region 5: 16,000–20,000 daltons.

For investigation of patterns of protein synthesis during early stages of adventitious bud development, 0-, 2-, 4-, and 8-day old cultures were studied. Results obtained from comparing ^{14}C-labeled protein samples of day zero bud culture with that of ^{3}H-labeled protein samples of corresponding 2-, 4-, and 8-day old cultures showed a significant activity in protein synthesis in region 5 (molecular weights ranging from 16,000 to 20,000 daltons). An increase in the synthesis of proteins at region 5 was detected as early as 2 days in culture, reached a maximum level at 4 days, and at 8 days returned to a level similar to that of 2-day-old cultures. The observed changes in protein synthesis as a function of time in culture suggest that this new cellular activity was initiated by *in vitro* stimuli.

To determine whether the consistent increase in protein synthesis is associated with an early stage of morphogenesis leading to bud formation or is just a growth regulator–induced nonmorphogenetic growth phenomenon, the pattern of protein synthesis of the bud culture was compared with that of other cultures. Comparison of the distribution of ^{14}C-leucine labeled proteins of bud culture with that of ^{3}H-leucine labeled proteins of bud-callus, callus, and control cultures showed that bud culture consistently exhibited increased protein synthesis at region 5 (16,000–20,000 daltons).

To determine whether differences in the pattern of protein synthesis between bud culture and callus culture were caused by the presence or absence of BAP, cultures maintained on a nutrient medium with a fixed concentration of BAP (5 μM) but with varying concentrations of NAA (5η M, 5 μM, and 30 μM) were used. As NAA concentrations increased from 5 η M to 30 μM, the frequency of morphogenesis decreased from a maximal level (bud culture) to zero (callus culture). Results obtained again showed an increase in synthesis of low molecular weight proteins for bud culture.

Composition of newly synthesized proteins was also influenced by variations of plant materials; however, for all materials, low molecular weight protein synthesis was associated with bud cultures.

The synthesis of low molecular weight (16,000–20,000 daltons) proteins probably is associated with an early stage of adventitious bud development; that it is not a growth regulator induced, nonmorphogenetic growth phenomenon is shown by the experimental findings obtained from cotyledon cultures initiated by different combined concentrations of growth regulators. The high concentration of NAA used for initiation of bud-callus and callus cultures (100-fold higher than that for bud culture) was extremely effective in stimulating active cellular growth but did not enhance low molecular weight protein synthesis. Furthermore, results from the control culture maintained without growth regulators showed that the increase of low molecular weight protein synthesis observed for bud culture was an induced phenomenon under stringent control.

The finding that low molecular weight proteins specific for budforming cultures appears as early as 2 days is, to our knowledge, the earliest biochemical event associated with morphogenesis detected to date. It is possible that the *in vitro* stimulus for this event, which involved derepression of all necessary genes required for initiating the developmental processes and the consequent cellular metamorphosis, is the detection in the cell of specific gene products (i.e., newly synthesized proteins) derived from translation of messages transcribed from these activated genes. Identification of these low molecular weight proteins is necessary if the nature of the proteins and their roles in morphogenesis are to be understood.

SUMMARY

In this laboratory significant progress has been made in continued development of *in vitro* techniques to meet the requirements for mass clonal propagation of forest tree species. Methods for regeneration of plantlets from Douglas fir cotyledon have been established. Development of methods for culture of tissue pieces on the surface of a fabric tissue support with a liquid nutrient medium facilitates periodic changes of culture conditions to meet the requirement for each successive developmental stage without involving transfer of cultured tissues. The histogenesis and biochemical mechanisms of adventitious bud development *in vitro* have been studied. Adventitious buds were shown to originate from cells of the hypodermis and their development involved 4 distinguishable stages: (1) meristemoid stage, (2) bud primordium stage, (3) early, and (4) late stage of adventitious bud development. Synthesis of low molecular weight proteins (16,000–20,000 daltons) was detected as early as 2 days in culture and subsequently reached a maximum level at day 4. The association of these low molecular weight proteins with bud-forming cotyledon cultures is consistent with their involvement in the morphorgenetic process leading to adventitious bud development.

ACKNOWLEDGMENTS

Research was supported by a grant from the Weyerhaeuser Corporation, Tacoma, Washington. I wish to express my gratitude to Dr. G. Doyle Daves, Jr., for his support in this research.

LITERATURE CITED

Allen, G. S., and T. N. Owens. 1972. The life history of Douglas fir. Information Canada, Ottawa.

Ball, E. 1950. Differentiation in a callus culture of *Sequoia sempervirens*. Growth 14:295–325.

Baur, P. S., and C. H. Walkinshaw. 1974. Fine structure of tannin accumulations in callus cultures of *Pinus elliotti* (slash pine). Canad. J. Bot. 52:615–19.

Bonga, J. 1974. Vegetative propagation: Tissue and organ culture as an alternative to rooting cuttings. N. Z. J. For. Sci. 4:253–60.

Brown, C. L. 1976. Forests as energy sources in the year 2000: What man can imagine, man can do. J. For. 74:7–12.

Campbell, R. A., and D. J. Durzan. 1975. Induction of multiple buds and needles in tissue cultures of *Picea glauca*. Canad. J. Bot. 53:1652–56.

Cheah, K. T., and T. Y. Cheng. 1978. Histological analysis of adventitious bud formation in cultured Douglas fir cotyledon. Amer. J. Bot. (in press).

Cheng, T. Y. 1975. Adventitious bud formation in culture of Douglas fir (*Pseudotsuga menziesii* [Mirb.] Franco). Plant Sci. Lett. 5:97–102.

Cheng, T. Y. 1976a. Tissue culture techniques in tree improvement. Industr. For. Assoc. Tree Improv. Newsl. 28:2–7.

Cheng, T. Y. 1976b. Vegetative propagation of western hemlock (*Tsuga heterophylla*) through tissue culture. Plant and Cell Physiol. 17:1347–50.

Cheng, T. Y. 1977. Factors affecting adventitious bud formations of cotyledon culture of Douglas Fir. Plant Sci. Lett. 9:179–87.

Cheng, T. Y., and T. H. Voqui. 1977. Regeneration of Douglas fir plantlets through tissue culture. Science 198:306–7.

Constabel, F. 1963. Tannins in tissue cultures of *Juniperus communis* L. Planta Med. 11:417–21.

Dhindsa, R. S., and R. E. Cleland. 1975. Water stress and protein synthesis. I. Differential inhibition of protein synthesis. Plant Physiol. 55:778–81.

Durzan, D. J., and R. A. Campbell. 1974. Prospects for the mass production of improved stock of forest trees by cell and tissue culture. Canad. J. For. Res. 4:151–74.

Eppig, J. J., and R. A. Eckhardt. 1976. Protein labeling patterns in oocytes of *Xenopus laevis*. Differentiation 6:97–103.

Hasegawa, P. M., T. Yasuda, and T. Y. Cheng. 1978. Specific proteins associated with morphogenesis *in vitro* of Douglas fir cotyledon. (Submitted.)

Isikawa, H. 1974. *In vitro* formation of adventitious buds and roots on the hypocotyl of *Cryptomeria japonica*. Bot. Mag. Tokyo 87:73–77.

Jorgenson, E. and D. Balsillie. 1969. Formation of heartwood phenols in callus tissue of red pine (*Pinus vasinosa*). Canad. J. Bot. 47:1015–16.

Konar, R. N., 1972. Tissue and cell culture of pines and allied conifers. USDA PL 480.

Konar, R., and R. Magmani. 1974. Tissue culture as a method for vegetative propagation of forest trees. N. Z. J. For. Sci. 4:279–90.

Konar, R. N., and Y. P. Oberoi. 1965. *In vitro* development of embryoids on the cotyledons of *Biota orientalis*. Phytomorphology 15:137–40.

Rediske, J. H. 1974. The objective and potential for tree improvement. *In* F. T. Ledig (ed.), Toward the future forest: Applying physiology and genetics to the domestication of trees. Pp. 3–18. Yale University, New Haven.

Shashoua, V. E. 1976. Identification of specific changes in the pattern of brain protein synthesis after training. Science 193:1264–66.

Skoog, F., H. Q. Hamzi, and A. M. Szweykowska. 1967. Cytokinins: Structure/activity relationships. Phytochemistry 6:1169–92.

Sommer, H. E. 1975. Differentiation of adventitious buds on Douglas-Fir embryos *in vitro*. Proc. Intern. Plant Prop. Soc. 25:125.

Sommer, H. E., C. L. Brown, and P. P. Kormanik. 1975. Differentiation of plantlets in long leaf pine (*Pinus palustris* Mill.) tissue cultures *in vitro*. Bot. Gaz. 136:196–200.

Terman, S. A. 1970. Relative effect of transcription-level and translation-level control of protein synthesis during early development of the sea urchin. Proc. Nat. Acad. Sci. USA 65:985–92.

Thomas, M. J., E. Duhoux, and J. Vazart. 1977. *In vitro* organ initiation in tissue cultures of *Biota orientalis* and other species of the *Cupressaceae*. Plant Sci. Lett. 8:395–400.

Winton, L. L., and S. A. Veihagen. 1977. Shoots from Dougls-fir cultures. Canad. J. Bot. 55:1246–50.

ANTONIO NATAL GONÇALVES, MARCOS A. MACHADO,
L. S. CALDAS, W. R. SHARP, AND
HELLÁDIO DO AMARAL MELLO

Tissue Culture of *Eucalyptus*

27

INTRODUCTION

The rapid growth rate of *Eucalyptus* species and the desirable qualities of its timber, which is dense, hard, and durable, are among the factors that led to the introduction of eucalypts from Australia and its neighboring islands into many other parts of the world. Even in areas where no industrial plantations exist, various *Eucalyptus* species are valued as ornamentals and shade and shelter trees. In addition to hardwood lumber production, eucalypts have been used as poles and sleepers, for short fiber pulp and paper, soft and hard fiberboard, and charcoal for pig iron manufacture. Essential or volatile oils, tannins, and kinos are extracted from some species, and others provide raw material for the production of methyl alcohol, acetic acid, rayon, and cellophane. Finally, some species of *Eucalyptus* have been used as sources of food and fodder (Penfold and Willis, 1961).

The considerable size attained by some eucalypts and their regenerative capacity have also contributed to the widespread use of eucalypts in commercial plantations. In Brazil alone, 1.5 million hectares have been planted.

The list of *Eucalyptus* species published by Chippendale (1976) recognized 445 species plus some subspecies, varieties, and 115 interspecific

Antonio Natal Gonçalves, Department of Forestry, ESALQ, University of São Paulo, Piracicaba, São Paulo, Brazil.

hybrids, many of which had received specific names in the past. Since 1,200 different names have been variously applied to eucalypt taxons, the difficulties in systematic studies of this large genus can be appreciated. Many different classification schemes have been developed, based on the characteristics of leaves, bark, oils, buds, fruits, anthers, and cotyledons. More recently, other characteristics such as polyphenol composition and genetic analysis, which may indicate evolutionary relationships, have also been taken into consideration (Pryor and Johnson, 1971). An understanding of the taxonomic relationships is of importance in breeding programs and in the selection of planting material, as can be seen in the case of the interspecific hybrids, and further studies may modify the status of many species recognized now (Pryor and Johnson, 1971).

Whereas interspecific hybrids are fairly readily formed once geographical and other barriers to breeding are removed, individual eucalypts may show some degree of self-fertility (Hodgson, 1976). This is usually associated with various undesirable consequences of inbreeding depression, and, as a rule, *Eucalyptus* species are considered outcrossing. The barriers to self-fertilization in *Eucalyptus grandis* include protandry, selective fertilization, and reduced seed yield after self-pollination, which may be due to infertility (Hodgson, 1976).

The vast majority of eucalypts are evergreen, with only a few species being deciduous (Penfold and Willis, 1961; Pryor, 1976). At maturity, they vary from small shrubs or mallees, less than 3 m high, to the giant *Eucalyptus regnans*, over 100 m high. Each individual eucalypt passes from a juvenile stage to an adult stage during its life cycle; during the transition, leaf anatomy and morphology, rooting potential, and fiber and wood characteristics are modified. The leaf crown of the eucalypts is generally built up rapidly, due to the presence of naked buds that develop a number of branch orders in a few weeks. Even if the crown is partially or totally destroyed, the reserve system of accessory and dormant buds permits the elaboration of more foliage in a comparatively short time.

Most species of *Eucalyptus* present lignotubers. These structures, consisting of a mass of vegetative buds and associated vascular tissue with substantial food reserves, appear in the first year of growth and afterward are covered by the secondary growth of the stem. The buds produce juvenile leafy shoots under some conditions (after fire, for example); thus, they are important in regeneration. However, some of the most important commercial *Eucalyptus* species do not develop lignotubers.

Breeding programs involving *Eucalyptus* could be greatly improved by the use of vegetative propagation of selected material. The classical techniques of grafting and the rooting of cuttings have not been generally satisfactory because adult material rarely demonstrates the ability to form adventitious roots and graft incompatibility is seen in a relatively large percentage of grafts (Davidson, 1977). For this reason, vegetative propagation through tissue culture techniques has been studied as a possible alternative to classical methods (Fossard et al., 1974; Kitahara and Caldas, 1975; Gonçalves, 1975). Interest in the extension of these methods and development of others that could be applied to facilitate the vegetative propagation of *Eucalyptus* gave rise to the studies reported here.

CALLUS CULTURE OF EUCALYPTUS

Callus information in tissue cultures of *Eucalyptus* has been reported in a number of species (table 1). Different plant parts that have been shown to develop callus include seeds, hypocotyls, cotyledons, seedling roots, stem segments, petioles, leaf blades, apical shoots, lignotubers, anthers, bark explants, and pollen grains. Leaf blades, anthers, and bark explants

TABLE 1

EUCALYPTUS SPECIES USED IN CALLUS CULTURE

Species	References
Eucalyptus alba Reinw. ex Bl.	Kitahara and Caldas, 1975; Gonçalves, 1975
E. bancroftii (Maid.) Maid.	Fossard, 1974; Fossard et al., 1974; Lee and Fossard, 1974
E. camaldulensis Dehnh.	Jacquiot, 1964; Cronshaw, 1965; Sussex, 1965; Piton, 1969; Gonçalves, 1975
(*E. rostrata* Schlecht.)	Fine, 1968
E. citriodora Hook.	Aneja and Atal, 1969
E. cladocalyx F. Muell.	Jacquiot, 1964
E. dunnii Maid.	Gonçalves, unpublished
E. gomphocephala DC.	Jacquiot, 1964
E. grandis Hill ex Maid.	Fossard, 1974; Fossard et al., 1974; Gonçalves, 1975; Kitahara and Caldas, 1975
E. gunnii Hook. f.	Jacquiot, 1964
E. laevopinea R. T. Bak.	Fossard, 1974; Fossard et al., 1974
E. maculata Hook.	Gonçalves, unpublished
E. melliodora A. Cunn. ex Schau.	Fossard, 1974; Fossard et al., 1974
E. nicholii Maid. & Blakely	Fossard et al., 1974
E. nova-anglica Deane & Maid.	Winton, 1972
E. obliqua L'Herit.	Blake, 1972
E. robusta Sm.	Gonçalves, 1975
E. saligna Sm.	Gonçalves, 1975
E. tereticornis Sm.	Jacquiot, 1964; Gonçalves, 1975
E. × *trabuti* (hybrid: *botryoides* × *camaldulensis*)	Marcavillaca and Montaldi, 1964
E. urnigera Hook. f.	Fossard et al., 1974
E. urophylla S. T. Blake	Gonçalves, unpublished
E. viminalis Labill.	Blake, 1972

Nomenclature follows Chippendale, 1976; names in parentheses are considered by Chippendale to be synonyms.

including the cambium formed callus rapidly in almost all explants placed in culture in our work. Preference has been given to explants from adult trees that have already demonstrated their silvicultural characteristics.

The relative ease of sterilization of these explants also contributed to their high success rate. Surface sterilization was accomplished by washing in detergent, then treating for 20 min with a 20 percent solution of commercial hypochlorite bleach followed by three rinses in sterile distilled water. After this treatment, no contamination was encountered in greenhouse-grown leaf blade explants, and all cultures formed callus. There was 20 percent contamination of field-grown explants, but the remaining 80 percent of the cultures developed callus (Gonçalves, 1975).

A series of nutrient media have been tested with varying degrees of success, including those of Murashige and Skoog (1962), White (1963), Linsmaier and Skoog (1965), Tulecke et al. (1965), Nash and Davies (1972), Sommer et al. (1975), and the test of different component concentrations by Fossard (1974) and Lee and Fossard (1974). The best callus development from leaf blades in our work was noted on the medium of Nash and Davies (1972) and from anthers on White's medium containing 10 percent coconut milk (CCM), 10 mM KNO_3, and 1 mg/l 2,4-dichloro-phenoxyacetic acid (2,4-D). It may be significant that both these media have relatively low salt concentrations as compared with those of Murashige and Skoog's medium, for example.

Eucalyptus callus will grow well on chemically defined media (Fossard, 1974; Gonçalves, 1975), although such additives as casein hydrolysate and coconut milk, in particular, stimulate growth to an appreciable extent. For example, hypocotyl segments of *Eucalyptus grandis* seedlings grown on Nash and Davies (1972) medium increased fresh weight from 533.1 mg to 840.7 mg and dry weight from 16.9 to 49.9 mg when CCM was added at 10 percent v/v.

Liquid culture of *Eucalyptus* cell suspensions has been successful (Sussex, 1965; Gonçalves, 1975), but the majority of the work to date has been done with solid media. Gonçalves (1975) demonstrated the viability and growth of cells from suspensions plated on solid media. Roots were also formed in these cultures. Therefore, it is possible to imagine a system in which the multiplication of the cells could be done on a large scale in liquid medium before a differentiation stage on solid medium.

Media Components and Callus Growth

Increasing sucrose concentrations from 0.0 to 40.0 g/l caused an increase in dry weight of the *Eucalyptus grandis* leaf blade callus formed on Nash and Davies medium (table 2). At 50 g/l the dry weight was less than

TABLE 2

SUCROSE CONCENTRATIONS AND EUCALYPTUS CALLUS

Sucrose g/l	Fresh Weight (mg)	Dry Weight (mg)	Callus Characteristics
0.0	14.7	2.41	Dark brown compact callus
5.0	441.0	12.2	Dark brown compact callus
10.0	960.6	26.2	Brown-green friable callus
15.0	1,083.1	35.3	Pale green friable callus
20.0	1,275.0	52.7	Pale green-white friable callus
25.0	2,008.0	79.8	Pale green-white friable callus with roots
30.0	1,569.0	79.5	Pale green-white friable callus with roots
35.0	1,242.5	85.7	Pale green-white friable callus with roots
40.0	1,727.5	110.0	Pale green-white friable callus with roots
50.0	1,170.4	87.0	Pale green friable callus

Explants from leaf blades of adult *Eucalyptus grandis* were placed in Nash and Davies' (1972) medium with sucrose concentrations as noted in the table. All treatments had 20 replicates and were placed to grow under a 12-hour photoperiod for 60 days. Dry weight determination was after 24 hours at 110°C (Gonçalves, 1975).

the maximum. A consistent increase in fresh weight was also noted with increase in sucrose from 0 to 25 g/l, after which a certain variability in fresh weight as a function of increased sucrose was observed. These results are compatible with those of Lee and Fossard (1974), who found 20 g/l optimal for *Eucalyptus bancroftii* stem callus, and those of Fossard (1974), who noted an increase in fresh weight of stem callus of three *Eucalyptus* species when explants were grown on medium with 30 g/l ($9 \times 10^{-2}M$) sucrose as compared with growth on 10 g/l. However, when these same treatments were tested on callus subcultures, Fossard did not detect any significant differences in fresh weight increase. In another of our series of experiments, with five serial subcultures of *Eucalyptus grandis* anther callus on White's medium plus 10 percent CCM, 1 mg/l 2,4-D, and an additional 10 mM KNO_3, distinctly better growth occurred at 20 g/l sucrose than at 5 or 10 g/l.

The dark brown coloration of the callus at lower sucrose concentrations and greening at higher concentrations was not foreseen based on results indicating polyphenol synthesis limited by carbohydrate availability (Davies, 1972) and sucrose inhibition of callus greening (Edelman and Hanson, 1972).

Nitrogen supply also greatly affects callus growth, as does the form of nitrogen added. However, 2.5 mM NO^-_3 was approximately as effective as 15.0 mM nitrate in supporting growth of *E. grandis* leaf blade callus on Nash and Davies' (1972) medium (table 3). Ammonium alone as the nitrogen source gave consistently lower fresh weight values than those encountered at comparable nitrate concentrations, and no stimulatory effects of ammonium on dry weight were noted.

Tests of different nitrate concentrations with anther callus indicated that

TABLE 3

NITROGEN SUPPLY AND LEAF BLADE CALLUS GROWTH
(Fresh and Dry Weights [mg])

NH$_4^+$ mM	NO$_3^-$ mM				
	0.0	0.5	2.5	7.5	15.0
0.0	15.4	486.4	1019.0	1206.4	1190.4
	(1.6)	(30.9)	(49.9)	(48.3)	(45.6)
2.5	197.4	166.4	229.4	904.4	1244.4
	(11.6)	(6.6)	(8.6)	(29.6)	(38.1)
7.5	155.4	282.4	555.4	1349.4	981.0
	(7.6)	(12.9)	(30.1)	(48.6)	(32.1)
15.0	404.4	245.4	569.4	386.4	1569.4
	(12.6)	(6.1)	(19.6	(16.6)	(39.6)

Leaf blade explants from adult *Eucalyptus grandis* were cultured on Nash and Davies' (1972) medium with the specified concentrations of nitrate (NaNO$_3$) and/or ammonium (NH$_4$Cl). Fresh and dry weights (the latter after 24 hours at 110°C) were measured after 60 days' growth on a 12-hour photoperiod. Dry weights are given in parentheses. No consistent differences were observed in cultures on the same treatments grown in the dark (Gonçalves, 1975).

growth was optimal on medium (White's supplemented as noted above) to which 10 mM KNO$_3$ was added, whereas 20 mM KNO$_3$ was supra-optimal and led to greater tissue darkening. The importance of a nitrogen supplement was verified in experiments comparing growth on a two-week transfer schedule with that of cultures on a four-week schedule (fig. 1). After the 10 mM KNO$_3$ supplement was introduced into the medium, a much more rapid growth rate was sustained on the two-week transfer schedule than on the four-week schedule.

Lee and Fossard (1974) found that stem and lignotuber callus of *Eucalyptus bancroftii* almost invariably showed greater fresh weight increase at higher auxin (IAA, NAA, NOA, and 2,4-D) concentrations, as well as with higher cytokinin concentrations (BAP and kinetin). The *Eucalyptus grandis* leaf blade callus demonstrated this same tendency with some hormones: kinetin stimulated dry weight increase at 0.5 mg/l, but no additional effect with higher concentrations (to 8.0 mg/l) was observed. Increasing concentrations of the auxins IAA (indole-acetic acid) and NAA (naphthalene-acetic acid) were accompanied by increases in fresh weight and dry weight up to concentrations of 20 mg/l and 8 mg/l respectively, in the presence or absence of kinetin to 1.0 mg/l. Although 2,4-D stimulation of growth occurred at concentrations from 1.0 to 8.0 mg/l in the absence of kinetin, this auxin effect was observed only at 0.5 mg/l in the presence of kinetin (0.5 to 8.0 mg/l). A similar interaction between cytokinin and 2,4-D was observed by Lee and Fossard (1974).

Callus Darkening

The darkening of explants, callus, and media used in tissue culture,

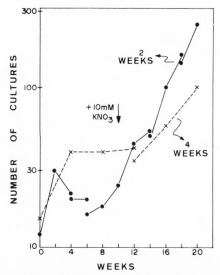

Fig. 1.   A supplement of 10 mM KNO₃ to White's medium containing 10% CCM and 1 mg/l 2,4-D was followed by significantly more rapid growth of *Eucalyptus grandis* anther callus. At each transfer, the callus was subdivided into sections with a fresh weight of approximately 50 mg that were used to inoculate fresh cultures. Thus, the increase in the number of cultures from one transfer to the next indicates the increase in callus mass during that period. A decrease in the number represents losses due to contamination or the elimination of cultures that did not grow. More rapid growth occurred on the two-week transfer schedule than on the four-week schedule.

usually ascribed to oxidation of phenols, is often observed with *Eucalyptus*. An apparent inverse correlation between the quantity of black exudate formed and growth of callus was reported by Fossard (1974), and darker areas of slash pine tissue cultures were found to have a lower respiratory rate (Hall et al., 1972). Several attempts were made to reduce this darkening using treatments to decrease phenol concentrations, inhibit polyphenol oxidase (PPO, an enzyme that reacts with O₂ and phenols to produce quinones), or simply serve as antioxidants.

Boron, through its effect on the pentose-phosphate shunt, can influence the quantity of phenols present in plant tissues (Epstein, 1972). In a series of treatments, concentrations of H₃BO₃ from 0.2 to 102.4 mg/l were added to Nash and Davies' (1972) medium without affecting the fresh or dry weight of *Eucalyptus grandis* leaf blade callus. The callus was pale green on all treatments with the exception of the highest boron concentration, on which the callus was brown. At intermediate and higher concentrations of H₃BO₃, callus formed mainly on opposite sides of the explant rather than all around it.

Many compounds have been used to inhibit the darkening of plant tissues *in vitro*, such as tyrosine (Reinert and White, 1956), polyvinyl pyrrolidone (Anderson and Sowers, 1968), and cysteine (Yeoman, 1973).

Vernon and Straus (1972) found that IAA stimulated the activity of PPO in tobacco tissue cultures (inducing the synthesis of this enzyme or some intermediate), whereas 2,4-D decreased its activity (possibly acting at the level of repression of PPO synthesis, but not directly on the enzyme). A similar effect of 2,4-D was found by Kovacs and Faludi (1974) in potato tissues, but only at much higher concentrations.

The effects of these various treatments on fresh weight and PPO activity of *Eucalyptus grandis* anther callus corroborate the reported effect of 2,4-D on reducing PPO activity (table 4); however, the presence of polyvinyl pyrrolidone, tyrosine, or cysteine was inhibitory to growth and did not reduce PPO activity below the control levels. The effect of 2,4-D both in increasing fresh weight and in decreasing PPO levels as compared to IAA is

TABLE 4

EUCALYPTUS ANTHER CALLUS GROWTH AND DARKENING INHIBITORS

TREATMENTS	2,4-D (1 MG/L)		IAA (2 MG/L)	
	Fresh Weight	PPO	Fresh Weight	PPO
PVP (5 g/l)	88 mg	12.3	66 mg	17.0
Tyrosine (40 mg/l)	127	8.9	70	14.1
Cysteine (10 mg/l)	129	12.4	68	16.0
Control	165	8.7	134	12.5
Average	127	10.6	84	14.7

significant at the 95 percent level. The effects of the inhibitors on growth and PPO activity were not significant. Although the lower PPO level under the influence of 2,4-D is correlated with increased fresh weight, the same inverse relation was not observed when the different inhibitors are compared.

Callus originated from stamens of *Eucalyptus grandis*, with fully developed pollen grains in the anthers, was placed on 10 ml of White's medium plus 10 percent CCM plus 10 mM KNO_3 and 1 mg/l 2,4-D or 2 mg/l IAA. Other additives were used in the concentrations indicated. After 100 days, the tissue was weighed, then homogenized in a chilled mortar and pestle with 3.0 ml of phosphate buffer, 0.1 M pH 7.0. The extract was filtered and its volume completed to 5.0 ml. The extract was added (1.5 ml) to an equal volume of $Na_2S_2O_5$ (0.1%), an inhibitor of PPO activity, and the reaction mixture maintained at 30°C for a period of 20 min before reading the absorbance at 450 nm in a Gilford 300-T spectrophotometer. The units of enzyme activity were calculated by comparing the absorbance of this tube with the absorbance of another similar tube in which the $Na_2S_2O_5$ was replaced by 5.0 mM catechol, pyrogallol, or other substrate.

A third tube containing extract and deionized water was used to measure the presence of endogenous substrates in the extract. The difference in absorbance at 450 nm between the substrate tube and that with $Na_2S_2O_5$, per gram fresh weight, is equivalent to one unit of enzyme activity.

A certain fraction of the darkening of the solution in this PPO assay was a nonenzymatic reaction between the added phenolic substrate and endogenous phenols present in the extract (fig. 2). After boiling the extract for 12 min and testing the reaction, 38 percent of the reaction in the presence of water remained, whereas only 18 percent of the reaction with pyrogallol persisted with the boiled extract (fig. 2, left). Time course studies of the reaction show that the absorbance in the presence of meta-bisulfite remains essentially unchanged from 5 to 35 min after mixing, whereas it increases in a linear fashion in both water and pyrogallol. The extract from *Eucalyptus grandis* anther callus is almost inactive with tyrosine as substrate, but highly active with pyrogallol (fig. 2, right).

Since no improvement in growth or reduction of tissue darkening was observed as a result of the treatments with inhibitors, and since a great deal of variation in growth occurred between replicates, the relationship between tissue darkening and growth of *Eucalyptus grandis* anther callus

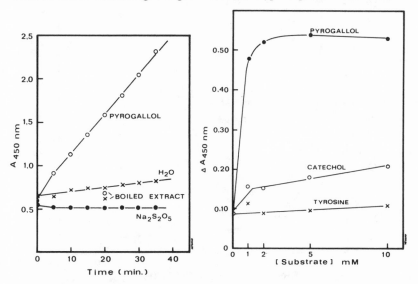

Fig. 2. (Left) Activity of extracts of *Eucalyptus grandis* anther callus in assays for PPO, as described in table 4, showed that only 18% of the activity with pyrogallol as substrate remained after boiling the extract for 12 min. Pyrogallol concentration was 10 mM. (Right) The extract from *Eucalyptus grandis* callus has a much higher activity with pyrogallol as substrate than with catechol and almost no activity with tyrosine.

was reexamined. Flasks of White's medium with 10 percent CCM, 10 mM KNO₃, and 1 mg/1 2,4-D added were inoculated with 150 mg of *Eucalyptus grandis* anther callus, each inoculum being farily homogeneous in coloration but with different degrees of discoloration from flask to flask. Of the 150 mg inoculum, 100 mg was used to measure the initial activity of PPO and the other 50 mg were left to grow for 50 days before the determination of final PPO activity and fresh weight.

No correlation was detected between the initial tissue darkness (as measured by the absorbance of the inoculum extract at 450 nm with metabisulfite) and subsequent growth. Although a wide range of PPO activities was represented among the different inocula, no correlation with subsequent growth was noted. The same was true for endogenous substrates (phenols).

On the other hand, there was a strong inverse relationship between the final activity of PPO with pyrogallol as substrate and the growth of the callus (fig. 3). Those flasks with the least growth had the highest activity at

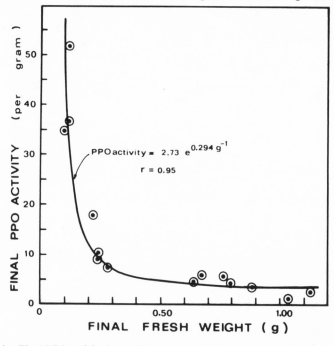

Fig. 3. Final PPO activity in replicate cultures of *Eucalyptus grandis* anther callus as a function of the final fresh weight. The initial inocula (50 mg each) differed in coloration and PPO activity. Each point represents the fresh weight attained by a single culture (after 50 days' growth on White's medium with 10% CCM, 10 mM KNO₃ and 1 mg/l 2,4-D) and its respective PPO activity.

the end of the growth period; the PPO activity was lowest in those cultures that had the greatest increases in fresh weight. However, an analysis of the actual values indicated that the initial PPO activity per gram had been reduced to the final activity by a factor that was of the same order of magnitude as the increase in fresh weight of the tissue (table 5). A possible

TABLE 5

EUCALYPTUS ANTHER CALLUS GROWTH AND PPO ACTIVITY

Initial PPO Activity of Replicate	$\frac{PPO_i}{PPO_f}$	Final Fresh Weight (mg)	$\frac{Fr. Wt._f}{Fr. Wt._i}$
39.2	11.8	878	17.6
47.8	31.8	1030	20.6
49.8	1.0	110	2.2
53.2	7.4	280	5.6
54.8	19.4	1130	22.6
55.3	1.5	110	2.2
59.0	10.2	760	15.2

Selected data from seven replicate cultures of *Eucalyptus grandis* anther callus grown for 50 days on White's medium with 10% CCM, 10 mM KNO_3, and 1 mg/1 2,4-D. Initial inoculum was 50 mg, and initial PPO activity was measured in a 100 mg sample of the same callus with similar aspect. Ratios of initial PPO activity (PPO_i) to final PPO activity (PPO_f) and final fresh weight to initial fresh weight were approximately the same, for instance, if the fresh weight increased 10 times, the final PPO activity was one-tenth the initial ($PPO_i / PPO_f = 10$).

explanation of the correlation between final PPO activity and fresh weight would be that little synthesis of PPO was occurring in the cells of *Eucalyptus grandis* anther callus under these conditions and the final activity represents the dilution of the enzyme by growth. Another possibility would be that PPO is limited in its distribution to more superficial cell layers within the callus and would increase as a function of the area of the callus while fresh weight increases the volume also (Henshaw, personal communication).

The lack of any correlation between growth and initial darkness or endogenous substrates of the inoculum suggests that the quinones already present and the endogenous phenols do not exert any appreciable effect on the growth. These substances may be compartmentalized, as suggested by Hall et al. (1972), in the vacuole or restricted to cells where their influence does not reach the actively growing cells in the callus. Neighboring cells do, in fact, demonstrate striking differences in tannic content (fig. 4). A group of nonlignified, dividing cells may be surrounded by other cells completely filled with tannins and with heavily lignified walls. The heterogeneity observed in cytological characteristics of the cells is certainly a demonstration of physiological or metabolic heterogeneity as well. The values of initial darkness, endogenous phenols, and PPO activity are averages for the whole mass of callus. However, considerable variation in growth could

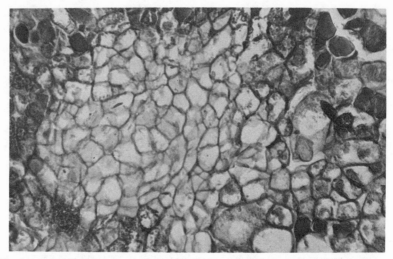

Fig. 4.   Callus from *Eucalyptus grandis* anther grown on White's medium with 10% CCM, 10 mM KNO₃, and 1 mg/l 2,4-D added showed a great deal of heterogeneity in the cytological characteristics from one part of the callus to another. This heterogeneity may help to explain the lack of correlation between initial PPO activity, tissue darkness, and endogenous substrate levels (of PPO) and subsequent callus growth. Callus was fixed in Navaschin's fixative and stained with safranin-fast green.

result from the same initial values depending on the distribution of the phenols and enzyme within the mass.

Finally, it is possible that there is no correlation because some other factor in the medium or callus is limiting to growth.

MORPHOGENESIS IN CALLUS CULTURES

Regeneration of roots has been observed in *Eucalyptus* callus cultures under certain conditions: callus from stem sections of *Eucalyptus grandis* formed roots on medium containing 2,4-D at 2×10^{-5} M and kinetin (Fossard, 1974); however, Lee and Fossard (1974) did not obtain root regeneration on any of a large number of auxin and cytokinin treatments of *Eucalyptus bancroftii* stem and lignotuber callus. Aneja and Atal (1969) had reported root formation in callus from roots of *Eucalyptus citriodora* seedlings at 1 mg/l NOA, and *Eucalyptus alba* hypocotyl callus formed roots at 1 mg/l IAA and 0.5 or 1.0 mg/l 2,4-D (Kitahara and Caldas, 1975). Sussex (1965) observed root formation in *Eucalyptus camaldulensis* only in the absence of 2,4-D or CCM.

Leaf blade callus from *Eucalyptus grandis* formed roots in a small number of the treatments with IAA and kinetin on Nash and Davies' (1972) medium and at concentrations of NAA from 1.0 to 8.0 mg/l in the presence of 0.5 mg/ 1 kinetin / Gonçalves, 1975). Root formation also occurred at the higher NAA concentrations in the absence of kinetin and at 1.0 mg/ 1 kinetin.

The results in table 6 show the growth and root formation that occurred

TABLE 6

ROOT FORMATION AND GROWTH OF CALLUS VERSUS 2,4-D AND KINETIN
(Fresh Weight [mg])

Kinetin	2,4-D mg/1						
mg/1	0.0	0.5	1.0	2.0	4.0	6.0	8.0
0.0	178.4	173.0	438.0	156.5	691.0		878.0
0.5	898.8	1,545.4	1,591.4	1,620.4	1,652.4	1,853.6	1,524.0
1.0	832.9	1,783.8	1,614.4	1,416.4	1,278.4	1,743.4	1,365.4
2.0	1,396.8	1,297.6	1,516.6	2,017.8	1,521.2	1,551.6	3,737.4
4.0	1,209.0	1,269.0	1,197.8	2,340.4	1,237.8	1,530.8	1,318.4
6.0	1,721.8	816.6	1,706.2	1,337.0	1,396.2	2,579.4	962.4
8.0	1,418.4	955.4	1,064.4	1,694.4	1,588.4	1,842.4	1,552.6

Leaf blade callus from *Eucalyptus grandis* was grown on Nash and Davies' (1972) medium with the hormone concentrations indicated in the table. Growth was measured after 60 days on a 12-hour photoperiod (Gonçalves, 1975).

in *Eucalyptus grandis* leaf blade callus. Roots were formed at almost all kinetin concentrations from 0.5 to 8.0 mg/ 1 when this was the only hormone in the medium, and at 2,4-D concentrations from 2.0 to 8.0 mg/ 1 when this hormone was used alone. In the treatments with both hormones, root formation occurred at progressively lower kinetin concentrations when the 2,4-D concentration was increased; that is, roots formed on callus at low 2,4-D plus high kinetin or high 2,4-D and low kinetin. No roots were formed in cultures maintained in the dark.

Of all the treatments tested, best root formation occurred on medium with 1.0 mg/ 1 IBA and no kinetin.

Activated charcoal, which has been used to absorb some toxic or inhibitory substances formed by callus or explants (Anagonostakis, 1974), inhibited callus formation from explants of different organs of *Eucalyptus* species when added at 8.0 g/ 1. Since *Eucalyptus* seeds planted on medium containing charcoal usually developed normally whereas those on medium without charcoal developed callus, the charcoal may be absorbing the exogenous hormones present in the nutrient medium (Gonçalves, 1975).

Shoot formation (fig. 5) from hypocotyl callus of *Eucalyptus alba* occurred on medium containing 1 mg/ 1 IAA (Kitahara and Caldas, 1975) and from lignotuber callus of *Eucalyptus citriodora* (Aneja and Atal, 1969). Since no shoots were formed from callus on the numerous combinations of auxins and cytokinins tested (Lee and Fossard, 1974; Gonçalves, 1975), more emphasis has been given to organ culture in attempts at vegetative propagation.

Fig. 5. Shoot and root formation on hypocotyl callus of *Eucalyptus alba* cultured on White's medium plus 10% CCM and 1 mg/l IAA after four months in culture (Kitahara and Caldas, 1975).

ORGAN CULTURE OF EUCALYPTS

Aside from the root cultures of *Eucalyptus camaldulensis* established by Bachelard and Stowe (1963) on medium containing CCM, and those of *Eucalyptus grandis* on minerals and sucrose (Cresswell and Fossard, 1974), the majority of organ cultures of *Eucalyptus* have used nodal segments or shoot apices in tests of rooting potential and bud development. A number of species have been tested (table 7).

Occasional root formation on cultured stem segments of *Eucalyptus × trabuti* was observed by Marcavellaca and Montaldi (1964). Subsequently, consistent rooting and axillary bud development in nodal cultures from

TABLE 7

EUCALYPTUS SPECIES TESTED IN ORGAN CULTURE

Species	Organ	References
Eucalyptus alba Reinw. ex Bl.	node	Gonçalves, 1975
E. bancroftii (Maid.) Maid.	shoot apex	Cresswell and Fossard, 1974
E. camaldulensis Dehnh.	root	Bachelard and Stowe, 1963
	node	Gonçalves, 1975
E. deglupta Bl.	shoot apex	Cresswell and Fossard, 1974
E. dunnii Maid.	node	Gonçalves, unpublished
E. ficifolia F. Muell.	node	Fossard, 1977
E. grandis Hill ex Maid.	node	Cresswell and Fossard, 1974; Cresswell and Nitsch, 1975; Gonçalves, 1975
	shoot apex	Cresswell and Fossard, 1974
E. maculata Hook.	node	Gonçalves, unpublished
E. polyanthemos Schau.	branch segment	Bachelard, 1969
E. robusta Sm.	node	Gonçalves, 1975
E. saligna Sm.	node	Gonçalves, 1975
E. tereticornis Sm.	node	Gonçalves, 1975
E. urophylla S. T. Blake	node	Gonçalves, unpublished

Nomenclature follows Chippendale, 1976.

Eucalyptus grandis seedlings was reported (Cresswell and Fossard, 1974). The presence of some leaf tissue on the explant stimulated rooting in this case, whereas further work with nodes taken from older plants (above node No. 14), demonstrated that the presence of part of the petiole permitted axillary bud development without inhibiting root formation (Cresswell and Nitsch, 1975). The auxin IBA at a concentration of 1×10^{-6} M was used to stimulate rooting of these explants.

This technique of rooting nodal explants was successfully extended to *Eucalyptus camaldulensis* and *Eucalyptus robusta*, which developed roots and shoots, as did *Eucalyptus grandis*, on Nash and Davies' (1972) medium with 1.0 mg/1 IBA (Gonçalves, 1975). The same treatment was not effective with *Eucalyptus alba* (shoot and callus development alone were noted), *Eucalyptus saligna* (callus only), or juvenile material from *Eucalyptus grandis* (shoot only). Growth in the dark slowed bud development but did not prevent either rooting or bud development as compared to growth on a 12-hour photoperiod. Soaking the nodal explants in gibberellic acid (GA$_3$ 100 mg/1) for two hours before placing them in culture was not noticeably effective in inducing a reversion to a more juvenile state, which could be accompanied by increased rooting capacity. This pretreatment did, however, inhibit shoot development in *Eucalyptus camaldulensis, Eucalyptus alba,* and juvenile *Eucalyptus grandis* (Gonçalves, 1975). Substitution of IBA by 1.0 mg/1 2,4-D plus 0.5 mg/1 kinetin in these experiments led to root development in only one species and treatment and shoot development in another.

Nodal culture seems to be a very promising technique for vegetative propagation of adult eucalypts if it can be applied to a larger number of species and a fairly high percentage of successful transplants to the field achieved.

FUTURE APPLICATIONS

The exploration of tissue culture techniques for vegetative propagation of selected trees, for inbreeding, for regeneration of individuals with mutants either spontaneously arising or induced *in vitro* (including polyploids), for somatic cell hybridization and embryo culture as well as various other purposes, has been raised by several authors (Bonga, 1977; Durzan and Campbell, 1974; Haissig, 1965). Many of these objectives are desirable in *Eucalyptus* species, as, for instance, the vegetative propagation on a scale of selected trees for seed orchards. However, since a greatly reduced seed yield and increased deformities have been noted in inbred *Eucalyptus* (Hodgson, 1976), a seed orchard might be composed of a mixed population of individuals of several or many selected clones.

Hybrids of Eucalyptus that currently appear promising in terms of growth and tree form, wood quality, disease resistance, and adaptation to diverse ecological conditions, could be propagated through tissue culture techniques. In addition, characterization of new hybrids with respect to nutritional demands and the lignification problems observed in some crosses could be studied *in vitro*, where metabolism is more easily controlled, influenced, and observed than in field conditions.

The ability to regenerate plants from callus cultures or cell suspensions would permit the maintenance of germplasm banks in the form of tissue cultures, which require relatively little space and permit a rapid multiplication of the material when desired.

Finally, basic studies of metabolism involving the biosynthesis of essential oils, tannins, lignification, and fiber formation, could be undertaken *in vitro*. Aspects of juvenility and its relation to the production of rooting inhibitor(s) could also be studied and characterized in tissue cultures.

ACKNOWLEDGMENTS

Grants from the Conselho Nacional de Desenvolvimento Cientifico e Tecnologico (CNP$_q$) and the Fundacao de Amparo a Pesquisa do Estãdo de São Paulo (FAPESP) were received by A.N.G. during the course of this work. We would also like to thank Henrique V. Amorim and G. G. Henshaw for helpful discussions and suggestions.

LITERATURE CITED

Anagnostakis, S. L. 1974. Haploid plants from anthers of tobacco—enhancement with charcoal. Planta 115:281-83.

Anderson, R. A., and J. A. Sowers. 1968. Optimum conditions for bonding of plant phenols to insoluble polyvinyl pyrrolidone. Phytochemistry 7:293-301.

Aneja, S., and C. K. Atal. 1969. Plantlet formation in tissue cultures from lignotubers of *Eucalyptus citriodora* Hook. Current Science (India) 38:69.

Bachelard, E. P. 1969. Studies on the formation of epicormic shoots on eucalypt stem segments. Aus. J. Biol. Sci. 22:1291-96.

Bachelard, E. P., and B. B. Stowe. 1963. Growth in vitro of roots of *Acer rubrum* L. and *Eucalyptus camaldulensis* Dehnh. Physiol. Plantarum 16:20-30.

Blake, T. J. 1972. Studies on the lignotubers of *Eucalyptus obliqua* L'Herit. III. The effects of seasonal and nutritional factors on dormant bud development. New Phytol. 71:327-34.

Bonga, J. M. 1977. Applications of tissue culture in forestry. *In* J. Reinert and Y. P. S. Bajaj (eds.), Applied and fundamental aspects of plant cell, tissue, and organ culture. Pp. 93-108. Springer-Verlag, Berling.

Chippendale, G. M. 1976. *Eucalyptus* nomenclature. Aust. For. Res. 7:69-107.

Cresswell, R. J., and R. A. de Fossard. 1974. Organ culture of *Eucalyptus grandis* Austr. For. 37:55-69.

Cresswell, R., and C. Nitsch. 1975. Organ culture of *Eucalyptus grandis* L. Planta 125:87-90.

Davidson, J. 1977. Problems of vegetative propagation of *Eucalyptus*. Third World Consultation on Forest Tree Breeding, Canberra.

Davies, M. E. 1972. Polyphenol synthesis in cell suspension cultures of Paul's Scarlet Rose. Planta 104:50-65.

Durzan, D. J., and R. A. Campbell. 1974. Prospects for the mass production of improved stock of forest trees by cell and tissue culture. Canad. J. For Res. 4:151-74.

Edelman, J., and A. D. Hanson. 1972. Sucrose suppression of chlorophyll synthesis in carrot-tissue cultures. J. Exp. Bot. 23:469-78.

Epstein, E. 1972. Mineral nutrition of plants: Principles and perspectives. John Wiley & Sons, New York.

Fine, M. 1968. The control of tracheary element formation in *Eucalyptus* tissue cultures. Ph.D. thesis, Yale University.

Fossard, R. A. de. 1974. Tissue culture of *Eucalyptus*. Austr. For. 37:43-54.

Fossard, R. A. de, C. Nitsch, R. J. Cresswell, and E. C. M. Lee. 1974. Tissue and organ culture of *Eucalyptus*. N. Z. J. For. Sci. 4:267-78.

Gonçalves, A. N. 1975. The growth and developmental physiology of *Eucalyptus* in cell and tissue culture systems. M. S. thesis, Ohio State University, Columbus.

Haissig, B. E. 1965. Organ formation *in vitro* as applicable to forest tree propagation. Bot. Rev. 31:607-26.

Hall, R. H., P. S. Baur, and C. H. Walkinshaw. 1972. Variability in oxygen consumption and cell morphology in slash pine tissue cultures. For Sci. 18:298-307.

Hodgson, L. M. 1976. Some aspects of flowering and reproductive behaviour in *Eucalyptus grandis* (Hill) Maiden at J.D.M. Keet Forest Research Station: 3. Relative yield, breeding systems, barriers to selfing, and general conclusions. S. Afr. For. 99:53-58.

Jacquiot, C. 1964. Structure of excised roots or of organs formed *de novo* from cambial tissue of trees grown in culture. Rev. Cytol. Biol. Veg. 27:319-22.

Kitahara, E. H., and L. S. Caldas. 1975. Shoot and root formation in hypocotyl callus cultures of *Eucalyptus*. For. Sci. 22:242–43.

Kovacs, E. I., and B. Faludi. 1974. Effect of 2,4-D on the polyphenol oxidase activity of isolated potato tissues. Acta Agron. Sci. Hung. 22:335–42.

Lee, E.C.M., and R.A. de Fossard. 1974. The effects of various auxins and cytokinins on the *in vitro* cultures of stem and lignotuber tissues of *Eucalyptus bancroftii* Maiden. New Phytol. 73:707–17.

Linsmaier, E. M., and F. Skoog. 1965. Organic growth factor requirements of tobacco tissue cultures. Physiol. Plantarum 18:100–127.

Marcavillaca, M. C., and E. R. Montaldi. 1964. Cultivo "in vitro" de tejidos de eucalipto. IDIA Supl. 12:62–64.

Murashige, T., and F. Skoog. 1962. A revised medium for rapid growth and bioassays with tobacco tissue cultures. Physiol. Plantarum 15:473–97.

Nash, D. T., and M. E. Davies. 1972. Some aspects of growth and metabolism of Paul's Scarlet Rose cell suspensions. J. Exp. Bot. 23:75–91.

Penfold, A. R., and J. L. Willis. 1961. The eucalypts: Botany, cultivation, chemistry, and utilization. Leonard Hill (Books), London.

Pryor, L. D. 1976. The biology of eucalypts. Institute of Biology's Studies in Biology No. 61. Edward Arnold, London.

Pryor, L. D., and L. A. S. Johnson. 1971. Classification of eucalypts. Australian National University Press, Canberra.

Reinert, J., and P. R. White. 1956. The cultivation in vitro of tumor tissues and normal tissues of *Picea glauca*. Physiol. Plantarum 9:177–89.

Sommer, H. E., C. L. Brown, and P. P. Kormanik. 1975. Differentiation of plantlets in longleaf pine (*Pinus palustris* Mill.) tissue cultured *in vitro*. Bot. Gaz. 136:196–200.

Sussex, I. M. 1965. The origin and morphogenesis of *Eucalyptus* cell populations. *In* P. R. White and A. R. Grove (eds.), Proceedings of the international conference on plant tissue culture. McCutchan Publishing Co., Berkeley, California.

Tulecke, W., R. Taggaret, and L. Colavito. 1965. Continuous culture of higher plant cells in liquid media. Contrib. Boyce Thompson Inst. 23:33–46.

Vernon, S. L., and J. Straus. 1972. Effects of IAA and 2,4-D on polyphenol oxidase in tobacco tissue cultures. Phytochemistry 11:2723–27.

White, P. R. 1963. The cultivation of animal and plant cells. 2d ed. Ronald Press, New York.

Winton, L. 1972. Bibliography of somatic callus cultures from deciduous trees. Genetics and Physiology Notes No. 17. Institute of Paper Chemistry, Appleton, Wis. 19 pp.

Yeoman, M. M. 1973. Tissue (callus) culture techniques. *In* H. E. Street (ed.), Plant tissue and cell culture. Blackwell, Oxford.

M. R. SÖNDAHL AND W. R. SHARP

Research in *Coffea* spp. and Applications of Tissue Culture Methods

28

INTRODUCTION

It is not known exactly when coffee was initially used in Africa as a stimulatory beverage. The history of coffee begins with its transference from the high land of Southwest Ethiopia to Yemen (Arabian Peninsula) during the thirteenth century. From Yemen, the British took coffee to Ceylon and the Dutch to Java. Later, seeds were taken to the Botanical Gardens of Amsterdam and Paris and from these gardens *Coffea* seeds were brought to South America (Surinam and Cayenne). Plantations were established north of Brazil (Belem) in 1727 and one century later Brazil became the leading world coffee producer. Despite the fact that Africa is the major center of germplasm variability of the genus *Coffea*, the increase in importance of commercial *Coffea* plantations on this continent was only observed in the 1950s after the establishment of a policy of price stabilization supported by Brazil. The autogamous and perennial characteristics of *Coffea arabica*, the most important cultivated *Coffea* species, have led to the development of homogenous plantations all over the world. This is especially true in South and Central America, where all of the plantations were historically derived from a single source. This uniformity is a potential risk to the economy of coffee plantations, since the outbreak of any severe disease endangers all plantations in a given area.

M. R. Söndahl, Department of Genetics, Agronomic Institute, Campinas, São Paulo, Brazil; W. R. Sharp, Department of Microbiology, Ohio State University, Columbus, Ohio 43210.

Coffee trees are usually 2–4 m high and the cherries produced by lateral branches carry two seeds each. The coffee beverage is prepared by percolation of ground and roasted seeds. Plantations in Brazil are established from 6-month seedlings grown in nurseries (under 50% field irradiation). A plastic bag (ca. 12 × 20 cm) filled with rich soil is used for greenhouse seedlings. Young plants are transplanted to the field during the rainy season (December to January). Hills containing two seedlings each are spaced at intervals of 4.0 × 2.5 m (Mundo Novo) or 3.5 × 2.0 m (Catuai) giving 1,000 and ca. 1,600 hills/ha, respectively. In Brazil, all plantations exist under full sunlight, i.e., without shade trees. It has been shown that the vegetative growth (Franco, 1963) and productivity (Moraes, 1963) of young plants are optimal under full sunlight. It is known that at high leaf temperatures (40°C) caused by direct irradiation there is no net assimilation because of high respiratory rates, high CO_2 concentrations in the mesophyll cells, and closure of the stomates. However, only about 50% of the leaves of an adult plant are exposed to sun at a given time, because of the mutual shading of leaves and multiple angles of incidence of sunlight; therefore, leaf temperatures and other deleterious consequences of direct irradiation are reduced. The temperature of illuminated leaves is generally between 10 and 15°C above air temperature, but 1 to 3°C below air temperature when leaves are in the shade (Buttler, 1977). It seems that the high leaf density of an adult plant (ca. 3,000 dm^2) provides enough leaf surface area for maximum photosynthetic assimilation under optimal conditions. Another disadvantage of the shading practice, in addition to restriction of full irradiation, is the competition for water between the shade trees and *Coffea* plants during the dry season. Franco (1963) states that an additional 250 mm water deficit occurs during the period April to September due to the transpiration of shade trees in Campinas, Brazil.

Coffee cherries can be harvested in two ways: (a) individual selection of ripe cherries or (b) stripping the branches after the cherries have been dried at the end of the season. Depending upon whether mature cherries or dried cherries are used, the preparation of coffee seeds follows a so-called *wet* or *dry* process, respectively. As a consequence of harvesting and preparation there is a clear distinction in the classification of coffee quality. When cherries are used, the exocarp is removed to liberate the seeds, which are coated with a gelatinous-pectic substance (mesocarp) around the seed coat or parchment (endocarp), and this pectic material is removed by fermentation in water tanks. When the cherries are allowed to dry on the trees, the fermentative process involves the participation of assorted microorganisms under different conditions of humidity and temperature. The extent to which this fermentative process affects the final beverage quality is not yet fully understood.

After preparation, the *Coffea* seeds are called "green coffee" and submitted to different classifications, according to the country of origin. In Colombia, for instance, Colombian Mild coffee is further classified by the altitude where the plants were grown and also the seed grading. Other Latin American countries (arabic coffee producers), Costa Rica, El Salvador, Guatemala, Honduras, Mexico, Nicaragua, and Panama, have adopted a similar grading system as Colombia. In African countries (mainly Robusta producers) the classification system is based on the geographic area and species. In Kenya and Tanzania (Colombia Mild producers), coffee classification is based on seed size and cup quality. In Brazil (unwashed arabic producer) and some African countries, including Angola, Cameroun, and Ivory Coast, coffee seeds are classified according to the number of physical defects and cup quality. At present, the Brazilian system considers 7 types of green coffee classified from 2 to 8 depending on the number of defects found in 300 grams of green coffee. Type 2 has practically no defects, whereas type 8 has 360 defects in 300 grams. Sieves with different different numbers of meshes per inch are used to determine seed size. For instance, grade 17 coffee seeds will pass through sieves of 17/64″, but are retained on sieves of 16/64″. For most markets, cup quality is also required; freshly prepared infusion is classified as riado (0), rio (1), hard (2), softish (3), soft (4), and strictly soft (5). Roasting is of great importance in maintaining good cup quality and the roasting intensity varies from country to country. The cup quality standard in the United States requires a light roasting process, whereas in Brazil a heavy roasting process is adopted. The speed of the roasting process is also an important determinant of coffee quality. High temperature (200°C) and vigorous agitation permit the achievement of good roasting grade in short time (5–10 min). During roasting, seeds become dark as sugar caramelizes and the Maillard reaction occurs between reducing sugars and amino acids. Extensive research efforts have been undertaken with green and roasted coffee seeds to characterize the chemical constituents contributing to coffee aroma and flavor. However, no real clue to the chemical and biochemical aspects of a good coffee quality is yet available. This aspect could be very important, since breeding and selection programs for better flavor are still hindered by the lack of such parameters.

CLASSIFICATION OF THE GENUS COFFEA

The family Rubiaceae includes more than 500 genera and about 8,000 species. The genus *Coffea* was established in 1735 by Linnaeus, who described the species *C. arabica* in 1753. This genus probably consists of more than 70 species, but if all the species described are accepted, the total would be over 100 species. However, it is believed that many interspecific hybrids have been classified as species and a critical systematic review in this genus is much needed. Chevallier (1947) divided the genus *Coffea* into four sections: Paracoffea (13 species), Argocoffea (11 species), Mascaro-coffea (18 species), and Eucoffea (24 species). All economically important species belong to the section Eucoffea, which is further divided into five subsections: Erytrocoffea (*C. arabica, C. canephora, C. congensis*, and *C. eugenioides*); Pachycoffea (*C. liberica, C. dewevrei*, and 2 other species), Melannocoffea, Nanocoffea, and Mozambicoffea. The genus *Coffea* has a tropical and subtropical distribution in Africa, and a few species are present in Asia. Four species have been raised in commercial plantations: *C. arabica* (ca. 70% plantations), *C. canephora* (ca. 26% plantations), *C. liberica* and *C. dewevrei* cv. Excelsa (together ca. 4% plantations), *C. arabica* is the only self-pollinated and tetraploid ($4\times = 44$) species known so far in the genus; it provides a superior beverage quality and is typically a species of high lands (up to 2,000 m). *C. canephora* cv. Robusta is a diploid ($2\times = 22$) species, outcrossed and typically growing in low and humid regions.

MORPHOLOGY

C. arabica, the first species to be described, consists of a tree of 2–4 m high, bearing a single cylindric stem (orthotropic shoots) with opposite lateral branches (plagiotropic shoots) at each leaf axil of the stem and a long main root. The leaves of the orthotropic shoot are opposite with 1/4 phyllotaxy, and the plagiotropic shoots twist in such a way as to place the leaves in the same orientation. The lateral branches are very long and flexible and frequently have secondary and tertiary branches, all derived from leaf axils. In the axils of the first nodes of orthotropic shoots arrested orthotropic buds are present. After the 9–11th node, plagiotropic buds, which are situated above the orthotropic buds, are also differentiated at the leaf axils. In the leaf axils of plagiotropic branches, only buds for lateral branches and flower buds are formed. Orthotropic buds never differentiate at the lateral branches. This has a practical limitation for the vegetative propagation of *Coffea* plants, since only orthotropic shoots can be used for rooting. When the apical bud is removed, two orthotropic shoots are produced at the uppermost leaf axil and a few additional shoots are obtainable if the main axis is bent. Flower buds only develop in the leaf axes of lateral branches formed during the previous growing season. A minimum juvenile period (ca. 3–4 mo) at each node is apparently required for flower bud differentiation, since leaf axils formed late in the growing season do not bear flowers during the forthcoming inductive period. A practical implication of this observation is that the number of productive leaf axils per season is dependent on the extent of lateral branch elongation achieved in the previous vegetative period. One should select for *Coffea* cultivars with rapid vegetative growth (maximum utilization of fertilizer, water, irradiation at optimum temperature, and photoperiod) and higher number of leaf axils (or nodes) per unit length of orthotropic and plagiotropic shoots. Precocious secondary and tertiary lateral branch formation would also be advantageous. The adult leaves of cultivated coffee have a shiny dark green color on the adaxial (upper) side and a dull light green color on the abaxial side. The stomates are present only on the lower epidermal surface. The average measurements of closed stomates and their frequencies per unit of leaf area are listed in table 1 (Söndahl et

TABLE 1

STOMATE MEASUREMENTS AND FREQUENCY PER MM^2 IN
FOUR CULTIVARS OF C. ARABICA AND ONE OF C. CANEPHORA

Cultivars of Coffee	Stomate Length (μm)	Stomate Width (μm)	No. Stomates per mm^2
C. arabica, cv. Mundo Novo	15.9 ($\pm$0.37)	24.7 ($\pm$0.76)	197.6 ($\pm$4.1)
C. arabica, cv. Catuai	16.7 ($\pm$0.33)	26.7 ($\pm$0.33)	206.6 ($\pm$4.9)
C. arabica, cv. 1130–13	16.7 ($\pm$0.33)	26.5 ($\pm$0.40)	219.9 ($\pm$4.3)
C. arabica, cv. H 6586-2	17.0 ($\pm$0.50)	28.2 ($\pm$0.84)	185.5 ($\pm$3.1)
C. canephora, cv. Guarini	14.6 ($\pm$0.65)	22.2 ($\pm$0.45)	282.5 ($\pm$5.1)

al., 1976). The flower buds develop as glomerules (short inflorescences) containing 2–19 flowers per axil (Carvalho and Monaco, 1963). The ovaries are inferior, bilocular bearing one single anatropous ovule attached to a central placenta. The styles are attached to the corolla and have about the same length as the corolla. The anthers open longitudinally and 5 anthers per flower bud are found in *C. arabica, Coffea* cherries are drupe with a red or yellow exocarp when ripe, with a gelatinous-pectic mesocarp and a hard endocarp (parchment) covering each seed. Underneath the parchment there is a thin pellicule (silver skin), which ontogenetically is derived from the perisperm (Dedecca, 1957). The seeds are convex with one flat side filled by a hemicellulosic endosperm containing one small embryo (fig. 1).

GROWTH PHYSIOLOGY

Seeds

The physiology of *Coffea* seeds was recently reviewed by Valio (1976), who confirmed that the optimum temperature for germination of coffee seeds is 30°C in darkness and that the seeds remain viable for long periods (8 mo) when stored at a high seed moisture content (40%). This ideal seed

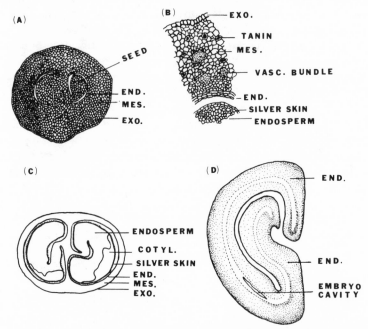

Fig. 1.   Coffee fruit development (after Dedecca, 1957). (A) Cross section of a recently fecundated ovary (30×): two locules bearing one fecundated ovule (seed) each; the ovary epidermis is represented by a one-layer thickness of cells (exoderm), underneath which appears an extensive region of parenchymatous cells (mesoderm) and a layer of cells representing the endodermis. (B) Cross section of a mature cherry (80×): single layer exoderm derived from the external ovary epidermis; exoderm cells are small with thin walls bearing stomates; mesoderm is formed by about 20 layers of large parenchymatous cells, some of which are filled with tanin; 2–3 concentric series of vascular bundles are also found in the mesoderm; 5–7 layers of small cells form the endodermis, which constitutes the parchment of a seed in a mature fruit; a thin layer (ca. 70 μm thick) covers the external surface of the endosperm (silver skin); it is formed from crushed dead elongated cells that are derived ontogentically from the perisperm. (C) Cross section of a dry cherry (16×): mesoderm is greatly reduced; endodermis (parchment) surrounds the seed, underneath which the silver skin covers the endosperm. (D) Cross section of a seed presenting the embryo cavity and the hemicellulosic endosperm.

moisture content for the storage of seeds is at odds with the findings of Bachi (1958), who observed that a 10% seed moisture content provided a high germination frequency after 21 months of storage. It seems important that the moisture content of seeds does not reach values lower that 8–9%. This effect is not fully understood yet, but may be due to physical damage of the embryo caused by contractions of the hemicellulose fibers compressing the embryo chamber (see fig. 1D) and/or to stimulation of synthesis of ABA-like substances and irreversible enzymatic denaturations.

Valio (1976) found that ABA (10^{-6}–10^{-4} M) inhibits both seed germination and embryo growth, whereas GA_3 (10^{-6}–10^{-4} M) only inhibits seed germination. It was postulated that GA inhibition is probably indirect, i.e., through induction of enzymatic synthesis in the endosperm. Kinetin (10^{-6}–10^{-4} M) enhances germination of seeds—especially old seeds with a low moisture content—and it also slightly reverses some of the inhibitory effect caused by GA_3 and ABA. Based on 80% methanol extracts of seeds and subsequent fractionation, it was concluded that high germination is related to low concentrations of endogenous GA- and ABA-like substances and high concentrations of endogenous cytokinin-like substances.

Vegetative Growth

Growth of *Coffea* is mainly affected by variations in the conditions of temperature, light (intensity and duration), disponibility of nutrients, and water. Extensive literature exists dealing with nutrition of *Coffea* (Malavolta, 1963; Malavolta and Moraes, 1963; Malavolta, 1975; Carvajal et al., 1969), but critical studies of nutrient uptake among different genotypes are still unavailable. Preliminary experiments (Söndahl, unpublished) have demonstrated that there is a great deal of variability concerning rate of absorption of Zn (between *C. arabica* cv. Mundo Novo and Amarelo Botucatu) and ability of Fe absorption under high pH (7.2). A gradient of Fe-deficiency was observed among seven *C. arabica* cultivars growing in nutritive solution in greenhouses. Data do not exist concerning the interaction of genotypes under varying water conditions.

Went (1957) reported that a thermoperiod of 30°C day/23°C night was optimum for vegetative growth of *C. arabica* seedlings and that 23°C day/17°C night for 1.5-year coffee plants. Franco (1958) reported 26°C day/20°C night as optimum for growth of young *C. arabica* plants. Nunes et al. (1968) found 25°C day/20°C night optimum for vegetative growth of 1-year-old coffee plants and 20°C for maximum net photosynthesis at a saturant light intensity of 0.11 cal.cm^{-2} min^{-1}. They also claimed that net photosynthesis would decrease about 10% for each degree above 24°C in the experimental conditions used. The effect of temperature in the root system and its effect on the water absorption was studied by Franco (1963). A thermoperiod regime of 26°C day/20°C night provided best growth and the temperature extremes of 23 and 33°C promoted transpiratory values of 80% and 100%, respectively. Below 23°C or above 33°C the capacity of water absorption by the roots is reduced seriously. High temperature (above 45°C) or extremely low temperatures (-2°C) cause necrosis of the base of the stem (girdling) in field and laboratory conditions. Low temperatures (3°C or lower) for a minimum of 6 hours can also cause discoloration and bleached areas of young coffee leaves, apparently due to a permanent destruction of the chloroplasts (Franco, 1963). The tolerance of *C. arabica* coffee genotypes already present in the *Coffea* germplasm collections to low and high temperatures has not been explored. Such studies would certainly open up the possibilities for geographical expansion of plantations.

The effect of light intensity on growth and productivity of plantations has been discussed previously. Stimulatory effects on vegetative growth caused by long photoperiods have been described by Piringer and Borthwick (1955) in 18-month-old *C. arabica* cvs. Naranjos (T2P3-4-209 El Salvador) and Bourbon Vermelho (B-43341 Brazil.) The length of lateral branches and the number of nodes were significantly increased (5% level) in plants under 14 and 18 hours in comparison to plants under 9 and 12 hours of light. Since these data were obtained using limited germplasm, the photoperiodical effect on growth of potted plants was further studied with two highly productive *C. arabica* cultivars (Catuai Amarelo and Mundo Novo-Acaia) and three hybrids of *C. arabica* cv. Mundo Novo: (1) H-5157, S.6 Cioicia from Ethiopia; (2) H-5202, KP 228 from Tanzania; and (3) H-3613, S. 964 from India. Eight replicates of each cultivar were used in a random block distribution (Söndahl et al., 1977). A normal field illumination period of 11–14 hours (Campinas) and an 18-hour illumination period (supplemented with incandescent light, 150 ft-c) were adopted for all experiments. A great deal of variation in growth was observed throughout the 4–13-month growth period among these highly diversified types of *Coffea* germplasm.

The length of lateral branches and leaf area were enhanced by an 18-hour light period in two progenies (M. Novo-Acaia and H-3613), but lowered or unchanged in the remaining three progenies. However, when dry weight and growth rate were measured, only the progeny M. Novo-Acaia presented increased values at an 18-hour light period, whereas the others had unchanged or lower values. In summary, M. Novo-Acaia was the most positively affected by an increase in day length, Catuai Amarelo and the other progenies were negatively affected or unchanged. Perhaps these data have practical implications concerning the regularity of vegetative growth at different latitudes. These results point out that variability of vegetative growth (leaf area, length of lateral branches, dry weight, and growth rate) due to interaction of *Coffea* germplasm and photoperiod can be expected.

To complement the study on the effects of photoperiod on vegetative growth and flowering, another experiment was designed with coffee plants submitted to normal day length at Campinas (11-14 hr) and an 18-hour light period. The *C. arabica* cultivars, Mundo Novo, Catuai Amarelo, Bourbon Vermelho, Geisha, and Semperflorens at the cotyledonary stage (ca. 2 mo old), were transplanted to soil (0.5 m apart) in a random block distribution with 8 replicates of each cultivar and grown up to 33 months (Monaco, Medina, Söndahl, and Miranda, 1977). All *C. arabica* cultivars grown under conditions of extended light period (18 hr) for 15 months were characterized by a 1.4-1.9 increase in height, a 1.8-2.4 increase in lateral branches, and a 1.7-2.8 increase in leaf number as compared to the normal field day length. This stimulatory effect of the 18-hour photoperiod on vegetative growth among 28-month-old plants was not as characteristic as at the 15-month stage. This can be attributed to mutual shading and competition among plants due to high density of plants at the later age. The cultivar Catuai Amarelo was characterized by a higher number of lateral branches and leaves at all stages of development. Interactions of vegetative growth and photoperiod will be discussed further in respect to flowering.

Water Relations

Measurements of transpiration in plants during a 1-year period reveal an average daily water loss of 6.3 $g/dm^2/day$ (Franco and Inforzato, 1950). Since the experimental coffee plants have a surface area of 31.46 m^2 (one side only, since only the lower surface has stomates), a total of 7,273 liters of water/year (ca. 20 liters/day) is transpired. In an unshaded coffee plantation where plants are spaced at 3.5 × 3.5 m, the total water loss is equivalent to about 600 mm of rainfall (Franco, 1963). However, if the water lost by coffee plant transpiration and soil surface loss and weed transpiration (evapotranspiration) are taken into account, the total water loss is equivalent to 840 mm of rainfall in the conditions of Kenya (Pereira, 1957). This required amount of yearly rainfall would be even higher if shade tree plantations were to be considered. Moreover, it should be noted that water precipitation is not considered. Moreover, it should be noted that water precipitation is not evenly distributed among the months of the year. This fact would impose some water deficits in the coffee plantations. In a 3.5 × 3.5 m plantation in Campinas, Brazil, a total water deficit of 25 mm of rainfall was found during the months of May, July, and August. If 10.5 × 10.5 m shading trees (Ingazeiro, *Cassia strobilacea*) were adopted, this deficit would total to 274.8 mm of rain among the months of April, May, June, July, August, and September (Franco, 1963). In both cases, the water deficits were estimated by considering only transpiration values and not the total water loss given by evapotranspiration, but these values are enough to stress the point that plants have periods of water deficit during some months of the year due to irregular distribution of rain and/or insufficient total water precipitation, commonly found in marginal plantation regions.

Interesting results have been presented by Browning and Fisher (1975), suggesting stimulatory vegetative growth after water deficit periods. Trees that were exposed to 12 weeks of drought produced approximately 70% more new lateral branches than irrigated trees. They also observed that *C. arabica* plants (irrigated and unirrigated) in the conditions of Kenya were severely stressed in the afternoon and were unable to recover during the night from daytime stress. This indicates that the evaporative demand of the atmosphere plays a larger part than soil moisture in determining the water status of a coffee plant. The authors explain their findings on the ability of ABA to increase the permeability of plant cells to water (Glinka, 1973). Abscisic acid synthesis would be stimulated under water stress, and thus would counteract the high root resistance to water uptake. After rehydration, a stimulatory effect on the synthetic processes for cell expansion and division associated with or without a hormonal balance shift may be postulated.

Increased water efficiency and resistance to drought are two desirable characteristics to be selected in improved *Coffea* cultivars to allow for the geographical expansion of cultivation of *Coffea*. Such research should be taken with a broader approach, that is utilizing coffee plants with smaller and reduced numbers of leaves. Certainly photosynthetically more efficient *Coffea* leaf surfaces in terms of mg CO_2 fixed per unit of leaf area should be selected for in order to avoid a drop in productivity per plant.

Flowering, Fruit Set, and Maturation

In *C. arabica* coffee plantations growing in Brazil the flowering process can be clearly separated into two phases: induction period and anthesis.

Induction period. *C. arabica* cultivar Bourbon Vermelho was initially identified as a short-day plant (SDP) by Franco (1940) and later confirmed by Piringer and Borthwick (1955) and Went (1957). The critical photoperiod for this cultivar was found to be 13–14 hours. More flowers were formed as the photoperiod was reduced to 12 hours, 10 hours, and 8 hours, and so Bourbon Vermelho would be a quantitative SDP. The photoperiodic effect in the cultivated *C. arabica* cultivars in Brazil (M. Novo, Catuai, Bourbon) is recognized because of the synchronization of flower bud differentiation occurring in late March to early April (beginning of fall). However, there is evidence that the induction of flower bud formation in coffee plants is not a simple photoperiodic phenomenon (Monaco, Medina, Söndahl, and Miranda, 1977).

Experiments with *C. arabica* cultivars (Mundo Novo, Catuai Amarelo, Bourbon Vermelho, Geisha, and Semperflorens) in soil conditions, spaced at 0.5×0.5 m in random blocks with 8 replicates under two photoperiodic conditions were evaluated under nursery conditions in Campinas, Brazil. This study was initiated with 1- to 2-month-old seedlings and grown up to 33-month-old plants. Normal day lengths (11-14 hr) and long photoperiodic conditions (18 hr) were provided to two separated sets of plants. It was found that, under an 18-hour light period, not only do all plants set flowers but also they set more flowers. This trend was observed in plants at 22 and 33 months old. At later stages, an increase in the frequency of flowers occurs apparently as a consequence of increased vegetative growth because of a higher number of nodes per branch. As a matter of fact, it is a known practice to produce premature flowering by inducing rapid growth of coffee plants with heavy fertilization. At 22 months, the intensity of flower bud formation under an 18-hour light period was found to be most pronounced in Catuai Amarelo (5.8), followed by Mundo Novo (4.3), Geisha (3.9), Bourbon Vermelho (1.9), and Semperflorens (1.8). Semperflorens flower buds were detected throughout the year under both light regimes as a normal characteristic of photoperiodic insensitivity of this cultivar (Monaco, Medina, Söndahl, and Miranda, 1977). In conclusion, the cultivars used in these studies appear to be day-neutral plants, the vegetative growth induced by long days (18 hr) seems to be correlated with an increase in the frequency of flower bud induction. The results of this study point out the necessity of furthering our knowledge of flower bud induction in *Coffea*, especially the interaction with vegetative growth. Similar conclusions have been discussed by Cannell (1972) in Kenya where induction of flowers occurs in photoperiods of less than 12 hours.

Growth and anthesis. After induction, *Coffea* flower buds grow for about 2 months, until they reach a size of about 5 mm in length, and then become quiescent (ca. 2 months old). At this time plants are regularly subjected to a water deficit on account of the yearly dry season. Later, the buds resume growth until they reach a length of 8-10 mm when they assume a pale yellow coloration. Anthesis, along with synchronous growth of the buds, will occur after a rainfall, and this will lead to the first blooming of the season 7 days later. Normal flowering in *C. arabica* under South Indian conditions was found to depend on the number of leaves at each node and as the total number on a branch, the starch index of the wood, and the nutritional condition of the plants (Gopal, Raju, Venkataramanan, and Janardham, 1975).

There is some controversy in the literature concerning the effect of rain on the anthesis of coffee buds. Flower buds are covered with a gum-like substance exuded by glandular hairs. It has been proposed that this gum-like substance may have some role in the flowering mechanism. It could act as an effective barrier to transpiration and perhaps liberate active growth substances after being dissolved by blossom showers in South India (Gopal, Venkataramanan, and Rathna, 1975). Mess (1957) suggested that a slight, rapid drop in the air temperature along with a washing effect were the principals factors controlling anthesis. However, Franco (unpublished results) was unable to induce anthesis of dormant flower buds by cooling treatments. Piringer and Borthwick (1955) and Alvim (1960) stressed the necessity of a dry period for anthesis of coffee buds to occur. Alvim (1958) demonstrated that the spraying of gibberellic acid (GA3) induced anthesis of quiescent coffee buds within 10 days, and that this effect was more pronounced with flower buds under water stress. Söndahl (unpublished) demonstrated this effect again with *C. arabica* cv. Catuai in Campinas when GA3 sprays were applied in late July. Alvim (1964) proposed the term "hydroperiodism" to such a plant-water relationship where anthesis occurs under water stress after a rainfall. Van der Veen (1968) proposed that a minimum water deficit was necessary for induction of flowering in hydroponic *Coffea* cultures; blooming occurred after the water potential was raised. Alvim (1973) hypothesized that water stress is necessary to release the flower buds from the dormant phase and that the increase in water potential (high air humidity of rain) provided the necessary conditions for further development. The physiological process of anthesis of buds became more clear after Magalhães and Angelocci (1976) measured the leaf potential threshold (−12 bars) below which anthesis will occur in response to watered roots of potted plants. No anthesis was observed in plants kept with a leaf potential higher than −11 bars. It was concluded that irrigation releases flower buds from dormancy in response to a sudden reversal of the negative water potential gradient established between flower buds and subtending leaves, causing an influx of water into buds due to a rapid increase in leaf water potential.

What happens to the endogenous hormone balance during water deficits and subsequent rise in water potential is still not completely clear. Annual and daily fluctuations of gibberellins were found in *C. arabica* leaves and fluctuations also occurred during the drought period and 6 days after plants were watered (Humphrey and Ballantyne, 1974). High levels of abscisic acid have been reported in dormant coffee flower buds and its endogenous level may contribute to the dormancy period (Browning et al., 1973). Critical experiments demonstrating shifts in the endogenous concentrations of gibberellins, abscisic acid, and perhaps other growth hormones during drought treatment and afterwards are necessary to clarify the control of anthesis in flower buds.

Fruit set and maturation. A few hours after anthesis of flower buds, the anthers will open up longitudinally. In *C. arabica* plants, after fecundation the ovaries will develop slowly and full-size green *Coffea* cherries will be present 5–6 months later. The maturation process will take 2 months or more, depending on the ecological conditions and coffee cultivar. This process of fruit set, growth, and maturation progresses according to the sequence of flowering and so brings about another problem in this crop: the presence of green and mature fruits side by side. At Campinas, at least three flowering periods are normally observed between September and November, with the first one being the most important; the number of flower buds opened at each flowering decreases as the season progresses. It is interesting to mention that, in many cases, not all of the flower buds of a single coffee inflorescence open during a single flowering period. It is common to see that, in the same inflorescence, 3–5 buds will open at the first blooming, another 2–3 buds at the second, and finally 1–2 buds at the third blooming period. To alleviate this problem, GA3 has been sprayed to inhibit "early fruit setting" in Kenya (Cannell, 1971), or Ethrel has been applied on new fruits to cause abscission (Adnikinju, 1975).

The lack of synchrony in growth and maturation of coffee cherries poses a special problem to the development of mechanical harvesting procedures. It has been suggested that the application of Ethylene liberated by Ethrel may help to achieve uniform maturation in coffee cherries (Rodrigues and Molero, 1970; Browning and Cannell, 1970; Monaco and Söndahl, 1974; Oyebade, 1976). With unshaded trees at Campinas, Brazil, it was concluded that Ethylene would produce a beneficial uniformity of ripening in coffee fruits if applied 2–3 weeks before the first major harvest. If Ethylene is applied earlier than this, it may cause pronounced abscission of young fruits, and it may also cause fruits to appear to be ripe (red or yellow color of exoderm) when their endosperms are not fully developed.

In order to evaluate the effect of ripening caused by Ethylene in the beverage quality, an experiment was conducted in Campinas (table 2). Ethrel was applied twice, in April and in May of 1972 (Söndahl et al.,

TABLE 2

AVERAGED DATA OF A NUMERICAL SCALE OF CLASSIFICATION OF CUP COFFEE
QUALITY BY THREE INDEPENDENT EXPERTS
(A, B, AND C) FROM COFFEE PLANTS TREATED WITH ETHREL

CONCENTRATION (PPM)	APRIL 5				MAY 6			
	A	B	C	Average	A	B	C	Average
0					3.7	4.4	4.0	4.0
500	2.1	2.1	2.0	2.1	3.7	4.3	4.0	4.0
1000	2.0	2.0	2.0	2.0	3.1	4.1	3.4	3.6
2000	2.0	2.3	2.1	2.2	3.6	4.1	4.0	3.9

1975). The cup quality of the beverage prepared with mature coffee fruits (harvested 15 days after Ethylene treatments) from the first application (April) was classified "hard" (grade 2), because of the high percentage of immature seeds; the beverage prepared from mature fruits from the second application (May) was classified as "soft" (grade 3.6–4.0). It was concluded that Ethrel could be used to bring about uniform maturation in coffee cherries without degrading the quality of the beverage only after the fruits were fully developed (mature endosperm). Ethrel applications induce acceleration of metabolic reactions leading only to maturation of the exoderm in cherries of *Coffea*.

Applications of In Vitro Cultures to Growth Physiology

Growth of excised *Coffea* embryos *in vitro* would provide a means for further study of embryo-endosperm relationships during aging of seeds. The excising of embryos at various times following fertilization would permit one to devise defined cultural conditions for embryo development *in vitro* following interspecific crosses and to establish a protocol for successful embryo culture.

In vitro culture of lateral branch explants (nodes and internodes) would provide an elegant and well-controlled system to approach the phenomenon of coffee flower bud differentiation as was done with tobacco (Wardell and Skoog, 1969). Culture of lateral branches with subtending leaves and arrested flower buds *in vitro* could be used to provide several water potential values in experiments where tissue extracts were sampled for endogenous concentrations of growth hormones. Additional *in vitro* experiments could be designed pertaining to the effects of exogenously applied hormones.

Development *in vitro* of excised lateral branches bearing flower buds and of excised coffee fruits could provide a technical advantage in the study of nutritional and hormonal balances during the development of coffee fruits, the objective being the synchronization of floral induction, anthesis, growth, and maturation of coffee berries.

As an extension of the expected natural variability concerning nutrient uptake and efficiency at standard thermoperiod and water supply, *in vitro* techniques could be utilized to induce and/or select out mutant cell lines with either more efficient permease systems and/or more tolerance to high levels of aluminum and acid soils. Natural or induced mutant cells could conceivably be isolated by a convenient selective method from suspension cultures growing at a concentration density of 10^7 cells/ml. Furthermore, cell suspensions or calluses growing in media of varying osmotic pressures could also lead to the isolation of mutant cells that could be used for the induction of drought-resistant plantlets. At the same time, one could select for cold hardy plants, since cold hardiness is generally associated with tolerance to drought.

PHOTOSYNTHESIS

The positive correlation between vegetative growth and productivity has been used as a selection criterion for nursery-grown seedlings in coffee-breeding programs (Carvalho, personal communication). Comparative studies of dry matter production and photosynthetic rates under both controlled laboratory conditions and in-field conditions may provide important information about the productivity and ecological adaptability of young *Coffea* plants. Earlier knowledge about the genotype potential of new cultivars would save time in breeding programs of coffee and other perennial species.

Net photosynthesis showed properties of heterosis in several examples of single crosses of inbred maize (Heichel and Musgrave, 1969). Heterosis was also found in photosynthetic activity and increased dry weight accumulation of single and double crosses of maize hybrids in which photosynthesis was well-correlated with increases in grain yield (Fousova and Avratov-scukovan, 1967). In the case of rice selection, two of the high-yielding cultivars had as one of the parents a variety in which leaves showed the highest rate of net photosynthesis among 50 rice cultivars tested (Chandler, 1969). In wheat, a comparison between two cultivars with 30% grain yield difference, a 30% greater assimilation and translocation was found in the more productive cultivar (Lupton, 1969). Among twelve soybean cultivars, average grain yields in the field were generally higher in cultivars with higher photosynthetic rates, but there were exceptions (Curtis et al., 1969). Although there is generally an excellent correlation between net photosynthesis and productivity, this is not necessarily an essential condition, since it can be counteracted by other factors. For such as the occurrence of high rates of respiration, poor root growth, excessive leaf shading, low total leaf area, low stomatal numbers, high stomatal resistances (partially or fully closed most of the time), lower photosynthetic rates at the time of seed production, or poor translocation of photosynthates. Certainly in breeding programs utilizing photosynthetic rates as a parameter for the selection of improved cultivars, previous correlations between the photosynthates of isolated leaves and crop productivity should be established. Moreover, these measurements should be taken in environmental conditions as close as possible to the natural field conditions of the crop to be evaluated. Obviously, field measurements would appear to be advantageous for such work where the proper overall environmental conditions (light, temperature, water and nutrient availability, etc.) of a particular geographical region are met. However, the screening of thousands of plants derived from several crosses in a long-term project requires controlled conditions to allow standard comparisons. Perhaps field tests of the few hybrids selected as having higher photosynthetic rates would later compensate for factors not taken into account during mass screening under controlled conditions.

Net photosynthetic rates in coffee plants measured to date are very low compared with other crops (table 3). Variations have been found among

TABLE 3

NET PHOTOSYNTHESIS OF SOME CULTIVATED C3 PLANTS

Species	Net Photosynthesis $mgCO_2\ dm^{-2}\ h^{-1}$	Reference
Tobacco	16–21	Zelitch, 1971
Sugar beet	24–28	Zelitch, 1971
Wheat	17–31	Zelitch, 1971
Bean (5 varieties)	13.7–18.5	Zelitch, 1971
Soybean (65 varieties)	12–43	Zelitch, 1971
Gossypium spp. (26 species)	24–40	Zelitch, 1971
Maple	6	Zelitch, 1971
Dogwood (*Cornus florida*)	7	Zelitch, 1971
Coleus (14 cultivars)	2.8–13.3	Rouhani and Khash-Khui, 1977
Oak	10	Rouhani and Khash-Khui, 1977
Oak	31 ± 3	Leopold and Kriedemann, 1975
Hibiscus	23 ± 5	Leopold and Kriedemann, 1975
Coffea arabica (2 cultivars)	5.6–7.0	Nunes et al., 1969
Coffea canephora	5.0	Nunes et al., 1969
Coffea arabica cv. Bourbon	6.1	Tio, 1962
Coffea arabica (field)	0.7–4.5	Nutman, 1937
Coffea arabica (4 cultivars)	5.4–7.4	Sondahl et al., 1976
C. canephora cv. Guarini	4.0	Sondahl et al., 1976

selected cultivars and the highest net photosynthesis was found in *C. arabica* cultivars (Nunes et al., 1969; Söndahl et al., 1976). A comparison of CO_2 fixation among 5 cultivars in controlled conditions demonstrated that the 2 most productive coffee cultivars, *C. arabica* cv. Catuai and M. Novo, have the highest photosynthetic rates in single leaves, respectively 7.4 and 6.4 $mgCO_2\ dm^{-2} \cdot h^{-2}$ (Söndahl et al., 1976). The *C. Arabica* cultivar 1130–13, which has averaged yields comparable to Bourbon Vermelho, presented 5.4 $mgCO_2 \cdot dm^{-2} \cdot h^{-1}$. Carvalho et al. (1961) compared the productivity of several *Coffea* cultivars and reported that, in relation to *C. arabica* cv. typica, the cultivars Bourbon Vermelho and Mundo Novo were 53% and 150% more productive. Probably one very important aspect of productivity in *Coffea* is the photosynthetic activity of the green fruits during the long growth period (5–6 months). Stimultaneous measurements of CO_2 uptake and respiration of green coffee berries and leaves would clarify the relative contribution of net photosynthesis in green cherries and subtending leaves in relation to productivity. Changes in the oxygen uptake during development of coffee fruits were reported by Cannell (1971), but no data on the photosynthetic activity were recorded.

When the effects of O_2 concentrations (Warburg effect) on the CO_2 fixation were studied, it was found that within the range of 5–52% O_2 the fixation of CO_2 was more altered in *C. canephora* cv. Guarini (37% drop) than in *C. arabica* cv. Catuai (28% drop; Söndahl et al., 1976). Ribulose 1,5-diphosphate carboxylase (RuDP) has been proposed as an enzyme with dual functions, that is a carboxylating activity, whereby CO_2 is fixed through the Calvin cycle, and an oxygenase activity, whereby phospho-glycolate is produced contributing to photorespiration (Ogren and Bowes, 1971; Ryan and Tolbert, 1975; Laing et al., 1974, 1975; Jensen and Bahr, 1977). It is possible that not all photorespired glycolate is a consequence of RuDP oxygenase activity (Kelly et al., 1976; Oliver and Zelitch, 1977). The different comportment between coffee cultivars suggests a different degree of O_2 competition at the oxygenase site of RuDP or different regulatory properties upon different O_2 tensions. In higher plants, RuDP is an enzyme commonly found with 8 large subunits (containing the sites for CO_2 fixation) coded by chloroplastic genes and 8 small subunits (sites of Mg^{++} attachment and regulation) coded by nuclear genes (Kelley et al., 1976). It is interesting to mention that there is a ploidy difference between the coffee plants studied and perhaps the results observed were a consequence of gene dosage as discussed in tall fescue by Randall et al. (1977).

The variations in photosynthetic rates among coffee cultivars and a possible difference in the RuDP carboxylase-oxygenase activities suggest the possibility of exploiting variations already available in *Coffea* germplasm or to induce nuclear or chloroplastic mutants for more efficient coffee cultivars.

Because CO_2 is the most important limiting factor of photosynthesis in field crops, and 90–95% of the dry weight of plants is derived from photosynthetically fixed CO_2, methods of increasing its fixation or decreasing its losses by respiration should take high priority in plant breeding programs seeking higher yields.

There are many ways of approaching the objective of obtaining plants with higher net photosynthesis. It seems that all plants have similar photosynthetic machineries for converting electromagnetic energy into chemical energy; that is, 8 photons per mol O_2 evolved per 4 mol ATP per 2 mol $NADPH + H^+$ (Nobel, 1974; Zelitch, 1975). Maximal rates of photosynthesis are only slightly higher in maize than in tobacco at saturated CO_2 and high irradiance (Zelitch, 1975). This means that all the differences found in net photosynthesis among plants come from metabolic interconversions. Thus, higher net photosynthesis can be approached in two ways: (1) selection of mutants with more efficient CO_2-trapping machinery (for instance, RuDP has a very low substrate affinity), or (2) selection of mutants with reduced respiratory activities (lower photorespiration and/or lower cyanide-insensitive dark respiration).

Cell Cultures and Photosynthesis

The use of cell cultures can provide large populations of cells within a short time and reduce the space usually required for seedlings or adult plants; they are also very suitable for induction of mutation and selection.

The increasing success of growing plant cells and tissues photoautotrophically (removal of sugar with growth at the expense of light energy and CO_2) would certainly open new opportunities to select for mutant cell lines with increased efficiency of CO_2 fixation (table 4). The survivors of plant cells treated with a high O_2 concentration may allow for selection of Warburg effect insensitive plants or for cell lines with increased intracellular concentrations of metabolites that exert feedback control in the glycolate synthesis or oxidation. In fact, Oliver and Zelitch (1977) reported that glycolate synthesis was inhibited and net photosynthesis increased when tobacco leaf discs were floated on solutions of glutamate, aspartate, phosphoenolpyruvate, and glyoxylate. This is evidence that genetic-regulated metabolic pools of intermediates of photorespiration could be used to produce more efficient plants. Resistant tobacco *car-1-1* plants have been raised from carboxin resistant cells and it remains to be seen if these mutants still bear cyanide-sensitive mitochondria (Polacco and Polacco, 1977).

GENETICS

Since *Coffea arabica* is the only tetraploid and self-pollinated species in the genus *Coffea*, the utilization of other morphological and metabolical variabilities found in other species of *Coffea* are hindered. Several *C. arabica* mutants have already been identified (table 5), and these have been used intensively in plant breeding programs for *C. arabica* at the Agronomic Institute, Campinas, Brazil, which holds one of the most complete *C. arabica* mutant collections of the world (Carvalho, 1958; Monaco and Carvalho, 1969).

A detailed description of some of these *C. arabica* mutants has been provided by Carvalho (1958). The mutants listed below are of special interest:

Angustifolia, recessive factor *ag ag* modifies leaf shape and reduces the leaf area to ca. 25% of *C. arabica* cv. M. Novo; ag_1 and ag_2 segregate 9:7 (double recessive alleles), but ag_3 segregates independently.

TABLE 4

Growth Conditions, Chlorophyll Content, and CO_2 Fixation of Photoautotrophic Plant Cells and Fixation of CO_2 in Leaves

Species	Type of Culture	CO_2 Growth Condition	Chlorophyll Content ($\mu g/g$ fr. w.)	Optimum CO_2 Fixation ($\mu mol CO_2/mg$ Ch/h)	Reference
Photoautotrophic cell cultures					
N. tabacum var. Sansun	Liquid	1–5%	4,200	170	Bergmann, 1967
Ruta graveolens	Solid	1%	· · ·	· · ·	Corduan, 1970
N. tabacum	Liquid	2%	61	670	Chandler et al., 1972
Carrot	Solid	1%	50	91	Hanson and Edelman, 1972
Haploid N. tabacum JWB su/su	Liquid	1%	14.6	297	Berlyn and Zelitch, 1975
Chenopodium rubrum	Liquid	1%	80 ± 4	185 ± 9	Husemann and Barz, 1976
Leaves					
N. tabacum JWB su/su	· · · ·	0.03%	3,000	250	Zelitch, 1971
N. tabacum	· · · ·	Satur.	· · ·	500	Zelitch, 1971
Coffea arabica cv. Catuai	· · · ·	0.03%	2,720	175.7	Sondahl, 1974
Coffea canephora	· · · ·	0.03%	3,180	80.4	Sondahl, 1974

TABLE 5

MORPHOLOGICAL AND BIOCHEMICAL MUTANTS OF
C. ARABICA L. VAR. TYPICA CRAMER

Abramulosa	Minutifolia
Abyssinica	Mokka
Angustifolia	Mucronata
Anomala	Murta
Anormalis	Nana
Bourbon	Palidoviridis
Calycanthema	Pendula
Caturra	Polyorthotropica
Cera	Polysperma
Columnaris	Purpurascens
Crassinervia	São Bernardo
Crispa	San Ramon
Erecta	Semi-erecta
Goiaba	Semperflorens
Laurina	Volutifolia
Macrodiscus	Xanthocarpa
Maragogipe	

Caturra, dominant factor *Ct Ct* gives shorter internodes resulting in a reduction in plant size.

Cera, yellowish endosperm color instead of green due to the presence of the recessive *ce* factor. The genotype *Ce ce ce* is green due to the xenia effect.

Erecta, lateral branches have tendency to grow orthotropically due to the presence of a dominant allele *Er*.

Laurina, recessive allele *lr* brings about an intensive pleiotropic effect characterized by reduced leaves, secondary branching with short internodes, long fruits, and pointed seeds containing one-half the caffeine content (0.6%) of *C. arabica* cv. Bourbon (1.2%).

Nana, internodes are so small that the plant only attains a few inches after several years. Plants with the genotype *tt Na Na* are known as Bourbon, *tt Na na* as Murta, and *tt na na* as Nana.

Purpuracens, the recessive pleiotropic factor *pr* results in a brownish or purple-green coloration of young and mature leaves, flowers have a pink coloration, and the exocarp of young fruits have purple stripes.

San Ramon and *San Bernardo*, characterized by short internodes due to a dominant and independent factor.

Semperflorens, the pleiotropic recessive factor *sf* confers a flowering pattern all year round, with short internodes and lateral branches.

Xanthocarpa, mature fruits appear with yellow exocarp due to incomplete recessive factor *xc*.

C. arabica mutant plants bearing one or more single genes for resistance against coffee leaf rust (*Hemileia vastatrix*) have been described (Bettencourt and Carvalho, 1968). At present there are six (and most certainly three others) single dominant genes responsible for resistance to *H. vastatrix* (Monaco and Carvalho, 1975). It is known that SH_1, SH_2, SH_4, and SH_5 genes belong to *C. arabica*, SH_3 to *C. liberica*, and SH_6 *C. canephora* and to interspecific hybrids between *C. arabica* and *C. canephora* (Timor Hybrid and Icatu).

Breeders have special interest in the absence of caffeine in *Coffea* of the sections Paracoffea and Mascarocoffea. Charrier and Berthaud (1975) report that, despite the fact that caffeine content is influenced by environmental factors, the character is genetically controlled by polygenes, whereby the caffeine content of the coffee reflects the average caffeine content of the parents. Within Paracoffea, *C. bengalensis* is characterized as a caffeine-free species.

The genetic relationships of 10 species of the genus *Coffea*, section Eucoffea, representing four subsections, were studied by Carvalho and Monaco (1968). The following species were cross-pollinated artificially: *C. arabica, C. canephora, C. congensis,* and *C. eugenioides* (subsection Erythrocoffea); *C. liberica* and *C. Dewevrei* (subsection Pachycoffea); *C. stenophylla* (subsection Melanocoffea); *C. racemosa, C. salvatrix,* and *C. kapakata* (subsection Mozambicoffea, fig. 2). The species within the same

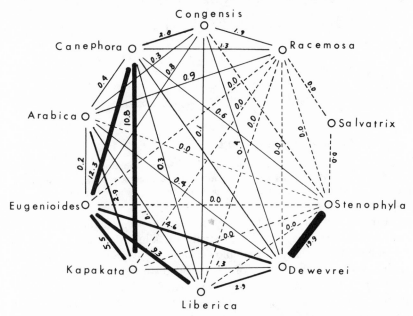

Fig. 2. Genetic relationship of ten species of *Coffea*, section Eucoffea. The numbers represent the percentages of plants obtained per 100 fecundated ovaries.

subsection usually crossed more readily, but a few exceptions were found: *C. kapakata* was shown to hybridize easily with species of Erythrocoffea and the hybrids were relatively fertile. On the other hand, *C. stenophylla* produced vigorous hybrids with *C. dewevrei*. These two species seem to behave as bridges between the respective subsections. In several cases, sympatric species provide fertile hybrid combinations, indicating the existence of special isolating mechanisms for prevention of the extinction of a fertile cross resulting between species. *C. dewevrei* and *C. stenophylla* occupy remote taxonomical positions and are characterized by fertile F_1's. Considering the relationship of the tetraploid *C. arabica* with other diploid species of the same subsection, Erythrocoffea, it was found that better seed set was obtained when *C. arabica* was used as the female parent. Seed set was higher in crosses using artificially induced tetraploid forms of the diploid species *C. canephora, C. congensis,* and *C. eugenioides,* which are shown to have many characteristics indicating their participation in the origin of *C. arabica* (fig. 3). *C. arabica* is described as being native to the high lands of Southwest Ethiopia (Carvalho, 1946; Meyer, 1965). The

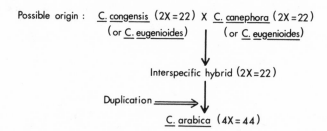

Fig. 3. Possible origin of the allotetraploid *C. arabica* species. *C. canephora, C. congensis,* and *C. eugenioides* have many morphological characteristics in common and produce relatively high percentages of hybrid plants. *C. arabica* is the only tetraploid and autofertile species (90% autogamy) in the genus *Coffea.*

appearance and survival of this single tetraploid species in the genus *Coffea* agrees with the concept of high plasticity and a high degree of invasion of the polyploid species into new ecological areas. Polyploidy is also believed to increase with the degree of ecological disturbances (Morton, 1966; Johnson and Packer, 1965; Grant, 1963). Allotetraploids offer advantages in seed crops in contrast to autotetraploids on account of increased heterozygosity and fertility (only bivalents are formed). The autogamous character of *C. arabica* (90%) allows this species to perpetuate recessive mutations within the population.

Research pertaining to the origin of *C. arabica* could be approached by cytological techniques as proposed schematically in figure 4 in a way

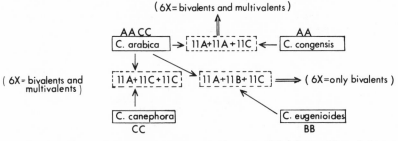

Fig. 4. Theoretical cytological scheme for studying the origin of *C. arabica*. It is proposed that the *C. arabica* genome (AA CC) is derived from *C. congensis* (AA) and *C. canephora* (CC). Full lines enclose the actual species, dashed lines denote artificial triploid hybrids, and double-line arrows represent the colchicine treatments in production of hexaploids.

similar to the studies made in cotton (Phillips, 1963; Frywell, 1971; Edwards et al., 1974). In this theoretical scheme, it is hypothesized that *C. arabica* was derived from a hybrid of *C. canephora* vs. *C. eugenioides*, since *C. arabica* is hypothetically presented with A and C genomes. However, only the detailed study of the meiotic comportment of the hexaploid plants derived from the colchicine-treated triploids will depict the actual constitution of the interspecific hybrid that gave rise to *C. arabica*. Artificial *C. arabica* plants could then be generated from the established parents for meiotic analysis. Obviously, the success of this phylogenetic study would be dependent on the succesful utilization of the interspecific hybrids and subsequent chromosome doubling for production of flowering hexaploids and tetraploids. Techniques for chromosome doubling in cultured tissues, zygotes, young embryos, seeds, seedlings, and plants have been reviewed by Jensen (1974). A new technique of colchicine treatment for coffee seedlings has been reported by Monaco et al. (1975) in which a high percentage of polyploids were evaluated in regard to leaf phenotypic (33–58%) and stomatal counts (23–25%) of the first pair of fully expanded leaves (table 6). A decrease in stomatal number was consistent with an increase in ploidy level as reported by Franco (1939) with *C. arabica* plants having 22, 33, 44, 55, 66, and 88 chromosomes. Mendes (1947) compared the efficiency of three methods of colchicine treatments in *Coffea* tissues and obtained good results following a 24-hour immersion of cuttings in 0.25–0.50% colchicine solutions and subsequent grafting.

Another possible way to study the origin of *C. arabica* would be through the analysis of Fraction I protein (RuDP) from leaves as revealed by isoelectrical focusing. This approach has been used to determine the identity of the interspecific hybrid that gave rise to the commercial tobacco

TABLE 6

Percentage of Possible Polyploids Induced by Colchicine Application
through Glass Microneedles

Dosage per Plant (μg)	Scoring (%)		Stunt (%)
	Phenotypic	Stomata*	
0	0.0	0.0	5.0
50	33.3	23.3	15.0
100	43.0	23.1	27.5
200	50.0	25.0	27.5
400	57.9	15.8	55.0

* Stomate countings deviating from the control plants

plant (Sakano et al., 1974; Gray et al., 1974) and the origin of polyploid wheats (Cheng et al., 1975). As reported in the previous chapter, RuDP carboxylase is an enzyme bearing several large and small subunits with the large subunits being coded by the chloroplast genome and the others by the nuclear genome. The use of the isoelectric focusing technique should allow for the identification of the female parental cytoplasmic contributor of the original cross.

Autogamy in *C. arabica* was estimated to be about 90% in the conditions of Campinas, Brazil (Carvalho and Monaco, 1963). Coffee plants with green endosperm (wild type) were grown in the vicinity of plants with yellow endosperm (mutant *cera*). This mutant was utilized because of the Xenia effect in *Ce ce ce* endosperm. Wind and insects are found to be the principal agents for pollination in *C. arabica*. Allogamy of other species of the genus *Coffea* has been suggested to be under control of S alleles of the gametophytic type (Monaco and Carvalho, 1969).

The phylogenetic study of *Coffea* species has been hindered by difficulties in regard to the importing of wild species of the genus. Difficulties are not only the availability of necessary funding for collecting and maintaining large collections of plants, but also the danger of introducing new pathogens or new physiological races of existing pathogens from geographical centers of *Coffea* variability to the commercial plantations.

Applications of Tissue Culture in Genetics and Breeding

Germplasm of valuable cultivars could be stored and exchanged in aseptic flasks utilizing the techniques of freeze preservation as discussed in this book by Dr. Y. P. S. Bajaj. This is a very important aspect to be developed with *Coffea* tissues because of the short period of coffee seed viability (ca. 6 mo), the danger of extinguishing wild ancestors (natural mutants, etc.), and the high cost of maintaining large tree collections in the field.

In vitro fertilization has potential for the hybridization of selfing of species, especially in crossing species with different flower time regimes. Chemical treatments such as calcium could also be utilized in an attempt to increase the degree of success of *in vitro* fertilization. Related to this topic is the utilization of pollen preservation described for *C. arabica* (Walyaro and Van der Vossen, 1977). This would allow the possibility for collecting, storing, and shipping pollen material for greenhouse or *in vitro* pollination. A controlled growth system for the culture of ovaries, placental tissues, isolated ovules, and pollen grains would perhaps allow studies of how to circumvent autoincompatibility. Such work might provide information pertaining to the maintenance of preferentially viable ovules and at the same time disrupt pollen grain viability by means of exogenous hormone treatments for the development of an emasculation protocol suitable to field conditions. Treatments with chemical mutagens could be applied to asceptic isolated pollen prior to *in vitro* pollination prior to androgenesis.

Embryo culture, as discussed in this volume by Dr. Norstog and by Raghavan (1976, 1977), is a promising tool for allowing researchers to nurture the embryos resulting from interspecific and intergeneric crosses which are accompanied by endosperm malfunction and abortion shortly after fertilization. Such crosses have potential for developing new cultivars having genes for disease and pest resistance, desirable vegetative characteristics, and drought and cold resistance, and would certainly justify long-term research projects in raising immature *Coffea* embryos. Isolated embryos in cultue can also be utilized for the study of host-parasite interactions (resistance to toxins, etc.) in aseptic conditions under controlled environmental conditions. Dual cultures using parabiotic chambers for the growth of *Coffea* tissue cultures and coffee leaf rust (*Hemileia vastatrix*), an obligate parasite, can provide a continuous source of sterile uredospores for testing resistance-susceptible genotypes in controlled environmental conditions. For these purposes and other physiological studies, high-frequency somatic embryos and coffee plantlets generated from the culture of leaf explants (Staritsky, 1970; Söndahl and Sharp, 1977 a,c) can be used.

The culture of endosperm (triploid tissue) can be used in *Coffea* genetics or breeding programs to generate triploid plants for genetic, morphological, or cytological analysis or as an intermediate step to produce fertile hexaploid plants following colchicine treatment. Attempts to cultivate *Coffea* endosperm (Keller et al., 1972) and *Coffea* perisperm-endosperm (Monaco, Söndahl, Carvalho, Crocomo, and Sharp, 1977) have been described, but plantlet differentiation has not yet been accomplished. In fact, in a recent review, Johri and Bhojwani (1977) reported that in the so-called autotrophic taxa (*Euphorbiaceae* and *Gramineae*), plantlets have been successfully induced from cultured endosperm for only three species.

A comprehensive review of the methods for haploid production in higher plants and utilization in research and plant breeding has been published (Kasha, 1974). Three general methods for producing haploid plants are described: (1) parthenogenesis and androgenesis, (2) chromosome elimination, and (3) *in vitro* culture of haploid tissues. Techniques for culturing haploid tissues are reviewed in this volume by Dr. V. Raghavan and Dr. N. Sunderland and are also found in Sunderland (1973, 1974), Nitsch (1974, 1977) and Reinert and Bajaj (1977).

Few of these techniques have been successfully applied in *Coffea* spp. to date (Monaco, Söndahl, Carvalho, Crocomo, and Sharp, 1977), despite their enormous potential in plant breeding. Genome fixation and gene duplication through diploidization of haploid plantlets of *C. canephora* or other allogamous diploid species would allow for increased homogeneity (fig. 5). Haploid culture, diploidization, and plant regeneration allow for the development of homozygous diploids containing both recessive and

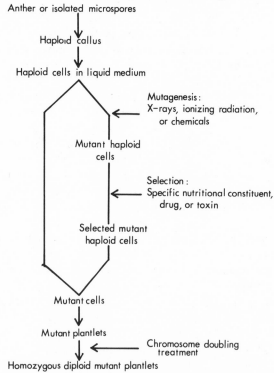

Anther or isolated microspores

Haploid callus

Haploid cells in liquid medium

Mutagenesis:
X-rays, ionizing radiation,
or chemicals

Mutant haploid
cells

Selection :
Specific nutritional constituent,
drug, or toxin

Selected mutant
haploid cells

Mutant cells

Mutant plantlets

Chromosome doubling
treatment

Homozygous diploid mutant plantlets

Fig. 5. A model for genome fixation through haploid diploidization in *C. canephora* and other heterozygous species of *Coffea* and/or obtainment of homozygous mutant *Coffea* plants.

dominant alleles and hence providing plant material for genetic analysis and subsequent selection for desirable commercial characteristics.

Isolation of self-incompatible clones with different self-incompatible alleles in *C. canephora* or other diploid species has been discussed by Monaco, Söndahl, Carvalho, Crocomo, and Sharp (1977). Application of the scheme depicted in figure 5 would be of special interest in *C. arabica*. Plants of this autogamous species have at least a 90% chance of being homozygous (propagation by seeds) with mutant types being easily distinguished by the population. Duplication of desirable recessive or dominant genes perhaps will produce valuable new commercial cultivars. Obtainment of flowering dihaploid plants from *C. arabica* would be another way to circumvent the ploidy barrier for crosses with the diploid species. Dihaploid tissues could be used as a source for protoplast liberation for fusion experiments with somatic protoplasts of 2X species of *Coffea* in which 4X coffee plants could conceivably be produced via shoot organogenesis or somatic embryogenesis. The mixoploidy of haploid callus cultures could be used for plantlet generation of different ploidy levels. Such plantlets produced from *C. canephora* and other 2X species have practical applications. Finally, the advantages of *Coffea* haploids are obvious for induction of mutations and selection because of the immediate expression of recessive alleles without the complication of heterozygosity. Any mutant locus could then be selected independent of dominance or recessiveness.

Increase of variability and plasticity have high priority in cultivated *C. arabica* plants that comprise a highly homogenous population (Carvalho, 1946; Monaco, 1977). It should be easy to induce mutant cell lines of dihaploid or somatic cells to undergo plantlet regeneration (fig. 6). Uptake of isolated organelles (chloroplasts or mitochondria) in protoplast preparations is another approach in regard to genetic engineering with the expectation that foreign organelles carrying desirable genomes will divide and perpetuate in the new protoplast environment. In the event that male sterility of diploid *Coffea* is found to be due to cytoplasmic genes, as is the case with corn cms-T cytoplasm, cytoplasmic gene transfer will be a practical approach for production of multilineal hybrids in *C. arabica*. Transformation provides another alternative for genetic modification in *Coffea*. The success of transformation requires the fulfillment of at least four steps: (1) homologous or heterologous DNA uptake by recipient cells; (2) replication of this DNA; (3) gene expression (transcription and translation); and (4) transmission of this DNA through offspring (stability). Despite the theoretically unlimited possibilities of this technique to increase the gene pool and to transfer polygenic characteristics, several technical difficulties are hindering further experiments: (1) availability of

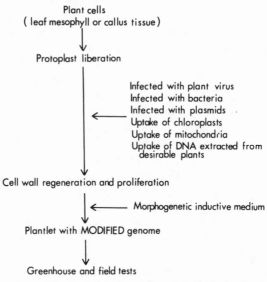

Plant cells
(leaf mesophyll or callus tissue)

Protoplast liberation

Infected with plant virus
Infected with bacteria
Infected with plasmids
Uptake of chloroplasts
Uptake of mitochondria
Uptake of DNA extracted from
    desirable plants

Cell wall regeneration and proliferation

Morphogenetic inductive medium

Plantlet with MODIFIED genome

Greenhouse and field tests

Fig. 6.   A model for modification of a plant cell genome through protoplast liberation, uptake of subcellular particles, or DNA and subsequent plantlet regeneration.

stable markers; (2) state of purification of DNA preparations and DNAse activity of recipient cells; (3) presence or absence of suppressor genes; (4) expression of the desirable traits after DNA was taken up and stabilized; (5) possibility of generating an imbalanced genome. Somatic hybridization could lead to genetic modification of cultivated coffee plants assuring that protocols for protoplast isolation, growth, and plant regeneration are available and biochemical and morphological markers are used (fig. 7).

An overall scheme of genetic manipulation through tissue culture methods is presented in figure 8. Several technical problems must be overcome before large-scale development of new cultivars is feasible. The last barrier to practical application in agriculture of cultured plant cells is the possibility and frequency of the regenerative step. The list of protocols for plant regeneration from callus, suspension, haploid, and protoplast cultures of cultivated crops must be enlarged. Continuous emphasis on high-frequency regenerative protocols is the only way of assuring the utilization of the potentially unlimited possibilities of genetic manipulation of cell cultures in contrast to those through the sexual route. In the case of *C. arabica*, a high frequency of somatic production is obtainable from cultured leaf callus tissues. The utilization of such regenerative protocols in cell suspension cultures and isolated regenerative protoplasts of *Coffea* cells will provide the impetus for resolving the technical problems in regard

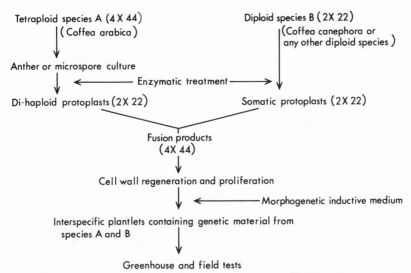

Fig. 7. A model for interspecific cell hybridization in *Coffea* utilizing dihaploid protoplasts of *C. arabica* (2X 22) with somatic protoplasts of *C. canephora* (2X 22) or any other diploid species producing a tetraploid hybrid (4X 44).

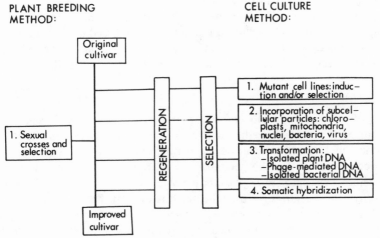

Fig. 8. Overall scheme for genetic manipulation utilizing plant cells or protoplasts in culture followed by plantlet regeneration.

to the utilization of genetic manipulation at the cellular and molecular levels to achieve improved *Coffea* cultivars.

Vegetative propagation through tissue culture techniques would have practical application in the case of *C. canephora* and other diploid species where the fixation of germplasm might be justified. Coffee plants do not root very easily, the frequency of rooting success varies throughout the year (Söndahl and Medina, unpublished), and only orthotrophic shoots will give rise to normal plants. Considering these limitations and the fact that somatic embryogenesis has already been described in *C. canephora* from stem tissues (Staritsky, 1970) and from leaf tissues (Söndahl and Sharp, 1977a), it seems worthwhile to perfect this technique for *C. canephora* and other highly heterozygous diploid species. Attempts at rooting *C. canephora* stems with one subtending leaf (Boudrand, 1974) and *C. canephora* leaves pretreated with NAA (Nsumbu and Boharmont, 1977) have been described.

MAJOR DISEASES AND PESTS IN COFFEA

The number of severe diseases and pests in coffee plants is small, despite the extensive cultivated area and the homogeneity of plantations. The major pathological problem is fungal diseases of the leaves (coffee leaf rust) and fruits (coffee berry disease). Among insects in Brazil, the coffee borer can cause extensive damage if recommended cultural practices are not adopted. Leaf miners have recently caused considerable damage in some coffee-growing areas in Brazil, apparently after copper sprays were used against the leaf rust. Also, root nematode damage is of increasing economic importance to Brazil. Cultural practices and breeding programs are already under way to circumvent this problem. Bacterial and viral diseases are not of economic importance to coffee plantations in Brazil.

Diseases

Coffee leaf rust belongs to the order Uredinales, family Pucciniaceae, genus *Hemileia*. Altogether, there are about 40 species of *Hemileia*, attacking 11 families of flowering plants, and 16 of these occur on Rubiacea. It seems that this genus *Hemileia* originated in Africa and is host specific. Two species, *H. vastatrix* and *H. coffeicola*, attack the genus *Coffea*; *H. vastatrix* is by far the most widely disseminated species (Monaco, 1977). Outside Africa, *H. vastatrix* was first detected in Ceylon in 1869 and soon thereafter in Java. Productivity in Ceylon was reduced from 450 to 200 kg/ha, and within 20 years coffee plantations were abandoned and replaced by tea (Meyer, 1965; Monaco, 1977). In 1970, coffee leaf rust, which previously had not been present in coffee plantations of Latin America, was detected on the eastern coast of Brazil (Bahia State). This dissemination may have been caused by the wind transporting coffee rust spores across the Atlantic Ocean (Bowden et al., 1971). Cupric fungicides are currently being used in Brazil to control *H. vastatrix*, and they account for about 20% of the expense associated with coffee production. Several mutant plants have already been identified in the *C. arabica* population with "vertical resistance" to one or a few physiological races of *H. vastatrix*, due to the presence of single dominant genes (Bettencourt and Carvalho, 1968; Carvalho and Monaco, 1972; Monaco and Carvalho, 1975). Utilization of coffee plants resistant to commonly found races of this fungus will favor proliferation of other physiological races by selection pressure. It has been demonstrated that the *C. arabica-H. vastatrix* system fits the Flor hypothesis of the gene-for-gene relationship (Flor, 1955, 1965; Bettencourt and Carvalho, 1968). A continual research program is required for identification of new sources of vertical resistance against new races of *H. vastatrix*. The genetics of resistance by single genes can be attributed to independent gene action or to gene interactions in which complementary gene action or modifier genes are involved (Williams, 1975). In 1968 there were six known dominant genes in coffee plants that conferred resistance to *H. vastatrix* (Bettencourt and Carvalho, 1968), and now nine major genes have been identified (Monaco, 1977). The number of possible physiological races of the fungus after all the combinations are calculated is enormous, but only 30 races have been identified so far. There are three strategies proposed to deal with gene-for-gene relationships. One proposed by Person (1966, 1967) is a "temporal discontinuity," where unusual genes for resistance are removed from the population; another is a "spatial discontinuity" (Knott, 1972) restricting the geographic distribution of genes for resistance; the last is the "multiline cultivars" (Borlaug et al., 1969), in which multilines are constructed, containing varying mixtures of isogenic lines, each with genes for resistance. The best option for perennial crops such as *Coffea* spp. would be to develop "horizontal resistance" or "nonspecific resistance" against *H. vastatrix*, which is considered to be under polygenic control. This type of resistance is expressed widely against all races of a pathogen, but with varying degrees of restriction of individual races, depending mainly on race aggressiveness and ecological and nutritional conditions (Nelson, 1972; van der Plank, 1968; Robinson 1976). Complex interactions between numerous genes may work to retard penetration of the pathogen, to slow its growth, and to limit its sporulation (Williams, 1975).

Another serious fungal disease occurring in fruits of coffee plants is *Colletrotrichum coffeanum*, commonly called coffee berry disease (CBD). This disease is the major threat for *C. arabica* production in Kenya and other countries of eastern Africa. CBD is not found in coffee plantations in Latin America. In Brazil, *Colletrotrichum* spp. is found in dead coffee branches, presumably as an after-infection, associated with physiologically and nutritionally imbalanced "die-back." The anthracnose of green and ripening coffee fruits (CBD) can cause crop losses of 50% and over in years favorable to severe CBD epidemic (Griffiths et al., 1971). The disease can be controlled by fungicide spraying, which may account for up to one-third of the field costs of plantation management; this is prohibitively expensive for small farmers. Methods of preselection for plants resistant to CBD have been reported (van der Vossen et al., 1976) as part of a breeding program for resistance to *C. coffeanum* at the Coffee Research Station in Kenya.

Falling-off is a fungal disease caused by *Rhizoctonia solani*, which commonly attacks young seedings in nurseries and, on rare occasions, one-year-old field plants in Brazil. In both cases the infection is established in nurseries (at an early or late stage) favored with excessive shading and humidity. Routine spraying with 0.3% copper will prevent the disease (Bergamin and Abrahão, 1963).

Pests

The coffee berry borer (*Hypothenemus hampei*) infects coffee fruits by perforating the seeds, causing consequent loss of weight and lowering of the commercial classification of the final product (green coffee). This pest can be controlled by harvesting all coffee fruits in order to decrease the initial population prior to the next season, or by applying 1% BHC in powder when the infection occurs in more than 5% of the fruits (Bergamin and Abrahão, 1963). To our knowledge, there is no research program aimed toward developing cultivars resistant to this pest.

Leaf mining is caused by larvae of *Perileucoptera coffeella* (coffee leaf miner). The eggs are deposited on the upper leaf surface and, after eclosion, the larvae enter the leaf cuticle and establish themselves in the leaf mesophyll. The visible damage results in necrosis of the leaf blade area. Leaf miner infection is commonly higher after a long period of drought (Bergamin and Abrahão, 1963). However, after the adoption of regular copper sprays to protect coffee plantations against coffee leaf rust, it is suspected that a biological imbalance is set up. High incidences of this pest have been detected lately in some coffee-growing areas in Brazil (Parra, personal communication).

Root Nematodes

The increased incidence of root nematodes in coffee plantations in Brazil was detected in the 1960s. So far, there are four species described attacking *C. arabica* roots (*Meloidogyne coffeicola, M. exigua, M. incognita*, and *Pratylenchus* sp.) and one species in *C. canephora (M. incognita)*. In order of importance, *M. coffeicola* is of special interest, since it can cause the death of a coffee plant within a few years; *M. exigua* is especially severe in plants growing in sandy soils, and is widely distributed in all coffee-growing areas; *M. incognita* is potentially a species of great importance, since it attacks both *C. arabica* and *C. canephora* and also is a parasite of several crops and weeds; *Pratylenchus* sp. has a restricted area of dissemination in the state of São Paulo. Germplasm resistant to *M. exigua* and *Pratylenchus* sp. has been identified in some plants of *C. canephora*; grafting of *C. arabica* on seedlings derived from such plants is recommended (Moraes and Franco, 1973).

Applications of Plant Tissue Cultures

In vitro culture methods could be extremely valuable for research dealing with fungus and nematode diseases in *Coffea*. *H. vastatrix* is a specific pathogen of coffee plants and, so far, no synthetic medium exists for asceptic culture of this fungus. Callus tissues of coffee or coffee plantlets growing in asceptic conditions could potentially provide a suitable environment for establishment of asceptic cultures of *H. vastatrix*. Reactions of resistance-susceptibility in different clones could be easily assayed in controlled physical and chemical environments. Cultivation of coffee cells in association with *H. vastatrix* would permit one to explore the possibility of selecting mutant cell lines completely immune or partially resistant to the coffee rust. Reconstitution of coffee plants from those selected coffee cells (Söndahl and Sharp, 1977a,c) would eventually give rise to new sources of *C. arabica* plants with vertical or horizontal resistance to *H. vastatrix*. Selection pressure would be exerted upon millions of cells (potentially convertible to plants), exploring the naturally occurring variability after chemical or physical mutagenic treatment. It is worth noting that selection for resistance to *H. vastatrix* could be conducted at the diploid level, since the single genes conferring resistance are dominant.

The induction of gene mutation for resistance in the host plant has been applied in barley, wheat, and oats, but the release of a new resistant variety has only been reported so far for Verticilium Wilt-resistant peppermint (Murray, 1971; Williams, 1975).

Toxins have been used as agents of selective pressure to screen oat mutants resistant to *Helminthosporium victoriae* (Konzak, 1956; Wheeler and Luke, 1955), sugarcane resistant to *H. sacchari* (Wheeler et al., 1971; Lim et al., 1971), and germinating seeds of cereals pretreated with mutagenics against purified filtrates of *Helminthosporium sativum* (Dutreq, 1977).

Pathotoxins have been used as a means of selection on *in vitro* protoplast cultures of tabacco (Carlson, 1973) and corn cells (Gengenbach and Green, 1975; Gengenbach et al., 1977). Haploid tobacco protoplasts have been mutagenized with EMS (ethyl methyl sulfoxide) and subsequently exposed to MSO (methionine sulfoximine), a methionine analog. Since the natural toxin of Pseudomonas tabaci is a methionine analog, and MSO elicits the same chlorotic effects, it can be used as a detective agent. MSO-resistant survivors were diploidized and subsequently regenerated into plants. Analysis of the F2 generation demonstrated the presence of a single mutant locus among plants recovered as resistant to *P. tabaci* (Carlson, 1973). Selection of toxin-resistant A 619 (cms-T) callus tissues of corn following four sub-cultures (160 days) or medium containing toxin were stable for 127 days in the absence of toxin by Gengenbach and Green (1975). Corn plants regenerated from resistant cell lines isolated after the fifth selection cycle were toxin resistant, and 52 out of 65 were fully male fertile. Progeny tests indicated that resistance to the toxin was inherited only through the female parent (Gengenbach et al., 1977).

A classical demonstration of expression of disease-resistant genes in both plants and callus tissues was reported by Helgenson et al. (1976). These workers used homozygous-resistant and homozygous-susceptible tobacco plants to *Phytophthora parasitica* race O (black shank of tobacco) and tested for resistance and susceptibility in clonal cuttings and pith callus cells from the F1, F2, and F3 generations. In each case, plants yielding resistant cuttings gave rise to resistant callus and plants yielding susceptible cuttings gave rise to susceptible callus.

In the study of host-parasite relationships, the presence of a toxin has not always been characteristic. In other cases, production of phytoalexins by the host plant has been associated with plant resistance to the infection of pathogens. Accumulation of antifungal substances in *C. arabica* leaves associated with resistant reaction of SH_1 and SH_2 loci against *H. vastatrix* has been reported by Rodrigues and Medeiros (1975). Histopathological studies of the hypersensitivity reaction on leaves of resistant coffee plants demonstrate that there is an induced enlargement of mesophyll cells which causes a crushing of the fungal mycelium (Rijo, 1972). No correlations between chlorogenic acid content and coffee rust resistance in *C. arabica* plants bearing major genes for resistance to *H. vastatrix* have been found (Carelli et al., 1974).

Parabiotic cultures should provide a means for selection of *in vitro* cell lines capable of high-level synthesis of phytoalexins in response to the fungal stimulus and eventually allow for production of enough tissue for the physical and chemical characterization of the fungal-stimulated substance(s).

The development of multiline cultivars is the strategy to be adopted in which the major genes for rust resistance in coffee plants are involved. Plants of *C. arabica* carrying major genes for resistance have been crossed with the most productive *C. arabica* cultivars. Homozygous plants were used in backcrosses and subsequent F3 and F4 generations were developed. The individually selected lines were mechanically mixed to form the composite cultivar Iarana (Monaco, 1977). Since vertical resistance to leaf rust in coffee plants is conditioned by dominant genes, there is a possibility of exploiting commercial hybrid production in *C. arabica* as soon as a source of male sterility is incorporated in this species. *C. canephora* and *C. eugenioides* are probable sources of nuclear and cytoplasmic male sterility, respectively, as discussed by Monaco (1977). *In vitro* techniques could help master this achievement by isolating *C. canephora* clones containing different self-incompatibility alleles in haploid cultures or by uptake experiments with *C. eugenioides*, in which isolated organelles are incorporated into *C. arabica* protoplasts.

The use of root cultures or callus tissues in association with plant nematodes is reviewed by Dr. Krusberg in this volume. Certainly after establishment of such associations, techniques for induction of mutations and selection of root nematode-resistant cell lines can be devised.

TISSUE CULTURE IN COFFEA

Review

A recent review of methods, source of explants, and current progress of cultured *Coffea* tissues has been published (Monaco, Söndahl, Carvalho, Crocomo, and Sharp, 1977). This article provides supplemental data to that covered in the previous review. Table 7 summarizes the available current work with *Coffea* tissue cultures. The list of cultured *Coffea* species and cultivars undergoing morphogenesis has been increased to four species (*C. arabica, C. canephora, C. congensis,* and *C. dewevrei*) and five cultivars of *C. arabica* (Bourbon, Mundo Novo, Catuai, Laurina, and Purpuracens). Explants of these cultivars originated from either leaf or stem explants and were grown on solid media. There is no report to date on the success of morphogenic control with haploid tissue. Also, there is no published report on the isolation, growth, and differentiation of *Coffea* protoplasts. Progress on these aspects of *Coffea* tissue culture is needed before present *in vitro* techniques can be applied to the improvement of cultivated *Coffea*.

Coffea cells from *C. arabica* cv. El Salvador in suspension cultures were used to analyze caffeine and chlorogenic acid contents (Buckland and Townsley, 1975) and to compare unsaponifiable lipids in green beans (van der Voort and Townsley, 1975). The maximum caffeine content of a 20- to 26-day-old suspension culture was found to be 0.038% on a dry weight basis, which is substantially less than the 1.15% caffeine content found in green beans. However, two observations should be taken into account in

TABLE 7

Current Success in the Culture of Tissues of the Genus Coffea

Success	Species	Explant	Author
Undifferentiated callus tissue			
Haploid	*C. arabica*	Anther	Sharp et al., 1973
	C. liberica	Anther	Söndahl et al., 1972
	C. racemosa	Anther	Söndahl et al., 1972
	C. canephora	Anther	Söndahl et al., 1972
Somatic	*C. arabica*	Stem	Staritsky, 1970
	C. arabica	Stem	Frischknecht et al., 1977
	C. liberica	Stem	Staritsky, 1970
	C. stenophyla	Perisperm	Monaco et al., 1974
	C. bengalensis	Leaf	Söndahl and Sharp, 1977b
	C. racemosa	Leaf	Söndahl and Sharp, 1977b
	C. salvatrix	Leaf	Söndahl and Sharp, 1977b
	C. liberica	Leaf	Söndahl and Sharp, 1977b
Triploid	*C. arabica*	Endosperm	Keller et al., 1972
Suspension	*C. arabica*	Stem	Townsley, 1974
Embryo culture	*C. Dewevrei* cv. *excelsa*	Embryo	Colona, 1972
	C. canephora	Embryo	Colona, 1972
"*Organoids*"	*C. arabica*	Leaf	Herman and Hess, 1975
High frequency somatic embryos	*C. canephora*	Stem	Staritsky, 1970
	C. arabica cv. Bourbon	Leaf	Söndahl and Sharp, 1977a
	C. arabica cv. Catuai	Leaf	Söndahl and Sharp, 1977c
	C. arabica cv. Laurina	Leaf	Söndahl and Sharp, 1977c
	C. arabica cv. M. Novo	Leaf	Söndahl and Sharp, 1977c
	C. arabica cv. Purpuracens	Leaf	Söndahl and Sharp, 1977c
	C. Dewevrei cv. *excelsa*	Leaf	Söndahl and Sharp, 1977c
	C. canephora cv. Kouillou	Leaf	Söndahl and Sharp, 1977c
	C. congensis	Leaf	Söndahl and Sharp, 1977c

interpreting the reasons for this low caffeine concentration in the suspension cultures: (1) caffeine synthesis appears to be produced in increasing amounts as the cell population increases, and (2) caffeine is water soluble and diffuses into the liquid medium. The maximum chlorogenic acid level found in the 20- to 25-day-old coffee suspension cells was 1.3% on a dry weight basis, as compared to 6.5% found in green beans. When the chlorogenic acid content of 10 *C. arabica* cultivars and 9 other diploid *Coffea* species was evaluated (Carelli et al., 1974), it was found that the chlorogenic acid concentrations of *C. arabica* fluctuated within the range of 7.13-8.17% and those within the diploid species fluctuated between a minimum of 2.70% (*C. salvatrix*) and a maximum of 10.30% (*C. canephora* cv. Robusta). The analysis of unsaponifiable lipids in suspension cultures of coffee cells demonstrated the presence of three sterols, B-sitosterol, stigmasterol, and campesterol, as the major components that are also present in coffee beans as a minor fraction. In addition, coffee cells also contained two diterpenoid alcohols, cafestol and kahweol, as a minor component of their unsaponifiable lipids. However, in coffee beans, cafestol and kahweol represent the major fraction of unsaponifiable lipids and sterols play a minor role. The presence of the two diterpenoid alcohols in coffee cell suspension cultures suggests that coffee cells maintain their ability to synthesize unique compounds present in parent coffee plants (van der Voort and Townsley, 1975).

Solid cultures of *C. arabica* cv. Bourbon Vermelho stem explants were established for analysis of caffeine production *in vitro* and possible commercial utilization of synthesis of such alkaloids (Frischknecht et al., 1977). The formation of caffeine was parallel with the increase of callus dry weight, and production remained at a fairly constant rate per unit of callus dry weight. When caffeine was added to the medium and the caffeine in the tissues was concentrated in excess of 900–1,000 $\mu g/$ ml tissue, callus growth as well as caffeine synthesis was inhibited. It seems that caffeine biosynthesis is closely correlated with the growth process in both plant and callus tissues. The theoretical dry weight percentage (tissue plus agar medium contents) of caffeine in cultured coffee tissues was 1.0–1.6%. This amount is very similar to values found in Brazil for *C. arabica* cv. Bourbon Vermelho, Caturra, and Mundo Novo of 0.8–1.2% (Carvalho et al., 1965). The caffeine content on a dry weight basis of *C. arabica* selections of Madagascar averaged 1.15 ± 0.03% and ranged from 0.58 to 1.90% among 867 genotypes tested (Charrier and Berthaud, 1975). The levels of caffeine synthesis *in vitro* were considered relatively high when compared with the generally low *in vitro* production of secondary plant substances (Frischknecht et al., 1977).

Growth and Morphogenesis of Cultured Leaf Explants

Methods. Mature leaves of lateral branches of *Coffea* spp. were surface-sterilized in 1% sodium hypochlorite for 15–30 min and rinsed three times in sterile double distilled water. Leaf explants of about 7 mm^2 were cut, excluding the midvein, margins, and apical and base portions of the leaf blade. Sections were cut in a sterile saline-sugar medium devoid of agar and growth regulators and were immediately plated onto 10 X 100 mm petri dishes containing a solid saline-sugar medium for 36 hours in the dark (fig. 15A). This preexperimental incubation period was found very useful in selecting viable leaf explants and eliminating any contaminated explant material. The abaxial side of the leaf explants (which was clearly distinguished by a dull pale green coloration, in contrast to the shiny dark green coloration of the adaxial side) was always placed up. French square bottles were charged with 10 ml of autoclaved basal medium (BM) containing Murashige and Skoog (MS) inorganic salts, 30 μM thiamine-HCL, 210 μM L-cysteine, 550 μM meso-inositol, 117 mM sucrose, and 8–10 g/l Difco-agar. In primary culture, a conditioning medium containing a combination of kinetin (20 μM) and 2,4-dichlorophenoxyacetic acid, 2,4-D (5 μM) was used, and the bottles were incubuated in the dark at 25 ± 1°C for 45–50 days. The composition of this "conditioning medium" was found ideal for high-frequency embryogenesis (60% of the flasks) in leaf explants of *C. arabica* cv. Bourbon (Söndahl and Sharp, 1977a).

Secondary cultures were established under conditions of a 12-hour light period at 24–28°C by subculturing 45- to 50-day-old tissues onto an "inducing medium" containing half-strength MS organic salts, except KNO$_3$, which was added twice concentrated, 58.4 mM sucrose, kinetin (2.5 μM), and Naphthalene acetic acid, NAA (0.5 μM). The same medium without kinetin was used in solid or liquid cultures for further subculture of somatic embryo and plantlet development. Callus proliferation was studied in other *Coffea* species with diallel experiments using kinetin and 2,4-D within a concentration range of 2–20 μM during primary culture. Callus was subcultured onto the "inducing medium" used for Bourbon. Callus proliferation was evaluated just before subculturing according to the following grade scale: 1 (6.0 mg), 2 (103 mg), 3 (650 mg), 4 (1,340 mg), and 5 (2,600 mg).

Materials. A mature *C. arabica* cv. Bourbon plant (ca. 6 years old) and 1.5- to 2-year-old plants of *C. arabica* cv. Bourbon, Catuai, Laurina, Mundo Novo, Purpuracens, *C. bengalensis, C. canephora* cv. Kouillou, *C. congensis, C. Dewevrei* cv. excelsa *C. liberica, C. racemosa,* and *C. salvatrix* were used as sources of leaf explants.

Growth of C. arabica cv. Bourbon. Callus proliferation of cultured leaf explants of *C. arabica* cv. Bourbon was studied in diallel experiments with a single cytokinin, kinetin, and several auxins (Indole-3-acetic acid (IAA), Indole butyric acid (IBA), NAA, and 2,4-D). Evaluations were made with 50- to 70-day-old primary cultures.

Kinetin (0, 4.5, 9.0, 18, 36, and 72 μM) and IAA (0, 6.0, 12, 24, 48, and 96 μM) at all possible combinations were proven to be ineffective for callus proliferation (fig. 9).

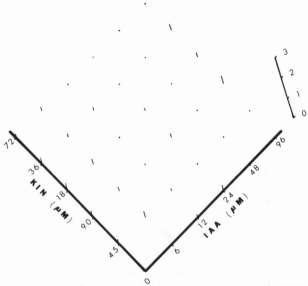

Fig. 9. Callus growth induced by diallelic interactions of Kinetin (KIN) and Indole-acetic acid (IAA) after seven weeks in darkness. Average data from 10–15 replicate treatments.

All 36 combinations of kinetin (0, 4.5, 9.0, 18, 36, and 72 μM) and IBA (0, 5.0, 10, 20, 40, and 80 μM) were ineffective in the promotion of substantial proliferation. The highest growth index, 1.4, was obtained with a kin/IBA concentration ratio (c.r.) of 4.5/20 μM (fig. 10).

Kinetin (0, 4.5, 9.0, 18, 36, and 72 μM) and NAA (0, 5.5, 11, 22, 44, and 88 μM) promoted excellent callus proliferation. Growth indexes of 2.9 and 2.80 were obtained with kin/NAA c.r. of 9/22 μM and 9/44 μM, respectively (fig. 11). A lower range of kinetin concentrations (0, 0.22, 0.45, 0.9, 1.8, and 3.6 μM) and NAA (0, 0.22, 0.45, 0.9, 1.8, and 3.6 μM) was tested (fig. 12). A growth index of 2.4 was observed at the kin/NAA c.r. of 0.45/2.0 μM, which coincides with the 1:2 or 1:4 optimum c.r. found in the previous experiment.

Combinations of kinetin (0, 4.5, 9.0, 18, 36, and 72 μM) and 2,4-D (0, 4.5,

9.0, 18, 36, and 72 μM) gave the highest growth indexes at the kin/2,4-D c.r. of 9/4.5 μM (3.3), 18/4.5 μM (3.0), and 18/9 μM (2.9), as seen in figure 13. A lower range of concentrations was also tested with kinetin (0, 0.22, 0.45, 0.9, 1.8, and 3.6 μM) and 2,4-D (0, 0.22, 0.45, 0.9, 1.8, and 3.6 μM) in order to tabulate information pertaining to the minimal and maximal

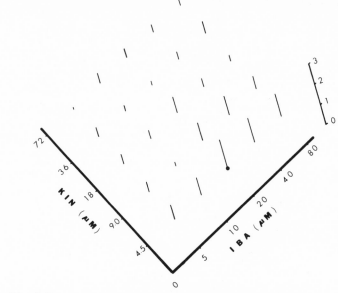

Fig. 10.   Callus growth induced by diallelic interactions of Kinetin (KIN) and Indole-butyric acid (IBA) after eight weeks in darkness. Average data from 10–15 replicate treatments.

concentration tolerances of coffee leaf tissues (fig. 14). The maximum growth index of 3.0 was found with kin/2,4-D c.r. of 1.8/3.6 μM and 3.6/3.6 μM.

It is interesting to note that out of 216 treatments tested, a broad range of combinations of kinetin with NAA or 2, 4-D were effective to promote callus proliferation. This means that in this coffee leaf system, no sharp sensitivity exists pertaining to callus growth within a wide range of growth regulator concentrations. Even though each experimental treatment was designed in such a way that subsequent concentrations doubled the previous concentration, the response of growth presented smooth transitions (figs. 11 and 13). In all growth regulator interactions tested, the necessity of having both a cytokinin and an auxin source was clearly demonstrated. If a minimum growth index of 1.0 is arbitrarily chosen, 2.0 μM of NAA, 0.9 μM of 2,4-D, and 0.45 μM of kinetin are required to achieve a minimum growth index. In the case of NAA, a ratio of about

1:2.5 or 1:5 was necessary for good callus proliferation. However, with 2,4-D, ratios of 5:1 and 2:1 were effective at high concentratons (more that 4.5 μM), whereas ratios of 1:2 and 1:1 were effective at lower range concentrations (less than 3.6 μM). The higher the auxin concentration the more friable the callus became; inversely, the higher the kinetin concentration the harder was the appearance of the callus tissues.

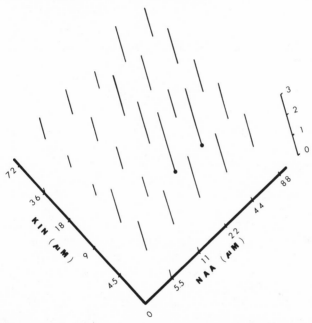

Fig. 11. Callus growth induced by diallelic interactions of Kinetin (KIN) and Naphthaleneacetic acid (NAA) at a range of high concentrations after eight weeks in darkness. Average data from 10–15 replicate treatments.

In another diallel experiment, six concentrations of sucrose were tested with six concentrations of MS inorganic salts and it was found that half-strength MS salts with 60 mM sucrose provided the best growth. Concentation ratios of KNO_3 and NH_3NO_3 were also studied at six diallel combinations and the higher growth indexes were observed with KNO_3 at the 2X MS concentrations (18.8mM) and $NH_4 NO_3$ to 1.0 $\times$ MS concentration (20.6mM).

Growth of other C. arabica cultivars and some diploid species. A series of diallel experiments was designed to study the effectiveness of kinetin (2, 4, 6, 10, and 20 μM) and 2,4-D (2, 4, 6, 10, and 20 μM) in promoting callus proliferation of leaf explants of several *C. arabica* cultivars and diploid species (table 8). A great deal of variation in the extent of callus was observed among these cultures. In general, *C. arabica* required concentrations of 2,4-D lower than 6μM, *C. canephora* required a 1:5 kinetin/2,4-D ratio, and *C. bengalensis* tissues underwent poor growth at all concentrations tested.

Morphogenesis in C. arabica cv. Bourbon. Following transfer to the inducing medium, the massive parenchymous type of callus growth ceases and the tissues slowly turn brown. Two sequences of morphogenetic

TABLE 8

KINETIN AND 2,4-D CONCENTRATION COMBINATIONS PROMOTING OPTIMAL
CALLUS GROWTH INDEXES FROM CULTURED LEAF EXPLANTS OF FOUR TETRAPLOID
C. ARABICA CULTIVARS AND SEVEN DIPLOID SPECIES OF COFFEA

Plant Material	Kinetin (μM)	2,4-D (μM)	Growth Index
C. arabica cv. Catuai	4.6	1.8–3.6	3.0
C. arabica cv. Laurina	10.0	6.0	1.8
C. arabica cv. Mundo Novo	2.0	2.0	2.4
C. arabica cv. Purpuracens	10.0	2.0	2.5
C. bengalensis	6.0	10.0	0.8
C. canephora cv. Kouillou	2.0	10.0	2.8
C. congensis	2.0	4.0	2.6
C. Dewevrei cv. excelsa	10.0	6.0	2.0
C. liberica	10.0	2.0	2.9
C. racemosa	2.0	2.0	3.0
C. salvatrix	20.0	4.0	1.8

differentiation have been characterized in secondary cultures of leaf explants in *Coffea*: low frequency somatic embryogenesis (LFSE) and high frequency somatic embryogenesis (HFSE) (fig. 15C). Adopting the standard cultural protocol described for *C. arabica* cv. Bourbon, LFSE is observed after 13–15 weeks and HFSE after 16–19 weeks of secondary culture. LFSE has appeared 3–6 weeks before the visible cluster of HFSE in more that 5,000 cultured flasks (Söndahl and Sharp, 1977 a,b,c).

LFSE is defined by the appearance of isolated somatic embryos developing into normal green plantlets in numbers ranging from 1 to 20 per culture. The occurrence of HFSE follows a unique developmental sequence: a white friable tissue containing globular structures develops from the nonproliferating brown callus cell mass (fig. 15D); the globular structures appear to develop synchronously for a period of 4–6 weeks. The pro-embryogenic globular mass gives rise to somatic embryos (fig. 15E) and finally to a plantlet, but this latter developmental process lacks the synchrony of the earlier stage, probably because of nutrient competition. The size of this pro-embryogenic tissue varies, but on the average about 100 somatic embryos develop per cluster of globular tissue. A cytological section through one somatic embryo is presented in figure 15F (Sphalinger, Söndahl, and Sharp, unpublished results). To speed up development and increase the percentage of fully developed plantlets, it is advisable to excise

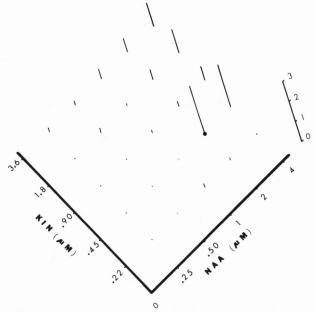

Fig. 12. Callus growth induced by diallelic interactions of Kinetin (KIN) and Naphthaleneacetic acid (NAA) at a range of low concentrations after 10 weeks in darkness. Average data from 10–15 replicate treatments.

the pro-embryogenic tissues and grow them under light conditions at 26° C in 5–10 ml of liquid inducing medium devoid of kinetin for 4–6 weeks (fig. 16A). After this period, the torpedo-shaped somatic embryos and young plantlets can be plated on saline-agar medium containing 0.5–1% sucrose in the presence of light (fig. 16B). Individual plantlets are removed from the agar medium, gently washed, and immediately transferred to Gif-pots inside a humid chamber. After a hardening period of 1–2 months, they can be exposed to normal atmospheric humidity and transferred to a greenhouse.

A minimum of 6–7 months is required for plantlet regeneration, if one considers the time interval from the inoculation of the leaf explant to the transfer of the HFSE plantlets into pots in a greenhouse. Considering that it takes 10–11 months for the development of young seedlings from fecundation, seed formation, and germination (September to July or August), the difference in length of time between somatic embryo and sexual embryo development is not significant. It is an open question whether the slow growth pattern interval between *Coffea* leaf tissue and plantlet development is inherent to the perennial habitat of *Coffea* or is a result of a suboptimum growth medium. The real importance of this morphogenetic protocol in *Coffea* is the potential for utilizing the genetic manipulation suggested in figures 6–9, and perhaps vegetative propagation, for a few extremely valuable heterozygous cross-pollinated plants.

The fact that HFSE is so effectively triggered (up to 60%) by the addition of 2,4-D in combination with kinetin during primary culture (Söndahl and Sharp, 1977a) is striking. Other sources of auxins (IBA and NAA) in combination with kinetin are not very effective in the induction of HFSE (10–20%). However, NAA in combination wtih kinetin induces LFSE in up to 60% of the cultured bottles. Moreover, the removal of 2,4-D and the

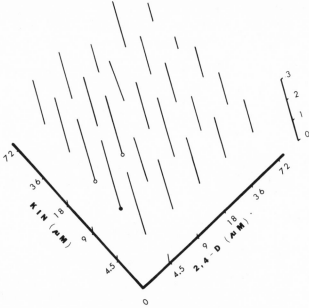

Fig. 13. Callus growth induced by diallelic interactions of Kinetin (KIN) and Dichlorophenoxyacetic acid (2,4-D) at a range of high concentrations after 10 weeks in darkness. Average data from 10–15 replicate treatments.

lowering of NAA and kinetin concentrations in the induction medium of the secondary cultures was found essential to the development of somatic embryos of HFSE in *Coffea* spp. The effective and preferential role of 2,4-D in triggering the developmental sequences leading to HFSE cannot be fully explained at present. Calluses induced by 2,4-D and NAA appear to be phenotypically similar upon visual observation. However, it is possible that 2,4-D induces the differentiation of a distinct population of cells according to one of the following phenomena: (1) inducing proliferation of a unique cell type from the original explant (cambium, secondary phloem, etc.); (2) lengthening the cell cycle of a certain cell population by interference with one or more control points of the cell cycle; or (3) removing a distinct population of cells from the cell cycle with a blockage at G1 or G2. More experimental work needs to be done in order to explain the effective physiological difference between 2,4-D and NAA in the induction of somatic embryogenesis in coffee leaf tissues. The presence of multiple cell populations has been described in developmental systems (Webster and Davidson, 1968; Friedberg and Davidson, 1971). Two subpopulations of cells were recognized at the early stages of lateral root formation in *Vicia faba*: a subpopulation at the central core, which was dividing actively, and

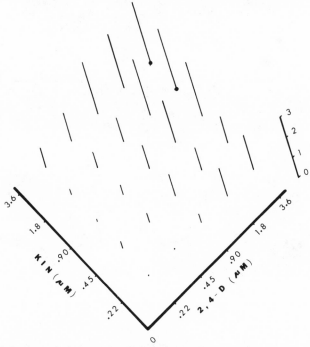

Fig. 14. Callus growth induced by diallelic interactions of Kinetin (KIN) and 2,4-dichlorophenoxyacetic acid (2,4-D) at a range of low concentrations after ten weeks in darkness. Average data from 10–15 replicate treatments.

an arrested subpopulation at the periphery. When the primordium emerged, the central core cells stopped dividing (arrested in G1 phase of cell cycle) and the peripheral cells divided actively for 48 hours, after which they stopped dividing and the central core cells resumed division again. The peripheral cells are the ones that contribute to the root cap and probably, changes in the mitotic cycle precede biochemical and morphological specialization. There is little information in the literature relating growth regulators and the cell cycle. Mac Leod (1968) reported that kinetin causes

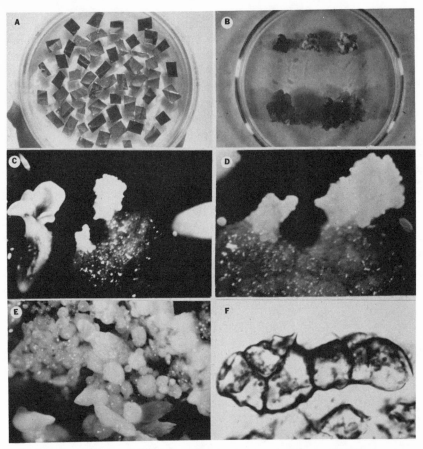

Fig. 15.   Culture of *Coffea* leaf explants. (A) Pre-experimental incubation of leaf explants (abadaxial side facing up) on saline-sugar medium for 36 hr in darkness (1X); leaf explants that retain the normal dull pale green coloration are transferred to the culture bottles and leaf explants changing to a pale dark green coloration are discarded. (B) Callus growth scale (1-5) utilized for evaluating diallelic experiments (1X). (C) Low- and high-frequency somatic embryo formation (LFSE and HFSE) *in vitro*; the pro-embryogenic tissue (HFSE) is observed proliferating on top of a brown callus tissue. A few plantlets have already differentiated (LFSE) (6X). (D) Emergence of the white globular proembryogenic tissue from the nonproliferating brown callus (10X). (E) Further development of the white proembryogenic tissue (6X). (F) Cross section of a somatic embryo (200X).

a blockage at the G1/S interphase and suggested that kinetin may affect the oxidation of carbohydrates by inhibiting glycolysis and, consequently, increasing carbohydrate oxidation through the pentose shunt. Neumann (1968) came to the same conclusion when he observed that carrot tissues with active cell division promoted by kinetin had a lower oxygen consump-

Fig. 16.   Development of *C. arabica* cv. Mundo Novo plantlets from somatic embryos and the colchicine doubling technique. (A) Torpedo shaped somatic embryos and young plantlets growing in liquid medium devoid of kinetin under illumination (1X). (B) Further development of plantlets on a saline-agar medium containing 0.5 or 1% sucrose in the presence of light (1X). (C) Growth of a plantlet in a Gif-pot which provides excellent conditions for root proliferation (0.5X). (E) Excised plantlets transferred to a vermiculite pot in the greenhouse (1X). (F) Close-up view of the colchicine treatment of an apical shoot meristem of *Coffea canephora* cv. Guarini seedling; the needle is calibrated for 10 μliter and filled with a 1% colchicine solution (100 μg colchicine per seedling) (1X).

tion rate than tissues growing without kinetin. He suggested that kinetin was an inhibitor of aerobic respiration in the carrot system. Differentiated cells from carrot explants can "differentiate" after a period of intensive cell division into embryo-competent carrot cells (Linser and Neumann, 1968). High rates of cell division are achieved by the addition in the nutrient medium of inositol, IAA, and kinetin (Linser and Neumann, 1968), or 2,4-D (Halperin, 1966), or coconut milk (Steward, 1963; Reinert, 1959). The rate of cell proliferation decreases following subculture into a 2,4-D free medium. The transcription and translation processes of the cellular subpopulations of the cultivated tissues are obviously differentially affected in regard to their phase intervals during the cell cycle on account of changed cytoplasmic and nuclear environments. As discussed by Linser and Neumann (1968), growth regulators added to the culture medium control morphogenesis indirectly or nonspecifically through the control of cell division and cell aging.

It is interesting to mention that aging of callus, omission of sucrose from the medium, or gamma irradiation are reported to stimulate embryogenesis in unfertilized ovule tissues of *Citrus sinensis* 'Shamouti' (Kochba and Button, 1974; Spiegel-Roy and Kochba, 1973). Baylis (1977) reported, in carrot cell suspension, a positive correlation between an increase in 2,4-D concentration and an increase of the mean generation time. It was also found that the mitotic duration increased and the mitotic index decreased with increasing 2,4-D concentrations (0.5–70 μM). These results were partially interpreted as a blockage or lengthening of G1 and G2 and lengthening of prophase and metaphase relative to anaphase and telphase. It is interesting to mention that, in the carrot system, somatic embryogenesis occurs only after removal of 2,4-D, and that concentrations as low as 0.1 μM inhibit embryo development (Baylis, 1977). Peaks in peroxidase activity were associated with the appearance of somatic embryos in orange ovular tissues and, following an isoenzyme analysis, the presence of a cathodic band was distinctly associated with the embryogenetic process (Kochba et al., 1977). Shortening of G1 and an increase in peroxidase activity were described by Gupta and Stebins (1969) in association with the developmental process in hooded primordia. It is possible that HFSE in *Coffea* leaf callus is induced by a resetting of enzyme patterns (peroxidases, etc.) during blockage of a certain population of cells in G1 or G2 caused by 2,4-D in primary culture. After subculture onto the induction medium, callus proliferation ceases and division only takes place in the previously arrested cells. It is also possible that 2,4-D alternatively disrupts the mitotic assembly and thus causes a shift in the polarity of the cell plate, leading to unequal cell division and/or cytokinesis at opposite angles, and hence to the differentiation of embryogenic cells.

CONCLUDING REMARKS

Coffee is a beverage derived from a perennial crop growing commercially only in tropical areas of the world and it contributes a significant share of the revenue generated by the national and international markets of developing nations. The money generated by this agricultural product in the international market (about 5 billion U.S. dollars annually) becomes strategically important among producing nations as a source of income for financing imports of modern industrial commodities and services.

Research with *Coffea* has been restricted to producing nations, primarily because of its tropical and perennial characteristics. A great deal of knowledge has been accumulated pertaining to *Coffea* in the areas of agronomy, genetics, morphology, physiology, and pathology. More information is needed in order to contribute to a definite systematic classification of the genus *Coffea* and to establish an urgently needed international collaboration for enlarging and preserving germplasm collections.

Attempts to utilize tissue culture methods for the improvement of coffee plants are very recent and still preliminary. The application of organ, cell, and protoplast technologies in *Coffea* spp. could be invaluable in many different areas of *Coffea* research. The fact that plant regeneration at high frequencies through somatic embryo formation has been observed so far in four species of *Coffea* (*C. arabica, C. canephora, C. congensis, and C. dewevrei*) brings us closer to the realization of applying the potentialities of *in vitro* techniques for the generation of improved *Coffea* cultivars. Some research projects that could be established in attempting to solve current problems related to the improvement of commerical coffee plants can be summarized as follows:

1. Embryo culture for growth physiology studies, to raise hybrids from interspecific crosses due to deficient abortive endosperms, to study host-parasite interactions, and finally to apply chemical or physical mutagens.

2. Culture of lateral branch explants containing 1–2 nodes to study differentiation and growth of flower buds. Culture of lateral branch explants with arrested flower buds and subtending leaves for the study of dormany and anthesis.

3. Induction and selection of mutant cell lines more tolerant to high levels of aluminum and low pH.

4. Induction and selection of cells with more efficient permease systems for nutrient uptake.

5. Induction and selection of mutant cells in varying osmotica and reconstitution of drought-resistant plantlets.

6. Induction and selection of cells and plantlet differentiation with improved resistance to low temperatures.

7. Induction and selection of cells with higher photosynthetic rates.

8. Freeze-preservation of germplasm of *Coffea* species and differential *C. arabica* clones for *Hemileia vastatrix* physiological races.

9. *In vitro* fertilization.

10. Pollen grains and ovule cultures seeking for hormonal emasculation.

11. Endosperm culture for generation of different ploidy levels.

12. Haploid plants of *Coffea* spp. for genetic analysis and isolation of self-incompatible alleles in diploid species of *Coffea*.

13. Generation of homozygous plants from haploid tissues of *C. canephora* or other allogamous diploid species of *Coffea*.

14. Obtainment of plantlets of several ploidy levels from cultivated haploid or somatic tissues of *Coffea* spp. Tetraploid plantlets generated from *C. canephora* tissues would be of valuable importance in breeding programs.

15. Increase of variability and plasticity in *C. arabica* seeking for sources of horizontal resistances, drought resistances, and so forth: (a) mutant cell lines and plant regeneration; (b) uptake of isolated chloroplasts and mitochondria by protoplast preparations; (c) transformation; and (d) interspecific hybridization between *C. arabica* and diploid species of *Coffea* by means of protoplast fusion (2 mult. dihaploid versus 2 mult. somatic protoplasts) and plantlet regeneration.

16. Vegetative propagation of *C. canephora* and other heterozygous diploid species of *Coffea*.

17. Induction and selection of disease-resistant cultivars: (a) isolation and *in vitro* cultivation of the pathogen onto chemically defined medium or in parabiotic vessels; and (b) selection of resistant coffee cell lines in toxin-isolated fractions or parabiotic cultures.

18. Study of resistance-susceptibility reactions in chemical and physical controlled conditions.

19. Transference of nuclear or cytoplasmic male sterility to *C. arabica* for production of multilineal hybrids to lessen selection pressure on pathogenic physiological races of *Hemileia vastatrix, Collectrotrichum coffeanum*, and other diseases.

20. Nematode studies by *in vitro* associations with root or callus cultures of *Coffea* spp.

REFERENCES

Adnikinju, B. A. 1975. A preliminary study of the influence of chemical sprays on fruit set and abscission in Robusta coffee (*Coffea canephora* Pierre ex Froehner). Turrialba 25(4):414–17.

Alvim, P. T. 1958. Use of growth regulators to induce fruit set in coffee (*Coffea arabica* L.). Turrialba 8:64–77.

Alvim, P. T. 1960. Moisture stress as a requirement for flowering of coffee. Science 132:354.

Alvim, P. T. 1964. Tree growth periodicity in tropical climates. *In* M. H. Zimmermann (ed.), Formation of wood in forest trees. Pp. 479–95. Academic Press, New York.

Alvim, P. T. 1973. Factors affecting flowering of coffee. *In* A. N. Srb (ed.), Genes, enzymes, and populations. Pp. 193–202. Plenum Press, London.

Bachi, O. 1958. Estudos sobre a conservacão de sementes. IV. Cafe. Bragantia 17:262–70.

Baylis, M. W. 1977. The effects of 2,4-D on growth and mitosis in suspension cultures of *Daucus carota*. Plant Sci. Lett. 8:99–103.

Bergamin, J. and J. Abrahão. 1963. Pragas e molestias do cafeeiro. *In* Cultura e adubacão do cafeeiro. Pp. 127–41. Inst. Bras. Potassa, São Paulo.

Bergmann, L. 1967. Wachustum grüner Suspensionskulturen von *Nicotiana tabacum* Var. Samsun mit CO_2 als Kohlenstoffquelle. Planta 74:243–49.

Berlin, M. B., and I. Zelitch. 1975. Photoautotrophic growth and photosynthesis in tobacco callus cells. Plant Physiol. 56:752–56.

Bettencourt, A. J., and A. Carvalho. 1968. Melhoramento visando a resistancia do cafeeiro a ferrugem. Bragantia 27(4):35–68.

Borlaug, N. E., J. A. Browning, and K. J. Frey. 1969. Multiline cultivars as a means of disease control. Ann. Rev. Phytopath. 7:355–82.

Boudrand, J. N. 1974. Le bouturage du cafeier canephora a Madagascar. Cafe Cacao The 18(1):31–47.

Bowden, J., P. H. Gregory, and C. G. Johnson. 1971. Possible wind transport of coffee rust across the Atlantic Ocean. Nature 229:500–501.

Browning, G., and M. G. R. Cannell. 1970. Use of 2-chloroethano-phosphonic acid to promote the abscission and ripening in fruit to *Coffea arabica* L. J. Hort. Sci. 45:233–32.

Browning, G., and N. M. Fisher. 1975. Shoot growth in *Coffea arabica* L. 11. Growth flushing stimulated by irrigation. J. Hort. Sci. 50:207–18.

Browning, G., G. V. Hoad, and P. Gaskin. 1973. Flower bud dormany in *Coffea arabica* L. I. Studies of gibberellin in flower buds and xylem sap and of abscisic acid in flower buds in relation to dormancy release. J. Hort. Sci. 48:29–41.

Buckland, E., and P. M. Townsley. 1975. Coffee cell suspension cultures. Caffeine and chlorogenic acid content. J. Inst. Canad. Sci. Technol. Aliment. 8(3):164–65.

Buttler, D. R. 1977. Coffee leaf temperatures in a tropical environment. Acta Bot. Neerl. 26(2):129–40.

Cannell, M. G. R. 1971. Changes in the respiration and growth rates of developing fruits of *Coffea arabica* L. J. Hort. Sci. 46:263–72.

Cannell, M. G. R. 1972. Photoperiodic response of mature trees of arabica coffee. Turrialba 22(2):198–206.

Carelli, M. I. C., C. R. Lopes, and L. C. Monaco. 1974. Chlorogenic acid content in species of *Coffea* and selections of *C. arabica*. Turrialba 24(4):398–401.

Carlson, P. S. 1973. Methionine sulfoximine-resistant mutants of tobacco. Science 180:1366–68.

Carvajal, F. F., A. Acevedo, and C. A. Lopes. 1969. Nutrient uptake by the coffee tree during a yearly cycle. Turrialba 19(1):13–20.

Carvalho, A. 1946. Distribuicão geografica e classificacão botanica de genero *Coffea* com referencia especial a especie arabica. Bull. Superint. Serv. Cafe 230:3–33.

Carvalho, A., 1958. Advances in coffee production technology. II. Genetics. Coffee and Tea Ind. Flav. Field 81:30–36.

Carvalho, A., and L. C. Monaco. 1963. Botanica e melhoramento. *In* Cultura e adubacão do dafeeiro. Pp. 45–58. Inst. Bras. Potassa, São Paulo.

Carvalho, A., and L. C. Monaco. 1968. Relaciones geneticas de especies seleccionadas de coffea. Cafe IICA 9(4):3–19.

Carvalho, A., and L. C. Monaco. 1972. Adaptacão e produtividade de cafeeiros portadores de fatores para resistencia a *Hemileia vastatrix.* Ciencia e Cultura 24(10):924–32.

Carvalho, A., H. J. Scaranari, H. Antunes, and L. C. Monaco. 1961. Melhoramento do cafeeiro. XXII. Resultados obtidos no ensaio de seleçoes regionais de Campinas. Bragantia 20:710–39.

Carvalho, A., J. S. Tango, and L. C. Monaco. 1965. Genetic control of the caffeine content of coffee. Nature 205:314.

Chadler, M. T., N. T. Marcac, and Y. Kouchkovsky. 1972. Photosynthetic growth of tobacco cells in liquid suspension. Canad. J. Bot. 50:2265–70.

Chandler, R. F. 1969. New horizons for ancient crop. 11th Intern. Bot. Congr. Symp. World Food Supply.

Charrier, A., and J. Berthaud. 1975. Variation de la teneur en cafeine cans le genre *Coffea.* Cafe Cacao The 19(4):251–64.

Chen, K., J. C. Gray, and S. G. Wildman. 1975. Fraction I protein and the origin of polyploid wheats. Science 190:1304–5.

Chevallier, A. 1947. Les cafeiers du globe. II. Systématique de caféiers et faux-caféiers, maladies et insects muisibles. Paul Le Chevalier, Paris.

Colonna, J. P., G. Gas, and H. Rabechault. 1971. Mise au point d'une méthode de culture *in vitro* d'embryons de caféiers. Application à deux variété de caféiers cultivés. Compt. Rend. 272:60–63.

Corduan, G. 1970. Autotrophe Gewebekulturen von *Ruta graveoleus* und deren $^{14}CO_2$ Markierungspordukte. Planta 91:291–301.

Curtis, P. E., W. L. Ogren, and R. H. Hageman. 1969. Varietal effects in soybean photosynthesis and photorespiration. Crop Sci. 9:323–27.

Dedecca, D. M. 1957. Anatomia e desenvolvimento ontogeneticode *Coffea arabica* L. var. typica Cramer. Bragantia 16(23):315–66.

Durtreq, A. J. E. 1977. *In vitro* selection of plants resistant or tolerant to pathogenic fungi. Intern. Symp. on the Use of Induced Mutations for Improving Disease Resistance in Crop Plants, IAEA/FAO, Vienna.

Edwards, G. A., J. E. Endrizzi, and R. Stein. 1974. Genome DNA content and chromosome organization in *Gossypium.* Chromosoma 47:309–26.

Flor, H. H. 1955. Host-parasite interaction in flax-rust—its genetics and other implications. Phytopathology 45:680–85.

Flor, H. H. 1965. Test for allelism of rust-resistance genes in flax. Crop Sci. 5:415–18.

Fousova, S., and N. Avratovscukovan. 1967. Hybrid vigor and photosynthetic rate of leaf disks in *Zea mays* L. Photosynthetics 1:3–12.

Franco, C. M. 1939. Relation between chromosome number and stomata in *Coffea.* Bot. Gaz. 100(4):817–27.

Franco, C. M. 1940. Fotoperiodismo em cafeeiro (*C. arabica*). Inst. Cafe Est. 27:1586–92.

Franco, C. M. 1958. Influence of temperature on growth of coffee plant. IBEC Research Institute No. 16. 24 pp.

Franco, C. M. 1963. Fisiologia do cafeeiro. *In* Cultura e adubação do cafeeiro. Pp. 59–76. Inst. Bras. Potassa, São Paulo.

Franco, C. M., and R. Inforzato. 1950. Quantidade de agua transpirada pelo cafeeiro cultivado ao sol. Bragantia 10:247–57.

Friedberg, S. H., and D. Davidson. 1971. Cell population studies in developing root primordia. Ann. Bot. 35:523–33.

Frischknecht, P. M., T. W. Baumann, and H. Wanner. 1977. Tissue culture of *Coffea arabica*. Growth and caffeine formation. Planta Medica 31:344–50.

Frywell, P. A. 1971. Phenetic analysis and phylogeny of the diploid species of *Gosypium* L. (Malvaceae). Evolution 25:554–62.

Gengenbach, B. G., and C. E. Green. 1975. Selection of T-cytoplasm maize callus cultures resistant to *Helminthosporium maydis* Race T pathotoxin. Crop Sci. 15:645–49.

Gengenbach, B. G., C. E. Green, and C. M. Donovan. 1977. Expression and inheritance of pathotoxin resistance in maize plants selected and regenerated from cell cultures. *In* this volume (abstr.).

Glinka, Z. 1973. Abscisic acid effect on root exudation related to increased permeability to water. Plant Physiology 48:103–5.

Gopal, N. H., K. I. Raju, D. Venkataramanan, and K. J. Janardham. 1975. Physiological studies on flowering in coffee under South Indian conditions. III. Flowering in relation to foliage and wood starch. Turrialba 25(3):239–42.

Gopal, N. H., D. Venkataramanan, and N. G. N. Rathna. 1975. Physiological studies on flowering in coffee under South Indian conditions. IV. Some physical properties and chromatographic assay of a gum-like substance exuded by flower buds. Turrialba 25(4):410–13.

Grant, V. 1963. The origin of adaptations. Columbia University Press, New York.

Gray, J. C., S. D. Kung, S. G. Wildman, and S. J. Sheen. 1974. Origin of *Nicotiana tabacum* L. detected by polypeptide composition of Fraction I protein. Nature 252:226–27.

Griffiths, E., J. N. Gibbs, and J. M. Waller. 1971. Control of coffee berry disease. Ann. Appl. Biol. 67:45–74.

Gupta, V., and G. L. Stebins. 1969. Peroxidase activity in hooded and awned barley at successive stages of development. Biochem. Genet. 3:15–34.

Halperin, W. 1966. Alternative morphogenetic events in cell suspensions. Amer. J. Bot. 53:443–52.

Hanson, A. D., and J. Edelman. 1972. Photosynthesis by carrot tissue cultures. Planta 102:11–25.

Heichel, G. H., and R. B. Musgrave. 1969. Relation of CO_2 compensation concentration to apparent photosynthesis in maize. Plant Physiol. 44:1724–28.

Helgenson, J. P., C. T. Haberlach, and C. D. Upper. 1976. A dominant gene conferring resistance to tobacco plants is expressed in tissue cultures. Phytopathology 66:91–96.

Herman, F. R. P., and G. J. Hass. 1975. Clonal propagation of *Coffea arabica* L. from callus culture. Hort. Sci. 10:588–89.

Humprey, D. M., and D. J. Ballantyne. 1974. Diurnal and annual flutuations of gibberellins in the leaves of *Coffea arabica* L. Turrialba 24(4):360–66.

Hüssemann, W., and W. Barz. 1972. Photoautotrophic growth and photosynthesis in cell suspension cultures of *Chenopodium rubrum*. Physiol. Plantarum 40:77–81.

Jensen, C. J. 1974. Chromosome doubling techniques in haploids. *In* K. J. Kasha (ed.), Haploids in higher plants. Advances and potential. Pp. 153–90. 1st Intern. Symp. Guelph. University of Guelph Press, Guelph, Ontario, Canada.

Jensen, R. G., and J. T. Bahr. 1977. Ribulose 1,5-biphosphate carboxylaseoxygenase. Ann. Rev. Plant Physiol. 28:379–400.

Johnson, A. W., and J. G. Packer. 1965. Polyploidy and environment in Arctic Alaska. Science 148:237–39.

Johri, R. M., and S. S. Bhojwani. 1977. Triploid plants through endosperm culture. *In* J. Reinert and Y. P. S. Bajaj (eds.), Applied and fundamental aspects of plant cell, tissue, and organ culture. Pp. 398–411. Springer-Verlag, Berlin.

Kasha, K. J. 1974. Haploids in higher plants. Advances and potential. 1st Intern. Symp. Guelph. University of Guelph Press, Guelph, Ontario, Canada.

Keller, H., H. Wanner, and T. W. Baumann. 1972. Caffeine synthesis in fruits and tissue cultures of *Coffea arabica*. Planta 108:339–50.

Kelly, G. J., E. Latzko, and M. Gibbs. 1976. Regulatory aspects of photosynthetic carbon metablism. Ann. Rev. Plant. Physiol. 27:181–205.

Knott, D. R. 1972. Using race-specific resistance to manage the evolution of plant pathogens. J. Environ. Qual. 1:227–31.

Kochba, J., and J. Button. 1974. the stimulation of embryoid development in habituated ovular callus from the 'Shamouti' orange (*Cirtus sinensis*) as affected by tissue age and sucrose concentration. Z. Pflanzenphysiol. 73:415–21.

Kochba, J., S. Lavee, and P. Spiegel-Roy. 1977. Differences in peroxidase activity and isoenzymes in embryogenic and non-embryogenic "Shamouti" orange ovular callus lines. Plant and Cell Physiol. 18:463–67.

Konzak. C. F. 1956. Induction of mutations for disease resistance in cereals. Genetics in Plant Breeding, Brookhaven Symp. in Biology 9:157.

Laing, W. A., W. L. Ogren, and R. H. Hageman. 1974. Regulation of soybean net photosynthetic CO_2 fixation by the interaction of CO_2 and O_2 and ribulose 1,5-diphosphate carboxylase. Plant Physiol. 54:678–85.

Laing, W. A., W. L. Ogren, and R. H. Hageman. 1975. Bicarbonate stabilization of ribulose 1,5-diphosphate carboxylase. Biochemistry 14:2269–75.

Leopold, A. C., and P. E. Kriedemann. 1975. Plant growth and development. McGraw-Hill Co., New York.

Lim, S. M., A. L. Hooker, and B. R. Smith. 1971. Use of *Helminthosporium maydis* Race T pathotoxin to determine disease reaction of germinating corn seed. Agron. J. 63:712.

Linser, H., and K. H. Neumann. 1968. Untersuchungen üben Beziehungen zwischen Zellteilung und Morphogenese bei Gewebekulturen von *Daucus carota*. I. Rhizogenes und Ausbildung ganger Pflanzen. Physiol. Plantarum 21:487–99.

Lupton, F. G. H. 1969. Estimation of yield in wheat from measurements of photosynthesis and translocation in the field. Ann. Appl. Biol. 64:363–74.

Mac Leod, R. D. 1968. Changes in the mitotic cycle in lateral root meristems of *Vicia faba* following kinetin treatment. Chromosoma 24:177–87.

Magalhães, A. C., and L. R. Angelocci. 1976. Sudden alterations in water balance associated with flower bud opening in coffee plants. J. Hort. Sci. 51:419–23.

Malavolta, E. 1963. Nutricão do cafeeiro. *In* Cultura e Adubação do Cafeeiro. Pp. 143-90. Inst. Bras. Potassa, São Paulo.

Malavolta, E. 1975. Nutrição mineral e adubação do cafeeiro. *In* E. Malavolta, H. P. Haag, F. A. F. Mello, and M. O. C. Brasil (eds.), Nutrição mineral e adubação de plantas cultivadas. Pp. 203-57. Livr. Pioneira Editora, São Paulo.

Malavolta, E., and F. R. P. Moraes. 1963. Resultadoes do ensaios de adubação. *In* Cultura e adubação do cafeeiro. Pp. 191-214. Inst. Bras. Potassa, São Paulo.

Mendes, A. J. T. 1947. Observações citologicas em *Coffea*. XI. Metodos do tratamento pela colchicina. Bragantia 7:221-30.

Mess, M. G. 1957. Studies on the flowering of *Coffea arabica* L. III. Various phenomena associated with the dormancy of coffee flower buds. Portugaliae Acta Biologica 5(1):25-44.

Meyer, F. G. 1965. Notes on wild *Coffea arabica* from Southwest Ethiopia with some historical considerations. Econ. Bot. 19:57-71.

Monaco, L. C. 1977. Consequences of the introduction of coffee rust into Brazil. Ann. N. Y. Acad. Sci. 287:57-71.

Monaco, L. C., and A. Carvalho. 1969. Coffee genetics and breeding in Brazil. Span. 12:2-5.

Monaco, L. C., and A. Carvalho. 1975. Resistencia a *Hemileia vastatrix* no melhoramento do cafeeiro. Ciencia e Cultura 27(10):1070-81.

Monaco, L. C., H. P. Medina, M. R. Söndahl, and M. A. L. Miranda. 1977. Efeito de dias longos no crescimento e florescimento de cultivares de *Coffea arabica*. Bragantia (in press).

Monaco, L. C., and M. R. Söndahl. 1974. Emprego do etileno na maturação de frutos de cafe. Pesq. Agropec. Bras. 9:135-37.

Monaco, L. C., M. R. Söndahl, and A. Carvalho. 1975. New technique for colchicine treatment of coffee seedlings. Turrialba 25(3):323-24.

Monaco, L. C., M. R. Söndahl, A. Carvalho, O. J. Crocomo, and W. R. Sharp. 1977. Applications of tissue cultures in the improvement of coffee. *In* J. Reinert and Y. P. S. Bajaj (eds.), Applied and fundamental aspects of plant cell, tissue, and organ culture. Pp. 109-29. Springer-Verlag, Berlin.

Moraes, F. R. P. 1963. Meio ambiente e praticas culturais. *In* Cultura e adulbação do cafeeiro. Pp. 77-126. Inst. Bras. Potassa, São Paulo.

Moraes, M. V., and C. M. Franco. 1973. Metodo expedito para enxertia em cafe. Bull. Inst. Bras. Cafe, GERCA.

Morton, J. K. 1966. The role of polyploidy in the evolution of a tropical flora. *In* C. D. Darlington and K. R. Lewis (eds.), Chromosomes today. Pp. 73-76. Plenum Press, New York.

Murray, M. J. 1971. Additional observations on mutation breeding to obtain verticillium-resistant strains of peppermint. *In* Mutation breeding for disease resistance. Pp. 171-95. International Atomic Energy Agency, Vienna.

Nelson, R. R. 1972. Stabilizing racial populations of plant pathogens by use of resistance genes. J. Environ. Qual. 1:220-27.

Neumann, K. H. 1968. Untersuchungen über Beziehungen zwischen Zellteilung und Morphogenese bei Gewebakulturen von *Daucus carota*. II. Respiration von Gewebekulturen Abhängigkeit von der Zellteilungsintensitat. Physiol. Plantarum 21:519-24.

Nitsch, C. 1974. Pollen culture. A new technique for mass production of haploid and homozygous plants. *In* K. J. Kasha (ed.), Haploids in higher plants. Advances and

Potential. Pp. 123–35. 1st Intern. Symp. Guelph. University of Guelph Press, Guelph, Ontario, Canada.

Nitsch, C. 1977. Culture of isolated microspores. *In* J. Reinert and Y. P. S. Bajaj (eds.), Applied and fundamental aspects of plant cell, tissue, and organ culture. Pp. 268–78. Springer-Verlag, Berlin.

Nobel, P. S. 1974. Introduction to biophysical plant physiology. W. H. Freeman & Co., San Francisco.

Nsumbu, N., and J. Bouharmont. 1977. Différentiation de racines et de tiges feuillées à partir de feuilles de *Coffea canephora*. Cafe Cacao The 21(1):3–8.

Nunes, M. A., J. F. Bierhuizen, and C. Ploegman. 1968. Studies of productivity of coffee. I. Effect of light, temperature, and CO_2 concentration on photosynthesis of *Coffea arabica*. Acta Bot. Neerl. 17(2):93–102.

Nunes, M. A., J. F. Beirhuizen, and C. Ploegman. 1969. Studies on productivity of Coffee. III. Differences in photosynthesis between four varieties of Coffee. Acta Bot. Neerl. 18(3):420–24.

Nutman, F. J. 1937. Studies of the physiology of *Coffea arabica*. I. Photosynthesis of coffee leaves under natural conditions. Ann. Bot. (NS) 1(4):353–67.

Ogren, W. L., and G. Bowes. 1971. Ribulose diphosphate carboxylase regulates soybeans photorespiration. Nature New Biol. 230(13):159–60.

Oliver, D. J., and I. Zelitch. 1977. Increasing photosynthesis by inhibiting photorespiration with glyoxylate. Science 196:1450–51.

Oyebade, T. 1976. Influence of pre-harvest sprays of ethrel on ripening and abscission of coffee berries. Turrialba 26(1):86–89.

Pereira, H. C. 1957. Field measurement of water use for irrigation control in Kenya. J. Agric. Sci. 49:459–66.

Person, C. 1966. Genetic ploymorphism in parasitic systems. Nature 212:266–67.

Person, C. 1967. Genetic aspects of parasitism. Canad. J. Bot. 45:1193–1204.

Phillips. L. L. 1963. The cytogenetics of *Gossypium* and the origin of new world cottons. Evolution 17:460–69.

Piringer, A. A., and H. A. Borthwick. 1955. Photoperiodic responses of coffee. Turrialba 5:72–77.

Polaco, J. C., and M. L. Polacco. 1977. Inducing and selecting valuable mutation in plant cell culture: A tobacco mutant resistant to carboxin. Ann. N. Y. Acad. Sci. 287:385–400.

Raghavan, V. 1976. Experimental embryogenesis in vascular plants. Academic Press, New York.

Raghavan, V. 1977. Applied aspects of embryo culture. *In* J. Reinert and Y. P. S. Bajaj (eds.), Applied and fundamental aspects of plant cell, tissue, and organ culture. Pp. 375–97. Springer-Verlag, Berlin.

Randall, D. D., C. J. Nelson, and K. H. Asay. 1977. Ribulose diphosphate carboxylase. Altered genetic expression in tal fescue. Plant Physiol. 59:38–41.

Rangaswamy, N. S. 1977. Applications of *in vitro* pollination and *in vitro* fertilization. *In* J. Reinert and Y. P. S. Bajaj (eds.), Applied and fundamental aspects of plant cell, tissue, and organ culture. Pp. 412–25. Springer-Verlag, Berlin.

Reinert, J. 1959. Über die Kontrolle der Morphogenese und die Induktion von Adventiven-bryonen und Gewebekulturen aus Karotten. Planta 53:318–33.

Reinert, J., and Y. P. S. Bajaj. 1977. Anther culture: Haploid production and its significance.

In J. Reinert and Y. P. S. Bajaj (eds.), Applied and fundamental aspects of plant cell, tissue, and organ culture. Pp. 251–67. Springer-Verlag, Berlin.

Rijo, L. 1972. Histophatology of the hypersensitive reaction t (tumefaction) induced on *Coffea* spp. by *Hemileia vastatrix* Berk and Br. Agronomia lusit. 33:427–31.

Robinson, R. A. 1976. Plant pathosystems. Springer-Verlag, Berlin.

Rodrigues, C. J., and E. F. Medeiros. 1975. Relationship between an phytoalexin-like response in coffee leaves (*Coffea arabica* L.) and compatibility with *Hemileia vastatrix* Berk and Br. Physiol. Plant Pathol. 6:35–41.

Rodrigues, C. J., and J. J. Molero. 1970. Ethrel: A potential coffee ripener. J. Agric. Univ. Puerto Rico. 54(4):689–90.

Rouhani, I., and M. Khash-khui. 1977. Variations in photosynthetic rates of fourteen coleus cultivars. Plant Physiol. 59:114–15.

Ryan, F. J., and N. E. Tolbert. 1975. Ribulose diphosphate carboxylase/oxygenase. III. Isolation and properties, J. Biol. Chem. 250:4229–33.

Sakano, K., S. D. Kung, and S. G. Wildman. 1974. Identification of several chloroplast DNA genes which code for the large subunit of *Nicotiana* Fraction I proteins. Molec. Gen. Genet. 130:91–97.

Sharp, W. R., L. S. Caldas, O. J. Crocomo, L. C. Monaco, and A. Carvalho. 1973. Production of *Coffea arabica* callus of three ploidy levels and subsequent morphogenesis. Phyton 31:67–74.

Söndahl, M. R. 1974. Incorporação de $^{14}Co_2$ em folhas intactas de cafeiro. M. S. thesis, University of São Paulo, Piracicaba.

Söndahl, M. R., O. J. Crocomo, and L. Sodek. 1976. Measurements of ^{14}C incorporation by illuminated intact leaves of Coffee plants from gas mixtures containing $^{14}CO_2$. J. Exp. Bot. 27(101):1187–95.

Söndahl, M. R., L. C. Monaco, L. C. Fazuoli, H. P. Medina and M. A. L. Miranda. 1977. Resposta de germoplasmas de *Coffea arabica* L. a benzil adenina e giberelina. Turrialba (in press).

Söndahl, M. R., L. C. Monaco, and W. R. Sharp. 1972. Unpublished results.

Söndahl, M. R., and W. R. Sharp. 1977a. High frequency induction of somatic embryos in cultured leaf explants of *Coffea arabica* L. Z. Pflanzenphysiol. 81:395–408.

Söndahl, M. R., and W. R. Sharp. 1977b. Interactions of cytokinin and auxin on growth and embryogenesis in cultured leaf explants of *Coffea*, 180. Intern. Conf. on Regulation of Develop. Processes in Plants, Halle.

Söndahl, M. R., and W. R. Sharp. 1977c. Growth and embryogenesis in leaf tissues of *Coffea*. Plant Physiol. 69(6):1.

Söndahl, M. R., A. A. Teixeira, L. C. Fazuoli, and L. C. Monaco. 1975. Efeito do etileno sobre o tipo e qualidade da bebida de cafe. Turrialba 24(1):17–19.

Spiegel-Roy, P. and J. Kochba. 1973. Stimulation of differentiation in orange (*Citrus sinensis*) ovular callus in relation to irradiation of the media. Rad. Bot. 13:97–103.

Staritsky, G. 1970. Embryoid formation in callus tissues of coffee. Acta Bot. Neerl. 19:509–14.

Steward, F. C. 1963. Totipotency and variation in cultured cells: Some metabolic and morphogenetic manifestations. *In* P. Maheshwari and N. S. Rangaswamy (eds.). Plant tissue and organ culture. Pp. 178–92. Int. Soc. Plant Morphol., Delhi.

Sunderland, N. 1973. Pollen and anther culture. *In* H. E. Street (ed.), Plant tissue and cell culture. Pp. 205–39. University of California Press, Berkeley.

Sunderland, N. 1974. Anther culture as a means of haploid induction. *In* K. J. Kasha (ed.), Haploids in higher plants. Advances and potential. Pp. 91–122. 1st Intern. Symp. Guelph. University of Guelph Press, Guelph, Ontario, Canada.

Tio, M. A. 1962. Effect of light intensity on the rate of apparent photosynthesis in coffee leaves. J. Agric. Univ. Puerto Rico. 46(3):159–66.

Townsley, P. M. 1974. Production of coffee from plant cell suspension cultures. J. Inst. Canad. Sci. Technol. Aliment. 7:79–81.

Valio, I. F. M. 1976. Germination of coffee seeds (*Coffea arabica* L. cv. Mundo Novo). J. Exp. Bot. 27(100):983–91.

van der Plank, J. E. 1968. Disease resistance in plants. Academic Press, New York.

van der Veen, R. 1968. Plant hormones and flowering in coffee. Acta Bot. Neerl. 17:373–76.

van der Voort, F., and P. M. Townsley. 1975. A comparison of the unsaponificable lipids isolated from coffee cell cultures and from green-coffee beans. J. Inst. Canad. Sci. Technol. Aliment. 8(4):199–201.

van der Vossen, H. A. M., R. T. A. Cook, and G. W. Murakaru. 1976. Breeding for resistance to coffee berry disease caused by *Colletrotrichum coffeanum* Noack (sensu Hindorf) in *Coffea arabica* L. I. Methods of preselection for resistance. Euphytica 25:733–45.

Walyaro, D. J., and A. M. Van der Vossen. 1977. Pollen longevity and artificial cross-pollination in *Coffea arabica* L. Euphytica 26:225–31.

Wardell, W. L., and F. Skoog. 1969. Flower formation in excised tobacco stem segments. II. Reversible removal of IAA inhibition by RNA base analogs. Plant Physiol. 44:1407–12.

Webster, P. L., and D. Davidson. 1968. Evidence from thymidine-^{3}H-labeled meristems of Vicia faba of two cell populations. J. Cell Biol. 39:332–38.

Went, F. W. 1957. The experimental control of plant growth. Chronica Botanica Co., Waltham, Mass.

Wheeler, H., and H. H. Luke. 1955. Mass screening for disease resistant mutants in oats. Science 128:1229.

Wheeler, H., A. S. Williams, and L. D. Young. 1971. *Helminthosporium maydis* t-toxin as an indicator of resistance to southern corn leaf blight. Plant Dis. Report. 55:667.

Williams, P. H. 1975. Genetics of resistance in plants. Genetics 79:409–19.

Zelitch, I. 1971. Photosynthesis, photorespiration, and plant productivity. Academic Press, New York.

Zelitch, I. 1975. Improving the efficiency of photosynthesis. Science 188:626–33.

Use of Tissue and Cell Culture Methods in Tobacco Improvement

29

Scientists having the responsibility of improving economic traits of tobacco (*Nicotiana tabacum* L.) are fortunate to have a wealth of fundamental scientific information derived from repeated use of tobacco as a model research system. Many early basic genetic, cytogenetic, and evolutionary principles were established from studies involving the genus *Nicotiana* (East, 1928; Goodspeed, 1954). The genus has also been successfully used in a wide range of *in vitro* tissue and cell culture investigations. For example, totipotency and the regeneration of plants from single cells in culture were demonstrated with tobacco (Vasil and Hildebrandt, 1965). Tobacco was also used extensively in the development of single cell cloning and suspension culture techniques (Bergmann, 1960). Although the first successful production of haploid embryos by *in vitro* culture of excised anthers was with *Datura innoxia* (Guha and Maheshwari, 1964), a large proportion of the subsequent anther culture efforts and successes have been recorded with *Nicotiana* species (Vasil and Nitsch, 1975). Finally, protoplasts were suitably prepared and fused to produce the first somatic or parasexual hybrid from two *Nicotiana* species (Carlson et al., 1972).

Tobacco is among the few plant species that can be classified as relatively easy to handle in an *in vitro* culture system. Organized plants can be obtained from either callus or cell suspension cultures of *N. tabacum* by manipulation of certain constituents of the culture medium and/or the

G. B. Collins and Paul D. Legg, Department of Agronomy, University of Kentucky, Lexington, Kentucky 40506.

cultural conditions. Isolated tobacco protoplasts can be stimulated to divide and plants can be recovered via an intermediate callus stage. Anther cultures of immature pollen grains yield large numbers of haploid plants. These haploid plants provide a valuable source of haploid cells and tissues for various types of *in vitro* studies, genetic experiments, selection studies, and the development of improved breeding lines. Tobacco was judged to be a promising species from a tissue culture standpoint but less desirable from a genetic point of view when Day et al. (1977) recently reviewed the area of plant manipulation via somatic cell genetic techniques.

This review will present an appraisal of the more important and more useful applications of cell and tissue culture methods in research efforts aimed at the improvement of tobacco as an economic crop species. Major emphasis will be given to the use of haploids and their derivatives obtained from anther or pollen culture as these materials have been used in tobacco genetics and breeding programs. Other tissue and cell culture techniques will be discussed to the extent that their applications have provided experimental materials useful toward tobacco improvement.

USES OF ANTHER-DERIVED HAPLOIDS

Haploid plants are obtained routinely by culturing anthers or isolated pollen from tobacco on a suitable, chemically defined culture medium. The most recent reviews of the techniques for successful anther and pollen culture have been provided by Reinert and Bajaj (1977) and Nitsch (1977), respectively. In previous review articles, we have enumerated some of the uses of anther-derived haploids and their derivatives in plant improvement (Collins, 1976; Collins and Legg, 1975).

In tobacco genetics and breeding efforts, haploids and their derivatives have been used rather extensively to date. In the following sections, we will deal with genetic analyses with haploids, selection at the haploid level, characteristics of doubled haploids, and breeding and varietal development using haploids.

Genetic Analyses with Haploids

When haploids and their derivatives are used to study genetic segregation for qualitative characters, the expected ratios are simpler because one is actually working with gametic ratios. For example, a monohybrid F_2 phenotypic ratio becomes a 1:1 ratio instead of the classical 3:1 ratio, and the F_2 ratio for duplicate factor inheritance becomes 3:1 instead of 15:1.

Several investigators have conducted inheritance studies with haploids or doubled haploids of tobacco. Nakamura et al. (1974c) used haploids to confirm that leaf shape in flue-cured tobacco is controlled by two genes.

From a cross between Hicks 2 and Coker 139, four types of leaf shape were distinguished in haploids and diploids. Thus, the genotype of the F_1 was identified as *PtptPdpd*, and would yield haploid plants with the genotypes *PtPd*, *Ptpd*, *ptPd*, and *ptpd*.

Three different plant color traits have been investigated using haploids. From a cross of Hicks 2 with a line bearing the *Ws* gene from *N. plumbaginifolia*, Nakata (1971) established the genotype of the F_1 to be $Ws_1ws_1 Ws_2ws_2$ by showing a 3 green to 1 albino segregation among haploid plants from the F_1. Ternovskii et al. (1975) studied the inheritance of a white color mutant by extracting haploids from a normal green line named Dubek 44F, a white color mutant named Belolist 600, and the F_1 hybrid. All haploids from Dubek 44F were normal green, haploids from Belolist 600 were all pale yellow, and 97 haploids from the hybrid were classified as 47 pale yellow and 50 normal green. The production of two classes of gametes with equal frequency from the hybrid showed that the character was controlled by a single gene. For the white burley character, Nakata and Kurihara (1972) verified the control by two genes when they observed a ratio of 3 green to 1 white in haploids from the F_1 hybrid between a normal green line, NFT 700, and Burley 21.

Segregation in haploids has been used also to study the disease reactions of tobacco to black shank, wildfire, and the tobacco mosaic virus. In five crosses between resistant and susceptible varieties, Nakata and Kurihara (1972) found haploids from the F_1 hybrids to segregate in a ratio of 1 resistant to 1 susceptible for both wildfire and tobacco mosaic. Collins et al. (1971) obtained haploids from the F_1 hybrids of L8 × Kentucky 16 and L8 × Burley 21. L8 is a breeding line with resistance to the common or race 0 of black shank, and the two varieties, Kentucky 16 and Burley 21, are susceptible to black shank. When haploids from the hybrids were tested with the race 0 black shank organism, significant deviations from the expected 1:1 ratio were observed with more haploids being susceptible than resistant. A larger number of susceptible to resistant progenies than expected was also observed in F_2 and backcross populations from crosses between L8 and susceptible varieties. The authors concluded that gametes carrying the resistance factor have an impaired function and do not always survive.

Haploids and doubled haploids of tobacco have been characterized for several quantitative traits such as vigor, yield, plant height, and leaf size; but they have not been used extensively in developing genetic information on multiple factor characters. Legg and Collins (1974) used biparental progenies from noninbred and doubled-haploid plants to estimate genetic variances for quantitative traits in a random-intercrossed population of burley tobacco. By using plants with different degrees of inbreeding, it was possible to obtain more information on genetic variances and functions of genetic variances. Additive genetic variances were important in the population for days to flower, plant height, leaf length, and number of leaves. Additive × additive epistatic variances were found for yield and total alkaloids. All estimates of dominance variance were nonsignificant.

Selection at the Haploid Level

Haploids are generally less vigorous and have smaller morphological features than diploids. Nakamura and Itagaki (1973) found haploids of *N. tabacum, N. otophora, N. paniculata, N. glutinosa*, and *N. rustica* to be about 1/4 to 1/2 smaller than the diploid counterparts in size of the plants, leaves, and flowers. Also, haploid plants had lighter-colored flowers and narrower leaves than the diploids. Burk et al. (1972) and Nakamura et al. (1974c) found that haploid plants flowered earlier than diploids. In chemical components, Izman et al. (1975) found the qualitative composition of alkaloids and volatile acids of haploids to be the same as that in the original variety. The quantitative content of the principal alkaloids and volatile acids in the haploids was appreciably lower in the haploids in comparison to the diploids. Such a reduction in alkaloid content appears to be characteristic of haploids and doubled haploids from standard varieties or hybrids, but does not relate to the ploidy level specifically. When Collins et al. (1974) compared haploid progenies derived from doubled haploids with the parental doubled haploids, alkaloid levels were remarkably similar (table 1). The aaBB line 58-1 was an exception complicated by the observation that the alkaloid content was higher in the haploid form at the first sampling, whereas the reverse situation was revealed at the second sampling.

TABLE 1

COMPARISON OF TOTAL ALKALOID CONTENT IN DOUBLED HAPLOID LINES AND IN HAPLOID PLANTS EXTRACTED FROM THE DOUBLED HAPLOIDS

Genotype	Pedigree	FLOWERING STAGE		4 WEEKS AFTER FLOWERING	
		# Plants	% Total Alkaloids	# Plants	% Total Alkaloids
aabb	50-1	16	0.08	10	0.22
ab	50-1	36	0.07	10	0.35
aaBB	58-1	25	0.41	10	1.89
aB	58-1	70	0.60	10	1.14
AAbb	26-2	19	1.03	10	2.51
Ab	26-2	50	1.09	10	2.63
AAbb	38-5	23	1.07	10	2.58
Ab	38-5	40	1.06	10	2.58

SOURCE: Collins et al., 1974. Adapted from Crop Science, Volume 14, pages 77–80, 1974, by permission of the Crop Science Society of America.

Although haploid plants are smaller and less vigorous than diploids, they can be evaluated in selection experiments and they will endure manipulations required to return them to the diploid condition. Thus, selection among haploids should be effective and advantageous for traits that have the same qualitative expression in haploids as in diploids. This would appear to be true for most qualitatively inherited traits, since expected phenotypes have been observed in populations of haploids following crosses between genetically different parents (Burk, 1970; Nakata, 1971; Nakata and Kurihara, 1972; Collins et al., 1971; Collins et al., 1974; Ternovskii et al., 1975).

Five cases are cited where selection among haploids has been conducted for qualitative traits. Mikoyan (1975) has initiated a program to select economically valuable variants in haploids derived from F_1 hybrids on the basis of morphological, chemical, and disease-resistant traits. Nakamura et al. (1974c) were able to derive four distinct leaf shape diploids by doubling selected haploids from the cross between Hicks 2 and Coker 139. Collins et al. (1974) established true breeding lines from the cross of a low alkaloid line, LA Burley 21, and Burley 21. Earlier genetic studies (Legg et al., 1969; Legg and Collins, 1971; Legg et al., 1970) had identified the genotypes of LA Burley 21 and Burley 21 to be *aabb* and *AABB*, respectively. From a total of 234 haploids from the F_1 hybrid, five were chosen to represent each of the four genotypic classes. Doubling the chromosome number of these selected haploids produced plants in each class that were comparable to conventionally derived lines. In a program designed to improve the variety MC 1610, Nakamura et al. (1974b,c) were able to reduce a population of 2016 haploids to 539 on the basis of selection for leaf size and plant type.

Nakamura et al. (1974c) conducted experiments to examine the efficiency of selection of haploid plants for certain qualitative traits including reaction to certain diseases, leaf shape, and alkaloid content. Screening was done with black shank, black root rot, powdery mildew, and wildfire. High correlations were found between reaction of haploids and diploids to each of the four diseases. It was concluded that screening for disease resistances in haploid plants is generally as effective as in diploid plants. Also, heritability values and haploid-diploid correlations for leaf shape and alkaloids were high enough to expect effective selection of haploid plants for these characters.

For quantitatively inherited traits, selection among haploid plants is complicated by the influence of environmental factors. Historically, plant breeders have had limited success in single-plant selection for complex characters. On the basis of a heritability study and diploid-haploid correlations, Nakamura et al. (1974c) concluded that selection among haploid plants would be relatively inefficient for several quantitative traits.

Low correlations between haploids and diploids for total nitrogen content, total soluble sugar level, total leaf number, and days to flower would make selection among haploids ineffective. Heritabilities for plant height and leaf length were low among haploids and would limit selection progress.

Characteristics of Doubled Haploids

Several experiments have been conducted that provide information on the vigor, uniformity, and general performance of doubled haploids relative to other diploid materials. Doubled haploid lines have been extremely uniform in most studies, but their vigor and other characteristics seem to vary with the material and the environmental conditions. Uniformity within doubled haploid lines has been reported for oriental varieties by Devreux and Laneri (1974); for flue-cured varieties by Burk et al. (1972), Burk and Matzinger (1976), and Nakamura et al. (1975); and for burley varieties by Collins et al. (1974). Also, Ternovskii et al. (1975) found restituted diploid lines produced from haploid tissue *in vitro* to be statistically more uniform under field conditions than the original inbred parent. Devreux and Saccardo (1972) observed significant differences among two isogenic lines for leaf and flower characteristics, but progeny of a given doubled haploid was uniform. An exception to higher uniformity among doubled haploid plants than among plants from other diploids was reported by Nakamura et al. (1974a,c). In a replicated test with 14 doubled haploid lines and selfed progeny of Hicks 2 and Coker 139, they found comparable variation among plants of the doubled haploid lines and the standard varieties.

Doubled haploid lines obtained from the same population, hybrid, or variety have shown considerable line-to-line variation, and they have been statistically different from the parental source in most cases. In a study of four alkaloid genotypes derived by doubling haploids from the LA Burley 21 × Burley 21 cross, Collins et al. (1973) found genotypic differences for six agronomic traits (table 2). When doubled haploid and conventionally derived lines of comparable alkaloid genotypes were compared, differences were generally statistically nonsignificant for the aabb, aaBB, and AAbb genotypes. For the AABB genotype, which has the highest level of alkaloids, the doubled haploid line flowered 3 1/2 days later, had leaves that were 3.5 cm shorter and 2.3 cm narrower, and yielded 20 grams per plant less than the conventionally derived line.

In a second experiment, Collins et al. (1973) compared the commercial variety, Burley 21, and two generations of doubled haploid lines from Burley 21. Except for leaf length and percentage of total alkaloids, large and significant differences were observed between Burley 21 and the two

TABLE 2

MEAN PERFORMANCE OF HAPLOID-DERIVED LINES FROM THE CROSS LA
BURLEY 21 × BURLEY 21 FOR SIX AGRONOMIC TRAITS

CHARACTER	ALKALOID GENOTYPE				TEST OF DIFFERENCES
	aabb	aaBB	AAbb	AABB	
Days to flower	71.9	69.1	71.0	70.1	**
Plant height (cm)	133.6	134.7	135.3	141.8	**
No. of leaves	21.0	21.2	21.3	21.6	*
Leaf length (cm)	65.7	64.7	64.7	64.9	**
Leaf width (cm)	28.3	29.8	30.4	30.1	**
Yield per plant (g)	147.8	159.2	157.2	163.6	**

*, ** Significant difference in means at P=.05 and .01, respectively.
SOURCE: Collins et al., 1973; reprinted with permission.

TABLE 3

COMPARISON OF BURLEY 21 AND DOUBLED HAPLOID LINES FROM BURLEY 21

Character	Burley 21	Doubled haploid Burley 21 lines		Burley 21 vs ($^1/_2$) (G_1+G_2)	G_1 lines	G_1 vs G_2
		G_1	G_2			
Days to flower	62.9	70.9	70.6	**	ns	ns
Plant height (cm)	148.0	137.6	135.8	**	ns	ns
No. of leaves	22.2	20.7	20.8	ns	ns	ns
Leaf length (cm)	65.6	65.3	64.1	**	ns	ns
Leaf width (cm)	27.7	29.4	29.3	**	ns	ns
Yield per plant (g)	182.6	164.5	166.0	**	ns	ns
Total alkaloids (%)	4.5	4.2	4.3	ns	ns	ns

** Significant difference in means at P=.01.
ns = Nonsignificant difference in means at P=.05; G_1—generation 1, G_2—generation 2.
SOURCE: Collins et al., 1973; reprinted with permission.

haploid-derived generations (table 3). All comparisons between the two generations of doubled haploids and among generation one lines were nonsignificant. A comparison of G_2 lines from the same G_1 showed that sister lines from G_1 line 4 differed for all six traits (table 4). Other comparisons were generally nonsignificant, and the authors concluded that a mutation or some cytological change had occurred in one of the G_2 lines derived from G_1 line 4.

Collins and Legg (1974) reported the results of evaluating 73 doubled haploid lines from a random-intercrossed population of burley tobacco. The population was developed by combining eight cultivars—Burley 1, Burley 21, Burley 37, Virginia B-29, Ky 10, Ky 12, Ky 16, and Ky 41A—into a single population by making two double crosses and then pollinating 100 plants from one double cross with a bulk of pollen from 100 plants of the other double cross. The resulting population was advanced two generations by pollinating 100 plants with a bulk sample of pollen from the same plants. In this third generation of intercrossing, a doubled haploid line was developed from each of 73 plants. These lines were evaluated in a replicated

TABLE 4

Comparison of Five Pairs of Generation 2 (G_2) Haploid-Derived
Lines for Six Agronomic Traits

	Days to Flower	Plant Height (cm)	No. Leaves	Leaf Length (cm)	Leaf Width (cm)	Yield/Plant (g)
Line 1						
Line 1	69.33	139.10	21.40	63.97	29.68	177.52
Line 2	68.83	138.43	21.13	66.78*	29.68	174.97
Line 2						
Line 1	72.33	128.92	20.68	64.23	29.55	160.62
Line 2	69.83*	138.22*	20.82	63.82	28.85	154.75
Line 3						
Line 1	70.17	135.88	20.93	65.03	29.47	175.13
Line 2	71.00	135.90	20.63	63.92	30.02	168.85
Line 4						
Line 1	74.00	123.77	19.12	61.30	27.87	147.067
Line 2	70.83*	135.75*	20.98*	63.82*	29.57*	167.65*
Line 5						
Line 1	70.33	141.42	21.20	63.25	28.65	166.50
Line 2	69.00	140.62	20.92	64.75	29.45	170.52
LSD .05	1.45	4.31	0.92	2.33	0.98	9.50

* Significant difference in means at P=.05.
Source: Collins et al., 1973; reprinted with permission.

TABLE 5

Performance of Doubled Haploid Lines from a Random-Intercrossed
Population of Burley Tobacco and of Six of the Eight Parental Cultivars

	Days to Flower	Plant Height (cm)	Leaf Length (cm)	Leaf Width (cm)	No. of Leaves	Yield Plant (g)	Total Alkaloids (%)
Doubled haploids							
Mean	71.9	147.6	72.1	33.8	20.0	188.0	4.1
Range	64.0–	140.1–	57.2–	27.0–	17.4–	111.7–	2.2–
	80.0	155.6	80.4	41.2	21.9	237.7	5.4
Burley 21	68.8	148.9	69.5	34.5	21.5	196.4	4.6
Burley 37	68.5	149.7	68.4	35.8	21.0	198.7	4.8
Va B-29	70.0	148.1	73.3	34.8	21.6	215.5	5.2
Ky 10	77.2	137.8	71.6	35.8	21.0	224.6	5.0
Ky 12	75.0	148.7	71.6	33.4	22.2	209.9	4.0
Ky 41A	70.0	140.9	73.4	37.7	21.2	208.5	3.5

Source: Collins and Legg, 1974; reprinted with permission.

test, and the results are given in table 5. In most all characters the range of values obtained exceeded the range found in six of the parental cultivars.

In Japan, Oinuma and Yoshida (1974) evaluated parental plants of Burley 21, Ky 10, Harrow Velvet, and three doubled haploid lines from each of the three varieties. Growth characters measured included flowering time, yield, and total alkaloid content (table 6). Analysis of variance indicated that the difference between entries was statistically significant in 19 out of 27 cases. Plant height, stem diameter, and leaf size had significant differences between entries in all three varieties. In the Ky 10 group, doubled haploid lines DH-4 and DH-6 flowered three to four days later and had four to five more leaves than the normal diploid line (ND). The yield of doubled haploid lines was comparable to the parental varieties except for DH-1, DH-4, and DH-7, which were lower than the varieties. Further testing in 1972 showed the reduced yield of DH-1 and DH-4 and the additional leaves on DH-6.

Also, Nakata (1971) found that a doubled haploid line of Shiroenshu, a burley-type variety, was not as vigorous as its parental variety, and the vigor was not recovered through repeated selfing.

Flue-cured tobacco doubled haploids from a given source have been characterized as being different and in most cases less vigorous than the parental material. Burk et al. (1972) evaluated 86 doubled haploids from the F_1 hybrid between LN 38 and NC 95. In yield, half of the lines were numerically lower than the lower parent, LN 38, and six numerically exceeded the higher parent, NC 95. There were 12 lines with alkaloid levels numerically above the 3.52 percent level of NC 95 and 13 lines below the 0.33 percent of LN 38. The authors felt that these extremes might be greater than one would observe in a normal F_2 population following a cross of two pure lines. After studying 150 doubled haploids from crosses of Hicks 2 × Coker 139, MC 1610 × Coker 139, and MC 1610 × Coker 254, Nakamura et al. (1975) seemed to reach the same conclusion. They reported that some doubled haploids had significantly lower yields or higher alkaloid content than the parental varieties, but many other lines had segregation patterns as expected in diploid hybrids. In another experiment, Nakamura et al. (1974c) compared the distribution of 77 doubled haploids from the F_1 hybrid of Hicks 2 × Coker 139 with the normal F_2 population. The distributions were comparable in days to flowering, plant height, leaf width, and leaf length. In leaf number and alkaloid content, doubled haploid lines were skewed toward large numbers and higher contents.

Experiments designed to compare the vigor of doubled haploids in flue-cured tobacco with standard varieties have been conducted by Kadotani and Kubo (1969), Nakamura et al. (1974c), and Burk and Matzinger (1976). Kadotani and Kubo (1969) found a doubled haploid line from the

TABLE 6

CHARACTERISTICS OF NORMAL DIPLOID AND DOUBLED HAPLOID LINES IN 1971

Variety	Line	Days to Flower	Plant Height (cm)	Stem Diameter (cm)	Total Leaf Number	Leaf Length (cm)	Leaf Width (cm)	Leaf Shape Index	Yield (kg/10a)	Total Alkaloids (%)
Burley 21	ND	74.3 a	158.3 c	3.29 b	39.7 a	74.3 c	32.5 b	0.438 a	232.4 b	2.44 a
	DH-1	73.5 a	136.4 a	3.13 a	41.5 a	69.2 a	29.7 a	0.430 a	199.2 a	2.72 a
	DH-2	73.2 a	144.1 b	3.29 b	39.7 a	72.4 b	32.4 b	0.447 a	240.5 b	2.35 a
	DH-3	74.2 a	155.6 c	3.35 b	39.7 a	75.3 c	33.4 b	0.444 a	243.0 b	2.25 a
Ky 10	ND	74.8 a	153.5 b	3.30 b	41.6 a	69.4 b	35.4 c	0.510 a	251.1 bc	2.41 b
	DH-4	80.3 b	135.9 a	3.13 a	46.7 b	64.8 a	32.6 a	0.528 a	230.6 a	2.02 a
	DH-5	77.8 ab	141.0 a	3.13 a	43.2 a	66.2 b	33.7 ab	0.510 a	237.4 ab	2.10 a
	DH-6	79.1 b	154.6 b	3.26 b	46.0 b	68.4 b	34.3 bc	0.501 a	259.7 c	1.95 a
Harrow Velvet	ND	71.8 a	133.8 b	2.93 b	43.7 a	60.8 b	28.4 b	0.467 a	201.1 b	2.15 ab
	DH-7	76.9 a	117.5 a	2.61 a	47.3 a	53.7 a	24.9 a	0.464 a	172.9 a	2.35 b
	DH-8	73.6 a	130.2 b	2.83 b	45.4 a	59.1 b	27.6 b	0.467 a	202.0 b	2.21 ab
	DH-9	73.6 a	129.1 b	2.82 b	45.4 a	60.2 b	27.8 b	0.463 a	206.7 b	1.94 a

ND = Normal diploid line; DH = Doubled haploid line.

Values within a column followed by the same letter are not significantly different at the 5% level.

SOURCE: Oinuma and Yoshida, 1974; reprinted with permission.

variety Bright Yellow to have similar flowers and narrower leaves than the normal diploid plants. Nakamura et al. (1974) compared doubled haploid lines and selfed progenies from single plants of Hicks 2 and Coker 139. The data given in table 7 indicate that the doubled haploid lines had almost the same mean values as their respective original varieties in all characters except number of leaves and total alkaloid content.

In a second test, Nakamura et al. (1974c) evaluated selfed progenies (S_2 and S_4 generations) of four doubled haploid lines and a doubled haploid line extracted from a doubled haploid. As shown in table 8, the S_2 and S_4 generations were generally comparable, and the selfed progenies had similar differences among them as did their parental doubled haploid lines. The doubled haploid from the doubled haploid, DH-2, differed from DH-2 in several traits.

Burk and Matzinger (1976) tested the performance of doubled haploids from an S_{15} line of Coker 139. The results presented in tables 9 and 10 include differences in the doubled haploids and between doubled haploids and Coker 139. Growth characteristics were generally reduced in the doubled haploids. For plant height, 45 or 46 doubled haploids were significantly shorter than Coker 139. More than 50 percent of the doubled haploids had fewer and narrower leaves, and flowered earlier than Coker 139. Yield and leaf length were significantly reduced in 18 of the 46 doubled haploids. Only six doubled haploids were significantly lower than Coker in grade index. The mean quantity of total alkaloids for doubled haploids was higher than for Coker 139, with 18 progenies exceeding, and none less than, Coker 139. Among the doubled haploid families, 6 exceeded the Coker 139 control for reducing sugars, but 40 did not differ significantly.

The question as to whether doubled haploids are less vigorous than completely homozygous lines derived by other methods is unresolved. Also, the presence of statistically significant differences among doubled haploids from the same inbred variety has not been adequately investigated. First, it is questionable whether lines developed by conventional inbreeding are completely homozygous. Second, the limited number of studies have produced inconsistent results.

Since doubled haploids are genetically pure, their sensitivity to environmental conditions might be higher than conventionally derived varieties. After testing seven doubled haploid lines and standard varieties at three locations in Japan, Nakamura et al. (1974c) concluded that the two types of material did not differ in response to environments. Oka et al. (1976) also found ten doubled haploid lines to have about the same agronomic adaptability over four different environments as seven ordinary varieties.

TABLE 7
CHARACTER MEANS OF DOUBLED HAPLOID LINES

Varieties	Lines	Days to Flower	Plant Height (cm)	Total No. of Leaves	Stem Diameter (cm)	Leaf Length (cm)	Leaf Width (cm)	Yield per Plant (g)	Total Alkaloids (%)
Hicks 2	DH-1	54.5 a	132.7 b	27.0 a	2.52 ab	50.4 ab	22.8 ab	60.8	2.97 abc
	DH-2	56.0 abc	123.4 ab	28.2 bc	2.63 b	49.1 ab	22.1 ab	56.6	3.22 cd
	DH-3	56.5 bc	137.2 b	28.7 bc	2.56 ab	50.4 ab	22.6 ab	66.9	2.82 abc
	DH-4	56.5 bc	117.9 a	28.4 bc	2.42 ab	47.8 a	21.3 a	50.5	3.46 d
	DH-5	57.3 cd	130.0 ab	29.0 c	2.34 a	47.6 a	21.4 a	50.6	2.81 abc
	DH-6	55.7 abc	131.4 ab	27.7 ab	2.58 ab	51.9 b	23.7 ab	63.6	2.73 ab
	DH-7	58.3 cd	131.5 ab	27.7 ab	2.50 ab	49.0 ab	24.3 b	60.7	3.14 bcd
	Average	56.4	129.1	28.1	2.51	49.5	22.6	59.2	3.02
	ND	55.5 ab	129.8 ab	27.1 a	2.50 ab	50.1 ab	22.5 ab	57.1	2.64 a
Coker 139	DH-8	64.1 a	118.8 abc	35.6 c	2.71 b	51.6 c	24.6 b	80.8 b	1.08 ab
	DH-9	64.7 a	120.5 bc	33.7 ab	2.63 b	51.4 bc	26.6 b	84.3 b	1.28 bc
	DH-10	67.9 b	110.9 a	37.1 d	2.42 a	48.3 ab	22.0 a	76.2 b	0.95 a
	DH-11	63.7 a	113.5 ab	32.8 a	2.52 ab	48.1 a	25.1 b	76.8 b	1.50 c
	DH-12	64.1 a	121.0 bc	34.4 b	2.60 ab	48.2 a	25.2 b	79.3 b	1.31 bc
	DH-13	64.4 ab	117.2 abc	34.8 bc	2.62ab	49.6 abc	25.3 b	83.7 b	1.12 ab
	DH-14	68.4 b	122.5 c	37.6 d	2.55 ab	49.6 abc	22.1 a	64.9 a	0.95 a
	Average	65.3	117.8	35.1	2.58	49.5	24.4	78.0	1.17
	ND	64.2 a	114.5 abc	34.0 b	2.51 ab	49.4 abc	24.5 b	81.7 b	1.01 a

Two values with different letters in the same column differ at 5% level after Duncan's multiple range test.

DH = Doubled haploid; ND = Normal diploid

SOURCE: Nakamura et al., 1974c; reprinted with permission.

TABLE 8

CHARACTER MEANS FOR PROGENIES OF DOUBLED HAPLOID LINES

Progeny	Generation	Line	Days to Flower	Plant Height (cm)	Number of Leaves	Leaf Length (cm)	Leaf Width (cm)	Total Alkaloids (%)
Selfed Progeny	S_2	DH-2	66.9 c	146.4 b	17.6 c	54.8	24.8 bc	4.73 abc
		DH-4	66.7 c	141.0 b	17.8 c	52.4	23.5 ab	5.44 cd
		DH-7	68.6 d	148.4 bc	17.2 bc	54.1	25.2 c	4.72 abc
	S_4	DH-2	66.1 bc	144.1 b	17.3 bc	54.7	24.6 abc	4.29 ab
		DH-4	66.8 c	138.5 b	17.4 bc	52.3	23.3 a	4.98 bc
		DH-7	68.5 d	147.2 bc	16.8 b	52.8	24.8 bc	4.64 abc
	Haploid-derived progeny DH-2		62.8 a	117.0 a	13.8 a	52.8	25.0 c	6.02 d
	Normal diploid		65.0 b	156.0 c	17.0 b	54.4	25.3 c	4.13 a
Selfed Progeny	S_2	DH-10	73.1	146.8	24.0 b	57.4	26.8 a	2.04
	S_4	DH-10	73.1	143.9	23.3 b	57.2	26.8 a	1.90
	Normal diploid		72.1	145.2	22.2 a	56.4	28.2 b	1.79

Two values with different letters in the same column differ at 5% level after Duncan's multiple range test.

SOURCE: Nakamura et al., 1974c; reprinted with permission.

Breeding and Varietal Development Using Haploids

Melchers and Labib (1970) were among the first to enumerate possible uses of anther-derived haploids in plant breeding. Their suggestions included induced mutagenesis, determination of genetic ratios, and the development of breeding lines with specific characteristics such as the combination of several dominant genes. From a practical standpoint, the tobacco breeder can immediately expect to make use of haploids and their derivatives mainly as a time-saving technique.

The breeder must deal with two kinds of traits, namely, those controlled by one or a few major genes or qualitative traits and those controlled by many genes with small effects or quantitative traits. The breeding objective with a qualitative trait is usually to incorporate one or a few desirable traits into an existing variety or breeding line. The backcross method is most often selected for this transfer and involves the successful completion of a series of backcrosses to the recurrent parent with maintenance of the desirable trait(s) by selection. Selfing is required when backcrossing is completed to stabilize the transferred gene(s) in the homozygous form. Six backcrosses and six to eight years are generally required to transfer a trait by the backcross method. At least three distinct potential advantages of the doubled haploid method can be identified. These include a reduction in the time required to complete the program, an increased probability of retaining the character under transfer, and more rapid stabilization of the transferred genetic material in homozygous form at the end of the backcross program. The time-saving and ease of manipulation features of using the haploid breeding method were demonstrated in the experiments carried out by Collins et al. (1974) in which four homozygous lines with different total alkaloid contents were produced from an F_1 hybrid that was heterozygous for two major loci controlling alkaloid content.

In breeding for quantitatively inherited traits, the doubled haploids enable the tobacco breeder to fix the genetic system of individual gametes in a testable, reproducible form. Again, a major advantage of the haploid method will be to reduce the time and effort required to generate reproducible genotypes for performance testing. The tobacco breeder must either identify or develop a genetically variable population whether he plans to use a conventional approach or the haploid method. The source populations might be a released variety or line, a random-intercrossed population, or a segregating generation of a cross or crosses involving two or more lines.

Assuming that the breeder will use a segregating generation of a population produced from a cross of genetically diverse lines or by random intercrossing, there are several decisions to be made as a part of the haploid method. Of course, the source population must contain the genetic variation needed for improvement. The first decision is one of when to sample the population via extracted haploids. The essential question is simply one of which generation will contain a reasonable number of recombinant gametes that will produce plants superior to the parental lines. Another decision to be made involves selection of a sampling method and determining how many haploid plants to obtain. Gene frequencies in the population and the success of getting desirable recombinations will greatly influence the number of haploids required. We have attempted to obtain a large number of doubled haploids and evaluate them in small nonreplicated plots with selection of the better lines for more extensive evaluation. This approach permits one to maximize the sampling of the genetic variability in the population and also to eliminate a considerable number of undesirable genotypes prior to critical evaluation tests.

The theoretical advantages of doubled-haploid over diploid selection methods are that (1) the variance on which selection operates, σ^2_A (additive genetic variance), is twice as much in doubled-haploid populations as in random-mated diploid populations, and (2) the dominance variance is eliminated from the doubled-haploid phenotypic variance for individuals and clones. Nei (1963) compared the probability of obtaining the desirable genotype in the F_2 by the haploid and diploid methods. Under the assumptions of no selection during haploid development, no linkage, and a high frequency of haploid development and doubling, Nei (1963) found the efficiency of the haploid method over the diploid method for self-fertilizing plants to be about 2^n when n is the number of segregating factors. Griffing (1975) considered the use of haploids or doubled haploids in a recurrent selection program. Statistical comparisons of the efficiency of selection made on haploids or doubled haploids versus selection made on regular diploids had considerable advantage (up to approximately six times as efficient) for the haploid procedures regardless of plant numbers. Some of the advantages disappear if long periods of time are required to extract haploids or obtain doubled plants from haploids. However, culturing and doubling times should not be a major consideration in using doubled haploids of tobacco.

Another method of using doubled haploids in breeding work might be to use them to divide a population into nonrandom groups. Griffing (1976) has reported that such a separation of populations will increse the efficiency of truncation selection. One way of forming nonrandom groups in a random-mating, heterozygous population would be to convert the population into a population of homozygous genotypes through extracting and doubling random haploids and forming nonrandom groups of the homozygous material.

Programs designed to develop new varieties for commercial production have resulted in one new Chinese variety (Cooperative Group of Haploid Breeding of Tobacco of Shangtung Institute of Tobacco and Peking Institute of Botany, Academia Sinica, 1974) and three improved breeding lines (Nakamura, 1974b,c). The Japanese program was undertaken to improve MC 1610, a leading variety developed from the cross between Hicks 2 and Coker 139. Anthers from the crosses Hicks 2 × Coker 139, MC 1610 × Coker 139, and MC 1610 × Coker 254 were used to produce 2,016 haploids. At the flowering stage, 539 haploids were selected on the basis of leaf shape and desirable plant type. Immersion of the inflorescence of these selected plants into 0.1 percent aqueous colchicine solution produced diploid shoots on 152 of the plants. These 152 doubled haploids were performance tested in 1972 for agronomic traits, chemical characters, and resistance to bacterial wilt, black shank, and black root rot. Further field testing was done on 19 selected lines in 1973 at three locations. Three lines that resembled MC 1610 in agronomic and chemical characteristics but had improved disease resistance were selected as promising breeding lines. The three lines produced cured leaves that were normally rich lemon to orange in color and medium in body with acceptable texture. All the lines were normal and mild in smoking quality.

Nakamura et al. (1974c) also reported that Oinuma had obtained a promising line from the burley tobacco variety Ky 10 by anther culture. This line had about 5 more leaves and lighter leaf color than Ky 10.

INTERSPECIFIC HYBRIDIZATION

Fusion of Protoplasts

The availability of large numbers of plant cell protoplasts ranks as a major contribution in the application of cell and tissue culture methods in crop improvement (Cocking, 1972). The utility of protoplasts in various types of studies has been discussed by many investigators, including those with specific reference to agricultue (Bajaj, 1974; Bottino, 1975). An up-to-date summary regarding protoplast hybridization has been provided by Bajaj (1977).

Interspecific hybridization has been involved extensively in the evolution of the genus *Nicotiana* (Goodspeed, 1954). Tobacco breeders have relied heavily on interspecific transfer of genetic material for such traits as disease resistance, insect resistance, chemical composition, male sterility, and plant morphological features in breeding improved varieties of tobacco (Chaplin and Burk, 1970). Many desirable species crosses in *Nicotiana* have not been possible with conventional crossing techniques. Somatic hybridization of protoplasts from *Nicotiana* species possessing traits that

TABLE 9

MEANS FOR MEASURED CHARACTERISTICS OF COKER 139 TOBACCO SOURCE VARIETY
DIHAPLOID PROGENY, AND RANGE OF DIHAPLOID PROGENY

	Yield kg/ha	Grade Index	Flower Days	Plant Height (cm)	No. of Leaves	Leaf Length (cm)	Leaf Width (cm)	Total Alkaloids (%)	Reducing Sugars (%)
Coker 139 S$_{16}$	2761	42.5	57.0	102.7	20.2	61.2	30.0	2.61	13.3
Dihaploid mean	2500	40.3	54.1	87.4	18.5	58.2	27.4	2.96	15.3
Dihaploid range									
High	2934	45.8	59.2	98.0	20.3	61.2	31.0	3.88	19.2
Low	1848	29.7	46.5	75.7	15.2	53.2	24.2	2.20	11.0
F-Tests									
Among DH	**	**	**	**	**	**	**	**	**
Coker 139 vs. DH	**	NS	**	**	**	**	**	**	NS

* Significant difference at 0.05 level of probability.

** Significant difference at 0.01 level of probability.

SOURCE: Burk and Matzinger, 1976. Copyright 1976 by the American Genetic Association. Reprinted with permission.

TABLE 10

Dihaploid Performance Compared with That of Coker 139 Parent Tobacco Cultivar

	Yield kg/ha	Grade Index	Flower Days	Plant Height (cm)	No. of Leaves	Leaf Length (cm)	Leaf Width (cm)	Total Alkaloids (%)	Reducing Sugars (%)
Dihaploids									
Greater*	0	0	0	0	0	0	0	18	6
Not different	28	40	22	1	10	28	17	28	40
Less*	18	6	24	45	36	18	29	0	0

* Significantly different at 0.05 level of probability.

SOURCE: Burk and Matzinger, 1976. Copyright 1976 by the American Genetic Association. Reprinted with permission.

would enhance the usability of *N. tabacum* are of specific interest. This approach appeared very promising as Carlson et al. (1972) reported the successful fusion of *N. glauca* and *N. langsdorffi* protoplasts followed by regeneration and characterization of the parasexual hybrid plants. Selection of the *glauca-langsdorffi* hybrid cells was possible because of differential growth of parental species cells and hybrid cells on a specific culture medium. Fusion of *N. glauca* and *N. langsdorffi* protoplasts has been repeated by Smith et al. (1976). An interesting additional observation in the latter study was that hybrid plants had high and variable chromosome numbers, i.e., 54 to 64 chromosomes instead of the 42 chromosomes expected.

Two other cases of parasexual hybridization have been reported in *Nicotiana* (Melchers and Labib, 1975; Gleba et al., 1975). However, both of the hybrids obtained were intraspecific types involving *N. tabacum* cells. In *Nicotiana*, it appears that cells from different species can be induced to fuse; and in those cases where the hybrid cells can be selected, hybrid plants can be regenerated. As Power and Cocking (1977) have pointed out in their recent review, the development of selection procedures has been limited by the lack of suitable biochemical mutants in higher plants. The achievement of various somatic hybrids between *Nicotiana* species and *N. tabacum* is of great practical importance for tobacco improvement; however, efficient transfer of genetic information by this method depends on the development of selection procedures for identifying hybrid fusion products.

Tissue and Organ Culture

Failure to obtain interspecific combinations between *N. tabacum* and several of the *Nicotiana* species involves either the production of hybrid seed that will not germinate or the young hybrid plants fail to grow beyond the seedling stage. The *in vitro* culture of embryos, ovules, or somatic tissues of the desired hybrids may facilitate the attainment of viable hybrid combinations.

Raghavan (1977) has thoroughly reviewed the use of embryo cultures in obtaining otherwise inviable hybrids in plants. He does not cite any examples of *Nicotiana* embryo culture for the purpose of obtaining hybrids.

The technique of *in vitro* pollination and fertilization has been extensively investigated, and applications of the technique include facilitation of difficult hybrid crosses (Rangaswamy, 1977). In *Nicotiana*, the technique has apparently been applied only to compatible intraspecific matings involving *N. rustica* (Rao, 1965) and *N. tabacum* (Dulieu, 1966). Both embryo culture and *in vitro* pollination methods are potentially useful in obtaining interspecific hybrids in *Nicotiana*.

Reed and Collins (1977) cultured fertilized ovules from three interspecific hybrids of *Nicotiana*. The crosses involved were *N. repanda* × *N. tabacum, N. stocktonii* × *N. tabacum,* and *N. nesophila* × *N. tabacum,* in which the purpose of the interspecific crosses was to transfer black shank fungus (*Phytophthora parasitica* var. *Nicotianae*) resistance to *N. tabacum.* Hybrid plants were obtained from all combinations although only the hybrids involving *stocktonii* and *nesophila* grew to maturity.

Lloyd (1975) generated callus cultures from the cotyledons of the ordinarily lethal *N. suaveolens* × *N. tabacum* hybrid in order to generate hybrid plants, thus transferring the brown spot (*Alternaria longipes*) resistance factors of *N. suaveolens* to *N. tabacum.* Large numbers of F$_1$ hybrid plants were produced succssfully in this manner, whereas hybrid seedlings produced from conventional crosses died at an early age.

The techniques of embryo culture, *in vitro* pollination and fertilization, ovule culture and callus culture, and regeneration of hybrid plants offer the potential for obtaining many interspecific hybrids in *Nicotiana* that have not previously been possible using sexual crossing techniques. The favorable environment provided by the *in vitro* culture system may permit the successful germination and growth of conventionally produced hybrid seeds that do not survive under standard germination and growth conditions.

ISOLATION OF ANEUPLOID CHROMOSOME TYPES

Aneuploid Plants from Anther Cultures

It is possible to obtain plants with aneuploid chromosome numbers when anthers from monosomic *N. tabacum* lines are cultured *in vitro* (Mattingly and Collins, 1974). Seven of the 24 possible nullihaploid (n – 1 =23) plants have now been isolated. A problem has arisen with regard to identification of nullihaploid progenies from anther cultures of a given monosomic family. Several phenotypically different n – 1 = 23 plants have been obtained from a single monosomic family. This result is apparently related to the phenomenon of monosomic shift in which a chromosome other than the one missing in the original monosomic plant is eliminated during meiosis.

Plants with n + 1 = 25 chromosomes are frequently encountered among the plants obtained in anther cultures from monosomic plants. Doubling the chromosome number of such plants yields tetrasomic lines. Crossing a tetrasomic (2 n + 2 = 50) plant with a normal disomic (2 n = 48) *N. tabacum* plant will give the trisomic (2 n + 1 =49) chromosomal condition. Once the problem of identifying the aneuploid chromsosome types is solved, the viable nullisomic, tetrasomic, and trisomic aneuploid lines of *N. tabacum* can be established. Such aneuploid types are useful in establishing linkage relationships, studying the control of chromosome pairing, studying evolutionary relationships in the genus, and achieving substitutions involving specific chromosomes.

Anthers from haploid, diploid, triploid, and tetraploid *N. tabacum* plants were cultured by Niizeki and Kita (1975). No plants were obtained from the haploid anthers, and plants obtained from diploid anthers were haploid with the exception of an occasional n + 1 = 25 chromosome plant or a diploid plant with 2 n = 46 chromosomes. Plants with chromosome numbers in the range of 26 to 41 and 43 to 49 were observed from the triploid and tetraploid anthers, respectively. The aneuploid chromosome numbers encountered in the plants obtained from the anther cultures were generally attributed to the irregular meiotic chromosomal segregation in the polyploid and haploid plants from which the anthers were taken.

Aneuploid Plants from Callus Cultures

The scientific literature dealing with the subject of chromosome numbers in plant tissue cultures is voluminous. One can easily draw the conclusion that variation in chromosome number is a phenomenon typical of most species cultured *in vitro*, expecially after the tissue has been repeatedly subcultured. Nuclear changes associated with cells and tissues in culture have been reviewed recently by Sunderland (1973), Sheridan (1975), and D'Amato, (1977).

The frequency and extent of aneuploidy increases in tobacco with culture age (Murashige and Nakano, 1966, 1967). Although aneuploid cells and tissues may have a lower regeneration potential than euploid tissues, as suggested by Murashige and Nakano (1967), there are several documented cases of the regeneration of aneuploid shoots and plants (Sacristan and Melchers, 1969). The isolation of aneuploid plant types from *in vitro* plant cultures is straightforward in *Nicotiana*. However, the identification of specific aneuploid types represents a more difficult and preplexing problem. The occurrence of polyploid cells in *in vitro* cultures has been recognized and suggested as a tool for obtaining plants with multiple chromosome sets (Murashige and Nakano, 1966). We have routinely used this method to produce doubled haploids in *Nicotiana* (Kasperbauer and Collins, 1972).

MUTANT SELECTION AT THE CELLULAR OR TISSUE LEVELS

Considerable enthusiasm has been generated among crop scientists with regard to the application of microbiological techniques in the induction and selection of genetic mutants in higher plant cells. Several review articles have been written on the subject (Chaleff and Carlson, 1974; Carlson and Polacco, 1975; Rice and Carlson, 1975; Redei, 1976; Chaleff and Carlson, 1976). In tobacco, unlike most crop species, regeneration of plants from selected cells or tissues is not a problem. However, the majority of the selected mutant types in tobacco have not involved traits of commercial importance to tobacco.

The first auxotrophic mutants isolated from tobacco included auxotrophs for amino acids, vitamins, and nucleic acid bases (Carlson, 1970). These mutants were leaky as judged by their continued slow growth on unsupplemented media. Nonetheless, these results stimulated research on the induction and isolation of other mutants. To date, mutants with resistance to L-threonine (Heimer and Filner, 1970), streptomycin (Maliga, Sz.-Breznovits, and Marton, 1973), 5-bromodeoxyuridine (Maliga, Marton, and Sz.-Breznovits, 1973), 8-azaguanine (Lescure, 1973), methionine sulfoximine (Carlson, 1973), 5-methyltryptophan (Widholm, 1972), fluorophenylalanine (Palmer and Widholm, 1975), DL-ethionine (Zenk, 1974), S-2-aminoethyl-L-cysteine and delta-hydroxylysine (Widholm, 1976), and cyclohexamide (Maliga et al., 1976) have been reported in *Nicotiana*. In addition, a strain with increased tolerance to the herbicide 2,4-D has been selected from *N. sylvestris* haploid cell suspension (Zenk, 1974). In the same report, Zenk reported the successful selection of a salt-tolerant *N. sylvestris* strain. Increased salt tolerance has also been achieved by using tissue culture methods in *N. sylvestris* (Dix and Street, 1975) and *N. tabacum* (Nabors et al., 1975). *N. sylvestris* lines have been isolated under low-temperature stress that have increased chilling resistance (Dix and Street, 1976).

The number and variety of mutants cited above suggest that tissue-cell culture techniques afford considerable potential in future work aimed at identifying useful mutants for commercial tobacco production. A word of caution is warranted in that most of the mutants mentioned have not been characterized as to their transmission and stability in successive sexual generations after isolation.

Helgeson et al. (1972, 1976) have used an *N. tabacum* tissue culture system to characterize the degree of black shank (*Phytophthora parasitica* var. *Nicotianae*) disease resistance in tissue from plants with different genotypes. The research established that the gene for black shank resistance was expressed in tissue cultures as well as in the intact plants.

In studies still in progress (unpublished observations), we have isolated presumed mutants of *N. tabacum* that are overproducers of nicotine. Ethylmethane sulfonate-treated regenerating haploid callus tissue was repeatedly subcultured on media containing inhibitory levels of nicotinic acid analogs. The tissues that grew on the screening media were eventually placed on media conducive to plant regeneration. Plants with a 4- to 5-fold increase in nicotine content over that observed in the plants providing the original explant tissue were obtained. The chromosome number of the selected lines was doubled and selfed seed obtained. These lines are currently being evaluated for nicotine content and agronomic traits under field conditions.

Potential applications of cellular selection techniques to induce and/or select mutants of value for commercial tobacco improvement are numerous. Establishment of genotypes of tobacco with resistance to specific diseases is an area of high priority. Selection in this case might be based on the use of the actual disease toxin or by using the disease organism in the *in vitro* screening operation. Alterations in the chemical composition of tobacco are desirable as levels of potentially harmful chemical constituents are reduced or removed. Also, tobacco quality is related to specific chemical constituents that possibly can be manipulated with the aid of cellular selection techniques. Unique morphological mutants of benefit in the development of tobacco varieties suitable for mechanized handling can be selected from among morphological variants derived from cell cultures.

UPTAKE OF DNA AND CELLULAR ORGANELLES BY PROTOPLASTS

One of the principal uses that has been advocated for protoplasts is to directly transfer genetic material from a donor source into a recipient protoplast. Also, the possibility of stimulating the uptake of organelles and microorganisms by protoplasts has been investigatd (Davey and Cocking, 1972; Davey and Power, 1975; Burgoon and Bottino, 1976).

Progress in the successful transfer of DNA or genetic material from various donor sources to higher plant protoplasts and/or cells is very preliminary. Hess (1977) states that successful gene transfer involves consideration of DNA uptake, expression, integration, and replication. To date, one can confidently conclude only that DNA can be taken up by protoplasts or cells. Evidence for expression, integration, and replication of exogenous DNA is tenuous and requires further evidence and substantiation.

Giles (1977) provides a similarly bleak picture for the potential of chloroplast transfer via protoplasts in the improvement of crops within the near future. He draws his conclusions from an extensive review of articles published on the subject, and his conclusions are based largely on the lack of evidence that chloroplast uptake by the protoplast actually occurs, plus a failure to demonstrate activity of the chloroplasts where uptake has been demonstrated.

The results of efforts to transfer genetic material and/or organelles into higher plant cells should not be viewed too negatively. The transfer of genes, organelles, or microorganisms into higher plant cells has not proceeded to the point of using such altered plant materials as improved forms of economic crop plants, but the research approach does have potential for making useful alterations in crop plants.

SUMMARY

Tissue and cell culture methods have been developed and refined for *Nicotiana tabacum* and many other *Nicotiana* species to the extent that cultural techniques including regeneration of plants from cells grown in suspension cultures and from protoplasts are possible. Many studies in *Nicotiana* have involved the use of callus cultures, suspension cultures, anther cultures, and protoplasts. Genetic information, improved breeding lines, specialized cytogenetic stocks, interspecific hybrid combinations, and several genetic mutants have been generated as a result of using *in vitro* tissue and cell culture techniques in *Nicotiana*. The potential contributions of tissue-cell culture methods in *Nicotiana* are numerous and include obtaining interspecific and intergeneric hybrids by protoplast fusion, direct transfer of desirable genes from various donor sources into *Nicotiana* protoplasts, and the uptake of organelles by protoplasts.

ACKNOWLEDGMENTS

The information summarized in this paper represents the cooperative efforts of the Department of Agronomy and the ARS, USDA. The paper (72-2-165) is submitted in connection with a project of the Kentucky Agricultural Experiment Station, and is published with the approval of the director.

LITERATURE CITED

Bajaj, Y. P. S. 1974. Potentials of protoplast culture work in agriculture. Euphytica 23:633–49.

Bajaj, Y. P. S. 1977. Protoplast isolation, culture, and somatic hybridization. *In* J. Reinert and Y. P. S. Bajaj (eds.), Applied and fundamental aspects of plant cell, tissue, and organ culture, Pp. 467–96. Springer-Verlag, Berlin.

Bergmann, L. 1960. Growth and division of single cells of higher plants *in vitro*. J. Gen. Physiol. 43:841–51.

Bottino, P. J. 1975. The potential of genetic manipulation in plant cell cultures for plant breeding. Rad. Bot. 15:1–16.

Burgoon, A. C., and P. J. Bottino. 1976. Uptake of the nitrogen fixing blue-green algae *Gloeocapsa* into protoplasts of tobacco and maize. J. Hered. 67:223–26.

Burk, L. G. 1970. Green and light-yellow haploid seedlings from anthers of sulfur tobacco. J. Hered. 61:279.

Burk, L. G., G. R. Gwynn, and J. F. Chaplin. 1972. Diploidized haploids from aseptically cultured anthers of *Nicotiana tabacum*. J. Hered. 63:355–60.

Burk, L. G., and D. F. Matzinger. 1976. Variation among anther-derived doubled haploids from an inbred line of tobacco. J. Hered. 67 (1976):382–84.

Carlson, P. S. 1970. Induction and isolation of auxotrophic mutants in somatic cell cultures of *Nicotiana tabacum*. Science 168:487–89.

Carlson, P. S. 1973. Methionine sulfoximine-resistant mutants of tobacco. Science 180:1366–68.

Carlson, P. S., and J. C. Polacco. 1975. Plant cell cultures: Genetic aspects of crop improvement. Science 188:622–25.

Carlson, P. S., H. H. Smith, and R. D. Dearing. 1972. Parasexual interspecific plant hybridization. Proc. Nat. Acad. Sci. USA 69:2292–94.

Chaleff, R. S., and P. S. Carlson. 1974. Somatic cell genetics of higher plants. Ann. Rev. Genet. 8:267–78.

Chaleff, R. S., and P. S. Carlson. 1976. *In vitro* selection for mutants of higher plants. *In* L. Ledoux (ed.), Genetic manipulations with plant material. Pp. 351–63. Plenum Publishing Corp., New York.

Chaplin, J. F., and L. G. Burk. 1970. Interspecific hybridization and gene transfer in *Nicotiana*: Problems and possible solutions. *In* Proceedings of the fifth international tobacco scientific congress. Pp. 59–67. Hamburg.

Cocking, E. C. 1972. Plant cell protoplasts—isolation and development. Ann. Rev. Plant Physiol. 23:29–50.

Collins, G. B. 1976. Use of anther-derived haploids and their derivatives in plant improvement programs. *In* R. H. Burris and C. C. Black (eds.), CO_2 metabolism and plant productivity. Pp. 359–84. University Park Press, Baltimore.

Collins, G. B., and P. D. Legg. 1974. The use of haploids in breeding allopolyploid species. *In* K. J. Kasha (ed.), Haploids in higher plants: Advances and potential. Pp. 231–47. University of Guelph, Guelph, Ontario, Canada.

Collins, G. B., and P. D. Legg. 1975. Prospects of using anther-derived haploids and tissue culture methods in a tobacco breeding program. Tob. Res. 1:35–44.

Collins, G. B., P. D. Legg, and C. C. Litton. 1973. The use of anther-derived haploids in *Nicotiana*. I. Isolation of breeding lines differing in total alkaloid content. Crop Sci. 14:77–80.

Collins, G. B., P. D. Legg, and C. C. Litton. 1973. The use of anther-derived haploids in *Nicotiana*. II. Comparison of doubled haploid lines with lines obtained by conventional breeding methods. Tob. Sci. 18:40–42.

Collins, G. B., P. D. Legg, C. C. Litton, and M. J. Kasperbauer. 1971. Inheritance of resistance to black shank in *Nicotiana tabacum* L. Canad. J. Genet. Cytol. 13:422–28.

Cooperative Group of Haploid Breeding of Tobacco of Shantung Institute of Tobacco and Peking Institute of Botany, Academia Sinica. 1974. Success of breeding the new tobacco cultivar "Tan-yuk No. 1." Acta Botanica Sinica 16:300–303.

D'Amato, F. 1977. Cytogenetics of differentiation in tissue and cell cultures. *In* J. Reinert and Y. P. S. Bajaj (eds.), Applied and fundamental aspects of plant cell, tissue, and organ culture. Pp. 343–57. Springer-Verlag, Berlin.

Davey, M. R., and E. C. Cocking. 1972. Uptake of bacteria by isolated higher plant protoplasts. Nature 239:455–56.

Davey, M. R., and J. B. Power. 1975. Polyethylene glycol-induced uptake of microorganisms into higher plant protoplasts: An ultrastructural study. Plant Sci. Lett. 5:269–74.

Day, P. R., P. S. Carlson, O. L. Gamborg, E. G. Jaworski, A. Maretzki, O. E. Nelson, I. M. Sussex, and J. G. Torrey. 1977. Somatic cell genetic manipulation in plants. Bioscience 27:116–18.

Devreux, M., and U. Laneri. 1974. Anther culture, haploid plant isogenic line, and breeding researches in *Nicotiana tabacum* L. *In* Polyploidy and induced mutations in plant breeding. Pp. 101–7. International Atomic Energy Agency, Vienna.

Devreux, M., and F. Saccardo. 1972. Ibridazione di linee isogeniche di tobacco ottenute da coltura *in vitro* de antere. Genet. Agraria 26:143–46.

Dix, P. J., and H. E. Street. 1975. Sodium chloride resistant cultured cell lines from *Nicotiana sylvestris* and *Capsicum annuum*. Plant Sci. Lett. 5:231–37.

Dix, P. J., and H. E. Street. 1976. Selection of plant cell lines with enhanced chilling resistance. Ann. Bot. 40:903–10.

Dulieu, H. L. 1966. Pollination of excised ovaries and culture of ovules of *Nicotiana tabacum* L. Phytomorphology 16: 69–75.

East, E. M. 1928. The genetics of the genus *Nicotiana*. Bibliogr. Genet. 4:243–318.

Giles, K. L. 1977. Chloroplast uptake and genetic complementation. *In* J. Reinert and Y. P. S. Bajaj (eds.), Applied and fundamental aspects of plant cell, tissue, and organ culture. Pp. 536–50. Springer-Verlag, Berlin.

Gleba, Yu., R. G. Butenko, and K. M. Sytnyak. 1975. Protoplast fusion and parasexual hybridization in *Nicotiana tabacum* L. Genetika 221:1196–98.

Goodspeed, T. H. 1954. The genus *Nicotiana*. Chronica Botanica, Waltham, Mass.

Griffing, B. 1975. Efficiency changes due to use of doubled-haploids in recurrent selection methods. Theor. Appl. Genet. 46(8):367–85.

Griffing, B. 1976. Selection in reference to biological groups. VI. Use of extreme forms of nonrandom groups to increase selection efficiency. Genetics 82:723–31.

Guha, S., and S. C. Maheshwari. 1964. *In vitro* production of embryos from anthers of *Datura*. Nature 204:497.

Heimer, Y. M., and P. Filner. 1970. Regulation of the nitrate assimilation pathway of cultured tobacco cells. Biochim. Biophys. Acta 215:152–65.

Helgeson, J. P., G. T. Haberlach, and C. D. Upper. 1976. A dominant gene conferring disease resistance to tobacco plants is expressed in tissue cultures. Phytopathology 66:91–96.

Helgeson, J. P., J. D. Kemp, G. T. Haberlach, and D. P. Maxwell. 1972. A tissue culture system for studying disease resistance: The black shank disease in tobacco callus cultures. Phytopathology 62:1439–43.

Hess, D. 1977. Cell modification by DNA uptake. *In* J. Reinert and Y. P. S. Bajaj (eds.), Applied and fundamental aspects of plant cell, tissue, and organ culture. Pp. 506–35. Springer-Verlag, Berling.

Izman, G. V., N. A. Sherstyanykh, and L. G. Astakhona. 1975. Analysis of tobacco haploids obtained by anther culture *in vitro* for the content of alkaloids and volatile acids. Soviet Genet. 11:312–15.

Kadotani, N., and T. Kubo. 1969. Studies on the haploid method of plant breeding by pollen culture. I. Diploidization of tobacco haploid by root culture and progeny test of diploid plant derived from pith culture. Jap. J. Breed. 19 (Suppl.) 2:125–26.

Kasperbauer, M. J., and G. B. Collins. 1972. Reconstitution of diploids from leaf tissue of anther-derived haploids in tobacco. Crop Sci. 12:98–101.

Legg, P. D., J. F. Chaplin, and G. B. Collins, 1969. Inheritance of percent alkaloids in *Nicotiana tabacum* L. I. Populations derived from crosses of low alkaloid lines with burley and flue-cured varieties. J. Hered. 60:213–17.

Legg, P. D., and G. B. Collins. 1968. Variation in selfed progeny of doubled haploid stocks of *Nicotiana tabacum* L. Crop Sci. 60:620–21.

Legg, P. D., and G. B. Collins. 1971. Inheritance of percent total alkaloids in *Nicotiana tabacum* L. II. Genetic effects of two loci in Burley 21 × LA Burley 21 populations. Canad. J. Genet. Cytol. 13:287–91.

Legg, P. D., and G. B. Collins. 1974. Genetic variances in a random-intercrossed population of burley tobacco. Crop Sci. 14:805–8.

Legg, P. D., G. B. Collins, and C. C. Litton, 1970. Registration of LA Burley 21 tobacco germplasm. Crop Sci. 10:212.

Lescure, A. M. 1973. Selection of markers of resistance to base-analogues in somatic cell cultures of *Nicotiana tabacum*. Plant Sci. Lett. 1:375–83.

Lloyd, R. 1975. Tissue culture as a means of circumventing lethality in an interspecific *Nicotiana* hybrid. Tob. Sci. 19:4–6.

Maliga, P., G. Lazar, Z. Svab, and F. Nagy. 1976. Transient cyclohexamide resistance in a tobacco cell line. Molec Gen. Genet. 149:267–71.

Maliga, P., L. Marton, and A. Sz.-Breznovits. 1973. 5-Bromodeoxyuridine-resistant cell lines from haploid tobacco. Plant Sci. Lett. 1:119–21.

Maliga, P., A. Sz-Breznovits, and L. Marton. 1973. Streptomycin-resistant plants from callus culture of haploid tobacco. Nature New Biol. 244:29–30.

Mattingly, C. F., and G. B. Collins. 1974. The use of anther-derived haploids in *Nicotiana*. III. Isolation of nullisomics from monosomic lines. Chromosoma 46:29–36.

Melchers, G., and G. Labib. 1970. Die bedentung haploider hoherer Pflanzen fur Pflanzen Physiologie und Pflanzenzuchtung. Ber. Dtsh. Bot. Ges. Bd. 83:129–50.

Melchers, G., and G. Labib. 1975. Somatic hybridization of plants by fusion of protoplasts. I. Selection of light resistant hybrids of "haploid" light sensitive varieties of tobacco. Molec. Gen. Genet. 135:277–84.

Mikoyan, A. I. 1975. Androhaploids and restitutional diploids induced *in vitro* as the initial material for the analysis and for the selection of characters in synthetic breeding. Genetika 11(9):8–14.

Murashige, T., and R. Nakano. 1965. Morphogenetic behavior of tobacco tissue cultures and implication of plant senescence. Amer. J. Bot. 52:819–27.

Murashige, T., and R. Nakano. 1966. Tissue culture as a potential tool in obtaining polyploid plants. J. Hered. 57:115–18.

Murashige, T., and R. Nakano. 1967. Chromosome complement as a determinant of the morphogenetic potential of tobacco cells. Amer. J. Bot. 54:963–70.

Nabors, M. W., A. Daniels, L. Nadolny, and C. Brown. 1975. Sodium chloride tolerant lines of tobacco cells. Plant Sci. Lett. 4:155–59.

Nakamura, A., and R. Itagaki. 1973. Anther culture in *Nicotiana* and characteristics of the haploid plants. Jap. J. Breed. 23:71–78.

Nakamura, A., R. Itagaki, and K. Kobayashi. 1974a. Studies on the haploid method of breeding by anther culture in tobacco. IV. Comparison between doubled haploid lines and normal diploid lines of the two tobacco varieties Hicks 2 and Coker 139. Iwata Tob. Exp. Sta. Bull. 6:29–34.

Nakamura, A., T. Yamada, N. Kadotani, and R. Itagaki. 1974b. Improvement of flue-cured tobacco variety MC 1610 by means of haploid breeding method and investigations of some problems of this method. *In* K. J. Kasha (ed.), Haploids in higher plants: Advances and potential. Pp. 277–78. University of Guelph Press, Guelph, Ontario, Canada.

Nakamura, A., T. Yamada, N. Kadotani, R. Itagaki, and M. Oka. 1974c. Studies on the haploid method of breeding in tobacco. Sabrao J. 6:107–31.

Nakamura, A., T. Yamada, M. Oka, Y. Tatemichi, K. Eguchi, T. Ayabe, and K. Kobayashi. 1975. Studies on the haploid method of breeding by anther culture in tobacco. V. Breeding of mild flue-cured variety F211 by haploid method. Iwata Tob. Exp. Sta. Bull. 7:29–39.

Nakata, K. 1970. Competition among pollen grains for haploid tobacco plant formation by anther culture. I. Analysis with leaf color character. Jap. J. Breed. 21:29–34.

Nakata, K. 1971. Anther culture of tobacco and doubling of the chromosome number of haploid plants. Symposium on anther culture held at Nagano Pref.:74–75.

Nakata, K., and T. Kurihara. 1972. Competition among pollen grains for haploid tobacco plant formation by anther culture. II. Analysis with resistance to tobacco virus (TMV) and wildfire diseases, leaf color, and leafbase shape characters. Jap. J. Breed. 22:92–98.

Nei, M. 1963. The efficiency of the haploid method of plant breeding. Heredity 18:95–100.

Niizeki, M., and F. Kita. 1975. Production of aneuploid plants by anther culture in *Nicotiana tabcum* L. Jap. J. Breed. 25:52–58.

Nitsch, C. 1977. Culture of isolated microspores. *In* J. Reinert and Y. P. S. Bajaj (eds.), Applied and fundamental aspects of plant cell, tissue, and organ culture. Pp. 268–78. Springer-Verlag, Berling.

Oinuma, T., and T. Yoshida. 1974. Genetic variation among doubled haploid lines of burley tobacco varieties. Jap. J. Breed. 24:211–16.

Oka, M., A. Nakamura, and T. Yamada. 1976. Studies on the haploid method of breeding in tobacco. Coresta Information Bull. 6th Intern. Tob. Scientific Congress, Tokyo. P. 114.

Palmer, J. E., and J. M. Widholm. 1975. Characterization of carrot and tobacco cell cultures resistant to p-florophenylalanine. Plant Physiol. 56:233–38.

Power, J. B., and E. C. Cocking. 1977. Selection systems for somatic hybrids. *In* J. Reinert and Y. P. S. Bajaj (eds.), Applied and fundamental aspects of plant cell, tissue, and organ culture. Pp. 497–505. Springer-Verlag, Berlin.

Raghavan, V. 1977. Applied aspects of embryo culture. *In*. J. Reinert and Y. P. S. Bajaj (eds.), Applied and fundamental aspects of plant cell, tissue, and organ culture. Pp. 374–97. Springer-Verlag, Berlin.

Rangaswamy, N. S. 1977. Application of *in vitro* pollination and *in vitro* fertilization. *In* J. Reinert and Y. P. S. Bajaj (eds.), Applied and fundamental aspects of plant cell, tissue, and organ culture. Pp. 412–45. Springer-Verlag, Berlin.

Rao, P. S. 1965. The *in vitro* fertilization and seed formation in *Nicotiana rustica* L. Phyton 22:165–67.

Redei, G.P. 1976. Induction of auxotrophic mutations in plants. *In* L. Ledoux (ed.), Genetic manipulations with plant material. Pp. 329–50. Plenum Publishing Corp., New York.

Reed, S. M., and G. B. Collins. 1978. Use of tissue culture methods to obtain interspecific hybrids in Nicotiana. *In* this volume (abstr.).

Reinert, J., and Y. P. S. Bajaj. 1977. Anther culture: Haploid production and its significance.

In J. Reinert and Y. P. S. Bajaj (eds), Applied and fundamental aspects of plant cell, tissue, and organ culture. Pp. 251–67. Springer-Verlag, Berlin.

Rice, T. B., and P. S. Carlson. 1975. Genetic analysis and plant improvement. Ann. Rev. Plant Physiol. 26:279–308.

Sacristan, M. D., and G. Melchers. 1969. The caryological analysis of plants regenerated from tumorous and other callus cultures of tobacco. Molec. Gen. Genet. 105:317–33.

Sheridan, W. F. 1975. Plant regeneration and chromosome stability in tissue cultures. *In* L. Ledoux (ed.), Genetic manipulation with plant material. Plenum Publishing Corp., New York.

Smith, H. H., K. N. Kao, and N. C. Combatti. 1976. Interspecific hybridization by protoplast fusion in *Nicotiana*. Confirmation and extension. J. Hered. 67:123–28.

Sunderland, N. 1973. Nuclear cytology. *In* H. E. Street (ed.), Plant tissue and cell culture. Pp. 161–90. University of California Press, Berkeley.

Ternovskii, M. F., G. V. Izman, and Yu. F. Sarychev. 1975. Androploids and restitutional diploids *in vitro* as the initial material for the analysis and selection of characters in synthetic breeding. Soviet Genet. 11:1079–85.

Vasil, I. K., and C. Nitsch. 1975. Experimental production of pollen haploids and their uses. Z. Pflanzenphysiol. 76:191–212.

Vasil, V., and A. C. Hildebrandt. 1965. Differentiation of tobacco plants from single, isolated cells in microcultures. Science 150:889–92.

Widholm. J. M. 1972. Cultured *Nicotiana tabacum* cells with an altered anthranilate synthetase which is less sensitive to feedback inhibition. Biochim. Biophys. Acta 261:52–58.

Widholm, J. M. 1976. Selection and characterization of cultured carrot and tobacco cells resistant to lysine, methionine, and proline analogs. Canad. J. Bot. 54:1523–29.

Zenk. M. H. 1974. Haploids in physiological and biochemical research. *In* K. Kasha (ed.), Haploids in higher plants: Advances and potential. Pp. 339–53. University of Guelph Press, Guelph, Ontario, Canada.

DONALD BOULTER AND OTTO J. CROCOMO

Plant Cell Culture Implications: Legumes

30

This paper describes the role and economic importance of legumes, reviews the work on legumes in culture, and illustrates how, in principle, culture methods can be of potential use in improving protein content and quality of legumes.

ROLE AND ECONOMIC IMPORTANCE OF LEGUMES

Plants are the major food supply of the world. Legumes are of special significance since they are high protein crops containing between 20 and 50% protein in the harvested seeds. They play a key role in developing countries, specifically augmenting the low protein content of the main starchy staples whether these are cereals, e.g., wheat and maize, root crops, e.g., cassava or yams, or starchy fruits, e.g., plantain.

With present population trends, the importance of these crops in the diet will increase. However, world cereal production is increasing much more rapidly than that of legumes, and this imbalance must be corrected. Thus, in the developing countries between 1952 and 1972, population increased 53 percent, food production 62 percent, but legumes only 40 percent. In Asia and the Far East, population increased 51 percent, food production 55 percent, and legumes 21 percent (FAO, 1971, 1973).

An important consideration is that cereals and legumes nutritionally complement one another; the low levels of sulphur amino acids in legumes

Donald Boulter, University of Durham, Botany Department, Durham, England DH1 3LE; Otto J. Crocomo, Department of Chemistry and CENA, ESALQ, University of São Paulo, Piracicaba-SP, Brazil.

are largely offset by their higher concentration in many cereals, and the legumes compensate for the low levels of lysine in cereals.

Legumes are less important in developed countries, where the main protein sources are of animal origin; but even so, many of these countries are looking for a legume that can replace soya as a source of textured protein products. Although it is relatively less efficient to produce protein from animals than plants, the increasing demand for meat and animal products means that the need for high protein feed supplements (e.g., legume meals) will continue to rise concurrently.

Worldwide, most legumes are used directly in human nutrition, but some, notably soya beans and peanuts, are also cultivated for their oil and/or as a source of animal feed (table 1). The world production figures for legumes are given in table 2; for additional information on world production and trade in legumes, see Hulse et al. (1975). Legumes add diversity to the agriculture of a country; they are important break crops that improve soil fertility, since by virtue of their association with Rhizobia they can fix nitrogen. In view of the energetically costly industrial process of manufacturing mitrogen fertiliser, legumes offer interesting possibilities both from the ecological and economic standpoints.

REVIEW OF LEGUME TISSUE CULTURE

One of the first problems to be overcome if plant cultures are to play an important role in plant improvement is the development of growth media suited to individual legume species. This development has proceeded from the use of undefined media to completely defined (minimal) media that support excellent growth; media for the induction of morphogenesis are also required. Hildebrandt et al. (1963), in an attempt to establish several varieties of plant tissues *in vitro*, included two varieties of *Phaseolus vulgaris*, navy and red kidney bean, in their experiments. Although excellent growth was obtained on a medium containing coconut milk, α-napthaleneacetic acid (NAA), and 2,4-dichlorophenoxyacetic acid (2,4-D), no chlorophyll formation occurred even though moderate growth was obtained without 2,4-D. Subsequently, Kant and Hildebrandt (1969) made some cytological observations in these chlorophyll-free *Phaseolus* cultures that revealed that they contained relatively high numbers of starch granules around the nuclei.

Because of the problem of obtaining chlorophyll synthesis and differentation in bean callus cultures, attention shifted to physiological experimentation with suspension cultures. Dougall (1964) and Lamport (1964) independently examined suspension cultures of higher plants, including *Phaseolus* varieties, from the point of view of growth energetics.

TABLE 1

LEGUMES OF MAJOR IMPORTANCE

Botanical Name	Common Name	Areas of Consumption
Cajanus cajan	Pigeon pea	India, Pakistan, Middle East, East Africa
Cicer arietinum	Chickpea	India, Pakistan
Lens esculenta	Lentil	Near East, North Africa, India, Central and South America
Vigna radiata (Phaseolus aureus)	Mung bean	South, Southeast, and East Asia, East Africa, India
Phaseolus lunatus	Lima bean	Tropical America, West Indies, Madagascar
Vigna mungo (Phaseolus mungo)	Black gram	India, Iran, East Africa, West Indies
Phaseolus vulgaris	Kidney bean	North, Central and South America, Mexico, East Africa
Pisum sativum	Green pea	Mainly temperate zones, parts of India and Africa
Vicia faba	Broad bean	Temperate zones, Near East, North Africa, Central and South America
Vigna unguiculata (Vigna sinensis) Oil Seeds, etc.	Cowpea	Asia, Tropical Africa, West Indies, China
Arachis hypogaea	Ground nuts	Asia, Africa, North, Central, and South America
Glycine max	Soya bean	North and Central America, Asia, Europe

SOURCE: Modified from Siegel and Fawcett, 1976.

TABLE 2

WORLD PRODUCTION OF MAJOR LEGUMES, 1972

Botanical Name	World Production (1,000 metric tons)	World Legume Production (%)
Glycine max	53,024	49.4
Arachis hypogaea	16,887	15.7
Phaseolus vulgaris	10,899	10.2
Pisum sativum	10,218	9.5
Cicer arietinum	6,718	6.3
Vicia faba	5,326	5.0
Cajanus cajan	1,720	1.6
Vigna unguiculata (Vigna sinensis)	1,260	1.2
Lens esculenta	1,182	1.1
World total	107,234	100

SOURCE: FAO, 1973; modified from Siegel and Fawcett, 1976.

It was apparent that *Phaseolus vulgaris* could easily be cultured in suspension with a high growth rate. Mehta (1966) defined a method for obtaining rapidly growing cells in suspension from macerated *Phaseolus* roots and then employed this system to study growth rates using mitotic activity as an index (Mehta et al., 1967). It was found that mitotic activity reached a peak during the first week or after 10 days of inoculation in fresh medium and fell markedly thereafter.

Suspension cultures were recognized as a system in which large scale experimentation could be conducted. Veliky and Martin (1970) developed a 67-V medium for fermentation-type experiments with *Phaseolus vulgaris*, and Gamborg et al. (1968) used a defined medium, B-5, to study the effect of different chemicals on growth of soya bean (*Glycine max*) cells. This medium was used in a phytostat developed by Miller et al. (1968) for the continuous culture and automatic sampling of soya bean cell suspensions.

Of primary importance was the development of a defined nutrient medium. Coconut milk has long been used as an essential requirement, and Liau and Boll (1970) found cell yield to be 300 percent greater when coconut milk was present; the addition of a mixture of amino acids to the same culture medium without coconut milk only increased growth by 50 percent.

Crocomo et al. (unpublished observations) have shown that a mixture of arginine, aspartic acid, and cysteine was better than casein hydrolysate as a source of organic nitrogen for root morphogenesis using explants of *Phaseolus vulgaris* in a defined medium. Mixtures of other amino acids did not enhance growth or differentiation.

However, Crocomo et al. (1967a) have recently reported that the addition of a bean seed extract successfully induced plantlet morphogenesis in *Phaseolus vulgaris*. The extract consisted of bean seeds homogenized with 67-V salts (Veliky and Martin, 1970) at concentrations ranging from 1/100 bean seed/ml to bean seed/ml. Seeds were soaked in aerated tap water prior to homogenization, and the extract was subsequently filtered through cheese cloth. The most striking and potentially significant results were obtained with extract from bean seeds soaked for two hours. At a concentration of 1/4 bean seed/ml, 2 plantlets were induced in 9 cultures.

In the search for a defined medium for growth and induction of morphogenesis, Crocomo et al. (unpublished observations) have shown that IAA and kinetin plus arginine, aspartic acid, and cysteine, enhance root morphogenesis better than that of NAA, IAA (indole-3-acetic acid), kinetin, and 2,4-D. In the latter medium, there was rapid callus development but little or no root formation.

Experiments based on the theory of organogenesis (Reinert, 1973) have lead to interesting and practical observations. This theory states that minimal quantitative changes in the ratio of certain components in the medium are decisive and determine whether organs are initiated or not. In this respect the use of 2,4-D has been shown to be more effective than IAA or NAA at equivalent concentrations (Liau and Boll, 1970; Gamborg, 1966). Witham (1968) found 2,4-D could stimulate growth of soya bean without the presence of a cytokinin, but that the addition of kinetin could stimulate growth to higher levels than 2,4-D alone. It was also found that when 2,4-D concentration was high, kinetin could be inhibitory.

Interaction between individual auxins and cytokinins at specific concentrations in the culture media regulate growth and root morphogenesis of *Phaseolus vulgaris* explants (Crocomo et al., 1976b). The highest growth index was observed in leaf explants, cultured on IAA-zeatin culture medium. Furthermore, callus from explants grown on the IAA-zeatin medium was characterized by the presence of a pronounced greening that was not present in the IAA-kinetin- or NAA-kinetin-treated explants.

With regard to the induction of root morphogenesis, the IAA-zeatin-containing medium was the best. Optimal root morphogenetic index occurred at IAA concentrations of 7.5–10.0 mg/l and at zeatin concentrations of 0.05 and 0.10 mg/l. A higher proportion of the roots induced by the IAA-zeatin treatment showed a geotropic response as compared with those induced with IAA-kinetin or NAA-kinetin. Concentrations of IAA (7.5–10.0 mg/l) and kinetin (0.4–1.0 mg/l), or NAA (10.0 mg/l) and kinetin (10.0 mg/l), were optimal for the induction of roots in other media that were tested. The mode of action of kinetin was studied in soya bean cultures by Blaydes (1966), and it was concluded that the action centered on RNA metabolism.

Jeffs and Northcote (1967) studied the influence of IAA and sugar on the induction of differentiation in the *Phaseolus* system. Using agar wedges containing IAA or sucrose inserted into callus blocks, it was possible to induce vascular differentiation and "meristematic regions." Gradients were found, but these apparently were the result of diffusion and not of active transport. Kinetin (0.1 mg/l) was found to stimulate cell division, and in the presence of IAA and sucrose, both cell division and cell differentiation were enhanced. A marked increase in phloem differentiation was noted in kinetin-treated callus blocks.

Conditions for morphogenesis have been described by Liau and Boll (1970), and Olieman-van der Meer et al. (1971) have studied adventitious root formation from cultivated epicotyls. Root formation was completely inhibited when sugar and auxin were not present in the culture medium and was also greatly reduced in the light. Liau and Boll (1970) inhibited root morphogenesis by changing the constituents of the inorganic fraction of the nutrient, but found no effect from altering the organic fraction. Differences in response were also noted in this work according to the origin (root, hypocotyl, or cotyledon) of the cultured cells. Such differences raise interesting questions with regard to totipotency and of the persistence of differentiation of cells in culture.

Work on plants other than legumes has shown that root morphogenesis can be induced in callus cell cultures by exogenous alkaloids, e.g., nicotine in *Nicotiana rustica* (Peters et al., 1974) where it has played a prominent role. Recently, Peters et al. (1976) reported an effect of exogenous nicotine in the culture medium on the control of growth and morphogenesis in

Phaseolus vulgaris tissue cultures; different concentrations of kinetin and IAA were used. This unequivocally establishes that nicotine possesses regulatory activity in species other than those in which it is synthesized. The removal of kinetin from the medium caused a drastic reduction in both callus cell growth and root induction (geotropic roots). Explants cultured on medium containing kinetin, IAA, and nicotine developed massive amounts of callus and aerial roots. Optimal amounts of root induction occurred at the 50 mg/ml nicotine concentration. Another alkaloid, caffeine, was also studied, but it had an inhibitory effect on both growth and root development.

Evans et al. (1976) worked with explants of the $Y_{11}Y_{11}$ genotype of *Glycine max* strain T219, grown on Murashige and Skoog's 1962 media (MSM) in an attempt to ascertain the optimal growing conditions of soya bean leaf tissues *in vitro* and to obtain some control over morphogenesis. These cultures were grown under normal conditions of 26–30°C with a 12 hr day of 250 ft-c. Optimal callus growth and inconsistent root morphogenesis were recorded when 10.75 μM NAA and 2.33μM kinetin were added to MSM. Maximal callus growth of 0.8 g fresh weight/culture was obtained with 3.8 mM $(NH_4)_2SO_4$ and 18.8 mM KNO_3 as nitrogen sources. By utilizing *diallel* tests, a high rooting medium (HRM) was identified that contained 100% MSM inorganic salts and 0.06M sucrose with 3.8 mM $(NH_4)_2SO_4$ and 18.8 mM KNO_3. When 50 leaf tissue explants on HRM were stored in the dark at a reduced temperature, all developed more roots than explants cultured on HRM in the light, but the roots were thinner and mostly aerial. On the other hand, when 16.26 μM nicotinic acid was added to HRM under normal conditions, all the roots from 17 explants were thicker and showed geotropic responses.

While callus cultures on agar have been used in attempts to study morphogenesis, suspension cultures have been utilized in the study of growth and biochemistry. *Phaseolus* suspension cultures have been a very good system for biochemical studies. Liau and Boll (1971) defined a 12-day growth cycle, in which cell division and appreciable synthesis of cellular material starts within 24 hours of innoculation. During the exponential phase, the cell number increased rapidly and clumps formed. However, this was not accompanied, directly, by increase in cell size and dry weight. Increase in dry weight was delayed by about 2 days, and cell enlargement started only when the log phase was more or less completed. With all cell types, cell growth frequently started by formation of protrusions, yeast-like budding of the outgrowth of papillae and filaments.

The same system was used by Liau and Boll (1972) to follow extracellular polysaccharide synthesis in *Phaseolus vulgaris*. It was observed that polysaccharide decreased in amount slightly during the log phase and increased at the onset of the stationary phase.

Suspension cultures have been used to study the amino acid composition of TCA-precipitated protein in *Phaseolus* cells by Gamborg and Finlayson (1969). Pathways in aromatic metabolism of *Phaseolus* have also been studied in such systems (Gamborg, 1966). It was found that the specific activities of enzymes extracted from suspension cell cultures were considerably higher than those obtained from differentiated tissues, suggesting that the possibility exists for the large-scale production and extraction of enzymes involved in the formation of aromatic compounds from cultures.

Manipulation of the nutrient composition in suspension cultures has made it possible to induce the biosynthesis of compounds not found in plant tissues *in vivo*. Thus, Veliky (1972) found that suspension cultures of *Phaseolus vulgaris* transformed tryptophan into the alkaloids harman and norharman, which are not usually found in *Phaseolus* plants.

Histological detection of differentiation generally depends upon the specific staining of a component localized in the cell type under consideration; quantitative biochemical estimations can examine either the amount of these components or the activity of enzymes believed to control their biosynthesis. The latter approach has the advantage that enzyme activity is a more sensitive indicator of a change in cellular development than the total amount of an end-product accumulated throughout the growth of the cell. The pattern of variation of activity could also suggest possible mechanisms for the molecular basis of differentiation.

Xylem vessels are generally detected by their heavily lignified walls, although the walls of other cells, such as extraxylary fibers, are lignified to some extent. Chemical analysis has shown that the lignin content of callus does rise when xylem is formed (Jeffs and Northcote, 1966); the activity of the enzyme phenylalanine ammonia lyase (PAL) has been correlated with xylogenesis.

Phloem vessels could be identified by the presence of paired pads of callose ($\alpha,\beta1\rightarrow3$ glucan) on the sieve plates. Callose is the major polysaccharide containing $\beta\ 1\rightarrow3$ linked glucose in dicotyledons such as *Phaseolus* spp. The precise mechanism of polysaccharide synthesis is unknown, but it seems likely that one control step is the transfer of the sugar moiety from the activated form of nucleotide-di-phosphate-sugar to the growing polysaccharide chain. The activity of UDP-glucose: $\beta1\rightarrow3$ glucan glycosyl transferase (callose synthetase) was therefore used as a marker for phloem formation by Haddon and Northcote (1975).

Two callus strains were developed from bean hypocotyl and grown on a defined maintenance medium supplemented with 2 mg/ l 2,4-D and 2 percent sucrose. Root initiation was observed in one strain and formation of nodules containing xylem and phloem in both strains after transfer to an induction medium supplemented with l mg/ l NAA, 0.2 mg/ l kinetin, and 3 percent sucrose.

Callus was transferred from maintenance medium after 3, 4, 5, and 6 transfers. The number of vascular nodules containing differentiated tissue after 21 days on induction medium decreased as the period the callus was kept in culture increased. No further differentiation could be induced after 6 transfers. The fall in the number of nodules formed was paralleled by a decrease in phenylalanine ammonia lyase (PAL) and $\beta 1\rightarrow 3$ glucan synthetase as measured, 21 days after transfer. Haddon and Northcote (1975) concluded that the ratio of activities of these enzymes in callus growing on an induction medium to those in callus growing on maintenance medium could be used as a measure of the amount of cell differentiation which had been induced.

Further studies by the same authors (Haddon and Northcote, 1976) showed that the activities of PAL and caffeate-O-methyl transferase in bean callus tissue were induced in a coordinated manner over a period of 21 days. Neither enzyme activity could be induced when differentiation was repressed by the addition of abscisic acid to the induction medium. PAL activity was more directly correlated with xylogenesis and nodule induction than with the accumulation of soluble phenols.

Peroxidase activity remained constant in callus grown on maintenance medium (150 μmol of peroxide oxidized $min^{-1}g^{-1}$fresh wt.) but fell rapidly in callus transferred to induction medium. By 6 days the activity had decreased to approximately 60% of the control value and it remained at this level (70 μmol min $^{-1}g^{-1}$ for at least a further 4 weeks. Gel electrophoresis of extracts from callus grown on either medium showed the presence of 3 main isoenzymes. The major band (Rp 0.65) was not observed in electrophoretic separations of extracts of bean hypocotyl.

Recently, Peters et al. (1977) reported the induction of haploid callus cells from the culture of isolated anthers of the Brazilian bean (*Phaseolus vulgaris* variety Bico de Ouro. The haploidy was confirmed using a Giemsa staining method. The percentage of anthers producing callus at 25° C was about the same, 57 percent, regardless of the presence of 2,4-D in the medium. However, the amount of callus produced was almost twice as much when the anthers were grown on medium plus 2,4-D. Almost the same distribution of haploids and diploids was observed among 40 cells. Polyploidy was observed in less that 3 percent of the cells in the form of near-hexaploidy at metaphase and near-tetraploidy at middle prophase.

IMPROVEMENT OF YIELD OF LEGUMES: USE OF CULTURE METHODS

The major problem for the plant breeder in his attempt to improve legumes is to maximize and stabilize yields. The constraints on yield are likely to be numerous and to vary with different legumes in different agro-ecological situations and from season to season.

However, rhizobial infection and activity may in many circumstances be crucial, since measurements of the nitrogen "fixed" by a particular legume can vary in the field situation by several hundred percent. Research directed toward an understanding of the factors affecting rhizobial infection and the efficient working of the symbiotic process are of considerable importance; the report by Holsten et al. (1971) of the establishment of symbiosis between Rhizobia and soya bean cells *in vitro* is of considerable significance, since it will now be possible to investigate these important problems in experimental systems subject to much greater control than that of the bacterial-root association.

The evidence available suggests that another constraint to improvement of yields is the lack of a sufficient genetic diversity of parental types in breeding programs. It would be a great advantage if the characteristics of one legume species could be incorporated into another, but it is not easy to make interspecific crosses often due to failure of fertilization. This is a possible way in which protoplast fusion might prove to be of considerable help to the breeder, although failure to fertilization is not the only development barrier and abortion due to genome incompatability or somato-plastic dysfunction may also occur.

For plant cultures to be used successfully in agriculture, it will be necessary to establish the following chain of events: the establishment of a callus, derivation of single cells from it, removal of cell walls, fusion of protoplasts, regeneration of callus, and differentiation of the latter into plantlets that can complete their life cycle. Some legumes have been put into callus culture; single-cell cultures have also been achieved and in some cases protoplasts produced from them (Arnison and Boll, 1974; Bingham et al., 1975; Nickell, 1956; Phillips, 1974a,b). Recently , Oswald et al. (1977), using soya and *Trifolium repens* (clover), have shown that it is possible to go from callus to single cell, back to callus, and thence to vegetative plantlets. It is likely, therefore that plant cultures will play an increasingly important role in the problem of recombining genetic information that would not be otherwise easily accomplished.

Legumes are susceptible to a variety of viral diseases, and it is possible that, apart from cases where there are overt symptoms, depression of yield may be due to this cause and be unsuspected. Kartha et al. (1972) have grown complete pea plants from shoot apical meristems, and this is a method that may have considerable importance in developing virus-free stocks.

Haploids are especially useful in studies on the induction and detection of mutations; they may also provide a fast and simple method for the production of homozygous plants and therefore have a potential role in breeding programs (Sunderland, 1973; Reinert and Bajaj, 1977). Before new genetic rearrangements deriving from recombinations can be exploited in crop production, the new genetic condition must be stabilized. With inbred species this is more easily accomplished than with outbred crops. Even with inbred species, eight or more generations may be needed before sufficiently homozygous populations are available. By the use of haploids, which on treatment with colchicine would double their chromosome number, it is potentially possible in two generations to go from an F_1 hybrid to a new variety. However, the breeder requires that his specific agronomically desirable materials shall be put into pollen culture and so on. Experience so far has shown that this is often a difficult and time-consuming operation that may not be successful even after one or two years' of experimentation. Basic research should therefore be directed toward finding, if possible, conditions for the general induction of haploid callus cells irrespective of their source. Furthermore, other methods exist for the production of haploids, for example, the use of chromosome elimination in barley varieties after pollination with *Hordeum bulbosum* (see Riley, 1974). Thus, the possibility exists that other methods of haploid production may sometimes prove more efficient.

IMPROVEMENT OF PROTEIN CONTENT AND QUALITY IN LEGUMES: USE OF TISSUE CULTURE METHODS

Increased supplies of legume protein could be achieved (1) by increasing the area planted, the present trend being a reduction due to the production of more efficient cereals; (2) by increasing yield—an improvement here might further offset the decrease noted in (1); or (3) by increasing protein content, yield remaining approximately the same. Table 3 lists the averge protein content of legumes of economic importance.

Generally speaking, there is a negative correlation between protein content and yield. The basic biochemical analysis of this phenomenon is discussed by Boulter and Gatehouse (1977); it has been concluded that in legumes it is not likely to be a serious impediment in present breeding programs (see Boulter, 1975).

The total protein of legume seeds is nutritionally limited by one or both of the sulphur amino acids methionine and cysteine. There are nutritional situations, for example, if the main staple is cassava or plantain, where

TABLE 3

Typical Protein Content of Legume Meals
(g/100g meal)

Botanical Name	g/100 g
Arachis hypogaea	25.6
Cajanus cajan	20.9
Cicer arietinum	20.1
Glycine max	38.0
Lens esculenta	24.2
Phaseolus lunatus	19.7
Phaseolus vulgaris	22.1
Pisum sativum	22.5
Vicia faba	23.4
Vigna mungo (Phaseolus mungo)	23.9
Vigna radiata (Phaseolus aureus)	23.0
Vigna unguiculata	23.4

Source: FAO (1971), "Amino acid content of foods," FAO Nutritional Studies No. 24, Rome and Boulter (unpublished observations)

sulphur amino acids are limiting in the diet as eaten, leading to inefficient use of dietary protein. Similarly in developed countries it is desirable that textured protein products of legumes should have a balanced amino acid profile. Consideration of the energetics of the biochemical reactions involved using the analysis of Penning de Vries et al. (1974) suggests that improving protein quality will have only a small initial effect on depressing yield.

At the moment breeding programs for improving legume protein content and quality screen the end-product, protein, of the complex sequence of events that includes nitrogen uptake, its incorporation into organic nitrogen compounds, transport, and finally the deposition of protein in the harvested parts. However, sufficient evidence is already available to suggest that, in different lines of a crop, various steps in this sequence of reactions and processes may be rate-limiting. In wheat, for instance, there is evidence that in some lines reduction of nitrogen in the leaves by the enzyme nitrate reductase may be the rate-limiting step (Hageman et al., 1976), and in other lines transport of nitrogen to the seeds is rate-limiting (Johnson et al., 1969). If further work substantiates these suggestions, improvements might be expected by combining parental types in which protein synthesis is limited by different steps; some of these might occur in parts of the plant quite remote from the site of protein deposition. Selection methods would be needed to identify these types. One possible approach would be to screen large numbers of cultures and lines for particular enzyme activities, and, as discussed by other speakers in this symposium, cell cultures have particular advantages for selection procedures.

Furthermore, it has been shown by the work of Carlson (1973) and Widholm (1974) that a useful characteristic selected in culture, can be maintained in the regenerated plant. Examples include not only selection for resistance to pathogenic toxins but also for improving nutritional status for particular amino acids by the selection of mutants with nonregulated enzymes involved in the biosynthesis of the amino acid concerned. The work of Widholm on the production of nonregulated tryptophan-synthesizing enzymes is particularly apposite, since, for example, it has been shown that in sheep the production of better-quality wool protein is limited by the availability of sulphur amino acids to the synthesizing system. However, there is no evidence that protein quality in legumes is dependent on the supply of particular amino acids.

Work to date on the quality of seed protein emphasizes the importance of the "storage" proteins (Boulter, 1976). Where quality mutants exist, as in some cereals, changes in regulating genes that affect the proportions of storage proteins and sometimes other proteins have been found to be responsible (Nelson, 1976). Improvements due to the production of new proteins or changed amino acid profiles of existing proteins have not been demonstrated, and are statistically less likely to occur than the former type of change.

The biochemical basis for improvements of this kind is the fact that individual storage proteins and other major seed proteins differ in their amino acid composition and hence nutritional status. Theoretically, therefore, selection for nutritionally important proteins using cultures might be expected to supplement existing screening methods. An example of this approach is illustrated by the work of Polacco (1977), who studied the assimilation of urea by cultured soya bean cells. Urease of soya bean is relatively rich in methionine, the nutritionally limiting amino acid of soya bean, as compared with the total protein of soya bean seeds (Bailey and Boulter, 1971). Urease occurs in a range of legumes but the levels vary, and soya bean has only about 10 percent of that found for example in jack bean. Selection of lines with high urease activity could be agronomically valuable if carried through into the adult plant since it might permit more efficient assimilation of urea fertilizer and also result in improved nutritional quality if reflected in higher levels of urease in the seeds (Polacco, 1977).

Polacco (1977) devised several selective systems to recover overproducing urease mutants as follows: (1) utilization of urea in the presence of urease inhibitors; (2) utilization of urea in the presence of known metabolizable repressors of urease production; and (3) utilization of urea in the presence of high levels of nitrate. These selective procedures may also produce in the case of (1) mutants with altered urease and in the case of (3) nitrate reductase constitutive mutants, i.e., those that produce nitrate reductase without the prior metabolism of a reduced nitrogen source. The long-range aim of work of this type is primarily to effect changes in seed protein composition by mutant selection in cultures; successful accomplishment of this final aim has yet to be demonstrated.

It has been suggested that the genetic potential of cell cultures is restricted and hence their usefulness as selection systems limited. However, it is more feasible that their limited biosynthesis is not intrinisically genetic but due to our lack of understanding of culture techniques that could lead to the complete expression of their biochemical potential. Although it has usually been reported that cultures of legume seed cotyledonary cells do not synthesize storage proteins, Muntz (personal communication) has found serological evidence for the production of storage proteins in a few cultures of *Vicia faba*, but quite sporadically, probably underlining our lack of control of the necessary environmental factors.

In considering whether or not characters selected in culture will be expressed in the seeds of a regenerated plant, nonexpression in the latter could be due to no transcription and/or translation of information or alternatively to the activity of a formed enzyme being inhibited by the internal environment. Possible analogous situations respectively might be cereal storage proteins genes that are not expressed in the embryo but only in the endosperm and *nif* genes that are only expressed in most organisms in low dissolved oxygen tensions.

The storage proteins of legumes, which are deposited in the cotyledons, have been separated into two fractions called legumin and vicilin. The former consists predominently of one protein with sedimentation characteristics of 12S. The second fraction contains several different proteins that have not been fully characterized (Derbyshire et al., 1976). Normally vicilin is synthesized prior to legumin at about one-third of the way through the seed development period (Millerd, 1975).

Young developing excised embryos in which storage protein synthesis has started can be cultured, and they continue to synthesize storage proteins. Millerd et al. (1975) have shown that if young developing pods of *Pisum* are detached prior to storage protein synthesis they can also be successfully cultured and the major storage proteins will be synthesized within the seeds. If, however, young seeds themselves are excised prior to storage protein synthesis, they synthesize only vicilin and not legumin, indicating the importance of a substance (or substances) presumably obtained from the pod tissue. Thus, organ culture could prove a useful experimental system for manipulating and elucidating the controls of protein synthesis in seeds. Another important use of embryo culture is in interspecific crosses, e.g., *Phaseolus vulgaris* and *Phaseolus acutifolius* are somatically compatible, but adverse interaction occurs in the endosperm and aborting embryos can be excised and cultured without difficulty.

ACKNOWLEDGMENTS

We thank Mrs. C. Webb and Mrs. A. Reis for help in preparing the manuscript and the Ministry of Overseas Development, U. K., and the Brazilian Council of Research (CNPQ) and the Brazilian Nuclear Energy Commission (CENA), Brazil, for financial support.

REFERENCES

Arnison, P. G., and W. G. Boll. 1974. Isoenzymes in cell cultures of bush bean (*Phaseolus vulgaris* cv. Contender). Canad. J. Bot. 51:2521–29.

Bailey, C. J., and D. Boulter. 1971. Urease of a typical seed protein of the Leguminosae. *In* J. B. Harbourne, D. Boulter, and B. L. Turner (eds.), Chemotaxonomy of the Leguminosae. Pp. 485–502. Academic Press, New York.

Bingham, E. T., L. V. Hurley, D. M. Kaatz, and J. W. Saunders. 1975. Breeding alfalfa which regenerates from callus tissue in culture. Crop. Sci. 15:719–21.

Blaydes, D. R. 1966. Interaction of Kinetin and various inhibitors in the growth of soybean tissue. Physiol. Plantarum 19:748–53.

Boulter, D. 1975. Breeding for protein yield and quality. Nature 256:168–69.

Boulter, D. 1976. Biochemistry of protein synthesis in seeds. *In* Genetic improvement of seed proteins. Pp. 231–50. National Academy of Science, Washington, D.C.

Boulter, D., and J. A. Gatehouse. 1977. The efficiency of plant protein synthesis in nature. 11th FEBS Meeting, Copenhagen. In press.

Bourgin, J. P., and J. P. Nitsch. 1967. Obtention de Nicotiana haploides à partir d-étamines cultivées *in vitro*. Ann. Physiol. Végét. 9:377–82.

Carlson, P. E. 1973. The use of protoplasts for genetic research. Proc. Nat. Acad. Sci. USA 70:598–602.

Crocomo, O. J., J. E. Peters, and W. R. Sharp. 1976a. Plantlet morphogenesis and the control of callus growth and root induction of *Phaseolus vulgaris* with the addition of a bean seed extract. Z. Pflanzenphysiol. 78S:456–60.

Crocomo, O. J., J. E. Peters, and W. R. Sharp. 1976b. Interactions of phytohormones on the control of growth and root morphogenesis in cultured *Phaseolus vulgaris* leaf explants. Turrialba 26(3):232–36.

Derbyshire, E., D. J. Wright, and D. Boulter, 1976. Legumin and vicilin, storage proteins of legume seeds. Rev. in Phytochemistry 15:3–24.

Dougall, D. K. 1964. A method of plant tissue culture giving high growth energetics. Exp. Cell Res. 33:438–44.

Evans, D. A., W. R. Sharp, and E. F. Paddock. Variations in callus proliferation in leaf tissue cultures of *Glycine max* strain T219. Phytomorphology 26:379–81.

Food and agricultural organisation of the United Nations. Production Year Book. 1970. 1971. FAO 24.

Food and agricultural organisation of the United Nations. Production Year Book. 1972. 1973. FAO 26.

Gamborg, O. L. 1966. Aromatic metabolism in plants: Enzymes of the shikimate pathway in suspension cultures of plant cells. Canad. J. Bot. 44:791–99.

Gamborg, O. L., and A. J. Finlayson. 1969. The amino acid composition of TCA-precipitated proteins and of total residues of plant cells grown in suspension culture. Canad. J. Bot. 48:1857–63.

Gamborg, O. L., R. A. Miller, and K. Ojima. 1968. Nutrient requirements of suspension cultures of soybean root cells. Exp. Cell Res. 50:151–58.

Guha, S., and S. C. Maheshwari. 1964. *In vitro* production of embryos from anthers of *Datura*. Nature 204:497.

Haddon, L. E., and D. H. Northcote. 1975. Quantitative measurement of the course of bean callus differentiation. J. Cell Sci. 17:11–26.

Haddon, L. E., and D. H. Northcote. 1976. Correlation of the induction of various enzymes concerned with phenylpropanoid and lignin synthesis during differentiation of bean callus (*Phaseolus vulgaris*). Planta 128:255–62.

Hageman, R. H., R. J. Lambert, D. Loussaert, M. Dalling, and L. A. Klepper. 1976. Nitrate and nitrate reductase as factors limiting protein synthesis. *In* Genetic improvement of seed proteins. Pp. 103–34. National Academy of Science, Washington, D.C.

Hildebrandt, A. C., J. C. Wilman, H. Johns, and A. J. Riker. 1963. Growth of edible chlorophyllous plant tissues *in vitro*. Amer. J. Bot. 50:248–54.

Holsten, R. D., R. C. Burns, R. W. F. Hardy, and R. R. Herbert. 1971. Establishment of symbiosis between *Rhizobium* and plant cells *in vitro*. Nature 232:173–76.

Hulse, J. H., B. Fawcett, and W. D. Daniels. 1975. Protein supplements: World production and trade. *In* J. T. Harapiak (ed.), Oilseed and pulse crops in western Canada. Calgary, Alberta, Western Co-operative Fertilizers Ltd. 1.

Jeffs, R. A., and D. H. Northcote. 1966. Experimental induction of vascular tissue in an undifferentiated plant callus. Biochem. J. 101:146–52.

Jeffs, R. A., and D. H. Northcote. 1967. The influence of IAA and sugar on the patterns of induced differentiatiion in plant tissue culture. J. Cell Sci. 2:77–88.

Johnson, V. A., P. J. Mattern, D. A. Whited, and J. W. Schmidt. 1969. Breeding for high protein control and quality in wheat. *In* New approaches to breeding for improved plant protein. Pp. 29–40. International Atomic Energy Agency, Vienna.

Kant, V., and A. C. Hildebrandt. 1969. Morphology of edible plant cells and tissue *in vitro*. Canad. J. Bot. 47:849–52.

Kartha, K. K., O. L. Gamborg, and F. Constabel. 1972. Regeneration of pea (*Pisum sativum* L.) plants from shoot apical meristems. Z. Pflanzenphysiol. Pp. 172–76.

Kasha, K. J., and K. N. Kao. 1970. High frequency haploid production in barley (*Hordeum vulgare*). Nature 225:874–75.

Lamport, D. T. A. 1964. Cell suspension cultures of higher plants: Isolation and growth energetics. Exp. Cell Res. 33:195–206.

Liau, D. F., and W. G. Boll. 1970. Callus and cell suspension culture of bush bean (*Phaseolus vulgaris*). Canad. J. Bot. 48:1119–30.

Liau, D. F., and W. G. Boll. 1971. Growth and patterns of growth and division of cell suspension cultures of bush bean (*Phaseolus vulgaris* cv. Contender). Canad. J. Bot.49:1131–39.

Liau, D. F., and W. G. Boll. 1972. Extracellular polysaccharide from cell suspension cultures of bush bean (*Phaseolus vulgaris* var. Contender). Canad. J. Bot. 50:2031–37.

Mehta, A. R. 1966. *In vitro* initiation and growth of root callus of *Phaseolus vulgaris*. Indian J. Exp: Biol. 4:187–88.

Mehta, A. R., G. C. Henshaw, and H. E. Street. 1967. Aspects of growth in suspension cultures of *Phaseolus vulgaris* and *Linum usitatissumum*. Indian J. Plant Physiol. 10:44–53.

Miller, R. A., J. P. Shyluk, O. L. Gamborg, and J. W. Kirkpatrick. 1968. Phytostat for continuous culture and automatic sampling of plant-cell suspension. Science 159:540–42.

Millerd, A. 1975. Biochemistry of legume seed proteins. Ann. Rev. Plant Physiol. 26:53–72.

Millerd, A., D. Spencer, W. F. Dudman, and M. Stiller. 1975. Growth of immature pea cotyledons in culture. Austr. J. Plant Physiol. 2:51–59.

Murashige, T., and F. Skoog. 1962. A revised medium for rapid growth and bioassay with tobacco tissue cultures. Physiol. Plantarum 15:473–97.

Nelson, O. E. 1976. Interpretative summary and review. *In* Genetic improvement in seed proteins. Pp. 383–91. National Academy of Science, Washington, D.C.

Nickell. L. G. 1956. The continuous submerged cultivation of plant tissue in single cells. Proc. Nat. Acad. Sci. USA 42:848–50.

Nitsch, C. 1974. La culture de pollen isolé sur milieu synthétique. Comptes Rendus D278:1031–34.

Olieman-van der Meer, A. W., R. L. M. Pierik, and S. Ruest. 1971. Effects of sugar, auxin, and light on adventitious root formation in isolated stem explants of *Phaseolus* and *Rhododendron*. Meded. Fac. Landbonwwet Rijksunic Genet. 36:511–18.

Oswald, T. H., A. E. Smith, and D.V. Phillips. 1977. Callus and plantlet regeneration from cell cultures of Ladino clover and soybean. Physiol. Plantarum 39:129–34.

Penning de Vries, F. W. T., A. H. M. Brunsting, and H. H. van Laars. 1974. Product requirements and efficiency of biosynthesis: A quantitative approach. J. Theor. Biol. 45:339–77.

Peters, J. E., O. J. Crocomo, and W. R. Sharp. 1976 Effects of caffeine and nicotine on the callus growth and root morphogenesis of *Phaseolus vulgaris* tissue cultures. Turrialba 86(4):337–41.

Peters, J. E., O. J. Crocomo, W. R. Sharp, E. F. Paddock, I. Tegenkamp, and T. Tegenkamp. 1977. Development of haploid callus cells from pollen of *Phaseolus vulgaris*. Phytomorphology 27:79–85.

Peters, J. E., P. H. L. Wu, W. R. Sharp, and E. F. Paddock. 1974. Rooting and metabolism of nicotine in tobacco callus cultures. Physiol. Plantarum 31:97–100.

Phillips, D. A. 1974a. Factors affecting the production of acetylene by *Rhizobium*-soybean cell associations *in vitro*. Plant Physiol. 53:67–72.

Phillips, D. A. 1974b. Promotion of acetylene reduction by Rhizobium-soybean cell associations *in vitro*. Plant Physiol. 54:654–55.

Polacco, J. C. 1977. Nitrogen metabolism in soybean tissue culture. Plant Physiol. 58:350–57.

Reinert, J., and Y. P. S. Bajaj. 1977. Anther culture-haploid production and its significance. *In* J. Reinert and Y. P. S. Bajaj (eds.), Applied and fundamental aspects of plant cell, tissue, and organ culture. Pp. 251–67. Springer-Verlag, Berlin.

Riley, R. 1974. Developments in plant breeding. Agric. Prog. 49:1–16.

Siegel, A., and B. Fawcett. 1976. Food legume processing and utilisation. International Research Centre, Ottawa, Ontario,Canada.

Sunderland, N. 1973. Pollen and anther culture. *In* H. Street (ed.), Plant tissue and cell culture. Pp. 205–39. Blackwell, Oxford.

Tulecke, W. 1953. A tissue derived from the pollen of *Ginkgo bilboa*. Science 117:599–600.

Veliky, I. A. 1972. Synthesis of carboline alkaloids by plant cell cultures. Phytochemistry 11:1405–6.

Veliky, I. A., and S. M. Martin. 1970. A fermenter for plant cell suspension cultures. Canad. J. Microbiol. 16:223–25.

Widholm. J. M. 1974. Cultured carrot cell mutants: 5-methyltryptophan resistant trait carried from cell to plant and back. Plant Sci. Lett. 3:323–30.

Witham, F. W. 1968. Effect of 2,4-dichlorophenoxyacetic acid on the cytokinin requirement of soybean cotyledon and tobacco stem pith callus tissues. Plant Physiol. 43:1455–57.

DEEPAK PENTAL AND JAMES E. GUNCKEL

Cereals

31

Although this paper reviews and reports on the principles and applications of cell and tissue cultures in the cereal grains, its central theme is concerned with a system for tissue manipulation (cf. *Triticum aestivum*) that will be highly responsive to microbial techniques for selection and isolation of desired mutants in order to redesign important crop species for improved yield and food quality.

I. APPLICATIONS

A. GENERAL PERSPECTIVES ON THE WORLD FOOD PROBLEM

It is very hard to reach any final conclusion on all of the ramifications of the world food situation, as there are views ranging from optimistic to very pessimistic (Sanderson, 1975). However, there is an emerging awareness in the United States about the food problem and both Science (Abelson, 1975) and Scientific American (Flanagan, 1976) have devoted full issues to the topic.

B. AGRICULTURAL RESEARCH IMPERATIVES

In the past two years a concensus has emerged—that agricultural research in the U.S. is lagging and that there is a necessity for exploring new innovative techniques to *augment* the returns from conventional plant breeding. The lack of new approaches was underscored by the National Research Council/Board of Agricultural and Renewable Resources (NRC/BARR) Study on World Food and Nutrition (1975) and the International Conference on Crop Productivity—Research Imperatives (1975), airing the views of some prominent researchers in the

Deepak Pental and James E. Gunckel, Botany Department, Rutgers University, New Brunswick, New Jersey.

field of agriculture. It was concluded that yields achieved by conventional breeding methods have tapered off and a redesigning of the important crop species should be undertaken, in terms of yield, fertilizer responsiveness, nutrition, nitrogen fixation, and photosynthetic activity.

C. REDESIGNING OF CROP SPECIES

A number of recent books and review articles have discussed the development of a number of innovative techniques for genome alterations and have reported on their potential and current applications for crop improvement (Bates and Deyoe, 1973; Bottino, 1975; Carlson, 1975; Carlson and Polacco, 1975; Chaleff and Carlson, 1975; Green, 1977; Kasha, 1974; Melchers, 1972; Nickell and Heinz, 1973; Reinert and Bajaj, 1977; Smith, 1974; Street, 1974, 1977a,b; Thomas and Davey, 1975; Vasil and Nitsch, 1975).

The desired alterations in the genome of crop plants can be achieved by *intercellular gene transfer* and by *somatic mutagenesis*. All of the terms used here are defined following Merril and Stanbro (1974).

II. INTERCELLULAR GENE TRANSFER

Intercellular gene transfer can be accomplished by somatic hybridization, transformation, and transgenosis. *Somatic hybridization* consists of using protoplasts from different genera, species, varieties, and cultivars and fusing them in sterile culture (this process can be compared to bacterial *conjugation*). *Transformation* is the uptake of soluble DNA/RNA to initiate transcription or translation of parts of the genome or genetic transformation of homologous or even heterologous DNA, or to initiate viral infection. *Transgenosis* is gene transfer by bacteriophage or virus, as in bacterial *transduction*. The term *transgenosis* is used rather than *transduction* to avoid premature assumption of chromosomal integration.

The potentialities of these techniques for intercellular gene transfer have been reviewed recently by Bottino (1975), Carlson and Polacco (1975), Chaleff and Carlson (1974), Holl (1975), Ledoux (1975), Merril and Stanbro (1974), Reinert and Bajaj (1977), and Hess (1977).

A. SOMATIC HYBRIDIZATION

Cocking (1972), Gamborg and Miller (1973), Street (1973a,b), Eriksson et al. (1974), and Bajaj (1977) reviewed techniques for the isolation, culture, fusion, and regeneration of protoplasts. The techniques are important for cell modification and somatic hybridization (Bajaj, 1974c).

The importance of isolated protoplasts lies in their capacity to take up macromolecules (Ohyama et al., 1972), viruses and bacteria (Davey and Cocking, 1972), and isolated chloroplasts (Potrykus, 1973), or plasmids, all leading to cell transformations. An especially intriguing possibility with bacterial uptake is to introduce the capacity for nitrogen fixation into new species (Davey, 1977). Under the proper conditions, isolated protoplasts in proximity may fuse, offering the possibility of wide crosses and overcoming crossability barriers (Bates and Deyoe, 1973). Isolated protoplasts also offer suitable material for cell biochemical studies.

   1. *Establishment of Large, Homogeneous Populations*
      of Plant Protoplasts
    a. *Protoplast isolation*

     Isolated plant protoplasts are derived from cells that have had their cell walls removed by mechanical or enzymatic methods. Mechanical isolation involves immersing the plant tissue in hypertonic sugar solution, causing plasmolysis, and then cutting the tissue to release the protoplasts. The deleterious effects of cell wall–degrading enzymes on cellular metabolism are avoided by this method.

     The first isolation of higher plant protoplasts with enzymes (Cocking, 1960) was performed with a cellulase obtained from culture filtrates of the fungus *Myothecium verrucaria*. Since that time, enzymes with a high degree of purity and activity have been available commercially. The major advantages of enzymatic over mechanical isolation are: (1) there is a larger yield of protoplasts per gram of tissue, (2) meristematic tissues can be used, and (3) no broken cells result as from cutting tissue (Ruesink, 1971), so there are no cell degradation products.

     Protoplasts have been isolated from leaves, pollen mother cells, microsporocytes, callus cell suspensions, and ovary tissue. For example, Power and Cocking (1970) treated the peeled lower epidermis of *Nicotiana* leaves in an enzyme mixture (0.5% macerozyme + 2% Onozuka cellulase in 13% sorbitol or mannitol at pH 5.4) and incubated 15–18 hr at 25 C. The incubated peels are teased, filtered, centrifuged, washed with 13% sorbitol solution and 20% sucrose solution, and concentrated by centrifugation. This method, with very little variation, is still used for isolating leaf protoplasts. Bajaj and Davey (1974) and Bajaj (1974a, b) detail the techniques for isolating pollen protoplasts. Pollen is squeezed out of the anther, sterilized with a 2% solution of sodium hypochlorite for 10 min, centrifuged into a pellet, and washed in distilled water. The pollen is incubated with various enzymes for 24–36 hours until released. Release of protoplasts from pollen mother cells and pollen tetrads is more simple and is accomplished in less than an hour.

Release of meiotic protoplasts from *Lilium* and *Trillium* was accomplished by Onozuka cellulase and macerozyme in modified White's medium plus sucrose. The protoplasts were cultured with yeast extract added to basal medium (Ito, 1973). Protoplasts from callus cell suspensions are readily obtained by treating with 2–4% Onozuka cellulase in 0.6M mannitol for 4–6 hr at 30–33 C (Schenk and Hildebrandt, 1969; Bajaj, 1977). Protoplasts have been isolated from nondifferentiating ovular callus of *Citrus sinensis* (Vardi et al., 1975). It is also possible to isolate intact vacuoles and chloroplasts of mature plant tissues (Wagner and Siegelman, 1975).

Isolation of protoplasts from cereals and other monocots is not as common as for dicots. Only the protoplasts of corn (Bawa and Torrey, 1971; Motoyoshi, 1972), oats (Ruesink and Thimann, 1965; Motoyoshi, 1972; Hall and Cocking, 1974; Kaur-Sawhney et al., 1976), rice (Deka and Sen, 1976; Maeda and Hagiwara, 1974; Somatic Hybridization Research Group, 1975), brome grass (Schenk and Hildebrandt, 1969), sugarcane (Maretzki and Nickell, 1973), barley (Taiz and Jones, 1971; Evans et al., 1972; Schaskolskaya et al., 1973; Dudits et al., 1976a; Koblitz and Saalbach, 1976), wheat (Evans et al., 1972; Bajaj and Davey, 1974), and rye (Evans et al., 1972) have been isolated. Cocking (1972) and Cocking and Evans (1973) reviewed the methods of protoplast isolation. Bhojwani et al. (1977) reviewed protoplast technology in relation to crop plants.

Robert Sotak in this laboratory evaluated most of the methods and enzymes used previously, except the nonspecific enzyme Xylonase, used to isolate protoplasts from sugar beet leaves (Nam et al., 1976). We conducted experiments on isolating callus cells and leaf protoplasts from *Triticum aestivum* cv. McNair 701 closely following the techniques of Evans et al. (1972).

Protoplasts were isolated in very large numbers from leaf tissue. When thin leaf sections were vacuum infiltrated for 10, 20, or 30 min at the beginning of enzyme incubation, the maximum yield was approximately 1.3×10^6 protoplasts/g leaf tissue in all cases. Vacuum infiltration for longer periods of time was detrimental to yield. A variety of enzyme mixtures were dissolved in 0.6 M sorbitol or mannitol as osmotic stabilizers, and 1% dextran sulfate was added to stabilize the membranes and facilitate cell separation (Takebe et al., 1968). The mixture was adjusted to pH 5.6 with 1% NaOH and buffered with Trizma maleate. Lower epidermis peels (Kaur-Sawhney et al., 1976) exposed mesophyll cells, which were added to the enzyme mixture at 30 C in the dark and incubated for 4 hr. Protoplasts were collected by centrifugation at 500 rpm for 10 min. The pellet was resuspended in 0.6 M sorbitol and centrifugation was repeated. Only normally appearing, intact, spherical protoplasts were counted in a Neubauer hemacytometer, and the yield was determined. Protoplast viability was determined by observing protoplasmic streaming or by staining with fluorescein diacetate, which is taken up by living protoplasts and converted to a bright green fluorescent derivative by enzymatic cleavage (Heslop-Harrison and Heslop-Harrison, 1970). Protoplast viability was also determined by measuring the rate of respiration with an oxygen electrode (Taiz and Jones, 1971).

Protoplast isolation was carried out with several different pectolytic and cellulolytic enzymes. *Macerozyme* (used at 0.5–2%) is described by the producers as a pectinase derived from *Rhizopus* species. Sigma *pectinase* is isolated from *Aspergillus* species. *Driselase* is described as a one-step enzyme, since it is a mixture of cellulase and pectinase derived from *Trichoderma* species. *Cellulysin* (2%) is a cellulase derived from *Trichoderma viride* and exhibits slight hemicellulose activity as well. *Onozuka R-10* (2%) is a cellulase derived from *Trichoderma* species. Macerozyme was selected as the pectinase over the Sigma pectinase. No hemicellulases were used.

Wheat callus was subjected to enzymatic treatment with Cellulysin, Onozuka R-10, and Driselase, but was resistant to even high concentrations (5%) of these enzymes after 18 hr incubation. Longer incubation time was impossible, since 18 hr treatment resulted in abnormal protoplasts. The few callus protoplasts isolated exhibited an extrusion of the nucleus, resulting in nonspherical protoplasts that did not survive in subculture. Callus age was not a factor in protoplast isolation, since 10-day-old callus released very few protoplasts.

Cellulysin, Onozuka R-10, and Macerozyme were successful in isolating leaf protoplasts, and Driselase was ineffective. Cellulysin had the most activity for wheat protoplasts, yielding 4.5×10^6 protoplasts per g from *young* leaves. Onozuka R-10 and Macerozyme yielded a maximum of 3.3×10^6 protoplasts per g for comparably aged leaves.

Isolation of wheat leaf protoplasts depended upon a number of factors:

(1) Leaf age and type of enzyme

Using 10-, 20-, and 40-day-old leaves, 10-day-old leaves treated with cellulysin yielded the greatest number of protoplasts, nearly twice that for other enzymes and older leaves. Onozuka R-10 and Macerozyme released the greatest number of protoplasts from 20- and 40-day-old leaves, but only about 0.5×10^6 cells/g more than cellulysin.

(2) Osmolarity

The optimum range of osmotic concentration was found to be 0.6–0.7 M mannitol or sorbitol.

(3) pH

The optimum pH range was 5.5–5.8.

(4) Photoperiod

Incubation in the dark resulted in less clumping of chloroplasts than in the light. In the dark, the chloroplasts formed a distinct peripheral boundary along the plasmalemma. Incubation in the light tended to cause clumping of the chloroplasts and thus decreased the yield.

(5) Temperature

The optimum temperature for incubation was 30 C, but in a range of 24–35 C.

(6) Ancillary chemical

The addition of potassium dextran sulphate to the enzyme mixture increased the yield of protoplasts and decreased protoplast damage (Takebe et al., 1968). It was found superior to divalent cations, such as Mg^{2+} or Ca^{2+}, in maintaining the integrity of the plasmalemma, preventing bursting and increasing protoplasmic streaming. A combination of $MgCl_2$ or $CaCl_2$ with potassium dextran sulphate resulted in the same clumping of chloroplasts as when potassium dextran sulphate was omitted and in lower yields. We did not evaluate the effect of pretreatment with a solution of 1 μg/ml cycloheximide or 5 μg/ml kinetin for 18 hr in the dark at 23 C, reported to reduce cellular debris, cause less damage to the protoplasts, and perhaps inhibit hydrolase activity (Kaur-Sawhney et al., 1976).

b.   *Protoplast culture*

(Narayanaswamy, 1977; Bajaj, 1977; Street, 1973b.)

Protoplasts have been cultured in one of three ways: (1) suspended in liquid medium in hanging drops (Bawa and Torrey, 1971), (2) in microchambers (Jones et al., 1960; Vasil and Vasil, 1973), or (3) by plating them onto agar nutrient medium (Bergmann, 1960; Nagata and Takebe, 1971; Dougall, 1973). Plating is the most used and most satisfactory method since, in most cases, either a nurse culture or neighboring cells are necessary for continued growth and division of isolated single cells (Hildebrandt, 1977).

All protoplast culture media are essentially either a modified White's (1939), on the one hand, or a modified Murashige and Skoog (1962) (M-S) medium on the other.

c. *Cell wall regeneration and cell division of protoplasts*

Pojnar et al. (1967) reported that continued survival of isolated protoplasts occurred only if they reverted back to their original cell structure. Protoplasts in culture either start to regenerate a cell wall very quickly or do not survive. Reinert and Hellmann (1971) observed that carrot protoplasts underwent several nuclear divisions without cytokinesis. After wall regeneration, *Nicotiana* protoplasts enlarge and elongate and the chloroplasts become organized (Bajaj, 1972, 1977). The cells start to divide in 2–3 days, and after 3 weeks light green colonies are visible. Once small colonies are formed, further growth is slowed unless the colonies are transferred to a mannitol-free medium. In other words, regeneration of a wall is not a prerequisite for nuclear division (Reinert and Hellmann, 1971), but is necessary for cytokinesis (Pojnar et al., 1967; Meyer and Abel, 1975b; Abo El-Nil and Hildebrandt, 1976). Meyer and Abel (1975a) suggest that the complete wall is the place of synthesis of the substance(s) necessary for the division activity of the cytoplasm.

Abo El-Nil and Hildebrandt (1976) attempted to quantify the progressive cell wall deposition on the surface of *Pelargonium hortorum* anther callus protoplasts. Cell wall regeneration was detected after 24 hr incubation and maximum deposition was achieved in 60 hr. Mitotic divisions started in 12 out of 174 reconstituted cells after 4–5 days incubation. After nuclear division, new cell wall material was deposited on both sides of the "middle lamella." Cytokinesis was completed after 15 hr. Binding (1966) found that only nucleated protoplasts regenerated a cell wall.

The biochemical growth requirements for isolated, cultured cells are quite analogous to the growth requirements in excised tissues and whole plants, and specifically as related to cell wall synthesis. The requirements for cell wall synthesis in whole plants are sources of sugar (Baker and Ray, 1965), auxin (Ray and Baker, 1965), protein synthesis (Nooden and Thimann, 1963), RNA and protein synthesis (Key, 1964), and mRNA (Key and Ingle, 1964). Zelcer and Galun (1976) followed the incorporation rates of radioactive precursors of DNA, RNA, and protein in newly isolated tobacco mesophyll protoplasts, leading to the onset of cell division in this system. Mäder, Meyer, and Bopp (1976) discuss the relationship of peroxidase isoenzymes to the dedifferentiation of isolated protoplasts.

The primary cell walls of six suspension-cultured monocots (wheat, oats, rice, sugarcane, Brome grass, and rye grass) were investigated (Burke et al., 1974). The monocot primary cell walls contain arabinoxylan as a major component (40%), plus 9-14% cellulose, 7-18% uronic acids, 7-17% protein, and less than 0.2% hydroxyproline. Mixed glucans were found only in the cell wall preparations of rye grass endosperm cells. For a more complete treatment of primary cell walls in culture see Albersheim (1974).

Aggregates of cells divide more often than free cells, resulting in an increase in size of cell clusters (Narayanaswamy, 1977). The density of plated cells is a key factor in determining cell division. In general, protoplast plating density must exceed $10^4/ml$ for induction of division, although lower plating densities are possible if either synthetic "conditioned" medium (Street, 1973 a,b) or the nurse culture technique (Hildebrandt, 1977; Vardi and Raveh, 1976) is used. A "conditioned" medium is one in which cells or tissues in culture release into that medium their own biosynthetic products, thereby assisting the growth of single cells or a few cells per unit volume.

d.   *Fusion of protoplasts*

This topic has been reviewed comprehensively by Bajaj (1977). He discusses protoplast fusion in two categories: (1) spontaneous fusion and (2) induced fusion.

Spontaneous fusion, usually in young leaf protoplasts, is possible when the cell wall is degraded, the plasmodesmata expand, and adjacent protoplasts, intraspecific only, fuse (Bajaj and Davey, 1974). As an example of spontaneous fusion, Motoyoshi (1972) enzymatically isolated protoplasts from callus cells of hybrid maize endosperm (subcultured every 30 days for 7 yr) and cultured them on liquid White's medium (with organics) plus yeast extract and 0.4 M mannitol, with or without 0.1 mg/l kinetin. In 3-6 days some of the protoplasts enlarged and budded. Nonsynchronous mitotic figures were detected in mononucleate and multinucleate protoplasts. About 50% of total protoplasts were multinucleate, suggesting that fusion of single protoplasts had occurred spontaneously during the protoplast isolation process.

Induced fusion requires an inducing agent to cause the protoplasts to adhere and the membranes to fuse locally and then more extensively until the protoplasts intermingle (Withers and Cocking, 1972). A wide range of additives have been used, especially dextran sulphate (Kameya, 1975), sodium nitrate (Powers et al., 1970), and polyethylene glycol = PEG (Kao and Michayluk, 1974). For induced fusion, Power et al. (1970) outlined the controlled conditions under which plant protoplasts can be induced to fuse. Oat and maize root protoplasts treated with sodium nitrite readily adhered and fused to form aggregates with the potential to provide new genetic combinations.

No comprehensive study on protoplast isolation, culture, and fusion in monocots has been made, so a few selected dicot studies are included here as a guide to the direction that future work on monocots might take.

Protoplasts of *Nicotiana glauca* and *N. langsdorffii* were isolated and fused as an example of interspecific hybridization (Carlson et al., 1972). They succeeded in producing a somatic hybrid of amphidiploid nature. That is, the somatic hybrid had 42 chromosomes and biochemical and morphological characters that were identical with those of the sexual amphidiploid *N. glauca langsdorffii* (2n = 42). Kameya (1975) developed haploid plants of tobacco from anther culture and induced callus. Cell suspensions from callus were cultured on a M-S medium with 0.1 mg/l 2,4-D for 10 days at 30 C and the cell suspension plated on nutrient agar, after straining out the protoplasts with a stainless steel mesh and washing in 10% mannitol solution. Carrot and *Senecio* protoplasts were prepared in the same way. Mixed populations of protoplasts (*Daucus carota* [root] – *N. tabacum* [leaf], *Senecio vulgaris* [leaf] – *N. tabacum* [leaf], *N. tabacum* [leaf] – *N. tabacum* [albino callus]) were cultured in M-S medium supplemented with kinetin (1.0 mg/l) and IAA (0.1 mg/l) at 25 C under constant illumination. Isolated protoplasts were suspended in 7.5% mannitol solution containing 15–20% dextrans. When mannitol concentration increased, the amount of dextran sulphate necessary for aggregation decreased. Protoplasts aggregated at random and divided to form colonies after about 2 weeks, but then stopped dividing and did not grow as colonies. Power et al. (1975) produced cell suspensions of crown gall callus of *Parthenocissus tricuspidata* and *Petunia hybrida* cv. Comanche. Protoplasts were resuspended in 13% mannitol solution with Heller's (1953) salt medium, centrifuged, and washed with 20% sucrose solution. Equal aliquots (0.5 ml) of the two protoplast suspensions were mixed together and resuspended in 7 ml of a fusion-induction solution—10.2% sucrose and 5.5% sodium nitrate, centrifuged twice, and held for 25 min at 25 C. The fusion-induction medium was replaced with 20% sucrose and maintained for 3 days at 25 C under constant illumination. Fusion aggregates were transferred to Nagata-Takebe (1971) (N-T) medium with 13% mannitol for 1 hr. The aggregates were broken up and plated on N-T medium with 10% mannitol and agar, and again subcultured on the same medium with 6.5% mannitol at 25 C in continuous light for 15 wk. Cell/callus were transferred to M-S medium without growth regulators and subcultured every 6–8 wk. The majority of calluses died after 10 wk culture, some differentiated roots, and a few recovered. The detection of petunia-specific peroxidase isoenzymes in cultured cells possessing *Parthenocissus* chromosomes has here provided a marker for identification of these calls. It is not clear whether the hybrid cells are cytoplasmic (cybrids) hybrids in which there is a mixture of cytoplasm of both genera with the nucleus of only one (*Parthenocissus*). They assumed the results were due to incompatibility and chromosome elimination. Izhar and Power (1977), using highly inbred lines of petunia, identified genetic variations in the hormone requirements of two species together with their F_1 and in backcrosses to the two parental lines. Only a few genes may control such growth requirement differences, and the different stages of protoplast development may be controlled by different genes. This suggests that it might be possible to breed individual plants possessing a desired genetic marker linked to a specific hormone requirement at the protoplast level. One might then select to recover a particular somatic hybrid following proto-plast fusion. This has not been too successful on tobacco hybrids (Carlson et al., 1972).

Constabel et al. (1976) fused protoplasts from soybean cell suspension cultures with those from leaves of two *Pisum sativum* varieties, four *Nicotiana* species, and *Colchicum autumnale*, and maintained them in a B-5 medium with additives. The heteroka-ryons were identified by the co-occurrence of chloroplasts from leaf primordia and leucoplasts and dense cytoplasm from soybean protoplasts. The heterokaryons all divided in 2–3 days. The percen-tage of heterokaryons obtained by intergeneric fusion varied from 3 (soybean–*N. langsdorffii*) to 23 (soybean–*Pisum sativum*).

Binding (1976) isolated leaf protoplasts of *N. tabacum* var. Xanthi and *Petunia* hybrids and cultured in modified N-T medium.

Free nuclei of *Petunia* were prepared by diluting the osmolarity of the protoplast suspension, centrifuging in 0.6 M sucrose, and resuspending in a solution of equal parts of 0.6 M sucrose and modified N-T medium. The liquid pericarp of tomato was separated into larger and smaller subprotoplasts by the same procedure as for *Petunia* nuclei. The fraction with smaller subprotoplasts was used for fusion experiments. This fraction was a mixture of subprotoplasts containing nuclei and chromoplasts, chromoplasts without nuclei, and no chromoplasts but with the nucleus embedded in a portion of cytoplasm. *Petunia* nuclei and protoplasts, tomato protoplasts, and tomato subprotoplasts were mixed in several combinations, agglutinated by PEG, and fused by treatment with calcium nitrate solution at pH 9. None of the tobacco-petunia protoplasts fused to form heterokaryons. Most of the heteroplastic cell clusters exhibited a segregation of two types of plastids. With tobacco protoplasts and *Petunia* nuclei, many nuclei entered the protoplasts after agglutination but no mitoses occurred. Tobacco protoplasts combined with tomato subprotoplasts to form about 10% symplasms. Symplasms regenerated cell walls but did not divide. Heterokaryotic protoplasts were visible in phase contrast. Lack of division in symplasms and fusion bodies indicated incompatibility and possibly also the segregation of plastids. Incompatibility in the *Petunia*-tomato system was probably due to the interaction of the tomato nucleus with the *Petunia* cytoplasm, since no tomato nuclei were present in multicellular regenerants.

Miller et al. (1971) have shown in somatic hybrids that the cytoplasms and nuclei fuse in interphase for soybean heterokaryons; Constabel et al. (1975) showed fusion in soybean-pea heterokaryons; Kao et al. (1974) demonstrated that protoplast fusions took place by association of nuclei during metaphase in heterokaryocytes of soybean-barley, soybean-pea, and soybean-corn; and Kartha et al. (1974) demonstrated the same for rapeseed-soybean. All showed a fusion of intergeneric protoplasts by means of PEG. The heterokaryons in all cases regenerated a cell wall and divided to a limited extent in appropriate media.

Gosh and Reinert (1976) investigated the behavior of the nuclei and the chromosomes during cell division of heterkaryons of *Petunia hybrida* and *Atropa belladonna*. In newly fused *Atropa* and *Petunia* "heterokaryons," two or more nuclei were randomly distributed in the protoplast. Twenty-four hr later anaphase nuclei were closely associated, probably on a common spindle, with each complement clearly distinguished. Continuous division of these unfused nuclei led to the formation of chimeral colonies "which could lead to hybrid callus and ultimately hybrid plants." There is nothing in this report to indicate that hybrid plants are actually produced.

Reinert and Gosch (1976) produced "heterons" of protoplasts from *Daucus carota* and *Petunia hybrida*, which divided continuously to form colonies of up to 15 cells. In liquid medium the fused protoplasts regenerated cell walls after 48 hr in culture; in 2-3 days they started to divide and formed chimeral colonies, which developed into calli "which could be induced to regenerate plantlets." Again, there is no evidence presented that plantlets actually were regenerated.

Dudits et al. (1976a) fused the protoplasts of *Daucus carota* with leaf protoplasts of *Hordeum vulgare* by treatment with PEG and cultured them in a modified B-5 medium. More than 50% of the parental carrot protoplasts formed cell groups, which in turn formed callus or embryoids, while barley parent protoplasts were unchanged or dead. Nonfused *barley* protoplasts never underwent mitosis. The heterokaryocytes (carrot and barley fused cytoplasms) contained one nucleus of either parent in 35-40% of the fusion products; the rest had more than one nucleus of one or both patents indicating multiple protoplast fusion. In 4-5% of the fused cells, barley chloroplasts were observed within the dense cytoplasm of carrot protoplasts, an indication of heteroplastic fusion. Homoplastic fusion products were also observed. The heterokaryocytes regenerated a cell wall within 24 hr; after three days of culture about 3% of the heterokaryocytes had divided at least once, and in 1 wk they formed small cell clusters. In general, nuclear division occurred synchronously following premitotic fusion of interphase nuclei, probably requiring the activation of barley by carrot protoplasts. Daughter cells of fusion products divided and formed cell clusters and then calli, indistinguishable from those derived from nonfused *carrot* protoplasts. It was not established if the synkaryons formed actually divided.

Dudits et al. (1976b) isolated carrot protoplasts from cell suspensions with an enzyme mixture and cultured on B-5 medium supplemented with 1 mg/l 2,4-D. Protoplasts were subcultured on a protoplast culture medium (Kao, 1975), regenerated cell walls and started to divide within 2 days, in 3-4 days formed cell clusters and small proembryo cell aggregates, and formed embryoids within 2 wk with the proper zeatin-NAA ratio, with a frequency of about

12%. Isolated protoplasts were also cultured and induced to fuse with PEG. About 30% of the isolated protoplasts treated with PEG formed homokaryons, and about 10% underwent mitosis and developed into embryoids. PEG treatments, then, do not affect the capacity of protoplasts to divide and form embryoids.

The latest word seems to be that the *intrageneric* fusion product, instead of being an amphidiploid heterokaryon, might best be described as a dikaryon or dikaryophase nucleus. By contrast, in *intraspecific* somatic cell hybrids, up to about half of the regenerated plants are normal amphidiploids and the remainder are some degree of polyploidy with associated aneuploidy (Melchers and Labib, 1974).

B. TRANSFORMATION

By uptake of soluble DNA: Holl et al. (1974) and Ledoux (1975).

The uptake of nucleic acids is now well established (Ledoux, 1975; Bhargava and Shanmugan, 1971; Hess, 1977), but their subsequent fate is not well established (decay of DNA and RNA, integration or no integration in the case of DNA). Isolated cell organelles, protoplasts, or cells and tissues in culture may act as receptors (Hess, 1977).

Ledoux and Huart (1972) reported the successful uptake of isolated, polymeric, labeled DNA by germinating barley. In general, most of the foreign DNA is broken down and reutilized, but in a few cases it escapes destruction and persists. Either the foreign DNA is kept free (outside the nucleus) (Hotta and Stern, 1971) or it is bound to the DNA of the recipient (Ledoux and Huart, 1972) by episome or exosome model, but the results in either case are essentially the same.

Ledoux and Huart (1972) incubated the seeds of barley in labeled DNA of *B. subtilis* and isolated a small amount of DNA that had a higher density than the bacterial or most DNA; that is, there was evidence for a duplex structure in the barley seedling root tip indicating integration and replication of the exogenous DNA. Ledoux et al. (1975) took this as evidence for an "episome" model. The foreign gene is integrated into the recipient DNA, like episomal DNA, through the homology of a few base pairs.

Fox and Yoon (1970), on the other hand, proposed an "exosome" model for exogenous DNA. The transplanted genetic information is associated with the original gene locus but is never integrated with the linear chromosomes structure. The exosome replicates in step with the chromosome and can be lost or transmitted with it at the time of cell division or meiosis. Glick and Majumdar (1972) suggested that the association of DNA in the exosome model might better be explained by a "loop" model in which both strands of the donor are covalently linked to the host DNA and there is then replication of the integrated material. A selective and mutually exclusive transcription of either allele may occur. The phenotypic result is the same as for the "exosome" model.

Ledoux (1975) notes that locus association is limited to a few plants like barley and *Arabidopsis thaliana*. A survey of *Helianthus, Pisum, Nicotiana, Trifolium, Medicago, Lycopersicon,* and *Sinapis* showed that no generalization was possible—foreign DNA may be degraded or destroyed, found intact or free, and integrated into a variety of tissues to varying degrees.

Ledoux's observations have been challenged by Kleinhofs (1975) and Kleinhofs et al. (1975), who could not confirm the evidence for integration or replication of exogenous DNA in pea, tomato, and barley reported by Ledoux's group. They concluded that contamination was what had been interpreted as evidence for integration. Clarification of these contradictory results is needed.

Hess (1975) reviewed earlier work in which a white-flowered petunia mutant was transformed to a red-flowered plant by treating seedlings of the white-flowered plant with DNA isolated from the leaves of the red-flowered plant. This work was vigorously challenged by Bianchi et al. (1974a,b), but countered by Hess (1977).

Watts et al. (1973) prepared *N. tabacum* var. White Barley protoplasts by the method of Motoyoshi et al. (1973) and traumatized by centrifugation and resuspension before inoculation. Inoculation was at 25 C or 0 C with virus (Cowpea chlorotic mottle, 1 $\mu g/ml$) and a polycation (e.g., poly-L-ornithine, 2 $\mu g/ml$) incubated together in buffered mannitol (unless the virus is positively charged). RNA degrades rapidly, possibly because it is not integrated into the host's metabolism and integration could not be demonstrated.

Hess (1975), in contrast to the views of Watts et al. (1973) and others, considered the *expression* of exogenous DNA in cultured cells to be convincing, and this gave more credance to uptake and integration. For example, Doy et al. (1973b) detected 5–10% of tomato calli (treated with $\phi 80$ plac[+], the bacterial gene for β-galactosidase) growing better than usual, even on 10% lactose media. In Hess's view the long duration of the experiments made it highly probable that the better growth was not due to stored carbohydrates but to higher β-galactosidase activity, but Hess has found (in petunia) that β-galactosidase activity can vary widely from plant to plant. If one accepts the exosome model, this change in activity could result from a selective and mutually exclusive transcription of either allele following replication of the integrated material.

The many attempts to transform plant cells with soluble DNA have produced few positive results (Hess, 1977).

C. TRANSGENOSIS

(Holl et al, 1974; Gresshoff, 1975; Hess, 1977.)

The term *transgenosis* describes the transfer and subsequent expression of exogenous DNA and carries no implication of a mechanism for integration. The advantage of bacteriophage transmission has been reviewed by Merril and Stanbro (1974). Use of the complete phage and the ability of phage genome to circulate in the cytoplasm may protect exogenous DNA from degradation by restricting enzymes of the host. Hence, Doy et al. (1973a) coined the term *transgenosis* to describe bacteriophage transmission of genetic information to plant cells when imprecise data for uptake and integration prevailed. Gresshoff (1975) recalled that many transmissions had been called "transformations" because of their analogy to bacterial systems in which donor DNA was used to confer a new phenotype to host cell. Actually, however, none of the prokaryotic-eukaryotic gene transfer systems have been investigated enough to justify the term *transformation* with its precise understanding of transcriptions and translations.

Ohyama et al. (1972) reported the uptake of *E. coli* DNA by protoplasts from cell suspension cultures of soybean, carrot, and *Ammu visnaga*. Up to 1% of the exogenous *E. coli* DNA supplied was incorporated in the acid-precipitable fraction during a 4-hr incubation.

Carlson (1973) reported that two T-3 specific enzymes were synthesized in barley after infection with coli phage T-3, which normally grows in *E. coli*. The S-adenosylmethionine-cleaving enzyme and a RNA-polymerase are not normally produced in *Hordeum*.

Doy et al. (1973a) reported the transfer of genes from the galactose and lactose operons of *E. coli* to haploid tomato cell callus. The cells were plated on a medium where either galactose or lactose was used as the sole carbon source, after infection with pgal$^+$ or ϕplac$^+$. The transformed cells and controls could not then grow on galactose or lactose as a carbon source.

Johnson et al. (1973) have reported the expression of *E. coli* lac genes carried on sycamore cells by phage λplac 5. Again, only the treated cells would grow on lactose medium.

In all these studies on transformations and transgenosis, the potential for successful gene transfer seems assured, since uptake of exogenous DNA into the cytoplasm is proved for protoplasts and is highly probable in cell and tissue culture (Hess, 1977). Isolated cell organelles, protoplasts, or cells and tissues in culture may act as receptors. The advantages of bacteriophage transmission (for animals) has been reviewed by Merril and Stanbro (1974).

In all these studies it is unclear whether the phage is integrated into the host chromosome, merely shows episomal maintanance, or is being fragmented and shows a temporary transgenetic state. The transferred genetic material may exist in the form of exosomes in the cytoplasm or be integrated into receptor DNA chromosomes (episome model). The phenotypic expression may reflect several alternatives (Hess, 1977). Only a few examples are available that deal with *integration* of exogenous DNA into the genome or plastome of the DNA receptor.

The *expression* of transferred genes, involving transcription and translation, seems to be more demonstrable in the *correction of mutations* than in the introduction of new mutations. An example of "correction" is the induction of anthocyanin synthesis in the white-flowered petunia (Hess, 1977).

We have little substantial data on the replication of exogenous DNA. Soyfer et al. (1976) injected DNA extracts from endospermal milk and leaves of normal distichous barley plants into the grains of hexastichous barley, as a milk-ripeness stage, carrying waxy mutation (defective synthesis of amylose). Upon injection of wild-type DNA, a certain fraction of plants showed changes toward wild type (normal starch synthesis). In the second generation (seeds) most plants returned to the recipient type. In the first generation with 99.5% of pollen wild type, the alterations were relatively stable, as only 1/5 to 1/6 of the plants returned to the original recipient type.

Concurrent with alterations of starch structure, alterations of the type of spike were observed. In the first generation, plants preserved a hexastichum spike, whereas in the second generation the plant with wild-type pollen developed donor spikes.

A general difficulty seems to have arisen in genetic transformations. In barley, potato, and other plants studied, the transformations do not extend through the seed generation, so many problems must be resolved before any real advances may be achieved in intercellular gene transfer.

An alternative to feeding soluble DNA, or using phage as a vector, is TIP (tumor-inducing principle) of *Agrobacterium tumefaciens*, which, after much speculation, seems to be a plasmid (or certain sequences of the plasmid).

Zaenen et al. (1974) isolated plasmids from crown gall–inducing strains of *Agrobacterium tumefaciens*; they were unable to find such plasmids in nonpathogenic strains. They concluded that plasmids present in crown gall–inducing bacteria could be TIP. Chilton et al. (1977) have shown the presence of multiple copies of a small part of the plasmid genome in a tobacco crown gall tumor line by DNA hybridization studies. Successful infection by virulent bacterium leads to:(1) hormonal autotrophy for cell proliferation in culture; (2) synthesis of unusual amino acid derivatives, e.g., octapine and nopaline; and (3) RNA transcripts of the foreign genetic information in the tumor cells (detected by Drummond et al., 1977). Thus, plant eukaryotic cells seem to be transformed by addition of new genetic information to the host DNA, which is stably maintained through subsequent cell divisions. The transmissible sequences of the virulence-plasmid could be used as a vector for future genetic manipulations in higher plants, offering a unique system of genetic transformation.

III. SOMATIC MUTAGENESIS

Somatic mutagenesis is the introduction of genetic variability into large, homogeneous populations of plant cells by exposure to chemical and physical mutagens. Conditions that select for defined mutant types may then be imposed upon this population of cells.

Somatic mutagenesis can be accomplished by the establishment of large, homogeneous populations of plant cells, mutagenesis, and selection.

Dominant and semidominant mutants are obtainable from cultures of diploid cells. If haploid tissue is used, direct selection of selfed, recessive mutants becomes possible. These possibilities have been reviewed by Smith (1974) and Nabors (1976).

A. ESTABLISHMENT OF LARGE, HOMOGENEOUS POPULATIONS OF PLANT CELLS

(Yamada, 1977 [diploids] and Clapham, 1977 [haploids].)

The first step in establishing a large, homogeneous population of plant cells is to obtain a suitable source of material. Traditionally, the pith tissues of the stem and the cambium layer of the root supply large amounts of homogeneous tissue. In the cereal grains, however, these areas are relatively small. The answer for cereals would seem to lie with inducing and maintaining callus cell cultures. Diploid callus cells might be obtained from cell and protoplast suspension cultures or nutrient agar cultures of germinating embryos. Haploid callus could derive from anther-pollen culture or from immature embryos by chromosome elimination.

Graebe and Novelli (1966) describe a practical method for large-scale plant cell cultures; King and Street (1973) and Street (1974) provide a comprehensive review of cell suspension culture techniques and Street (1977b) of the applications. The general cytology of cultured cells is given very comprehensive treatment by Yeoman and Street (1973). The chemical nature of the walls in suspension-cultured monocots has been surveyed by Burke et al. (1974).

1. Cell Suspension Cultures—Haploid and Diploid

In general, the media used first for cell suspension cultures were not fully defined, but included such things as coconut milk, yeast extract, or malt extract. Torrey and Reinert (1961) reported a synthetic medium for suspension cultures of certain dicots and Straus (1960) a medium for corn endosperm. Nickell and Maretzki (1969) developed a chemically defined medium (modified White's) for cell suspension cultures of sugarcane. Schenk and Hildebrandt (1972) developed a synthetic medium for monocots in general.

Sugarcane is the only monocot close to the cereal grains whose cell isolation, culture, and regeneration is fully elucidatd (Heinz et al., 1977).

Nickell and Maretzki (1969) and Nickell and Heinz (1973) isolated the cells of sugarcane in a modified White's medium in liquid culture in order to study sugarcane cell biochemistry (Higa, 1975) and specifically the production of various plant cell products and their potential usefulness. As another example of how cell suspensions can be used to study cell biochemistry, Yamaya and Ohira (1976) isolated a nitrate reductase–inactivating factor from a study of rice cells in suspension culture. Other studies have demonstrated physiologically active substances or secondary cell products from cell suspension cultures (Misawa et al., 1974; Butcher, 1977; Staba, 1977).

Heinz et al. (1969) reported that sugarcane cell suspension cultures (5 species) maintained in a yeast extract–enriched medium for more than 6 yr showed variability in chromosome number in regenerated plants, with a partial aneuploid series at the haploid and/or polyploid level associated with loss of potency.

In cereals, Kao et al. (1970) showed that in cell suspensions of *Triticum monococcum* and *T. aestivum* there were changes in chromosome numbers. *T. monococcum* ($2n = 14$) after 2–3 yr in culture had chromosome numbers of 15–17, with the majority at 16, or 27–30, $1/2$ at 28, $1/4$ at 29, and the other $1/4$ higher or lower. In *T. aestivum* cv. Thatcher ($2n = 42$), the chromosome numbers after long-term culture were 28, 32, 35, and 36. A number of dicentrics, fragments, or giant chromosomes were also observed. The implications of chromosome abnormalities will be discussed with diploid callus.

During the period of cell suspension culture there is a peak in mitosis at about 7 days, followed by a gradually falling frequency until complete cessation in 3 wk (Torret et al., 1962); the same cycle is followed when subculturing *Daucus carota* L., *Convolvulus arvensis* L., and *Haplopappus gracilis* (Nutt.) Gray.

Gamborg and Eveleigh (1968) studied cell suspensions in barley and wheat, primarily for the detection of glucanases.

The cytogenetics of cell culture and differentiation are thoroughly discussed by D'Amato (1977). Outside of genetic stability in meristems, and to a lesser extent in stem callus, somatic and germinal plant tissue and cell cultures show variation in ploidy, chromosome number (aneuploidy), and karyotype.

2. *Diploid Cell Populations*

a. *Callus from isolated protoplasts*

In general, regeneration of cell walls and division to form callus from cell protoplast suspensions are limited to a very few species, mostly dicots, at present: Cowpea, *Vigna* sinensis cv. Black Eye (Davey et al., 1974), cucumber, *Cucumis sativus* (Coutts and Wood, 1975, 1977), *Ranunculus sceleratus* (Dorion et al., 1975), *Petunia* sp. (Power et al., 1976), *Nicotiana* sp. (Banks and Evans, 1976), *Haplopappus gracilis* (Eriksson and Jonasson, 1969), oat, *Avena sativa* (Brenneman and Galston, 1975), *Phaseolus vulgaris* (Pelcher et al., 1974), *Convolvulus arvenis, Allium cepa* (Bawa and Torrey, 1971), rice, *Oryza sativa* (Somatic Hybridization Research Group, 1975; Deka and Sen, 1976), pea, *Pisum sativum* var. Little Marvel (Landgren and Torrey, 1973), barley, *Hordeum vulgare* cv. Diamant (Koblitz, 1976), *Lycopersicon esculentum*, and *L. peruvianum* (Zapata et al., 1977).

In many cases the differentiation is limited. For example (Davey et al., 1974), cowpea (*Vigna sinesis*) leaf mesophyll protoplasts divide undifferentiated on Schenk and Hildebrandt (1972) and M-S (1962) media that includes 1.0 mg/l 2,4-D and several auxin-cytokinin regimes. Shootlike structures developed on tobacco callus in Nagata and Takabe (1971) medium without mannitol and with 0.5 mg/l IAA and 0.5 mg/l 6-BAP (benzyl amino purine). Root production is prolific, especially with IAA or NAA (0.1–1.0 mg/l) and a corresponding cytokinin (6-BAP, 2iP, kinetin, 0–5.0 mg/l). Coutts and Wood (1975, 1977) cultured protoplasts from the cotyledons and first leaves of cucumber, and the callus induced in 6–10 weeks of culture formed roots but no shoots.

Brenneman and Galston (1975), after peeling the lower epidermis, isolated and cultured seedling mesophyll protoplasts of *Avena sativa* on a modified B-2 medium (Gamborg, 1970). Cell wall formation and a few divisions were reported, and then the protoplasts deteriorated after 5–6 wk in culture.

Deka and Sen (1976) isolated mesophyll protoplasts from the leaf sheaths of rice, *Oryza sativa* L., by treatment with 2% pectinase followed by 3% cellulase at pH 5.4 in 0.45 M mannitol. Callus cells were obtained by culturing stem tissue on S-H (1972) medium plus 0.5 mg/l 2,4-D, 0.5 mg/l IAA, and 0.1 mg/l kinetin. Callus protoplasts were isolated by treating with a mixture of 2% pectinase and 3% cellulase (Onozuka SS) in 0.45 M mannitol at pH 5.4 and 30 C. Protoplasts were cultured following the methods of Takebe et al. (1971). Mesophyll protoplasts were put under continuous light (2,000 lux), whereas the callus protoplasts were put in the dark. Regeneration of the cell wall took place after 24 hr in culture. The protoplasts changed from round to oval, the first cell division occurred after 4–5 days, and cell clusters formed. When calli reached 1.5–2 mm after 8 wk in culture, they were transferred to S-H medium with 0.5 mg/l 2,4-D, 0.5 mg/l IAA, 0.25 mg/l NAA, and 0.1 mg/l kinetin to induce differentiation at 26 C with alternating light (5,000 lux) and dark cycles of 12 hr each. The colonies grew actively, and after 2 wk some callus differentiated into healthy roots but no shoots. It is quite probable that this differentiation was from stem callus, but the paper does not state whether or not there was any difference in response of mesophyll or callus protoplasts.

Koblitz (1976) started with callus cultures of barley isolated in 1973 (cv. Diament) and isolated the protoplasts in a two-step procedure: (1) treatment with 5% Onozuka-Cellulase P 1500 and 2% Macerozyme mixture in 0.3 M sorbitol and mannitol at pH 5.7, and (2) treatment with 0.25% Driselase and 0.25% Penicillium enzyme. The isolated protoplasts were filtered and centrifuged for concentration. The protoplasts were then cultured in various concentrations and combinations of the treatment 2 mixture. After the cell wall regenerated, and after 6 days in culture, 10% of the protoplasts divided; after 12 days they had formed aggregates of about 15 cells, and after 20 days, 50 cells. After 18–21 days in culture, the callus clusters were plated on B-5 medium. Cell variants are reported but no differentiation.

The Somatic Hybridization Research Group (1975) isolated protoplasts from pollen callus of rice (*Oryza sativa* L. var Lian Kiang Mi Tsao). The callus was shredded and treated with two combinations of enzyme mixture: (1) 5% cellulase (Onozuka R-10), 1.5% pectinase, 0.2% potassium dextran sulphate, and 0.8 M mannitol at pH 5.4; and (2) 3–4% self-made cellulase, 1% macerozyme R-10, and 0.85 M mannitol adjusted to pH 5.7. The enzyme-callus mixture was incubated at 34 C for 2–3 hr and in a gyrorotary shaker (60–80 rpm) at the same temperature for 2 hr. The mixture was then filtered, resuspended, washed twice in sterile 0.75 M mannitol with 0.1 M $CaCl_2$, and centrifuged at 500 rpm for 2–3 min.

The protoplasts were inoculated in a medium containing the following, in mg/l: KNO_3 (2200), NH_4NO_3 (360), $CaCl_2$ (340), KH_2PO_4 (300), $MgSO_4 \cdot 7H_2O$ (185), $MnSO_4 \cdot 4H_2O$ (4.4), $ZnSO_4 \cdot 7H_2O$ (1.5), H_3BO_3 (1.6), KI (0.8), Fe-EDTA (5 ml of soln. containing 7.45 g/l Na-EDTA and 5.57 g/l $FeSO_4 \cdot 7H_2O$), glycine (2), thiamine·HCl (1), pyridoxine·HCl (0.5), nicotinic acid (0.5), 2,4-D (0.5), 6-benzyladenine (0.3), casein hydrolysate (200 or 400), mannitol (0.6M), and sucrose (10 g/l). The pH was adjusted to 5.8 with $1N$ NaOH and the medium sterilized by autoclaving. About $10^5 - 2 \times 10^5$ protoplasts were suspended in 1 ml of the medium and incubated in a sealed flask at 28 C under low light (10 hr per day).

The cells were generally sperical, with slightly vacuolated, dense cytoplasm and nucleus. After 1 wk in culture, some protoplasts deteriorated, some were unchanged, and others increased in size and became oval in shape. New cell walls were evident after 2 wk, and 4 conditions of cells then became evident: (1) plasmolysis occurred in protoplasts no longer viable, (2) irregluar wrinkles appeared in cells and the cytoplasm became thinner, (3) cells divided symmetrically into two hemispherical daughter cells and eventually degenerated, and (4) the cells underwent asymmetrical division. Only cells in condition 3 and 4 were capable of further division. The frequency of cells in the latter two conditions was 0.1%. After 3 wk in culture, some protoplasts formed clusters of 4–10 cells and more than 30 cells in 4 wk. Supplementing the medium with 400 mg/l casein hydrolysate produced a number of cells within one common cell wall. This study simply demonstrated how difficult it is to get isolated monocot protoplasts to divide.

The most comprehensive studies on callus from protoplast isolates have been done on *Petunia* (Frearson et al., 1973; Hayward and Power, 1975; Power et al., 1976; Banks and Evans, 1976) and are included here as possible guides to work on cereals.

Hayward and Power (1975) enzymatically isolated protoplasts of *Petunia parodi*. The seedling leaf protoplasts were isolated with 3 % Meicelase and 0.3% Macerozyme in 13% mannitol and inorganic salts at pH 5.8. The cells were washed with 13% mannitol and then resuspended in 21% sucrose solution plus inorganic salts, centrifuged, resuspended, and plated on nutrient medium (Frearson et al., 1973) at 25 C and 1000 lux. Cell colonies were transferred to new media and formed compact green callus that initiated shoots after 1 mo.

Shoots, excised from callus and transferred to M-S medium with 0.1 mg/l NAA and no cytokinins, developed adventitious roots in 2 weeks. Regenerated plants showed the normal chromosome number of 14.

Power et al. (1976) isolated, cultured, and induced callus formation and plantlet production in several *Petunia* species. Enzymatic incubation was in an enzyme mixture containing 3% Meicelase P = cellulase of *Trichoderma*. Callus was grown on Frearson et al. (1973) medium plus 0.5 μg/ml 2,4-D, 1.5 μg/ml NAA and 1.0 μg/ml 6-BAP. Organogenesis was stimulated by a wide variety of growth regulators, depending upon the *Petunia* species. Some calli produced shoots and roots on M-S medium with IAA + kinetin or zeatin or 6-BAP, or NAA or IAA + 6-BAP, or zeatin alone. Roots are induced by transferring shoots to M-S medium + NAA (0.1 μg/ml) and 0.35% agar.

Some of the factors to be evaluated for aiding in cell and protoplast isolation are: colchicine (Umetsu et al.,1975), reported to increase the degree of cell separation; gibberellin, which induced separation of cells in isolated endorsperm (Jacobsen et al., 1976); chelating agents as cell loosening agents that prevent the formation of cell aggregates (El Hinnawy. 1974); cyclohexamide and kinetin, which affect the yield, integrity, and metabolic activity of leaf protoplasts (Kaur-Sawhney et al., 1976); the critical influence of temperature in isolation and division (Zapata et al., 1977); the isolation of mesophyll (tobacco) protoplasts in salts rather than the traditional sugars for osmotic control (Meyer, 1974); growth substances (Power and Cocking, 1970); and different cell wall-digesting enzymes.

In summary, although the leaf mesophyll is a good source for protoplasts in cereals, the mesophyll protoplasts at best undergo only a few divisions. Concomitant with this is the fact that there is no report of callus induction from leaf mesophyll cells in cereals. Therefore, a better source of protoplasts might be from cell suspensions or callus derived from germinating embryo cells or stem and root explants—cells better adapted to culture conditions and able to give rise to callus.

b.   *Callus from tissue explants—and plant regeneration*

The factors that govern the success of tissue cultures, principally age and selection of tissue and the culture medium, will be reviewed briefly. After a suitable mineral salt medium has been chosen, the specific additives must be determined for each species.

In general, the earlier culture media (e.g., White's) were deficient in nitrogen and potassium salts for callus growth or for suspension cultures. This deficiency was compensated for by adding undefined organics such as yeast extract, various amino acids, casein hydrolysate, and coconut milk. In most cases the nitrogen-potassium requirement can be provided by increasing the concentration of inorganic salts, particularly nitrate and ammonium nitrogen, as well as sucrose and vitamins (Murashige and Skoog [M-S], 1962; Murashige, 1974; Gamborg et al. [B-5 medium], 1968; Eriksson [ER], 1965; Ohira et al., 1973; Schenk and Hildebrandt [S-H], 1972). Some species prefer one salt medium over the other, but any one of the above mineral salt media, with the exception of White's, may be used for callus culture. The principal difference between the M-S or ER media and the other preferred media is the ammonium nitrogen, which may be essential for some cells and tissues but may be inhibitory to others (Gamborg and Shyluk, 1970).

Yamada (1977) presents a table that summarizes some of the more common media used for induction and maintenance of callus from cereal plants. For the most part, the culture media are modifications of White (1939), Heller (1953), Murashige and Skoog (1962), Gamborg et al. (1968), and Schenk and Heldebrandt (1972). Gamborg et al. (1976) have presented an excellent analysis of these more common culture media.

For callus induction the cereals, as for most plants, require a strong auxin (2,4-D or 2,4,5-trichlorophenoxyacetic acid) to induce and maintain callus (Yamada et al., 1967; Carter et al., 1967). Cytokinins may or may not be beneficial. Narayanaswamy (1977) has reviewed some of the subtle interactions among auxins and cytokinins and between tissues and organs. Organogenesis *in vitro* from diploid callus cultures will be reviewed for the cereal grains and closely related species.

(1)   Maize (*Zea mays* L.)

Straus (1960) cultured corn endorsperm on synthetic medium (mineral salts, 2 % sucrose, vitamins, and 1.5×10^{-2} M asparagine). Viable callus was always obtained from the endosperm (excised 19 days after pollination) (LaRue, 1949). Mascarenhas et al. (1975a,b) initiated viable callus from embryos, which differentiated roots when subcultured. Green et al. (1974) excised corn embryos and implanted on a modified Linsmaier-Skoog (L-S) medium (2,4-D added and kinetin and m-inositol omitted). Calli formed on the seedling first node and occasionally from the radicle and coleoptile regions. Calli from 17 of 23 single crosses and 10 of 17 inbred crosses varied in frequency with genetic variability. Gresshoff and Doy (1973), using the germinated seeds of *Zea mays* var. B-48, excised the embryos, macerated them to a cell paste, and placed them on a fully defined growth medium (Gamborg and Eveleigh, 1968) plus 2.5 mg/l NAA and 0.1 mg/l kinetin in the dark. After 2 weeks callus growth doubled, and the callus was then subcultured on growth medium 2 (8 mg/l NAA, 0.01 mg/l kinetin) in the dark. Calli cells on medium 2 are large, highly vacuolated, and friable. The calli are next subcultured on medium 3 (basal + 2.5 mg/l NAA, 1.0 mg/l kinetin) in the dark. The callus cells become smaller and are closely packed and give rise to aerial and normal roots. Aerial roots, on touching the agar, form calli that serve as centers for renewed root formation. Transfer to medium 4 (basal + 0.1 mg/l NAA, 2.0 mg/l kinetin) induces shoots in 4-6 weeks with a 16-hr/8-hr light-dark cycle. When transferred to medium 5 (basal + 8 mg/l NAA, 0.1 mg/l kinetin) in continuous light, phallic horns developed. Protrusions on the horns developed into what were thought to be minute ovules. These fail to develop further.

Green and Phillips (1975) initiated maize callus from embryo scutellar tissues (18 days after pollination) on M-S inorganics plus vitamins, amino acids, sucrose, and 2 mg/l 2,4-D. Combinations such as 1 mg/l 2,4-D, 4 mg/l NAA, and 0.05 mg/l 2iP increase the efficiency of scutellar callus initiation. Callus was maintained on the same basal medium but with the 2,4-D concentration reduced to 0.25 mg/l. Plantlets formed on transfer to 2,4-D-free medium. After transfer to soil, 10-15% of the plants survived and grew normally. Sheridan (1975) contributed to the nutritional requirements for inducing and maintaining callus cultures of mutant maize with twice the normal amount of lysine and tryptophane.

Harms et al. (1976) reported on *Zea mays* L. (cv. 'Prior,' 'Inkrakorn,' and two experimental lines, M9473 and M0003) regeneration from callus induced in the mesocotyl region of maize seedlings. In contrast to Green and Phillips (1975), callus induction for the 4 cultivars was an average of 77% if the scutellum was placed downwards and 38% if placed upwards on the medium. 'Prior' and 'Inkakrorn' formed callus equally well in light or dark. Unless all organized parts of the mesocotyl region (vessels and scutellum remnants) are removed, the resulting browning reduces both callus initiation and proliferation. Using modified M-S, S-H, and C-H media, the addition of 2,4-D was necessary for callus induction (up to 15 mg/l) and maintenance

(2–2.5 mg/ 1), and kinetin (0.1–1 mg/ 1) had no effect. Reduced 2,4-D levels decreased callus growth rates but stimulated root formation. Root formation ranged from 8% on modified M-S medium containing 2.5 mg/ 1 2,4-D to over 50% on M-S medium lacking 2,4-D. At least four different types of roots were formed, all nonfunctional. For shoot induction, calluses on M-S plus 2.5 mg/ 1 2,4-D were transferred to M-S plus 0.25 mg/ 1 2,4-D and then 2,4-D–free medium. Shootlike structures and buds were observed, but these did not develop into shoots. However, 3 of 19 calluses of cv. 'Prior' on the 6th subculture formed green shoots on 2,4-D–free medium after 20 days in a 16/8 hr photoperiod of 3,000 lux at 24 C. Roots formed directly from the base of the shoots. Plantlets were transferred to potting compost in a 16/8 hr photoperiod of 10,000 lux at 27 C and grew very rapidly.

(2)   Rye (*Secale cereale* L.)

Carew and Schwarting (1958) obtained callus from the rye embryo and demonstrated that 2,4-D was superior to IAA for induction. Mullin (1970) obtained vigorously growing callus from a clone of excised roots of the cultivar Black Winter. Gamborg et al. (1970) point out the problem of chlorophyll-deficient plants derived from induced callus. Norstog (1956) grew rye grass endosperm *in vitro*.

(3)   Rice (*Oryza sativa* L.)

Ohira et al. (1975) studied the nutritional requirements of rice cell cultures. Starting with B-5 medium they excluded cobalt, iodine, pyridoxine-HCL, and m-inositol as nonessential by adjusting upward the concentration of NO_3- and NH_4- N, P and K to produce their R-2 medium. Cell growth of rice on R-2 medium was believed to be superior to growth on B-5, Heller's, M-S, or White's.

Furuhashi and Yatazawa (1964) induced callus from seeds, roots, and rice stem nodes. The frequency of aneuploid cells was 80% for root callus and 57% for node callus; seed callus was not scored. Yamada et al. (1967) induced rice callus with 2,4-D and cultured it on a synthetic medium.

Nishi et al. (1968) cultured callus derived from the roots of rice on the L-S (1965) medium plus 10^{-5} M 2,4-D. Subcultured on the same medium but without auxin and incubated in the light, the calli differentiated to form both roots and shoots, mostly diploid.

Maeda (1968) induced callus and shoot formation from the plumule of rice embryos and subsequently (1971) studied the histology, differentiation, and varietal differences.

Mascarenhas et al. (1975a,b) induced callus that formed only roots.

Schenk and Hildebrandt (1972) developed a superior nutrient medium for rice callus but required the addition of 2,4-D and pCPA to avoid dedifferentiation.

Yatazawa et al. (1967) induced callus in the roots of rice seedlings that could be cultured indefinitely on a modified Heller salt solution plus yeast extract and 2,4-D. No organ formation occurred.

Tamura (1968) was the first to report buds, formed *in vitro*, from the callus originated from the rice embryo of 10 different varieties. Callus-inducing medium was White's (salts) +2,4-D. The most effective organ-forming medium was a revised L-S (1965) medium + thiamine-HCL, m-inositol, IAA, kinetin, and sucrose. Visible buds were excised from the callus and subcultured on root-forming medium with the IAA level reduced (to 0.2 mg/l) and kinetin omitted.

Kawata and Ishihara (1968) succeeded in developing rice plants from callus cultures induced by 2,4-D from the seminal root. The callus had to be subcultured on a medium containing 1 mg/l IAA and no 2,4-D to obtain roots and/or buds.

Nishi and Mitsuoka (1969) isolated anthers and ovaries of rice aseptically for callus induction on L-S (1965) basal medium plus 2,4-D and no cytokinins. After 2–3 months' incubation, the callus was transferred to fresh medium without auxin for 2 months, and both roots and shoots formed. Diploid plants only formed from embryo or shoot nodule callus, but haploid, diploid, triploid, and pentaploid plants were derived from anther callus and tetraploids from ovary callus.

(4)   Oats (*Avena sativa* L.)

Webster (1966) established callus from germinating oat seeds grown on Heller's nutrient agar medium supplemented with IAA, NAA, and 2,4-D. The callus was maintained for 3 years.

Carter et al. (1967) also induced callus from germinating oat seeds on an L-S basal medium (optional constituents omitted) plus 2,4-D and no cytokinins. Callus formed from the meristematic zone adjacent to the root tip at low auxin concentrations and then differentiated large numbers of shoots and a few roots, with or without auxin.

Brennaman and Galston (1975) used root-hypocotyl tissue from macerated seedlings to produce callus of *Avena sativa* L. on

L-S medium plus 5 mg/l 2,4-D at 24 C and 18/6 hr photo-period. Before subculturing, the callus was separated from preformed plant tissue. Excised and subcultured, the callus formed roots on media low in IAA, NAA, or GA$_3$ and supplemented with cytokinins (BAP, 2iP, and kinetin). On a modified L-S medium, elongated, bright green nodules and meristemoids formed, independent of auxins or cytokinins. If 2,4-D was omitted, roots but not shoots formed.

Cummings et al. (1976) used the *immature* embryos of 25 oat genotypes for callus initiation on M-S or B-5 medium containing various concentrations of 2,4-D and 20 g sucrose at pH 6.0 and incubated at 28–30 C under 3,000 lux. After callus initiation all cultures and subcultures were grown on B-5 medium with 1 mg/l 2,4-D. B-5 without 2,4-D was used as organ-inducing medium. Regenerated plantlets were planted in sterile soil and grown in the greenhouse. Plant regeneration began with shoot initiation and then delayed root initiation. Seed set was variable and ranged from complete sterility in 22 of 133 plants to high viability in field plantings. Phenotype variability was observed in regenerated plants—from occasional albinos to variegated plants to awn and heading variability. Some differences carried over into the next generation and were associated with chromosome instability. Of the *mature* germinating embryos in 5 genotypes investigated, only Lodi genotype produced callus from which plants regenerated. Apical meristems of Portal seedlings proliferated calli, and many plants were regenerated but none were grown to maturity.

Lörz et al. (1976) produced callus in 15–20 days from the seedling hypocotyl of three varieties of *Avena sativa*— 'Flamingskrone,' 'Arnold,' and 'Tiger'—when cultured on S-H medium plus 0.5 mg/l 2,4-D, 2 mg/l pCPA, 2% sucrose, and 1.2% agar. Callus growth could be stimulated by increasing the 2,4-D concentration to 2 mg/l. Callus was subcultured every 4 wk, and at the 8th subculture organ induction was initiated by varying the hormone combinations and concentrations. Roots only formed with IAA-kinetin or NAA-BAP-GA$_3$, and roots and shoots on NAA (3×10^{-5}M), IAA (10^{-5} M), and BAP ($1.5 + 10^{-6}$ M), all cultured in 4,000 lux, 16/8 hr photoperiod, at 24 C. More than 50% of calli formed shoots (16/28), ranging from 1–5 shoots per callus. Calli developing shoots without roots were transplanted on hormone-free M-S medium where some developed roots and could be planted into planting compost.

(5)   Barley (*Hordeum vulgare* L.)

Yamaguchi et al. (1970) reported callus induction from the barley embryo.

Cheng and Smith (1975) established rapidly proliferating callus cultures from the apical meristem of 1-week-old seedlings of two cultivars of barley on an M-S medium supplemented with auxins and a cytokinin (2iP, not essential). The growth media were divided into a "callus-initiating" medium, a "maintenance" medium, and an "organ-inducing" medium. The basal medium was the same M-S salts, but the auxin differed. The callus-initiating medium contained 10 μM IAA and 15 μM 2,4-D, the callus-maintaining medium contained 10 μM IAA + 5–20 μM 2,4-D or 20–40 μM NAA or 20–40 μM pCPA. No growth substances were included in the organ-inducing medium, and the basal medium was diluted 1:1. After 3 consecutive transfers (every 3–4 weeks), 85% of the calli exhibited organogenesis. After 5 transfers, 60% of calli exhibited organogenesis. After 4–5 months the ability to produce shoots was lost. The restored plants maintained a constant diploid chromosome number and there were no aberrations.

(6)   Sorghum (*Sorghum vulgare* Pere)

Masteller and Holden (1970) reported the induction of sorghum callus from the seedling shoot and the development of plants from this callus, using the same medium as Carter et al. (1967) for oats.

Strogonov et al. (1968) cultured the seeds of sorghum on M-S salts plus thaimine, m-inositol, Ca-pantothenate, casein hydrolysate, 2,4-D, and kinetin. Calli formed on the roots or on tillering nodes of sorghum shoots. Root calli formed no buds; tillering node callus formed stem buds.

Mascarenhas et al. (1975a,b,) initiated viable callus of sorghum from germinating embryos, which when subcultured formed roots.

Gamborg et al. (1977) used immature embryos of sorghum (*Sorghum bicolor* L. Moench hybrid X4004), 12–18 days after pollination, grown on M-S salts plus Gamborg (1975) vitamins, 5 μM 2,4-D and 10–50 μM zeatin, and 30 g sucrose and 0.6% agar at pH 6.0, 2,000 lux and a photoperiod of 20/4 hr at 20 and 15 C respectively. The light and temperature regimes were fairly critical. Callus formed on 20–50% of the embryo explants in 4–6 wk and leafy shoots in 10 wk. Subcultured on the same basal medium but with only $5\mu M$ IAA as the growth hormone, roots and complete plantlets formed within 10 wk. The chromosome complement was the normal $2n = 20$. Plantlets transferred to vermiculite-sand-peatmoss and grown in the greenhouse produced both sterile and seed-bearing plants. This paper is rather unique in carefully testing and reporting on a number of culture variables.

(7)   Indian millets (*Paspalum scrobiculatum* L., *Eleusine coracana* Gaertn., *Pennisetum typhoideum* Pers.)

Rangan (1976) germinated grains of indian millets, and the mesocotyl from 5-day-old seedlings were excised and planted on M-S salts, supplemented with auxin plus malt extract, yeast extract, or coconut milk to augment growth of callus. Various cytokinins were useful to initiate shoot buds after withdrawal of auxin. Young shoot buds, excised and grown on M-S medium + NAA + coconut milk, developed adventitious roots, and complete plantlets were obtained in 4 weeks.

(8)   Millet (*Panicum miliaceum* L.)

Rangan (1974) established a diploid callus and regenerated plantlets from this callus.

(9)   Sugarcane (*Saccharum officinarum* L.)

Heinz et al. (1969) produced callus from internode parenchyma on White's medium (salts) plus 2,4-D. The callus was subcultured for 6 years on a liquid yeast-enriched shake medium with 0.5 mg/1 2,4-D. The chromosome number was stable in each of four parental clones ($2n = 122, 114, 114, 112$) but was variable in another ($2n = 108-128$).

(10)   Triticale (*Triticum* X *Secale* hybrid)

Triticale surpasses wheat in total protein as well as in the content of the essential amino acid lysine. In addition, it has the extreme hardiness and disease-resistance of rye. Schenk and Hildebrandt (1972) induced callus from the seed mesocotyl on S-H medium plus a high concentration of auxins (2,4-D and pCPA) and a low concentration of cytokinin in the dark.

(11)   Rye-grass (*Lolium perenne* L.)

Norstog (1956) grew rye-grass endosperm on White's medium fortified with yeast extract. The callus induced was unstable and did not differentiate. The callus was characterized by chromosomal aberrations—breakages.

(12)   Brome grass (*Bromus inermis* Leyss.)

Schenk and Hildebrandt (1972) initiated callus from the

mesocotyl of Canadian brome grass on S-H medium plus 2,4-D and pCPA. No plantlets differentiated.

(13) Wheat (*Triticum aestivum* L.)

(a) *Triticum aestivum* as experimental material

Most of the tissues of diploid plants (2n = 6x = 42) form callus rapidly. The diploid calli are cytologically more stable than most other monocot or dicot tissues. A very low percentage of shoot differentiation occur from diploid calli.

(b) Choice of wheat variety

The variety of *Triticum aestivum* that may be the best for culture work is Chinese Spring, for these reasons: (1) Chinese Spring is genetically the best known wheat variety (Sears, 1974); (2) there is a comprehensive paper on cell development in the anther, ovule, and young seed of this variety (Bennett et al., 1973); and (3) it was the variety originally described that can cross with *Hordeum bulbosum* (male) to produce haploid embryos by chromosomal elimination (Barclay, 1975).

(c) Previous work on *Triticum aestivum* diploid callus cultures.

Authors	Year	Salient Finding
Gamborg and Eveleigh	1968	Suspension cultures on defined medium —mineral salts, vitamins, 2,4-D.
Trione et al.	1968	Somatic cells of wheat in callus cultures retain the capacity to form roots.
Shimada et al.	1969	Best callus growth on 2,4-D (5 mg/ 1), supplementing a basal White's medium and grown in dark. Calli consisted of half eudiploid and half aneuploid cells, the majority with 42 ± 3 chromosomes. Six calli formed shoots.
Kao et al.	1970	Cell suspension cultures of *T. monococcum* and *T. aestivum* var. Thatcher callus on B5 medium + CH developed from root culture. Both species after 2 yr showed changes in chromosome number and abnormal karyotypes—accounting for lack of morphogenesis.
Fujii	1970	Callus production was low in *T. monococcum* (2x), absent in *T. durum* (4x),

T. aestivum (6x) and *T. spelta* (6x), and highest in *T. dicocioides* (4x) followed by *T. Aegilopoides* (2x).

Shimada	1971	Callus cells derived from root tips and subcultured for 2–4 yr were predominantly diploid. Wheat callus is more stable than tobacco callus.
Schenk and Hildebrandt	1972	Developed a superior nutrient solution for monocot callus, including 2,4-D and pCPA to avoid dedifferentiation.
Prokhorov et al.	1974	Used M-S medium with 2,4-D for callus induction. Calli were raised from normal embryos and one plant differentiated from the callus.
Mascarenhas et al.	1975 a,b	Induced callus from germinating embryos which differentiated roots when subcultured. Calli were stable, viable.
Nakai and Shimada	1975	Study of morphological, cytological, and biochemical characteristics of calli from hexaploid wheat.
Asami et al.	1975	Callus cells induced from embryos of *T. aestivum* var. Chinese Spring ($2n = 6x = 42$) and cultured 6 mo consisted of 97% normal diploid cells. In nulli-5B, tetra 5D, and ditel-5AL calli the chromosome number varied considerably.
Dudits et al.	1975	7.1% of calli induced from rachis and shoot on B-5 and T-medium supplemented with zeatin, IAA and 2,4-D differentiated into complete plants. TIBA influenced the frequency of shoots.
Udvardi et al.	1976	Optimal concentrations of different auxins for callus growth and correlations of auxin concentrations with levels of nucleolytic enzymes.
Baroncelli et al.	1977	Studies on genetic control of callus growth and root differentiation. 26 ditelocentric stocks of Chinese Spring were used. Group J chromosomes were of relevance in controlling cell proliferation in both roots and nodes. Root

formation may be controlled by genome D chromosomes.

Chin and Scott (1977) initiated root, embryo, and glume (3–5 cm florets) callus tissue of *Triticum aestivum* L. cv. Mengavi on a callus-inducing medium of basal salts plus 2,4-D (5 mg/1) and pCPA (1 mg/1) in 6 wk. Callus was subcultured on basal plus 2,4-D(2 mg/1)and pCPA (4 mg/1) every 4 wk, maintained under continuous light (200 lux) at 23 C. Under these conditions, root callus was white and very friable, embryo callus was yellowish and very compact, and the glume callus was light green. Eighteen of 28 root calluses (64%) formed roots in 7 days, and 3 of 42 (7%) formed shoots on primary callus. The 2–3 buds formed on each callus usually appeared before root primordia, and only 1–2 formed shoots. Glume callus formed no roots or shoots. Both root and embryo calluses decreased their root-forming capacity in subculture, and the capacity was completely lost by the 5th or 6th transfer. Concentrations from 0.05 to 2.0 mg/1 of 2,4-D, pCPA, NAA, IBA, and kinetin were tested for root-inducing efficiency. Whereas a single concentration of 2,4-D and pCPA could induce 20–70% calli with roots, only NAA (at 0.5 and 1.0 mg/1) could induce greater than 70% calli with roots. Kinetin and BAP were ineffective at any concentration.

c. *Embryoids* (Narayanaswamy, 1977.)

An alternative for cell protoplasts forming callus, as described earlier, is for cell colonies to initiate organ primordia *de novo* by forming an embryoid (Halperin, 1966).

Diffusion or physiological gradients of substances and resulting polarity of tissues along this diffusion gradient must occur in these cell clusters if organ formation or embryogenesis is to occur (Ross and Thorpe, 1973). 2,4-D in the medium is not conducive for either organogenesis or embryogenesis (Reinert and Backs, 1968). However, in general, the balance of hormone levels (auxin-cytokinin ratio) leads to organogenesis. An auxin must be present in the growth medium (Halperin, 1970). Cytokinins and gibberellins alone cause a partial or complete inhibition of embryogenesis. Halperin and Wetherell (1965) observed that ammonium ions and casein hydrolysate in low concentrations are stimulatory to embryogenesis in comparison to nitrate. Tazawa and Reinert (1969) reported that the addition of nitrogen, either inorganic (KNO_3, NH_4NO_3) or organic (amino acids and amides), initiates embryogenesis.

Callus cell cultures from the mesocotyl of Canadian brome grass (*Bromus inermis* Leyss, cv. Manchar) were grown and maintained for 2 yr on an agar medium containing mineral salts, sucrose, casein hydrolysate, vitamins, 0.5 mg/ l 2,4-D, 2 mg/ l pCPA, and 0.1 mg/ l kinetin (Gamborg et al. 1970). These same calli were grown in *liquid* B-5 medium (no casein hydrolysate) with 1 mg/ l 2,4-D as the only auxin at 28 C in the dark. When callus cells were sieved, washed, and incubated in 2,4-D–free medium, embryos were produced in 2 wk and plantlets within 1 mo. Embryoids formed as globular outgrowths along the surface of an irregular cell aggregate of an 8- to 10-day-old culture. A root-shoot axis developed first (the embryoid became polarized) and then the coleoptile. Several dozen plantlets were produced from the embryoids, all albino.

From a structural point of view, two kinds of callus cell aggregates could be recognized (Constabel et al., 1971). One type was irregular, was unorganized, varied in size, and frequently contained outgrowths linked by multicellular short stalks. These loosely arranged, irregular cell aggregates probably gave rise to globular or ellipsoidal aggregates of small cells surrounded by a distinct outer layer. Some of these globular aggregates appeared proembryolike. After 18 days the culture contained "tadpolelike" structures and irregular cell aggregates. The formation of true roots was initiated in the tadpolelike structures, and then a *white* coleoptile appeared and later the leaves. The cell aggregates with embryogenic potential differed from the other cells in being very rich in starch.

As a recent example of induced embryogenesis in dicots, Söndahl and Sharp (1976) grew *Coffea arabica* leaf tissue explants on a conditioning medium containing M-S salts, 3% sucrose, 2,4-D, and kinetin for 7 wk and then subcultured in the same formulation but with the inorganic salts at half-strength for 4 wk. Tissues were then cultured on an induction medium consisting of half-strength M-S salts, a doubling of the concentration of KNO_3, and the usual vitamins, kinetin, and NAA to produce high-frequency (50-60%) somatic embryo induction. Low-frequency somatic embryo development occurred if NAA was submitted for 2,4-D in the conditioning medium, or if NAA-substituted conditioning medium was further supplemented with coconut milk, casein hydrolysate, yeast extract, or leaf extracts of tobacco and Kalanchöe. Using this approach they currently have more than 2500 embryos in various stages of development and 729 normal-appearing plantlets, with the oldest being 19 mo. old.

Churchill et al. (1973) reported similar results for an orchid leaf-tip culture.

3. *Haploid Cell Populations*

Guha and Maheshwari (1964) first showed in *Datura innoxia* that haploids could be produced from anthers in culture. Niizeki and Oono (1968) extended this technique to rice, and subsequently many other investigators reported notable success with rice anther culture. Sunderland (1973, 1974), Nitsch (1974, 1975, 1977), Vasil and Nitsch (1975), and Reinert and Bajaj (1977) have published very comprehensive reviews of anther-pollen haploids. Clapham (1977) has reviewed progress in anther culture of cereals and has compared this with other methods of raising haploids, and discusses the use of haploids in plant breeding. DeFossard (1974) has put the production of haploids from pollen in perspective by suggesting a series of four partial, linked processes that might be treated experimentally.

Haploids may be produced from anther-pollen culture in one of four ways: (1) haploid callus from anther-pollen culture, (2) haploid embryoids from anther culture, (3) haploid embryoids from isolated pollen grown in liquid medium (Vyskot and Novak, 1976; Tyrnov and Khokhlov, 1976), or (4) haploid callus from immature embryos produced by the Bulbosum Method (Kasha and Kao, 1970).

a. *Haploid callus from anther-pollen culture*

Anthers are implanted on solid medium. After remaining dormant for a time, some of the pollen grains divide to form calli. Haploid plantlets may or may not differentiate from this haploid callus. A number of factors are limiting, but the most important is not the nutrient medium per se but the growth factor additives. Haploid, pollen-derived calli are cytogenetically unstable in many cases. For example, in *Oryza sativa*, haploids to pentaploids have been reported to differentiate from an originally haploid callus (Nishi and Mitsuoka, 1969; Niizeki and Grant, 1971). Haploid wheat anther-pollen callus is, however, cytogenetically stable (Wang et al., 1973; Ouyang et al., 1973; Craig, 1974). The induction frequency in wheat, as in all other cereals, is very low.

The induction and culture of haploid callus and organogenesis *in vitro* from haploid callus cultures will be reviewed for the cereal grains and closely related species.

(1)   Rice (*Oryza sativa* L.)

Niizeki and Oono (1968, 1971), using the nutrient medium of Blaydes (1966) supplemented with IAA, kinetin, 2,4-D. adenine sulfate, and yeast extract, singly or in combination, developed haploid rice plants from anther cultures. Anthers of 10 rice varieties, taken 1–2 days prior to heading, were plated on agar nutrient medium. After 3 wk the anthers turned black; after 4–8 wk callus formation occurred on basal plus 1–2 mg/1 IAA, 1–2 mg/1 kinetin, and 1–2 mg/1 2,4-D. Combinations of IAA and kinetin (2 mg/1 IAA + 2–4 mg/1 kinetin were optimum for organ formation, reminiscent of the chemical regulation of growth and organ formation described by Skoog and Miller (1957) for tobacco callus. After 4 wk on this medium, small plants with coleoptiles formed, and then leaves developed rapidly. Seven plants grew to maturity from Norin-20 and Toride-2 varieties. The plants were usually green and had normal morphology, but an albino plant rarely formed. Only 20 out of 3,500 anthers (0.57%) formed callus. Chromosome counts in callus squashes were $2n = 12$.

Niizeki (1968) extended the work of Niizeki and Oono to 10 rice varieties and 9 F_1 hybrids on Blaydes' medium plus 2,4-D and NAA to form anther callus. Shoots and roots were induced with a proper balance of IAA and kinetin.

Myint and deFossard (1974) induced haploid callus from rice anthers and the regeneration of rice plants. Anther cultures of 5–6 varieties of rice were induced to form callus, but the number of anthers induced was very low (0.21–2.20%). Haploid callus produced either albinos or green plants. Green plants grew well and set seed, but numerous infertile panicles were observed. Details are lacking.

Nishi and Matsuoka (1969) restored rice plants from anther and ovary culture (L-S medium + 10^{-5} M 2,4-D but without optional constituents and cytokinins). Haploid, diploid, triploid, and pentaploid callus was derived from anther culture, tetraploid from ovary callus.

Second Division, 3rd Laboratory, Institute of Genetics, Academia Sinica (1974) cultured rice pollen (*Oryza sativa* subsp. King) on Niizeki and Oono (1968) basal medium plus 2,4-D, IAA, and kinetin. Callus appeared in 20 days. Shoots and roots were induced on basal plus IAA and kinetin and no 2,4-D in about 20 days. When the callus-inducing medium was supplemented with yeast extract and nucleic acid hydrolysate, the frequency of induced callus raised significantly. When supplemented with coconut milk, lactalbumen hydrolysate, and nucleic acid hydrolysates, the callus had a greater capacity for root and shoot differentiation. Pollen plant ploidies varied from X to 4X. From 6 to 68% of the seedlings were yellow or white (albinos), probably due to temperature fluctuations and to different sucrose concentrations.

Yin et al. (1976) produced a new cultivar of rice by anther culture. Hybrids of F_1 and F_2 *Oryza sativa* subsp. King were used to produce pollen callus on a modification of the medium used by Niizeki and Oono (1968), the N_6 medium, raising the average of callus from 3.4% to 38%. Callus differentiated on medium devoid of 2,4-D and supplemented with IAA and kinetin, or no kinetin at all. The frequency of albinos was 50%. Modification of the nitrogen source and concentration of Mg and Fe did not affect the frequency of albinos. About 50% of the pollen plants are haploids, the rest diploids whose ploidy number increases with age of callus—spontaneous diploidization by endomitosis takes place during culture.

(2)   Triticale (*Triticum* × *Secale* hybrid)

Wang et al. (1973) induced callus from anthers on M-S salts plus thiamine (0.4 mg/ 1) and 6% sucrose, supplemented with 2–5 mg/ 1 2,4-D, 40–80 mg/ 1 RNA-nucleotide mixture, 1 mg/ 1 kinetin, and 15% coconut milk—singly or in combination. Only 5 of 8 strains formed callus (1.2–17.2%). For inducing shoot differentiation, M-S medium was supplemented with 0.2 mg/ 1 NAA, 0.5 mg/ 1 IAA, 1 mg/ 1 kinetin, 80 mg/ 1 nucleotide, and 15% coconut milk, all in various combinations, at 25–30 C under fluorescent light during the day time. M-S medium with 0.2 mg/ 1 NAA and 15% coconut milk was the best organ-inducing medium. Calli transferred after 27–32 days to differentiation-medium formed 90% plantlets, whereas after 45 days the induction was less. Of 42 amphidiploid plantlets, 29 were albino, 12 green, and 1 an albino-green chimera seedling.

Ono and Larter (1976) cultured anthers from three strains of *Triticale* on B-5 and Blaydes' basic medium plus various concentrations and combinations of sucrose, 2,4-D, IAA, kinetin, amino acids, and coconut milk, on an 18 hr photoperiod at 21 C. The highest callus response was 5.3% from the 'Rosper' strain cultured on B-5 and 10% sucrose plus 2 mg/ 1 IAA, kinetin, and 2,4-D. Substitution of 10% coconut milk for kinetin and IAA completely suppressed callus growth. Of 9,246 anthers cultured, 80 calluses were transferred to Blaydes' basic medium for differentiation, and 7 plantlets, all albino, were formed.

(3)   Aegilops (goat grass)

Kimata and Sakamoto (1972) cultured the anthers of 8 *Aegilops* species and one artificial amphidiploid (*Ae. caudata* × *Ae.*

umbellulata) on Miller's (1963) basal medium supplemented with 2.21 mg/1 2,4-D at pH 6.0 on agar. Callus formed only in anthers of the amphidiploid at 1% frequency. Transferred to basal medium minus 2,4-D plantlets formed in 2 wk, all albinos.

(4)   Barley (*Hordeum vulgare* L.)

Clapham (1971), on L-S or White-Heller medium, found that 28% of the anthers of *Hordeum vulgare* formed callus during a 16-hr light period at 22–27 C, with a growth factor additive of TIBA, IAA, or BAP. Differentiation into plantlets occurred; the majority were green but up to 40% were albino, and ploidy levels were n, 2n, and 4n.

Grunewaldt and Malepszy (1975) found that 30% of the anthers of *H. vulgare* cv. Vogelsanger Gold and Ortolan on L-S medium formed callus, which consisted of about 7% haploid cells and only 1% after 8 subcultures, and the number of diploid cells increased from 60 to 70% after 8 subcultures. Of nearly 1,000 plantlets, 4 were green and the rest albino. Three of the 4 green plants were haploid. The albino plants were mostly haploids and diploids, but also included some polyploids and aneuploids.

Dale (1975) cultured *H. vulgare* L. cv. Akka anthers on modified L-S medium plus p-aminobenzoic acid, IAA, 6-benzylaminopurine, and coconut milk. The majority of pollen grains stained with acetocarmine from the mid-binuclueat stage onwards. Only those cells that are nonstaining respond in culture to form pollen callus at 15–20% frequency, as observed also by Sunderland (1974). LaCour (1949) suggested that atypical grains are due to an atypical angle of the spindle apparatus and abnormal diffferentiation of cytoplasm.

Foroughi-Wehr et al. (1977) cultured anthers of 19 *H. vulgare* varieties on the medium of Clapham (1973) plus 1 mg/1 IAA and 1 mg/1 BAP, with cobalt, iron sulfate, and coconut milk omitted, and produced callus. Differentiation was induced by halving the concentration of IAA, BAP, and sucrose. The percentages of callus varied with the variety (0.8–14.2). The best result was 14 green plants out of 176 anthers plated. A preponderance of haploid cells was detected in callus; in the regenerated green plants there was a predominance of diploid cells, but some haploid and aneuploid cells were present. Green plants were produced from only 7 of the 19 varieties.

Jensen (1977) reported that callus from the scutellum of monoploid *H. vulgare* embryos on his C-17 medium formed mostly normal green monoploid plants.

The work of Grunewaldt and Malepszy (1975) points up the fact that chlorophyll deficiency and different levels of ploidy make the plantlets from barley anther callus of little use in the production of monoploids in plant breeding and genetics.

(5)   *Lolium* × *Festuca*

Nitzsche (1970) cultured the anthers of *Lolium multiflorum* (4X) X *Festuca arundinacea* (12 ×) (2n = 49) and formed callus that gave rise to 1 green plant and 7 white plants. The green plant showed 25 chromosomes.

(6)   Rye (*Secale cereale* L.)

Thomas et al. (1975) cultured the anthers of rye, and after 2 wk in culture developed two types—a firm white and a more flaccid brownish type. The microspores form multicellar structures, more from the white than the brown type. After about 8–9 wk the multicellular structures form callus and/or embryoids, varying in ploidy level from haploid to tetraploid.

(7)   Wheat (*Triticum aestivum*)

Picard and Buyser (1973) cultured *in vitro* wheat anthers and obtained callus and regenerated plants. The anther-callus–inducing medium consisted of the salts of Miller (1963), glycine (2×10^{-6}M), EDTA (0.5×10^{-4}M), 2,4-D ($7 \times 10^{-7} - 2 \times 10^{-5}$), 12 % sucrose, and 1% agar adjusted to pH 5.8–6.0. Callus extruded from the anther in 4–5 wk with a frequency of 0.5%. The callus consisted of rounded haploid cells with fragile membranes and a greenish-white color. They also described a series of stages—division of pollen nucleus, multinucleated cell clusters, and multicellular proembryos. Callus was transferred to the same or to two other media for organ induction. The other media were: (1) basal salts plus 12% sucrose and TIBA, and (2) basal salts plus 10^{-6} M IAA, 10^{-4} M BA, 10% coconut milk, and 10^{-4} M alanine. M-S and White's media were also tried and gave negative results. Callus formation was best on the initial callus-inducing medium with high sugar and 2,4-D, although the yield was so low that one cannot demonstrate a preference for medium 1 or 2. Medium 3, without sugar, did not form any callus. Calli were then transplanted to an organ-inducing medium with basal salts, 2% sugar, and a weak auxin (10^{-6}M IAA). After a few days, 3 of the 27 calluses obtained presented a different type of regeneration: (1) after 1 wk 6 small roots formed on the callus, (2) after 10 days an albino plantlet with two roots formed, and (3) a green plantlet formed quickly and in about 15 days possessed two leaves and two roots. Only the winter wheat callus from *T. aestivum* cv.

Champlain (male sterile cytoplasm) regenerated a normal plantlet.

HAPLOID CALLUS CULTURES (n = 3X = 21)

Picard and Buyser	1973	Division induced in the pollen of cultured anthers by low temperature treatment. No callus or embryoids formed.
Wang et al.	1973	Haploid callus from anther-cultured pollen. In M-S medium plus 6% sucrose, 2,4-D, and kinetin, 1% of pollen (26/2850) formed callus. 60% of induced calli differentiated plants, all with n = 21. Of 26 calli, 8 gave rise to green shoots, 6 produced albinos, 1 was a green-white chimera, 4 did not differentiate, 7 must have died.
Ouyang et al.	1973	Lactalbumen hydrolysate increases the percent of induced calli to 0.82–3.1, depending on the variety. Up to 75% of haploid calli differentiated into shoots, all haploid.
Chu et al.	1974	On M-S medium supplemented with many organics plus 2,4-D, 103 of 21,094 inoculated anthers produced pollen calli that formed plantlets. Anthers also formed embryoids and plantlets directly. Of 20 calli differentiating plantlets, 8 were green, 2 half-white and half-green, 8 white, 1 with purple pigment, and 2 partly green and partly white.
Picard et al.	1974	Increased percentage of anther calli by traumatisms and cold treatments.
Craig	1974	Used Ouyang et al. medium. One anther from 550 formed callus. Nine haploid plants differentiated from this callus.
Picard and Buyser	1975	Anthers from lower part of spike yield higher frequency of callus than those from upper ones after low temperature treatment.

Shimada and Makino	1975	Cultured *T. aestivum* var. Chinese Spring anthers on Blaydes' or L-S medium plus 2,4-D and IAA and/or kinetin plus 3% sucrose. Only one pollen callus was obtained from 1,341 anthers cultured; it was from a nulli-2A line. Many albino shoots differentiated from this callus. Some factor inhibiting callus induction is located on the B arm of chromosome 4A.

Heszky and Mesch (1976) cultured *Triticum* anthers from the cereals world collection on M-S salts supplied with casein hydrolysate, sucrose, coconut milk, NAA, 2,4-D and kinetin at a pH of 5.7–5.8. Pollen mitoses were observed on the 5th day, usually only in the vegetative nucleus but occasionally both cells divided in 4–10 pollen grains per anther. Further, only 20–40% of multicellular pollen grains developed callus or embryoids. Only 5–20% of haploids produced developed into mature haploid plants. Roots and then shoots developed from haploid callus. Only 11 of 66 varieties of *Triticum* investigated produced callus. The pollen callus of only one variety of *T. aestivum* (Halle 4848/37) formed a haploid plantlet and that one was albino.

Returning now to *Triticum aestivum* as experimental material: (1) A very low frequency of pollen callus is formed when anthers are implanted on specific media. The maximum frequency of callus formation reported for wheat is 3.1%, except for the filament callus of Shimada and Makino (1975). (2) Shoot differentiation from the haploid callus ($n = 3x = 21$) is much more frequent than from diploid callus. (3) Haploid plants differentiated from haploid callus are cytologically stable. (4) For anther-pollen cultures in most *Gramineae*, there is the general problem of chlorophyll-deficient plants from induced callus, chromosomal abnormalities, and unpredictable differentiation (Gamborg et al., 1970; Clapham, 1971, 1973, 1977).

b. *Haploid embryoids from anther-pollen culture*

An alternative for pollen cultures forming callus, as just described, is for cell colonies to initiate organ primordia *de novo* by forming an embryoid (Halperin, 1966). The cells that are to form an embryoid differ in their cytology from other cell aggregates, and this development involves the synthesis of rRNA and protein (Sussex, 1972). The literature on RNA and protein synthesis during pollen development has been reviewed by Heslop-Harrison (1972) and Mascarenhas (1971, 1975).

Anthers are implanted on solid medium. After remaining dormant for a time, some of the pollen inside the anther starts dividing and extrudes as callus masses that become polarized and give rise to embryoids. Direct regeneration of plants from embryoids would be highly desirable by virtually eliminating chromosomal aberrations and the formation of chimeras.

The potentiality for pollen to undergo embryogenesis cannot be understood without taking into account the developmental sequences (Wulff and Maheshwari, 1938) and biochemical events normal in pollen development. The most critical stage in embryogenetic development may be the period beginning just before microspore mitosis and ending soon after the division. Intense metabolic activity is seen in the microspore in the period immediately preceding and following microspore mitosis.

In addition to physical factors, such as light and temperature, the nutrient factors that initiate an essentially embryonal type of development have been studied extensively, albeit mostly in dicots. The nutrient factors that trigger embryogenesis are different from those that initiate callus formation. Reinert (1959) showed that an auxin-free medium and the addition of nitrogen compounds such as amino acids triggered embryogenesis. There was no qualitative difference between the capacity of oxidized and reduced nitrogen to induce embryos, demonstrating that it is the *concentration* and not the form of nitrogen in the medium that is important (Reinert, 1959; Reinert et al. 1967; Tazawa and Reinert, 1969). Stated simply, a somatic cell may form an embryo by an increased ratio of nitrogen to auxin (Reinert, 1973). This is contradictory to the view of Halperin and Wetherell (1965) that ammonium is a requirement that cannot be replaced by potassium nitrate or amino acids. Similarly, Steward (1963) demonstrated that coconut milk may provide conditions favorable for embryo formation, but embryogenesis can occur in a purely synthetic medium (Homès, 1967). Söndahl and Sharp (1977) demonstrated that a high nitrogen concentration is necessary for inducing embryogenesis, but there may be selection against some nitrogen sources.

Guha and Maheshwari (1964, 1967) investigated *Datura inoxia* and *D. stramonium* and were the first to demonstrate that pollen grains could be cultured to form embryoids. Using several basal media (e.g., Nitsch, 1951) plus 2% sucrose and 0.7% agar plus 10^{-6} M IAA and 10^{-6} M kinetin, the anthers callused vigorously, especially connective tissue. On basal salt medium with 15% coconut milk or kinetin, pollen tubes and embryoids formed, but no callus. The embryoids arise by division of pollen in mature anthers. Physical factors such as light and temperature affect embryoid development (Sopory and Maheshwari, 1976a). The addition of auxins, gibberellins, and cytokinins, especially BAP, enhance embryoid production (Sopory and Maheshwari, 1976b).

Guha et al. (1970) inoculated rice anthers into Blaydes' (1966) nutrient medium plus IAA, 2,4-D, kinetin, yeast extract, and coconut milk. After 2–3 wk some pollen grains enlarged 2–3 times their initial volume. In 4 wk there were divisions within the exine coat, which then ruptured and callus was extruded, resembling a globular embryo. These pollen embryoids in their early development are attached to the anther wall by a suspensorlike outgrowth. Detached embryoids were transferred to fresh medium without auxin, resumed growth, and gave rise to haploid seedlings. Of some 20 rice varities examined, the more primitive types (from Assam) gave the best regeneration, illustrating genotypic differences.

Gamborg et al. (1970) reported successful suspension cultures of brome grass (*Bromus inermis* Leyss.) anther callus on B-5 medium plus 2,4-D. The callus formed embryoids and 100% albino plants. The embryoids formed as globular outgrowths along the surface of irregular cell aggregates of an 8- to 10-day-old culture. A root-shoot axis developed first and then the coleoptile.

Butenko et al. (1967) noted that mineral salts should be present in high concentrations for embryo induction.

Zenkteler and Misiura (1974) reported androgenous embryoids from barley anther cultures, but there was no report on plant formation.

Embryogenesis in cultured anthers can follow any one of the following four pathways (Sunderland, 1974; Sunderland and Dunwell, 1974).

Pathway A. This is the principal pathway and is via divisions in the normally quiescent vegetative cell (nonparticipation of the generative cell), operative in anthers in which the gametophytic cells are already formed, but also in anthers inoculated before the first gametophytic mitosis. It has been confirmed in *Nicotiana tabacum* (Sunderland and Wicks, 1971), *Datura metel* (Iyer and Raina, 1972), *Hordeum vulgare* (Clapham, 1973), *Triticum aestivum* (Ouyang et al., 1973), *Triticale* (Wang et al., 1973; Sun et al., 1974), and *Capsicum annuum* (Kuo et al., 1973).

Pathway B. Microspores diverted into irregular mitosis result in two equal nuclei instead of a smaller, compact generative nucleus and a larger, diffuse vegetative nucleus. One of the two equal nuclei undergoes limited division but, like the generative cell in A, does not participate in the final product. Division of the second nucleus is delayed, but then the first division is accompanied by wall formation, and subsequent divisions form the pollen callus embryoids. This has been confirmed in *Datura innoxia* (Sunderland et al., 1974), *Triticum aestivum* (Ouyang et al., 1973), *Triticale* (Wang et al., 1973; Sun et al., 1974), *Nicotiana*, and *Capsicum annuum* (Kuo et al., 1973).

Pathway C. Both gametic cells are involved in this pathway. This is reported only in *Datura innoxia* (Sunderland et al., 1974).

Pathway D. The generative cell divides to form embryoids. This pathway is reported only in *Hyoscyamus niger* (Raghavan, 1976).

It may be stressed here that in all but two of the above cases, embryogenesis took place inside the anther. In two plants, *Datura innoxia* and *Nicotiana tabacum*, where *isolated* pollen gave rise to embryoids, pathway B is followed (Nitsch, 1974).

c. *Haploid embryoids from isolated pollen grown in liquid medium*

Nitsch (1974) has reported that when anthers of *Datura* and *Nicotiana* are given low temperature treatment and the pollen is suspended in a hormone-free medium supplemented with certain amino acids, a high frequency of haploid embryoids are formed that give rise to haploid plants.

d. *Haploid callus from immature embryos produced by the Bulbosum Method*

The most recent development in haploid production is the chromosome elimination method by Kasha and Kao (1970). Work on barley, the cereal grain with which there has been notable success in haploid production and on which chromosome elimination studies have focused, is reviewed by Jensen (1977).

Kasha and Kao (1970) made an interspecific cross between autotetraploid *Hordeum vulgare* (2n = 4x = 28) and tetraploid *H. bulbosum* (2n = 4x = 28), and the progeny were nearly all 14-chromosome diploid plants resembling cultivated barley. The induced diploid embryos developed vigorously for 10 days and then showed signs of aborting. The embryos were excised and grown on the B-5 medium of Gamborg et al. (1968). All 23 plants obtained were haploid. When crosses between the diploid forms of these species were carried out, haploid plants were obtained (n = 7).

A dihybrid potato was obtained from crossing the tetraploid

potato (*Solanum tuberosum*) with pollen from diploid *S. phureja*, the same as the *H. vulgare* × *H. bulbosum* cross (Hougas et al., 1958).

Kostoff (1943) produced haploid plants with only the characteristics of the male plant from crossing *Nicotiana tabacum* macrophylla × *N. langsdorffii*, probably by androgenesis in the embryo sac. Clausen and Lammerts (1929) described similar results from a *N. digluta* × *N. tabacum* cross.

The Bulbosum Method is based upon making an interspecific cross, with *H. vulgare* as the female and *H. bulbosum* as the male, to produce frequencies of 11–68.5% barley monoploids (Kasha, 1974b; Jensen, 1973, 1974b, 1977) and a much lower frequency of wheat haploids when *Triticum aestivum* is crossed with *H. bulbosum*. Barclay (1975) produced *T. aestivum* var. Chinese Spring haploids by crossing with *H. bulbosum* (male) and elimination of the male genome by some unknown method (see Kasha, 1974b, for possibilities) but controlled by chromosomes 2 and 3 of *H. vulgare* (Kasha, 1974b) in the case of *H. vulgare* × *H. bulbosum* crosses.

There are two stages in the production of haploids by the Bulbosum Method where nutrition is critical. The first stage is at fertilization and includes early embryo development on the plant. The second is during later embryo development on culture medium (Kasha, 1974b; Jensen, 1976). Treating pollinated florets with GA$_3$ improves seed set and embryo development. See also the effect of ABA on excised barley embryos (Norstog and Blume, 1974). Jensen (1973) detached tillers at pollination and placed them in Hoagland's solution, and this led to better embryo development.

Embryos left too long in the fruit may show brown spots or become infected. It is common practice to excise barley embryos at 13–15 days after pollination and culture the immature embryos on either a nutrient agar medium—B-5 (Gamborg et al., 1968), B^{II} (Norstog, 1973), or R-M-IS (Islam and Sparrow, 1974)—or a floating culture system (Jensen, 1975, 1977) with C-17 or C-21 media. The whole procedure for monoploid production is carefully worked out by Jensen (1977).

4. *Importance of Haploidy*

Haploid plants are sporophytes that have the gametophytic chromosome number. Haploids usually arise by parthenogenetic functioning of some component of the embryo sac or from cultured anthers and pollen (Riley, 1974; Magoon and Khanna, 1963).

Kimber and Riley (1963) and Magoon and Khanna (1963) defined a number of kinds of haploids: (1) euhaploids (chromosome number balanced), (2) aneuhaploids (chromosome number unbalanced, mostly derived from polyploids), (3) monoploids (haploids derived from diploid species), (4) polyhaploids (haploids derived from polyploid species), (5) disomic haploids (haploids with certain chromosomes additional to the normal complement), (6) nullihaploids or subhaploids (haploids lacking a number of the normal complement of chromosomes).

In considering haploids for breeding purposes or for viability in culture, it is important to realize that: (1) every locus is in the hemizygous condition, (2) haploids derived from outbreeding certain diploid species will expose deleterious alleles so the haploids will be inviable or abnormal, except in maize, and (3) haploids derived from inbreeding homozygous species or polyhaploids in which there is redundancy will not show morphological abnormalities and will be viable, except for barley (Riley, 1974).

Kimber and Riley (1963) and Magoon and Khanna (1963) also reviewed the literature on the origin of haploids. Chase (1969) reviewed monoploid and monoploid-derivatives of maize. Haploids may arise spontaneously or by induced parthenogenesis and androgenesis (Lacadena, 1974), or by parthenogenesis following interspecific hybridization (Rowe, 1974). Haploids may arise by isolation in marker crosses: (1) a seed parent carrying one or more recessive gene(s) is pollinated by pollen from the male parent carrying the dominant allele and the progeny with the dominant character are discarded, or (2) for detection of androgenic haploids the recessive gene markers are carried in the pollinator stock. Markers may be observed at the seedling stage, as an embryo, or as dry seed markers (Sarkar, 1974). Haploids may be induced by wide crossing and selection of twins, somatic reduction (Clapham, 1977), and alien cytoplasm (Johns and Harvey, 1974; Tsunewaki et al., 1976a,b). The frequency of haploids may be increased by physical and chemical treatments (Lacadena, 1974). Routes to haploidy have been described also by Sunderland (1973) and by deFossard (1974).

Nei (1963) examined the efficiency of the haploid method of plant breeding in comparison with the conventional diploid methods.

The inadequacy of multicullular diploid material for mutation research with higher plants has been well highlighted by Devereux and deNettancourt (1974). In the great majority of cases, the conventional methodology involves the exposure of dry seeds, growing parts of flowers or flower buds to mutagens, and the scoring of mutations during the next generation at the seedling level or in the populations of mature plants (Gaul, 1964; Binding, 1974).

The best solution to these inadequacies might be to use haploid tissues and to induce and screen mutants at the single cell level. The importance of haploid tissues in mutagenesis has been discussed by Binding (1974), Melchers (1974), Sadasivaiah (1974), Chaleff and Carlson (1974), Carlson and Polacco (1975), and Clapham (1977).

Kao and Puck (1975), working with mammalian cells, have underlined the importance of haploidy in raising auxotrophic lines. They have used CHO-K1 cell lines of Chinese hamster ovary tissue cultures. This line has 20 chromosomes instead of 22, and karyotype analysis shows that several chromosomal alterations and delations have occurred. They have concluded that almost all of their mutations are in the haploid loci of the genome.

The recently found exotic genotypes in corn and barley (Ingversen et al., 1973; Doll et al., 1974; Doll, 1975; Ingversen, 1975) that have high lysine content, though at the expense of carbohydrates, are due to recessive mutations (Sylvester-Bradley and Folkes, 1976). If that is any indication, haploid tissues have great importance for obtaining "nutritional" mutants.

B. Mutations in Haploid and Diploid Cell Populations

Exposure to chemical and physical mutagens (Binding, 1974) of cultures has been reviewed extensively, so it will not be dealt with here. See Konzak (1957), Matsumura (1964), Smith (1972), and Gaina and Valeva (1975) for comparative genetic effects of different physical mutagens in higher plants, and Howland and Hart (1977) for the radiation biology of cultured cells. For the action of chemical agents (or culture media), see Murashige and Nakano (1966), Kasha (1974a), Lacadena (1974), Konstantinov et al. (1976), and Clapham (1977).

C.   Selection Systems for Somatic Hybrids and
         Mutated Cells

(Power and Cocking, 1977.)

Genetic variability may be induced into large, homogeneous populations of plant cells by exposure to chemical (including exogenous DNA) or physical mutagens. Conditions that select for defined mutant types may then be imposed upon this population. Dominant and semidominant mutants are obtainable from cultures of diploid cells. If haploid tissue is used, direct selection of recessive mutants becomes possible. These possibilities have been reviewed by Smith (1974) and Nabors (1976).

Negrutiu et al. (1975) proposed *Arabidopsis thaliana* as a model system in somatic cell genetics. However, the *in vitro* culture phase must be kept short and differentiation of whole plants achieved quickly in order to allow genetic stability. Selection of mutants might be achieved by manipulating cultural conditions.

The process of somatic mutagenesis can be used to achieve different kinds of mutant lines such as auxotrophy; resistance to drugs, analogs, and herbicides; resistance to phytotoxins; resistance to certain physical conditions, as low temperature; elevated amino acid levels; and increased photosynthetic efficiency (see Zenk, 1974).

Carlson (1970) isolated auxotrophic mutants from mutagenized haploid cells of *Nicotiana tabacum*. The selective system depended upon differential incorporation of 5-bromodeoxyuridine (BUDR) into DNA of phototrophic cells dividing actively in minimal medium. Only auxotrophs that synthesize little or no DNA during the BUDR pulse survive subsequent illumination. Mutants resistant to antibiotics (Maliga et al., 1973a; Binding, 1972; Sung, 1976; Nakdimon and Goldner, 1976) and nucleic acid analogs (Maliga et al., 1973b) have been isolated. Auxotrophic or drug-resistant lines can serve as markers in somatic hybridization.

Cell lines resistant to low temperature have been isolated in *Nicotiana* and *Capsicum* (Street and Dix, 1976).

Widholm (1972, 1974) has shown that in higher plants the endogenous concentration of a specific metabolite can be increased by selecting for resistance to a structural analog of that metabolite. He has isolated cell lines of tobacco and carrot capable of growth in presence of normally inhibiting concentrations of 5-methyl tryptophan. Endogenous levels of free tryptophan in resistant lines of tobacco and carrot were 15 and 27 times higher, respectively, than the wild type. This resistant phenotype can be transmitted from callus to plant and back to callus and remain stable. Chaleff and Carlson (1975) have reported altered levels of the aspartate group of amino acids, using lysine analog, S-aminoethyl cysteine (SAEC), in rice callus tissue. However, the callus in this case failed to differentiate into plants, as it was too old or lacked inductive factors. Similar resistant lines have been reported in carrot by Pallet and Miflin (Norris and Lea, 1976).

The problem of photorespiration and plant productivity has been discussed by Zelitch (1971). He believes that somatic mutagenesis could be used to provide selection methods for superior CO_2 uptake by illuminated plant tissue cultures grown in the absence of an organic carbon source, with the hope that such a genotype would be expressed in the intact plants regenerated from this callus. Contrary to previous reports, Berlyn and Zelitch (1975) have shown that haploid tobacco cultures can be maintained on CO_2 for long periods of time. An alternative approach would be to select for decreased photorespiration in haploid tissue cultures grown on sucrose by obtaining mutants with decreased rates of glycolic acid synthesis.

Increasing the photosynthetic efficiency, and thereby the photosynthate, can have great implications in increasing general yield and amino acid content of cereal grains. Bhatia and Rabson (1976) have summarized the bioenergetic implications of changing cereal grain protein concentrations. Any simultaneous increases in grain protein concentration and grain yield are incompatible for energetic considerations. Opaque-2 maize and high lysin sorghum and barley show increased lysine contents in their endosperm at the expense of other endosperm proteins and reduced grain yield. The synthesis of additional protein or carbohydrate requires the availability of additional photosynthesis. This requirement can be met only by increasing the photosynthetic efficiency of the cereal plants.

IV.   OVERVIEW OF CULTURAL SYSTEMS FOR SOMATIC MUTAGENESIS: IDENTIFICATION OF PROBLEMS

A.   Induction and Differentiation of Callus

1.   *Age and Kind of Tissue*

There is great variability among cultivars in the ability to respond to callus and organ induction (Picard and Buyser, 1975; Heszky and Mesch, 1976). Excised roots of Hilgendorf wheat grew on White's medium plus 5 mg/1 tryptophan as the auxin source, glucose rather than sucrose, and thiamine as the only vitamin. Only 3 of 28 varieties of wheat roots grew on the tryptophan medium used for Hilgendorf cv. (Ferguson, 1967).

Germinating embryos are the best source of material for diploid callus and immature embryos by the Bulbosum Method for haploid callus. Meristems and, to a lesser extent, stem nodes proliferate readily and are genetically stable, whereas mature somatic tissues or germinal plant tissues proliferate much less readily and show variations in ploidy, chromosome number, and karyotype. As tissues and cultures age, there is a progressive loss in callus- and organ-forming capacity, correlated with higher chromosome numbers and more aneuploidy. Callus induction falls off rapidly with subculturing (e.g., Wang et al., 1973b; Grunewaldt and Malepszy, 1973). Plantlet induction frequency is usually quite low, and all reports of 1–2 plantlets derived per callus must be discounted as possibly derived from organized tissue in the explant.

2.   *Sequential Treatment*

Most cultures require the principle of sequential treatment established by Steward (1958, 1963). It is useful to think in terms of callus-initiating, maintenance, and organ-inducing media.

Callus initiation requires a basal salt medium supplemented with a strong auxin, such as 2,4-D or 2,4,5-T. Occasionally weaker auxins such as IAA, NAA, and pCPA are added as well, but with no clearly demonstratèd efficacy. Also, the induction medium may be fortified with nucleotide or coconut milk. Coconut milk gives slightly better results in some cases (e.g., Guha et al., 1970, for rice) but occasionally coconut milk is inhibitory (e.g., Ono and Larter, 1976, for *Triticale*).

Callus maintenance may be the same medium as for callus induction, or the medium may contain reduced concentrations of growth substances.

An organ-inducing medium usually requires either dilution of the 2,4-D concentration (Chin and Scott, 1976), sustitution of 2,4-D with a weak auxin (Wang et al., 1973; Picard and Buyser, 1975; Shimada and Makino, 1975), elimination of auxin entirely (Second Division, 1974, Kimata and Sakamoto, 1972), or the addition of an antiauxin (Picard and Buyser, 1973). Occasionally, no change in auxin concentration is required between callus induction and organ induction (Foroughi-Wehr et al., 1977; Guha et al., 1970; Heszky and Mesch, 1976; Gamborg et al., 1970). In one case cytokinin substituted for auxin produced pollen tubes and embryoids but no callus (Guha and Maheshwari, 1964, 1967).

Only occasionally is a balance of auxins and cytokinins helpful (Niizeki and Oono, 1968). For the most part, cytokinins do not seem to contribute anything to monocot cultures.

Successful callus cultures have been achieved with a spectrum of media ranging from White's on the one hand to M-S on the other. With such a wide range of concentrations, it is unlikely that the basal medium is critical to successful cell and callus cultures in the monocots.

3. *Sugar Concentration*

Monocot cultures may require higher concentrations, on the order of 10-15%, than dicot cultures (Picard and Buyser, 1975; Ono and Larter, 1976). A contributing factor may be the use of 2,4-D as the hormone source in nearly all cultures.

4. *The Use of 2,4-D as a Growth Hormone*

The classical view has been that exogenous 2,4-D introduces mitotic irregularities into tissue culture, most certainly into longterm ones at high concentrations (say above 5 $\mu g/ml$; cf. Melchers and Bergmann, 1959; Torrey, 1967). On the other hand, Singh and Harvey (1975) contend that the higher the exogenous concentrations of 2,4-D, the fewer the mitotic irregularities (bridges and fragments) in cultures, at least in short-term ones.

Bayliss (1977) suggested that increasing the concentration of 2,4-D lengthened the mean generation time, particularly the duration of prophase and metaphase when the mitotic spindle mechanism is assembled, and might thereby affect mitotis through malfunction of the spindle apparatus. He showed that at 2,4-D concentrations of 15–30 μg/ml, the amount of Feulgen-positive materials increased. On the other hand, the work of Soma (1968) on application of 2,4-D to the surface of the apical meristem suggests to us that, whereas exogenous 2,4-D may affect DNA cell content, its main effect will be on cell enlargement rather than on cell division.

Feung et al. (1976) reported that the major metabolites of 2,4-D–grown rice callus are carboxylic glucosides, and two-ring hydroxylated aglycones in corn callus cultures. In monocot cultures where the medium is rather routinely supplemented with 2,4-D as a growth hormone, this paper raises the possibility of a 2,4-D–induced depletion of sugar in the tissues as sugar metabolites accumulate in the medium. That is, sugar may be inhibitory *or* limiting, depending upon the cultivar, since the major metabolites of 2,4-D–grown cultures are sugars. Harada and Nikumi (1950) reported that cell wall metabolites will react to form colored products with Schiff reagents. It would be impossible, by staining methods alone, to distinguish between 2,4-D–induced sugar metabolites and, 2,4-D–induced nucleotide residues that are Feulgen positive.

At the least, the work reviewed herein that reports the same requirements of DNA/RNA, protein synthesis, and auxin, plus a sugar source for growth and cell wall synthesis for both protoplast cultures and intact plants, suggests that the 2,4-D situation may not be simple and requires a new look.

5. *Cytogenetics in Cell Cultures*

As reviewed by Narayanaswamy (1977), many cytological changes may occur in somatic cells *in vitro*. Callus cultures maintained by subculture may consist solely of cells that are tetraploid or octoploid, or even aneuploid (Norstog et al., 1969). As pointed out before, wheat callus is more stable than that of other *Gramineae*.

In the monocots, Kao et al. (1970), Shimada (1971), and Heinz et al. (1971) described cytological changes in somatic cells leading to polyploidy and aneuploidy, and tissues became mixaploid. Diploid cells may spontaneously become polyploid through endoduplication or selective stimulation (D'Amato, 1977), and this may occur regularly in long-term cell culture (Murashige and Nakano, 1966).

6. *Chlorophyll Deficiency*

The problem of chlorophyll-deficient and genetically unstable plants discussed in the text has not been adequately treated. High frequencies of albino plant differentiation has been observed from cali of *Hordeum* (Clapham 1971), *Triticale*, and *Oryza* (Sun et al. 1974). Clapham's (1973) EM studies showed that there were almost normal chloroplasts in the leaf cells of haploid albino barley, indicating that chlorophyll systhesis was not normal. In *Oryza* (Sun et al., 1974), however, lamellar development in plastids is arrested; plastids had DNA fibrils but lacked ribosomes, so could not carry out vital protein synthesis. The male line—the pollen-callus—haploid pathway is inferior because differentiation is unpredictable, ploidy status is variable, and albinos are formed. The female line, discussed in the text under "Haploid callus from immature embryos produced by the Bulbosum Method," has been genetically stable, and Jensen (1975) was able to show dramatic progress in monoploid production of normal, green barley plants in liquid, continuous-flow nutrient solution.

There is now a distinct possibility that variations may be determined by expression of the physiological condition at the time of treatment, as well as by the possibility that the chromosomes of *H. bulbosum* were not completely eliminated before differentiation occurred, that there are genotypic differences for the ability to produce haploids, or that cytoplasmic factors play a strong role in induction and diffferentiation.

7. *New Direction*

From the above discussion and the review of current literature, it follows that at least two of the innovations now needed to lay the groundwork for genome alteration in wheat by somatic mutagenesis are to (1) make the diploid callus differentiate more frequently, and (2) devise some way for obtaining haploid callus in high frequency, since it is reported to differentiate more readily than diploid callus.

B. ANTHER-POLLEN CALLUS

The basis for many, if not most, of the experimental treatments and procedures in pollen-anther cultures rests on some fundamental studies that have guided investigators who have achieved some success in pollen-anther culture: Brewbaker (1959, 1967, 1971); Walker and Dietrich (1961); Linskens (1964); Linskens and Schwauwen (1968); Kihara and Hori (1966); Vasil (1967); Pfahler (1967, 1968); Bennett et al. (1973); and Mascarenhas (1975).

Further advances in pollen-anther culture must be based upon the fundamental morphology and physiology of anther-pollen development (cf. Bennett et al., 1973). A cursory survey of some physiological and environmental factors that affect the developmental responses of anther-pollen cultures and that should be borne in mind and reevaluated are:

1. *Stage of Pollen Development*

Usually the uninucleate condition of pollen grains is most favorable for anther-pollen culture (Nitsch and Nitsch, 1969; Dale, 1975; Heberle and Reinert, 1977).

Isolated pollen behaves differently than pollen in anther culture, never producing plantlets at the uninucleate stage but regularly at the binucleate stage (Heberle and Reinert, 1977).

2. *Physiological Condition of the Donor Plant*

Sunderland (1971, 1974) and Dunwell (1976) have pointed out that much of the variability and abnormalities in anther-pollen culture derive from the poor physiological condition of the donor plant.

3. *Culture Medium*

As for diploid callus, a wide spectrum of culture media have been used to successfully culture anthers-pollen. A number of these media have been carefully compared (Gamborg et al., 1976).

The role of mineral elements should be reevaluated, particularly with regard to calcium ion concentration (Brewbaker and Kwack, 1971), pollen source, calcium and boron interactions (Pfahler, 1967, 1968), and temperature and calcium in pollen germination (Cook and Walden, 1967).

4. *Physical Factors*

Temperature (Sopory and Maheshwari, 1976a; Nitsch, 1974; Nitsch and Norreel, 1972; Sunderland, 1971; Picard and Buyser, 1975), temperature and photoperiod (Dunwell, 1976; Muraoka and Ohira, 1956; Kasperbauer, 1970), temperature and pH Vasil and Bose, 1959)—all affect the efficiency of anther-pollen cultures.

5. *Cytological Factors*

Dale (1975) described pollen dimorphism in developing anthers, which reflects a different potential for different pollen grains.

Mascarenhas (1966, 1971) noted the synthesis of mRNA and rRNA in the pollen grain prior to anther dehiscence. Jalouzot (1969) noted differences in RNA in the vegetative and generative nuclei of pollen grains (*Lilium candidum*).

Synthesis of DNA, RNA, and nuclear histones takes place at microspore mitosis (Linskens and Schrauwen, 1968). After microspore mitosis, DNA synthesis takes place in both the vegetative and generative nuclei and reaches a 2C level. However, there is much less RNA synthesis in the generative cell. Most of the RNA synthesis takes place in the vegetative cell, and the bulk of it is rRNA. Most of the rRNA, and possibly tRNA also, appears to be synthesized in the period 24 hours before and following microspore mitosis, after which the rRNA and tRNA genes are turned off (Mascarenhas, 1971, 1975).

Bose (1959) described the effect of gibberellin in causing the generative cell of pollen to divide frequently.

6. *Timing of Exogenous Hormone Application*

Hatcher (1945) reported that auxin synthesis in the developing anther began at the uninucleate stage of pollen development. It is quite likely that organ-induction is triggered by a subtle interaction of growth hormones that would be overridden by 2,4-D residual in the cells. We believe that the most important aspect of organ induction may well be the timing of the application of exogenous growth hormone relative to the normal production and concentration of endogenous growth hormone.

7. *Faulty Experimental Design*

Many investigations lack experimental design and do not provide adequate information to evaluate the contribution of the study. Lacking most often are data on induction efficiency, whether the plantlets differentiated were transferred to potting soil and survived, whether the plantlets are morphologically and cytogenetically aberrant, and whether the plantlets were green or albino, and in what numbers.

8. *Ancillary Chemicals*

The possibility that p-fluorophenylaianine can be used to maintain, or even to increase, the population of haploid cells in haploid plants derived from anther cultures (of tobacco) needs to be investigated further (Matthews and Vasil, 1976), although Corduan (1975) failed in *Hyoscyamus.*

C. Haploid Embryoids—Anther-Pollen Culture

Transcription and translation move according to a program as the pollen matures. Successful embryogenesis is possible only by interrupting a normal program and shifting the microspore to a sporophytic pattern of development. The first indication of a new pattern could be a mitotic division leading to two equal cells (pathway B), or active divisions in the vegetative cell (pathway A), the generative cell (pathway D), or both (pathway C). This summary on the production of haploid embryoids follows closely the series of partial, though linked, processes and experimental treatments suggested by de Fossard (1974).

Street and Withers (1974) and Kohlenbach (1977) have reviewed some of the basic aspects of differentiation and plant regeneration from cell and tissue culture, particularly the anatomy and cytology of embryogenesis in culture.

D. HAPLOID EMBRYOS

In haploid embryogenesis by the Bulbosum Method (Kasha and Kao, 1970), there are two critical stages—at fertilization and early embryo development, and later embryo development on culture medium (Norstog, 1961, 1970, 1973; Jensen, 1976, 1977). It is crucially important to combine the proper nutritional factors with particular cytological factors.

E. CYTOPLASMIC INHERITANCE

Cytoplasmic inheritance in wheat and *Aegilops* was discovered by Kihara in 1951. Since then a number of protoplasms have been introduced into wheat from alien species, and Fukasawa (1967) and Kihara and Tsunewaki (1967) showed that certain phenotypic effects of the alien cytoplasm remain constant in wheat for many generations, and that genetic differences of cytoplasm exist in some *Triticum* and *Aegilops* species.

Tsunewaki et al. (1976a) produced nucleus-cytoplasm hybrids in all possible combinations between the nuclei of 12 strains of common wheat and the cytoplasms of 22 species of *Triticum* and *Aegilops*. Consideration was given to individual cytoplasms as to their genetic effects on the character expression of common wheats, and 8 major plasma types were distinguished—causing such effects as growth depression, male sterility and haploid formation, anther malformation, winter killing, and so forth.

In wheat, Kihara and Tsunewaki (1962) found that *Triticum aestivum* strain Salmon produces haploids at a high frequncy (7.1–33.6%) when its cytoplasm is substituted by the cytoplasm of 15 species of *Aegilops* and 3 species of *Triticum* (all with 2n = 46). Three pollen parents of *T. aestivum*, Salmon, Chinese Spring, and Jones Fife, showed different haploid-inducing powers in the order of Fife Jones (22.0%), Salmon (18.3%), and Chinese Spring (12.3%). All the haploid-inducing cytoplasm except one (*Ae. umbellulata*) produced twin and triplet seedlings at high frequencies (2.8–14.3%) (Tsunewaki et al., 1976b).

Manipulation of cytoplasm offers an alternative to the Bulbosum Method for obtaining haploids. The fact that wheat callus is much more stable than other monocot callus might be due to wheat containing attenuated alien cytoplasm.

F. PROTOPLASTS

As already discussed, mesophyll protoplasts can be easily isolated in cereals, but show limited divisions, if any. Potrykus et al. (1977) used 80,000 variations in culture media but failed to elicit any divisions. Maybe an effective answer lies in preconditioning the cells—with different physiologically active substances, culturing in conditioned media (Street, 1973a), or using nurse culture techniques (Hildebrandt, 1977). This approach has been largely neglected in culturing monocots. Summarizing, it could be said that cultural systems in cereals need much more painstaking work before these could be amenable to genetic modification by somatic mutagenesis or transformation.

REFERENCES

Abelson, P. H. (ed.). 1975. Food: Politics, economics, nutrition, and research. Amer. Assoc. Adv. Sci., Washington, D. C. See also: Science 188:501-650.

Abo El-Nil, M. M., and A. C. Hildebrandt. 1976. Cell wall regeneration and colony formation from isolated single geranium protoplasts in microculture. Canad. J. Bot. 54:1530-34.

Albersheim, P. 1974. Structure and growth of wall of cells in culture. *In* H. E. Street (ed.). Tissue culture and plant science. Pp. 379-404. Academic Press, New York.

Asami, H., N. Inomata, and M. Okamoto. 1976. Chromosome variation in callus cells derived from *Secale cercale* L. with and without B-chromosome. Jap. J. Genet. 51:297-303.

Asami, H., T. Shimada, N. Inomata, and M. Okamoto. 1975. Chromosome constitution in cultured callus cells from four aneuploid lines of the homoeologous group 5 of *Triticum aestivum*. Jap. J. Genet. 50:283-89.

Bajaj, Y. P. S. 1972. Protoplast culture and regeneration of haploid tobacco plants. Amer. J. Bot. 59:647.

Bajaj, Y. P. S. 1974a. The isolation, culture, and ultrastructure of pollen protoplasts. *In* K. Kasha (ed.), Haploids in higher plants—advances and potential. Pp. 139-40. University of Guelph, Guelph, Ontario, Canada.

Bajaj, Y. P. S. 1974b. Isolation and culture studies on pollen tetrad and pollen-mother-cell protoplasts. Plant Sci. Lett. 3:93-99.

Bajaj, Y. P. S. 1974c. Potentials of protoplast culture work in agriculture. Euphytica 23:633-49.

Bajaj, Y. P. S. 1976. In vitro induction of haploid plants. *In* P. K. Evans (ed.), Towards plant improvement by in-vitro methods. Academic Press, New York.

Bajaj, Y. P. S. 1977. Protoplast isolation, culture, and somatic hybridization. *In* J. Reinert and Y. P. S. Bajaj (eds.), Applied and fundamental aspects of plant cell, tissue, and organ culture. Pp. 467-96, 563-77. Springer-Verlag, Berlin.

Bajaj, Y. P. S., and M. R. Davey. 1974. The isolation and ultrastructure of pollen protoplasts. *In* H. F. Linskens (ed.), Fertilization in higher plants. Pp. 73-80. North-Holland Publishing Co., Amsterdam.

Baker, D. B. 1965. Relation between effects of auxin on cell wall synthesis and cell elongation. Plant Phsyiol. 40:360-68.

Baker, D. B., and P. R. Ray. 1965. Direct and indirect effects of auxin on cell wall synthesis in oat coleoptile tissue. Plant Physiol. 40:345-52.

Banks, M. S., and P. K. Evans. 1976. A comparison of the isolation and culture of mesophyll protoplasts from several *Nicotiana* species and their hybrids. Plant Sci. Lett. 7:409-16.

Barclay, I. R. 1975. High frequencies of haploid production in wheat (*Triticum aestivum*) by chromosome elimination. Nature 256:410–11.

Barclay, I. R., K. W. Shepherd, and D. H. B. Sparrow. 1972. Control of chromosome elimination in *Hordeum vulgare–H. bulbosum* hybrids. Barley Genet. Newsl. 2:22–24.

Baroncelli, S., M. Buiatti, A. Bennici, G. Foroghi-Wehr, G. Mix, H. Craul, and B. Giorgi. 1977. Genetic control of callus growth in hexaploid wheat. Experientia 34 (in press).

Bates, L. S., and C. W. Deyoe. 1973. Wide hybridization and cereal improvement. Econ. Bot. 27:401–12.

Bawa, S. B., and J. G. Torrey. 1971. Budding and nuclear division in cultured protoplasts of corn, *Convolvulus*, and onion. Bot. Gaz. 132:240–45.

Bayliss, M. W. 1977. The effects of 2,4-D on growth and mitosis in suspension cultures of *Daucus carota*. Plant Sci. Lett. 8:99–103.

Bennett, M. D., and W. G. Hughes. 1972. Additional mitosis in wheat pollen induced by ethrel. Nature 240:566–68.

Bennett, M. D., M. K. Rao, J. B. Smith, and M. W. Bayliss. 1973. Cell development in the anther, the ovule, and the young seed of *Triticum aestivum* L. var. Chinese Spring. Proc. Roy. Soc. Lond. 266B:39–81.

Bergmann, L. 1960. Growth and division of single cells of higher plants *in vitro*. J. Gen. Physiol. 43:841–51.

Berlyn, M. B., and I. Zelitch. 1975. Photoautotrophic growth and photosynthesis in tobacco callus cells. Plant Physiol. 56:752–56.

Bhargava, P. M., and G. Shanmugam. 1971. Uptake of non-viral nucleic acids by mammalian cells. *In* J. N. Davidson and W. E. Cohn (eds.), Progress in nucleic acid research and molecular biology. Pp. 104–59. Academic Press, New York.

Bhatia, C. R., and R. Rabson. 1976. Bioenergetic considerations in cereal breeding for protein improvement. Science 194:1418–20.

Bhojwani, S. S., J. M. Dunwell, and N. Sunderland. 1973. Nucleic acid and protein contents of embryogenic pollen. J. Exp. Bot. 24:863–71.

Bhojwani, S. S., P. K. Evans, and E. C. Cocking. 1977. Protoplast technology in relation to crop plants: Progress and problems. Euphytica 26:343–60.

Bianchi, F., R. deBoer, and A. J. Pompe. 1974. An investigation of spontaneous reversions in a dwarf mutant of *Petunia hybrida* in connection with the interpretation of the results of transformation experiments. Acta Bot. Neerl. 23:691–700.

Bianchi, F., and H. G. Walet-Foederer. 1974. An investigation into the anatomy of the shoot apex of *Petunia hybrida* in connection with the results of transformation experiments. Acta Bot. Neerl. 23:1–6.

Binding, H. 1966. Regeneration und verschmelzung nackter laubmoos protoplasten. Z. Pflanzenphysiol. 55:305–21.

Binding, H. 1972. Selektion in Kallus-Kulturen mit haploiden zellen. Z. Pflanzenphysiol. 67:33–38.

Binding, H. 1974. Mutation in haploid cultures. *In* K. Kasha (ed.), Haploids in higher plants—advances and potential. Pp. 323–37. University of Guelph, Guelph, Ontario, Canada.

Binding, H. 1976. Somatic hybridization experiments in Solanaceous species. Molec. Gen. Genet. 144:171–75.

Binh, L. T., and K. H. Köhler. 1976. Untersuchungen über die DNA-Synthese während der

Frühphase der Kallusbildung bei *Zea mays*—Wurzeln. (English summary.) Biochem. Physiol. Pflanz. 170:189–200.

Blaydes, D. F. 1966. Interaction of kinetin and various inhibitors in the growth of soybean tissue. Physiol. Plantarum 19:748–53.

Bose, N. 1959. Effect of gibberellin on the growth of pollen tubes. Nature 184:1577–79.

Bottino, P. J. 1975. The potential of genetic manipulation in plant cell cultures for plant breeding. Rad. Bot. 15:1–17.

Bouharmont, J. 1974. Study of the first stages of pollen development in cultured anthers of barley. *In* K. Kasha (ed.), Haploids in higher plants—Advances and potential. P. 140. University of Guelph, Guelph, Ontario, Canada.

Brenneman, F. N., and A. W. Galston. 1975. Experiments on cultivation of protoplasts and calli of agriculturally important plants. 1. Oat (*Avena sativa* L.). Biochem. Physiol. Pflanz. 168:453–71.

Brewbaker, J. L. 1959. Biology of the angiosperm pollen grain. Indian J. Genet. Plant Breed. 19:121–33.

Brewbaker, J. L. 1967. The distribution and phylogenetic significance of binucleate and trinucleate pollen grains in the angiosperms. Amer. J. Bot. 54:1069–83.

Brewbaker, J. L. 1971. Pollen enzymes and isoenzymes. *In* J. Heslop-Harrison (ed.), Pollen: Development and physiology. Pp. 156–70. Butterworths, London.

Brewbaker, J. L., and B. H. Kwack. 1963. The essential role of calcium ions in pollen germination and pollen tube growth. Amer. J. Bot. 50:859–65.

Buiatti, M. 1977. DNA amplification and tissue cultures. *In* J. Reinert and Y. P. S. Bajaj (eds.), Applied and fundamental aspects of plant cell, tissue, and organ culture. Pp. 358–74, 442–64. Springer-Verlag, Berlin.

Burke, D., P. Kaufman, M. McNeil, and P. Albersheim. 1974. The structure of plant cell walls. VI. A survey of the walls of suspension-cultured monocots. Plant Physiol. 54:109–15.

Butcher, D. N. 1977. Secondary products in plant tissues. *In* J. Reinert and Y. P. S. Bajaj (eds.), Applied and fundamental aspects of plant cell, tissue, and organ culture. Pp. 668–93, 703–16. Springer-Verlag, Berlin.

Butenko, R. G., B. P. Strogonov, and J. A. Babaeva. 1967. Somatic embryogenesis in carrot tissue cultures under conditions of high salt concentration in the medium. Dokl. Akad. Nauk. SSSR 175:1179–81.

Carew, D. P., and A. E. Schwarting. 1958. Production of rye embryo callus. Bot. Gaz. 119:237–39.

Carlson, P. S. 1970. Induction and isolation of auxotrophic mutants in somatic cell cultures of *Nicotiana tabacum*. Science 168:487–89.

Carlson, P. S. 1973. The use of protoplasts for genetic research. Proc. Nat. Acad. Sci. USA 70:598–602.

Carlson, P. S. 1975. Crop improvement through techniques of plant cell and tissue culture. Bioscience 25:747–49.

Carlson, P. S., and J. C. Polacco. 1975. Plant cell cultures: Genetic aspects of crop improvement. Science 188:622–26.

Carlson, P. S., H. H. Smith, and R. D. Dearing. 1972. Parasexual interspecific plant hybridization. Proc. Nat. Acad. Sci. USA 69:2292–94.

Carter, O., Y. Yamada, and E. Takahashi. 1967. Tissue culture of oats. Nature 214:1029–30.

Chaleff, R. S., and P. S. Carlson. 1974. Somatic cell genetics of higher plants. Ann. Rev. Genet. 8:267–78.

Chaleff, R. S., and P. S. Carlson. 1975. Higher plant cells as experimental organisms. *In* R. Markham et al. (eds.), Modification of the information content of plant cells. Pp. 197–214. North-Holland Publishing Co., Amsterdam.

Chang, W. C., and P. L. Chin. 1976. Induction of rice plantlets from anther culture. Bot. Bull. Acad. Sinica 17:18–24.

Chase, S. S. 1969. Monoploid and monoploid-derivatives of maize (*Zea mays* L.). Bot. Rev. 35:117–67.

Cheng, T-Y., and H. H. Smith. 1975. Organogenesis from callus culture of *Hordeum vulgare*. Planta 123:307–10.

Chilton, M-D, M. H. Drummond, D. J. Merlo, D. Sciaky, A. L. Montoya, M. P. Gordon, and E. W. Nester. 1977. Stable incorporation of plasmid DNA into higher plant cells: The molecular basis of crown gall tumorigenesis. Cell 11:263–73.

Chin, J. C., and K. J. Scott. 1977. Studies on the formation of roots and shoots in wheat callus cultures. Ann. Bot. 41:473–81.

Chu, C-c., C-c. Wang, C-s. Sun, N-f. Chien, K-c. Yin, and C. Hsü. 1974. Investigations on induction and morphogenesis of wheat (*Triticum vulgare*). Acta Bot. Sinica 151–9.

Churchill, M. E., E. A. Ball, and J. Arditti. 1973. Tissue culture of orchids. I. Methods for leaf tips. New Phytol. 72:161–66.

Clapham, D. H. 1971. In vitro development of callus from the pollen of *Lolium* and *Hordeum*. Z. Pflanzenzüchtg. 65:285–92.

Clapham, D. H. 1973. Haploid *Hordeum* plants from anthers in vitro. Z. Pflanzenzüchtg 69:142–55.

Clapham, D. H. 1977. Haploid induction in cereals. *In* J. Reinert and Y. P. S. Bajaj (eds.), Applied and fundamental aspects of plant cell, tissue, and organ culture. Pp. 279–98, 331–40. Springer-Verlag, Berlin.

Clausen, R. E., and W. E. Lammerts. 1929. Interspecific hybridization in *Nicotiana*. X. Haploid and diploid merogony. Amer. Nat. 63:279–82.

Cocking, E. C. 1960. A method for isolation of plant protoplasts and vacuoles. Nature 187:962–63.

Cocking, E. C. 1972. Plant cell protoplasts—isolation and development. Ann. Rev. Plant Physiol. 23:29–50.

Cocking, E. C., and P. K. Evans. 1973. The isolation of protoplasts. *In* H. E. Street (ed.), Plant tissue and cell culture. Pp. 100–120. University of California Press, Berkeley.

Constabel, F., D. Dudits, O. L. Gamborg, and K. N. Kao. 1975. Nuclear fusion in intergeneric heterokaryons. Canad. J. Bot. 53:2092–95.

Constabel, F., R. A. Miller, and O. L. Gamborg. 1971. Histological studies on embryos produced from cell cultures of *Bromus inermis*. Canad. J. Bot. 49:1415–17.

Constabel, F., G. Weber, J. W. Kirkpatrick, and K. Pahl. 1976. Cell division of intergeneric protoplast fusion products. Z. Pflanzenphysiol. 79:1–7.

Cook, F. S., and D. B. Walden. 1967. The male gametophyte of *Zea mays* L. III. The influence of temperature and calcium on pollen germination and tube growth. Canad. J. Bot. 45:605–13.

Corduan, G., 1975. Regeneration of anther-derived plants of *Hyoscyamus niger* L. Planta 127:27–36.

Coutts, R. H. A., and K. R. Wood. 1975. The isolation and culture of cucumber mesophyll protoplasts. Plant Sci. Lett. 4:189–93.

Coutts, R. H. A., and K. R. Wood. 1977. Improved isolation and culture methods for cucumber mesophyll protoplasts. Plant Sci. Lett. 9:45–51.

Craig, I. L. 1974. Haploid plants (2n = 21) from in vitro anther culture of *Triticum aestivum*. Canad. J. Genet. Cytol. 16:697–700.

Cummings, D. P., C. E. Green, and D. D. Stuthman. 1976. Callus induction and plant regeneration in oats. Crop Sci. 16:465–70.

Dale, P. J. 1975. Pollen dimorphism and anther culture in barley. Planta 127:213–20.

D'Amato, F. 1977. Cytogenetics of differentiation in tissue and cell cultures. *In* J. Reinert and Y. P. S. Bajaj (eds.), Applied and fundamental aspects of plant cell, tissue, and organ culture. Pp. 343–57, 442–64. Springer-Verlag, Berlin.

Davey, M. R. 1977. Bacterial uptake and nitrogen fixation. *In* J. Reinert and Y. P. S. Bajaj (eds.), Applied and fundamental aspects of plant cell, tissue, and organ culture. Pp. 551–62, 563–77. Springer-Verlag, Berlin.

Davey, M. R., E. Bush, and J. B. Power. 1974. Cultural studies of a dividing legume leaf protoplast system. Plant Sci. Lett. 3:127–33.

Davey, M. R., and E. C. Cocking. 1972. Uptake of bacteria by isolated higher plant protoplasts. Nature 239:455–56.

Davies, D. R. 1974. Chromosome elimination in interspecific hybrids. Heredity 32:267–70.

Deka, P. C., and S. K. Sen. 1976. Differentiation in calli originated from isolated protoplasts of rice (*Oryza sativa* L.) through plating technique. Molec. Gen. Genet. 145:239–43.

Devreux, M., and D. deNettancourt. 1974. Screening mutations in haploid plants. *In* K. J. Kasha (ed.), Haploids in higher plants—advances and potential. Pp. 309–22. University of Guelph, Guelph, Ontario, Canada.

Doll, H. 1975. Genetic studies of high lysine barley mutants. *In* Barley Genetics III. Pp. 541–46. Proc. 3rd Int. Barley Genet. Symp., Garching.

Doll, H., B. Koie, and B. O. Eggum. 1974. Induced high lysine mutants in barley. Rad. Bot. 14:73–80.

Dorion, N., Y. Chupeau, and J. P. Bourgin. 1975. Isolation, culture, and regeneration into plants of *Ranunculus scleratus* L. leaf protoplasts. Plant Sci. Lett. 6:325–31.

Dougall, D. K. 1973. Dilution plating and nutritional considerations. *In* P. F. Kruse, Jr., and M. K. Patterson, Jr. (eds.), Plant cells. Pp. 261–64. Academic Press, New York.

Doy, C. H., P. M. Gresshoff, and B. G. Rolfe. 1973a. Biological and molecular evidence for the transgenosis of genes from bacteria to plant cells. Proc. Nat. Acad. Sci. USA 70:723–26.

Doy, C. H., P. M. Gresshoff, and B. G. Rolfe. 1973b. Time course of phenotypic expression of *E. coli* gene Z following transgenosis in haploid *Lycopersicon esculentum* cells. Nature New Biol. 244:90–91.

Drummond, M. H., M. P. Gordon, E. W. Nester, and M. D. Chilton. 1977. Foreign DNA of bacterial plasmid origin is transcribed in crown gall tumors. Nature 269:535–36.

Dudits, D., K. N. Kao, F. Constabel, and O. L. Gamborg. 1976a. Fusion of carrot and barley protoplasts and division of heterokaryotes. Canad. J. Genet. Cytol. 18:263–69.

Dudits, D., K. N. Kao, F. Constabel, and O. L. Gamborg. 1976b. Embryogenesis and formation of tetra- and hexaploid plants from carrot protoplasts. Canad. J. Bot. 54: 1063–67.

Dudits, D., G. Nemet, and Z. Haydu. 1975. Study of callus growth and organ formation in wheat (*Triticum aestivum*) tissue cultures. Canad. J. Bot. 53:957–63.

Dunwell, J. M. 1976. A comparative study of environmental and developmental factors which influence embryo induction and growth in cultured anthers of *Nicotiana tabacum*. Envir. Exp. Bot. 16:109–18.

Dunwell, J. M., and N. Sunderland. 1974a. Pollen ultrastructure in anther cultures of *Nicotiana tabacum*. I. Early stages of culture. J. Exp. Bot. 25:353–61.

Dunwell, J. M., and N. Sunderland. 1974b. Pollen ultrastructure in anther cultures of *Nicotiana tabacum*. II. Changes associated with embryogenesis. J. Exp. Bot. 25:363–73.

Ehrenberg, L. 1971. Higher plants. *In* A. Hollaender (ed.), Chemical mutagens. 2:365–82. Plenum Publishing Corp., New York.

El Hinnawy, J. M. 1974. Chelating compounds as cell wall–loosening agents in cell suspension cultures of *Melilotus alba* Desr. Z. Pflanzenphysiol. 71:207–19.

Eriksson, T. 1965. Studies on the growth requirements and growth measurements of cell cultures of *Haplopappus gracilis*. Physiol. Plantarum 18:148–51.

Eriksson, T., H. T. Bonnet, K. Glimelius, and A. Wallin. 1974. Technical advances in protoplast isolation, culture, and fusion. *In* H. E. Street (ed.), Tissue culture and plant science. Pp. 213–29. Academic Press, New York.

Eriksson, T., and K. Jonasson. 1969. Nuclear division is isolated protoplasts from cells of higher plants grown *in vitro*. Planta 89:85–89.

Evans, P. K., A. G. Keates, and E. C. Cocking. 1972. Isolation of protoplasts from cereal leaves. Planta 104:178–81.

Ferguson, J. D. 1967. The nutrition of excised wheat roots. Physiol. Plantarum 20:276–84.

Feung, C-s., R. H. Hamilton, and R. O. Mumma. 1976. Metabolism of 2,4-dichlorophenoxyacetic acid. IV. Identification of metabolites in rice root callus tissue cultures. J. Agric. Food Chem. 24:1013–15.

Flanagan, D. 1976. Twelve articles on food and agriculture. Sci. Amer. 235:30–196.

Foroughi-Wehr, B., G. Mix, H. Gaul, and H. M. Wilson. 1977. Plant production from cultured anthers of *Hordeum vulgare* L. Z. Pflanzenzüchtg. 77:198–204.

Fossard, R. A. de. 1974. Summation: Methods for producing haploids. *In* K. J. Kasha (ed.), Haploids in higher plants—Advances and potential. Pp. 145–50. University of Guelph, Guelph, Ontario, Canada.

Fox, A. S., and S. B. Yoon. 1970. DNA-induced transformation in *Drosophila*: Locus-specificity and the establishment of transformed stocks. Proc. Nat. Acad. Sci. USA 67:1608–55.

Frearson, E. M., J. B. Power, and E. C. Cocking. 1973. The isolation, culture and regeneration of *Petunia* leaf protoplasts. Devel. Biol. 33:130–37.

Fujii, T. 1970. Callus formation in wheat anthers. Wheat Info. Serv. 31:1–2.

Fukasawa, H. Constancy of cytoplasmic property during successive backcrosses. Amer. Natur. 101:41–46.

Furuhashi, K., and M. Yatazawa. 1964. Indefinite culture of rice stem node callus. (In Japanese.) Kagaku 34:623.

Gaina, L. V., and S. A. Valeva. 1975. An investigation of the effect of mutagens on wheat varieties and species of different ploidy levels. Trans. from Genetika 10(1):7–14. Article UDC 575.24:633.11. Plenum Publishing Corp., New York.

Gamborg, O. L. 1970. The effects of amino acids and ammonium on the growth of plant cells in suspension culture. Plant Physiol. 45:372–75.

Gamborg, O. L. 1975. Callus and cell culture. *In* O. L. Gamborg and L. R. Wetter (eds.), Plant tissue culture methods. Pp. 1–10. National Research Council, C.N.R.S., Ottawa, Ontario, Canada.

Gamborg, O. L., F. Constabel, and R. A. Miller. 1970. Embryogenesis and production of albino plants from cell cultures of *Bromus inermis*. Planta 95:355–58.

Gamborg, O. L., and D. E. Eveleigh. 1968. Culture methods and detection of glucanases in suspension cultures of wheat and barley. Canad. J. Biochem. 46:417–21.

Gamborg, O. L., and R. A. Miller. 1973. Isolation, culture, and uses of plant protoplasts. Canad. J. Bot. 51:1795–99.

Gamborg, O. L., R. A. Miller, and K. Ojima. 1968. Nutrient requirements of suspension cultures of soybean root cells. Exp. Cell Res. 50:151–58.

Gamborg, O. L., T. Murashige, T. A. Thorpe, and I. K. Vasil. 1976. Plant tissue culture media. In Vitro 12:473–78.

Gamborg, O. L., and J. P. Shyluk. 1970. The culture of plant cells with ammonium salts as the sole nitrogen source. Plant Physiol. 45:598–600.

Gamborg, O. L., J. P. Shyluk, D. S. Brar, and F. Constabel. 1977. Morphogenesis and plant regeneration from callus of immature embryos of sorghum. Plant Sci. Lett. 10:67–74.

Gaul, H. 1964. Mutations in plant breeding. Rad. Bot. 4:155–232.

Gengenbach, B. G., and C. E. Green. 1975. Selection of T-cytoplasm maize callus cultures resistant to *Helminthosporium maydis* Race T. Pathotoxin. Crop Sci. 15:645–49.

Giménez-Martin, G., M. C. Risueno, and J. F. López-Sáez. 1969. Generative cell envelope in pollen grains as a secretion system, a postulate. Protoplasma 67:223–35.

Glick, J. L., and A. Majumdar. 1972. A "loop" model for integration of donor DNA into host DNA for Eukaryote cells. J. Theor. Biol. 36:503–12.

Gosch, G., and J. Reinert. 1976. Nuclear fusion in intergeneric heterokaryotes and subsequent mitosis of hybrid nuclei. Naturwiss. 63:534–35.

Graebe, J. E., and G. D. Novelli. 1966. A practical method for large-scale plant tissue culture. Exp. Cell Res. 41:501–20.

Green, C. E. 1977. Prospects for crop improvement in the field of cell culture. Hort. Sci. 12:131–34.

Green, C. E., and R. L. Phillips. 1975. Plant regeneration from tissue cultures of maize. Crop Sci. 15:417–21.

Green, C. E., R. L. Phillips, and R. A. Kleese. 1974. Tissue cultures of maize (Zea mays L.): Initiation, maintenance, and organic growth factors. Crop Sci. 14:54–58.

Gresshoff, P. M. 1975. Theoretical and comparative aspects of bacteriophage transfer and expression in eukaryotic cells in culture. *In* L. Ledoux (ed.), Genetic manipulations with plant material. Pp. 539–49. Plenum Publishing Corp., New York.

Gresshoff, P. M., and C. H. Doy. 1973. *Zea mays*: Methods for diploid callus culture and the subsequent differentiation of various plant structures. Austr. J. Biol. Sci. 26:505–8.

Grunewaldt, J., and S. Malepszy. 1975. Beobachtungen an Antherenkallus von *Hordeum vulgare* L. Z. Pflanzenzüchtg. 75:55–61.

Guha, S., R. D. Iyer, N. Gupta, and M. S. Swaminathan. 1970. Totipotency of gametic cells and the production of haploids in rice. Curr. Sci. 39:174–76.

Guha, S., and S. C. Maheshwari. 1964. *In vivo* production of embryos from anthers of *Datura*. Nature 204:497.

Guha, S., and S. C. Maheshwari. 1967. Development of embryoids from pollen grains of *Datura* in vivo. Phytomorphology 17:454–61.

Hall, M. D., and E. C. Cocking. 1974. The response of isolated *Avena* coleoptile protoplasts to indole-3 acetic acid. Protoplasma 19:225–34.

Halperin, W. 1966. Alternative morphogenetic events in cell suspensions. Amer. J. Bot. 53:443–53.

Halperin, W. 1970. Embryos from somatic plant cells. *In* H. A. Padykula (ed.), Control mechanisms in the expression of cellular phenotypes. Intern. Soc. Cell Biol. Symposia. 9:169–91. Academic Press, New York.

Halperin, W., and D. F. Wetherell. 1965. Ammonium requirement for embryogenesis in vitro. Nature 205:519–20.

Harada, T., and Z. Nikumi. 1950. Structure of lignin. J. Agr. Chem. Soc. Japan 23:415–21.

Harms, C. T., H. Lörz, and I. Potrykus. 1976. Regeneration of plantlets from callus cultures of *Zea mays* L. Z. Pflanzenzüchtg. 77:347–51.

Hatcher, E. S. J. 1945. Studies in the vernalization of cereals. IX. Auxin production during development and ripening of the anther and carpel of spring and winter rye. Ann. Bot. 9:235–66.

Hayward, C., and J. B. Power. 1975. Plant production from leaf protoplasts of *Petunia parodi*. Plant Sci. Lett. 4:407–10.

Heberle, E., and J. Reinert. 1977. Factors of haploid production by isolated pollen cultures. Naturwiss. 64:100–101.

Heinz, D. J., M. Krishnamurthi, L. G. Nickell, and A. Maretzki. 1977. Cell tissue and organ culture in sugarcane improvement. *In* J. Reinert and Y. P. S. Bajaj (eds.), Applied and fundamental aspects of plant cell, tissue, and organ culture. Pp. 3–17, 207–48. Springer-Verlag, Berlin.

Heinz, D. J., G. W. P. Mee, and L. G. Nickell. 1969. Chromosome numbers of some *Saccharum* species hybrids and their cell suspension cultures. Amer. J. Bot. 56:450–56.

Heller, R. 1953. Recherches sur la nutrition minérale des tissus végétaux cultivés in vitro. Ann. Sci. Nat. Bot. Biol. Veg. 14:1–223.

Heslop-Harrison, J. 1968. Synchronous pollen mitosis and the formation of the generative cell in massulate orchids. J. Cell Sci. 3:457–66.

Heslop-Harrison, J. (ed.). 1971. Pollen: Development and physiology. Butterworths, London.

Heslop-Harrison, J. 1972. Sexuality of angiosperms. *In* F. C. Steward (ed.), Plant physiology—A treatise 6 C:133–89. Academic Press, New York.

Heslop-Harrison, J., and Y. Heslop-Harrison. 1970. Evaluation of pollen viability by enzymatically induced fluoroescence; Intracellular hydrolysis of fluorescein diacetate. Stain Tech. 45:115–20.

Hess, D. 1975. Uptake of DNA and bacteriophage into pollen and genetic manipulation. *In* L. Ledoux (ed.), Genetic manipulations with plant material. Pp. 519–37. Plenum Publishing Corp., New York.

Hess, D. 1977. Cell modification by DNA uptake. *In* J. Reinert and Y. P. S. Bajaj (eds.), Applied and fundamental aspects of plant cell, tissue, and organ culture. Pp. 506–35, 563–77. Springer-Verlag, Berlin.

Hess, D., and G. Wagner. 1974. Induction of haploid parthenogenesis in *Mimulus luteus* by in vitro pollination with foreign pollen. Z. Pflanzenphysiol. 72:466–68.

Heszky, L., and J. Mesch. 1976. Anther culture investigations in cereal gene bank collection. Z. Pflanzenzüchtg. 77:187–97.

Higa, A. 1975. The changes in nucleic acid and protein accumulation during the batch cultivation of sugarcane cells. Plant Cell Physiol. 16:247–56.

Hildebrandt, A. C. 1977. Single cell culture, protoplasts, and plant viruses. *In* J. Reinert and Y. P. S. Bajaj (eds.), Applied and fundamental aspects of plant cell, tissue, and organ culture. Pp. 581–97, 636–46. Springer-Verlag, Berlin.

Holl, F. B. 1975. Innovative approaches to genetics in agriculture. Canad. J. Genet. Cytol. 17:517–24.

Holl, F. B., O. L. Gamborg, K. Ohyama, and L. Pelcher. 1974. Genetic transformation in plants. *In* H. E. Street (ed.), Tissue culture and plant science: 1974. Pp. 301–27. Academic Press, New York.

Homès, J. 1967. Induction de plantules dans des cultures in vitro de tissus de carotte. Compt. Rend. Soc. Biol. 161:730–32.

Hotta, Y., and H. Stern. 1971. Uptake and distribution of heterologous DNA in living cells. *In* L. Ledoux (ed.), Informative molecules in biological systems. Pp. 176–86. North-Holland Publishing Co., Amsterdam.

Hougas, R. S., S. J. Peloquin, and R. W. Ross. 1958. Haploids of the common potato. J. Hered. 49:103–6.

Howland, G. P., and R. W. Hart. 1977. Radiation biology of cultured plant cells. *In* J. Reinert and Y. P. S. Bajaj (eds.), Applied and fundamental aspects of plant cell, tissue, and organ culture. Pp. 731–56, 778–89. Springer-Verlag, Berlin.

Ingversen, J. 1975. Structure and composition of protein bodies from wild-type and high-lysine barley endosperm. Heriditas 81:69–76.

Ingversen, J., B. Køie, and H. Doll. 1973. Induced seed protein mutant of barley. Experientia 29:1151–52.

Inomata, N., M. Okamoto, and H. Asam. 1976. Behaviour of unstable telocentric chromosomes in cultured callus cells of Chinese Spring wheat. Jap. J. Genet. 51:223–28.

International Conference on Crop Productivity—Research Imperatives. 1975. Proceedings, Harbor Spring, Michigan, Oct. 20–24, 1975. Michigan Agr. Exp. Sta., East Lansing; Charles F. Kettering Foundation, Yellow Springs, Ohio.

Islam, R., and D. B. H. Sparrow. 1974. Production of haploids in barley. Barley News. 17:40–42.

Ito, Michio. 1973. Studies on the behavior of meiotic protoplasts. I. Isolation from microsporocytes of liliaceous plants. Bot. Mag. Tokyo 86:133–41.

Iyer, R. D., and S. K. Rama. 1972. The early ontogeny of embryoids and callus from pollen and subsequent organogenesis in anther cultures of *Datura metel* and rice. Planta 104:146–56.

Izhar, S., and J. B. Power. 1977. Genetical studies with *Petunia* leaf protoplasts. I. Genetic variation to specific growth hormones and possible genetic control on stages of protoplast development in culture. Plant Sci. Lett. 8:375–83.

Jacobsen, H. J. 1976. Genotypic and environmental influences on germination and callus induction in *Hordeum distichum*. Biochem. Physiol. Pflanzen 169:453–60.

Jacobsen, J. V., E. Pressman, and N. A. Pyliotis. 1976. Gibberellin-induced separation of cells in isolated endosperm of celery seed. Planta 129:113–22.

Jalouzot, R. 1969. Différenciation nucléaire et cytoplasmique du grain de pollen de *Lilium candidum*. Exp. Cell Res. 55:1–8.

Jensen, C. J. 1973. Production of monoploids in barley. Barley Genet. Newsl. 3:23–24.

Jensen, C. J. 1974a. Chromosome doubling techniques in haploids. *In* K. J. Kasha (ed.), Haploids in higher plants—advances and potential. Pp. 151–90. University of Guelph, Guelph, Ontario, Canada.

Jensen, C. J. 1974b. Production of monoploids in barley—a progress report. *In* Polyploidy and induced mutations in plant breeding. Pp. 169–79. Proceedings of the FAO/IAEA and Eucarpia Meeting, Bari, Italy, 2–10 Oct. 1972. International Atomic Energy Agency, Vienna.

Jensen, C. J. 1975. Barley monoploids and doubled monoploids: Techniques and experience. Barley Genet. 3:316–45.

Jensen, C. J. 1976. Embryo culture as an implement in hybridization, haploid production, and in genetical and developmental studies. Newsl. Int. Assn. Plant Tissue Culture, No. 18, July, 1976.

Jensen, C. J. 1977. Monoploid production by chromosome elimination. *In* J. Reinert and Y. P. S. Bajaj (eds.), Applied and fundamental aspects of plant cell, tissue, and organ culture. Pp. 299–330, 331–40. Springer-Verlag, Berlin.

Johns, W. A., and B. L. Harvey. 1974. The effects of *Hordeum bulbosum* L. cytoplasm on *H. vulgare* L. *In* K. J. Kasha (ed.), Haploids in higher plants—advances and potential. Pp. 276–77. University of Guelph, Guelph, Ontario, Canada.

Johnson, C. B., D. Grierson, and H. Smith. 1973. Expression of plac5 DNA in cultured cells of a higher plant. Nature New Biol. 244:105–6.

Johri, B. M., and S. S. Bhojwani. 1977. Triploid plants through endosperm culture. *In* J. Reinert and Y. P. S. Bajaj (eds.), Applied and fundamental aspects of plant cell, tissue, and organ culture. Pp. 398–411, 442–64. Springer-Verlag, Berlin.

Jones, L. E., A. C. Hildebrandt, A. J. Riker, and J. H. Wu. 1960. Growth of somatic tobacco cells in microculture. Amer. J. Bot. 47:468–75.

Kameya, T. 1975. Induction of hybrids through cell fusion with dextran sulfate and gelatin. Jap. J. Genet. 50:235–46.

Kao, F. T., and T. T. Puck. 1975. Mutagenesis and genetic analysis into Chinese hamster auxotrophic cell markers. Genetics 79:343–52.

Kao, K. N. 1975. A method for fusion of plant protoplasts with polyethylene glycol. *In* O. L. Gamborg and L. R. Wetter (eds.), Plant tissue culture methods. Pp. 22–27. National Research Council, C.N.R.S., Ottawa, Ontario, Canada.

Kao, K. N., F. Constabel, M. R. Michayluk, and O. L. Gamborg. 1974. Plant protoplast fusion and growth of intergeneric hybrid cells. Planta 120:215–27.

Kao, K. N., and M. R. Michayluk. 1974. A method for high frequency intergeneric somatic fusion of plant protoplasts. Planta 115:355–67.

Kao, K. N., R. A. Miller, O. L. Gamborg, and B. L. Harvey. 1970. Variations in chromosome number and structure in plant cells grown in suspension cultures. Canad. J. Genet. Cytol. 12:297–301.

Kartha, K. K., O. L. Gamborg, F. Constabel, and K. N. Kao. 1974. Fusion of rapeseed and soybean protoplasts and subsequent division of heterokaryocytes. Canad. J. Bot. 52:2435–36.

Kasha, K. J. (ed.). 1974a. Haploids in higher plants—advances and potential. University of Guelph, Guelph, Ontario, Canada.

Kasha, K. J. 1974b. Haploids from somatic cells. *In* K. J. Kasha (ed.), Haploids in higher plants—advances and potential. Pp. 67–91. University of Guelph, Guelph, Ontario, Canada.

Kasha, K. J., and K. N. Kao. 1970. High frequency haploid production in barley (Hordeum vulgare L.). Nature 225:874–76.

Kasperbauer, M. J. 1970. Photo- and thermo-control of flowering in tobacco (Nicotiana tabacum L.). Agron. J. 62:825–27.

Kaur-Sawhney, R., M. Rancillac, B. Staskawicz, W. R. Adams, Jr., and A. W. Galston. 1976. Effect of cycloheximide and kinetin on yield, integrity, and metabolic activity of oat leaf protoplasts. Plant Sci. Lett. 7:57–67.

Kawata, S., and A. Ishihara. 1968. The regeneration of rice plant, *Oryza sativa* L. in the callus derived from the seminal root. Proc. Japan Acad. 44:549–53.

Kermicle, J. L. 1974. Origin of androgenetic haploids and diploids induced by the indeterminate gametophytic (ig) mutation in maize. *In* K. J. Kasha (ed.), Haploids in higher plants—advances and potential. P. 137. University of Guelph, Guelph, Ontario, Canada.

Key, J. L. 1964. Ribonucleic acid and protein synthesis as essential processes for auxin-induced cell elongation. Plant Physiol. 39:365–70.

Key, J. L., and J. Ingle. 1964. Requirement for the synthesis of DNA-like RNA for growth of excised plant tissue. Proc. Nat. Acad. Sci. USA 52:1382–88.

Kihara, H. 1951. Substitution of nucleus and its effect on genome manifestations. Cytologia 16:177–93.

Kihara, H., and T. Hori. 1966. The behavior of nuclei in germinating pollen grains of wheat, rice, and maize. Der Züchter 36:145–50.

Kihara, H., and K. Tsunewaki. 1962. Use of an alien cytoplasm as a new method of producing haploids. Jap. J. Genet. 37:310–13.

Kihara, H., and K. Tsunewaki. 1967. Genetic principles applied to the breeding of crop plants. *In* A. Brink (ed.), Heritage from Mendel. Pp. 403–18. University of Wisconsin Press, Madison.

Kimata, M., and S. Sakamoto. 1972. Production of haploid albino plants of *Aegilops* by anther culture. Jap. J. Genet. 47:61–63.

Kimber, G., and R. Riley. 1963. Haploid angiosperms. Bot. Rev. 90:480–531.

King, P. J., and H. E. Street. 1973. Growth patterns in cell cultures. *In* H. E. Street (ed.), Plant tissue and cell culture. Botanical Monographs Vol. 11, Pp. 269–337. University of California Press, Berkeley.

Kleinhofs, A. 1975. DNA-hybridization studies of the fate of bacterial DNA in plants. *In* L. Ledoux (ed.), Genetic manipulations with plant material. Pp. 461–77. Plenum Publishing Corp., New York.

Kleinhofs, A., F. C. Eden, M. D. Chilton, and A. J. Bendich. 1975. On the question of the integration of exogenous bacterial DNA into plant DNA. Proc. Nat. Acad. Sci. USA 72:2748–52.

Koblitz, H. 1974. Methodische Aspekte der Zell- und Gewebezüchtung bei Gramineen unter besonderer Berücksichtigung der Getreide. Kulturpflanze 22:95–157.

Koblitz, H. 1976. Isolierung und Kultivierung von Protoplasten aus Calluskulturen der Gerste. (English summary.) Biochem. Physiol. Pflanzen 170:287–93.

Koblitz, H., and G. Saalbach. 1976. Callus culture from apical meristem of barley. Biochem. Physiol. Pflanzen 170:97–102.

Kohlenbach, H. W. 1977. Basic aspects of differentiation and plant regeneration from cell and tissue cultures. *In* W. Barz, E. Reinhard, and M. H. Zenk (eds.), Plant tissue culture and its bio-technological application. Pp. 355–66. Springer-Verlag, Berlin.

Konzak, C. F. 1957. The genetic effects of radiation on higher plants. Quart. Rev. Biol. 32:27–45.

Kostoff, D. 1943. Cytogenetics of the genus *Nicotiana*. Pp. 1–1071. States Printing House, Sofia, Bulgaria.

Krishnamurt, M. 1976. Isolation, fusion, and multiplication of sugarcane protoplasts and comparison of sexual and parasexual hybridization. Euphytica 25:145–50.

Kuo, J. S., Y. Y. Wang, N. F. Chien, S. S. Kuo, M. L. Kung, and H. C. Hou. 1973. Investigations on the anther culture in vitro of *Nicotiana tabacum* L. and *Capsicum annuum* L. Acta Bot. Sinica 15:37–50.

Lacadena, J. R. 1974. Spontaneous and induced parthenogenesis and androgenesis. *In* K. J. Kasha (ed.), Haploids in higher plants—advances and potential. Pp. 13–32. University of Guelph, Guelph, Ontario, Canada.

LaCour, L. F. 1949. Nuclear differentiation in the pollen grain. Heredity 3:319–37.

Lai, K. L., and L. F. Liu. 1976. On the isolation and fusion of rice protoplasts. Agric. Assoc. China 93:1–9.

Landgren, C. R., and J. G. Torrey. 1973. The culture of protoplasts derived from explants of seedling pea roots. *In* Protoplastes et fusion de cellules somatique végétales. Colloques Intern. C.N.R.S. 212:281–89. Institut National de la Recherche Agronomique, Paris.

Lange, W. 1971. Crosses between *Hordeum vulgare* L. and *H. bulbosum* L. I. Production, morphology, and meiosis of hybrids, haploids, and dihaploids. Euphytica 20:14–29.

LaRue, C. D. 1949. Cultures on the endosperm of maize. Amer. J. Bot. 34:585–86.

Ledoux, L. (ed.). 1975. Genetic manipulations with plant material. Plenum Publishing Corp., New York.

Ledoux, L. 1975. Fate of exogenous DNA in plants. *In* L. Ledoux (ed.), Genetic manipulations with plant material. Pp. 479–98. Plenum Publishing Corp., New York.

Ledoux, L., and R. Huart. 1972. Fate of exogenous DNA in plants. *In* L. Ledoux (ed.), Uptake of informative molecules by living cells. Pp. 249–76. North-Holland Publishing Co., Amsterdam.

Ledoux, L., R. Huart, M. Mergeay. P. Charles, and M. Jacobs. 1975. DNA mediated genetic correction of thiamineless *Arabidopsis thaliana*. *In* R. Markham, D. R. Davies, D. A. Hopwood, and R. W. Horne (eds.), Modification of the information content of plant cells. Pp. 67–89. North-Holland Publishing Co., Amsterdam.

Linskens, H. F. (ed.). 1964. Pollen physiology and fertilization. North-Holland Publishing Co., Amsterdam.

Linskens, H. F., and J. Schrauwen. 1968. Quantitative nucleic acid determinations in the microspore and tapetum fractions of lily anthers. Proc. Koninkl. Nederl. Akad. v. Wetenschappen-Amsterdam 71C:267–79.

Linsmaier, E. M., and F. Skoog. 1965. Organic growth factor requirements of tobacco tissue cultures. Physiol. Plantarum 18:100–127.

Liu, M. C., and W. H. Chen. 1974. Isolation and fusion of protoplasts from sugarcane young leaves. *In* K. J. Kasha (ed.), Haploids in higher plants—advances and potential. Pp. 141–42. University of Guelph, Guelph, Ontario, Canada.

Liu, M. C., and W. H. Chen. 1976. Tissue and cell culture as aids to sugarcane breeding. I. Creation of genetic variation through callus cultures. Euphytica 25:393–403.

Lörz, H., C. T. Harms, and I. Potrykus. 1976. Regeneration of plants from callus in *Avena sativa* L. Z. Pflanzenzüchtg. 77:257–59.

Mäder, M., Y. Meyer, and M. Bopp. 1976. Zellwand regeneration und Peroxidase-Isoenzym-Synthese isolieter Protoplasten von *Nicotiana tabacum*. Planta 129:33–38.

Maeda, E. 1968. Subculture and organ formation in the callus derived from rice embryos *in vitro*. Proc. Crop Sci. Soc. Japan 37:51–58.

Maeda, E. 1971. Growth of rice callus derived from the embryo under subculture conditions. Proc. Crop Sci. Soc. Japan 40:141–49.

Maeda, E., and T. Hagiwara. 1974. Enzymatic isolation of protoplasts from the rice leaves and callus cultures. Proc. Crop Sci. Soc. Japan 43:68–76.

Maeda, I. M. 1976. Effect of auxin concentration on the callus induction from the various organs of rice seedlings. Proc. Crop Sci. Soc. Japan 45:545–57.

Magoon, M. L., and K. R. Khanna. 1963. Haploids. Caryologia 16:191–235.

Maliga, P. S., A. Breznovitz, and L. Marton. 1973a. Streptomycin-resistant plants from callus culture of haploid tobacco. Nature New Biol. 244:29–30.

Maliga, P. S., L. Marton, and A. Sz-Breznovits. 1973b. 5-Bromodeoxyuridine-resistant cell lines from haploid tobacco. Plant Sci. Lett. 1:119–21.

Maretzki, A., and L. G. Nickell. 1973. Formation of protoplasts from sugarcane cell suspensions and the regeneration of cell cultures from protoplasts. *In* Protoplastes et fusion de cellules somatique végétales. Colloques Intern. C.N.R.S. 212:51–63. Institut National de la Recherche Agronomique, Paris.

Mascarenhas, A. F., M. Pathak, R. R. Hendre, D. D. Ghugale, and V. Jagannathan. 1975a. Tissue culture of maize, wheat, rice, and sorghum. IV. Studies of organ differentiation in tissue cultures of maize, wheat, and rice. Indian J. Exp. Biol. 13:116–19.

Mascarenhas, A. F., M. Pathak, R. R. Hendre, and V. Jagannathan. 1975b. Tissue culture of maize, wheat, rice, and sorghum. I. Initiation of viable callus and root cultures. Indian J. Exp. Biol. 13:103–7.

Mascarenhas, J. P. 1966. Pollen tube growth and ribonucleic acid synthesis by vegetative and generative nuclei of *Tradescantia*. Amer. J. Bot. 53:563–69.

Mascarenhas, J. P. 1971. RNA and protein synthesis during pollen development and tube growth. *In* J. Heslop-Harrison (ed.), Pollen: Development and physiology. Pp. 201–23. Butterworths, London.

Mascarenhas, J. P. 1975. The biochemistry of angiosperm pollen development. Bot. Rev. 41:259–315.

Mascarenhas, J. P., and P. R. Bell. 1969. Protein synthesis during germination of pollen: Studies on polyribosome formation. Biochim. Biophys. Acta 179:199–203.

Masteller, V. J., and D. J. Holden. 1970. The growth of and organ formation from callus tissue of sorghum. Plant Physiol. 45:363–64.

Matsumura, S. 1964. Complete report of "mechanism of mutagenesis." Jap. J. Genet. 39:83–198.

Matthews, P. S., and I. K. Vasil. 1976. The dynamics of cell proliferation in haploid and diploid tissues of *Nicotiana tabacum*. Z. Pflanzenphysiol. 77:222–36.

McLaren, A. D., W. A. Jensen, and L. Jacobson. 1960. Absorption of enzymes and other proteins by barley roots. Plant Physiol. 35:549–56.

Melchers, G. 1972. Haploid higher plants for plant breeding. Z. Pflanzenzüchtg. 67:19–32.

Melchers, G. 1974. Haploid research in higher plants. *In* K. J. Kasha (ed.), Haploids in higher plants—advances and potential. Pp. 393–401. University of Guelph, Guelph, Ontario, Canada.

Melchers, G., and L. Bergmann. 1959. Untersuchungen an Kulturen von haploiden Geweben von *Antirrhinum majus*. Ber. d. Bot. Ges. 71:459–73.

Melchers, G., and G. Labib. 1974. Somatic hybridization of plants by fusion of protoplasts. I. Selection of light resistant hybrids of "haploid" light sensitive varieties of tobacco. Molec. Gen. Genet. 135:277–94.

Merril, C. R., and H. Stanbro. 1974. Intercellular gene transfer. Z. Pflanzenphysiol. 72:371–88.

Meyer, Y. 1974. Isolation and culture of tobacco mesophyll protoplasts using saline medium. Protoplasma 81:363–72.

Meyer, Y., and W. O. Abel. 1975a. Importance of the wall for cell division and in the activity of the cytoplasm in cultured tobacco protoplasts. Planta 123:33–40.

Meyer, Y., and W. O. Abel. 1975b. Budding and cleavage division of tobacco mesophyll protoplasts in relation to pseudo-wall and wall formation. Planta 125:1–13.

Mezentsev, A. V., R. G. Butenko, and N. A. Rodionova. 1976. Production of isolated protoplasts from mesophyll of perennial grasses and barley. Sov. Plant Physiol. 23:431–35.

Miller, C. O. 1963. Kinetin and kinetin-like compounds. *In* H. F. Linskens and M. V. Tracey (eds.), Moderne Methoden der Pflanzen-analyse. 6:194–202. Springer-Verlag, Berlin.

Miller, R. A., O. L. Gamborg, W. A. Keller, and K. N. Kao. 1971. Fusion and division of nuclei in multinucleated soybean protoplasts. Canad. J. Genet. Cytol. 13:347–53.

Misawa, M., K. Sakato, H. Tanaka, M. Hayashi, and H. Samejima. 1974. Production of physiologically active substances by plant cell suspension cultures. *In* H. E. Street (ed.), Tissue culture and plant science: 1974. Pp. 405–32. Academic Press, New York.

Motoyoshi, F. 1972. Protoplast isolation from callus cells of maize endosperm. Exp. Cell Res. 68:452–56.

Motoyoshi, F., J. B. Bancroft, J. W. Watts, and J. Burgess. 1973. The infection of tobacco protoplasts with cowpea chlorotic mottle virus and its RNA. J. Gen. Virol. 20:117–93.

Mullin, M. 1970. Tissue culture of some monocotyledonous plants. Austr. J. Biol. Sci. 23:473–77.

Muraoka, Y., and K. Ohori. 1956. Developmental responses of tobacco plants under various temperatures and photoperiods. Proc. Crop Sci. Soc. Japan 25:104–6.

Murashige, T. 1974. Plant propagation through tissue cultures. Ann. Rev. Plant Physiol. 25:135–66.

Murashige, T., and R. Nakano. 1966. Tissue culture as a tool in obtaining polyploid plants. J. Heredity 57:115–18.

Murashige, T., and R. Nakano. 1967. Chromosome complement as a determinant of the morphogenetic potential of tobacco cells. Amer. J. Bot. 54:963–70.

Murashige, T., and F. Skoog. 1962. A revised medium for rapid growth and bioassays with tobacco tissue cultures. Physiol. Plantarum 15:473–97.

Myint, A., and R. A. de Fossard. 1974. Induction of haploid callus from rice anthers and regeneration of plants. *In* K. J. Kasha (ed.), Haploids in higher plants—advances and potential. P. 139. University of Guelph, Guelph, Ontario, Canada.

Nabors, M. W. 1976. Using spontaneously occurring and induced mutations to obtain agriculturally useful plants. Bioscience 26:761–68.

Nagata, T., and I. Takebe. 1971. Plating of isolated tobacco mesophyll protoplasts on agar medium. Planta 99:12–20.

Nakai, Y., and T. Shimada. 1975. In vitro culture of wheat tissues. II. Morphological, cytological, and enzymatic variations induced in wheat callus by growth regulators, adenine sulfate, and casein hydrolysate. Jap. J. Genet. 50:19–31.

Nakata, K., and M. Tanaka. 1968. Differentiation of embryoids from developing germ cells in anther culture of tobacco. (Japanese with English summary.) Jap. J. Genet. 43:65–71.

Nakdimon, U., and R. Goldner. 1976. Effects of streptomycin on diploid tobacco callus cultures and the isolation of resistant mutants. Protoplasma 89:83–89.

Nam, L-s, B. Landová, and Z. Landa. 1976. Isolation of protoplasts from sugar beet leaves. Biol. Plantarum 18:389–92.

Narayanaswamy, S. 1977. Regeneration of plants from tissue cultures. In J. Reinert and Y. P. S. Bajaj (eds.), Applied and fundamental aspects of plant cell, tissue, and organ culture. Pp. 179–206, 207–48. Springer-Verlag, Berlin.

National Research Council/Board of Agriculture and Renewable Resources. 1975. NRC (Barr) Study on world food and nutrition: Enhancement of food production for the United States.

Negrutiu, I., F. Beeftink, and M. Jacobs. 1975. *Arabidopsis thaliana* as a model system in somatic cell genetics. Plant Sci. Lett. 5:293–304.

Nei, M. 1963. The efficiency of haploid method of plant breeding. Heredity 18:95–100.

Nickell, L. G., and D. J. Heinz. 1973. Potential of cell and tissue culture techniques as aids in economic plant improvement. In A. M. Srb (ed.), Genes, enzymes, and populations. Pp. 109–28. Plenum Publishing Corp., New York.

Nickell, L. G., and A. Maretzki. 1969. Growth of suspension cultures of sugarcane cells in chemically defined media. Physiol. Plantarum 22:117–25.

Niizeki, H. 1968. Induction of haploid plants from anther culture. Jarg: Japan. Agric. Res. Quart. 3:41–45.

Niizeki, H., and K. Oono. 1968. Induction of haploid rice plant from anther culture. Proc. Jap. Acad. 44:554–57.

Niizeki, H., and K. Oono. 1971. Rice plants obtained by anther culture. In Cultures de tissus de plantes. Colloques Intern. C.N.R.S. 193:251–57. Institut National Recherche Agronomique, Paris.

Niizeki, M., and W. F. Grant. 1971. Callus, plantlet formation, and polyploidy from cultured anthers of lotus and *Nicotiana*. Canad. J. Bot. 49:2041–51.

Nishi, T., and S. Mitsuoka. 1969. Occurrence of various ploidy plants from anther and ovary culture of rice plants. Jap. J. Genet. 44:341–46.

Nishi, T., Y. Yamada, and E. Takahashi, 1968. Organ redifferentiation and plant restoration in rice callus. Nature 219:508–9.

Nishi, T., Y. Yamada, and E. Takahashi. 1973. The role of auxins in differentiation of rice tissues cultured *in vitro*. Bot. Mag. Tokyo 86:183–88.

Nitsan, J., and A. Lang. 1965. Inhibition of cell division and cell elongation in higher plants by inhibitors of DNA synthesis. Devel. Biol. 12:358–76.

Nitsch, C. 1974. Pollen culture—a new technique for mass production of haploid and homozygous plants. In K. J. Kasha (ed.), Haploids in higher plants—advances and potential Pp. 123–35. University of Guelph, Guelph, Ontario, Canada.

Nitsch, C. 1975. Single cell culture of an haploid cell: The microspore. *In* L. Ledoux (ed.), Genetic manipulations with plant material. Pp. 297–310. Plenum Publishing Corp., New York.

Nitsch, C. 1977. Culture of isolated microspores. *In* J. Reinert and Y. P. S. Bajaj (eds.), Applied and fundamental aspects of plant cell tissue, and organ culture. Pp. 268–78, 331–40. Springer-Verlag, Berlin.

Nitsch, C., and B. Norreel. 1972. Factors favoring the formation of androgenetic embryos in anther culture. *In* A. M. Srb (ed.), Gene, enzymes, and populations. Pp. 129–44. Plenum Publishing Corp., New York.

Nitsch, J. P. 1951. Growth and development in vitro of excised ovaries. Amer. J. Bot. 38:566–77.

Nitsch, J. P., and C. Nitsch. 1969. Haploid plants from pollen grains. Science 163:85–87.

Nitzsche, W. 1970. Herstellung haploider Pflanzen aus *Festuca-Lolium*-bastarden. Naturwiss. 57:199–200.

Nitzsche, W., and L. Hennig. 1977. Frachtknotenkultur bei Gräsern. (English summary.) Z. Pflanzenzüchtg. 77:80–82.

Nooden, L. D., and K. V. Thimann. 1963. Evidence for a requirement for protein synthesis for auxin-induced cell enlargement. Proc. Nat. Acad. Sci. USA 50:194–200.

Norris, R. D., and P. J. Lea. 1976. The use of amino acid analogues in biological studies. Sci. Prog. 63:65–85.

Norstog, K. 1956. Growth of rye-grass endosperm in vitro. Bot. Gaz. 117:253–59.

Norstog, K. 1961. The growth and differentiation of cultured barley embryos. Amer. J. Bot. 48:876–84.

Norstog, K. 1970. Studies on the survival of very small embryos in culture. Bull. Torrey Bot. Club 94:223–29.

Norstog, K. 1973. New synthetic medium for the culture of premature barley embryos. In Vitro 8:307–8.

Norstog, K., and D. Blume. 1974. Abscissic acid promotion of development of excised immature barley embryos. *In* K. J. Kasha (ed.), Haploids in higher plants—advances and potential. P. 142. University of Guelph, Guelph, Ontario, Canada.

Norstog, K., W. E. Wall, and G. P. Howland. 1969. Cytological characteristics of ten-year-old rye-grass endosperm tissue cultures. Bot. Gaz. 130:83–86.

Novak, F. J., and B. Vyskot. 1975. Karyology of callus cultures derived from *Nicotiana tabacum* L. haploids and ploidy of regenerants. Z. Pflanzenzüchtg. 75:62–70.

Ogura, H. 1976. The cytological chimeras in original regenerants from tobacco tissue cultures and in their offsprings. Jap. J. Genet. 51:161–74.

Ohira, K., L. Ojima, and A. Fujiwara. 1973. Studies on the nutrition of rice cell culture. I. A simple, defined medium for rapid growth in suspension culture. Plant Cell Physiol. 14:1113–21.

Ohira, K., K. Ojima, M. Saigusa, and A. Fujiwara. 1975. Studies on the nutrition of rice cell culture. II. Microelement requirements and the effects of deficiency. Plant Cell Physiol. 16:73–82.

Ohyama, K., O. L. Gamborg, and R. A. Miller. 1972. Uptake of exogenous DNA by plant protoplasts. Canad. J. Bot. 50:2077–80.

Ohyama, K., and J. P. Nitsch. 1972. Flowering haploid plants obtained from protoplasts of tobacco leaves. Plant Cell Physiol. 13:229–36.

Ono, H., and E. N. Larter. 1976. Anther culture of Triticale. Crop Sci. 16:120–22.

Ouyang, T. W., H. Hu, C. C. Chuang, and C. C. Tseng. 1973. Induction of pollen plants from anthers of *Triticum aestivum* L. cultured *in vitro*. Sci. Sinica 16:79–95.

Pelcher, L. E., O. L. Gamborg, and K. N. Kao. 1974. Bean mesophyll protoplasts: Production, culture, and callus formation. Plant Sci. Lett. 3:107–11.

Pfahler, P. L. 1967. In vitro germination, and pollen tube growth of maize (Zea mays L) pollen. I. Calcium and boron effects. Canad. J. Bot. 45:839–45.

Pfahler, P. L. 1968. In vitro germination and pollen tube growth of maize (Zea mays) pollen. II. Pollen source, calcium, and boron interactions. Canad. J. Bot. 46:235–40.

Picard, E. 1973. Influence de modifications dans les corrélations internes sur le devenir du gametophyte male de *Triticum aestivum* L. in situ et en culture in vitro. C. R. Acad. Sci. Paris 277D:777–80.

Picard, E., and J. deBuyser. 1973. Obtention de plantules haploides de *Triticum aestivum* L. à partir de culture d'anthères in vitro. C. R. Acad. Sci. Paris 277D: 1463–66.

Picard, E., and J. deBuyser. 1975. Nouveaux résultats concernant la culture d'anthères in vitro de blé tendre (*Triticum aestivum* L.). Effets d'un choc thermique et de la position de l'anthère dans l'épi. C. R. Acad. Sci. Paris 281D: 127–30.

Picard, E., J. deBuyser, and J. Bozza. 1974. Production of haploid plants by anther culture of wheat. *In* K. J. Kasha (ed.), Haploids in higher plants—advances and potential. P. 143. University of Guelph, Guelph, Ontario, Canada.

Pojnar, E., J. H. M. Willison, and E. C. Cocking. 1967. Cell wall regeneration by isolated tomato fruit protoplasts. Protoplasma 64:460–80.

Potrykus, I. 1973. Transplantation of chloroplasts into protoplasts of *Petunia*. Z. Pflanzenphysiol. 70:364–66.

Potrykus, I., H. Lörz, and C. T. Harms. 1977. On some selected problems and results concerning culture and genetic modification of higher plant protoplasts. *In* W. Barz, E. Reinhard, and M. H. Zenk (eds.), Plant tissue culture and its bio-technological application. Pp. 323–34. Springer-Verlag, Berlin.

Power, J. B., and E. C. Cocking. 1970. Isolation of leaf protoplasts: Macro-molecule uptake and growth substance response. J. Exp. Bot. 21:64–70.

Power, J. B., and E. C. Cocking. 1977. Selection systems for somatic hybrids. *In* J. Reinhert and Y. P. S. Bajaj (eds.), Applied and fundamental aspects of plant cell, tissue, and organ culture. Pp. 497–505, 563–77. Springer-Verlag, Berlin.

Power, J. B., S. E. Cummins, and E. C. Cocking. 1970. Fusion of isolated plant protoplasts. Nature 225:1016–18.

Power, J. B., E. M. Frearson, C. Hayward, and E. C. Cocking. 1975. Some consequences of the fusion and selective culture of *Petunia* and *Parthenocissus* protoplasts. Plant Sci. Lett. 5:197–207.

Power, J. B., E. M. Frearson, D. George, P. K. Evans, S. F. Berry, C. Hayward, and E. C. Cocking. 1976. The isolation, culture, and regeneration of leaf protoplasts in the genus *Petunia*. Plant Sci. Let. 7:51–55.

Prokharov, M. N., L. K. Chernova, and B. V. Folin-Koldakov. 1974. Growing wheat tissues in culture and the regeneration of an entire plant. Dokl. Akad. SSSR 214:472–75.

Raghavan, V. 1976. Role of generative cell in androgenesis in henbane. Science 191: 388–89.

Raghavan, V. 1977. Applied aspects of embryo culture. *In* J. Reinert and Y. P. S. Bajaj (eds.), Applied and fundamental aspects of plant cell, tissue, and organ culture. Pp. 375–97, 442–64. Springer-Verlag, Berlin.

Raman, K., and R. I. Greyson. 1977. Graft unions between floral half-meristems of differing genotypes of *Nigella damascens* L. Plant Sci. Lett. 8:367–73.

Rangan, T. S. 1974. Morphogenetic investigations on tissue cultures of *Panicum miliaceum*. Z. Pflanzenphysiol. 72:456–59.

Rangan, T. S. 1976. Growth and plantlet regeneration in tissue cultures of some Indian millets: *Papsalum scrobiculatum* L., *Eleusine coracana* Gaertn., *Pennisetum typhoideum* Pers. Z. Pflanzenphysiol. 78:208–16.

Rangaswamy, N. J. 1977. Applications of in vitro pollination and in vitro fertilization. *In* J. Reinert and Y. P. S. Bajaj (eds.), Applied and fundamental aspects of plant cell, tissue, and organ culture. Pp. 412–25, 442–64. Springer-Verlag, Berlin.

Ray, P. M., and D. B. Baker. 1965. The effect of auxin on synthesis of oat coleoptile cell wall constituents. Plant Physiol. 40:353–60.

Reinert, J. 1959. Über die Kontrolle der Morphogenese und die Induktion von Adventiveembryonen an Gewebekulturen aus Karotten. Planta 53:318–33.

Reinert, J. 1973. Aspects of organization—organogenesis and embryogenosis. *In* H. E. Street (ed.), Plant tissue and cell culture. Pp. 338–55. Blackwell, Oxford.

Reinert, J., and D. Backs. 1968. Control of totipotency in plant cells growing in vitro. Nature 220:1340–41.

Reinert, J., and Y. P. S. Bajaj (eds.). 1977. Applied and fundamental aspects of plant cell, tissue, and organ culture. Springer-Verlag, Berlin.

Reinert, J., and Y. P. S. Bajaj. 1977. Anther culture: Haploid production and its significance. *In* J. Reinert and Y. P. S. Bajaj (eds.), Applied and fundamental aspects of plant cell, tissue, and organ culture. Pp. 251–67, 331–40. Springer-Verlag, Berlin.

Reinert, J., and G. Gosch. 1976. Continuous division of heterokaryons from *Daucus carota* and *Petunia hybrida* protoplasts. Naturwiss. 63:534.

Reinert, J., and S. Hellmann. 1971. Mechanism of formation of polynuclear protoplasts from cells of higher plants. Naturwiss, 58:419.

Reinert, J., M. Tazawa, and S. Semonoff, 1967. Nitrogen compounds as factors of embryogenesis in vitro. Nature 216:1213–14.

Riley, Ralph. 1974. The status of haploid research. *In* K. J. Kasha (ed.), Haploids in higher plants—advances and potential. Pp. 3–9. University of Guelph, Guelph, Ontario, Canada.

Ross, M. K., and T. A. Thorpe. 1973. Physiological gradients and shoot initiation in tobacco callus cultures. Plant Cell Physiol. 14:473–80.

Rowe, P. R. 1974. Methods of producing haploids: Parthenogenesis following interspecific hybridization. *In* K. J. Kasha (ed.), Haploids in higher plants—advances and potential. Pp. 43–52. University of Guelph, Guelph, Ontario, Canada.

Ruesink, A. W. 1971. Protoplasts of plant cells. Meth. Enzymol. 23A:197–209.

Ruesink, A. W., and K. V. Thimann. 1965. Protoplasts from the *Avena* coleoptile. Proc. Nat. Acad. Sci. USA 54:56–64.

Sadasivaiah, R. S. 1974. Haploids in genetic and cytological research. *In* K. J. Kasha (ed.), Haploids in higher plants—advances and potential. Pp. 355–86. University of Guelph, Guelph, Ontario, Canada.

Sanderson, F. H. 1975. The great food fumble. Science 188:503–9.

Sarkar, K. R. 1974. Genetic selection techniques for the production of haploid plants. *In* K. J. Kasha (ed.), Haploids in higher plants—advances and potential. Pp. 33–41. University of Guelph, Guelph, Ontario, Canada.

Scandalios, J. G., and J. C. Sorensen. 1977. Isozymes in plant tissue culture. *In* J. Reinert and Y. P. S. Bajaj (eds.), Applied and fundamental aspects of plant cell, tissue, and organ culture. Pp. 719-30, 778-89. Springer-Verlag, Berlin.

Schaskolskaya, N. D., G. N. Sacharovskaya, and E. V. Sacharova. 1973. The optimal conditions for isolation and incubation of barley mesophyll protoplasts. *In* Protoplastes et fusion cellules somatiques végétales. Colloques Intern. C.N.R.S. 212:93-98.

Schenk, R. U., and A. C. Hildebrandt. 1969. Production of protoplasts from plant cells in liquid cultures using purified commercial cellulases. Crop Sci. 9:629-31.

Schenk, R. U., and A. C. Hildebrandt. 1972. Medium and techniques for induction and growth of monocotyledonous and dicotyledonous cell cultures. Canad. J. Bot. 50:199-204.

Sears, E. R. 1974. The wheats and their relatives. *In* R. C. King (ed.), Handbook of genetics. Pp. 59-93. Plenum Publishing Corp., New York.

Sharp, W. R., R. S. Raskin, and H. E. Sommer. 1972. The use of nurse culture in the development of haploid clones in tomato. Planta 104:357-61.

Sheridan, W. F. 1975. Tissue culture of maize. I. Callus induction and growth. Physiol. Plantarum 33:151-56.

Shimada, T. 1971. Chromosome numbers in cultured pith tissue of tobacco. Jap. J. Genet. 46:235-41.

Shimada, T., and T. Makino. 1975. In vitro culture of wheat. III. Anther culture of the A genome aneuploids in common wheat. Theor. Appl. Genet. 46:407-10.

Shimada, T., T. Sasakuam, and K. Tsunewaki. 1969. In vitro culture of wheat tissues. I. Callus formation, organ redifferentiation, and single cell culture. Canad. J. Genet. Cytol. 11:294-304.

Singh, B. D., and B. L. Harvey. 1975. Does 2,4-D induce mitotic irregularities in plant tissue cultures? Experientia 31:785-87.

Skoog, F., and C. O. Miller. 1957. Chemical regulation of growth and organ formation in plant tissues cultivated in vitro. Symp. Soc. Exp. Biol. 11:118-30.

Smith, H. H. 1972. Comparative genetic effects of different physical mutagens in higher plants. *In* Induced mutations and plant improvement. Pp. 75-93. International Atomic Energy Agency, Vienna.

Smith, H. H. 1974. Model systems for somatic plant cell genetics. Bioscience 24:269-76.

Soma, K. 1968. The effect of direct application of 2,4-D to the shoot apex of *Phaseolus vulgaris*. Phytomorphology 18:305-24.

Somatic Hybridization Research Group and Cytobiochemistry Research Group. Peking Institute of Botany, Academia Sinica. 1975. Isolation and culture of rice protoplasts. Sci. Sinica 18:779-89.

Söndahl, M. R., and W. R. Sharp. 1977. High frequency induction of somatic embryos in cultured leaf explants of *Coffea arabica* L. Z. Pflanzenphysiol. 81:395-408.

Sopory, S. K., and S. C. Maheshwari. 1976a. Development of pollen embryoids in anther cultures of *Datura innoxia*. I. General observations and effects of physical factors. J. Exp. Bot. 27:49-57.

Sopory, S. K., and S. C. Maheshwari. 1976b. Development of pollen embryoids in anther cultures of *Datura innoxia*. II. Effects of growth hormones. J. Exp. Bot. 27:58-68.

Soyfer, V. N., N. A. Kartel, N. M. Chekalin, Y. B. Titov, K. K. Cieminis, and N. V. Turbin. 1976. Genetic modification of the waxy character in barley after an injection of wild-type exogenous DNA. Analysis of the second seed generation. Mutat. Res. 36:303-10.

Staba, E. J. 1977. Tissue culture and pharmacy. *In* J. Reinert and Y. P. S. Bajaj (eds.), Applied and fundamental aspects of plant cell, tissue, and organ culture. Pp. 694–702, 703–16. Springer-Verlag, Berlin.

Steward, F. C. 1963. Carrots and coconuts: Some investigations on growth. *In* P. Maheshwari and N. S. Rangaswamy (eds.), Plant tissue and organ culture—a symposium. Pp. 178–97. International Society of Plant Morphologists, Delhi.

Steward, F. C., M. O. Mapes, A. E. Kent, and R. D. Holsten. 1958. Growth and organized development of cultured cells. I. Growth and division of freely-suspended cells. Amer. J. Bot. 45:693–703.

Straus, J. 1960. Maize endosperm tissue grown in vitro. III. Development of a synthetic medium. Amer. J. Bot. 47:641–47.

Street, H. E. 1973a. Cell (suspension) cultures—techniques. *In* H. E. Street (ed.), Plant tissue and cell culture. Pp. 59–99. University of California Press, Berkeley.

Street, H. E. 1973b. Single-cell clones. *In* H. E. Street (ed.), Plant tissue and cell culture. Pp. 191–204. University of California Press, Berkeley.

Street, H. E. (ed.). 1974. Tissue culture and plant science. Academic Press, New York.

Street, H. E. 1977a. Engineering with plant cells. Interdisciplinary Sci. Rev. 2:62–74.

Street, H. E. 1977b. Applications of cell suspension cultures. *In* J. Reinert and Y. P. S. Bajaj (eds.), Applied and fundamental aspects of plant cell, tissue, and organ culture. Pp. 649–67, 703–16. Springer-Verlag, Berlin.

Street, H. E., and P. J. Dix. 1976. Selection of plant cell lines with enhanced chilling resistance. Ann. Bot. 40:903–10.

Street, H. E., and L. A. Withers. 1974. The anatomy of embryogenesis in culture. *In* H. E. Street (ed.), Tissue culture and plant science. Pp. 71–100. Academic Press, New York.

Strogonov, B. P., E. I. Komizerko, and R. G. Butenko. 1968. Culturing of isolated grasswort, sorghum, sweetclover, and cabbage tissues for a comparative study of their salt resistance. Sov. Plant Physiol. 15:173–77.

Sun, C. S., C. C. Wang, and C. C. Chu. 1974. Cell division and differentiation of pollen grains in *Triticale* anthers cultured in vitro. Sci. Sinica 17:47–51.

Sunderland, N. 1971. Anther culture: A progress report. Sci. Prog. 59:527–49.

Sunderland, N. 1973. Pollen and anther culture. *In* H. E. Street (ed.), Plant tissue and cell culture. Pp. 205–39. University of California Press, Berkeley.

Sunderland, N. 1974. Anther culture as a means of haploid induction. *In* K. J. Kasha (ed.), Haploids in higher plants—advances and potential. Pp. 91–122. Fniversity of Guelph, Guelph, Ontario, Canada.

Sunderland, N., G. B. Collins, and J. M. Dunwell. 1974. The role of nuclear fusion in pollen embryogenesis of *Datura innoxia*. Planta 117:227–41.

Sunderland, N., and J. M. Dunwell. 1974. Pathways in pollen embryogenesis. *In* H. E. Street (ed.), Tissue culture and plant science. Pp. 141–67. Academic Press, New York.

Sunderland, N., and F. M. Wicks. 1971. Embryoid formation in pollen grains of *Nicotiana tabacum*. J. Exp. Bot. 22:213–16.

Sung, Z. R. 1976. Mutagenesis of cultured plant cells. Genetics 81:51–57.

Sussex, I. M. 1972. Somatic embryos in long-term carrot tissue cultures: Histology, cytology, and development. Phytomorphology 22:50–58.

Sylvester-Bradley, R., and B. F. Folkes. 1976. Cereal grains: Their protein components and nutritional quality. Sci. Prog. 63:241–63.

Taiz, I., and R. L. Jones. 1971. The isolation of barley-aleurone protoplasts. Planta 101:95–100.

Takebe, I., Y. Otsuke, and S. Aoki. 1968. Isolation of tobacco mesophyll cells in intact and active state. Plant Cell Physiol. 9:115–24.

Tamura, S. 1968. Shoot formation in calli originated from rice embryo. Proc. Japan Acad. 44:544–48.

Tanaka, M., and K. Nakata. 1969. Tobacco plants obtained by anther culture and the experiment to get diploid seeds from haploids. (English summary.) Jap. J. Genet. 44:47–54.

Taylor, A. R. D., and J. D. Hall. 1976. Some physiological properties of protoplasts isolated from maize and tobacco tissue. J. Exp. Bot. 27:389–91.

Tazawa, M., and J. Reinert. 1969. Extracellular and intracellular chemical environments in relation to embryogenesis in vitro. Protoplasma 68:157–73.

Thomas, E., and M. R. Davey. 1975. From single cells to plants. Wykeham Publ., London; Springer-Verlag, New York.

Thomas, E., F. Hoffmann, and G. Wenzel. 1975. Haploid plantlets from microspores of rye. Z. Pflanzenzüchtg. 75:7–14.

Torrey, J. G. 1959. Experimental modification of development in the root. *In* D. Rudnick (ed.), Cell Organism and milieu. Pp. 189–222. Ronald Press, New York.

Torrey, J. G. 1967. Morphogenesis in relation to chromosomal constitution in long-term plant tissue cultures. Physiol. Plantarum 20:265–75.

Torrey, J. G., and J. Reinert. 1961. Suspension cultures of higher plant cells in synthetic media. Plant Physiol. 36:483–91.

Torrey, J. G., J. Reinert, and N. Merkel. 1962. Mitosis in suspension cultures of higher plant cells in synthetic medium. Amer. J. Bot. 49:420–25.

Torrey, J. G., and Y. Shigemura. 1957. Growth and controlled morphogenesis in pea root callus tissue grown in liquid media. Amer. J. Bot. 44:334–44.

Townsend, C. O. 1884. Der einfluss des zellkerns auf die bildung der zellhaut. Jahrb. Wiss. Bot. 30:484–510.

Trione, E. J., L. E. Jones, and R. J. Metzger. 1968. In vitro culture of somatic wheat callus tissue. Amer. J. Bot. 55:529–31.

Tsuji, S., and K. Tsunewaki. 1976. Genetic diversity of the cytoplasm in *Triticum* and *Aegilops*. III. On the origin of the cytoplasm of two hexaploid *Aegiops* species. Jap. J. Genet. 51:149–59.

Tsunewaki, K., Y. Mukai, T. R. Endo, S. Tsuji, and M. Murata. 1976a. Genetic diversity of the cytoplasm in *Triticum* and *Aegilops*. V. Classification of 23 cytoplasms into 8 plasma types. Jap. J. Genet. 51:175–91.

Tsunewaki, K., Y. Mukai, T. R. Endo, S. Tsuji, and M. Murata. 1976b. Genetic diversity of the cytoplasm in *Triticum* and *Aegilops*. VI. Distribution of the haploid-inducing cytoplasms. Jap. J. Genet. 51:193–200.

Tyrnov, V. S., and S. S. Khokhlov. 1976. Androgenesis in angiosperms. Transl. from Genetika 10(9):154–67. Article UDC 575:581.16:575.42. Pp. 1183–93. Plenum Publishing Corp., New York.

Udvardy, J., B. Sivok, and G. Nemet. 1976. Effect of napthylacetic acid, 2,4-5-trichlorophenoxyacetic acid and 3,6-dichloro-o-anisic acid on nucleolytic enzymes in callus cultures from wheat root. Z. Pflanzenphysiol. 78:33–40.

Umetsu, N., K. Ojima, and K. Matsuda. 1975. Enhancement of cell separation by colchicine in cell suspension cultures of soybean. Planta 125:197–200.

Vardi, A., and D. Raveh. 1976. Cross-feeder experiments between tobacco and orange protoplasts. Z. Pflanzenphysiol. 78:350–59.

Vardi, A., P. Spiegel-Roy, and E. Galun. 1975. Citrus cell culture: Isolation of protoplasts, plating densities, effect of mutagens, and regeneration of embryos. Plant Sci. Lett. 4:231–36.

Vasil, I. K. 1967. Physiology and cytology of anther development. Biol. Rev. 42:327–73.

Vasil, I. K., and C. Nitsch. 1975. Experimental production of pollen haploids and their uses. Z. Pflanzenphysiol. 76:191–212.

Vasil, V., and I. K. Vasil. 1973. Growth and cell division in isolated plant protoplasts in microchambers. *In* Protoplasts et fusion cellules somatiques végétales. Colloques Intern. C.N.R.S. 212:139–49.

Vyskot, B. M., and F. J. Novak. 1976. Induced androgenesis *in vitro*—a new method of obtaining haploid plants. Transl. from Genetika 11(1):135–45. Article UDC 576.356.52. Pp. 103–9. Plenum Publishing Corp., New York.

Wagner, G. J., and H. W. Siegelman. 1975. Large-scale isolation of intact vacuoles and isolation of chloroplasts from protoplasts of mature plant tissues. Science 190:1298–99.

Walker, G. W. R., and J. F. Dietrich. 1961. Abnormal mircrosporogenesis in *Tradescantia* anthers cultured with sucrose-deficiency and kinetin-supplementation. Canad. J. Genet. Cytol. 3:170–83.

Wang, C-c, C-c Chu, Ss Sun, S-h Wu, K-c Yin, and C. Hsü. 1973a. The androgenesis in wheat (*Triticum aestivum*) anthers cultured in vitro. Sci. Sinica 16:218–22.

Wang, Y. Y., C. S. Sun, C. C. Wang, and N. F. Chien. 1973b. The induction of the pollen plantlets of Triticale and Capsicum annuum anther culture. Sci. Sinica 16:147–51.

Watts, J. W., D. Cooper, and J. M. King. 1975. Plant protoplasts in transformation studies: Some practical considerations. *In* R. Markham, D. R. Davies, D. A. Hopwood, and R. W. Horne (eds.), Modification of the information content of plant cells. Pp. 119–31. North-Holland Publishing Co., Amsterdam.

Webster, J. M. 1966. Production of oat callus and its susceptibility to a plant parasitic nematode. Nature 212:1472.

White, P. R. 1939. Potentially unlimited growth of excised plant callus in artificial nutrient. Amer. J. Bot. 26:54–64.

Widholm, J. M. 1972. Cultured *Nicotiana tabacum* cells with an altered anthranilate synthetase which is less sensitive to feedback inhibition. Biochim. Biophys. Acta 261:52–58.

Widholm, J. 1974. Cultured carrot cell mutants: 5-methyltryptophan-resistance trait carried from cell to plant and back. Plant Sci. Lett. 3:323–30.

Withers, L., and E. C. Cocking. 1972. Fine structural studies on spontaneous and induced fusion of higher plant protoplasts. J. Cell Sci. 11:59–75.

Wulff, H. D., and P. Maheshwari. 1938. The male gametophyte of angiosperms. J. Indian Bot. Soc, 17:117–40.

Yamada, Y. 1977. Tissue culture studies on cereals. *In* J. Reinert and Y. P. S. Bajaj (eds.), Applied and fundamental aspects of plant cell, tissue, and organ culture. Pp. 144–59, 207–48. Springer-Verlag, Berlin.

Yamada, Y., K. Tanaka, and E. Takahashi. 1967. Callus induction in rice, *Oryza sativa* L. Proc. Japan Acad. 43:156–60.

Yamaguchi, H., H. Tokuda, and M. Fukazawa. 1970. Growth of barley cells in shaking culture. Jap. J. Breed. 20:160–64.

Yamauchi, F., M. Hashimoto, and I. Nishiyama. 1976. Cytogenetics of the F_2 progeny of a highly sterile hybrid *Avena longiglumis* × *A. strigosa*. Jap. J. Genet. 51:109–13.

Yamaya, T., and K. Ohira. 1976. Nitrate reductase inactivating factor from rice cells in suspension culture. Plant Cell Physiol. 17:633–41.

Yatazawa, M., K. Furuhashi, and M. Shimizu. 1967. Growth of callus tissue from rice-root in vitro. Plant Cell Physiol. 8:363–73.

Yeoman, M. M. 1973. Tissue (callus) cultures—techniques. *In*: H. E. Street (ed.), Plant tissue and cell culture. Pp. 31–58. University of California Press, Berkeley.

Yeoman, M. M., and H. E. Street. 1973. General cytology of cultured cells. *In* H. E. Street (ed.), Plant tissue and cell culture. Pp. 121–60. University of California Press, Berkeley.

Yin, Kuang-chu, et al. 1976. A study of the new cultivar of rice raised by haploid breeding method. Sci. Sinica 19:227–41.

Yudin, R. F. 1974. Means of increasing the efficiency of the haploid method of producing homozygous lines in maize. *In* K. J. Kasha (ed.), Haploids in higher plants— advances and potential. Pp. 138–39. University of Guelph, Guelph, Ontario, Canada.

Zaenen, I., H. van Larebeke, M. Teuchy, M. van Montagu, and J. Schell. 1974. Supercoiled circular DNA in crown gall inducing *Agrobacterium* strain. J. Molec. Biol. 86:109–127.

Zaitlin, M., and R. N. Beachy. 1974. Protoplasts and separated cells: Some new vistas for plant virology. *In* H. E. Street (ed.), Tissue culture and plant science: 1974. Pp. 265–85. Academic Press, New York.

Zapata, F. J., P. K. Evans, J. B. Power, and E. C. Cocking. 1977. The effect of temperature on the division of leaf protoplasts of *Lycopersicon esculentum* and *Lycopersicon peruvianum*. Plant Sci. Lett. 8:119–24.

Zelcer, A., and E. Galun. 1976. Culture of newly isolated tobacco protoplasts: Precursor incorporation into protein, RNA, and DNA. Plant Sci. Lett. 7:331–36.

Zelitch, I. 1971. Photosynthesis, photorespiraton, and plant productivity. Academic Press, New York.

Zenk, M. H. 1974. Haploids in physiological and biochemical research. *In* K. J. Kasha (ed.), Haploids in higher plants—advances and potential. Pp. 339–53. University of Guelph, Guelph, Ontario, Canada.

Zenkteler, M., and E. Misiura. 1974. Induction of adrogenic embryos from cultured anthers of *Hordeum, Secale,* and *Festuca*. Biochem. Physiol. Pflanzen 165:337–40.

Zenkteler, M., E. Misiura, and A. Ponitka. 1975. Induction of androgenetic embryoids in the *in vitro* cultured anthers of several species. Experientia 31:289–91.

K. K. KARTHA AND O. L. GAMBORG

Cassava Tissue Culture—Principles and Applications

32

INTRODUCTION

Cassava (*Manihot esculenta* Crantz), a vegetatively propagated tuber crop, belongs to the family Euphorbiaceae. The genus *Manihot* consists of two sections, the Arboreae and Fruticosa. The former contains tree species and the latter cassava. Although opinions differ as to the origin, it is believed that cassava was first cultivated either in Brazil, Venezuela, or Central America (Rogers, 1965). Cassava, however, can be successfully grown in zones ranging from latitudes 30° north and south and at elevations of up to 2,000 m. It has an exceptional ability to withstand temperature variations of 18 to 35°C, precipitation of 50 to 500 mm and a soil pH range of 5 to 9.

Cassava is one of the 12 to 15 important food crops of the world and is considered as a subsistence crop for people in the tropical areas. In addition to its ability to tolerate drought conditions, cassava is also capable of growing in poor soils and is relatively resistant to weeds and pests. Unlike other.major crops, cassava is not season-bound in the sense that it can be planted and harvested during any time of the year or, if desired, can be left unharvested for considerable periods of time. The latter feature establishes cassava as a risk-aversion crop and a security against famine for people living close to subsistence levels in tropical countries.

K. K. Kartha and O. L. Gamborg, National Research Council of Canada, Prairie Regional Laboratory, Saskatoon, Saskatchewan, Canada.

USES OF CASSAVA

The most important use of cassava is as human food. It is also being used for industrial purposes, mostly as a source of starch, and for animal feed supplement.

It is estimated that around 300 million people in the tropical areas of the world depend on cassava as a major carbohydrate source (Nestel, 1973). Cassava is produced in more than 80 countries, but two-thirds of the world's production takes place in Brazil, Indonesia, Zaire, Nigeria, and India. Current annual production of cassava on a global basis exceeds 92 million tons, of which 55 million tons are used for human food. Cassava productivity in terms of calories per unit of land per unit of time is significantly higher than other staple food crops (de Vries et al., 1967). Coursey and Haynes (1970) report that cassava can produce 250×10^3 cal/ha/day as compared with 176×10^3 for rice and 110×10^3 for sorghum.

Nutritionally, cassava is inferior to other crops. Though it is low in calcium and protein, it is rich in phosphorous, iron, and the vitamin B complex (Schewerin, 1970). Fortification of cassava flour with soy protein isolate or soy grits has been sought as a means of increasing its nutrient status. Extensive studies of this type have been carried out in Brazil, where cassava flour is widely used in the production of composite flours as substitutes for wheat flour. People in the tropics who depend heavily on cassava as a major dietary item usually supplement it with either fish or legumes to achieve a protein balance.

The form in which cassava is consumed varies considerably and is linked to personal preferences and traditional dietary habits. For example, in Africa it is mostly consumed as a vegetable or in the form of pastes or mashes made from cassava flour. In East Africa consumption of leaves or pastes made from fermented roots is popular, whereas in West Africa it is consumed in the form of "gari" made from dried, grated, fermented cassava tubers. It is, however, significant to note the considerable increase in consumption of nonwheat flours and starches, particularly of cassava origin, as total or partial substitutes for wheat flour in bread making. In North America and Europe cassava starch is used as a gel and thickener in convenience foods.

Although the food requirements of more than 300 million people are derived from cassava, fatal cases of poisoning due to ingestion of large quantities of cassava by humans and animals are not uncommon. The toxicity of cassava is caused by the presence of the cyanogenic glycoside, linamarin, together with small amounts of closely related lotaustralin. These substances are hydrolyzed by the endogenous enzyme linamarase to liberate hydrogen cyanide (HCN) (Coursey, 1973). Fortunately, under

normal growing conditions, the enzymes and the substrate are compartmentalized and do not come into contact. But contact might occur
subsequent to the destruction of structural integrity caused by either
mechanical injury or post-harvest deterioration. For detailed information
concerning the chronic cassava toxicity the reader may refer to Nestel and
MacIntyre (1973).

Factors affecting the cyanogenic content of cassava have been studied in
detail by de Bruijn (1973). Environmental conditions influence the cyanogenic glycoside content, although different genotypes behave differently to
changing ecological conditions. Nitrogen fertilization as well as drought
conditions increase the glycoside content, whereas potassium and farm-
yard manure decrease it. Shading young plants increases the glycoside
content in the leaves, but decreases it in the roots. In De Bruijn's opinion,
glycoside content of a clone appears to be positively correlated to the water
content of leaves and tuberous roots. Moreover, the enzyme activity was
found to be highest in very young expanding leaves and is very high in the
peel fraction of tuberous roots as opposed to the low activity in the tuber.
However, the following guide to toxicity based on mg HCN per kg of fresh
peeled tuber was adopted by Koch (1933), Bolhuis (1954), and de Bruijn
(1971): less than 50 mg HCN, innocuous; 50–100 mg HCN, moderately
poisonous; and over 100 mg HCN, dangerously poisonous. The traditional
processing procedure, which involves boiling, baking, roasting, or fermentation has been found to be adequate for eliminating the toxic component.

A wide variety of industrial products can be made from cassava, which is
regarded as the cheapest known source of starch. Cassava starch contains
only 17% amylose as opposed to 22 and 27% for potato and corn starch,
respectively. Cassava starch is exceptionally suitable for sizing paper or
fiber with greater tensile strength (Ayres, 1972) and is also suitable in the
manufacture of glucose, alcohols, dextrins, and starch-based adhesives
(Park and Lima, 1973; Teixeira, 1964; Evans and Wurzburg, 1967).

The United States, at present, is the principal consumer of starch and
imports around 90,000 tons per year. Phillips (1974) estimates that the
Canadian demand for cassava starch could range from 44 to 46 million
pounds. The projected 1980 world demand for cassava starch is 20 to 447%
greater than 1970 levels, which means that the collective demand for
cassava starch in the 1970s will grow at a compound annual rate of 2 to 16%
(Phillips, 1974).

Cassava's low price and high energy content have been primarily
responsible for its increasing use as an animal feed supplement. Cassava
mixed with appropriate quantities of protein supplements such as soybean
meal have produced cheaper feeds than those which are cereal-based. The
nutrient content of cassava could also be increased and attempts in that

direction are already under way in Canada where Gregory (1974), using an *Aspergillus* sp. that grows at 50°C and pH 3.5 on liquid cassava substrate, developed a process that permits the production of an animal feed of 15% protein at a relatively low cost. Such an approach, if found practical, will have very useful applications in cassava-producing countries.

The main users of cassava for animal feed are Germany, Holland and Belgium. In the last decade the importation of cassava to the European Economic Community (EEC) has more than tripled and 80 to 90% of the world market is supplied by Indonesia and Thailand. The projected 1980 demand for cassava in the compound feed for EEC may be 246 to 634% greater than the 1970 demand (Phillips, 1974).

DISEASES OF CASSAVA

Like other crops, cassava is susceptible to a wide range of pathogens. Recently Lozano and Booth (1976) have published a comprehensive edition on various diseases of cassava dealing with the nature and type of the etiological agents, epidemiology, and control. They point out that, on a global basis, cassava bacterial blight (*Xanthomonas manihotis*) is considered to be one of the most devastating diseases, which might result in total loss of yield. Fortunately, effective control measures to combat the pathogen have already been developed at the CIAT (Centro Internacional de Agricultura Tropical), Colombia.

There are few viral diseases reported on cassava, viz. mosaic, common mosaic, brown streak, vein mosaic, and a latent virus; cassava mosaic disease is of most serious concern in Africa and India.

Cassava Mosaic Disease

This disease has been found to occur in East and West Africa, Madagascar, Java, and India (Menon and Raychaudhuri, 1970). Estimates of loss in yield due to the disease range from 20 to 90% (Lefebre, 1935, Jennings, 1960; Doku, 1965; Beck, 1971; Hahn and Howland, 1972).

African and Indian cassava mosaic disease differs from the Brazilian cassava common mosaic in that the former is: (1) present in many widely used cultivars and readily disseminated by vegetative propagation; (2) not readily transmissible either manually or mechanically; and (3) transmitted in the field by the "sweet potato" white fly *Bemisia tabaci*. Furthermore, the exact nature of the causative agent has not been conclusively proved to be a virus. Therefore, the etiology of the disease is uncertain.

Cassava mosaic disease, as the name implies, causes mosaic symptoms accompanied by crinkling, distortion, downward curling, and reduction in size of leaf laminae. The pathogen has a very narrow host range and is confined to *Manihot* spp. However, under experimental conditions, the disease has been transmitted to Cucumber (*Cucumis sativus*) through *Bemisia tabaci* (Menon and Raychaudhuri, 1970). Although transmission from cucumber to cucumber was achieved, it is not certain whether transmission from cucumber to cassava is possible.

Several attempts have been made to characterize the causative agent. Recently Bock and Guthrie (1976) reported the isolation of two serologically related but distinct viruses from cassava infected with brown streak and mosaic. Despite repeated attempts, they were unable to infect cassava with either virus. Peterson and Yang (1976) observed signs of infectivity on *Nicotiana clevelandi* by the inoculum taken from mosaic-diseased cassava. However, until confirmatory and conclusive evidences are found, the African and Indian cassava mosaic disease cannot be categorically classified as virus disease.

Attempts are being made at IITA, Nigeria, and CTCRI, India, to produce varieties resistant to mosaic disease. According to Hahn and Howland (1972), no resistance has been found either in African *M. esculenta* local cultivars nor in those grown from Latin American seeds. Eradication of the white fly vector as a means of controlling the spread of the disease on a large scale is considered impractical. An effective approach providing control of the disease would then be the production of disease-free stock plants by tissue culture techniques.

ELIMINATION OF CASSAVA MOSAIC DISEASE BY
MERISTEM CULTURE

Shoot apical meristem culture is becoming increasingly popular for the asexual propagation of cultivated plants and the elimination of systemic viral infections from vegetatively propagated crops. In a plant that has been systemically invaded by a virus, all cells are not uniformly infected and in most cases the shoot meristems are generally virus-free (Limasset and Cornuet, 1949). This significant observation led Morel and Martin (1952) to develop meristem culture techniques in order to produce virus-free plants. Since then, meristem culture alone or in combination with heat treatment has been successfully used to eliminate viral pathogens from a wide range of plant species (for a detailed review see Hollings, 1965; Quak, 1972, 1977). At present, plant virus disease control by chemotherapeutic agents has met with limited success. An alternate approach involves the production and distribution of large numbers of virus-free plants by meristem culture in order to dilute the amount of inoculum in a given locality. A prerequisite to such an approach would be the availability of a procedure of *in vitro* techniques to produce plants in high frequency to satisfy the demand. The next step involves the maintenance of the foundation virus-free stock isolated and fully protected against reinfection since they are not immune to the virus. Also, one must ensure that only the progeny propagated from the foundation stock is exposed to hazards of cultivation.

As mentioned earlier, mosaic disease poses serious problems to cassava cultivation. The vegetative propagation of the crop, intensive monoculture, and the lack of resistant varieties promote the spread of the disease to devastating proportions and impede the exchange of germplasm between cassava-growing countries.

Under the terms of a contract with the International Development Research Centre we have developed a reproducible technique to regenerate whole plants from the shoot apical meristems and adapted the procedures to permit the elimination of the mosaic disease in cassava plants. For details of the experiments, the reader may refer Kartha et al., 1974; Kartha and Gamborg, 1975, and Gamborg and Kartha, 1976.

The Technique

The materials used for this purpose are young buds produced on mature stakes. Dormant diseased stakes of Indian and Nigerian cassava, cultivars Kalikalan and Ogunjobi, respectively, were cut in sections with two nodes each. The upper cut ends were sealed with paraffin and the sections were planted in pots in a greenhouse where the cuttings sprouted in 5 to 7 days.

In other experiments, cuttings from diseased stakes were planted in vermiculite in pots and grown in a growth cabinet at constant 35°C, 16-hour photoperiod (4,000 lx from banks of cool-white fluorescent lamps) and 70% relative humidity. After 30 days growth, meristem tips were isolated and cultured. The control experiments consisted of cuttings originating from the same diseased stakes but grown under greenhouse conditions (21°C, 14-hour photoperiod, and 40 to 50% relative humidity).

Within 5 to 7 days, the diseased cuttings sprouted and the foliage showed typical mosaic symptoms. As the plants grew older, the symptoms became severe and caused crinkling, distortion, and reduction in size of leaf laminae (fig. 1).

The meristem culture medium consisted of micro and macro elements according to Murashige and Skoog (1962), vitamins as in B_5 medium (Gamborg et al., 1968), and 2% sucrose. Before adding Difco Bacto-Agar (0.6%) the pH of the medium was adjusted to 5.7. The growth hormones benzyladenine (BA), naphthaleneacetic acid (NAA), and gibberellic acid (GA_3) were added after the agar was dissolved at molar concentrations of

5×10^{-7}, 10^{-6}, and 10^{-7}, respectively. Aliquots of 2.5 ml of medium were individually dispensed into 10×2.5 cm pyrex test tubes, the tubes were plugged with absorbent cotton and autoclaved at $1.46 \text{ kg}/\text{cm}^2$ for 20 minutes. The medium was left to cool and solidify at ambient temperature.

Meristems were dissected in a laminar flow cabinet equipped with sterile air circulation. The shoot apices were sterilized by immersing in 70% ethanol for 1 minute followed by three washings in sterile, distilled water. Shoot meristem domes measuring approximately 0.2 to 0.5 mm were carefully excised from the shoot apices with razor blades chipped to form fine scalpels and mounted on steel holders. The dissection was performed under a Wild-M5 stereo microscope. The meristems thus removed were planted on the agar media and the tubes stoppered, sealed with Parafilm and incubated in a growth cabinet programmed to provide light intensity of 3000 lx, a light and dark cycle of 16/8 hours, 26°C, and 70% relative humidity.

The response of the meristem tips to culture conditions was noticeable within 3 days in the form of considerable swelling at the basal cut portion leading to the development of a callus. The shoot differentiation was observed within 7 to 10 days, followed by root development (fig. 2). Only explants exceeding 0.2 mm in length formed either callus or callus with roots. The regeneration potential (number of plants differentiated from the total number of meristems cultured) of meristem tips exceeding 0.2 mm was as high as 90 to 95% in both Indian and Nigerian cultivars. The meristem tips regenerated into whole plants within 26 days (fig. 3). At this stage they were transferred to pots of vermiculite and the plants were grown either in a greenhouse or in a growth room. The plants were frequently observed for symptom expression for a period of 6 months.

In a population of 135 plants regenerated in several experiments from 150 meristems (size 0.4 mm) of mosaic-diseased cultivar Kalikalan, 60% showed no symptoms of mosaic even after 6 months. The mortality of plants subsequent to potting averaged 4 to 5%. In subsequent experiments, when the meristem size exceeded 0.4 mm (0.5 to 0.8 mm), all the resulting plants exhibited mosaic symptoms. In a test on Nigerian cultivar Ogunjobi, out of 42 plants regenerated from 45 meristems, 40 were symptom-free. The healthy regenerated plants had deep green foliage and grew vigorously (fig. 4).

Transmission experiments were conducted by grafting the scions from regenerated plants onto healthy but susceptible CIAT, Colombia, cultivars, Llanera, and Colombia #800 at monthly intervals for 6 months. Control experiments consisted of grafting scions from diseased plants. No visible symptoms were noticeable on the stock plants grafted with healthy

Figs. 1-5. (1) Cassava plant exhibiting typical mosaic symptoms. Note the crinkling, distortion, downward curling, and reduction in size of leaf laminae. (2) Meristem differentiating into shoot and roots. (3) A five-week-old mosaic disease-free plant regenerated from meristem. (4) (a) Cutting from diseased plant grown in a greenhouse at 21°C; and (b) in a growth cabinet at 35°C for 30 days (note the disappearance of mosaic symptoms and increased vegetative growth in [b]). (5) A meristem-derived mosaic disease-free plant.

plants regenerated from meristems. On the other hand, the control plants using diseased scions exhibited typical mosaic symptoms on the newly formed axillary shoots within 21 to 28 days.

Heat Treatment and Meristem Culture

Cuttings from diseased stakes of cultivars Kalikalan and Ogunjobi, when grown at 35°C (constant), 16-hour photoperiod (4000 lx), and 70% relative humidity, exhibited vigorous growth as compared to similar plants grown under greenhouse conditions. Masking of mosaic symptoms on the young leaves was apparent from day 15, and no symptoms were visible on the new leaves produced by day 30 (fig. 4b). The control plants of the same age derived from the same cutting grown at 21°C developed severe mosaic symptoms with extensive distortion and reduction in size of leaf laminae (fig. 4a). In plants with foliage free of mosaic symptoms, typical mosaic symptoms appeared within 7 to 10 days when they were transferred from the 35°C growth cabinet to the greenhouse at 21°C.

Meristems were cultured from the plants grown at 35°C after day 30, which were still maintained under growth cabinet conditions, and plants regenerated. The regeneration potential averaged 90 to 95% when the explant size exceeded 0.2 mm, thus confirming results of other experiments. Mosaic symptoms could not be detected in any plants regenerated from meristems of up to 0.8 mm in length. Plantlets developed from larger meristems (0.9 to 1 mm) developed mosaic symptoms. Fifty plants of each cultivar have been produced and grown to adult plants by the combined 35°C growth and meristem culture technique. Transmission experiments carried out as described earlier showed no sign that the mosaic disease agent was present in the regenerated plants (fig. 5). However, in the absence of the white fly vector *Bemisia* spp, insect transmission studies were not done. Transmission studies were not attempted on other herbaceous plants.

The experiments clearly indicate that symptoms of the cassava mosaic disease prevalent in India and Nigeria could be eliminated by meristem culture alone or in combination with heat therapy. The success and frequency of producing symptom-free plants appear to be governed by the size of the meristems (maximum size 0.4 mm). Apparently, the cassava mosaic disease agent is present in shoot apical regions of a diseased plant below 0.4 mm. However, when the diseased stakes were grown under a higher temperature (35°C), masking of symptoms leading to their total disappearance was observed, and meristem tips up to 0.8 mm were free of the pathogen. The higher temperature apparently favored plant growth and may have retarded invasion and multiplication of the causative agent. Plant hormonal balance may be a determining factor by stimulating plant growth and providing adverse conditions for the multiplication and invasion of the pathogen.

The absence of symptoms on plants grown at 35°C is not due to inactivation of the pathogen, since the symptoms appeared on the foliage after growth at a lower temperature. Chant (1959) reported that the cassava mosaic "virus" had been inactivated from the Nigerian cassava treated at 35–39°C for periods of 28–42 days. From 18 surviving plants he obtained four plants that remained healthy after 42 days at 39°C, and the rest were infected. In our experience cassava plants grown above a temperature of 36–37°C for periods exceeding 20 days develop spindly growth and premature senescence. Moreover, the regeneration potential of meristems from such plants was as low as 2–5%. Our results, therefore, suggest that the causative agent of cassava mosaic disease present in Indian and Nigerian plants is not inactivated by thermotherapy but is eliminated only by meristem culture.

Employing stem-tip culture techniques, Berbee et al. (1973) succeeded in producing cassava plants free of leaf distortion symptoms. High-frequency plant regeneration from shoot apical meristems has also been reported from Taiwan (Liu, 1975).

Mosaic disease-free cassava plants produced by meristem culture techniques are not immune to reinfection by the same or other causative agents, although the technique makes it possible to produce mosaic disease-free plants in large numbers and provide a foundation stock from which progenies can be propagated and distributed for cultivation. Further vegetative propagation of such stock is possible using shoot tip cuttings in humid chambers and employing rooting hormones (Wholey and Cock, 1974).

The meristem culture operation requires relatively simply facilities and a modest working area. The propagation of healthy stock plants requires a larger area. Such an approach may be economically established in a region in which vegetative growth can occur throughout the year. Other requirements are a location isolated from commercial cassava production and the absence of vectors transmitting the causal agent of mosaic. Plant quarantine stations may be in a position to initiate projects of this type. The plant quarantine division of the East African community in Nairobi is already in the process of introducing the meristem culture technique for cleaning up the cassava germplasm within the region and also the materials imported from other countries.

OTHER POTENTIAL TISSUE CULTURE APPROACHES

In recent years considerable progress has been made in the propagation of plants through tissues other than meristems (Murashige, 1974). Plant regeneration has been achieved from excised leaf sections, eg. tomato (Kartha et al., 1976), rapeseed (Kartha and Gamborg, unpublished) and potato (Gamborg, unpublished), hypocotyl segments of flax (Gamborg and Shyluk, 1976), immature embryos of maize (Green and Phillips, 1975) and sorghum (Gamborg et al., 1977). The method of regenerating plants from excised leaf sections has recently been applied to tobacco to produce mosaic virus-free plants (Murakishi and Carlson, 1976). Plant cells can be grown on a solidified nutrient agar as callus or in liquid culture as suspended aggregates (Gamborg, 1975). Under appropriate conditions, such cells, depending upon the species, could be induced to undergo differentiation, organ development, and whole plant regeneration. Such systems would be valuable for: (1) mass propagation of plants, (2) genetic improvement of crops, (3) recovery of disease-free plants, and (4) production of pharmaceuticals (Murashige, 1974). An example of plant regeneration from somatic cells on a large scale is the production of plants from sugar cane callus and selection of lines resistant to Fiji disease (Krishnamurthi and Tlaskal, 1974). Similar plant regeneration has not been achieved with cassava callus. When methods become available, one could envision selecting lines that are resistant to various pathogens, and also screening for cyanide levels and eventually producing low-cyanide cassava plants.

It has been proposed that the most efficient, practical, and economical attack on the mosaic disease problem is breeding for resistance (Beck, 1971; Jennings, 1972), and some degree of success has been achieved in introducing resistance into cultivated varieties (Beck, 1971; Bock and Guthrie, 1976). The germplasm of most varieties is derived from *Manihot esculenta*, which apparently possesses no resistance. Moreover, cassava is a monoecious crop and some varieties are male sterile, thereby rendering sexual crossing difficult. One of the wild species, *M. glaziovii*, is resistant to cassava mosaic. Crosses have been made between *M. esculenta* and *M. glaziovii* but natural barriers lower the efficiency of the process (Beck, 1971) and resistance in the progeny may not persist (Hahn and Howland, 1972). The most recent innovation of *in vitro* methods is the production of somatic hybrids between different plant genera and families by fusion of protoplasts (Gamborg et al., 1974). Advances in the technology have reached the stage where predictions can be made about its potential use in plant breeding programs. Somatic hybridization could permit expanding the genetic base far beyond what is now possible by conventional methods. It would allow for new and desirable crosses and thus increase the scope for crop improvement (Gamborg and Kartha, 1976).

The continuing search for higher-yielding varieties of crop plants with pathogen resistance necessitates the availability of germplasm resources. For crop species that are vegetatively propagated, such as cassava, preservation of germplasm implies handling and maintaining a large number of genotypes, thus taxing heavily the manpower resources and land area. At present, large collections of cassava are maintained at research stations (CIAT; IITA; CTCRI, etc.), which demand considerable space and time. Moreover, under natural growing conditions, they are subject to attacks of disease pathogens and insect pests resulting in heavy losses. An alternate but effective way of overcoming such problems is controlled freezing and low-temperature storage of isolated cassava meristems. Such an approach would be extremely beneficial in the exchange and preservation of germplasm.

The survival of shoot tips of carnations after freezing to −196°C, the subsequent recovery of whole plants (Seibert, 1976), and the available information on successful attempts to freeze-preserve plant cells (Nag and Street, 1973, 1975a, 1975b; Bajaj and Reinert, 1977) are positive indications to the effect that such an approach may be feasible with cassava.

SUMMARY

Cassava (*Manihot esculenta* Crantz), a vegetatively propagated tuber crop, is one of the 12 to 15 important food crops. Around 300 million people in the tropical areas of the world depend on cassava as a major carbohydrate source. Cassava productivity in terms of calories is significantly higher than that of other staple food crops. Cassava starch is regarded as the cheapest known source of starch. A wide variety of industrial products, such as paper, glucose, alcohol, dextrins, and adhesives, can be made from cassava starch. Cassava is also used as an ingredient in animal feed.

A major factor contributing to low yield and handicapping rapid expansion of cassava production is cassava mosaic disease, which reduces yield by as much as 20 to 90%. The presence of this disease not only reduces yield, but also inhibits the movement of germplasm for breeding purposes. No resistance against the pathogen has been found either in African or Latin American cultivars. As an alternate approach to control the disease, a meristem culture technique has been developed to produce mosaic disease-free cassava plants in high frequency from cassava stakes infected with mosaic disease of Nigerian and Indian origin. The principles, methodology, and details of the experiments are presented, along with potential and possible applications of other tissue culture approaches for cassava crop improvement.

ACKNOWLEDGMENTS

The tissue culture work presented in this paper was carried out with the aid of a grant from the International Development Research Centre, Ottawa, Canada. We are extremely thankful to Drs. L. R. Wetter and L. E. Pelcher for reviewing this manuscript and offering valuable suggestions. Grateful appreciation is extended to Dr. C. Lozano (CIAT, Columbia) for providing us with excellent literature on cassava. Finally, our thanks are also due to Mr. A. S. Lutzko for preparing the photographic plates.

LITERATURE CITED

Ayres, J. C. 1972. Manioc: The potential exists for increased use of this tropical plant and its products. Food Technol. 26:128-32.

Bajaj, Y. P.S., and J. Reinert. 1977. Cryobiology of plant cell cultures and establishment of gene-banks. *In* J.Reinert and Y. P. S. Bajaj (eds,), Applied and fundamental aspects of plant cell, tissue, and organ culture. Pp. 757-77. Springer-Verlag, Berlin.

Beck, B. D. A. 1971. The breeding goals in cassava breeding program in West Africa. Ford Foundation, Lagos, Nigeria. 5 pp.

Berbee, F. M., J. G. Berbee, and A. C. Hildebrandt. 1973. Induction of callus and virus-symptomless plants from stem tip cultures of cassava. In Vitro 8:421.

Bock, K. R., and E. J. Guthrie. 1976. Recent advances in research on cassava viruses in East Africa. *In* B. L. Nestel (ed.), African cassava mosaic: Report of an interdisciplinary workshop, Muguga, Kenya, 19-22 February 1976. Pp. 11-26. Intern. Devel. Res. Centre. IDRC-071ᵉ.

Bolhuis, G. G. 1954. The toxicity of cassava roots. Neth. J. Agric. Sci, 2:176-85.

Chant, S. R. 1959. A note on the inactivation of mosaic virus in cassava (*Manihot utilissima* Pohl) by heat treatment. Emp. J. Exp. Agric. 27:55-58.

Coursey, D. G. 1973. Cassava as food: Toxicity and technology. *In* Chronic cassava toxicity: Proceedings of an interdisciplinary workshop. London, England. 29-30 January 1973. Pp. 27-36. Intern. Devel. Res. Centre. IDRC-010ᵉ.

Coursey, D. G., and P. H. Haynes. 1970. Root crops and their potential as food in the tropics. World Crops 22:261-65.

de Bruijn, G. H. 1971. Etude de caractère cyanogénétique du manioc. Veeman & Zonen, Wageningen.

de Bruijn, G. H. 1973. The cyanogenic character of cassava (*Manihot esculenta*). *In* Chromic cassava toxicity: Proceedings of an interdisciplinary workshop, London, England, 29-30 January 1973. Pp. 43-48. Intern. Devel. Res. Centre.IDRC-010ᵉ.

de Vries, C. A., J. D. Ferwerda, and M. Flach. 1967. Choice of food crops in relation to actual and potential production in the tropics. Neth. J. Agr. Sci. 15:241-48.

Doku, E. V. 1965. Breeding for yield in cassava. I. Indices of yield. Ghana J. Sci. 5:42-59.

Evans, R. B., and O. B. Wurzburg. 1967. *In* R. L. Whistler, et al. (eds.), Starch chemistry and technology vol. 2. Academic Press, New York.

Gamborg, O. L. 1975. Callus and cell culture. *In* O. L. Gamborg and L. R. Wetter (eds.), Plant tissue culture methods. Pp. 1-10. National Research Council of Canada, Ottawa, Canada.

Gamborg, O. L., F. Constabel, L. C. Fowke, K. N. Kao, K. Ohyama, and K. K. Kartha. 1974. Protoplasts and cell culture methods in somatic hybridization in higher plants. Canad. J. Genet. Cytol. 16:737-50.

Gamborg, O. L., and K. K. Kartha. 1976. *In vitro* techniques in the control of cassava mosaic disease. *In* B. L. Nestel (ed.), African cassava mosaic: Report on an interdisciplinary workship, Maguga, Kenya, 19-22 February 1976. Pp. 30-35. Intern. Devel. Res. Centre. IDRC-071ᵉ.

Gamborg, O. L., R. A. Miller, and K. Ojima. 1968. Nutrient requirement of suspension cultures of soybean root cells. Exp. Cell Res. 50:151-58.

Gamborg, O. L., and J. P. Shyluk. 1976. Tissue culture, protoplasts, and morphogenesis in flax. Bot. Gaz. 137:301-6.

Gamborg, O. L., J. P. Shyluk, D. S. Brar, and F. Constabel. 1977. Morphogenesis and plant regeneration from callus of immature embryos of sorghum. Plant Sci. Lett. (in press).

Green, C. E., and R. L. Phillips. 1975. Plant regeneration and tissue culture of maize. Crop Sci. 15:417–21.

Gregory, K. F. 1974. Enrichment of cassava with single-cell protein. *In* Papers for Canadian cassava review. 27–28 January 1975. Intern. Devel. Res. Centre, Ottawa, Canada.

Hahn, S. K., and A. K. Howland. 1972. Breeding for resistance to cassava mosaic. Proceedings of the IITA/IDRC Cassava Mosaic Workshop, IITA, Ibadan, Nigeria. Pp. 37–39.

Hollings, M. 1965. Disease control through virus-free stock. Ann. Rev, Phytopath. 3:367–96.

Jennings, D. L. 1960. Observations on virus diseases of cassava in resistant and susceptible varieties. I. Mosaic disease. Emp. J. Exp. Agric. 28:261–70.

Jennings, D. L. 1972. Breeding for resistance to cassava viruses in East Africa. Proceedings of the IITA/IDRC Cassava Mosaic Workshop, IITA, Ibadan, Nigeria, pp. 40–42.

Kartha, K. K., and O. L. Gamborg. 1975. Elimination of cassava mosaic disease by meristem culture. Phytopathology 65:826–28.

Kartha, K. K., O. L. Gamborg, F. Constabel, and J. P. Shyluk. 1974. Regeneration of cassava plants from apical meristems. Plant Sci. Lett. 2:107–13.

Kartha, K. K., O. L. Gamborg, J. P. Shyluk, and F. Constabel. 1976. Morphogenetic investigations on *in vitro* leaf culture of tomato (*Lycopersicon esculentum* Mill. cv. starfire) and high frequency plant regeneration. Z. Pflanzenphysiol. 77:292–301.

Koch, L. 1933. Cassaveselectie. Veeman & Zonen, Wageningen.

Krishnamurthi, M., and J. Tlaskal. 1974. Fiji disease-resistant *Sacharum officinarum* var. Pindor sub-clones from tissue culture. Proc. Intern. Soc. Sugarcane Tech. 15.

Lefebre, P 1935. Quelques considerations sur la mosaique du manioc. Bull. Agri. du Congo Belge. 26:442–47.

Limasset, P., and P. Cornuet. 1949. Recherche de virus de la mosaique du tabac (*Marmor tabaci*, Holmes) dans les méristèmes des plantes infectées. C. R. Acad. Sci. Paris 228:1971–72.

Liu, M. C. 1975. The *in vitro* induction of callus and regeneration of cassava plants from shoot apical meristems. Taiwan Sugar, September–October 1975, pp, 171–77.

Lozano, J. C., and R. H. Booth. 1976. Diseases of cassava (*Manihot esculenta* Crantz). Technical Bulletin-Series DE-5, Centro Internacional de Agricultura Tropical (CIAT), Cali, Colombia.

Menon, M. R., and S. P. Raychaudhur. 1970. Cucumber: A herbaceous host of cassava mosaic virus. Plant Dis. Report. 54:34–35.

Morel, G., and C. Martin. 1952. Guérison de dahlias atteints d'une maladie à virus. C. R. Acad. Sci. Paris 235:1324–25.

Murakishi, H. M., and P. S. Carlson. 1976. Regeneration of virus-free plants from dark-green islands of tobacco mosaic virus–infected tobacco leaves. Phytopathology 66:931–32.

Murashige, T. 1974. Plant propagation through tissue culture. Ann. Rev. Plant Physiol. 25:135–66.

Murashige, T., and F. Skoog. 1962. A revised medium for rapid growth and bioassays with tobacco tissue culture. Physiol. Plantarum 15:473–97.

Nag, K. K., and H. E. Street. 1973. Carrot embryogenesis from frozen cultured cells. Nature 245:270–72.

Nag, K. K., and H. E. Street. 1975a. Freeze preservation of cultured plant cells. I. The pretreatment phase. Physiol. Plantarum 34:254–60.

Nag, K. K., and H. E. Street. 1975b. Freeze preservation of cultured plant cells. II. The freezing and thawing phases. Physiol. Plantarum 34:261–65.

Nestel, B. L. 1973. Current utilization and future potential for cassava. *In* B. L. Nestel and R. MacIntyre (ed.), Chronic cassava toxicity: Proceedings of an interdisciplinary workshop, London, England, 29–30 January 1973. Pp. 11–26. Intern. Devel. Res. Centre. IDRC–010^e.

Nestel, B. L., and R. MacIntyre (eds.). 1973. Chronic cassava toxicity: Proceedings of an interdisciplinary workshop, London, England, 29–30 January 1973. Intern. Devel. Res. Centre. IDRC–010^e.

Park, Y. K., and D. C. Lima. 1973. Continuous conversion of starch to glucose by an amyloglucosidase-resin complex. J. Food Sci. 38:358–59.

Peterson, J. F., and A. F. Yang. 1976. Characterization studies of cassava mosaic agents. pp. 17–26. *In* B. L. Nestel (ed.), African cassava mosaic: Report of an interdisciplinary workshop, Muguga, Kenya, 19–22 February 1976. Intern. Devel. Res. Centre. IDRC–071^e.

Phillips, T. P. 1974. Cassava utilization and potential markets. Intern. Devel. Res. Centre. IDRC–020^e. P. 182.

Quak, F. 1972. Review of heat treatment and meristem-tip culture as methods to obtain virus-free plants. 10th Intern. Hort. Cong. Proc. 3:12–25.

Quak, F. 1977. Meristem culture and virus-free plants. *In* J. Reinert and Y. P. S. Bajaj (eds.), Applied and fundamental aspects of plant cell, tissue, and organ culture. Pp. 598–615. Springer-Verlag, Berlin.

Rogers, D. J. 1965. Some botanical and ethanological considerations of *Manihot esculenta*. Econ. Bot. 19:369–77.

Schewerin, K. H. 1970. Apuntes sobre la yuca y sus origenes (Notes on cassava and its origin). Tropical Root and Tuber Crops Newsletter 3:4–12.

Seibert, M. 1976. Shoot initiation from carnation shoot apices frozen to –196°C. Science 191:1178–79.

Teixeira, C. G. 1964. Aguardente de mandioca (A liquor distilled from cassava). Agronomico 16:9–10.

Wholey, D. W., and J. H. Cock. 1974. Rooted shoots for physiological experiments with cassava. Trop. Agric. (Trinidad) 52:187–89.

DONALD K. DOUGALL

Factors Affecting the Yields of Secondary Products in Plant Tissue Cultures

33

Plants are used as sources of many compounds and mixtures of compounds that are useful to man. These compounds, which are referred to as secondary products, include drugs, flavors, enzymes, essential oils, and colorings. Secondary products do not include compounds, such as amino acids or nucleotides, that are intermediates in the metabolism of cells essential for survival. In addition, secondary products do not include compounds that are ubiquitous in plants, such as cellulose or sucrose. Specific secondary products are not widely distributed in the plant kingdom; each is restricted to a limited number of species. When considered collectively, the secondary products from plants form a very heterogeneous group of compounds in terms of their chemistry and their biological activity. Because of their accumulation in plants, these compounds have sometimes been thought of as end products or waste products of metabolism. However, this does not seem to be valid in many cases because the compounds are also degraded in plants.

The quantity of a plant species available for extraction of a particular secondary product may be limited by a variety of factors. These factors include natural restriction of the geographical area in which the plant grows and political decisions such as closed frontiers or warfare. The availability of plant material may also be affected by transportation problems, difficulties of collecting plants of a species that grow at low population densities, and quarantine requirements. In some cases an

Donald K. Dougall, W. Alton Jones Cell Science Center, Old Barn Road, Lake Placid, New York 12946.

economical chemical synthesis of the compound of interest ensures its supply. In others the complexity of the molecule is such as to preclude an economical chemical synthesis, and so a biological source is the only reasonable alternative.

Plant tissue culture has provided an alternative to whole plants as a biological source of useful compounds. This alternative arose as a possibility in the period 1950–55 as a result of demonstrations that plant cells could be grown suspended in liquid medium, as is done with microorganisms, and commercial use of microorganisms to produce useful compounds such as penicillin. At that time it was thought that cells of the plant that produces the desired compound could be grown outside the plant to increase the quantity of tissue from which to extract the compound of interest. Since the recognition of the possibility, there has been continued development of the data and relevant ideas. My objective is to describe to you the current state of these ideas. To do so I will divide the presentation into five parts:

1. Summary of data on the yields of specific secondary products obtained in plant cell cultures.

2. Demonstrations that the capacity for synthesis of secondary products can be retained in cell cultures giving low yields.

3. Evidence showing that within plant cell cultures cells may differ in their ability to produce specific secondary products.

4. Evidence showing that the yields of specific secondary products are influenced by the conditions of culture.

5. A hypothesis describing the way that the yield of secondary products changes as cells multiply in culture. Some possible limitations on this hypothesis will be outlined.

YIELD OF SECONDARY PRODUCTS IN PLANT CELL CULTURES

The yield of a secondary product, expressed per unit weight of tissue, is used as a crude comparison of the productivity of cell cultures with that of whole plants. If the yield approaches or exceeds that of the whole plant, then there is some basis for hope that cell cultures can become a realistic alternative to whole plants.

Tissue cultures from sixteen species of plants produced a variety of specific secondary products at levels approximating or exceeding those found in the whole plant. These examples are shown in table 1. The chemical types of compounds listed in table 1 are diverse. Many of the species listed are sources of compounds used in medicine. In the cases of the tissue cultures of *Ruta graveolens* and *Andrographis paniculata,* each yielded a series of novel compounds.

Tissues from a wide range of species in addition to those shown in table 1 have been grown in culture and then examined for the presence of compounds characteristic of that species. In many cases the compounds characteristic of the species of origin could be demonstrated in the cells grown in tissue culture, but at low yields. This work has been reviewed by Butcher (1977), Staba (1969, 1977), Puhan and Martin (1971), and Constabel et al. (1974).

CAPACITY FOR SYNTHESIS IS RETAINED

The sixteen examples in table 1 are a sufficient number to indicate that it is possible to obtain high yields of specific compounds in tissue culture. The question now is, What are the requirements for the maintenance of yield? We can assume that the tissue used to initiate the cultures had the capacity to produce the compound of interest. After the multiplication of the tissue in culture, the yield of compound declined. Is the synthesis of the specific compound turned off but the capacity retained, or is the capacity for synthesis lost from the cells in culture? This question has been answered in a number of cases by examining the specific compounds of interest in cell cultures and in plants regenerated from these cell cultures. These studies were performed with cell cultures of *Coptis japonica* (Ikuta et al., 1975), of eleven members of the family *Papaveraceae* (Ikuta et al., 1974), of *Scopolia parviflora* (Tabata et al., 1972), of *Nicotiana tabacum* (Tabata et al., 1971), of *Digitalis purpurea* (Hirotani and Furuya, 1977), and of *Datura innoxia* (Hiraoka and Tabata, 1974). In all cases the cell cultures had low levels of alkaloids. The alkaloids present in the cell cultures were also qualitatively different from those in the species of origin. The plants regenerated from the cell cultures recovered the patterns and levels of alkaloids characteristic of the species. Thus, it appears the capacity for synthesis of specific compounds is usually retained during culture.

CELLS WITHIN A CULTURE MAY VARY IN YIELD
OF SECONDARY PRODUCTS

The study of Hiraoka and Tabata (1974) with *Datura innoxia* suggested that altered cell lines develop in cell cultures and that a selection for "normal" cells may operate during the regeneration of plants. They showed that the range of chromosome numbers in cells in culture was wider than the range in the plants regenerated from the cultures; that the plants were not chimeras, indicating that each had been derived from a single cell; and that the regenerated plants displayed differences in the pattern of appearance of alkaloids during development.

Two additional studies provide evidence for wide differences in yield of

TABLE I

Compounds Isolated from Plant Tissue Cultures in Yields Approaching or Exceeding Those Found in Whole Plants

Compound	Species	Yield from Tissue Culture	Yield from Plant	Reference
Biscoclaurine alkaloids	Stephania cepharantha	10–22.9 mg/g d w Total alkaloids	Tuber 8.2 mg/g d w	Akasu et al., 1976
Nicotine	Nicotiana rustica	0.291% d w	Aerial part 1.0 mg/g d w	Tabata and Hiraoka, 1976
Serpentine	Catharanthus roseus	0.5% d w	Root 0.25% of d w Leaf 0.35% of d w	Döller et al., 1976
Anthraquinones	Morinda citrifolia	900 μmoles/g d w	Root 110 μmoles/g d w	Zenk et al., 1975
Anthraquinones	Cassia tora	0.334% of f w	0.209% of d w of seeds	Tabata et al., 1975
Glutamine	Symphytum officinale	924 μmoles/g d w	Leaf 3.5 μmoles/g d w	Tanaka et al., 1974
Diosgenin	Dioscorea deltoidea	26 mg/g d w		Kaul et al., 1969
Diosgenin	Trigonella occulta	0.37% d w	Seeds 0.32% d w	Jain et al., 1977
Gitogenin	Trigonella occulta	0.14% d w	Seeds 0.04% d w	Jain et al., 1977
Tigogenin	Trigonella occulta	0.05% d w	Seeds 0.01% d w	Jain et al., 1977
Ubiquinone	Nicotiana tabacum	0.5 mg/g d w	Leaf circa 16 μg/g d w	Ikeda et al., 1975 Ikeda et al., 1976
Thebaine	Papaver bracteatum	130 μg/g d w	Leaves 1,400 μg/g d w Roots 3,000–3,500 μg/g d w	Kamimura et al., 1976
Proteinase inhibitors	Scopolia japonica	41.0 mg/g d w	Stem 12.5 mg/g d w Leaf 24.1 mg/g d w Root 37.1 mg/g d w	Misawa et al., 1975
Ginseng saponins	Panax ginseng Panax quinquefolium	0.38% f w 0.41% f w	0.3–3.3% f w	Jhang et al., 1974
Phenolics	Acer pseudoplatanus	132 mg/g d w	Buds 72 mg/g d w Bark 32 mg/g d w Root 14 mg/g d w	Westcott and Henshaw, 1976
Flavanols	Acer pseudoplatanus	68 mg/g d w	Buds 38 mg/g d w Bark 15 mg/g d w Root 7 mg/g d w	Westcott and Henshaw, 1976
Leuco anthocyandin	Acer pseudoplatanus	305 OD units/g d w	Buds 45 OD units/g d w Bark 12 OD units/g d w Root 8 OD units/g d w	Westcott and Henshaw, 1976

TABLE 1—*Continued*

Compound	Species	Yield from Tissue Culture	Yield from Plant	Reference
Coumarins	*Ruta graveolens*	Isopimpinellin 1.2–1.6 mg/g d w	Not present	Steck et al., 1971
		Rutamarin 0.1–0.2 mg/g d w	Traces only	
		Rutacultin 0.4–0.6 mg/g d w	Not detected	
Alkaloids	*Ruta graveolens*	Kokusaginine edulinine	Not present	Steck et al., 1971
Paniculides A, B, C (sesquiterpene lactones)	*Andrographis paniculata*	Paniculides A,B,C	Not found in plants	Butcher and Connolly, 1971
Serpentine plus Ajmalacine	*Cathranthus roseus*	1.3% d w	0.26% d w	Zenk et al., 1977

specific compounds between clonal lines derived from tissue cultures (Tabata and Hiraoka, 1976; Zenk et al., 1977). In each of these studies, some of the clonal lines gave yields as high as those obtained in the whole plant and were stable on further culture. In addition, a *p*-fluorophenylalanine-resistant subline selected from cultures of *Nicotiana tabacum* has been shown to accumulate phenolic compounds to a level ten times higher than that accumulated by the parent culture (Berlin and Widholm, 1977).

Clonal variation in yield of secondary products would provide the basis of an explanation for low yields in much of the published literature. In many of these cases, the methods used for detection of the compounds of interest were not sensitive. Large quantities of tissue had to be used to establish the presence of the compound. As a consequence, tissue was grown through many generations to provide the amounts needed for detection of the compounds. During this time there would be substantial opportunity for selection of low-yielding cell lines.

Two hypotheses can be proposed to account for the sixteen cases of tissue cultures that produce high yields of specific compounds (table 1). These hypotheses are:

1. High-yielding sublines were selected during culturing of the tissues.
2. The individual plants from which these cultures were established had the capacity to retain the yield of specific compounds in tissue culture.

Although there is data consistent with the first hypothesis, there appears to be no evidence in the literature to directly confirm the second hypothesis. However, Zenk et al., (1977), have shown that if the yields of serpentine or ajmalacine in tissue cultures established from high-yielding and low-yielding plants of *C. roseus* are compared, then, on the average, tissue cultures from high-yielding plants give higher yields of alkaloids than do the tissue cultures from low-yielding plants.

This observation suggests that in attempts to develop cell cultures giving high yields of specific compounds, cultures should be initiated from individual plants that have high yields of the desired compounds. The evidence already presented further suggests that clonal lines of cells with high yields of the desired compound should be selected from the cultures established from plants with high yields.

Selection for high yield in clonal lines from tissue culture has only been performed by measurement of yield in clones and propagation of those with high yield (Tabata and Hiraoka, 1976; Zenk et al., 1977). In both cases clones with high and stable yields have been recovered. With this procedure, the ability to select clones with desired characteristics is limited. If the frequency of the desired subline in the population is low, a large number of clones has to be examined to ensure that the desired subline will appear at least once. If the frequency of high-yielding clones decreases with increasing passage number as can be inferred from much of the experience that shows decreasing yields of specific compounds in cell cultures, then cloning and measurement of yields should be performed at relatively early passages of the cultures. Selected lines should be recloned and reselected for high yield at intervals to ensure that the characteristics are maintained.

Two general methods of cloning plant cell cultures have been described. The first involves the isolation and nurturing of single cells (microculture). The most widely used microculture method is that of Muir et al. (1958), in which single cells are grown separated from a relatively large piece of cultured tissue by filter paper. Here the larger piece of tissue modifies the environment of the single cell and allows it to grow. The second general method is to localize single cells by plating them in a suitable medium containing agar and allowing them to grow (Bergmann, 1960). The conditions required to achieve plating efficiencies greater than 70 percent with tobacco cell cultures were established by Gibbs & Dougall, (1965). Kao and Michayluk (1975) grew cells from *Vicia hajastana* from inocula containing 5 cells/ml. Recently, Schulte and Zenk (1977) have described a method for replicate plating of plant cells using a nylon screen to transfer a replica from one plate to another.

CULTURE CONDITIONS AFFECT SECONDARY PRODUCT YIELDS

The evidence dealt with in the preceeding sections points to two factors that will increase the yield of specific compounds in cell cultures. Another factor that needs to be considered is the effect of media composition and culture conditions on the yields of desired compounds. Evidence for alteration of yield in cell cultures by the conditions of culture is shown in table 2. It is clear that the medium composition and culture conditions can play a key role. Yields of specific compounds in cell culture may be altered by growth substances, nitrogen sources, carbon sources, illumination, and several other variables.

Growth Substances

Historically, the highest priority for investigation has been given to growth substances because (a) these are required for growth of cultures, (b) altering the concentrations of growth substances in the culture medium initiates organized structures such as shoots or roots in some plant cell cultures, and (c) in some cases there appears to be an association between production of secondary compounds and formation of organized struc-

TABLE 2

Effects of Medium Components on Secondary Products in Plant Tissue Culture

Compound(s)	Species	Range of Levels	Factors that Affect Yield	Reference
Nicotine	*Nicotiana tabacum*	0–0.25% d w	Type and concentration of auxin	Furuya et al., 1971
Nicotine	*Nicotiana tabacum*	0–0.25% d w	Concentration of cytokinin (kinetin)	Tabata et al., 1971
Carboline alkaloids	*Peganum harmala*	0–850 nanomoles/g f w	Concentration of auxin (2,4-D) Concentration of phosphate Type of nitrogen source Time of culture	Nettleship and Slaytor, 1974
Thebaine	*Papaver bracteatum*	0.1–130 μg/g d w	Concentration of auxin Concentration of kinetin Coconut milk	Kamimura et al., 1976
Serpentine	*Catharanthus roseus*	0.01–0.5% d w	Type and concentration of auxin Light or dark	Döller et al., 1976
Serpentine Ajmalacine	*Catharanthus roseus* *Catharanthus roseus*	0–0.8% d w 0–1.0% d w	Type and concentration of auxin Presence of tryptophan (precursor)	Zenk et al., 1977
Berberine Jatrorrhizine	*Coptis japonica* *Coptis japonica*	432–774 μg/g f w 459–912 μg/g f w	Concentration of auxin (2,4-D) Concentration of cytokinin (kinetin)	Ikuta et al., 1975
Anthocyanin	*Haplopappus gracilis*	0–20 mg/g d w	Types and concentration of auxin Types and concentration of cytokinin	Constabel et al., 1971
Anthocyanins	*Rosa* sp.	0–40 absorbance units/30 ml culture	Concentration of auxin Light or dark	Davies, 1972 a
Anthocyanin	*Populus* sp.	0.1–2.4 mg/100 ml culture	Concentration of kinetin Light or dark Type of auxin Initial pH of medium Sugar type and concentration Period of growth	Matsumoto et al., 1973

TABLE 2—Continued

Compound(s)	Species	Range of Levels	Factors that Affect Yield	Reference
Anthocyanin	*Haplopappus gracilis*	0–6 mg/d w	Light or dark Concentration of auxin (2,4-D) Time of exposure to light	Strickland and Sunderland, 1972
Anthraquinones	*Morinda citrifolia*	0–900 μmoles/g d w	Type and concentration of auxin Type and concentration of carbohydrates Type and concentration of nitrogen source	Zenk et al., 1975
Anthraquinones	*Cassia tora*	0.025–0.334% of f w	Concentration of auxin Concentration of kinetin	Tabata et al., 1975
Polyphenols	*Rosa* sp.	150–300 μmoles/30 ml culture	Type and concentration of auxin Concentration nitrate	Davies, 1972a Davies, 1972b
Phenolic acids	*Daucus carota*	54–1070 μg/d w	Light or dark Period of growth Concentration of auxin	Sugano et al., 1975
Chlorogenic acid	*Haplopappus gracilis*	14–30 mg/g d w	Light or dark Concentration of auxin (2,4-D) Time of exposure to light	Strickland and Sunderland, 1972
Phenolics	*Acer pseudoplatanus*	0.33–13.3 mg/g f w	Concentration of 2,4-D Sucrose, ethylene, and nitrogen source	Westcott, 1976 Westcott and Henshaw, 1976
Phenolics	*Rosa* sp.	0.9–13.4 μg/50 ml	Concentration of sugar Concentration of nitrate	Amorim et al., 1977
Daidzin	*Glycine max*	>2-fold increase	Concentration of cytokinin Concentration of auxin Concentration of malonic acid	Miller, 1969
Lipid content and composition	*Glycine max*	0.06–0.12% f w linoleic acid 4–18% of total fatty acids	Presence or absence of auxin, cytokinin, gibberellin	Stearns and Morton, 1975
Ubiquinone	*Nicotiana tabacum*	200–600 μg/g d w	Concentration of auxin (2,4-D) Concentration of cytokinin	Ikeda et al., 1976
Chlorophyll	*Daucus carota*	12–55 μg chl/g f w	Type of sugar	Edelman and Hanson, 1971

TABLE 2—Continued

Compound(s)	Species	Range of Levels	Factors that Affect Yield	Reference
Chlorophyll	*Spinacia oleracea*	24–416 µg/g d w	Ethylene concentration CO_2 concentration O_2 concentration	Dalton and Street, 1976
Steroids Diosgenin	*Solanum xanthocarpum* *Dioscorea deltoidea*	0.32–0.55 mg/g d w 0.77–2.6 mg/g d w	Types of auxin Cholesterol concentration Yeast extract concentration Concentration of auxin (2,4-D)	Heble et al., 1971 Kaul et al., 1969
Glutamine	*Symphytum officinale*	0–200 mg/g d w	Concentration of auxin (2,4-D) Concentration of kinetin Nitrogen source	Tanaka et al., 1974
Plasmin inhibitor	*Scopolia japonica*	0.1–1.5 mg/ml culture	Concentration of auxin (2,4-D) Concentration of cytokinin (kinetin) Nitrogen source	Misawa et al., 1975

tures. The types of growth substances required by plant cell cultures fall into two classes, auxins and cytokinins. Auxins are required for growth of cultures of normal plant cells. Examples of auxins are: indoleacetic acid (IAA), naphthaleneacetic acid (NAA), and 2,4-dichlorophenoxyacetic acid (2,4-D). Cells from some plant species require cytokinins in addition to auxins for growth. Examples of cytokinins are: 6-benzyladenine, 6-furfuryladenine and 6-isopentenyladenine. The specific auxin or cytokinin used, as well as the concentration of each, alters both growth of cells and yield of specific compounds. These compounds are used in the concentration range of 0.1–10 mg/l.

Nitrogen Sources

Some plant tissue cultures will grow on nitrate as the sole nitrogen source, others require a mixture of nitrate and ammonium ions, and some require glutamine. Other variations include the presence or absence of casein hydrolysate. In some cases the nature of the nitrogen source and its concentration affects the yields of specific compounds.

Carbon Source

Sucrose is the carbon and energy source for heterotrophic growth of plant cell cultures that gives the highest growth rate. Cultures from different species may also achieve highest growth rate on other carbon sources such as mannose, galactose, or glucose. Both the sugar and its concentration affect the yield of secondary products in some cases.

Illumination

Despite the fact that plant cell cultures grow heterotrophically, growth of cultures in light or dark alters the yield of secondary products in several cases. It is to be noted that autotrophic growth of plant cell cultures are rare and then only at low growth rate.

Miscellaneous

In a limited number of cases, medium pH, phosphate concentration, gas type and concentration (O_2, CO_2, or ethylene), and the addition of biologicals (e.g., yeast extract or coconut water) have led to alterations in yield of secondary products.

HYPOTHESIS

The preceeding survey of the literature leads to the following hypothesis to explain the widely reported observation that in plant cell cultures the yields of secondary products are low:

1. The culture conditions were not optimal for secondary product accumulation.
2. The conditions of maintenance of the cultures allowed the development and selection of low-yielding cell lines.

Two sets of comments need to be made about these hypotheses. The first is that the culture conditions for maximal accumulation of a specific secondary product may not be the optimal conditions for maintenance of other characteristics of a cell line (Zenk et al., 1977). Second, there are essentially no data available to show the rate of appearance of low-yielding cell lines in cultures. Further, there are essentially no data on the factors in the culture conditions that influence the rate of appearance of low-yielding cell lines in culture.

If cell cultures that maintain high yields of specific compounds are desired, then the following studies should be included.

1. The yield of desired compounds from individual plants within the species producing the compounds should be examined because high-yielding individuals are likely to give high-yielding cell cultures.
2. The cultures should be cloned, and the yields in individual clones measured. If the frequency of high-yielding clones declines with increased passaging of cultures, then high-yielding clones can be selected and propagated.
3. The media and culture conditions giving maximum yield should be determined so that efficient production can be achieved.

This approach has been successfully applied to cell cultures of *Catharanthus roseus* to obtain stable cultures giving high yields of serpentine and ajmalacine by Zenk et al. (1977). Although the one coherent example and the additional data are sufficient to lead to the hypothesis, they are not sufficient to establish it firmly. There are two additional ideas that should be considered. The first of these is that, in some cases, secondary product accumulation only occurs in specific morphological structures such as the oil glands in mint or the lactifers of rubber trees. The second of these ideas is that the reactions that lead to secondary products do not have an absolute specificity for substrates, so that if suitable analogs of the substrates are available then analogs of the product will also be formed.

The idea that some secondary products can only accumulate in specific morphological structures is the outcome of some experience with plant tissue cultures (Hiraoka and Tabata, 1974; Ikuta et al., 1974; Ikuta et al.,

1975; Tabata et al., 1972; Tabata et al., 1971, and Hirotani and Furuya, 1977) and of the accumulation of some secondary products in particular parts of plants. It implies that in cell cultures accumulation can only occur if there are specific structures present in the cultures. The idea further implies that the genetic information required to accumulate the compound is part of the information required for the construction of the morphological structure. This seems unlikely on the basis of knowledge of gene specificity. In cases where a close association between morphology and secondary product accumulation occurs in tissue culture, it seems more likely that the culture conditions that allow the genes for morphology to be expressed coincidentially allow the genes for secondary product accumulation also to be expressed. If so, then it should be possible to find conditions in which secondary product accumulation occurs in the absence of specific morphological structures. The compounds in table 1 are examples of such situations. A rational approach to identifying conditions of culture that allow secondary product accumulation to occur in unorganized cell cultures is unlikely at this time because knowledge of the control of gene activity in eucaryotic cells is fragmentary at best. An empirical approach may be productive. If bona fide cases of an obligatory relationship between morphology and secondary product accumulation exist, it would be very difficult to prove the relationship.

When secondary products accumulate in a cell, three conditions must exist within the cells: the rate of degradation must be lower than the rate of synthesis of the compounds; reactions for the synthesis must be possible; and the intermediates in synthesis must be present. There appear to be cases where some of the intermediates in the synthesis of secondary products are present in limiting amounts in cell cultures. Freeman et al. (1974) showed that onion tissue cultures that have no odor contain the enzyme alliinase, required for odor production, but lack the precursors for the odor compounds. Steck et al. (1973) and Boulanger et al. (1973) showed that on feeding the precursor 4-hydroxy-2-quinolone to cell cultures of *Ruta graveolens* large increases in quinoline alkaloids were obtained. Margna (1977) has discussed the possibility that the level of the substrate phenylalanine rather than the level of phenylalanine-ammonia lyase, the first enzyme in the pathway from phenylalanine to flavonoids, lignin, and related phenylpropanoids, controls the rate of accumulation of these compounds. These situations indicate either that early enzymes in the biosynthetic pathways are present at inadequate levels or that intermediates are diverted to other uses within the cell. These possibilities reduce to problems of gene activity or nutrition. These observations open the further possibility that plant cell cultures can be used to biotransform precursors or intermediates to desired end products. These possibilities have been reviewed by Steck and Constabel (1974), Reinhard (1974), Alfermann et al. (1977), and Stohs (1977).

In addition to possibly increasing production of desired compounds, biotransformation offers an opportunity to explore the specificity of some of the reactions of biosynthesis. Steck and Constabel (1974) fed 4-methyl- or 7-methyl-7-hydroxycoumarin to *Ruta graveolens* cultures and showed that these were converted into the 4-methyl- or 7-methyl-analogs of the compound obtained from 7-hydroxycoumarin. Alfermann et al. (1977) showed that cell cultures of *Digitalis lanata* will biotransform the unnatural β-methyl digitoxin into β-methyl digoxin, which is used in cardiac therapy. Hahlbrock (1977) has summarized their work on the specificity of enzymes of lignin and flavanoid biosynthesis and concluded that the enzymes have limited but not absolute substrate specificities. Thus in terms of substrate specificities of the reactions of secondary product synthesis we can anticipate that they are not absolute but are likely to be limited. This opens opportunities for the synthesis of novel modifications of secondary products.

The question of substrate specificity in the synthesis of secondary products has also been raised as a result of many observations that compounds that are closely related in structure occur in plants. These groups of compounds could be the result of a series of closely related substrates being available to a series of partially specific reactions. However, this is not the only possibility, and the explanation of their presence is not available.

CONCLUSION

There have been major developments in the achievement of accumulation of secondary products in plant cell cultures, and there is reason to be hopeful that commercial utilization of these possibilities is near at hand. A basic strategy for the achievement of high yields of secondary products can now be described. In addition, the synthesis of desired compounds, either natural or novel, by feeding of precursors is a real possibility.

LITERATURE CITED

Akasu, M., H. Itokawa, and M. Fujita. 1976. Biscoclaurine alkaloids in callus tissues of *Stephania cepharantha*. Phytochemistry 15:471–73.

Alfermann, A. W., H. M. Boy, P. C. Döller, W. Hagedorn, M. Heins, J. Wahl, and E. Reinhard. 1977. Biotransformation of cardiac glycosides by plant cell cultures. *In* W. Barz,

E. Reinhard and M. H. Zenk (eds.), Plant tissue culture and its bio-technological application. Pp. 125–41. Springer-Verlag, Berlin.

Amorim, H. V., D. K. Dougall, and W. R. Sharp. 1977. The effect of carbohydrate and nitrogen concentration on phenol synthesis in Paul's Scarlet Rose cells grown in tissue culture. Physiol. Plantarum 39:91–95.

Bergmann, L. 1960. Growth and division of single cells of higher plants *in vitro*. J. Gen. Physiol. 43(4):841–51.

Berlin, J., and J. M. Widholm. 1977. Correlation between phenylalanine ammonia lyase activity and phenolic biosynthesis in *p*-fluorophenylalanine-sensitive and -resistant tobacco and carrot tissue cultures. Plant Physiol. 59:550–53.

Boulanger, D., B. K. Bailey, and W. Steck. 1973. Formation of edulinine and furoquinoline alkaloids from quinoline derivatives by cell suspension cultures of *Ruta graveolens*. Phytochemistry 12:2399–2405.

Butcher, D. N. 1977. Secondary products in tissue cultures. *In* J. Reinert and Y. P. S. Bajaj (eds.), Applied and fundamental aspects of plant cell, tissue, and organ culture. Pp. 668–93. Springer-Verlag, Berlin.

Butcher, D. N., and J. D. Connolly. 1971. An investigation of factors which influence the production of abnormal terpenoids by callus cultures of *Andrographis paniculata* Nees. J. Exp. Bot. 22(71):314–22.

Constabel, F., O. L. Gamborg, W. G. W. Kurz, and W. Steck. 1974. Production of secondary metabolites in plant cell cultures. Planta Med. 25:158–65.

Constabel, F., J. P. Shyluk, and O. L. Gamborg. 1971. The effect of hormones on anthocyanin accumulation in cell cultures of *Haplopappus gracilis*. Planta 96:306–16.

Dalton, C. C., and H. E. Street. 1976. The role of the gas phase in the greening and growth of illuminated cell suspension cultures of spinach (*Spinacia oleracea* L.). In Vitro 12(7):485–94.

Davies, M. E. 1972a. Polyphenol synthesis in cell suspension cultures of Paul's Scarlet Rose. Planta 104:50–65.

Davies, M. E. 1972b. Effects of auxin on polyphenol accumulation and the development of phenylalanine ammonia-lyase activity in darkgrown suspension cultures of Paul's Scarlet Rose. Planta 104:66–67.

Döller, Von G., A. W. Alfermann, and E. Reinhard. 1976. Production of indole alkaloids in tissue cultures of *Catharanthus roseus*. Planta Med. 30(1):14–20.

Edelman, J., and A. D. Hanson. 1971. Sucrose suppression of chlorophyll synthesis in carrot callus cultures. Planta 98:150–56.

Freeman, G. G., R. J. Whenham, I. A. MacKenzie, and M. R. Davey. 1974. Flavour components in tissue cultures of onion (*Allium cepa* L.). Plant Sci. Lett. 3:121–25.

Furuya, T., H. Kojima, and K. Syōno. 1971. Regulation of nicotine biosynthesis by auxins in tobacco callus tissues. Phytochemistry 10:1529–32.

Gibbs, J. L., and D. K. Dougall. 1965. The growth of single cells from *Nicotiana tabacum* callus tissue in nutrient medium containing agar. Exp. Cell Res. 40:85–95.

Hahlbrock, K. 1977. Regulatory aspects of phenylpropanoid biosynthesis in cell cultures. *In* W. Barz, E. Reinhard and M. H. Zenk (eds.), Plant tissue culture and its biotechnological application. Pp. 95–122. Springer-Verlag, Berlin.

Heble, M. R., S. Narayanaswami, and M. S. Chadha. 1971. Hormonal control of steroid synthesis in *Solanum xanthocarpum* tissue cultures. Phytochemistry 10:2393–94.

Hiraoka, N., and M. Tabata. 1974. Alkaloid production by plants regenerated from cultured cells of *Datura innoxia*. Phytochemistry 13:1671-75.

Hirotani, M., and T. Furuya. 1977. Restoration of cardenolide-synthesis in redifferentiated shoots from callus cultures of *Digitalis purpurea*. Phytochemistry 16:610-11.

Ikeda, T., T. Matsumoto, and M. Noguchi. 1975. Formation of ubiquinone by tobacco plant cells in suspension culture. Phytochemistry 15:568-69.

Ikeda, T., T. Matsumoto, and M. Noguchi. 1976. Effects of nutritional factors on the formation of ubiquinone by tobacco plant cells in suspension culture. Agric. Biol. Chem. 40(9):1765-70.

Ikuta, A., K. Syōno, and T. Furuya. 1974. Alkaloids of callus tissues and redifferentiated plantlets in *Papaveraceae*. Phytochemistry 13:2175-79.

Ikuta, A., K. Syōno, and T. Furuya. 1975. Alkaloids in plants regenerated from *Coptis* callus cultures. Phytochemistry 14:1209-10.

Jain, S. C., H. Rosenberg, and S. J. Stohs. 1977. Steroidal constituents of *Trigonella occulta* tissue cultures. Planta Med. 31:109-11.

Jhang, J. J., E. J. Staba, and J. Y. Kim. 1974. American and Korean ginseng tissue cultures: Growth, chemical analysis, and plantlet production. In Vitro 9(4):253-59.

Kamimura, S., M. Akutsu, and M. Nishikawa. 1976. Formation of thebaine in the suspension culture of *Papaver bracteatum*. Agric. Biol. Chem. 40(5):913-19.

Kao, K. N., and M. R. Michayluk. 1975. Nutritional requirements for growth of *Vicia hajastana* cells and protoplasts at very low population density in liquid media. Planta 126:105-10.

Kaul, B., S. J. Stohs, and E. J. Staba. 1969. Dioscorea tissue cultures. III. Influence of various factors on diosgenin production by *Dioscorea deltoidea* callus and suspension cultures. Lloydia 32(3):347-59.

Margna, U. 1977. Control at the level of substrate supply: An alternative in the regulation of phenylpropanoid accumulation in plant cells. Phytochemistry 16:419-26.

Matsumoto, T., K. Nishida, M. Noguchi, and E. E. Tamaki. 1973. Some factors affecting the anthocyanin formation by *Populus* cells in suspension culture. Agric. Biol. Chem. 37(3):561-67.

Miller, C. O. 1969. Control of deoxyisoflavone synthesis in soybean tissue. Planta 87:26-35.

Misawa, M., H. Tanaka. O. Chiyo, and N. Mukai. 1975. Production of a plasmin inhibitory substance by *Scopolia japonica* suspension cultures. Biotechnol. Bioeng. 17(3):305-14.

Muir, W. H., A. C. Hildebrandt, and A. J. Riker. 1958. The preparation, isolation, and growth in culture of single cells from higher plants. Amer. J. Bot. 45:589-97.

Nettleship, L., and M. Slaytor. 1974. Adaption of *Peganum harmala* callus to alkaloid production. J. Exp. Bot. 25(89):1114-23.

Puhan, Z., and S. M. Martin. 1971. The industrial potential of plant cell culture. Prog. Ind. Microbiol. 9:13-39.

Reinhard, E. 1974. Biotransformations by plant tissue cultures. *In* H. E. Street (ed.), Tissue culture and plant science. Pp. 433-59. Academic Press, New York.

Schulte, U., and M. H. Zenk. 1977. A replica plating method for plant cells. Physiol. Plantarum 39:139-42.

Staba, E. J. 1969. Plant tissue culture as a technique for the phytochemist. Recent Adv. Phytochem. 2:75-106.

Staba, E. J. 1977. Tissue culture and pharmacy. *In* J. Reinert and Y. P. S. Bajaj (eds.), Applied and fundamental aspects of plant cell, tissue, and organ culture. Pp. 694–716. Springer-Verlag, Berlin.

Stearns, E. M., Jr., and W. T. Morton. 1975. Effects of growth regulators on fatty acids of soybean suspension cultures. Phytochemistry 14:619–22.

Steck, W., B. K. Bailey, J. P. Shyluk, and O. L. Gamborg. 1971. Coumarins and alkaloids from cell cultures of *Ruta graveolens*. Phytochemistry 10:191–94.

Steck, W., and F. Constabel. 1974. Biotransformations in plant cell cultures. Lloydia 37(2):185–91.

Steck, W., O. L. Gamborg, and B. K. Bailey. 1973. Increased yields of alkaloids through precursor biotransformations in cell suspension cultures of *Ruta graveolens*. Lloydia 38(1):93–95.

Stickland, R. G., and N. Sunderland. 1972. Production of anthocyanins, flavonols, and chlorogenic acids by cultured callus tissues of *Haplopappus gracilis*. Ann. Bot. 36(146):443–57.

Stohs, S. J. 1977. Metabolism of steroids in plant tissue cultures. *In* W. Barz, E. Reinhard, and M. H. Zenk (eds.), Plant tissue culture and its bio-technological application. Pp. 142–71. Springer-Verlag, Berlin.

Sugano, N., R. Iwata, and A. Nichi. 1975. Formation of phenolic acid in carrot cells in suspension cultures. Phytochemistry 14:1205–7.

Tabata, M., and N. Hiraoka. 1976. Variation of alkaloid production in *Nicotiana rustica* callus cultures. Physiol. Plantarum 38:19–23.

Tabata, M., N. Hiraoka, M. Ikenoue, Y. Sano, and M. Konoshima. 1975. The production of anthraquinones in callus cultures of *Cassia tora*. Lloydia 38(2):131–34.

Tabata, M., H. Yamamoto, N. Hiraoka, and M. Konoshima. 1972. Organization and alkaloid production in tissue cultures of *Scopolia parviflora*. Phytochemistry 11:949–55.

Tabata, M., H. Yamamoto, N. Hiraoka, Y. Marumoto, and M. Konoshima. 1971. Regulation of nicotine production in tobacco tissue culture by plant growth regulators. Phytochemistry 10:723–29.

Tanaka, H., Y. Machida, H. Tanka, N. Mukai, and M. Misawa. 1974. Accumulation of glutamine by suspension cultures of *Symphytum officinale*. Agric. Biol. Chem. 38:987–92.

Westcott, R. J. 1976. Changes in the phenolic metabolism of suspension cultures of *Acer pseudoplatanus* L. caused by the addition of 2-(chloroethyl) phosphonic acid (CEPA). Planta 131:209–10.

Westcott, R. J., and G. G. Henshaw. 1976. Phenolic synthesis and phenylalanine ammonia-lyase activity in suspension cultures of *Acer pseudoplatanus* L. Planta 131:67–73.

Zenk, M. H., H. El-Shagi, H. Arens, J. Stöckigt, E. W. Weiler, and B. Deus. 1977. Formation of the indole alkaloids serpentine and ajmalicine in cell suspension cultures of *Catharanthus roseus*. *In* W. Barz, E. Reinhard and M. H. Zenk (eds.), Plant tissue culture and its bio-technological application. Pp. 27–43. Springer-Verlag, Berlin.

Zenk, M. H., H. El-Shagi, and U. Schulte. 1975. Anthraquinone production by cell suspension cultures of *Morinda citrifolia*. Planta Med. 79–101.

Y. P. S. BAJAJ

Establishment of Germplasm Banks through
Freeze-Storage of Plant Tissue Culture
and Their Implications in Agriculture

34

INTRODUCTION

The cryobiology of animal and human cells has made great advance-
ments. The revival of larvae, caterpillars (Scholander et al., 1953), and
entire insects (Asahina and Aoki, 1958) at super low temperatures has been
recorded. The storage of animal sperms in liquid nitrogen and their
subsequent use for artificial insemination is now routine, and the sperm
banks have done a great service in animal breeding programs. Various
appliances like cryostats, cryomicrotomes, and cryomicroscopes have been
specially designed and are used in medical research for cryosurgery and
organ transplantation. Recently, "Cryotoriums" (unlike crematoriums)
have been experimented with for the indefinite preservation and immortal-
izing of dead bodies. As a matter of fact, some people have expressed their
desire to be "Cryotombed." These paradoxical practices certainly open up
new vistas in experimental biology. With this rapid progress in the area of
animal cryobiology, suspended animation and the delay or temporary
arrest of the aging process do not seem to be beyond reach any more.

The freeze-storage of plant cells at super low temperatures is a rather
recent development but with abundant potential (see Bajaj, 1976a). The
limited work done on the freeze-storage of plant tissue culture during the

Y. P. S. Bajaj, Tissue Culture Laboratory, Department of Plant Breeding, Punjab
Agricultural University, Ludhiana, Punjab, India.

last few years (see Bajaj and Reinert, 1977) and the observation that entire plants can be regenerated from cell suspensions (Bajaj, 1976b), meristem tip (Seibert, 1976), and pollen embryos (Bajaj, 1976a, 1977a,b) frozen at $-196°C$, has greatly enhanced the hope that this technology could well be a meaningful tool for the preservation of important and rare genomes, especially of plants that are about to become extinct.

Plant cell cultures on repeated subculture and serial propagation have a tendency to show variations in nuclear condition, chromosome alteration, and genetic changes (see D'Amato, 1977). This situation is further aggravated in long-term cultures, which show cells with various levels of ploidy. To avoid periodic subculture, methods like mineral oil overlay for the conservation of bacteria (Scherf, 1943) and fungi (Buel and Weston, 1947), though, have proved to be useful but do not seem to be of much practical importance for higher plants (Caplin, 1959). At present no standard method for ensuring genetic stability in the long-term storage of plant tissue cultures is available. The establishment of germplasm banks by freeze-storage would ensure genetic uniformity of the material and should be seriously considered for the vegetatively propagated plants.

METHODOLOGY AND FREEZERS

The detailed methodology for the freeze preservation of plant cells, tissues, and organs, and the various freezers employed have been described earlier (Bajaj, 1976a; Bajaj and Reinert, 1977). A large variety of liquid nitrogen freezers, both automatic and manually (fig. 1) operated, is available. In figures 2–4 are shown the relatively newer and improved models. Freezers R201/200 and R202/200 can not only freeze large numbers of cultures by regulated rates of cooling but can also thaw them at $35°C$.

Although, as I said above, the freeze preservation of plant tissue culture is a relatively recent development, the technique has been considerably refined, and the progress made is tremendous. Following is a general survey of the work done on seedlings, callus cell suspensions, somatic embryos, meristem tips, excised anthers, and pollen embryos.

Seedlings

It has been known that seeds and spores are the two plant materials that can be stored under ordinary dry conditions for a long time. They also survive freezing in the dry state, but are killed if stored after soaking (Adams, 1905). The seeds of several plant species have shown normal germination after being placed under conditions of germination for two days followed by vacuum drying for eight days, and then subjected to

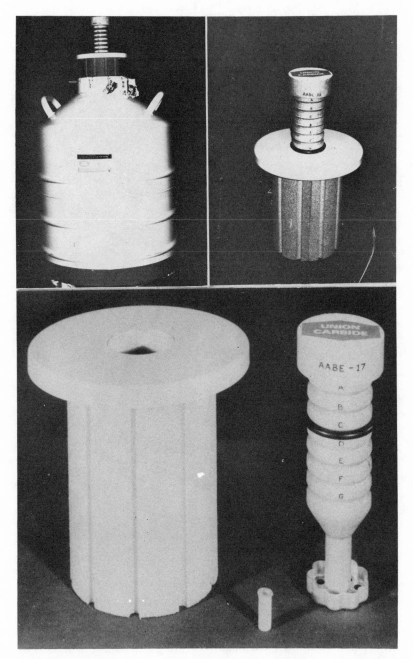

Fig. 1. Various parts of the manually operated (Union Carbide Corporation) Freezing Unit LR 33 with Biological Freezer-6. (Photograph courtesy of Messer Griesheim GMBH, West Berlin 12.)

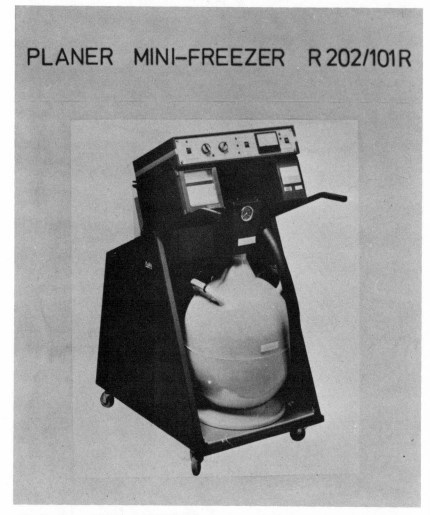

Fig. 2. Planer Mini-Freezer R202/101R (courtesy of Planer Products Ltd., Sunbury-on-Thames, England). A portable, modular, low-running-cost freezer is an improvement over model R202, and allows thawing of the specimen at 25°C.

$-192°C$ and $-269°C$ (Becquerel, 1932). Similarly, Nemmer and Luyet (1954) and Sun (1958) revived desiccated pea (*Pisum sativum*) seedlings. The decotylated embryos (7–12 mm) were pretreated by drying to final water content of 40.1 percent and then treating with liquid nitrogen for one minute. After freezing, the seedlings were quickly thawed in tap water and grown in White's nutrient agar. This work shows that root meristem even with normal water content survived freezing though the rest of the embryo

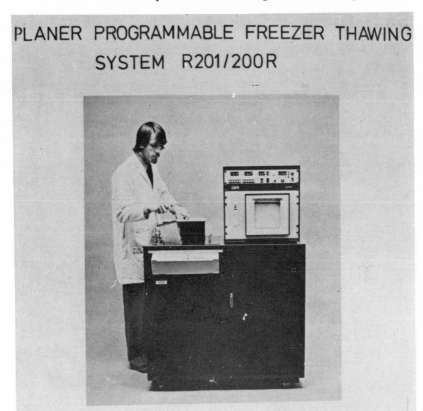

PLANER PROGRAMMABLE FREEZER THAWING SYSTEM R201/200R

Fig. 3. Planer Mini-Freezer R 202/200 R (courtesy of Planer Products Ltd., Sunbury-on-Thames, England). In addition to freezing, the specimen can be thawed at 35°C.

died. A strong correlation between the water content and the tissue survival was thus observed. Root tips survived exposure to liquid nitrogen only when they had a moisture content of 27–40 percent. The roots failed to survive when the moisture content was less than 14.2 percent. The root meristem with normal water content (without desiccation) was not injured by immersion in liquid nitrogen for one minute, but all other portions of the seedlings were injured and killed. In large desiccated seedlings only the stem tips survived after exposure to liquid nitrogen. The survival of root and shoot apices is interesting; these cells are non-vacuolated, thin-walled, young, and relatively less differentiated, and are thus closer to actively growing cell suspensions.

Callus Cell Suspensions

Cell suspensions of a number of plant species have been successfully

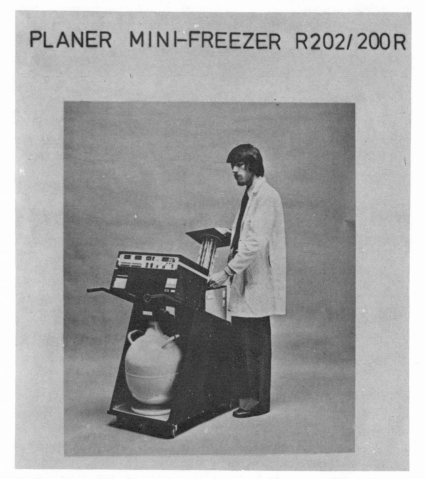

Fig. 4.   Planer Programmable, Freezing and Thawing Equipment System R201/200R (courtesy of Planer Products Ltd. Sunbury-on-Thames, England). A highly sophisticated, push-button programming with alarm warning system against power failure or misuse. Programmed thawing at 35°C.

employed for freeze-storage; some notable ones are *Acer pseudoplatanus* (Sugawara and Sakai, 1974), *Atropa belladonna* (Nag and Street, 1975), *Chrysanthemum morifolium* (Bannier and Steponkus, 1972, 1976), *Datura stramonium* (Bajaj, 1976b), *Daucus carota* (Latta, 1971; Nag and Street, 1973; Dougall and Wetherell, 1974; Bajaj and Reinert, 1975; Finkle et al., 1975; Bajaj, 1976b), *Glycine max* (Bajaj, 1976b), *Haplopappus ravenii* (Hollen and Blakely, 1975), *Ipomoea* (Latta, 1971), *Linum usitatissimum* (Quatrano, 1968), *Nicotiana tabacum* (Bajaj, 1976b), *Populus euramerica-*

na (Sakai and Sugawara, 1973), and *Prunus cerasus* (Tumanov et al., 1968). However, entire plants have been obtained so far from callus cultures of tobacco (fig. 5) and carrot (fig. 6).

These studies have demonstrated that the revival depends, to a great extent, on growth phase of cell suspension. Periodically transferred, young cell suspensions containing small clumps of cells in an exponential phase of growth are better equipped to withstand ultra-low temperatures as compared with old cultures.

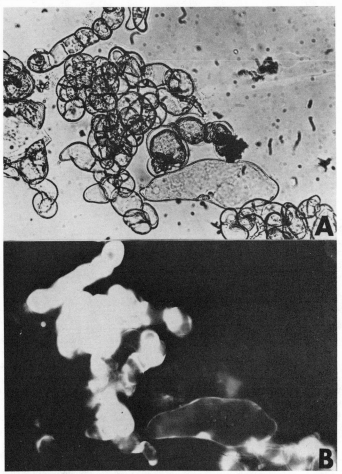

Fig. 5A–B. Cell suspension of *Nicotiana tabacum* frozen at the rate of 2°C min. and stored for 14 months in liquid nitrogen. Note the survival (fluorescence with fluorescein diacetate) of small, highly cytoplasmic clumps of cells and death of a large isolated cell. (A) Photographed in tungsten light; (B) in ultraviolet light. (After Bajaj, 1976b.)

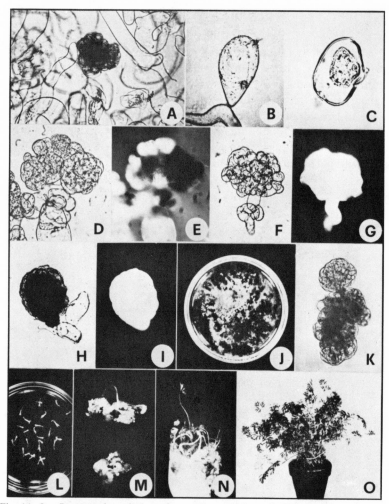

Fig. 6A–O. Regeneration of plants from cell suspensions of *Daucus carota* subjected to super-low temperatures. (A,B) Free cells from a suspension just before (A) and after freezing (B) in liquid nitrogen; note the plasmolysis of cell in (B). (C) 12-day-old suspension containing vacuolated free cells and an actively growing clump of cells. (D) A clump of cells (in tungsten light) taken from a culture frozen at –196°C for 3 months. (E) Same, in the ultraviolet light; note the partial survival of the clump (cryoprotectant DMSO 7%, cooling rate 2°C/min, thawing at 37°C. (F) A clump from cultures frozen at –20°C for a week. (G) Same, in ultraviolet light; note the complete survival of cells. (H,I) A proembryo along with free cells subjected to –20°C (1 h) and –70°C (30 min), and then immersed in liquid nitrogen (cryoprotectant DMSO 5%, thawing at 37°C); note the complete survival of the proembryo, and the death of single cells. (J–L) Various stages in the differentiation of embryos obtained from cell suspensions frozen in liquid nitrogen; the cultures were raised in a liquid medium. (M,N) Differentiation of plantlets in a 2,4-D-free agar solidified medium, 3(M) and 6 weeks (N), after transfer. (O) Plants of carrot obtained from cell suspensions stored in liquid nitrogen for 3 months (6 weeks after transfer to pot from cultures in [N]). (After Bajaj, 1976b.)

The induction of acclimation and hardening has attracted attention because of the possibility of obtaining "frost-resistant," or at least frost-tolerant, plants from cold-acclimated tissue cultures. In addition, tissue cultures offer a suitable tool for the study of the mechanism of cold acclimation at a cellular level.

Interest in this area started when Tumanov et al. (1968) demonstrated that cherry (*Prunus cerasus*) callus tissues grown on a medium containing high sucrose (10–20 percent) showed cold-resistance when first hardened at 2°C, and then subjected to subfreezing temperature of –30°C. Later, it was reported (Stephonkus and Bannier, 1971; Bannier and Steponkus, 1976) that the unacclimated callus of *Chrysanthemum morifolium* could withstand cooling down to –6.6°C only, whereas callus acclimated for six weeks at 4.5°C exhibited complete survival down to –16.1°C. Likewise, cell cultures of *Nicotiana sylvestris* and *Capsicum annuum* subjected to –3°C for 21 days have been claimed to show enhanced chilling resistance (Dix and Street, 1976).

The work of Sakai and Sugawara (1973) on the revival of poplar calluses masses subjected to liquid nitrogen in the absence of a cryoprotectant is commendable. They induced acclimation by first varying the day (12°C) and night (0°C) temperature, and then cooled the callus in 5°C steps at daily intervals to –30, –70, –120 and –196°C. This opens up the possibility that, by carefully manipulated prefreezing, the use of a cryoprotectant can be dispensed with.

This is an area that needs to be explored extensively, and work on these lines should be extended to other systems to see if the plants regenerated from the acclimated callus retain this tolerance to low temperatures.

Somatic Embryos

Carrot embryos obtained from cell suspensions have positively responded to freezing (Dougall and Wetherell, 1974; Bajaj, 1976b). The differentiating cell suspensions (growing in an auxin-free medium) undergoing embryogenesis and filtered through a nylon mesh were employed for freezing studies. Young globular embryos (fig. 7A,B) invariably withstood slow (1–3°C/min) freezing. In cultures they resumed growth (fig. 7C,D) after a lag phase of 3–4 months, and normal-looking plants (fig. 8A,B) were regenerated from such embryos stored for up to eight months in liquid nitrogen.

The embryo survival, however, dropped with advance in maturity. In embryos with fully developed cotyledons, occasionally a few cells survived that had a tendency to form callus.

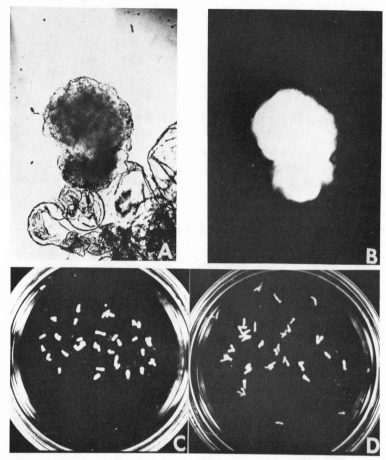

Fig. 7A–D.   Growth and development of carrot somatic embryos subjected to –196°C. (A) A proembryo along with cell suspension frozen at the rate of 2°C/min and stored for 1 year in liquid nitrogen, photographed in tungsten light. (B) Same, photographed in ultraviolet light after staining with fluorescein diacetate; note the survival of the embryo and death of the callus cells. (C,D) Frozen somatic embryos after 3 and 4 months (lag phase) in culture; note growth and development.

Meristem Tip

The meristem tip and shoot apices appear to be the ideal material for freeze preservation of vegetatively propagated plants, as they possess a number of advantages over callus cells:

1.   they are genetically very stable;

2.   the cells in the apices are small, highly cytoplasmic and thinwalled like the actively growing cell suspensions in an exponential phase;

Fig. 8A–B.   Regeneration of carrot plants from somatic embryos. (A) Control. (B) Plant obtained from frozen embryos (cooled at the rate of 2°C/min in the presence of DMSO 7% and thawed at 37°C) stored for 8 months in liquid nitrogen.

3.   in some cases it is difficult to regenerate an entire plant from callus, but the shoot apices are rather easy to grow on simple media;

4.   it is also an excellent material for the storage of pathogen-free stocks.

So far the survival of meristem tips of carnation (Seibert, 1976) and potato (Bajaj, 1977c) subjected to –196°C has been observed. In both these cases the shoot tips withstood direct immersion in liquid nitrogen. The potato cultures mostly underwent a lag period of two or more weeks, after which they showed elongation, and unfolding of the leaves. In ten weeks 3–4 cm-long shoots were formed. The percentage of survival was low, but it could certainly be improved by regulating the rate of cooling.

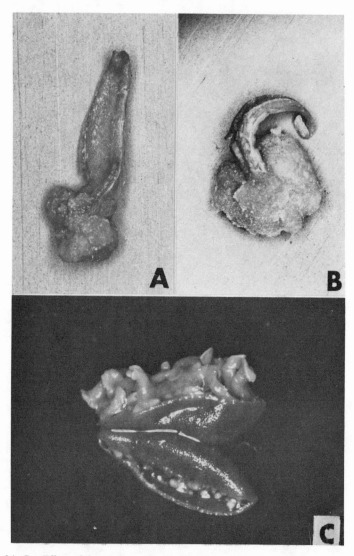

Fig. 9A–C. Effect of freezing on 4-week-old cultured anthers of *Petunia hybrida*. (A) Proliferated anther subjected to freezing. (B) Same, after 6 months in culture; note the increase in callusing. (C) Four-week-old culture of tobacco anther undergoing andro-genesis; note the protruding pollen embryos. (After Bajaj, 1978.)

In view of the above-mentioned advantages, the shoot apices should be preferred over other tissues for freeze preservation.

Anthers

The following three freezing experiments were conducted with the anthers of *Atropa, Nicotiana,* and *Petunia* (Bajaj, 1978):

1. freshly excised anthers from the flower buds;
2. anthers cultured for 3–4 weeks;
3. transversely cut halves of cultured anthers.

Freshly excised whole anthers of tobacco and *Atropa* could not withstand slow regulated cooling (1–3°C/min); but in certain cases, after sudden and brief immersion in liquid nitrogen followed by rapid warming at 37°C, occasional survival was observed. Such anthers, though, did not show any visual sign of survival for two to three months in cultures; however, occasionally localized callusing or multicellular pollen were observed.

The subsequent experiments were conducted on anthers (Nitsch and Nitsch, 1969; Zenkteler, 1971) that were first cultured for three to four weeks to induce callusing (fig. 9A)/androgenesis (fig. 9C) and then subjected to freezing. The results are summarized in tables 1–3. *Petunia* anthers excised from the flower buds and cultured for four weeks on a basal nutrient agar medium supplemented with NAA (0.5 ppm) + benzyladenine (0.5 ppm) in some cases showed proliferation (fig. 9A). As seen in table 1, treatment with dimethyl sulfoxide (DMSO) combined with washing at 37°C for 8 minutes reduced callusing from 44 percent to 24 percent. After freezing only 5 percent of the anthers continued to proliferate (fig. 9B), as compared with 44 percent in the controlled sample. In tobacco (table 2) androgenesis was induced in only 1.5 percent of the anthers, as compared with 81.1 percent in the control, and also the number of plants per anther was considerably reduced. Mostly, the anther tissue became soft and

TABLE 1

EFFECT OF DMSO AND FREEZING ON
FOUR-WEEK-OLD CULTURED ANTHERS OF PETUNIA HYBRIDA

Experiment	No. of Anthers Cultured	No. of Anthers Callused	Survival Percentage
1. Control (untreated)	85	38	44.7
2. Treated with 7% DMSO	60	15	25.0
3. Treated with 7% DMSO, frozen at the rate of 2°C/min and thawed at 37°C	95	5	5.2

NOTE: Ampules containing anthers were gradually lowered in the liquid nitrogen cylinder and then immersed for one minute.

SOURCE: After Bajaj, 1978.

TABLE 2

RESPONSE OF NICOTIANA TABACUM WHOLE ANTHERS (FOUR WEEKS AFTER
CULTURE) SUBJECTED TO VARIOUS TREATMENTS AND RECULTURED ON
AGAR-SOLIDIFIED MEDIUM

Experiment	No. of Anthers Cultured	No. of Growing Anthers	Survival Percentage	Total No. of Plantlets	No. of Plants per Anther
1. Untreated (control)	85	69	81.1	890	12.8
2. Treated with 7% DMSO	70	54	77.1	530	9.8
3. Treated with 7% DMSO and warmed at 37°C for 10 min	66	45	68.1	390	8.6
4. Treated with DMSO, cooled at the rate of 2°C/min to -196°C, and thawed at 37°C	130	2	1.5	7	3.5

TABLE 3

GROWTH RESPONSE OF TRANSVERSELY CUT TWO HALVES OF FOUR-WEEK-OLD
CULTURED ANTHERS OF NICOTIANA TABACUM SUBJECTED TO COLD
TREATMENT AND RECULTURED IN AGITATED LIQUID MEDIUM

Treatment	No. of Anthers Cultured	No. of Growing Anthers	Survival Percentage	Total No. of Plantlets	No. of Plants per Anther
1. Control (untreated)	45	30	66.6	623	20.7
2. Frozen at the rate of 2°C/min to -196°C	60	4	6.6	10	2.5

SOURCE: Bajaj, 1978; reprinted with permission.

spongy, and the pollen embryos inside the anther aborted—perhaps due to
the toxic substances produced by the degenerating anther. From the
foregoing, it appeared that if more pollen embryos were exposed but not
removed from the anther it might improve their chances of survival. An
experiment was, therefore, designed to cut the anther into two transverse
halves (to expose the pollen embryos) and then freeze them. By doing so,
more of pollen embryos survived and the number of plantlets per anther
was considerably increased (table 3).

Pollen Embryos

The procedure for the isolation, freezing, and culture of pollen embryos
is schematically represented in figure 10. The frozen-thawed pollen embryos of *Atropa* and *Nicotiana* cultured in drops of 250–500 μl of the
liquid synthetic medium were able to revive and regenerate plants (Bajaj,
1976a; 1977a,b). In addition to the method of cooling, the survival
percentage was strongly influenced by the stage of embryo development
(table 4) and the cryoprotectant (figs. 13, 14).

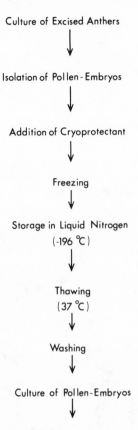

Culture of Excised Anthers

Isolation of Pollen-Embryos

Addition of Cryoprotectant

Freezing

Storage in Liquid Nitrogen
(-196 °C)

Thawing
(37 °C)

Washing

Culture of Pollen-Embryos

Regeneration of Haploid Plants

Fig. 10. Schematic representation showing the methodology for the freeze-preservation of pollen embryos in liquid nitrogen and the subsequent regeneration of haploid plants. (After Bajaj. 1976a.)

Stage of pollen embryo development. Globular embryos (fig. 11A,B) withstood freezing better (table 4) and showed 31 percent viability, in contrast to early and late heart-shaped embryos, which showed 9 percent and 2 percent survival respectively. Higher survival of globular embryos is attributed to their highly cytoplasmic, thin-walled, and nonvacuolated nature of cells as in young meristems and actively growing cell suspensions.

In cultures the pollen embryos showed a lag period that was linearly proportional to their age and the extent of freeze injury. The lag period varied from 2 to 4 weeks for globular embryos, and from 4 to 6 weeks for heart-shaped ones. Younger embryos continued to undergo androgenesis and formed haploid plants (fig. 11C,D), whereas late heart-shaped and

TABLE 4

EFFECT OF ULTRA-LOW TEMPERATURE (-196°C) ON POLLEN EMBRYOS
OF NICOTIANA TABACUM

Stage of Embryo Development	Survival Percentage	Lag Period
1. Globular	31	2-4 wks
2. Early heart-shaped	9	4-6 wks
3. Late heart-shaped	2	4-6 wks
4. Fully differentiated	0	. . . . .

SOURCE: After Bajaj, 1977b.

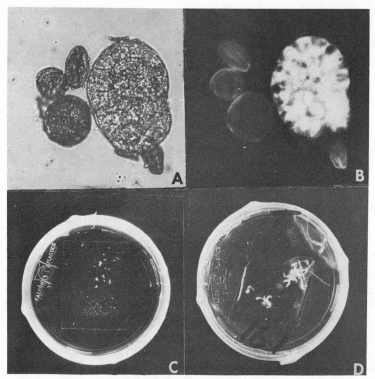

Fig. 11A–D. Survival of pollen embryos of *Atropa belladonna* and *Nicotiana tabacum*
subjected to -196°C in the presence of 5% DMSO. (A) Pollen suspension of *Atropa
belladonna* photographed in tungston light, (B) in Ultraviolet light; note the fluorescence
of pollen embryos in (B) after staining with fluorescein diacetate (from Bajaj, 1976).
(C) Pollen embryos of *Nicotiana* obtained from 4-week-old cultured anthers and frozen
at 196°C (at the time of culture). (D) Regeneration of pollen plantlets from pollen
embryos stored in liquid nitrogen for 7 months.

differentiated embryos showed a long lag phase and only partially survived.
They had a tendency to show callusing (fig. 12A) or produced abnormal
and malformed plants with multiple shoots (fig. 12B).

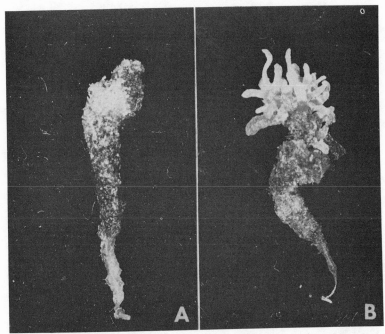

Fig. 12A-B. Growth response of late heart-shaped pollen embryos (frozen at the rate of 1°C/min in the presence of 15% sucrose and stored for 5 months in liquid nitrogen) 3 and 4 months after culture respectively. Note the proliferation and irregular growth in (A) and the regeneration of multiple shoots from the disturbed (freeze-injured) growing point in (B). Such embryos did not show any fluorescence with fluorescein diacetate staining; however, 2-3 months (lag phase) after culture they resumed growth. (Bajaj, 1978; reprinted with permission.)

Effect of cryoprotectants. The pollen-embryo survival was to a great extent affected by the nature of the cryoprotectant and its concentration (figs. 13, 14). The effect of various concentrations of DMSO ranging from 2-10 percent, as shown in figure 13, reveals that 7 percent gave the optimum survival value of 36 percent, whereas in the absence of DMSO the pollen embryos died. Ten percent and higher concentrations of DMSO decreased the percentage of survival. An experiment was also conducted (fig. 14) to compare the effects of DMSO, sucrose, and glycerol. The results obtained with 15 percent glycerol (survival 34 percent) were comparable to those of 7 percent DMSO (survival 33 percent).

GENERAL GUIDELINES AND PRECAUTIONS

"Freezing-Storage-Thawing-Culture" is a complicated multiple event, and a fault at any stage can result in lethal injury to the system. The detailed

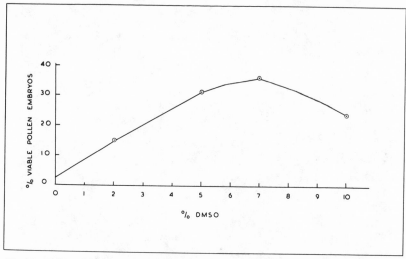

Fig. 13.   Effect of various concentrations of dimethyl sulfoxide on percentage of viability of pollen embryos frozen at the rate of 2°C/min and stored in liquid nitrogen (data based on fluorescein diacetate staining).

account of these aspects has already been given earlier (Bajaj and Reinert, 1977); following, however, are some of the general guidelines that have emerged from work on the cryobiology of plant tissue cultures and that can be helpful in successfully increasing the revival of cells and in the correct determination of their viability.

1. Nature and age of the culture
2. **Physiological state of the plant**
3. Nature and concentration of the cryoprotectant
4. Method and rate of cooling
5. Storage temperature
6. Method of thawing
7. Parameter for cell survival
8. Induction of hardiness and cold acclimation

Selection of the Material

Highly cytoplasmic, non-vacuolated thin-walled cells in small aggregates from periodically transferred and actively growing young suspensions are able to withstand freezing much better than relatively older cultures containing large and thick-walled single cells or callus masses.

In addition, high density of the cells or biomass per ampule (with relation to volume of the medium) is an important criterion for cell survival. The

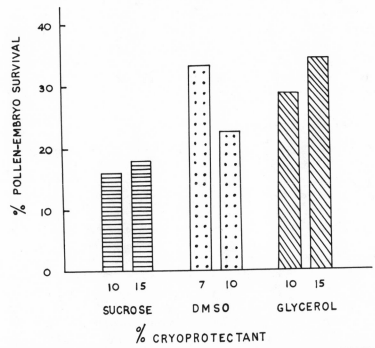

Fig. 14. Comparison of the effect of various cryoprotectants (i.e., sucrose, DMSO, and glucose) on percentage of viability of frozen pollen embryos (data based on fluorescein diacetate staining). (Bajaj, 1978; reprinted with permission.)

ampules containing thick and packed suspensions often show higher cell survival compared with those wtih low density (Bajaj, 1976b).

The cell cultures should be frozen at an optimum time in their growth cycle, i.e., the exponential phase of their growth. This also avoids the possibility of preserving only a selected portion of the cell population (Peterson and Stulberg, 1964).

The survival response of young seedlings, somatic embryo, and pollen embryos at various stages of development is comparable to those of cell suspensions. Young globular embryos showed higher survival when compared with heart-shaped and fully differentiated embryos. This is attributed to their highly cytoplasmic, thin-walled, and non-vacuolated nature of cells as in young meristems. Therefore, young embryos should be a preferred material for freeze-storage studies.

Physiological State of the Plant

It is not only the age and nature of the cultures but also the change of season and physiological state of the plant that affect their response to low temperatures. It has been recorded that due to seasonal variations there is a change in the soluble proteins (Siminovitch and Briggs, 1949), an increase in the level of sugars (Heber, 1958; Parker, 1959) and polyhydric alcohols (Sakai, 1960) that is generally believed to increase the cold resistance of plants. In this connection cultures of frozen shoot apices obtained from plants at various seasons might throw some light on this aspect.

Nature and Concentration of the Cryoprotectant

The long-term freeze-storage of cells though theoretically ideal for germplasm banks, at the same time requires caution particularly in the choice and concentration of the cryoprotectant. The limited cytogenetic studies on frozen animal cells have not shown any chromosomal variations, yet there exists the possibility of genetic erosion—especially when long treatments with strong cryoprotectants are involved. Dimethyl sulfoxide, though it fulfills all the criteria of a good solvent and has proved to be an excellent anti-freeze, in a rather high concentration, i.e., 10–15% (which is normally used), interferes with the RNA and protein synthesis in tissue cultures (Bajaj et al., 1970). In view of these observations, it is advisable to give more stress on the use of a mixture of glycerol, sucrose, lower concentrations of DMSO, and other mild cryoprotectants that are not harmful. Perhaps a combination of two or more chemicals would be more suitable.

Second, the cryprotectant should be added to the cultures gradually and at some interval, as sudden mixing causes plasmolysis (Towill and Mazur, 1976).

Method of Cooling and Freezing Injury

Intracellular freezing is almost instantaneous "flashing" (Meryman, 1974), and the injury caused is irreversible and lethal. Once caused, it cannot be reversed by cryoprotectants. Thus, the method and the velocity of cooling should be such that it does not cause intracellular freezing and crystal formation. To prevent freezing injury, cooling must be slow enough to let all freezable water flow out of the cell during cooling (Mazur 1970). This can either be achieved by prefreezing the tissue cultures (Sakai and Sugawara, 1973; Sugawara and Sakai, 1974) or by a slow regulated rate of cooling. In all the plant cell cultures successfully revived, a slow rate of cooling has been employed that varies between 1–5°C/min, though the optimum appears to be 2°C/min (see Latta, 1971; Dougall and Wetherell, 1974; Henshaw, 1975; Nag and Street, 1975; Bajaj, 1976a,b).

The data obtained on the effect of direct immersion in liquid nitrogen are inconsistent. Animal cells subjected to sudden immersion in liquid nitrogen mostly result in immediate lethality, which could be due to the denaturation of proteins (Brandts, 1967). Sakai (1971) revived cortical tissue of mulberry from sudden exposure to liquid nitrogen, when rewarmed rapidly; however, no survival was observed when warmed slowly in air at 0°C. Similarly, pollen embryos occasionally survived after sudden exposure of one to five minutes in liquid nitrogen followed by rewarming rapidly at 37°C. However, the embryos shriveled when thawed at a room temperature and did not show any sign of survival with fluorscein diacetate (FDA), though in a few cases sporadic callus at places was observed. Likewise, Seibert (1976) reported the survival of shoot apices of carnation suddenly frozen in liquid nitrogen and thawed at 35°C.

The freezing injury could be due to a number of interdependent factors, i.e., intracellular freezing, dehydration of the cells, increase in the concentrations of solutes (Lovelock, 1953), formation of disulfide bonds (Levitt, 1962), and also to mechanical stress and rupture of the organelle membrane. According to the "Minimum-Cell-Volume" theory (see Meryman, 1974), there is a strong correlation between cell volume reduction and the damage. It is believed that the damage is due to the water loss and the reduction in volume rather than to any absolute ionic or solute concentrations that are responsible for the hypertonic injury associated with freezing (Meryman, 1974). If reduction in volume is the main reason for injury, then by controlling the volume it would be possible to avoid intracellular freezing injury. If it is so, then one should be able to build mechanical resistance against injury simply by preventing cell shrinkage (to a point of critical volume) with the help of gradients. This concept is very interesting and needs to be further explored.

Storage Temperature

Another equally important point for long-term preservation is the storage temperature. When specimens are not stored at a sufficiently low temperature, an additional injury to the cells may be caused. For the storage of cells, -20°C is not a suitable temperature, as metabolic processes continue (see White, 1963), and cause protein denaturation (Brandts, 1967), which results in lethality. Other changes in concentration of solutes, variation in pH (Van den Berg, 1959), and water migration in frozen cells have also been recorded. Likewise, Grant and Alburn (1965) reported that certain enzymatic and non-enzymatic reactions proceed more rapidly at -18°C than at 1°C. Mitochondria are also known to be affected by storage temperatures that are not low enough. Dimethyl sulfoxide, for instance, reduced but did not prevent changes in the permeability of rat-liver mitochondria at -15°C (Lusena, 1965), or freezing damage in tomato mitochondria when stored at -18°C (Dickinson et al., 1967). In my experience, tobacco cells frozen for 6 months at –20°C did not show any growth response; however, cultures stored for 13 months at –196°C revived and resumed growth after a lag phase of 5 months. Likewise, in pollen embryos stored at –20°C, the viability was nil at the end of 3 months, whereas cultures stored at –196°C revived. So, it is obvious that extensive changes do occur at –20°C, and thus it is not an efficient temperature for long-term storage.

As a matter of fact, although at –196°C there is no metabolic activity, in animal cell cultures enzymes like aldolases and lactic acid dehydrogenase isozymes (Peterson and Stulberg, 1964; Heber, 1968) have been observed to show variations. In view of these observations, it might be worthwhile to investigate whether there is any difference in the cultures stored in liquid nitrogen (–196°C) and in helium (–269°C).

Method of Thawing

Although in a few instances (Dougall and Wetherell, 1974; Sugawara and Sakai, 1974) no difference in the survival of cell suspensions thawed at a room temperature, or at 30°C–40°C was observed, the importance of rapid thawing has been fairly emphasized (Sakai, 1971; Nag and Street, 1975; Bajaj, 1976b).

Rapid thawing (35°C–40°C) of the cultures is recommended because it avoids freeze injury by preventing recrystalization or growth of the ice crystals, and thus results in higher survival.

Parameters for Cell Survival

With regard to the viability of frozen cells, it is pertinent to point out that staining methods alone may not be sufficient to give accurate information about their survival. There are always some cells that, although giving a positive staining reaction immediately after thawing, later die in cultures. There are also cells that are partially injured and in a state of cold-shock, and do not give a positive reaction. In addition, with fluorescein diacetate (Widholm, 1972) the cells sometimes show partial or light fluorescence as compared with the control, and it is not possible to decide upon their viability. It has been observed (Bajaj, 1976b) that tobacco frozen-thawed cells showed mild fluorescence or gave a negative staining reaction; however, in cultures after a lag phase of 3–4 months they started to grow.

In view of the erratic response with stains, various growth parameters, i.e., mitotic index, cell number, cell culture volume, fresh and dry weights, and plating efficiency combined with the refined technique of triphenyl tetrazolium chloride (Steponkus and Lamphear, 1967; Towill and Mazur, 1975) should be the major yardstick for the evaluation of cell survival.

Induction of Hardiness and Frost Resistance

It is known that the conditions that favor hardening of the cells and tissues are the ones that retard growth. Sakai and Yoshida (1968) reported that gardenia (*Gardenia grandiflora*) and cabbage (*Brassica oleracea capitata*) leaves into which sugar was introduced showed a considerable rise in freezing resistance. Similarly, seedlings (Tumanov and Trunova, 1963) and callus (Tumanov et al., 1968) are known to show increase in hardiness after the incorporation of sugar. In addition, Levitt (1956) reported that heavy nitrate fertilizers decrease the ability of plants to harden, and plants native to severe cold climates show higher resistance to freezing than those growing in warm climates (Sakai and Okada, 1971).

Acclimation is also dependent on the age of the culture. Tumanov et al., (1968) reported that young cherry-stem tissue cultures exhibited the greatest acclimation capacity. Likewise, 10-day-old cultures of *Chrysanthemum* (Bannier and Steponkus 1976) showed highest acclimation; and with an increase in period of subculture, there was a gradual decrease in hardiness, and after 59 days the callus was unable to acclimate.

Thus, from the above-mentioned studies it emerges that to induce acclimation and hardiness the following points should be taken into consideration:

1. tissue cultures should be grown on a low nitrate and high sucrose media;
2. cultures should be raised from rather hardy cultivars grown at low temperatures;
3. actively growing and periodically subcultured young cultures should be used;
4. the change to different low-temperature regimes should be periodical and gradual (Sakai and Sugawara, 1973).

PROSPECTS OF FREEZE-STORAGE AND ITS
IMPLICATIONS IN AGRICULTURE

Isolated cells, tissues, and organs are the ideal systems for studying not only the basic and fundamental aspects of the biology of freezing and the mechanics involved, but they offer a number of applied possibilities in agricultural research as well (fig. 15). The regeneration of the entire plant from isolated cells and tissues frozen in liquid nitrogen has demonstrated the practical utility and potentials of the technology of freeze-storage for the preservation of germplasm. Some of the most obvious applications and potentials of freeze-storage of plant culture are:

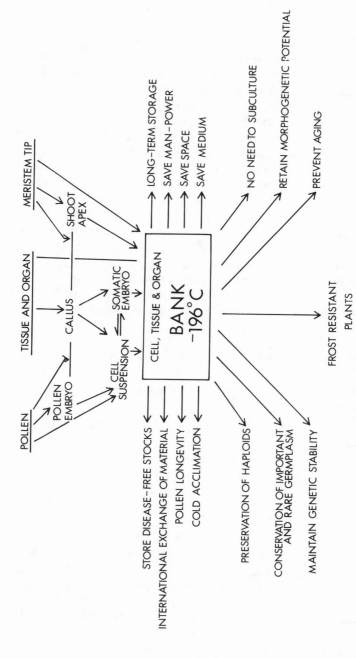

Fig. 15. Diagrammatic representation showing the freeze-storage of plant cell, tissue, and organs in liquid nitrogen and their prospects establishment of germplasm banks. (Bajaj, 1977c; reprinted with permission.)

1. preservation of genetic uniformity;
2. conservation of rare and important germplasm;
3. maintenance of unstable (haploid) cultures;
4. storage of disease-free material;
5. retention of morphogenetic potential;
6. avoid periodical subculturing;
7. save manpower, media, and storage space;
8. slow metabolism and prevention of aging;
9. international exchange of material;
10. induction of cold acclimation;
11. regeneration of frost-resistant (mutant) plants;
12. pollen storage for longevity and incompatible crossing.

The phenomena like chromosome aberration, mutations, and change of ploidy, though undesirable for maintaining uniformity of the clones, can, nevertheless, be picked up as variables and can be usefully incorporated into the breeding programs. A practical example is that of sugarcane (Heinz and Mee, 1971; Liu and Chen, 1976). Such variables, which do not occur in nature, could then be frozen and kept indefinitely. This would also be a repository for prototype strains of cells as reference stocks. New cell culture lines and mutants can be banked and used in response to research needs.

In addition, the haploids are of great importance for the induction of mutations and homozygous plants; however, they are highly unstable in culture and revert to diploid state in a relatively short period of time. In this connection their preservation as stem tips (Seibert, 1976) and pollen embryos (Bajaj, 1976a, 1977a,b,) would be highly rewarding.

Another equally important area would be the induction of hardiness through cold acclimation (Tumanov et al., 1968; Steponkus and Bannier, 1971; Bannier and Steponkus, 1976; Dix and Street, 1976) and the selection of "Cold Resistant Mutant" cell lines that could then be differentiated into "Frost Resistant Plants."

Pathogen-free stocks of important germplasm could be frozen, revived, and vegetatively propagated when desired. This would be an ideal method for the international exchange of such materials.

One of the most profitable areas from the agricultural point of view is the storage of pollen (Nath and Anderson, 1975). This would solve some of the problems often encountered in the incompatibility and longevity of pollen viability.

The revival of mouse embryos frozen at $-196°C$ and $-269°C$ and their subsequent growth and development into adults on transplanting to the uterus of a foster mother (Whittingham et al., 1972) has relevance in incompatible plants. In this connection the storage or transplantation of embryos in incompatible plants (where they normally abort and empty seeds are formed) would be worth attempting.

Emphasis should be placed especially on vegetatively propagated crop plants, and freeze-storage should be extended to explants from underground stems, roots, leaves, and so on.

PLEA FOR THE ESTABLISHMENT OF "GERMPLASM BANKS"

The existing germplasm banks for plants are customarily evidenced by seed germplasm collections. Although seeds offer an effective storage system for the genetic material, they are usually variable and the plants are not true to the mother plant. At present no method for the long-term preservation of vegetatively propagated plants is available.

Plant tissue cultures are the ideal material for germplasm banks. This system, which offers the ultimate in vegetative propagation, has the proven ability to regenerate entire plants from isolated cells and protoplasts. The plants thus regenerated possess certain advantages over the sexual method of reproduction. The vegetative method is not only faster but the daughter plants obtained are "true to type" to the mother of known desirable traits.

Although animal genetic material has been preserved by the freeze-storage method for about two decades, plants, which offer a wide range of possibilities, have been ignored, possibly due to the keeping quality of the seeds. The storage of plant material offers advantages over animal cells, i.e., in addition to being totipotent, there is no embargo on genetic engineering, and no social, moral, or legal complications are involved in generating clones or making unusual crosses. Now that the technology of freeze-preservation of plant cells is developed and refined, it is high time that stress should be given to extend it to crop plants.

With the ever-enthusiastic search for obtaining new cultivars and varieties of plants for producing more food, and the rapid increase in their number, it is becoming not only difficult but at times impossible to maintain or preserve some of the stocks that at present are not needed for breeding purposes. Thus, some of the germplasm that may not seem to be so important today but might in future be needed is ignored or lost. This is especially true for plants that are vegetatively propagated. It is for the preservation of such materials that are threatened with extinction that the "Germplasm Banks" should be established. These banks should form a part of various bureaus of germplasm that would be responsible for the storage, maintenance, distribution, and exchange of important and rare germplasm stocks at national and international levels.

The quest of agriculture for obtaining new cultivars for producing more food and the rapid multiplication of the plants requires the utilization of freeze-preservation methods for the storage of the desirable genomes.

LITERATURE CITED

Adams, J. 1905. The effect of very low temperature on moist seeds. Sci. Proc. Roy. Soc. Dublin 2:1-6.

Asahina, E., and K. Aoki. 1958. Survival of intact insects immersed in liquid oxygen without any antifreeze agent. Nature 182:327-28.

Bajaj, Y. P. S. 1976a. Gene preservation through freeze-storage of plant cell, tissue, and organ culture. Acta Horticulturae 63:75-84.

Bajaj, Y. P. S. 1976b. Regeneration of plants from cell suspensions frozen at -20, -70, and -196°C. Physiol. Plantarum 37:263-68.

Bajaj, Y. P. S. 1977a. Survival of *Atropa* and *Nicotiana* pollen-embryos frozen at -196°C. Curr. Sci. 46:305-7.

Bajaj, Y. P. S. 1977b. Regeneration of plants from pollen-embryos frozen at ultra-low temperature: A method for the preservation of haploids. *In* Proc. Intern. Palynological Conf. Lucknow (in press).

Bajaj, Y. P. S. 1977c. Initiation of shoots and callus from potato-tuber sprouts and axillary buds frozen at -196°C. Crop Improv. 4:48-53.

Bajaj, Y. P. S. 1978. Effect of super-low temperature on excised anthers and pollen-embryos of *Atropa, Nicotiana,* and *Petunia.* Phytomorphology 28 (in press).

Bajaj, Y. P. S., V. S. Rathore, S. H. Wittwer, and M. W. Adams. 1970. Effect of dimethyl sulfoxide on zinc[65] uptake, respiration, and RNA and protein metabolism in bean (*Phaseolus vulgaris*) tissues. Amer. J. Bot. 57:794-99.

Bajaj, Y. P. S., and J. Reinert. 1975. Regeneration of plants from cells subjected to super-low temperatures. Abstr. Intern. Bot. Congr. P. 278. Leningrad.

Bajaj, Y. P. S., and J. Reinert. 1977. Cryobiology of plant cell cultures and establishment of gene-banks. *In* J. Reinert and Y. P. S. Bajaj (eds.). Applied and fundamental aspects of plant cell, tissue, and organ culture. Pp. 757-77. Springer-Verlag, Berlin.

Bannier, L. J., and P. L. Steponkus. 1972. Freeze preservation of callus cultures of *Chrysanthemum morifolium.* Ramat. Hort. Sci. 7:194.

Bannier, L. J., and P. L. Steponkus. 1976. Cold acclimation of *Chrysanthemum* callus cultures. J. Amer. Soc. Hort. Sci. 101:409-12.

Becquerel, P. 1932. La reviviscence des plantules deséchées soumise aux actions du vide et des très basses températures. C. R. Acad. Sci. Paris 194:2158-59.

Berg, I., Van den. 1959. Arch. Biochem. Biophys. 84:305-15.

Brandts, J. F. 1967. Heat effects on proteins and enzymes. *In* A. H. Rose (ed.). Thermobiology. Academic Press, New York.

Buel, C. B., and W. H. Weston. 1947. Application of the mineral oil conservation method to making collection of fungus cultures. Amer. J. Bot. 34:555-61.

Caplin, S. M. 1959. Mineral oil overlay for conservation of plant tissue cultures. Amer. J. Bot. 46:324-29.

D'Amato, F. 1977. Cytogenetics of differentiation in tissue and cell cultures. *In* J. Reinert and Y. P. S. Bajaj (eds.). Applied and fundamental aspects of plant cell, tissue, and organ culture. Pp. 343–57. Springer-Verlag, Berlin.

Dickinson, D. B., M. J. Misch, and R. E. Drury. 1967. Dimethyl sulfoxide protects tightly coupled mitochondria from freezing damage. Science 156:1738–39.

Dix, P. J., and H. E. Street. 1976. Selection of plant cell lines with enhanced chilling resistance. Ann. Bot. 40:903–10.

Dougall, D. K., and D. F. Wetherell. 1974. Storage of wild carrot cultures in the frozen state. Cryobiology 11:410–15.

Finkle, B. J., Y. Sugawara, and A. Sakai. 1975. Freezing of carrot and tobacco suspension cultures. Plant Physiol. 56 (Suppl.) :80.

Grant, N. H., and H. E. Alburn. 1965. Fast reactions of ascorbic acid and hydrogen peroxide in ice: A presumptive early environment. Science 150:1589–90.

Heber, H. 1968. Freezing injury in relation to loss of enzyme activities and protection against freezing. Cryobiology 5:188–201.

Heber, U. 1958. Ursachen der Frostresistenz bei Winterweizen I. Die Bedeutung der Zucker für die Frostresistenz. Planta. 52:144–72.

Heinz, D. J., and G. W. P. Mee. 1971. Morphologic, cytogenetic, and enzymatic variation in *Saccharum* species hybrid clones derived from callus tissue. Amer. J. Bot. 58:257–62.

Henshaw, G. G. 1975. Technical aspects of tissue culture storage for genetic conservation. *In* Crop genetic resources for today and tomorrow. Intern. Biol. Programme. Pp. 349–57. Cambridge University Press, Cambridge.

Hollen, L. B., and L. M. Blakely. 1975. Effects of freezing on cell suspensions of *Haplopappus ravenii*. Plant Physiol. 56 (Suppl.) :39.

Latta, R. 1971. Preservation of suspension cultures of plant cells by freezing. Canad. J. Bot. 49:1253–54.

Levitt, J. 1956. Hardiness of plants. Academic Press, New York.

Levitt, J. 1962. A sulfhydryl-disulfide hypothesis of frost injury and resistance in plants. J. Theoret. Biol. 3:355–91.

Liu, Ming-Chin, and Wen Huei Chen. 1976. Tissue and cell culture as aids to sugarcane breeding. I. Creation of genetic variation through callus culture. Euphytica 25:393–403.

Lovelock, J. E. 1953. The haemolysis of human red blood-cells by freezing and thawing. Biochim. Biophys. Acta 10:414–26.

Lusena, C. V. 1965. Release of enzymes from rat liver mitochondria by freezing. Canad. J. Biochem. 43:1787–98.

Mazur, P. 1970. Cryobiology: The freezing of biochemical systems. Science 168:939–49.

Meryman, H. T. 1974. Freezing injury and its prevention in living cells. Ann. Rev. Biophys. Bioeng. 3:341–63.

Nag, K. K., and H. E. Street. 1973. Carrot embryogenesis from frozen cultured cells. Nature 245:270–72.

Nag, K. K., and H. E. Street. 1975. Freeze preservation of cultured plant cells. II. The freezing and thawing phases. Physiol. Plantarum 34:261–65.

Nath, T., and J. O. Anderson. 1975. Effect of freezing and freeze-drying on the viability and storage of *Lilium longiflorum* L. and *Zea mays* L. pollen. Cryobiology 12:81–88.

Nemner, M. W., and B. J. Luyet. 1954. Survival of dehydrated pea seedlings. Biodynamica 7:193-211.

Nitsch, J. P., and C. Nitsch. 1969. Haploid plants from pollen grains. Science 163:85-87.

Parker, J. 1959. Seasonal variations in sugars of conifers with some observations on cold resistance. For. Sci. 5:56-63.

Peterson, W. D., Jr., and C. S. Stulberg. 1964. Freeze preservation of cultured animal cells. Cryobiology 1:80-86.

Quatrano, R. S. 1968. Freeze-preservation of cultured flax cells utilizing dimethyl sulfoxide. Plant Physiol. 43:2057-61.

Sakai, A. 1960. The frost-hardening process of woody plants. VIII. Relation of polyhydric alcohols to frost hardiness. Low Temp. Sci. Ser. B. 18:15-22.

Sakai, A. 1971. Some factors contributing to the survival of rapidly cooled plant cells. Cryobiology 8:225-34.

Sakai, A., and S. Okada. 1971. Freezing resistance of conifers. Silvae Genetica 20:91-96.

Sakai, A., and Y. Sugawara. 1973. Survival of poplar callus at super low temperatures after cold acclimation. Plant & Cell Physiol. 14:1202-4.

Sakai, A., and S. Yoshida. 1968. The role of sugar and related compounds in variation and freezing resistance. Cryobiology 5:160-75.

Scholander, P. F., W. Flagg, R. J. Hock, and I. Irving. 1953. Studies on the physiology of frozen plants and animals in Arctic. J. Cell. Comp. Physiol. 42:1-56.

Seibert, M. 1976. Shoot initiation from carnation shoot apices frozen to -196°C. Science 191:1178-79.

Scherf, A. F. 1943. A method of maintaining *Phytomonas sepedonica* for long periods without transfer. Phytopathology 33:330-32.

Siminovitch, D., and D. R. Briggs. 1949. The chemistry of the living bark of the black locust tree in relation to frost hardiness. I. Seasonal variations in protein content. Arch. Biochem. 23:8-17.

Steponkus, P. L., and L. J. Bannier. 1971. Cold acclimation of plant tissue cultures. Cryobiology 8:386-87.

Steponkus, P. L., and F. O. Lanphear. 1967. Refinement of the triphenyl tetrazolium chloride method of determining cold injury. Plant Physiol. 42:1423-26.

Sugawara, Y., and A. Sakai. 1974. Survival of suspension-cultured sycamore cells cooled to the temperature of liquid nitrogen. Plant Physiol. 54:722-24.

Sun, C. N. 1958. The survival of excised pea seedlings after drying and freezing in liquid nitrogen. Bot. Gaz. 119:234-36.

Towill, L. E., and P. Mazur. 1975. Studies on the reduction of 2,3,5-triphenyl tetrazolium chloride as a viability assay for plant tissue cultures. Canad. J. Bot. 53:1097-1102.

Towill, L. E., and P. Mazur. 1976. Osmotic shrinkage as a factor in freezing injury in plant tissue cultures. Plant Physiol. 57:290-96.

Turmanov, I. I., R. G. Butenko, and I, V. Ogolevets. 1968. Application of isolated tissue culture technique for studying hardening process in plant cells. Fiziol. Rast. 15:749-56.

Tumanov, I. I., and T. L. Trunova. 1963. The first phase of frost hardening in the dark of winter plants on sugar solutions. Fiziol. Rast. 10:176-88.

White, P. R. 1963. The cultivation of animal and plant cells. Ronald Press, New York.

Whittingham, D. G., S. P. Leibo, and P. Mazur. 1972. Survival of mouse embryos frozen to 196°C and -269°C. Science 176:411-14.

Widholm, J. M. 1972. The use of fluorescein diacetate and phenosafranine for determining viability of cultured plant cells. Stain Technol. 47:189-94.

Zenkteler, M. 1971. In vitro production of haploid plants from pollen grains of *Atropa belladonna*. Experientia 27:1087.

YURY YU. GLEBA

Nonchromosomal Inheritance in Higher Plants as Studied by Somatic Cell Hybridization

35

EVIDENCE FOR BIPARENTAL TRANSMISSION OF EXTRANUCLEAR
GENETIC DETERMINANTS DURING SOMATIC CELL FUSION:
CYTOPLASMIC HYBRID (CYBRID) PLANTS FROM FUSED PROTOPLASTS

Hybrid plant production by means of protoplast fusion has been confirmed
in several laboratories. Carlson et al. (1972) were the first to demonstrate
that hybrid plants can be obtained by fusion of somatic cells. In the work of
Melchers and Labib (1974) it was shown for the first time that this new
technique can also be used to cross haploid plants, that is, plants that
cannot be crossed by sexual technique. Also, a generally applicable method
for hybrid selection based on the use of chlorophyll deficient mutants was
proposed in this work.

　　In our work we have tried to answer the important question of whether
genetically novel plants can be produced by protoplast fusion. Nonsexual
plant hybridization is of interest insofar as it enables the production of
plants that are genetically distinct from those obtained by a typical sexual
cross.

　　A reminder of the general characteristics of sexual crosses will facilitate
an understanding of the advantages of exploring nonsexual hybridization.
Firstly, sexual crosses are restricted to phylogenetically related species.
Attempts to overcome this restriction by protoplast fusion have so far been
unsuccessful in yielding whole hybrid plants (Melchers, 1977), but a wide

Yury Yu. Gleba, Institute of Botany, Academy of Sciences of Ukrainian S.S.R., Kiev,
USSR.

range of hybrids has been produced at the cellular level (Constabel et al., 1976; Kao, 1977; Wetter, 1977; Gosch and Reinert, 1976). Secondly, sexual hybridization produces, so to say, regular progeny from regular parents in a regular manner. This is the case because sexual hybridization involves special mechanisms governing transmission of parental genes to the progeny. Nuclear genes are inherited according to Mendelian laws, whereas extranuclear genes in most higher plants have strict uniparental maternal inheritance. It has become evident that eucaryotic plant cells contain several genophores in addition to a nucleus. A genetic approach to the problem of the role played by nonchromosomal genes in plants has been hampered by the fact that it is impossible to obtain, by sexual crosses, plants that are heterozygous for extranuclear genes. Recently it has been demonstrated that, during somatic hybridization of animal cells (Dawid et al., 1974), mitochondrial DNAs are inherited biparentally and that cytoplasmic heterozygosity can be revealed after a prolonged period of culturing the hybrid cell lines. The emphasis in our experiments has been on investigating somatic cell fusion in higher plants as a possible model system that can perform a different type of gene transmission: in particular, another mode of extranuclear inheritance.

Four different types of experiments have been performed. In all types, leaf protoplasts from two different mutants or species were fused and cultured en masse until the plants had regenerated. Presumed hybrid forms were then selected by means of genetic markers and analyzed. It should be mentioned that such a system inevitably includes not only the products of hybridization (protoplast fusion, nuclear fusion), but also products resulting from the segregation and elimination of genetic elements during subsequent multiplication of the fusion products.

In the first type of experiment (Gleba, 174; Gleba et al., 1975 a,b), we used plastome and genome chlorophyll deficiencies in tobacco, *Nicotiana tabacum* L., as markers that permit us to follow the fate of parental plastomes and genomes during hybridization. The first mutant used was described by von Wettstein and Eriksson (1964) and Wildman et al. (1973). It is a plastome mutant since the variegation has strict maternal inheritance, and cells with mixed plastids have been demonstrated. Variegation is also retained in androgenetic haploids (Nilsson-Tillgren and von Wettstein-Knowles, 1970; Gleba et al., 1974). Wong-Staal and Wildman (1973) have demonstrated that this mutation is correlated with an alteration in plastid DNA: using hybridization of DNAs from mutant and normal plastids, a nonhomologous region corresponding to about 1,000 base pairs was revealed. Reverse mutations are extremely rare or absent (Wildman et al., 1973). For experimental purposes, periclinal chimaeras of w-w-g type carrying only defective plastids in apical layers I and II were selected. We can be certain of obtaining genetically pure mutant material by using a suspension of protoplasts from the white marginal areas of the leaves. Starting from these protoplasts, thousands of white plantlets have been regenerated (Shvidkaya and Gleba, 1974) and green or variegated plants have never been found. In most later experiments, however, we used aseptically grown albino plants of homoplastidic line w-w-w- from plastome mutant (P^-) as a source of protoplasts. The second mutant, known as a "sulphur" tobacco, is genome semidominant (Burk and Menser, 1964). Seedlings carrying two mutant alleles (Su/Su) are yellow and die within 14 days of germination in the absence of sugar. Heterozygotes Su/su are yellow-green and viable. Direct and reverse mutations are frequent: both yellow and green spots are often seen on leaves of heterozygotes. We have regenerated several hundreds of plants out of protoplasts from sterile-grown yellow seedlings and they all were yellow (Shvidkaya and Gleba, 1974).

In hybridization experiments, a mixed suspension containing $1-3 \times 10^6$ of (su/su) (P^-) and $2-5 \times 10^5$ of (Su/Su) (P^+) protoplasts was used. The mixture was treated with polyethylene glycol (PEG) to induce fusion (method after Kao and Michayluk, 1974) and then cultured for plant regeneration (Gleba et al., 1974; Gleba, 1974). A similar untreated suspension was cultured as the control. Four independent experiments were performed (fig. 1).

The theoretical basis of this first type of experiment was as follows. Although the mutations belong to different genetic systems, they are both expressed as defects in photosynthesis. Consequently, in hybrid cells obtained by protoplast fusion, a genetic complementation must occur. Therefore, plants produced from hybrid cells must be capable of photosynthesis. Use of the plastome mutation on the one hand, and of the genome semidominant one on the other, permit identification of at least two classes of nonsexual hybridization. (A) If fusion of two (or more) protoplasts is accompanied by fusion of their nuclei, and if plastids are really transmitted biparentally through fusion, then after regeneration yellow-green variegated plants will arise. The yellow-green color indicates the presence of both defective and normal nuclear genomes, and variegation indicates the presence of both defective and normal plastomes. These plants can be defined as true hybrids. (B) If for any reason fusion of nuclei does not occur in fused cells, then the nuclei may segregate to different progeny cells. However, in such a case, integration of cytoplasm and the formation of organelle mixtures will take place, such that any regenerants obtained will be green (absence of defective genome), and probably variegated. The same

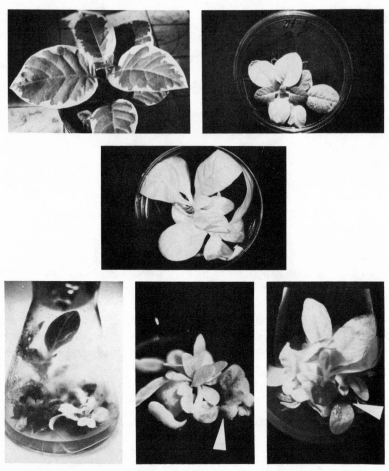

Fig. 1. Protoplast fusion and parasexual hybridization in *Nicotiana tabacum* L.: (upper left) periclinal chimaera of w-w-g type obtained from plastome chlorophyll deficient variegated mutant (*P⁻*), and used as protoplast source; (upper right) pure homoplastidic line of the same mutant; (middle) genome homozygote chlorophyll deficient (*Su/Su*) mutant used as second parent; (lower left) photosynthesizing plant regenerated from fused protoplasts, green variegated regenerant; (lower middle) the same, yellow-green variegated (arrow) regenerant; (lower right) the same, yellow-green regenerant.

kind of green variegated plants will also be found if genomal/chromosomal segregation takes place during the first mitoses in true hybrid cells. The term "cybrid" (Cocking, 1977) will be appropriate for such hybrid plants.

Electron microscopic analysis revealed a significant number of hetero-plasmic fusions in the cell population after PEG treatment (Sytnik et al., 1976). In contrast to the experiments of Giles (1973, 1974), we have not been able to detect any direct genetic complementation in the fused cells, even though thorough fluorescent microscopic analysis has been performed. In control experiments we have regenerated thousands of plantlets, but no photosynthesizing ones were ever found. In the experimental variant, organogenesis was rather poor. However, among the plants obtained from fused protoplasts, 35 photosynthesizing plantlets have been isolated. They were of three classes; yellow-green variegated, green variegated, and yellow-green, summarized in table 1. Most of the plants

TABLE 1

NUMBER OF PHOTOSYNTHESIZING PLANTS REGENERATED
FROM FUSED PROTOPLASTS

| EXPERIMENT NUMBER | PHOTOSYNTHESIZING PLANTS, NUMBER | |
	Experiment	Control
1	$P^+/P^-,Su/su$ 2	0
	$P^+/P^-,su$ 3	0
	$P^+,Su/su$ 9	0
2	$P^+/P^-,Su/su$ 1	0
	$P^+/P^-,su$ 3	0
	$P^+,Su/su$ 0	0
3	$P^+/P^-,Su/su$ 4	0
	$P^+/P^-,su$ 7	0
	$P^+,Su/su$ 1	0
4*	$P^+/P^-,Su/su$ 2	0
	$P^+/P^-,su$ 2	0
	$P^+,Su/su$ 0	0

* Albina plants (homoplastidic line w-w-w) of the plastome mutant were used as a source of protoplasts in this experiment.

demonstrated morphological abnormalities typical for polyploid and aneuploid forms, and were polyploid. One hexaploid was found. Different ploidy levels in the same plant as well as chromosome segregation were observed in many cases (table 2). Yellow-green variegated plants did not survive to the flowering stage. Green variegated plants were mostly triploid. Some of the variegated plants tested transmitted the character to F_1 progeny after pollination (table 3a).

For the following five reasons we conclude with certainty that the photosynthesizing plants obtained are nonsexual hybrids originating as a consequence of protoplast fusion. (1) Both alleles of each of two different genes located initially in the different donor plants are present in the yellow-green variegated regenerants. Two independent mutations would be required to explain this if nonsexual hybridization did not occur. (2) The presence of heteroplastidic cells in the starting material of the plastome mutant can be excluded in all four experiments. (3) Photosynthesizing plants were also found in an experiment where a pure homoplastidic line of plastome mutant plants had been used (experiment 4). (4) The variegation character was transferred to sexual progeny, which is impossible if plants

TABLE 2

VARIATIONS IN CHROMOSOME NUMBERS WITHIN THE DIFFERENT
CELLS AND DIFFERENT ROOTS OF THE PLANTS PRODUCED
OUT OF FUSED PROTOPLASTS

Plant No.	Root No.	Metaphases Counted	Chromosome Number (±2 chms)	Metaphases with Segregation
H1a	1	6	34, 44, 52, 58, 60, 69	
H1b*	1	2	90, 94	
H1c	1	3	52, 62, 64	
	2	11	60, 61, 63, 63, 65, 69, 69, 74 75, 76, 92	
	3	4	69, 69, 70, 75	
	4	4	54, 59, 62, 73	
	5	4	63, 65, 69, 90	
	6	6	64, 64, 66, 68, 70, 80	
H1d	1	1	59	1
	2	12	38, 39, 49, 51, 54, 55, 60, 61, 68, 76, 78, 106	
H1e	1	4	77, 85, 88, 92	
	2	5	33, 35, 42, 48, 50	
	3	4	59, 64, 72, 123	
H2a	1	9	68, 68, 71, 71, 72, 81, 88, 96, 105	4
	2	9	67, 72, 77, 78, 80, 82, 87, 89, 98	
	3	3	78, 81, 86	
	4	8	58, 64, 68, 69, 70, 75, 79, 97	
H2b	1	11	45, 54, 56, 59, 65, 70, 70, 73, 86, 87, 114	11
	2	1	90	
	3	15	57, 63, 66, 67, 70, 74, 75, 78, 79, 80, 82, 91, 84, 85, 85	
H4a*	1	2	85, 89	
H4b*	1	4	128, 152, 155, 156	

* Yellow-green variegated plants; others are green variegated.

TABLE 3

ANALYSIS OF THE F_1 PROGENY OF GREEN VARIEGATED
PLANTS REGENERATED OUT OF FUSED PROTOPLASTS

PLANT NUMBER AND TYPE OF CROSS PERFORMED	NUMBER OF SEEDLINGS WITH PHENOTYPES		
	Green	Variegated	White
A. Cross P^+, Su/Su + P^-, su/su			
(H1b) selfed	4	8	6
(H2a) selfed	4	0	2
H1c* × Samsun (green)	30	12	5
H2a* × Samsun (green)	28	0	2
H2b* × Samsun (green)	23	0	1
H4c* × Samsun (green)	49	19	27
B. Cross P^+_{chl}, C_{ms} + P^-_{chl}, $C_{wild\ type}$			
H10b × Samsun (green)	55	4	15
H11a × Samsun (green)	20	11	5

* Mother plants.

produced were chimaeras. (5) The control experiments in all cases lacked photosynthesizing plants among regenerants.

The results obtained indicate that: (1) through protoplast fusion, both parental plastomes are transferred to the nonsexual progeny; (2) at least in some cases, protoplast fusion is accompanied by nuclear fusion, yielding a true hybrid cell; and (3) in some cases, nuclear fusion does not necessarily follow fusion of the protoplasts, or genomal/chromosomal segregation takes place during mitotic divisions in the fusion product.

Thus, genetic variability of the somatic plant cells arising from the hybridization process as well as culture conditions permit us to synthesize new genetic combinations not possible before. Unfortunately, it was not possible to support genetic data with biochemical confirmation at the intraspecific level, so at this stage we had to terminate this type of experiment.

In the second type of experiment (Gleba et al., 1977), parent sources were the above-mentioned plastome mutant (homoplastidic line $P_{chl.}$) and a cytoplasmic male-sterile analog (cms-analog) of *N. tabacum* that carries cytoplasm from *N. debneyi*. Cms-analog was obtained by Grebenkin and Ternovsky (unpublished) through hybridization between *debneyi* as female and *tabacum*. The character has persisted unmodified through 13 successive backcrosses, and is therefore a non-Mendelian trait, apparently inherited as a cytoplasmic system contributed from *debneyi*, an Australian species. No gene-fertility restorers were found among different varieties of tobacco. The character is expressed as total absence of stamens, i.e. in flower unisexuality. As of the present, the following remain to be established: (1) whether the character is controlled by a single gene *and* single genophore, and (2) in which genophore(s) the cms-gene(s) are localized. Use of a tobacco analog with foreign cytoplasm enabled us to perform biochemical analysis of the hybridization process. Chen et al. (1975) analyzed Fraction I protein in similar cms-tobacco carrying cytoplasm of an unidentified Australian species using isoelectrofocusing of polypeptides. Differences in the large subunit composition were observed, which indicated that the large subunit of the analog was also contributed from Australian tobacco. We have been able to demonstrate the same for five cms-analogs (with cytoplasms from *N. debneyi*, three independently derived lines including the one used in this hybridization assay, *N. magalosiphon*, and *N. gossei* (Komarnitsky et al., in preparation). Thus, the crossing we performed was interspecific with regard to the extranuclear genetic determinants. Fraction I protein has been proved earlier to be an ideal marker for analysis in hybridization experiments, though it has not yet been rigidly confirmed that extranuclear genes controlling the large subunit are themselves localized in plastome. Maternal chlorophyll deficiency as mentioned above, was shown to be due to mutation in plastome.

So, three different extranuclear characters have been followed in this cross: plastome chlorophyll deficiency, cytoplasmic male sterility, and the polypeptide composition of the large subunit from Fraction I protein. This second type of experiment was based on the following considerations. Variegated plants obtained from this cross should be hybrids (cybrids) if: (1) variegation is maintained through sexual crosses, and is therefore not due to chimaericity, and (2) biochemical analysis reveals the presence of both *tabacum* and *debneyi* large subunits in Fraction I protein, which cannot be the case if variegation is a consequence of mutation rather than hybridization.

At present, we have performed 5 independent experiments, each of them up to plant regeneration. So far, 8 variegated plants (fig. 2) have been found in 3 different dishes. Two plants have reached flowering stage, and it was possible to demonstrate that variegation is transferred to sexual progeny (table 3b). From the leaves of all 8 variegated plants obtained from fusion, calluses have been initiated. In all cases after redifferentiation we have been able to find, in addition to pure green and white, some variegated plantlets. These findings can be used as a strong argument against the chimaeral nature of variegation produced in fusion experiments. In two plants that have been grown to flowering stage, variegation rapidly disappeared, seemingly because of more rapid multiplication of green plastids. Flowers of these two plants were completely male sterile, a fact that will be discussed later.

Biochemical analysis of the polypeptide composition of crystalline Fraction I protein from two variegated plants has been performed using isoelectrofocusing in PAA (methods after Chan et al., 1972; Kung et al., 1974). The large subunit of the protein, which is coded for by extranuclear genes, contains polypeptides of both *tabacum* and *debneyi* types (fig. 2). These experiments, though based on limited available material, provide clear evidence that by protoplast fusion it is possible to obtain plants heterozygous for extranuclear genetic determinants even at the interspecific level. We would like to reinforce that these plants cannot be produced by a normal sexual cross, and therefore hybrids obtained from somatic cell fusion probably have not existed before either in nature or in the breeders' test fields.

The third type of experiment dealt with crossing a plastome albino mutant of tobacco and another species, *N. debneyi*. These experiments were designed to duplicate the second type of cross, but under more complicated circumstances; namely, in the background presence of both *tabacum* and *debneyi* nuclei. In three experiments we have regenerated

Fig. 2. Cytoplasmic hybrids (cybrids) *Nicotiana tabacum + N. debneyi* obtained by protoplast fusion: (upper left) albino plants of plastome (*P⁻*) mutant used as protoplast source; (upper right) cytoplasmic male sterile analog of tobacco, carrying cytoplasm from *N. debneyi* and used as another protoplast source; (lower left) variegated plant regenerated from fused protoplasts; (lower right) analysis of Fraction I protein by isoelectric focusing; (a) plastome mutant of tobacco; (b) cms-analog with cytoplasm of *N. debneyi*; (c) variegated plant obtained from fused protoplasts; (d) variegated plant regenerated by tissue culture technique from vareigated leaf of the (c) plant; (e) green plant regenerated by tissue culture technique from variegated leaf of the (c) plant.

thousands of plantlets from fused protoplasts, but no variegated plant has so far been detected.

The fourth type of experiment is also in progress at the present time. It is a cross between a plastome albino mutant on one hand, and an extrachromosomal streptomycin-deficient mutant obtained by Maliga et al. (1973, 1975) on the other. The purpose for this cross will be clarified in the next chapter.

Results presented here are in good agreement with those obtained by other workers. Analysis of the polypeptide composition of Fraction I protein from the first parasexual hybrid derived from the fusion of protoplasts of *Nicotiana glauca* and *N. langsdorffii* (Carlson et al., 1972) have shown the presence of only *glauca* large subunits (Kung et al., 1975). Recently, however, Smith et al. (1976) have repeated this type of fusion experiments and regenerated 23 mature hybrid plants with the same species combination. The polypeptide composition of the hybrid's Fraction I protein has been analyzed in Prof. S. Wildman's laboratory as well (S.G. Wildman, personal communication; Chen et al., 1976). From the total of 20 plants analyzed, 12 had only *langsdorffii* plastome expressed, whereas 7 had only *glauca* plastome expressed. In one case, the plant possessed both *glauca* and *langsdorffii* large subunit polypeptides. It should be stressed that our experimental system is more efficient in investigating the problems of extrachromosomal genetics because it is based on the use of extrachromosomal markers (plastome chlorophyll deficiency) for hybrid (cybrid) detection. There is one more important experiment on cybrid induction by protoplast fusion that has been published recently. In the work of Belliard et al. (1977), two varieties of tobacco have been used, namely cytoplasmic male sterile tobacco with cytoplasm from *N. debneyi* and sessile leaves (monogenic Mendelian character), and normal varieties Samsun and Xanthi having petiolated leaves. In two nonsexual crosses, the authors were able to find as many as ten nuclear hybrids (as determined by leaf morphology) and several dozens of plants with altered flower morphology and different degrees of expression of the male sterility character. The latter is best explained in terms of biparental transmission of cytoplasmic genes.

GENETIC ANALYSIS BY MEANS OF NONSEXUAL HYBRIDIZATION

Hybridization by protoplast fusion can be used like conventional sexual crossing to study the nature and organization of genetic material in plant cells. Two possibilities are being tested in our laboratory. Both involve the reinstallation of heterozygosity of material by protoplast fusion in cases where sexual crossing is ineffective or impossible.

In many experiments, particularly on mutagenesis, one is not able to analyze a major part of the mutants obtained because they are lethal at the whole organism level and therefore cannot be maintained and/or analyzed by crossing. Fusion of protoplasts from these mutant plants with protoplasts from known mutants can help by reinstalling heterozygosity: (1) to characterize mutants, and (2) to keep mutations in question alive and available for further studies by sexual crosses. The following experiments illustrate. Three spontaneously arising chlorophyll-deficient spots on leaves of tobacco, var. Samsun, and one periclinal chimaera of unknown origin having a chlorophyll deficiency in the apical layer III (g-g-w) of var. Xanthi, have been regenerated by tissue culture techniques into albino plantlets. Protoplasts from leaves of these plantlets were extracted and hybridized with mesophyll protoplasts of sulphur genome semidominant homozygote (*Su/Su*). Plantlets have been regenerated from protoplasts and characterized as photosynthesizing or not. Table 4 is a summary of the results, from which one can conclude that: (1) mutations 1–3 are recessive, localized in the genome, and complement with gene *Su*, and (2) mutation 4 (out of a periclinal chimaera) is an extrachromosomal mutation, which does not complement with gene *Su*. Apart from these findings, we now have these mutations in the form of normal plants available for further experiments.

TABLE 4

PHENOTYPICAL CHARACTERISTICS AND NUMBER OF
PHOTOSYNTHESIZING PLANTS PRODUCED FROM FUSION OF
SU/SU PROTOPLASTS WITH PROTOPLASTS FROM NEWLY ISOLATED
CHLOROPHYLL DEFICIENT TOBACCO MUTANTS

Experiment Number	Plant Used	Phenotypical Characteristics	Number of Plants Produced
1	"1"	Yellow-green	13
2	"2"	Yellow-green	8
3	"3"	Yellow-green	3
4	"4"	Yellow-green variegated	3
		Green variegated	7
		Yellow-green	2

Analysis of the Extrachromosomal Genes Using Cosegregation Tests

Once we can establish heterozygosity for extranuclear genetic determinants, we are also able to follow the segregation process of cytoplasmic genes during multiplication of the cells originating from double heterozygotes, i.e. heterozygotes for two different cytoplasmic genes. If two characters that are coded for by extranuclear genes segregate independently, they are supposed to be controlled by different genophores and vice versa. This cosegregation analysis provides an instrument to characterize gene linkage groups outside the nucleus. We have used cybrids *N. tabacum* + *N. debneyi*, which are heterozygous for genes controlling plastome chlorophyll deficiency and polypeptide composition of the large subunit from Fraction I protein, to attack this question. Calluses were originated from variegated leaves of cybrids and, after two months of unorganized growth, plantlets were regenerated. Pure green and variegated regenerants were used for biochemical analysis of the Fraction I protein composition. The data are given in fig. 2. Variegated leaves again contained large subunit polypeptides of both species, whereas pure green (dechimaerized) plantlets had only *debneyi* type polypeptides expressed. The evident cosegregation of the two characters further confirms a prediction that polypeptides of the large subunit of Fraction I protein are coded for by a plastome. As far as we know, these data provide the first direct evidence of genetical linkage of extrachromosomal genes in higher plants.

Another important observation from studies with *tabacum* + *debneyi* cybrids is that they are completely male sterile. This can be explained as a consequence of the selection in favor of genophores that code for the male sterility character, and in this case, the genophore in question is obviously not a plastome, or more precisely, not only a plastome. Another explanation is also possible: it may be that in the expression of the male sterility character, which is a product of complex morphogenic events itself, some sort of dominance takes place in the heterozygous population of genophores.

More hybrid plants are required to prove these inferences definitely, and experiments are in progress. We hope that experiments of the types reported here will finally permit us to assign each of the extrachromosomal genes involved (coding for chlorophyll deficiency, male sterility, polypeptide composition of the large subunit of Fraction I protein, and streptomycin resistance) to one or more discrete genophores.

ACKNOWLEDGMENTS

I am grateful to Prof. William R. Sharp for the invitation to participate in the Colloquium. I thank Profs. Raisa G. Butenko and Konstantin M. Sytnik for scientific training, and Drs. I. Komarnitsky, A. Burgutin, N. Piven, L. Frolova, and L. Shvidkaya for their assistance and cooperation. Suggestions and criticism on the manuscript were made by Prof. G. Melchers and Dr. N. Umiel. Dr. William L. Cairns corrected the English version of the manuscript.

LITERATURE CITED

Belliard, G., G. Pelletier, and M. Ferault. 1977. Fusion de protoplastes de *Nicotiana tabacum* à cytoplasmes différents: Étude des hybrides cytoplasmiques néo-formés. C. R. Acad. Sci. Paris 284D:749–52.

Burk, L. G., and H. A. Menser. 1964. Dominant aurea mutation in tobacco. Tob. Sci. 8:101–4.

Carlson, P. S., H. H. Smith, and R. D. Dearing. 1972. Parasexual interspecific plant hybridization. Proc. Nat. Acad. Sci. USA 69:2292–94.

Chan, P. H., K. Sakano, and S. G. Wildman. 1972. Crystalline Fraction 1 protein: Preparation in large yield. Science 176:1145–46.

Chen, K., S. Johal, and S. G. Wildman. 1976. Phenotypic markers for chloroplast DNA genes in higher plants and their use in biochemical genetics. *In* J. H. Weil and L. Bogorad (eds.), Nucleic acids and protein synthesis in plants. Plenum Press, Strasburg (in press).

Chen, K., S. D. Kung, J. C. Gray, and S. G. Wildman. 1975. Polypeptide composition of Fraction 1 protein from *Nicotiana glauca* and from cultivars of *Nicotiana tabacum*, including a male sterile line. Biochem. Genet. 13:771–78.

Cocking, E. C. 1977. Uptake of foreign genetic material by plant protoplasts. Intern. Rev. Cytol. 48:323–43.

Constabel, F., G. Weber, J. W. Kirkpatrick, and K. Pahl. 1976. Cell division of intergeneric protoplast fusion products. Z. Pflanzenphysiol. 79:1–7.

Dawid, I. B., I. Horak, and H. G. Coon. 1974. The use of hybrid somatic cells as an approach to mitochondrial genetics in animals. Genetics 78:459–71.

Giles, K. L. 1973. Attempts to demonstrate genetic complementation by the technique of protoplast fusion. Colioques Intern. C. N. R. S. 212:485–95.

Giles, K. L. 1974. Complementation by protoplast fusion using mutant strains of maize. Plant and Cell Physiol. 15:281–85.

Gleba, Y. Y. 1974. Culture of isolated protoplasts of tobacco, *Nicotiana tabacum L.*, as model for genetic engineering. Thesis. Kiev, USSR.

Gleba, Y. Y., R. G. Butenko, and K. M. Sytnik. 1975a. Protoplast fusion and parasexual hybridization in *Nicotiana tabacum L.* Dokl. Acad. Nauk USSR 221:1196–98.

Gleba, Y. Y., N. M. Piven, I. K. Komarnitsky, and K. M. Sytnik. 1977. Cytoplasmic hybrids (cybrids) *Nicotiana tabacum* + *N. debneyi* obtained by protoplast fusion. Dokl. Acad. Nauk USSR (in press).

Gleba, Y. Y., L. G. Shvidkaya, R. G. Butenko, and K. M. Sytnik. 1974. Culture of isolated protoplasts. Sov. Plant Physiol. 21:598–605.

Gleba, Y. Y., K. M. Sytnik, and R. G. Butenko. 1975b. Genetic consequences of protoplast fusion in *Nicotiana tabacum L.* Abstr. Fourth Int. Symp. Yeast and Other Protopl. (abstr. 83). Nottingham, England.

Gosch, G., and J. Reinert. 1976. Nuclear fusion in intergeneric heterokaryocytes and subsequent mitosis of hybrid nuclei. Naturwiss. 11:534.

Kao, K. N. 1977. Chromosomal behaviour in somatic hybrids of soybean—*Nicotiana glauca*. Molec. Gen. Genet. 150:225–30.

Kao, K. N., and M. R. Michayluk. 1974. A method for high-frequency intergeneric fusion of plant protoplasts. Planta 115:277–94.

Kung, S. D., J. C. Gray, S. G. Wildman, and P. S. Carlson. 1975. Polypeptide composition of Fraction 1 protein from parasexual hybrid plants in the genus *Nicotiana*. Science 187:353–55.

Kung, S. D., K. Sakano, and S. G. Wildman. 1974. Multiple peptide composition of the large and small subunits of *Nicotiana tabacum* Fraction I protein ascertained by fingerprinting and electrofocusing. Biochim. Biophys. Acta 365:138–47.

Maliga, P., A. Sz.-Breznovits, and L. Marton. 1973. Streptomycin resistant plants from callus culture of haploid tobacco. Nature New Biol. 244:29–30.

Maliga, P., A. Sz.-Breznovits, L. Marton, and F. Joo. 1975. Non-Mendelian streptomycin-resistant tobacco mutant with altered chloroplasts and mitochondria. Nature 255:401–2.

Melchers, G. 1977. Microbial techniques in somatic hybridization by fusion of protoplasts. *In* B. R. Brinkley and Keith R. Porter (eds.), International cell biology 1976–1977. Rockefeller University Press, Boston (in press).

Melchers, G., and G. Labib. 1974. Somatic hybridisation of plants by fusion of protoplasts. I. Molec. Gen. Genet. 135:277–94.

Melchers, G., and M. D. Sacristan. 1977. Somatic hybridisation of plants by fusion of protoplasts. II. *In* La culture des tissus et des cellules des végétaux. Pp. 169–77. Masson, Paris.

Nilsson-Tillgren, T., and P. von Wettstein-Knowles. 1970. When is the male plastome eliminated? Nature 227:1265–66.

Shvidkaya, L. G., and Y. Y. Gleba. 1974. Plating of mesophyll protoplasts isolated from different photosynthetic mutants of *Nicotiana tabacum L.* Abstr. 3rd Intern. Congr. Plant Tissue Cell Culture (abstr. 195) Leicester, England.

Smith, H. H., K. N. Kao, and N. C. Combatti. 1976. Interspecific hybridization by protoplast fusion in *Nicotiana*. J. Hered. 67:123–28.

Sytnik, K. M., V. A. Sidorov, N. V. Belitser, and Y. Y. Gleba. 1976. Ultrastructural study of the initial stages of protoplast fusion in tobacco, *Nicotiana tabacum L.* Dokl. Acad. Nauk USSR 226:945–46.

von Wettstein, D., and G. Eriksson. 1964. The genetics of chloroplasts. *In* Genetics today. Proc. Eleventh Intern. Congr. Genet., The Hague, 1963. 3:591–612. Pergamon Press, Oxford.

Wetter, L. R. 1977. Isoenzyme patterns in soybean-*Nicotiana* somatic hybrid cell lines. Molec. Gen. Genet. 150:231–35.

Wildman, S. G., C. Lu-Liao, and F. Wong-Staal. 1973. Maternal inheritance, cytology, and macromolecular composition of defective chloroplasts in a variegated mutant of *Nicotiana tabacum*. Planta 113:293–312.

Wong-Staal, F., and S. G. Wildman. 1973. Identification of a mutation in chloroplast DNA correlated with formation of defective chloroplasts in a variegated mutant of *Nicotiana tabacum*. Planta 113:313–26.

Subprotoplasts and Organelle Transplantation

36

INTRODUCTION

Plant cells are complex biological systems in which the organelles are integrated, forming a functional unit. The metabolism is steered by the genetic information of several genophores that are located in the various cell organelles.

Recombination of the genetic information of higher plants is possible only by sexual events. The variety of combinations is limited by sexual incompatibilities, apomixis, sterility of hybrids, and by the fact that mostly the male parent contributes the nucleus only to his offspring. The sexual barriers possibly may be bypassed in experiments on protoplast fusion, organelle transplantation, and gene transfer. This paper will be concerned with transfer of cell organelles.

Before discussing experiments, features and hypotheses on this branch of parasexual hybridization, it seems reasonable to consider some information on extrakaryotic inheritance in plants. No complete survey of our knowledge will be presented here. Papers on this topic have been reviewed and discussed earlier (Laughnan and Gabay, 1975; Sager, 1975; Tilney-Bassett, 1975; Wildman et al., 1975). Arnold (1975) has reviewed papers on plastid genetics and, recently, a review article on mitochondrial genetics has been published by Michaelis (1976). Furthermore, nonchromosomal inheritance in somatic cell hybridization is the topic of the paper by Gleba

Horst Binding, Institute of Botany, University of Kiel, Kiel, West Germany.

et al. in this volume. My intention in the first part of this paper is to demonstrate the complex and complicated interrelationships between the cell organelles and how far away we are from a complete understanding of the genetic information contributed by the plastids and mitochondria to the functions of the plant cell.

SOME GENETIC ASPECTS OF CELL ORGANELLES

Early information on the participation of the plasmone in the differentiation of plants, which emerged from the analysis of reciprocal crosses, has been compiled by Correns (1937). Only the papers on reciprocal hybridization in the Funariaceae by von Wettstein (1928) and in *Saxifraga* by Melchers (1935) shall be mentioned here. Detailed investigations on reciprocal hybrids in *Epilobium* (Michaelis, 1965a) and in *Oenothera* (Stubbe, 1971) revealed a lot of evidence for a sensible balance between the genome and plasmone. Experiments on hybridization and mutation have shown that most of the genetic information in plant cells is located in the nucleus. Relatively few functions are known that are steered by plastid and mitochondrial genes. Almost nothing is known on the genetic potencies of the other cell organelles.

The Plastome

In the late nineteenth century, plastids have been recognized as self-replicating cell organelles (Schmitz, 1882). The occurrence of autonomous genetic information in plastids has been concluded from observations that altered plastid characters are stable in cells containing two types of plastids (van Wisselingh, 1920) and, most convincingly that the differentiation pattern of plastids was retained even after recurrent combination with foreign genetic background (Renner, 1934). The survival of chloroplasts outside the cell appeared to be possible but limited (Giles and Sarafis, 1971, 1972).

Papers on biochemical genetics of plastids have been reviewed by Arnold (1975). The information for most of the functions of the plastids is located in the nucleus. The corresponding mRNA is translated by cytoplasmic ribosomes (Harris et al., 1973; Wang et al., 1974) or by plastid ribosomes (Hovenkamp-Obbema and Stegwee, 1974; Jennings and Ohad, 1973; Preddie et al., 1973). Synthesis of RNA in higher plant plastids has been found by Caritt and Eisenstadt (1973a,b) and Detchon and Possingham (1973). The cistrons for plastid rRNA are located in the plastid genome. The formation of ribosomal proteins of the plastids appeared to be directed by plastid and nuclear genes as was concluded from the existence of Mendelian and non-Mendelian erythromycin resistance of the chloroplasts (Davidson et al., 1974). The synthesis of lamellar proteins is performed by plastid ribosomes (Herrmann, 1972; Herrmann et al., 1974). It has been assumed that most of the non–Mendelian chlorophyll deficiencies are attributable to defects in the systems that form the lamellar structure. Plastids are presumed to be the sites of tryptophan synthesis. Grosse (1976) detected high amounts of tryptophan-synthesizing enzymes in isolated ethioplasts of pea. He suggested that the mutants with an altered anthranilate synthetase (Widholm, 1972a,b) are located in the plastome. Maliga et al. (1973) proved the maternal inheritance of streptomycin resistance in tobacco. The discrimination between the different extrakaryotic genophores was not possible.

A great deal of information has been gained on the Fraction I protein of *Nicotiana* that represents the main fraction of soluble chloroplast proteins (Wildman et al., 1975). This protein fraction consists of large subunits that are coded by the chloroplast DNA and small subunits coded by nuclear genes. The proteins of different higher plant species have different electrophoretic behavior (Börner et al., 1976).

Observations on the induction of plastid mutations by nuclear genes (Rhoades, 1943; Michaelis, 1968; Potrykus, 1970; Redei, 1973; Redei and Plurad, 1973; Epp, 1973) may be taken as criteria that a nuclear coded enzyme is affected that is involved in repair mechanisms of the plastid genophores.

The Chondriome

Most detailed information on mitochondrial genetics has been gained in yeast (Michaelis, 1976). It has been possible to map loci for drug resistance and for recombination on the mitochondrial genome by using recombinants and extrakaryotic petite mutants (Michaelis et al., 1973). The mitochondrial genophores of higher plants are known to bear the genes for the mt-rRNA species. Furthermore, it is suggested that they contain their own tRNAs.

Evidence on mutation in plant mitochondria has been reported for an acriflavin-induced mutant of *Chlamydomonas* that forms small colonies. This character is inherited in a biparental non–Mendelian manner reminiscent of the petite mutants in yeast (Alexander et al., 1974). A series of investigations points to altered mitochondria as the basis for cytoplasmic male sterility.

Cytoplasmic Male Sterility

Michaelis (1931) detected the influence of special plasmone-genome combinations in *Epilobium* reducing the pollen fertility. The phenomenon of male sterility in combinations of a special plasmotype with special nuclear genes has been reported later in a lot of other plants (Frankel and Galun, 1977). The character was obtained by interspecific hybridization or arose spontaneously by mutation in the plasmone. Male fertility can be restored in most cases by a single dominant nuclear gene.

Male sterility has developed as an important factor in hybrid production in crops. The applicability of this breeding method bears some limitations. One of these limitations is instability of the sterility which has been reported by Barnes and Garboucheva (1973). Other disadvantages of male sterile clones are based on pleiotropic effects. Doig et al. (1975) observed preharvest sprouting of wheat induced by the cytoplasmic male sterility derived from *Triticum timopheevi*. The cytoplasm of *Aegilops* in combination with the wheat genome induced abnormal seed formation leading to twins and Haploids (Tsunewaki et al., 1974).

A widely known pleiotropic effect is the connection of male sterility with sensitivity to *Helminthosporium maydis* toxin in corn. A large number of independently arisen cytoplasmic male sterile mutants has been isolated, but it has not been possible to break the linkage. It must be concluded that male sterility and resistance are expressions of one principle that is probably located in the mitochondria. This suspicion is supported by the investigations of Peterson et al. (1975). They found that the activities of the inner membrane of the mitochondria were altered by the association of the *Helminthosporium* toxin. Gregory et al. (1977) isolated mitochondria from male sterile and male fertile corn and subjected them to digitonin treatment. Marked differences in the sensitivity to digitonin were found by measuring the respiratory activities.

Mitochondrial Complementation and Heterosis

Increased vigor of hybrid maize and wheat has been attributed to complementation of genetically different mitochondria (McDaniel and Sarkissian, 1966; 1968; Sarkissian and Srivastava, 1967; McDaniel, 1969; 1972; Zobl et al., 1972; Sage and Hobson, 1973). The authors suggested to take the activities *in vitro* and the values of complementation of the mitochondria of different clones as criteria for plant breeding. Sarkissian (1972) and Ellis et al. (1973), however, called the reliability of this method into question. Hanson et al. (1975) established that the respiration rates of mitochondria from vigorous plants were higher than the values from weaker plants, but that these differences depended on nuclear genes.

Conclusions with Respect to Organelle Transfer

Cytoplasmically inherited characters play an increasing role in plant breeding. But there are still some handicaps that restrict the applicability of breeding methods involving cytoplasmic characters. A relatively small number of genes is known to be located on the genophores of chloroplasts and mitochondria. Most of the functions of the cell organelles seem to be directed by nuclear genes. The collaboration between the information provided by the genophores of the cell organelles is in a balance that is impaired in interspecific hybrid combinations. Hence, important objectives in experiments on organelle transfer are analyses of extrakaryotic factors with respect to (1) the biochemical function in the cell metabolism, (2) the gene localization and (3) the fate of cell organelles in nuclear or cytoplasmic hybrids. Knowledge of these matters will make possible the selection of a suitable experimental approach to obtain a desired organelle transfer product.

In the second part of this paper, parasexual transfer of cell organelles and chromosomes will be discussed and compared to phenomena that are known from sexual hybridization.

ORGANELLE TRANSFER

Uptake of Cell Organelles

Transfer by cell fusion. The integration of foreign cell organelles arises spontaneously as a consequence of organelle segregation and elimination in sexual and parasexual hybrids. From this point of view, protoplast fusion can be taken as a tool of organelle transfer.

Transfer by subprotoplast fusion. Subprotoplasts are particles that are normally surrounded by a plasmalemma and contain only parts of the protoplast of a cell. There are found subprotoplasts with or without the nucleus (fig. 1) and with or without plastids. Subprotoplasts, therefore, can be taken as carriers of selected cell organelles (Binding, 1976; Binding and Kollmann, 1976).

Subprotoplasts can be obtained easily in large amounts from nearly or fully ripe berries of some solanaceous species. They are also formed during plasmolysis, especially in prosenchymatic cells and if ionic solutions are used for plasmolysis (Binding, 1966). Subprotoplasts of this type have been isolated from pith tissue of etiolated stems from shoot cultures of *Petunia*, tobacco, and *Solanum* species by the dissection method of Klerker (1892). Furthermore, subprotoplasts are formed by budding in protoplast and cell cultures. Subprotoplast formation during the incubation of protoplasts in fusion-inducing agents can be obtained if the sediment of agglutinated protoplasts is stirred. The protoplasts stick together so strongly that parts of one protoplast are extruded while they remain in contact with another protoplast (fig. 2).

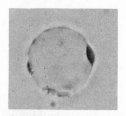

Figs. 1 and 2. (1) Subprotoplasts of dihaploid *Solanum tuberosum* containing three chloroplasts but no nucleus. The subprotoplast has been isolated from a stem of a shoot culture after plasmolysis by 0.25 m $Ca(NO_3)_2$. (2) Protoplasts of *Vicia faba* (light) and *Petunia* (dark) agglutinated. A *Petunia* protoplast was sticking to the *Vicia* protoplast but drifted apart leaving behind a small portion containing one chloroplast (next to the *Vicia* nucleus).

Subprotoplasts fuse under the same conditions as protoplasts, but they are sometimes less stable than protoplasts. That is the reason why we use a fusion method that is slightly modified from the PEG/high pH/calcium method (Keller and Melchers, 1973; Kao and Michayluk, 1974; Binding, 1974; Schieder, 1974). The protoplasts and subprotoplasts are sedimented in the enzyme incubation mixture, resuspended in 0.2–0.3 M $Ca(NO_3)_2$ solution at pH 6, poured together and centrifuged, again. The protoplast pellet is resuspended in 0.2 ml of the solution. Then, 0.5 ml of a PEG solution (40–50%) of a pH value of 10 adjusted by KOH is added gently, lifting up the protoplast-subprotoplast suspension. The protoplasts and subprotoplasts settle down onto the PEG solution forming a dense layer. After an incubation time of about 30 min, 10 ml of the $Ca(NO_3)_2$ solution or of culture media are added. Fusion is allowed to continue for another 30 min.

Organelle transplantation. The classical method for the transfer of cell organelles into foreign cells is organelle transplantation. Cell organelles are isolated from donor cells and implanted into the receptor protoplast. Organelle uptake is induced by the same agents as protoplast fusion. Transplantation of cell organelles of higher plants has been achieved in a few laboratories. Recent reviews have been prepared by Potrykus and Lörz (1976) and Giles (1976).

Implantation of plant nuclei into plant protoplasts has been obtained by Potrykus and Hoffmann (1973) and Binding (1976). Chromosome transplantation has not been reported. Isolated chloroplasts have been transplanted into protoplasts of the same species in *Petunia* (Potrykus, 1973). Algal chloroplasts have been incorporated into carrot protoplasts (Bonnet and Eriksson, 1974), into fungal protoplasts (Vasil and Giles, 1975), and into mouse cells (Nass, 1969). Chloroplast uptake in *Nicotiana*, which has been described by Carlson (1973), has been drawn into question by Potrykus (1973).

Transplantation of plant mitochondria has not been investigated, though the experience on the function of mitochondria *in vitro* suggests interesting results to be awaited from mitochondria transplantation.

Integration of Cell Organelles

Complete integration of the transferred cell organelle into the plant cell is necessary for the achievement of heritable recombination. "Complete integration" means that the transferred organelle replicates within the receptor organism. Up to now clearcut proof of the replication of transplanted organelles and their transmission to the daughter cells in higher plants is not available in the literature. Some ideas on the consequences of the combination of heterogenic organelles in plant cells have been revealed by investigations in sexual hybridization and fusion of protoplasts and subprotoplasts. Integration and segregation of heterogenic cell organelles depend on several factors that will be mentioned in the following part.

Segregation and Elimination of Cell Organelles in Hybrid Cells

Physical and statistical factors. Michaelis (1965b) followed up the fate of different plastid types in mixed cells of *Epilobium*. He established that, in some cases, the segregation of the plastids in the daughter cells followed statistical principles. A segregation was accelerated if plastids of one parent or sister plastids remained in near neighborhood during cytokinesis. A similar situation may have led to the segregation of plastids in the somatic interspecific hybrids of *Nicotiana* (Smith et al., 1976). The unequal distribution of the plastids of *Petunia hybrida* and *Vicia faba* in a bicellular regenerant of a fusion product (fig. 3) also may be taken as evidence for the beginning of segregation at random.

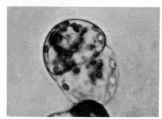

Fig. 3.   *Vicia faba* + *Petunia hybrida* hybrid cell divided once after three days of culture. The upper cell contained about 30 *Vicia* chloroplasts (dark), the lower cell only 9.

Destruction of cell organelles. Direct fast elimination has not been observed in parasexual recombinants. Elimination of the chromosomes of the male parent in the early development of plant embryos was described for the first time by Pace (1913). Recent investigations of this phenomenon have been discussed by Jensen in session 2 of this colloquium. Genophores of plastids normally degrade in zygotes of *Chlamydomonas*. Following the idea of Davies (1974), the selective elimination of the genophores of one parent in the early development of the hybrid may be explained by specific restriction enzymes.

Loss of cell organelles and chromosomes. Information on the fate of chromosomes in somatic hybrids has been gained in the soybean + tobacco system (Kao, 1976; 1977) and recently in a system of *Vicia faba* + *Petunia hybrida* (Binding and Nehls, 1978). Asynchrony of the division of cell organelles may lead to the loss of one organelle type. In fusion products of *Vicia* + *Petunia*, for example, frequently no nuclear fusion took place. In these cells, sometimes the nucleus of *Petunia* divided but the *Vicia* nucleus remained undivided. As a consequence, one daughter cell had lost the *Vicia* nucleus at this first cell division.

Asynchrony of chromosome replication is probably one of the mechanisms that caused the loss of chromosomes during mitosis of hybrid nuclei.

Some of the features observed by Kao in the soybean + tobacco system are explained by impaired chromosome replication. Similar observations have been made in the *Vicia* + *Petunia* hybrids. Single *Vicia* chromosomes in nuclei of *Petunia* are often heteropycnotic (fig. 4). Abnormal spindle formation (fig. 5) or failure of chromosomes to attach to the spindle apparatus are also important factors in chromosome elimination. Eliminated chromosomes are either left in the cytoplasm and degenerate or form micronuclei (fig. 6). Polyploid and aneuploid cells are also formed frequently in somatic hybrid tissue of *Vicia* + *Petunia* (fig. 7). This phenomenon has also been observed in somatic hybrids of tobacco (Melchers and Sacristán, 1977) and *Datura* (Schieder, 1977).

Up till now chloroplast elimination in somatic hybrids, which is probably under physiological control, has not been reported, although it is well known in plants with mixed plastid population. Hagemann and Scholze (1974), for example, observed retarded division of one chloroplast type and consequently elimination of this chloroplast type. The formation of large starch grains by tobacco chloroplasts in cells that contained only one or a few tobacco plastids but a higher number of *Petunia* chloroplasts may indicate that the tobacco plastids were affected by a factor of *Petunia* (Binding, 1976).

Organelle and chromosome elimination may be due partly to different replication rates inherent in their own replication systems. It is also very

Figs. 4 and 5.   (4) Two nuclei of a *Vicia faba* + *Petunia hybrida* hybrid callus each containing one heterochromatic *Vicia* chromosome. (5) Three-polar spindle in a *Vicia faba* + *Petunia hybrida* hybrid callus. Only *Vicia* chromosomes could be detected in this cell.

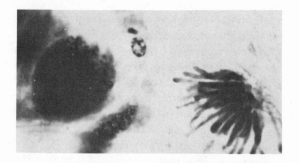

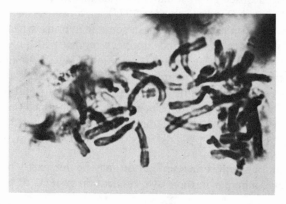

Figs. 6 and 7.   (6) A micronucleus between a normal nucleus (right) and a telophase pole of another—tetraploid—cell (left) of a *Vicia faba* + *Petunia hybrida* hybrid callus. The discernible chromosomes and the large nucleus are of *Vicia*-type. (7) Metaphase plate of a cell containing about 24 *Vicia* chromosomes and at least one fragment. The cell was found in a 50-day-old callus of a *Vicia faba* + *Petunia hybrida* hybrid cell.

likely as concluded from the knowledge on the implications of the genome and plasmone, that the combination of unrelated cell organelles produces impairment of the balance of cell metabolism, or that some kind of incompatibility hinders the normal development of the transferred organelles.

Some observations are probably explained by incompatibility of the newly combined systems. *Petunia* + tomato fusion hybrids were able to divide only if a small subprotoplast of tomato had been fused to a *Petunia* protoplast. Tobacco + tomato protoplasts never did undergo mitosis (Binding, 1976). Regeneration in somatic hybridization experiments in the combinations of tobacco + *Petunia* (Melchers, 1976), *Datura innoxia* + *Petunia hybrida* and *Datura innoxia* + *Nicotiana silvestris* (Schieder, 1977) failed, too. The origin of callus with hybrid character with respect to isozyme pattern but with uniparental chromosomes from a *Petunia* +

Parthenocissus hybridization experiment is also explained by some kind of organelle segregation (Power et al., 1975).

DISCUSSION AND OUTLOOK ON THE APPLICABILITY
OF ORGANELLE TRANSFER

It has been shown that several methods are available by which organelle transfer can be obtained: sexual and somatic hybridization and subsequent organelle elimination, subprotoplast fusion and organelle transplantation. Sexual hybridization is limited to sexually compatible plants. The advantage of somatic hybridization by protoplast fusion is that cell organelles of unrelated cells are easily and reliably combined in the fusion body. Organelle elimination may end up in the formation of cells that contain a complete set of organelles of one parent except that one organelle type is replaced by a foreign one. The end product of organelle elimination cannot be predicted. Organelle segregation, competition of cell organelles, incompatibility, and impairment of the balance in cell metabolism are factors that are hardly controlled by experimental conditions.

The ideal method for organelle transfer might be the transplantation of isolated organelles into plant protoplasts. The modest success in organelle transplantation experiments may be due to experimental problems that will probably be overcome in the near future. In the meantime, subprotoplast fusion seems to be a useful alternative (Binding, 1976).

Organelle transfer is supposed to be a suitable method for the investigation of the inheritance and expression of extrakaryotic genes.

The applicability of organelle transfer to plant breeding is still open to speculation. Transfer of complete nuclei in related plants can be obtained by cell fusion. This type of organelle transfer is mainly useful in breeding of highly heterozygous crops, as for example, of potato (Melchers, personal communication; Binding et al., 1978).

Transfer of nuclei in unrelated plants may be reasonable if subsequent chromosome elimination and rearrangements take place ending up in cell lines that contain the whole genetic information of the crop but only one or a few genes of the nucleus donor plant.

Transfer of chloroplasts or mitochondria may be an appropriate method in plant breeding if organelles with high metabolic activities are discovered in wild plants and should be introduced into related cultivars. In addition, male sterile combinations could be constructed, probably, by organelle transfer.

These examples demonstrate that organelle transfer might be useful in plant breeding. Another question is if and in which cases experimental success may be expected.

Transfer of the complete information of a special type of cell organelle into crops will probably be successful only in combinations of near-related plants where compatibility of the newly combined systems is more likely than in unrelated combinations.

Fusion products of far-related plants are able to form calluses (Kao et al., 1974; Kartha et al., 1974). Chromosome elimination and breakage have been observed in somatic hybrid cells (Kao, 1977; Binding and Nehls, 1978). Even in these combinations perhaps plant regeneration may occur if the procedures end up with strains that contain only a small portion of the genetic information of one parent but all or nearly all of the information of the other parent. If a selective pressure could be applied to regenerating hybrid calluses, it should be possible to select for strains that are comparable to gene transfer products.

The application of organelle transfer methods in plant breeding is only possible if the plants can be regenerated from isolated protoplasts. However, protoplasts of most of the crops do not regenerate after isolation. The dependence of success in plant regeneration from protoplasts on a good culture medium is demonstrated in potato. Melchers (personal communication) regenerated dihaploid plants from protoplasts that had been isolated from cell cultures in the KM medium (Kao and Michayluk, 1975). Vigorous callus formation in this medium was also obtained using potato protoplasts from shoot cultures (Binding et al., 1978) and protoplasts of other species of the *Solanaceae* (Binding and Nehls, 1977; Nehls, 1978; Schieder, 1977).

LITERATURE CITED

Alexander, N. J., N. W. Gillham, and J. E. Boynton. 1974. The mitochondrial genome of *Chlamydomonas*: Induction of minute colony mutations by acriflavin and their inheritance. Molec. Genet. 130:275–90.

Arnold, C. G. 1975. Extrakaryotic inheritance. Progr. Bot. 37:259–68.

Barnes, D. K., and R. A. Garboucheva. 1973. Intra-plant variation for pollen production in male sterile and fertile alfalfa. Crop Sci. 13:456–59.

Binding, H. 1966. Regeneration und Verschmelzung nackter Laubmoosprotoplasten. Z. Pflanzenphysiol. 55:305–21.

Binding, H. 1974. Fusionsversuche mit isolierten Protoplasten von *Petunia hybrida* L. Z. Pflanzenphysiol. 72:422–26.

Binding, H. 1976. Somatic hybridization experiments in solanaceous species. Molec. Gen. Genet. 144:171–75.

Binding, H., and R. Kollmann. 1976. The use of subprotoplasts for organelle transplantation.

In D. Dudits, G. L. Farkas, and P. Maliga (eds.), Cell genetics in higher plants. Pp. 191-206. Akadémiai Kiadó, Budapest.

Binding, H., and R. Nehls. 1977. Regeneration of isolated protoplasts to plants in *Solanum dulcamara* L. Z. Pflanzenphysiol. 85:279-80.

Binding, H., and R. Nehls. 1978. Somatic cell hybridization of *Vicia faba* + *Petunia hybrida*. Molec. Gen. Genet. (in press).

Binding, H., R. Nehls, O. Schieder, S. K. Sopory, and G. Wenzel. 1978. Regeneration of mesophyll protoplasts isolated from shoot cultures of dihaploid clones of *Solanum tuberosum* L. Physiol. Plantarum 43:52-54.

Bonnet, H. T., and T. Eriksson. 1974. Transfer of algal chloroplasts into protoplasts of higher plants. Planta 120:71-79.

Börner, T., G. Jahn, and R. Hagemann. 1976. Electrophoretic mobility of the subunits of Fraction I protein of higher plant species. Biochem. Physiol. Pflanz. 169:179-81.

Caritt, B., and J. M. Eisenstadt. 1973a. RNA-synthesis in isolated chloroplasts: Characterization of the newly synthesised RNA. FEBS-Lett. 36:116-20.

Caritt, B., and J. M. Eisenstadt. 1973b. Synthesis *in vitro* of high-molecular-weight RNA by isolated *Euglena* chloroplasts. Europ. J. Biochem. 36:482-88.

Carlson, P. S. 1973. The use of protoplasts for genetic research. Proc. Nat. Acad. Sci. USA 70:598-602.

Correns, C. 1937. Nicht mendelnde Vererbung. Gebr. Borntraeger, Berlin.

Davidson, J. N., M. R. Hanson, and L. Bogorad. 1974. An altered chloroplast ribosomal protein in ery-M1 mutants of *Chlamydomonas reinhardi*. Molec. Gen. Genet. 132:119-29.

Davies, D. R. 1974. Chromosome elimination in interspecific hybrids. Heredity 32:267-70.

Detchon, P., and J. V. Possingham. 1973. Chloroplast ribosomal nucleic acid synthesis in cultured spinach leaf tissue. Biochem. J. 136:829-36.

Doig, R. L., A. A. Dore, and D. F. Rogers. 1975. Preharvest sprouting in bread wheat (*Triticum aestivum*) as influenced by cytoplasmic male-sterility derived from *T. timopheevi*. Euphytica 24:229-32.

Ellis, I. R. S., C. J. Bunton, and J. M. Palmer. 1973. Can mitochondrial complementation be used as a tool in breeding hybrid cereals? Nature 241:45-47.

Epp, M. D. 1973. Nuclear gene-induced plastome mutations in *Oenothera hookeri*. I. Genetic analysis. Genetics 75:465-483.

Frankel, R., and E. Galun. 1977. Pollination mechanisms, reproduction, and plant breeding. Springer-Verlag, Berlin.

Giles, K. L. 1976. Chloroplast uptake and genetic complementation. *In* J. Reinert and Y. P. S. Bajaj (eds.), Applied and fundamental aspects of plant cell, tissue, and organ culture. Pp. 536 50. Springer-Verlag, Berlin.

Giles, K. L., and V. Sarafis. 1971. On the survival and reproduction of chloroplasts outside the cell. Cytobios 4:61 74.

Giles, K. L., and V. Sarafis. 1972. Chloroplast survival and division *in vitro*. Nature New Biol. 236:56 58.

Gleba, Y. Y. 1977. Nonchromosomal inheritance in higher plants as studied by somatic cell hybridization. *In* this volume.

Gregory, P., V. E. Gracen, O. C. Yoder, and N. A. Steinkraus. 1977. Differential effects of digitonin on mitochondria isolated from male-sterile cytoplasms of corn. Plant Sci. Lett. 9:17–21.

Grosse, W. 1976. Enzyme der Tryptophan-Synthese in Etioplasten von *Pisum sativum* L. Z. Pflanzenphysiol. 80:463–68.

Hagemann, R., and M. Scholze. 1974. Struktur und Funktion der genetischen Information in den Plastiden. VII. Vererbung und Entmischung genetisch unterschiedlicher Plastidensorten bei *Pelargonium zonale* Ait. Biol. Zbl. 93:625–48.

Hanson, W. D., D. E. Moreland, and C. R. Shriner. 1975. Correlation of mitochondrial activities and plant vigor with genotypic backgrounds in maize and soybeans. Crop Sci. 15:62–66.

Harris, E. H., J. F. Preston, and J. M. Eisenstadt. 1973. Amino acid incorporation and products of protein synthesis in isolated chloroplasts of *Euglena gracilis*. Biochemistry 12:1227–34.

Herrmann, F. H. 1972. Chloroplast lamellar proteins of the plasmid mutant en:viridis-1 of *Antirrhinum majus* having impaired photosystem II. Exp. Cell Res. 70:452–53.

Herrmann, F. H., D. Martorin, K. Timopheev. T. Börner, A. B. Rubin, and R. Hagemann. 1974. Structure and function of the genetic information in plastids. IX. Studies on primary reactions of photosynthesis in plastome mutants. Biochem. Physiol. Pflanz. 165:393–400.

Hovenkamp-Obbema, R., and D. Stegwee. 1974. Effect of chloramphenicol on the development of proplastids in *Euglena gracilis*. I. The synthesis of ribulosediphosphate carboxylase, NADP-linked glyceraldehyde-3-phosphate dehydrogenase and aminolaevulinate dehydratase. Z. Pflanzenphysiol. 73:430–38.

Jennings, R. C., and I. Ohad. 1973. XII. The influence of chloramphenicol on chlorophyll fluorescence and chlorophyll organisation in green cells of a mutant of *Chlamydomonas reinhardi* y-1. Plant Sci. Lett. 1:3–9.

Jensen, C. J. 1977. *In vitro* embryology and its application in genetics. (Unpublished.)

Kao, K. N. 1976. Cytological studies on plant heterokaryocytes.—Nuclear behavior. *In* D. Dudits, G. L. Farkas, and P. Maliga (eds.), Cell genetics in higher plants. Pp. 149–52. Akadémiai Kiadó, Budapest.

Kao, K. N. 1977. Chromosomal behavior in somatic hybrids of soybean–*Nicotiana glauca*. Molec. Gen. Genet. 150:225–35.

Kao, K. N., F. Constabel, M. R. Michayluk, and O. L. Gamborg. 1974. Plant protoplast fusion and growth of intergeneric hybrid cells. Planta 120:215–27.

Kao, K. N., and M. R. Michayluk. 1974. A method for high-frequency intergeneric fusion of plant protoplasts. Planta 115:355–67.

Kao, K. N., and M. R. Michayluk. 1975. Nutritional requirements for growth of *Vicia hajastana* cells and protoplasts of a very low population density in liquid media. Planta 126:105–10.

Kartha, K. K., O. L. Gamborg, F. Constabel, and K. N. Kao. 1974. Fusion of rapeseed and soybean protoplasts and subsequent division of heterokaryocytes. Canad. J. Bot. 52:2435–36.

Keller, W. A., and G. Melchers. 1973. The effect of high pH and calcium on tobacco leaf protoplast fusion. Z. Naturforschung 28C:737–41.

Klerker, J. 1892. Eine Methode zur Isolierung lebender Protoplasten. Öfvers. Vet.- Akad. Förhdl. 9:463–74.

Laughnan, J. R., and S. J. Gabay. 1975. An episomal basis for instability of S male sterility in maize and some implication for plant breeding. *In* C. W. Birky, Jr., P. S. Perlman, and T. J. Byers (eds.), Genetics and biogenesis in mitochondria and chloroplasts. Pp. 331–52. Ohio State University Press, Columbus.

McDaniel, R. G. 1969. Mitochondrial heterosis in barley. *In* R. A. Nilan (ed.), Barley genetics. 2:323–36. Washington State University Press.

McDaniel, R. G. 1972. Mitochondrial heterosis and complementation as biochemical measures of yield. Nature New Biol. 236:190–91.

McDaniel, R. G., and J. V. Sarkissian. 1966. Heterosis: Complementation by mitochondria. Science 152:1640–42.

McDaniel, R. G., and J. V. Sarkissian. 1968. Mitochondrial heterosis in maize. Genetics 59:465–75.

Maliga, P., A. S. Breznowits, and L. Marton. 1973. Streptomycin resistant plants from callus cultures of tobacco. Plant Sci. Lett. 1:119–21.

Melchers, G. 1935. Über reziprok verschiedene Merkmals ausbildung in F₁ der Kreuzung *Saxifraga adscendens* L. × *S. tridactylites* L. unter Berücksichtigung des Entwicklungsstadiums. Z. ind. Abstam.-u. Vererb. 69:263.

Melchers, G. 1976. Kombination somatischer und konventioneller Genetik für die Pflanzenzüchtung. Vortrag Ges. Dt. Naturforsch. u. Ärzte. Stuttgart.

Melchers, G., and M. D. Sakristán. 1977. Somatic hybridization of plants by fusion of protoplasts. II. The chromosome numbers of somatic hybrid plants of different fusion experiments. *In* R. J. Gautheret (ed.), Recueil de travaux dediés à la memoire de G. Morel. Masson, Paris.

Michaelis, G. 1976. Extrakaryotic inheritance. Progr. Bot. 38:205–17.

Michaelis, G., E. Petrochilo, and P. P. Slonimski. 1973. Mitochondrial genetics. III. Recombined molecules of mitochondrial DNA obtained from crosses between cytoplasmic petite mutants of *Saccharomyces cerevisiae*: Physical and genetic characterization. Molec. Gen. Genet. 123:51–65.

Michaelis, P. 1931. Die Bedeutung des Plasmas für die Pollenfertilität reziprok verschiedener *Epilobium*-Bastarde. Ber. dt. Botan. Ges. 49:96.

Michaelis, P. 1965a. II. The occurrence of plasmone-differences in the genus *Epilobium*. (A historic survey.) Nucleus 8:93–108.

Michaelis, P. 1965b. Genetische, entwicklungsgeschichtliche und cytologische Untersuchungen zur Plasmavererbung. IV. Beschleunigte Plastiden-Umkombination infolge unvollständiger Durchmischung der Plastiden vor der Zellteilung. Flora 156A:1–19.

Michaelis, P. 1968. Beiträge zum Problem der Plastiden-Abänderungen. IV. Über das Plasma- und Plastidenabänderungen auslösende, isotopen (^{32}P)–induzierte Kerngen mp₁ von Epilobium. Molec. Gen. Genet. 101:257–306.

Nass, M. M. K. 1969. Uptake of isolated chloroplasts by mammalian cells. Science 165:1128–31.

Nehls, R. 1978. Isolation and regeneration of protoplasts from *Solanum nigrum* L. (In press.)

Pace, L. 1913. Apogamy in Atamosco. Bot. Gaz. 56:376.

Peterson, P. A., R. B. Flavell, and D. H. P. Barratt. 1975. Altered mitochondrial membrane activities associated with cytoplasmically-inherited disease sensitivity in maize. Theoret. Appl. Genet. 45:309–14.

Potrykus, I. 1970. Mutation und Rückmutation extrachromosomal vererbter Plastidenmerkmale von *Petunia*. Z. Pflanzenzücht. 63:24–40.

Potrykus, I. 1973. Transplantation of chloroplasts in protoplasts of *Petunia*. Z. Pflanzenphysiol. 70:364–66.

Potrykus, I., and F. Hoffmann. 1973. Transplantation of nuclei into protoplasts of higher plants. Z. Pflanzenphysiol. 69:287–89.

Potrykus, I., and H. Lörz. 1976. Organelle transfer into isolated protoplasts. *In* D. Dudits, G. L. Farkas, and P. Maliga (eds.), Cell genetics in higher plants. Pp. 183–90. Akadémiai Kiadó, Budapest.

Power, I. B., E. M. Frearson, C. Hayward, and E. C. Cocking. 1975. Some consequences of the fusion and selective culture of *Petunia* and *Parthenocissus* protoplasts. Plant Sci. Lett. 5:197–207.

Preddie, D. L., E. C. Preddie, A. M. Guerrini, and T. Cremona. 1973. Two isoaccepting species of tryptophan transfer ribonucleic acid from *Chlamydomonas reinhardii*. Canad. J. Biochem. 51:951–53.

Redei, G. P. 1973. Extrachromosomal mutability determined by a nuclear gene locus in *Arabidopsis*. Mutat. Res. 18:149–62.

Redei, G. P., and S. B. Plurad. 1973. Heriditary structural alterations of plastids induced by a nuclear mutator gene in *Arabidopsis*. Protoplasma 77:361–80.

Renner, O. 1934. Die pflanzlichen Plastiden als selbständige Elemente der genetischen Konstitution. Ber. Math.-phys. Kl. d. Sächs. Akad. d. Wiss. (Leipzig) 86:241.

Rhoades, M. M. 1943. Genic induction of an inherited cytoplasmic difference. Proc. Nat. Acad. Sci. USA. 29:327–29.

Sage, G. C. M., and G. E. Hobson. 1973. The possible use of mitochondrial complementation as an indicator of yield heterosis in breeding hybrid wheat. Euphytica 22:61–69.

Sager, R. 1975. Patterns of inheritance of organelle genomes: Molecular basis and evolutionary significance. *In* C. W. Birky, Jr., P. S. Perlman, and T. J. Byers (eds.), Genetics and biogenesis in mitochondria and chloroplasts. Pp. 252–67. Ohio State University Press, Columbus.

Sarkissian, I. V. 1972. Mitochondria polymorphism and heterosis. Z. Pflanzenzüchtg. 67:53–54.

Sarkissian, I. V., and H. K. Srivastava. 1967. Mitochondrial polymorphism in maize. II. Further evidence of correlation of mitochondrial complementation and heterosis. Genetics 57:843–50.

Schieder, O. 1974. Fusionen zwischen Protoplasten von *Sphaerocarpos donnellii* Aust.-Mutanten. Biochem. Physiol. Pfl. 165:433–35.

Schieder, O. 1977. Hybridization experiments with protoplasts from chlorophyll-deficient mutants of some solanaceous species. Planta 137:253–57.

Schmitz, F. 1882. Die Chromatophoren der Algen. Verh. Naturhist. Ver. preuss. Rheinlande 40.

Smith, H. H., K. N. Kao, and N. C. Combatti. 1976. Intergeneric hybridization by protoplast fusion in *Nicotiana*. J. Hered. 67:123–28.

Stubbe, W. 1971. Origin and continuity of plastids. *In* J. Reinert and H. Ursprung (eds.), Origin and continuity of cell organelles. Pp. 65–81. Springer-Verlag, Berlin.

Tilney-Bassett, R.A. E. 1975. Genetics of variegated plants. *In* C. W. Birky, Jr., P. S. Perlman, and T. J. Byers (eds.), Genetics and biogenesis of mitochondria and chloroplasts. Pp. 268–308. Ohio State University Press, Columbus.

Tsunewaki, G., R. T. Endo, and Y. Mukai. 1974. Further discovery of alien cytoplasms inducing haploids and twins in common wheat. Theoret. Appl. Genet. 45:104–9.

Vasil, I. K., and K. Giles, 1975. Induced transfer of higher plant chloroplasts into fungal protoplasts. Science 190:680.

Wang, W.-Y., W. L. Wang, J. E. Boynton, and N. W. Gillham. 1974. Genetic control of chlorophyll biosynthesis in *Chlamydomonas*. Analysis of mutants at two loci mediating the conversion of protopophyrin-1x to magnesium protoporphyrin. J. Cell Biol. 63:806–23.

Wettstein, F. V. 1928. Über plasmatische Vererbung und über das Zusammenwirken von Genen und Plasma. Ber. dt. Bot. Ges. 46:32.

Widholm, J. M. 1972a. Anthranilate synthetase from 5-methyl-trytophan-susceptible and -resistant cultured *Daucus carota* cells. Biochim. Biophys. Acta 279:48–57.

Widholm, J. M. 1972b. Cultured *Nicotiana tabacum* cells with an altered anthranilate synthetase which is less sensitive to feedback inhibition. Biochim. Biophys. Acta 261:52–58.

Wildman, S. G., K. Chen, J. C. Gray, S.D. Kung, P. Kwanyuen, and K. Sakano. 1975. Evolution of ferredoxin and Fraction I protein in the genus *Nicotiana*. *In* C. W. Birky, Jr., P. S. Perlman, and T. J. Byers (eds.), Genetics and biogenesis in mitochondria and chloroplasts. Pp. 309–29. Ohio State University Press, Columbus.

Wisselingh, C. van 1920. Über Variabilität und Erblichkeit. Z. indukt. Abstam.-u. Vererbungsl. 22:65–126.

Zobl, R., G. Fischbeck, F. Keydel, E. Latzko, and G. Spark. 1972. Complementation of isolated mitochondria from several wheat varieties. Plant Physiol. 50:790–91.

The Transfer of Nitrogen-fixing Ability to Nonleguminous Plants

37

INTRODUCTION

Since the end of the last war, when the major fertilizer-manufacturing companies saw the tremendous potential of disused munitions factories for the production of fixed nitrogen, there has been an increase in both the production and use of nitrogenous fertilizers in North America. This has shown in increased production in many crops, notably the cereals. In 1973 the sudden realization that oil is not evenly distributed around the globe, and yet represents a prime energy resource, caused what has now become known as the "energy crisis." There are several other reasons for the increased price of fertilizer nitrogen in North America, but in third world countries the increase in energy costs has been responsible for most of the increased cost.

Since the world's population is expected to double within the next 30 years, there is inevitably going to be an increased demand for food. However, the ever-increasing cost of fossil fuels, and energy of all kinds, mitigates against the production of the necessary fertilizers to increase, or even maintain, present crop yields. To maintain present yields a much greater reliance will have to be placed on biological nitrogen fixation to maintain adequate nitrogen supplies for cropping. Nitrogen is only one parameter affecting plant growth and yield, but it is an important one,

K. L. Giles, Plant Physiology Division, Department of Scientific and Industrial Research, Palmerston North, New Zealand.

which might be satisfactorily supplied by biological systems. Presently only the legumes have symbiotic nitrogen-fixing activity associated with their roots. They fix as much as 26–32% of their total nitrogen requirements during their life cycle. Changes in management, utilizing the legumes in rotations, may allow us to use some of this biologically fixed nitrogen with nonlegume crops in alternate rotation. However, in order to increase crop productivity in third world countries, where management systems are not as well developed as in some western countries, and to maintain production in these highly developed agronomic societies a preferable system would be to create a nitrogen-fixing symbiosis with the agronomically important species. The symbiosis could take several forms and currently there are several apparently rational experimental approaches to the transfer of nitrogen-fixing ability to plants other than the legumes.

1. Experiments to induce the uptake of *Rhizobium* into protoplasts of nonleguminous plants in the hope that modification of the *Rhizobium* and/or host protoplast might occur and allow the formation of nitrogen-fixing tissues and, eventually, plantlets.

2. The fusion of protoplasts from leguminous and nonleguminous species in the hope of creating a hybrid cell line capable of forming an association with *Rhizobium* strains, probably after the elimination of much of one of the genomes of the fusion partners.

3. The fusion of *Rhizobium* bacteroid-containing cells from the nodules of leguminous plants with leaf protoplasts from nonleguminous plants.

4. The transfer of the *nif* genes from *Rhizobium* to *Agrobacterium tumifaciens*. The *nif* genes are those concerned with the production of enzymes used in the biological fixation of nitrogen. Their transference to a more widely distributed bacterium capable of forming tumors on many nonleguminous plants might induce the fixation of nitrogen within these tumors.

5. The transference of *nif* genes from *Rhizobium*, or other nitrogen-fixing species, directly to the higher plant by phage-mediated transfer of the genetic information or by DNA uptake by protoplasts.

6. The uptake of nitrogen-fixing blue-green algae by protoplasts in the hope of creating some form of nitrogen-fixing organelle within the cell, comparable with the systems in *Gunnera, Macrozamia* etc.

7. The uptake of free-living nitrogen-fixing bacteria such as *Azotobacter* and *Klebsiella*.

CHOICE OF A SUITABLE VECTOR

Mycorrhizal Fungi

To date nitrogen fixation has been found only in association with prokaryotic cells. To attempt the transfer of this process to eukaryotic cells directly, species by species, would be a massive job. Because in many agronomic species it is difficult, if not impossible, to culture protoplasts into whole plants, we took advantage of the almost ubiquitous association between soil fungi and the roots of higher plants in an attempt to create a nitrogen-fixing mycorrhizal association. Host specificity in the mycorrhizal fungi is not fully understood, but it is apparent that most mycorrhizal fungi are capable of of forming associations with several plant species. Ectomycorrhizal associations have been shown to be responsible for the bulk of phosphate uptake by woody plants, which are usually deficient or sparsely supplied with root hairs (Slankis, 1971). Other elements (calcium, magnesium, etc.) have also been implicated in mycorrhizal associations. Ericoid endomycorrhizae have been implicated in the transfer of nitrogen compounds to their hosts (Stribley and Read, 1974). Thus to ask a mycorrhizal fungus to transport nitrogen compounds to the higher plants is not asking the impossible, though it does represent a probable change of metabolic priorities. One of the advantages of using fungi is that their protoplasts are usually easy to grow into new mycelia. This allows one to use a very large number of protoplasts with a fair certainty of being able to produce 70–80% of them as reverted mycelia within a very short time after plating out.

Although most of the herbaceous agronomic species appear to favor the incorporation of a vesicular-arbuscular fungus of the *Endogone* type, woody species are usually associated with ectomycorrhizal fungi. The ectomycorrhizal fungi can be cultured *in vitro* in sterile culture, whereas the *Endogone* species necessitate rather more complicated treatments for their maintenance in the laboratory, and cannot yet be cultured aseptically *in vitro*.

The mycorrhizal fungus used in these experiments was a culture of *Rhizopogon* sp. supplied to us by the Forestry Research Institute, Rotorua. The particular isolate used had been originally isolated by Dr. Chu from the roots of *Pinus radiata* seedlings and has since been reassociated with seedlings showing that it had maintained its mycorrhizal habit in culture. The specific name of the cultures used could not be ascertained, since they had not been derived from fruiting bodies of the fungus but from plant roots. It is known that *Rhizopogon* species normally form associations with the roots of seedling pine trees of many species and maintain a mycorrhizal association with them until the tree is between five and six years old when this genus is usually replaced by other fungi. Preliminary work with mycelium of this fungus showed that protoplasts could be produced by enzymatic digestion of the cell walls with enzymes derived from *Trichoderma viride* as described by De Vries and Wessels (1972). The enzyme preparation was used at a concentration of 2 mg protein per ml. The $MgSO_4$ (0.5 M) osmotic stabilizer was dissolved in 0.05 M Na-maleate buffer, pH 5.8. Small pieces of mycelium were washed in this solution and 50–100 μg dry weight were incubated in 100 μl of the enzyme-supplemented osmoticum at room temperature. Eight- to ten-hr incubation was sufficient to release most protoplasts. Yields were variable, however, between 2×10^4 and 1×10^6 per ml. The protoplasts were washed from the mycelium debris by floatation of the vacuolate protoplasts.

The Oxygen Problem

Probably what promises to be the most difficult problem associated with the transfer of nitrogen fixation to eukaryote species is the oxygen sensitivity of the major enzyme concerned, nitrogenase. The extreme sensitivity of all nitrogenases to oxygen (Eady et al., 1972) means that aerobic nitrogen-fixing organisms must have physiological mechanisms that protect the functional enzyme. Heterocystous blue-green algae, such as *Anabaena* and *Nostoc*, restrict nitrogenase activity to heterocysts that lack the oxygen-evolving photosystem 2 (Steward, 1971). Fungi, into which we envisaged introducing free-living nitrogen-fixing bacteria, are fortunately free from the problems of photosynthetic oxygen evolution. However, it was considered wise to use members of the *Azotobacter* group, for their known methods of nitrogenase protection. The enzyme can change to a conformationally protected form that, although relatively inactive, is insensitive to oxygen. Protection within these cells is also afforded by respiration, which scavenges oxygen from the functional enzyme (Postgate, 1974). The species finally chosen was *Azotobacter vinelandii* and the strain was capable of reducing 1,500 nM of acetylene per mg per hr in pure culture. The culture of *Azotobacter vinelandii* was grown on Ashby's medium. (This was made up as follows: 3.2 mM K_2HPO_4; 2 mM $MgSO_4.7H_2O$; 0.1 mM Fe $SO_4.7H_2O$; 0.1 mM $MnSO_4.2H_2O$; 0.04 mM $Na_2 MoO_4$; 1.7 mM NaCl; 5 mM $CaCO_3$; 5 g glucose; solidified with 1.2% agar per liter pH 8). The bacteria were subcultured every five days to fresh medium.

Induced Uptake of the Bacteria and Selection of Active Strains

Uptake of bacterial cells by the fungal protoplast was induced by the

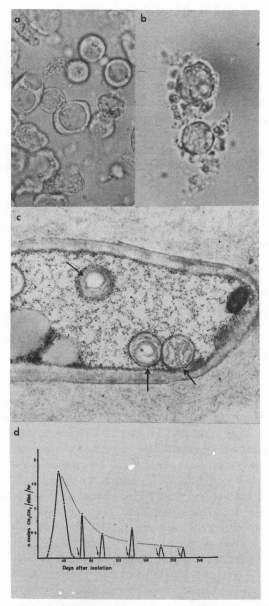

Fig. 1. (a) Protoplasts of *Rhizopogon* sp. suspended in 0.5 M MgSO₄. × 600. (b) Adhesion of *Azotobacter* cells around protoplasts of *Rhizopogon* sp. after addition of polyethylene glycol (PEG). × 600. (c) Electron micrograph of a hypha of the modified *Rhizopogon* sp. Included within the hypha are three L-forms of *Azotobacter* in cross-section (arrows). × 3600. (d) Acetylene reduction activity of the modified *Rhizopogon* strain 3. The arrows indicate transfer to fresh medium. (Giles and Whitehead, 1975; reproduced by courtesy of the Faculty Press, 88 Regent Street, Cambridge, England.)

presence of 20% w/v polyethylene glycol (PEG) molecular weight 6000, as described by Bonnett and Erikssen (1974) for the uptake of chloroplasts into higher plant protoplasts (fig. 1). Half a milliliter of a suspension of fungal protoplasts in 0.5 M $MgSO_4$ at a concentration of 2.5×10^5 per ml was mixed with 0.5 ml of the suspension 2.5×10^5 per ml vegetative cells of a three-day-old culture of *Azotobacter*, also suspended in 0.5 M $MgSO_4$, and 1 ml of 40% w/v PEG was added. Ten minutes after mixing, the suspension was transferred to an 8 μm Millepore filter and washed free of PEG with 0.5 M $MgSO_4$ supplemented nitrogen-deficient Hagem's medium (9.4 mM NH_4Cl; 3.4 mM KH_2PO_4; 2 mM $MgSO_4.7H_2O$; 1 ml/1 ml per liter of 1% $FeCl_3$; 5 g glucose/liter; 5 g malt extract/liter). Microscopic examination of the protoplasts after this washing indicated that most of the bacteria had been washed from the protoplasts. Three hours later, allowing time for uptake, any adhering bacteria still outside the protoplasts were lysed with a 500 μg/ml lysozyme solution. The protoplasts were not visibly affected by the lysozyme wash, since most had by this time become osmotically insensitive with the beginnings of formation of a new cell wall (Giles and Whitehead, 1976).

After the lysozyme wash the protoplasts were resuspended in 0.5 ml of liquid Hagem's medium still supplemented with 0.5 M $MgSO_4$ and spread on the surface of nitrogen-deficient Hagem's medium solidified with 1.2% agar. In nitrogen-deficient Hagem's medium KCl replaced NH_4Cl and the malt extract was replaced by 0.5 mg/1 nicotinic acid and 0.1 mg/1 pyridoxine HCl and 0.1 mg/1 thiamine-HCl. Benzylpenicillin at a concentration of 400 mg/ml was added to this medium. Tests had shown that this concentration of benzylpenicillin lysed the strain of *Azotobacter* used, without affecting the regeneration or reversion of the fungal protoplasts. The protoplasts varied widely in size (20–80 μm) and most of them were vacuolate. Treatment with PEG caused an immediate clustering of bacterial cells around the protoplasts, though it seemed to cause little fusion between the protoplasts. There was a tendency for the fungal protoplasts to adhere though no fusion occurred during cell wall regeneration whether or not PEG was present.

Within 22 hours of plating on Hagem's medium containing combined nitrogen, 70–80% of the unmodified fungal protoplasts (i.e., those not exposed to bacteria) had regenerated a cell wall, developed two or three yeast-like buds, and finally reverted to a normal hypha. On nitrogen-deficient medium 10–15% of the unmodified spheroplasts regenerated cell walls and some of them had occasional budding but none of the 5×10^6 protoplasts plated out reverted to mycelial growth. The PEG treatment given to induce bacterial uptake had no noticeable effect on the level of protoplast reversion on Hagem's medium containing nitrogen. Slightly higher rates of regeneration and reversion could be obtained using liquid culture medium, 85–95% reversion, but this method greatly complicated the isolation of individual colonies for subculturing.

Out of 7.5×10^6 *Azotobacter*-treated fungal protoplasts that were plated onto nitrogen-deficient medium, five colonies were initially isolated that were capable of reversion and growth. These colonies were isolated during the course of several experiments indicating the repeatability of the system: the control fungus on nitrogen-containing medium grew as a big fluffy mat mainly on the surface of the agar. The modified fungal mycelium had a similar growth form in the presence of nitrogen, but on deficient medium grew as a sparse mycelium under the agar surface. The mycelium of the control fungus on deficient medium hardly grew off the nitrogen-containing inoculum, and formed a yellow pigment that diffused into the medium. This was absent, or at least markedly reduced, in the cultures of the modified fungus on nitrogen-deficient medium (fig. 2).

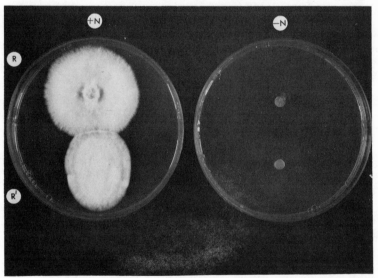

Fig. 2. Growth of *Rhizopogon* strains on +N and –N media. Control wild-type *Rhizopogon* (R) growing as a thick mat on +N Hagem's medium fails to grow on –N medium. Modified nitrogen-fixing *Rhizopogon* (R') growing in a similar manner to the control fungus on +N medium, but as a sparse mycelium under the agar surface on –N medium. All the colonies are 3 weeks old from the same sized inoculum in 9 cm Petri dishes.

Forty-eight hours after the original plating out, the very small modified colonies were transferred under a binocular microscope to fresh medium lacking the penicillin, and 96 hours later they were again washed with 500 μg/ml lysozyme solution. The cultures were then transferred to fresh nitrogen-deficient medium and subjected to rigorous contamination tests.

Portions of the mycelium were used to inoculate liquid Ashby's medium, to favor the growth of any contaminating *Azotobacter*. Seventy-two hours after inoculation and incubation at 25° C, samples were taken and plated on Ashby's medium solidified with 1.2% agar. No colonies of *Azotobacter* were found and no acetylene reduction activity was detected. Noninoculated controls gave negative results, whereas *Azotobacter*-inoculated flasks yielded many colonies that were actively reducing acetylene after 48 hours. Portions of the modified mycelium left in a liquid medium for up to 3 weeks gave negative results. The control *Rhizopogon* mycelium yielded two contaminants, one gram-positive rod and a gram-positive coccus, neither of which was capable of reducing acetylene. These two contaminants were not found in the modified fungal cultures.

Acetylene Reduction Assays of Nitrogenase Activity

The acetylene reduction assays for nitrogenase activity were all performed using a Perkin-Elmer gas chromatograph. A 30×0.15 cm column containing 0.35 g of Poropak T was used at an oven temperature of 75° C. The column was maintained at a pressure of 68.95 kPa nitrogen and oxygen and 137.9 kPa of hydrogen; a 10% atmospheric acetylene concentration was used in the assay vials and 0.2 ml samples were taken at appropriate intervals. Strict controls were run in conjunction with all tests to ascertain that there was no trivial reason for the acetylene reduction activities noted. At no stage was the wild-type fungus found to produce ethylene in culture and all acetylene gas samples were assayed for ethylene content. The ratio of acetylene to ethylene was used in the calculation of acetylene reduction activity.

Because the modified fungal hyphae tended to grow through the agar, rather than on the surface, it proved impossible to weigh the hyphae. Acetylene reduction measurements were made using five standard discs (5 mm diameter) of the agar and mycelium in each assay vial.

During the first 9 days in culture none of the five strains isolated reduced acetylene. Seventeen days after isolation, acetylene reduction was detected in four of the strains; the fifth showed no activity until day 19. The level of activity increased until days 29 and 30 and then decreased until, on day 55, there was no detectable acetylene reduction in any of the cultures. Upon transferring the cultures to fresh medium 61 days after isolation, there was a second peak of activity lasting 6–7 days, peaking on day 3 in all cultures. This reappearance of acetylene reduction was repeated on days 93, 154, 182, 212, 253, 296, 340, and until 1.5 years after the initial isolation of the cultures.

After 138 days in culture protoplasts were again produced by enzymatic digestion of the fungal wall. The protoplasts were plated on nitrogen-deficient medium and 65–70% reverted to mycelial growth. Acetylene reduction activity was detected after 72 hours and lasted for an additional 6 days. Samples of the unmodified *Rhizogogon* mycelium at no time either reduced acetylene or produced detectable ethylene during the course of these experiments.

Azotobacter vinelandii grown on Ashby's nitrogen-deficient medium at pH 8 had a peak of acetylene reduction activity, 24–36 hours after plating out, of 1,200–1,500 nM CH2 = CH2/mg dry weight per hour. When the samples were moved to fresh medium 60 days after the initial transfer to Hagem's medium there was no reappearance of activity. These results are in contrast to the activity detected in the fungus, which appeared after 17 days in culture on Hagem's medium and was repeatable on transfer to fresh medium. The pH optimum for acetylene reduction by *Azotobacter* was pH 8, whereas in the fungus, optimum reduction occurred near pH 4 and the pH optimum for growth of *Rhizopogon* is pH 3.8. Both these findings support the contention that the acetylene reduction detected was not due to contaminating bacteria alone.

Supplementation of the nitrogen-deficient Hagem's medium with 2.5 mg/l Na_2MoO_4 increased the rate of fixation by the fungus twofold, from 2.42 to 5.9 nM CH2 = CH2/disc/hr on day 112. Higher concentrations did not appreciably raise this level; the lower ones were not as effective.

LOCALIZATION OF THE INTRODUCED BACTERIA

Optical microscopy revealed that the *Azotobacter*-treated strains capable of nitrogen fixation had obvious inclusions within the hyphae and these inclusions were lacking in the control mycelium. Electron microscopy showed that these inclusions were, in fact, deposits of an electron transparent storage deposit. The control fungus had no such deposits within its hyphae, but similar deposits were found in the cysts of *Azotobacter*, and these were subsequently identified as poly-beta-hydroxybutyric acid (PHB) (Lemogne and Girard, 1943). Tests were carried out to discover whether the product was present in the modified strains capable of acetylene reduction using the spectrophotometric method of Law and Slepecky (1961). It was found that PHB was present in the modified fungal hyphae, but that none was detectable in cultures of the control *Rhizopogon*.

Poly-beta-hydroxybutyrate (PHB) Reserves

Stevenson and Socolofsky (1966) have shown that cells of the genus *Azotobacter* develop specialized resting cysts and deposit intracellular reserves of PHB. They propose that the accumulation of large amounts of PHB by *Azotobacter vinelandii* reflects an unbalanced growth condition that promotes cyst formation. They suggest that cells assimilate exogenous carbon faster than they can fix the molecular nitrogen necessary for conversion into nitrogenous cell components, and as a result the cells accumulate large amounts of the nonnitrogenous material, PHB. The buildup of PHB within the fungal hyphae indicates that not only the information for nitrogenase necessary for the acetylene reduction activity was transferred, but also that for the synthesis and accumulation of PHB itself. The PHB accumulation in fungal hyphae may reflect inability of the fungus to fix nitrogen at a rate comparable to its rate of carbon assimilation, thus creating imbalanced growth and triggering PHB synthesis. This could explain the intermittent nature of the activity in the modified fungus. The intermittent activity may, of course, also have been due to removal by the fungus of some toxic environment because of the accumulation in old media of extracellular toxins.

Under the electron microscope PHB appeared to build up within the fungal cytoplasm, being membrane bound in some cases but certainly never within apparent bacterial cells. In fact, no intact bacterial cells were ever found within the mycelium fungus by electron microscopy. However, small spherical bodies measuring between 0.4 and 1 μm in diameter were detected under the electron microscope. It was at first thought that these structures represented modified mitochondria of the fungus, but it would now appear that they probably represent modified L-forms of *Azotobacter*. The bodies, which appear circular under the electron microscope, consist of a series of concentric membranes surrounded by a double membrane, which often is associated with a large concentration of polyribosomes. Concentric membranes are reminiscent of the arrangement within vegetative cells of *Azotobacter*. Considerable effort has been put into discovering whether these structures in fact represent L-forms of the bacteria within the fungus.

Fluorescent Antibody Location of Bacterial Proteins and Membranes

Since the fate of the introduced bacterial information was at first obscure, it became of interest to locate the position and state of that information within the fungal hypha. Preliminary electron microscope data had shown the appearance of small, circular, membrane-bound bodies within the hypha of the modified strains. These bodies have already been described, and it was suggested that they may represent modified mitochondria or L-forms of the bacteria after the transgenotic event (Giles and Whitehead, 1975, 1976). Fluorescent antibodies were prepared against bacterial proteins and membranes in order to test whether the bacterial proteins could be demonstrated *in situ* within the fungal hypha. Because at no time under the electron microscope had intact bacterial cells been seen in the fungal hyphae, lysed bacterial cells were used as the antigens. Bacterial cells were frozen and thawed several times in phosphate buffer and treated with 0.6 mg/ml benzylpenicillin (pH 6.8) and the cell wall materials were spun down by centrifugation (300 g for 3 min). The proteins and membrane fragments were dialysed to a final concentration of 3 mg protein per ml. These proteins and membranes were injected, using Freund's adjuvant, into white laboratory bred rabbits, 1 ml being given in six separate injections. Four injections were subcutaneous on the back, two intramuscular on the thighs. They were administered twice over a period of two months. An additional month was allowed for the antibodies to build up and the rabbits were then bled either from the ear or by heart puncture.

Salting-out of the globulins. A solution of 3.25 M ammonium sulfate (analytical reagent) was used to precipitate the globulins. This precipitation was repeated three times in order to free the suspension of any grossly observable hemoglobin. The protein was then dialysed at 4° C against frequent changes of 0.85% NaCl until sulfate was no longer detectable in the saline. About 16–20 hours of continous dialysis was necessary. Protein content was then assayed.

Conjugation procedure. Solutions were warmed to 25° C before starting the conjugation procedure. Ten ml of the globulin solution at a concentration of 1% protein were added to 4 ml of 0.15 M phosphate buffer pH 9, and 0.05 mg of fluorescine-isothiocyanate (FITC) was added to the protein in 4 ml of 0.1 phosphate buffer pH 8. The pH was adjusted to 9 with 0.1 N NaOH and the volume made up to 20 ml with 0.1 M NaCl. The mixture was incubated at 25° C for 21 hours and then run through a Sephadex G 25 column using phosphate buffered saline as the eluant. The first yellow band traveling down the column was collected. The volume of this band was brought back to 20 ml by negative dialysis against 10% PEG (MW 6,000). Small quantitites of the conjugate were then frozen and maintained at –22° C.

Staining procedures. Lysed cells of *Azotobacter vinelandii* were placed on clean microscope slides and dried at 60° C in an oven. The slide was placed in 100% acetone for 5 min, removed and air dried. One drop of fluoresceine antibody conjugate was placed on each smear. The slide was then placed in a vessel over water and kept in the dark for 30 min. It was then removed and washed in phosphate-buffered saline, which was gently stirred for another 30 min. The bottoms of the slides were wiped and the slide completely dried at 60° C in an oven. A drop of glycerol was used as a mounting fluid and the cover slip was sealed with nail varnish. Slides were maintained at 4° C in the dark until examined.

As has been previously reported (Giles and Whitehead, 1975) the modified fungal strains capable of nitrogen fixation tend to grow in the agar substrate and small blocks of this were used as samples. Blocks were placed on the stubs of a cryomicrotome and 20 μm sections of agar and fungal hyphae were cut with a steel knife. These slides were then treated in the same way as the *Azotobacter* smears, except that sections were placed on gelatine subbed slides to maintain firm contact during processing. In order to minimize the nonspecific autofluorescence within the fungal hyphae, it was necessary to pre-stain the fungal sections with a Rhodamine B bovine serum albumin conjugate. This was done in the same way as described for the FITC antibody and preceded the fluoresceine staining. The agar sections had to be very thoroughly washed, since there was a tendency for the fluorescent conjugate to stain the agar nonspecifically. In most cases the washing period was extended to one hour in gently stirred phosphaste-buffered saline. Prepared slides were kept at 4° C in the dark until examined.

Microscopy and photography. A Zeiss photomicroscope II was used in conjunction with an HB200 illuminator. The KP500 and KP630 filters were used in the incident light excitation and barrier filters KP500 and BG38 were used to view the fluorescence. The fluorescence was photographed with a Polaroid film ASA rating 3000. Nomarski interference optics were used in the bright field examination of fungal hyphae. PAN film (ASA 50) was used for bright field photography.

Controlled preparations of *Azotobacter* cells that had been pretreated with penicillin to disrupt the cell wall structures gave strong fluorescence under the microscope when stained with the fluorescent antibody conjugate. When the cell walls were not broken, the fluorescence was poor or nonexistent, suggesting that the antibodies were indeed made mainly against internal proteins and membranes. Control *Rhizopogon* hyphae without treatment with Rhodamine B did have some slight autofluorescence, but this was completely eliminated by the use of Rhodamine B-BSA conjugate prior to staining with a fluorescent antibody. Hyphae from the modified strain of *Rhizopogon* had areas of intense fluorescence, which were contained entirely within the hyphae. The points could be seen under Nomarski differential interference microscopy as bodies within the fungal cytoplasm. The structures from which fluorescence emanated appeared either as single spheres or in some cases larger multiple collections of spheres. Sometimes both types were carried within a single cell. The multiple bodies were made up of small spherical structures similar in size to the individual spheres found in cytoplasm.

In a previous description of this system (Giles and Whitehead, 1975, 1976) sectoring within the fungal hyphae was blamed for the apparent loss or lowering of acetylene reduction activity in the fungal strains. In the present hyphae it became apparent that sectoring did in fact occur, since some large areas of mycelium consisted of hyphae that were completely devoid of fluorescence, whereas others had a very high percentage of cells with fluorescent bodies included.

It did not pass our notice that the occurrence of these fluorescent bodies within hyphae was often associated with a swelling of the hypha. Hyphae containing large numbers of fluorescent bodies were considerably larger in diameter (7–8 μm) than hyphae lacking such bodies (3–4 μm). In some hyphae isolated fluorescent bodies occurred and these bodies seemed to be osmotically stable. Cutting through swellings during the preparation of the slides allowed the fluorescent body to escape into the medium without any apparent damage. As yet, attempts to culture these isolated bodies have proved unsuccessful.

The cross reaction between fluorescent antibodies and structures within hyphae of the modified forms of *Rhizopogon* is an indication of the existence of bacterial proteins and/or membranes within fungal hyphae. The strains used in these localization experiments had been isolated two years previously, and subcultured at least ten times prior to being stained, indicating that replication and division of these bacterial bodies must have occurred. The fluorescence appears to be associated with discrete bodies or collections of bodies within hyphae. Average diameter of the fluorescent speres is 1.18 μm, whereas under the electron microscope these spheres were 0.41 μm in diameter. This difference in size could be caused by shrinkage during processing of fungal cytoplasm, a notoriously difficult subject for electron micrographic fixation, or it could be caused by the thin sectioning of the spheres. It seems likely that the spherical bodies described previously (Giles and Whitehead, 1976) as being associated with the appearance of acetylene reduction activity in the fungal hyphae represent some type of L-form of *Azotobacter* lying within the fungal cytoplasm. It is known that these structures are surrounded by a double membrane, the outer one of which is possibly of fungal origin. Attempts to conjugate ferritin with the *Azotobacter* antibodies in order to continue electron micrographic localization and identification of the bacterial proteins are currently underway.

THE REASSOCIATION OF THE MODIFIED STRAINS OF
RHIZOPOGON WITH THE ROOTS OF PINUS RADIATA SEEDLINGS

As was explained earlier, the decision to use mycorrhizal fungus for the transfer of nitrogen fixation had been made on the basis that it allowed one to reassociate the fungus with several different higher plant species. *Rhizopogon* has been shown to be mycorrhizal with roots of almost all species of pine seedlings at least during their first 5 to 6 years of growth. It was of obvious interest to discover whether the nitrogenase activity found *in vitro* was activated once in association with plant roots and whether any of the nitrogen fixed was transferred to the host plant.

The nitrogen and phosphorous levels in the medium are of prime importance during both mycorrhizal reassociation and the later symbiosis (Richards, 1965, Slankis, 1971). High levels of these elements tend to depress the level of fungal reassociation with host roots. The activity of modified strains capable of acetylene reduction and nitrogen fixation could have been a barrier to reassociation because of the levels of fixed nitrogen. Coupled with this, since the whole bacterial genome had been introduced, other factors present in the modified strains may have proved inhibitory to reassociation with the host plant.

In the legume nodule system, light increases the rate of nitrogen fixation, presumably because of the level of photosynthate available as an energy source, and tests were carried out to ascertain whether similar effects occurred in this association.

The five most active strains of *Rhizopogon* in nitrogen fixation assays were grown on nitrogen-deficient Hagem's medium. Pine seeds obtained from the Forest Research Institute, Rotorua, were surface sterilized and planted into sterile pumice and vermiculite. Seedlings were grown under sterile conditions with Hoagland's medium applied every 2 days for a period of 3 months. After 3 months, 200 seedlings were selected for their similarity and size and were potted up into sterile pumice and vermiculite and divided using a double technique into 8 lots of 25. Two batches of 25 (A, B) were inoculated with cultures of the modified *Rhizopogon* strains as follows (Giles and Whitehead, 1977).

Samples of the five different strains of fungus were diced and pieces placed in close association with seedling roots mixed throughout the potting medium. Five seedlings were inoculated with each of the five strains. Because of the modified nature of these mycorrhizae it was considered judicious to maintain them for the first 3 months in association with the plant roots, in screw-top jars, modified to allow for passage of sterile air in and out of them. Two further batches of 25 plants each (C, D) were inoculated with the wild-type cultures of *Rhizopogon* that had been grown on full Hagem's medium. Another 50 seedlings were potted up in the sterile potting mix but were not inoculated (E, F). Seedling trees batches A–E received modified Hoagland's nutrient containing 10% of the combined phosphorous (1.5 ppm) and no combined nitrogen or molybdic acid; calcium chloride replacing calcium nitrate and potassium chloride replacing potassium nitrate in the solution. Trees in batches B and D were given 2 ml of 0.005 g/ 1 molybdic acid every two days. Seedlings in batch F received full Hoagland's medium and acted as a non-inoculated, non-deprived control. Apart from batches A and B, seedling trees were maintained in a greenhouse under controlled temperatures of 25° C day, 15° C night.

Precautions taken in the full containment of trees to be inoculated with a modified strain of the fungus proved to be justified. All 10 seedlings inoculated with strain 1 of the fungus succumbed within the first few weeks of the association and died. It is not known whether their death was coincidental or whether it was directly due to the non-controlled growth of mycorrhiza within the plant tissues. Portions of leaves and stems of these young seedlings showed that there were traces of intercellular and intracellular fungi that superficially bore resemblances to the nitrogen-fixing strain. Reassociation of the wild-type fungus with roots can lead to death of seedlings if they are very young. The remaining four strains of the modified fungus formed apparently normal mycorrhizal relationships and rootlets with the seedlings after a 3-month period deprived of nitrogen. Examination of the roots of batches E and F showed no signs of mycorrhizal relationships after the 3-month period. Neither of these batches had been inoculated. Batch F, which had been receiving full Hoagland's medium during the period, had no signs of chlorosis or browning but batch E trees were much smaller in size and the needles were severely yellowed. Upon receiving full Hoagland's nutrient after their 3-month deprivation, batch E recovered fully and none died. Twenty-one plants in batch C and 19 in batch D had mycorrhizal reassociation. These batches had received inoculations of the wild-type mycorrhiza and had received nitrogen-deficient, phosphorus-low Hoagland's during the course of the 3-month test period. The plants had a fluffy white mantle around many doubly-bifurcating rootlets and the production of a Hartig net indicating normal mycorrhizal relationships. Batch D, which had received added molybdenum during the course of the nutrient deprivation, had no greater signs of reassociation than did C (121 ± 12 mycorrhizal rootlets per plant in D compared with 130 ± 15 in C). Batches A and B, which had been inoculated with a modified *Rhizopogon*, also had double bifurcating rootlets, though the mantle was usually much less prominent and in some cases only superficial. Fourteen plants in batch A and 12 in batch B, which had received a molybdenum supplement during the course of the deprivation, had a higher, though not statistically significant, rate of mycorrhizal association than A (128 ± 17 mycorrhizal rootlets in plant B compared with only 115 ± 10 in A).

Acetylene Reduction Activity of the Mycorrhizal Rootlets

Because of the low levels of acetylene reduction detected, stringent controls were necessary to maintain reproducibility in the acetylene reduction assay. Trees were tested for activity using the Perkin-Elmer gas chromatograph as described previously. Seedlings were contained in 200 ml bottles fitted with screw tops and were maintained at 25° C in a water bath. The background level of ethylene in the acetylene used was measured regularly and allowance was made for it in measuring the amounts of ethylene produced. New screw caps were used, and this helped to keep the loss of acetylene from the system to a minimum over the 6-hour assay period. Ethylene production was calculated from the ethylene to acetylene ratios and no ethylene reproduction by the tissue was detected in the absence of acetylene.

All seedlings were assayed for acetylene reduction activity. Seedlings inoculated with modified strains 2–5 in batches A and B had detectable acetylene reduction in 60% of the reassociated trees. The activity in B was about 20–25% higher than activity in A. No activity was recorded in seedlings lacking well-formed mycorrhizal rootlets. No activity was recorded at all in batches C–F. In table 1 the effects on acetylene reduction activity using mycorrhizal rootlets, small pieces of root attached to the rootlets, or using the whole seedling in the assay vial are evident. The mycorrhizal rootlets had low but detectable activity, whereas when rootlets were left attached to pieces of root only about 2.5–4 cm long, there was a much higher activity expressed as dry weight of mycorrhizal rootlets. The

TABLE 1

ACETYLENE REDUCTION ACTIVITY OF MODIFIED
FUNGAL STRAINS 1–5 IN BATCHES A AND B

Batch	Strain	Acetylene Reduction Activity nmoles C_2H_4/ g DW Mycorrhizal Rootlet/ hr	Mg/ g Nitrogen Dry Weight of Roots and Shoots
A	1	. . .	. . . .
	2	2.64	. . . .
	3	3.27	. . . .
	4	2.51	. . . .
	5	2.83	940 ± 25
B	1	. . .	. . . .
	2	3.38	. . . .
	3	4.20	. . . .
	4	2.45	. . . .
	5	3.67	. . . .
C,D	. . . . . .	0	820 ± 39
E	. . . . . .	0	808 ± 45
F	. . . . . .	0	1,040 ± 12
B$_2$	reisolated	0.77*	

Plants had 0.5 g to 2 g DW of mycorrhizal rootlet each. Strain 1 killed the seedlings it became associated with and no assays could be performed. All other figures represent the average for the group of five. Strain B$_2$ was reisolated from the seedling roots and its activity is expressed per mg dry weight of mycelium. Nitrogen content is based on an assay of 50 plants in batches A–D and on 25 plants in batches E and F. Variability is indicated by the standard deviation. Acetylene reduction assays were performed in 15 ml vials with a background sensitivity of 0.5 nm h^{-1} g^{-1}.

* Mg DW mycelium.

activity was much enhanced when the mycorrhizal rootlets were left attached to the intact plant, but varied depending on whether the experiment was performed in the light or dark. In the light whole plants had an activity of up to 45.6 nmoles of acetylene per gram mycorrhizal rootlets per hour, whereas in the dark only about 1/50 of this activity was recorded. This effect could be manipulated in the light and dark over a period of at least 60 hours.

To ascertain that there was not a trivial explanation for the acetylene reduction activity recorded in asociation with the plant, a series of controls was run in parallel with all the measurements made. As well as seedlings associated with a modified fungal strain, seedlings reassociated with the wild-type fungus similar to the seedlings in batch C were also assayed. Seedlings unassociated with the mycorrhizae were assayed with acetylene production both with and without acetylene in the asssay vial. Seedlings killed by exposure to 110° C for 6 hours were also used as controls. Seedlings reassociated with a modified strain were assayed after removal of the mycorrhizal rootlets in order to assure that this was the site primarily responsible for nitrogenase activity. All these controls gave no ethylene production above the sensitivity limits of the system, calculated for the whole plants in 200 ml flasks as being 7 nm $CH_2-CH_2/hr/g$.

Nitrogen Content

The analyses of shoots and roots of control and inoculated plants showed that controls receiving full Hoagland's solution had a higher nitrogen content than trees under nitrogen-deficient conditions whether they were inoculated with wild-type or modified strains of *Rhizopogon*. However, those trees under nitrogen-deficient conditions inoculated with a modified *Rhizopogon* capable of acetylene reduction had an intermediate level of nitrogen in both roots and shoots compared with those inoculated solely with the wild-type, indicating that some incorporation of the fixed nitrogen may well have occurred (table 2). Mycorrhizal rootlets were included in total nitrogen measurements so that some of the recorded nitrogen would have been fungal in origin. Phosphorous levels were the same in all treatments, suggesting that phosphate uptake by the modified strains was not impaired (table 3).

Electron Microscopy

Thin-sectioning for electron microscopy revealed the presence of fungal hyphae in the intercellular spaces of the cortex of seedling roots and closely adhering to the outside walls of the epidermal cells of the root. The fungi seemed identical ultrastructurally with the hyphae from *in vitro* cultures.

TABLE 2

THE EFFECT OF USING MYCORRHIZAL ROOTLETS, PIECES OF ROOTS WITH
ROOTLETS, AND WHOLE PLANTS IN THE ACETYLENE REDUCTION ASSAY

	NMOLES C_2H_4/g DW MYCORRHIZAL ROOTLETS/HR	
	1	2
Rootlets	0.773	0.57
Roots plus rootlets	3.18	2.83
Whole plant (8–10 cm)		
Light	45.6	38.2
Dark	1.2	1.65

Results are shown for two plants, 1 and 2; the whole plants were tested
in both continuous light and dark. The light-dark effect was reversible (see
fig. 1). Upon removal of the mycorrhizal rootlets, the activity fell to below
the sensitivity of the detection method used, indicating that the nitrogenase
activity was associated with them.

TABLE 3

ANALYSIS OF TOTAL PHOSPHORUS

Batch		Total P ug/g DW
A,B	Shoots	585
	Roots	515
C,D	Shoots	575
	Roots	519
E,F	Shoots	627
	Roots	601

Fifty plants in each of the treatments (AB—
modified mycorrhiza, CD—wild-type mycorrhiza,
and EF—no mycorrhiza, non-deprived control)
were dried, powdered, and mixed, and samples
were assayed. There was no significant difference
in phosphorus content between plants associated
with modified or wild-type mycorrhizae, though
both had a lower phosphorus content than non-
deprived controls. This indicates that phosphorus uptake by the modified and wild-type fungus
was comparable.

When comparing the ultrastructure of the roots inoculated with modified
and unmodified strains, it was found that hyphae of the former quite
frequently intruded into the intracellular spaces of the cortical cells of the
root. At no time was this noticed with the control *Rhizopogon* strain. The
cells into which the modified strain entered appeared dead, empty of
cytoplasm, and often contained massive tannin deposits. Whether the
death of the cells represented cause or effect is not known. Hyphae within
the cells were identical with the *in vitro* cultures containing small globules
of PHB and other spherical bodies in the cultures of the modified strains.
Hyphae re-isolated from the reassociated roots retained their normal
morphology and acetylene reduction activity. However, mycorrhizal
rootlets formed in association with the modified fungal strains seemed to

have a shorter life span than rootlets formed in association with the wild-type strain. There was a premature blackening of the mycorrhizal rootlets probably associated with the tannin deposits seen under the electron microscope. The mycorrhizal rootlets tended to be sloughed off after 3–4 months as opposed to 8–9 months with the wild-type fungal strain. It seems then that modification of the fungal strains causes them to induce premature senescence within the mycorrhizal rootlets and this may in some way be associated with the apparent pathogenic effect of Strain 1 in association with the plant roots.

However, despite the obvious difficulties associated with the system, the finding of acetylene reduction activity in plant roots reassociated with these modified fungal strains, and the increased nitrogen content of the reassociated plants, is an encouraging sign that this form of transfer of nitrogen fixation using genetically modified mycorrhizae may be successful. It would appear that the introduction of the intact bacterial cells into the cytoplasm of at least one mycorrhizal fungus can allow the creation of semi-stable organelle-like bodies within the mycelium of that fungus, capable of metabolic activity over and above that normally carried out by the fungus. This raises the possibility that other fungal hosts may be preferable to *Rhizopogon* in such transfers or that different free-living nitrogen-fixing bacteria or other organisms could be introduced. In either case the results support the contention that the genetic modification of these rhizosphere organisms, capable of forming symbiotic relationships with higher plants, might show them to be useful vectors for carrying introduced genetic information. Some of these possibilities will be discussed in conjunction with their associated problems in the next section.

DISCUSSION OF THE PROBLEMS AND POSSIBLE DEVELOPMENT
OF THIS FORM OF GENETIC MANIPULATION

Expression of the Genetic Information

The death of pine seedlings associated with the modified strains of *Rhizopogon* is cause for concern. That the metabolically modified strains should appear to have upset the subtle and stable relationship normally formed between the root and mycorrhizal fungi means that any metabolic changes introduced into these fungi would have to be tested extensively before there could be any field release. There has at no time been consideration of field trials with the modified *Rhizopogon* strains. The problem raises questions as to the applicability of using such a subtle relationship in the transfer of new genetic information to higher plant hosts. The successful reassociation of four of the strains with seedlings in such a manner that normal mycorrhizal rootlets were established, and the detection of acetylene reduction activity in conjunction with these rootlets, does give some support to the belief that this technique provides at least one way of transferring genetic information safely.

The time course for the expression of the nitrogenase gene, measured by acetylene reduction activity, is similar to that described for the expression of the Z gene of *Escherichia coli* in tomato cells (Doy et al., 1973c). In the case of nitrogenase, no activity was detected until 17 days after treatment of the protoplasts with *Azotobacter* cells, whereas in tomato cells 40 days elapsed before the expression of the Z gene was detected. The intermittent activity of the acetylene reduction activity in the modified *Rhizopogon* strains parallels the behavior of the β-glactosidase activity in transgenosed tomato cells (Doy et al., 1973a,b).

There is an ever-increasing body of evidence suggesting that bacterial DNA can be taken up by higher plant systems and the information carried on it can be expressed by the plant (Ledoux et al., 1971; Doy et al., 1973a). The present results suggest that some degree of stabilization of the genetic information for nitrogen fixation derived from bacteria is possible over extended periods within eucaryote cells provided sufficient selection pressure is present.

Endomycorrhizae

The ectomycorrhizae normally associated with tree species are replaced in the bulk of agronomically important plants, such as the cereals and legumes, with the endomycorrhizal vesicular-arbuscular (VA) fungi. These are usually classed under the general generic name of *Endogone*. It would seem logical that the next step in the transfer of nitrogen-fixing ability to the cereals and crop plants would be to investigate the possibility of the system described here being used in conjunction with these VA fungi. This poses considerable problems, since *Endogone* is presently impossible to culture away from plant roots, though it can be grown in sterile association with root systems *in vitro* (Mosse and Hepper, 1975). This allows development of hyphae in the root growth medium which would be accessible for manipulation. It would be possible to make cell wall preparations from these hyphae and to develop enzymes against them using a *Trichoderma viride* system, and hence to introduce bacteria into isolated protoplasts of the fungus. However, the next stage—the reinoculation of the root system with protoplasts of the fungus—would probably offer considerable resistance and difficulties. Although I believe that such a system should be the ultimate goal of the sort of technique that I have been describing, it would be simpler initially to concentrate on the Ericoid mycorrhizae, which also form intracellular associations but which can be cultured *in vitro*. The use of these fungi would give a lead in as to the behavior of modified VA fungi in root systems. Ericoid mycorrhizae from the roots of *Calluna vulgaris* have been grown *in vitro* by Pearson and Read (1973). We feel it should be possible to repeat everything we have done to date with *Rhizopogon*, and to reintroduce these fungi to the sterile roots of seedlings of *Calluna vulgaris*, or some other suitable Ericoid species, and watch for the development of the characteristic hair roots in which these fungi are found. A further factor that might be relevant to the use of these Ericoid mycorrhizae is that they have been associated with the transfer of nitrogen compounds from the soil to their host plants (Stribley and Read, 1974), so that they might represent a particularly suitable vehicle for the transfer of nitrogen-fixing bacteria and the transfer of the nitrogen products produced. The use of *Endogone* VA mycorrhizae must await the introduction of more effective techniques for their culture *in vitro* or in association with plant roots to allow the easy reinoculation of modified strains.

Derepressed Strains and Poplar Mycorrhizae

The use of derepressed bacterial strains for the nitrogen-fixing carrier offers potential for increasing the nitrogen-fixing potential of such a system. However, it must be borne in mind that not only would it fix more nitrogen but it would be expected to demand more energy from the plant. Thus the photosynthate reserves would be reduced, adversely affecting the growth of the host plant. It would seem more sensible, in order to create a harmonious symbiosis, to use conventional free-living repressed strains, which might only yield a percentage of the total nitrogen requirements of the plant, but which may act as self-regulating organelles—fixing nitrogen when it is needed and only when there is a sufficient carbon source available. The use of derepressed strains has still to be investigated.

We have repeated these experiments with mycorrhizal fungi isolated from the roots of the Lombardy poplar (*Populas nigra* L. cv "italica") and have managed to select strains of the modified fungus that contain PHB and also fluorescent bodies characteristic of the introduced *Azotobacter* after staining with fluorescent antibodies. Normally these bacteria are associated with small coils in the fungal hyphae, possibly suggesting that growth substances produced by the bacteria are associated with the formation of these coils, as perhaps the swellings in the fungal hyphae associated with *Azotobacter* in *Rhizopogon* may also represent some growth substance effect on the cell walls. These fungi, although possessing many of the characteristics expected of the nitrogen-fixing symbiosis, have as yet yielded no acetylene reduction activity. The selection sieve used to isolate the strains proved to be ineffective in this case, since the wild-type strains also are capable of growth on –N medium and it would appear that they are effective ammonia scavengers. Growing them in scrubbed air, clean of ammonia, very much limits their growth. However, even then they seem to be able to gather enough nitrogen from the agar substrate on which they are growing to show some growth. This has hindered the selection of the nitrogen-fixing strains and will probably be a factor to be taken into account in the choice of other mycorrhizal fungi for future work.

REFERENCES

Bonnett, H. T., and T. Erikssen. 1974. Transfer of algal chloroplasts into protoplasts of higher plants. Planta 120:71–79.

De Vries, O., M. H., and J. G. H. Wessels. 1972. Release of protoplasts from *Schizophylleum commune* by a lytic enzyme preparation from *Trichoderma viride*. J. Gen. Microbiol. 73:13–22.

Doy, C. H., P. M. Gresshoff, and B. G. Rolfe. 1973a. Biological and molecular evidence for the transgenosis of genes from bacteria to plant cells. Proc. Nat. Acad. Sci. USA 70:723–26.

Doy, C. H., P. M. Gresshoff, and B. G. Rolfe. 1973b. Transgenosis of bacterial genes from *Escherichia coli* to cultures of haploid *Lycopersicon esculentum* and haploid *Arabidopsis thaliana* plant cells. In J. Pollack and J. W. Lee (eds.), The biochemistry of gene expression in higher organisms. Pp. 21–37. Australian and New Zealand Book Co.

Doy, C. H., P. M. Greshoff, and B. G. Rolfe. 1973c. Time-course of phenotypic expression of *Escherichia coli* Z. gene following transgenosis in haploid *Lycopersicon esculentum* cells. Nature 244:90–91.

Eady, R. R., B. E. Smith, K. A. Cook, and J. R. Postgate. 1972. Nitrogenase of *Klebsiella pneumoniae*. Purification and properties of the component proteins. Biochem. J. 128:655–62.

Giles, K. L., and H. C. M. Whitehead. 1975. The transfer of nitrogen fixing ability to a eukaryote cell. Cytobios 14:49–61.

Giles, K. L., and H. C. M. Whitehead. 1976. Uptake and continued metabolic activity of *Azotobacter* within fungal protoplasts. Science 193:1125–26.

Giles, K. L., and H. C. M. Whitehead. 1977. Reassociation of a modified mycorrhiza with the host plant roots (*Pinus radiata*) and the transfer of acetylene reduction activity. Plant and Soil (in press).

Law, J. H., and R. A. Slepecky. 1961. Assay of poly beta hydroxybutyric acid. J. Bact. 82:33–36.

Ledoux, L., R. Huart, and M. Jacobs. 1971. Fate of exogenous DNA in *Arabidopsis thaliana*. Eur. J. Biochem. 23:96–108.

Lemoigne, M., and H. Girard. 1943. Réserves lipidiques beta-hydroxybutyriques chez *Azotobacter chroococcum*. Compt. Rend. 217:557–58.

Mosse, B., and C. Hepper. 1975. Vesicular-arbuscular mycorrhizal infections in root organ cultures. Physiol. Plant Path. 5:215–23.

Pearson, V., and D. J. Read. 1973. The biology of mycorrhiza in the *Ericaceae*. 1. The isolation of the endophyte and the synthesis of mycorrhiza in aseptic culture. New Phytol. 72:371–79.

Postgate, J. R. 1974. Prerequisites of biological nitrogen fixation. *In* A. Quispel (ed.), Biological nitrogen fixation. North-Holland Publishing Co., Amsterdam.

Richards, B. N. 1965. Mycorrhiza development of loblolly pine seedlings in relation to soil reaction and the supply of nitrate. Plant and Soil 22:187–99.

Slankis, V. 1971. Formation of ectomycorrhizae on forest trees in relation to light, carbohydrates, and auxins. *In* E. Hacskaylo (ed.), Proceedings of the First North American Conference. Pp. 151–67. USDA Forest Service, Washington, D. C.

Stevenson, L. H., and M. D. Socolofsky. 1966. Cyst formation and polyhydroxybutyric acid accumulation in *Azotobacter*. J. Bact. 91:304–10.

Steward, W. D. P. 1971. Physiological studies on nitrogen-fixing blue-green algae. *In* T. A. Lie and E. G. Mulder (eds.), Biological nitrogen fixation in natural and agricultural habitats. Plant and Soil (special volume).

Stribley, D. P., and D. J. Read. 1973. Some nutritional aspects of the biology of ericaceous mycorrhizas. *In* F. E. Sanders, B. Mosse, and P. B. Tinker (eds.), Endomycorrhizas. Pp. 195–208. Academic Press, New York.

Stribley, D. P., and D. J. Read. 1974. The biology of the mycorrhiza in the Ericaceae. IV. The effect of mycorrhizal infection on uptake of ^{15}N from labeled soil by *Vaccinium macrocanpon*. New Phytol. 73:1149–52.

Abstracts of Contributed Papers

Influence of Ammonium on Growth, Protein Synthesis, and Nitrogen Assimilating Enzymes in Suspension Cultures of Paul's Scarlet Rose

Suspension cultures of Paul's Scarlet rose were grown in the defined media that differed only in their inorganic nitrogen content. Both media possessed equal amounts of NO_3 (1,920 μmoles) but differed in that NH_4^+ (72.8 μmoles) was present in control medium, whereas, no NH_4^+ was present in the test medium. A comparison of fresh weight increases over a fourteen-day growth period showed that NH_4^+ caused a twofold stimulation in growth, and governed the pattern of development.

The presence of NH_4^+ in the medium had little influence on the uptake of NO_3^- but it had a pronounced influence on the assimilation of nitrogen from nitrate into amino acids. When cells were grown in medium containing NH_4^+, approximately 60 percent of the absorbed NO_3^- was utilized for amino acid synthesis, whereas only 30 percent of the absorbed NO_3^- was incorporated into amino acids of cells grown without NH_4^+. In both cultures the majority of the reduced nitrogen was in protein in which approximately 40 percent was recovered from the medium. Only trace amounts of soluble amino acids were present in the media.

A developmental study of nitrogen-assimilating enzymes showed that they reached their maximum activities on different days. The presence of ammonium enhanced the activities of nitrate reductase, glutamate dehydrogenase, and glutamate synthetase, but had no stimulatory influence on glutamine synthetase. When the accumulative activity of each of these enzymes over the fourteen-day growth period was compared with the amount of nitrogen entering amino acids it appeared that nitrate reductase was the limiting step.

J. S. Fletcher, Department of Botany and Microbiology, University of Oklahoma, Norman, Oklahoma 73069.

C. Y. HU AND E. C. SCRIVANI

Effect of Kinetin on Callus Morphogenesis from Stem Apex Explants of American Chestnut (*Castanea dentata*) Winter Buds

Dormant winter buds were dissected and the apexes were excised and cultured. Various combinations of growth regulators were added to the media to test callus induction potential. Results indicated that callus only developed when auxin was supplemented to the media. When kinetin was added in addition to auxin, there was a significant increase in the percentage of callus induction. The optimum concentration range of kinetin for increasing callus formation was 0.1 to 1.0 ppm. Distinct dimorphism was observed in the resulting calli. Without kinetin, callus growth was three-dimensional with randomly distributed growth center, resulting in calli with a nodular appearance. As kinetin was included, callus growth became primarily two-dimensional with growth concentrated at the periphery, resulting in flat and circular calli, resembling bacterial colonies. The effect of ethrel on callus development was also investigated. It was found that ethrel produced a more uniform development.

C. Y. Hu and E. C. Scrivani, Biology Department, William Paterson College, Wayne, New Jersey 07470.

A. D. KRIKORIAN AND R. P. KANN

Clonal Micropropagation of Daylilies

Callus induction of explants from 5 diploid cultivars of *Hemerocallis*, its ability to proliferate upon transfer or subculture, and its ability to undergo morphogenesis on various media have been investigated since 1974. Although explants from leaves, leaf bases, shoot apices, scapes, and flower petals have been tested, those derived from flower buds about 1.5–4 cm long have yielded the most consistent results. But explants that include the proximal half of the ovary with 1–2 mm of pedicel or the distal half of the ovary with ca. 2 mm of style provide choice starting material. The basal medium of Murashige and Skoog (B_{ms}) supplemented with N^6-benzylaminopurine (BAP) and 2,4-dichlorophenoxyacetic acid (2,4-D), each varied logarithmically from .001–10 mg/1, has proved to be a satisfactory medium. Relatively high levels of 2,4-D (e.g., 10 mg/1) in combination with relatively low levels of BAP (.001–.1 mg/1) foster proliferation usually within a month. Explants are then transferred once or twice prior to separation of the callus from the parent explant as a preliminary to subculture. Lower levels of 2,4-D (e.g., .001–.01 mg/1) with higher levels of BAP (e.g., .1–1 mg/1) also yield callus. (Equal amounts of 2,4-D and BAP are less satisfactory in inducing growth.) Growth cells (callus) initiated in this way are easily subcultured every thirty days provided relatively large (as large as .5–1 cm diameter) inocula are used. After an additional thirty days, portions of the callus may be transferred to media containing lower amounts of BAP and 2,4-D (ratio of cytokinin to auxin 10:1). Organized structures (shoots and roots) form, adventitiously, from such cultures within a few weeks and may be large enough for subdivision and transferral to large tubes containing 1/2 strength B_{ms} within thirty

A. D. Krikorian and R. P. Kann, Department of Biology, State University of New York, Stony Brook, New York 11794.

days. After an additional sixty days these plants can be transplanted to pots. To date, the minimal interval from excised explant to potting in the greenhouse has been ca. 135 days. A procedure that increases the plantlet yield but takes longer (ca. 180 days) involves the initial use of higher levels of BAP and 2,4-D.

To increase further the yield and to minimize the size of the inoculum needed to give organized growth, an additional step using liquid media has been introduced. Proliferations induced on explants in the usual way are subcultured initially on semi-solid media and later subdivided into 1–2 mm fragments for transfer to liquid. Three subcultures at thirty-day intervals using higher levels of auxin and cytokinin after the first transfer are usually necessary before cultures are adequately established in liquid. Such cultures may be maintained easily through repeated subculture and yield far greater numbers of small units (ca. 1–2 mm diameter) capable of organized development when transferred to semi-solid media.

S. R. LONG AND I. M. SUSSEX

Embryo Culture and Germination in *Phaseolus vulgaris*

Embryogenesis in higher vascular plants comprises the period during which the zygote develops into a multicellular embryo within the maternal tissue and is usually terminated by entry into a state of developmental arrest or dormancy. In germination the embryo resumes growth independently of the maternal plant. Embryogenesis and germination can be viewed as a developmental continuum, where control of the later stages is dependent upon functions carried out in the earlier stages. To investigate these controls, we have studied normal embryogenesis and germination in *Phaseolus vulgaris* and have compared it with precocious germination of cultured embryos, where there is no intervening dormant stage between embryogenesis and germination. We have used biochemical markers, and sensitivity to inhibitors and to water stress, to determine the stage during embryogenesis where the embryo becomes capable of germination. We have shown that in precocious germination, the embryo will complete, *in vitro*, steps characterizing embryogenesis before it begins germinative growth.

S. R. Long and I. M. Sussex, Department of Biology, Yale University, New Haven, Connecticut 06520.

Embryogeny of *Phaseolus coccineus*: The Functions of the Suspensor during Early Embryo Development

The suspensor is a specialized embryonic organ that connects the developing embryo to the maternal tissues. Classical interpretation has relegated the suspensor to a rather passive role in the development of the embryo. However, *in vitro* embryo culture experiments of *P. coccineus* clearly show that the presence of the suspensor greatly stimulates the growth of the embryo prior to the late heart stage of embryo during early embryogeny.

Detailed light and electron microscopic studies showed structural differentiation occurring in the suspensor at the proembryo stage with the formation of wall ingrowths, the appearance of densely stained plastids and smooth ER, a massive increase in the number of polysomes, and the development of polytene chromosomes in the nucleus. These specializations are not found in the embryo proper and, in fact, occur at the time when the suspensor is most needed, as shown by *in vitro* culture. These studies suggest that the suspensor may be a site for metabolism not occurring within the embryo proper and also the major uptake site of nutrients for the developing embryo. Therefore, the enzymatic contents of the suspensor and the embryo proper have been compared using micro-thin-layer gel electrophoresis and direct quantitative assays. Large differences in the quantity of certain enzymes involved in carbohydrate and amino acid metabolism have been found. Thus, one of the functions of the suspensor would appear to be related to the metabolism of these substances. By using C^{14}-labeled sucrose and 3-0-methyl glucose as the indicators of nutrient flow, the uptake patterns of these substances by the embryo can be followed. Results from these experiments indicate that the suspensor-

E. C. Yeung, Department of Biology, Yale University, New Haven, Connecticut 06520.

embryo junction is the major uptake site for the developing embryo and that the suspensor must play an important role during early embryogeny.

Hormonal Regulation of Differentiation in Leaf Discs of *Salpiglossis In Vitro*

Hormonal factors regulating adventitious roots, buds, and callus proliferation were investigated in leaf discs (9 mm dia.) of *Salpiglossis* ("splash") cultured on synthetic media. On basal medium (BM) containing Murashige and Skoog's minerals, vitamins, and sucrose, the leafs discs died within two weeks.

Napthalene acetic acid (NAA) (1 ppm) induced callus differentiation and profuse rooting. In the presence of 2,4-D, leaf discs produced a friable callus that was light yellow, compared with green callus produced on a NAA medium. In the presence of kinetin (1 ppm) and adenine (40 ppm), leaf discs produced a friable callus with isolated pink cell clusters. Leaf discs cultured on a medium containing BAP (1 ppm) and adenine (5 ppm) expanded and produced numerous adventitious buds. A combination of BAP and 2,4-D suppressed budding, and a compact white callus developed without organ differentiation.

Our results demonstrate that cultured leaf discs of *Salpiglossis* evoke defined morphogenetic responses to cytokinins and auxins. In addition to these studies, observations from suspension cultures and attempts to isolate protoplasts from mesophyll cells will be discussed.

K. Raman, Department of Plant Sciences, University of Western Ontario, London, Ontario, Canada.

C. SENGUPTA AND V. RAGHAVAN

Nucleic Acid Metabolism during Induction of Somatic Embryogenesis in Carrot

The regulation of nucleic acid and protein synthesis was investigated during division growth of carrot cells in a medium containing 2,4-dichlorophenoxyacetic acid (2,4-D) and kinetin and during their transformation into embryoids in a medium lacking the auxin.

Upon transfer of the cell suspension to a medium lacking 2,4-D, the rate of DNA synthesis as measured by ^{3}H-adenosine incorporation decreased within the very first hour of transfer. On the other hand, the rate of protein synthesis in the cells increased upon removal of auxin, and the increase was maintained for the first 12 hours, attaining a maximum at 1–2 hours of transfer to the auxin depleted medium. The rate of RNA synthesis as measured by precursor ATP pool method was high during the first 12 hours in the cells growing in the auxin-depleted medium, and attained a maximum at about 4–6 hours of transfer. However, as determined by continuous labeling experiments, the accumulation of protein and RNA in cells was lower during the early hours of growth in the auxin-depleted medium than in cells growing in a medium containing auxin. Electrophoresis of phenol-extracted RNA showed the synthesis of 16S and 14S RNA species within 3–4 hours of transfer to the auxin-depleted medium. Electrophoresis of RNA obtained from double-labeling experiment using ^{3}H-adenosine-labeled cells growing in 2,4-D omitted medium and ^{14}C-adenosine-labeled cells growing in auxin containing medium showed lower rates of rRNA and tRNA synthesis in the embryogenic cells as early as 24 hours of transfer. A whole array of RNA of different S values were also synthesized at this point.

C. Sengupta and V. Raghavan, Department of Botany, Ohio State University, Columbus, Ohio 43210.

The results suggest that the control of somatic embryogenesis in carrot cell suspension is post-transcriptional. The mRNA inducing embryogenesis is probably synthesized in the presence of auxin and is translated when the auxin is removed from the medium. The products of pre-transcribed RNA in turn may have a role in derepressing genes involved in embryogenesis.

L. D. SPIESS

Control of Development in Moss

The moss *Pylaisiella selwynii* grown in liquid inorganic salts medium is excellent for studying differentiation because of its small size, distinct stages in development, and well-documented responses to growth hormones. Some strains of *Agrobacterium* and *Rhizobium* initiate rapid gametophore development after physical attachment between a lipopolysaccharide portion of the bacterial envelope and a component of the plant cell wall. Other strains of *Agrobacterium* initiate abnormal gametophores, callus rhizoids, and tumors on gametophore "leaves." All these bacteria produce cytokinins and indoleacetic acid, but neither hormone alone, or in tested ratios, induces normal gametophores. Ethylene induces callus. Vitamin B_{12} and lysopine, an amino acid obtained from *Agrobacterium*-induced crown gall, produce some gametophores. Adenosine and guanosine added together in certain ratios initiate normal gametophores that elongate rapidly. Combinations of cAMP and cGMP give similar but less-pronounced results. Cyclic AMP has no effect on development but reduces the amount of callus initiated by cytokinins. Theophylline and other inhibitors or animal cAMP phosphodiesterase similarly reduce callus induction by cytokinins. Cyclic GMP, guanosine, and imidazole cause elongation of filament cells, suppression of branching, and some loss of chloroplasts. Imidazole and cGMP also suppress callus induction by cytokinins. The addition of indoleacetic acid, zeatin, and cAMP produce abnormal gametophore with "leaves" that revert to filamentous growth. Similar structures are induced by *Agrobacterium rubi*. Thus the moss-bacterial system is useful for analysis of the physiology of growth and morphogenesis in moss in which cyclic nucleotides play a role.

L. D. Spiess, Department of Biology, Lake Forest College, Lake Forest, Illinois 60045.

L. WEISMAN AND M. F. MORSELLI

Tetrazolium Dyes as Indicators of Metabolic Differentiation in *Acer saccharum* Callus Cultures

Massive root callus cultures of *Acer saccharum* Marsh (sugar maple) were incubated on agar medium containing tripenyl tetrazolium chloride, or nitro blue tetrazolium. These dyes are reduced to an insoluble colored formazan by enzymatic hydrogenation. The callus developed a discrete subsurface hemispherical cap of diffuse formazan, independent of tissue orientation during incubation. Fresh hand-cut callus tissue sections also developed the formazan layer corresponding to its position in the mother culture and independent of its orientation during incubation. Callus sections treated with mercuric chloride, an enzyme poison, were unable to reduce the tetrazolium dyes. We demonstrated that unreduced tetrazolium was present in all cells of incubated tissues, including those treated with $HgCl_2$ by chemical reduction with sodium dithionite. This formazan cap apparently delineates a cambium-like layer of metabolically active cells and suggests the use of tetrazolium dyes as indicators of cellular metabolic differentiation in callus cultures of sugar maple.

L. Weisman and M. F. Morselli, Department of Botany, University of Vermont, Burlington, Vermont 05401.

Using Embryo Culture to Reveal Factors Maintaining *Ilex* Rudimentary Embryos in Quiescent State

In vitro cultures of excised rudimentary embryos of *Ilex aquifolium, I. cornuta,* and *I. opaca* were used to test the validity of two hypotheses that have been proposed to explain the maintenance of these embryos at heart stage in mature seeds.

Testing of hypothesis I, the inability of rudimentary embryos to utilize nonreducing sugars that are found in *Ilex* endosperm, was carried out by supplementing the culture media with three nonreducing endosperm sugars (stachyose, raffinose, and sucrose) and one reducing sugar (glucose). Results indicated that all four sugars were sufficient to support embryonic growth, whereas sugar-free controls were not.

Testing of hypothesis II, the presence of chemical inhibitors in endosperm, was carried out by both culturing embryos alone and with *Ilex* endosperm adjacent to each cultured embryo. Only slight embryonic growth occurred in the endosperm-containing cultures, whereas most embryos in endosperm-free cultures developed into mature stage and germinated. Data obtained support hypothesis II. Ultrastructure comparisons were also made between *in vivo* quiescent and *in vitro* growing heart-shaped embryos.

C. Y. Hu, Biology Department, William Paterson College, Wayne, New Jersey 07470.

V. A. SIDOROV, YU. YU. GLEBA, AND K. M. SYTNIK

Culture of Protoplasts Isolated from *Arabidopsis thaliana* Heynh. Callus Tissue and Fusion of *Arabidopsis* Callus and Tobacco Mesophyll Protoplasts

Callus tissues of two races (Enkheim and Kijon) of *A. thaliana* on B_5 medium (Gamborg et al., 1969) were used as source of protoplasts. For protoplast isolation, tissue was digested with 10 ml of enzyme solution containing 0.2% "Onozuka R-10," 0.1% "Macerozyme R-10," 0.2% "Xylanase," and 0.4 M mannitol, at 25° C for 16–18 hours; 1 g of tissue yields 2 8.10^7 protoplasts. Protoplasts were separated quantitatively from cell clumps by filtering, washed from enzymes by centrifugation, and cultured at high densities (higher than 10^6 cells per ml) as liquid droplets in modified B_5 medium, containing 18 g per l of glucose instead of sucrose, 0.4 M mannitol, 7 mM $CaCl_2$, 0.3 ppm of kinetin, 1 ppm of NAA, and 0.3 ppm of 2,4-D. Cultures were maintained at 26° C under dim light (1–3 kilolux per 16 hours daily). First division was observed on the second day, second and third divisions on the third and fifth days. Plating efficiencies comprise from 30% to 80% (counting on the fifth day). By the third day cultures were diluted wtith equal volumes of fresh medium. Seven-day-old cultures were recollected by centrifugation and cultured on a solid B_5 medium. The process of cell colony regeneration from isolated protoplast has been followed by electron microscopy. Attempts to achieve plant regeneration were so far unsuccessful.

Using the method of treatment with polyethylene glycol (PEG) (Kao and Michayluk, 1974), fusion of protoplasts isolated from *Arabidopsis* callus and tobacco mesophyll has been achieved. Electron microscopic analysis

V. A. Sidorov, Institute of Botany, Academy of Sciences of the Ukrainian S.S.R., Kiev.

of the heteroplasmic fusion products indicates that mixing of cytoplasms takes place a few hours after PEG application. In three-hour cultures, fusion products are multinucleate. Our data, however, suggest that the fusion of nuclei takes place prior to first mitosis.

YU. YU. GLEBA, N. M. PIVEN,
J. K. KOMARNITSKY, AND K. M. SYTNIK

Analysis of the Parasexual Cytoplasmic Hybrids (Cybrids) of *Nicotiana tabacum* and *N. debneyi*, Obtained by Protoplast Fusion

After fusion of mesophyll protoplasts isolated from (1) plastome chlorophyll-deficient mutant of *N. tabacum* (von Wettstein, Eriksson, 1964) (aseptically grown albino plants of w-w-w type) and (2) cytoplasmic male sterile analog of *N. tabacum*, carrying cytoplasm of *N. debneyi*, six variegated plants have been regenerated. Biochemical analysis of the polypeptide composition of crystalline Fraction I protein using isoelectric focusing (methods after Kung et al., 1974) indicates the presence of large subunit polypeptides of both *tabacum* and *debneyi* types in the four isolates so far studied. After selection of pure green shoots from variegated ones, *tabacum* large subunit polypeptides in these disappear. Variegation is maintained during propagation through callus stage; therefore, simple chimericity as an explanation of the results is excluded. So far it has not been possible to induce flowering in the plants produced.

Results obtained indicate that (1) in hybridization by protoplast fusion both parental plastomes are transferred to parasexual progeny; (2) biparental transmission of extrachromosomal determinants holds also in interspecific crosses; and (3) gene(s) that code(s) for plastome chlorophyll deficiency and genes coding for polypeptide composition of Fraction I protein large subunit are probably linked.

Yu. Yu. Gleba, Institute of Botany, Academy of Sciences of the Ukrainian S.S.R., Kiev.

S. M. FLASHMAN AND P. FILNER

Selection and Characterization of Tobacco Cell Lines Resistant to Selenoamino Acids

Biosynthesis of cysteine and methionine in plants involves sulfate uptake, activation by ATP sulfurylase, reduction to carrier-bound sulfhydryl, and incorporation into the amino acids. The pathway is subject to feedback control by cysteine and methionine at the levels of sulfate uptake and ATP sulfurylase.

The toxic analogues selenocystine and selenomethionine have been used to select variant tobacco cell lines with alterations in their sulfur metabolism. Starting with 2×10^7 cells of the threnine-resistant XDR_6^{thr} cell line, three variant lines have been isolated. Two of these lines are resistant only to selenocystine. Sulfate utilization is apparently normally regulated in these lines, but they appear to have an increased rate of cystine uptake. The third variant line is apparently partially constitutive for sulfate utilization. It is resistant to both selenocystine and selenomethionine, but only when well-supplied with sulfur. Further, it is sensitive to the sulfate analogue selenate under conditions when normal cells are resistant due to repression of sulfate utilization. Studies are now in progress to measure ATP sulfurylase levels and sulfate uptake rates in this line, and to determine the pool sizes of sulfur compounds in all three variant cell lines.

S. M. Flashman, Department of Genetics, North Carolina State University, Raleigh, North Carolina 27607; P. Filner, Plant Research Laboratory, Michigan State University, East Lansing, Michigan 48824.

C. E. ISAACS AND K. W. FISHER

An *In Vitro* System for the Culture of *Solanum stenotomum* Cells and Selection of a Variant Resistant to 5-Methyltryptophan

Suspension cultures of *Solanum stenotomum* derived from callus were used for the selection of variants resistant to the tryptophan analogue 5-methyltryptophan (5MT). 5MT is a feedback inhibitor of the branch point enzyme anthranilate synthetase (AS). 5MT-resistant cell cultures were able to grow in the presence of high levels of the analogue, whereas the growth of 5MT-sensitive cultures was inhibited by 5MT.

Analysis of cells from some of the resistant clones revealed they possessed a greater AS specific activity than 5 MT-sensitive cells from which they were derived. AS from 5MT-resistant cells was less sensitive to feedback inhibition by 5MT and tryptophan than AS from 5MT-sensitive cells. Resistance to 5MT was determined to be a stable trait that was maintained when 5MT-resistant cultures were grown in the absence of the analogue. Callus tissue regenerated from 5MT-resistant suspension cultures maintained 5MT resistance and the higher level of AS specific activity.

Roots regenerated from 5MT-resistant callus had a lower specific activity of AS than the callus but a higher specific activity than roots regenerated from 5MT-sensitive callus. AS from 5MT-resistant roots was less sensitive to feedback inhibition by 5MT than AS from 5MT-sensitive roots. The level of AS activity in 5MT-resistant cultures was greater than any level of AS activity found in whole plant tissues. Thus, markers can be obtained in cultures of potato cells and then retained in differentiated plant tissue.

C. E. Isaacs and K. W. Fisher, Department of Biological Sciences, Douglass College, Rutgers University, New Brunswick, New Jersey 08903.

T. J. McCOY AND E. T. BINGHAM

Breeding and Selection of Diploid Alfalfa (*Medicago sativa* L.) That Regenerates Plants from Cells Grown in Suspension Culture

Alfalfa is an autotetraploid forage legume and leaf protein source in which a low frequency of genotypes will regenerate plants from cells maintained as solid callus. A diploid line capable of regenerating plants from cells grown in suspension was desired to initiate a variant selection program at the cellular level. Steps in accomplishing this goal were (1) selection of tetraploids that would regenerate from solid callus, (2) scaling their chromosome number down to the diploid level by maternal haploidy, (3) selection of two derived-diploids that would regenerate from solid callus, and, finally, (4) crossing these diploids and screening the progeny for ability to proliferate in suspension culture and retain the capacity to regenerate. This breeding and selection scheme was successful in producing a diploid line (HG2) that was exceptional in its ability to regenerate plants from cells grown in suspension culture. Ability to regenerate (96 percent of colonies) and the proportion of diploid plants (70 percent) both remained high after one transfer and a total of 21 days in suspension. Regeneration decreased with time in suspension to 68 percent at 42 days, 48 percent at 56 days, 3.5 percent at 77, and zero percent at 98 days. Kinetin concentration was a significant factor in increasing the ploidy level of cells in culture. The percentage of diploid cells after 14 days in suspension culture was 95 percent without kinetin, 81 percent with .5 mg/ l kinetin, 75 percent with 2 mg/ l kinetin, and 46 percent with 8 mg/ l kinetin. Ability to regenerate is under genetic control in alfalfa, and HG2 is being used to breed other diploids that regenerate. HG2 is also expected to provide the basic material for future experiments in variant selection.

T. J. McCoy and E. T. Bingham, Department of Agronomy and Plant Genetics, University of Wisconsin, Madison, Wisconsin 55108.

Y. E. CHU

Genetic Parameters of Soybean Cells Grown in Suspension Culture

Suspension cultures of *Glycine max* (L.) Merr. were grown at 22° and 33° C. The doubling times of dividing cells were 35 and 25 hours, respectively. G_2 was 6.2 and 6.7 hours, and S was 13.8 and 6.5 hours. G_1 was calculated as 13 and 10 hours, respectively. These values were determined by labeling cells with ^{3}H-thymidine and measuring the appearance of radioactive mitotic figures. Treatment with 5-fluorodeoxyuridine (FudR) inhibited DNA synthesis, and, as a result, cells accumulated in S. Such cells were viable and, upon removal of the FudR, proceeded synchronously into mitosis. Treatment with 5-bromodeoxyuridine, following FudR synchronization, sensitized the cells to white light. Thus, cells capable of synthesizing DNA could be killed. Twenty to thirty percent of the cells in suspension cultures growing at 22° or 33° dividing (Q) cells were able to synthesize RNA and protein at a reduced rate.

Y. E. Chu, Lilly Research Laboratories, Eli Lilly and Company, Greenfield, Indiana 64140.

W. A. KELLER, K. C. ARMSTRONG,
I. A. DE LA ROCHE, AND J. LACAPRA

Production and Utilization of Haploids from *Brassica* Species

Embryogenesis and plant regeneration have been obtained in anther cultures of *Brassica campestris, B. napus*, and *B. oleracea*. Factors influencing embryogenesis included medium composition (especially carbohydrates and growth regulators) and culture temperature. Elevated sucrose levels (optimum 9–11 percent) were necessary for induction of embryogenesis. Auxins were also essential, with a combination of NAA and 2,4-D each at 0.1 mg/l being effective for all three species. A stimulation of embryogenesis in *B. campestris* was obtained with the auxin-like compound, benazolin (10 mg/l). Culture temperature had a dramatic effect on embryogenesis. Short-term, high-temperature (35° C) treatments increased the frequency of embryogenic anthers as well as the number of embryos produced per anther. In *B. napus* the maximum number of embryos was obtained by continuous culture at 30° C. The majority of embryos obtained were abnormal, lacking the ability to develop directly into plantlets. Plant regeneration was, however, achieved by culturing hypocotyl explants of anther-derived embryos on media supplemented with benzyladenine. Haploids have been identified in anther-derived regenerates of *B. campestris* and *B. napus*. *Brassica* haploids might be readily utilized to solve several breeding problems such as modification of oil characteristics in rapeseed. Cotyledons of anther-derived embryos of *B. campestris* were individually analyzed and found to have a fatty acid composition similar to that of seeds. Such sampling techniques could be utilized for efficient selection of individuals with desirable seed oil characteristics.

W. A. Keller, Ottawa Research Station, Agriculture Canada, Ottawa, Ontario, Canada.

D. W. MEINKE

Embryo Lethal Mutants in *Arabidopsis*: A Genetic Approach to Plant Embryogeny

Our present understanding of plant embryo development is based primarily on descriptive, experimental, and biochemical studies. An alternative genetic approach is to study mutants that block or alter specific stages of embryo development, resulting in aborted seeds. *Arabidopsis thaliana* (Cruciferae) is an ideal plant for this type of analysis because of its short generation time, large number of seeds, self-fertilization, and ability to distinguish unfertilized ovules and aborted and normal seeds in developing fruits (siliques). Furthermore, embryogeny in *Arabidopsis* is essentially identical to that of *Capsella*, a favorite plant for descriptive and experimental studies. In the present study, mutagenized seeds (0.05–0.50% EMS, 8 hrs) were grown into mature plants, and the first 5 siliques were screened for the presence of aborted seeds. Since an average silique contains about forty seeds and the normal abortion frequency is less than 1 percent of fertilized ovules, it was possible to clearly identify plants that appeared to be segregating for recessive embryo lethal mutations. Phenotypically normal, mature seeds were then removed from segregating siliques of these plants, and the nature of the mutation was studied in heterozygotes identified in the next generation. Since a high mutagen dose appeared to cause multiple mutations in many plants, initial work has centered on five mutants isolated from a population of 300 plants grown from 0.1% EMS-treated strain "Columbia" seeds. These mutants segregate as single Mendelian recessives in the siliques of heterozygotes, and each mutant arrests embryo development at a characteristic stage: heart (1), globular (1), and preglobular (3). The histology and development of the mutant embryos have been studied to more accurately determine the time and site of any

D. W. Meinke, Department of Biology, Osborn Memorial Laboratory, Yale University, New Haven, Connecticut 06520.

abnormalities, and *in vitro* embryo rescue experiments have been designed that may aid in understanding the cause of lethality in the late-aborting mutants.

A. WERNER AND M. F. MORSELLI

Ploidy Level: A Factor in the Development of *Acer saccharum* and *Acer rubrum* Stem Tissue Cultures

Data from the literature on woody plants' tissue cultures suggest that the ploidy level of the cultured species in its natural condition may be a factor affecting the rate of growth of the culture and/or its successful differentiation *in vitro*. As part of a project aimed at achieving differentiation of *Acer saccharum* Marsh tissue culture, callus from one-year-old twigs of two maple species, *A. saccharum* (sugar maple, $2n = 2X = 26$) (1) and *A. rubrum* (red maple, $2n = 6X = 78$, or $2n = 7X = 91$, or $2n = 8X = 104$), were established simultaneously to compare their growth and/or level of differentiation. *A. rubrum* had a higher percentage of callus production and exhibited more vigorous growth. Several types of defined media were tested. Hormone treatments to induce differentiation are presently being applied to stem callus of both species, and results to date will be reported.

A. Werner and M. F. Morselli, Department of Botany, University of Vermont, Burlington, Vermont 05401.

G. B. GENGENBACH, C. E. GREEN, AND C.M. DONOVAN

Expression and Inheritance of Pathotoxin Resistance in Maize Plants Selected and Regenerated from Cell Cultures

Texas male-sterile cytoplasm (cms-T) maize (*Zea mays* L.) is susceptible to *Helminthosporium maydis* (Nisikado and Miyake) race T and its pathotoxin, whereas nonsterile (N) cytoplasm maize is resistant. Callus cultures, initiated from immature embryos of a cms-T genotype, were both susceptible to the pathotoxin and capable of plant regeneration. Pathotoxin-resistant cell lines were selected by a sublethal enrichment procedure where cms-T callus were grown for several selection cycles (subculture transfers) in the presence of progressively higher concentrations of toxin. Periodically during the selection process, plants were regenerated from the cms-T cultures to determine their susceptibility of resistance to the pathotoxin.

Plants regenerated after four cycles of selection with pathotoxin were male sterile and toxin susceptible as shown by a leaf bioassay. All plants regenerated from resistant cell lines isolated after the fifth selection cycle, however, were toxin resistant and 52 of 65 were fully male fertile. The 13 partially or completely "male-sterile" resistant plants either produced a limited amount of starch-filled pollen or did not resemble cms-T plants in tassel morphology. Leaf bioassay tests on progeny from regenerated resistant plants indicated that resistance to the pathotoxin was inherited only through the female. The male-fertility trait also was not transmitted through the pollen. Progeny from regenerated resistant plants inoculated with *H. maydis* race T spores showed good agreement between lesion length and resistance to the pathotoxin. This indicated that plant resistance to the pathogen was closely associated with the toxin resistance obtained through cell culture selection.

G. B. Gengenbach, Department of Agronomy and Plant Genetics, University of Minnesota, St. Paul, Minnesota 55108.

C. E. GREEN AND C. M. DONOVAN

Regeneration of Monoploid Plants from Tissue Cultures of Maize

The gene indeterminant gametophyte (*ig*) causes the frequency of monoploid individuals in maize to be increased to about 3 percent. The gene R-navajo (*R-nj*) can be used to identify those monoploids by the absence or presence of anthocyanin pigmentation in the embryo. Crosses involving W23 *ig R-nj* females × A188 or Black Mexican Sweet males were used to generate paternal monoploid embryos for tissue culture. Immature embryos were isolated from ears of these crosses 11–13 days after pollination and placed on MS medium containing 2,4-D for callus initiation. The embryos were exposed to continuous fluorescent light immediately after isolation for 3–6 days to stimulate anthocyanin development in the scutellum. Embryos that remained colorless after light treatment represented presumptive monoploids of paternal origin. These were selected for continued callus initiation. Embryos with a colored scutellum were discarded because they represented diploids or maternal monoploids. Callus cultures capable of plant regeneration were initiated from 16 presumptive monoploid embryos.

Root tips collected from cultures or regenerated plants were treated with colchicine and fixed in Farmer's solution. The chromosome number of these roots was determined by preparing squashes, staining with propiono carmine, and cytological analysis of cells with metaphase figures. Cultures derived from three embryos produced roots with 10 chromosomes per cell indicating that these cultures had monoploid origins. Currently, 21 plants have been regenerated from monoploid cultures, and six of these have grown to maturity. These plants had roots with 10 chromosomes and the

C. E. Green and C. M. Donovan, Department of Agronomy and Plant Genetics, University of Minnesota, St. Paul, Minnesota 55108.

reduced stature, narrow leaves, and sterility characteristic of maize monoploids.

J. McD. STEWART and C. L. HSU

In Ovulo Embryo Culture as a Means to Interspecific *Gossypium*

The technique for cotton ovule culture developed by Beasley and Ting was examined as a means for embryo culture. Ovules on the basal medium (BT) developed embryo axes but no cotyledon. A few embryos germinated and under optimum conditions grew to mature plants. BT with phytohormone supplement (BTP) produced no viable embryos; however, addition of NH_4^+ to BTP allowed full embryo development and germination. These plantlets were easily grown to maturity.

The technique was used to culture *G. hirsutum* (4X) ovules fertilized by *G. arboreum* (2X). BT supplemented with NH_4^+ produced about a 50 percent yield of embryos with small cotyledons, but these embryos could be cultured to mature flowering plants. $BTP + NH_4^+$ produced more vigorous embryos, but the tendency for the ovules to callus reduced the yield of embryos to 10 percent. Hybrid plants, although sterile, were vigorous and had many morphological characteristics intermediate between the two parents.

J. McD. Stewart and C. L. Hsu, U.S. Department of Agriculture, Agriculture Research Service and Department of Plant and Soil Science, University of Tennessee. Knoxville, Tennessee 37916.

Embryo Culture of *Cucumis metuliferus* and the Interspecific Hybrid with *C. melo*

Embryo culture was studied as a means to circumvent difficulties in crossing *Cucumis metuliferus* (African horned cucumber) with *C. melo* (muskmelon). Hybrid embryos from this interspecific cross failed to develop when *C. metuliferus* was used as the female. Fruit set did not occur in the reciprocal cross. Embryos from self-pollinated *C. metuliferus* were in the globular stage at nine days after pollination and in the early heart-shaped stage by the tenth day. The torpedo shape was reached by the twelfth day, and the embryos were mature by the fifteenth day. Development of selfed early heart- and heart-shaped stages grew rapidly on Murashige and Skoog's culture medium supplemented with 3 percent sucrose and 0.1 mg/1 IAA, 0.001 mg/1 kinetin, 0.4 mg/1 thiamin. About 23 percent of the embryos developed into whole plants. Less than 1 percent of globular embryos showed limited growth and only when they were left in contact with the endosperm. None grew when they were excised from the embryo sac. Hybrid embryos from the cross *C. metuliferus* (⊕) × *C. melo* (o) took three to five days longer to develop to comparable stages. They were smaller and did not grow beyond the early heart-shaped stage even in fifty-day-old fruit. They were reared in culture with varying degrees of success. One was raised to a whole fruiting plant, but the progeny had characteristics of the maternal plant only. We believe the embryo in this instance may have been apomictic. Successful hybridization of the African horned cucumber with the muskmelon should lead to improvements in the cultivated muskmelon because it has good resistance to the southern root-knot nematode, watermelon mosaic virus I, squash mosaic virus, and powdery mildew.

G. Fassuliotis, U.S. Vegetable Laboratory, U.S. Department of Agriculture, Agricultural Research Service, Charleston, South Carolina 29407.

D. CRESS, P. JACKSON, AND K. G. LARK

DNA Replication in Soybean

DNA replication in cell suspension cultures of soybean strain SB-1 has been studied. Strain SB-1 has a DNA content of 2 *pico* gm/cell corresponding to 36–38 chromosomes/cell.

Radioactive DNA precursors (thymidine or 5-bromodeoxyuridine) are taken up by cells or by "properly prepared" protoplasts with equal efficiency. The initial uptake is linear over a period of two hours and results in a semiconservative product. Isolation of large DNA from protoplasts is successful if nuclease action is avoided. This has been done in two ways: separation and subsequent lysis of nuclei; or inhibition of nucleases by appropriate physical and chemical treatment of lysing protoplasts. An interesting endonuclease reduces the DNA to a size of 200–800 nucleotides and may represent a restriction endonuclease. Newly synthesized DNA appears to be relatively resistant to this endonuclease action.

DNA is synthesized in small pieces that are gradually joined together. Incorporation of density label confirms previous autoradiographic observations that these small pieces are arranged in tandem arrays. Velocity sedimentation indicates a size of ca. 15μ for the repeating tandem unit. In the presence of FudR, replication is slowed and tandem units do not get joined into the larger pieces that can be observed in the absence of this inhibitor. However, some joining occurs. Incorporation of BudR into DNA yields a completely substituted DNA with a heterogeneous density indicative of variations in AT/GC ratio.

The results indicate that the general aspects of replication are similar to those observed in other eukaryotes such as yeast, mammalian cells, or

D. Cress, Department of Biology, University of Utah, Salt Lake City, Utah 84112.

drosophila and agree with previous work on plant cells. Some implications of these results will be discussed, i.e., aspects that may be uniquely important for future genetic or genetic engineering studies:

1. The nature of the endonuclease, the range of substrates that it can attack, and its intracellular location.
2. The use of BudR labeling to isolate particular fractions of the soybean genome corresponding to a particular AT/GC content.
3. The possible use of FudR to limit replication (and hence density substitution) to clusters of neighboring genes that can be isolated and purified according to density and length criteria.

Z. S. WOCHOK AND M. A. EL-NIL

Comparison of *In Vitro* Developmental Responses of Wild and Full-Sib Families of Douglas Fir

Douglas-fir cotyledons from wild seedlings have been successfully cultured, and produced adventitious shoots *in vitro*. Some of these have rooted to form plantlets. In this study a comparison has been made of the developmental response of cultured cotyledons from wild seedlings and several selected full-sib seedlings. Cotyledons from wild-type seedlings were cultured on different media. The number of bud primordia that developed varied both on a treatment basis and among seedlings, and ranged from 21 to 264 primordia per seedling. Analyses were made of shoot development in terms of the percentage of explants forming one or more shoots per explant. The responses of several full-sib and wild cotyledons on four media were compared. The data indicate a variability among the experimental group.

Experimental data suggest that cotyledons selected from seedlings two to four weeks after germination produce more shoots than cotyledons from older seedlings. Proper selection of plant material for tissue culture has been shown by other investigators to be of primary importance in the responsiveness of the tissue to treatment, and this system appears to be no exception.

Z. S. Wochok and M. A. El-Nil, Forestry Research Center, Weyerhaeuser Company, Centralia, Washington 98531.

T. A. THORPE, T. GASPAR, AND M. TRAN THANH VAN

Changes in Isoperoxidases during Differentiation of Cultured Tobacco Epidermal Layers

Explants of floral branches of tobacco (*Nicotiana tabacum* L., cv. Wisc. 38) consisting of 6–8 layers of epidermal and subepidermal cells differentiate in semi-solid culture into flowers or vegetative buds or roots or become unorganized into callus through interactions of auxin, cytokinin, and sucrose. Changes in isoperoxidases were examined by starch gell electrophoresis during these different morphogenetic programs. There was an increase in total specific peroxidases in all programs compared with the explant, with the appearance of new isoperoxidases. Though several bands were common to each program, there were qualitative and quantitative differences both in anodic and cathodic peroxidases during differentiation. The possible significance of these changes will be discussed.

T. A. Thorpe, Department of Biology, University of Calgary, Calgary, Alberta, Canada.

S. M. REED AND G. B. COLLINS

Use of Tissue Culture Methods to Obtain Interspecific Hybrids in *Nicotiana*

In an attempt to produce interspecific hybrids of *Nicotiana* that had not been obtained previously, fertilized ovules of the desired hybrids were cultured *in vitro* on a chemically defined agar medium. The hybrid combinations of interest were those of *N. tabacum* (2n = 48), *N. stocktonii* (2n = 48) and *N. nesophila* (2n = 48).

Ovaries of the hybrid crosses were collected at various intervals after pollination. After surface sterilization and removal of the ovary wall, the fertilized ovules were transferred aseptically to vials containing an appropriate culture medium. Hybrid plants that germinated and grew on the culture medium were either rooted or sectioned into small pieces and placed on a medium that favored the proliferation of a callus. Plants that were regenerated from such callus were induced to root for subsequent transfer to pots in the greenhouse.

Plants were obtained from all three hybrid combinations when *N. tabacum* was used as the pollen parent. No plants of the *N. repanda* × *N. tabacum* cross survived to maturity. *Nicotiana stocktonii* × *N. tabacum* and *N. nesophila* × *N. tabacum* plants survived to maturity and flowered. Meiotic analyses completed on each of the hybrids revealed a range of 0–4 bivalents, with a mode of 1 bivalent. Stainable pollen was not observed from either of the hybrid combinations. The hybrid plants were intermediate in appearance between the two parents. The hybrid plants are presently being used in an attempt to transfer race 1 black shank resistance to commercial tobacco.

S. M. Reed and G. B. Collins, Department of Agronomy, University of Kentucky, Lexington, Kentucky 40506.

E. D. EARLE, D. YORK, AND V. E. GRACEN

Response of Maize Mesophyll Protoplasts to the Host-Specific Toxin Produced by *Helminthosporium maydis* Race T

High yields of mesophyll protoplasts ($1-5 \times 10^6$ / g F. W.) were isolated from maize (*Zea mays* L., inbred W64A) leaves after 2–4 hours of incubation in 2 percent Cellulysin and 1 percent Driselase. Many of these protoplasts survived for at least 7 days in the dark in medium containing only 0.5 M sorbitol and 10 mM $CaCl_2$ 2 H_2O. Survival of protoplasts from plants with male-fertile (N) cytoplasm was unaffected by toxin produced by the fungus *Helminthosporium maydis* race T. Low concentrations of toxin (1/5 the level that caused 50 percent inhibition of seedling root growth) destroyed all protoplasts from plants with Texas male-sterile (T) cytoplasm. Damage was usually apparent within 18 hours and by 1–2 days, all toxin-treated T protoplasts collapsed. Toxin-treated T protoplasts survived longer when cultured in the light. Thirty minutes' exposure to toxin followed by washing was sufficient to destroy T protoplasts. Electron microscopy showed that 30–60 minutes' exposure to toxin severely damaged mitochondria of treated protoplasts. The rapid and sensitive interaction between T protoplasts and *H. maydis* race T toxin could be used as an assay for toxin activity and in studies of the physiological and ultrastructural basis for toxin action. It may also facilitate selection of protoplasts that have acquired toxin resistance through genetic manipulations such as protoplast fusion or mitochondrial transfer.

E. D. Earle, Department of Plant Breeding and Biometry, New York State College of Agriculture and Life Sciences, Cornell University, Ithaca, New York 14853.

W. R. KRUL and J. F. WORLEY

Formation of Adventitious Embryos in Callus Cultures of "Seyval"

Callus tissue of "Seyval" (Seyve-Villard 5-276) (*Vitis* sp.), a French hybrid grape, formed adventitious embryos when transferred from a medium with 2,4-dichlorophenoxyacetic acid (2,4-D) to medium containing 1-naphthaleneacetic acid (NAA). Embryos began to turn green and develop into apparently normal vines when placed on a medium free of hormones and vitamins in the light. Histological evidence indicated that plants derived from callus originated from embryo-like structures presumably derived from single cells and not from plantlets of excised buds. Secondary embryoids formed on primary embryoids, and tertiary embryoids occasionally formed on secondary embryoids. Outgrowth of secondary and tertiary embryos was enhanced when the apex of the parent plant was removed. More than fifty vines and several hundred adventitious embryoids were obtained from approximately one cubic centimeter of callus. An evaluation of the first season's field performance of vines derived from callus will be discussed.

W. R. Krul and J. F. Worley, Cell Culture and Nitrogen Fixation Lab, U. S. Department of Agriculture, Agricultural Research Service, Beltsville, Maryland 20705.

R. H. SMITH AND H. J. PRICE

Gossypium Species in Culture

Environmental and culture conditions have been defined by us for initiating callus from hypocotyl explants of several *Gossypium* species. The medium consists of the Murashige and Skoog inorganic formulation with 100 mg/ 1 myo-inostol, 0.4 mg/ 1 thiamine · HCl, 2 mg/ 1 IAA, 1mg/ 1 kinetin, 3 percent (w/v) glucose, solidified with 0.6 percent agar. Callus initiation is best under high light intensity (9000 lux) at 30°C. Only modifications of the phytohormone combinations and concentrations were necessary to obtain vigorous green callus subcultures that were free of the discoloration commonly reported in the literature. Hormone requirements for prolonged subculture of six species are as follows: *G. anomalum*, 10.0 mg/ 1 2iP and 1.0 mg/ 1 IAA; *G. arboreum*, 0.5–1.0 mg/ 1 BA and 2.0 mg/ 1 NAA, or 5.0–10.0 mg/ 1 2iP and 1.0 mg/ 1 NAA; *G. armourianum* 1.0 mg/ 1 2iP and 5.0 mg/ 1 IAA; *G. hirsutum*, 1.0 mg/ 1 2iP and 5.0 mg/ 1 IAA, or 1.0 mg/ 1 2iP and 0.1 mg/ 1 NAA; *G. klotzschianum* and *G. raimondii*, 1.0 mg/ 1 2iP and 0.1 mg/ 1 NAA. We have obtained suspension cultures derived from *Gossypium* callus and are currently selecting for mutants, e.g., those resistant to high concentrations of 5-methyltryptophan. Organogenesis is also being examined.

R. H. Smith and H. J. Price, Department of Plant Sciences, Texas A&M University, Texas Agricultural Experiment Station, College Station, Texas 77843.

K. N. PANDEY AND P. S. SABHARWAL

In Vitro Induction of Plant Tumors by Gamma Irradiation

Callus tissues were derived from inflorescence segments of *Haworthia mirabilis* Haw. Stock cultures were maintained on Murashige and Skoog's agar medium, supplemented with 1.5mg/1 NAA and 1.5 mg/1 kinetin. The callus tissues were irradiated with ^{60}Co-rays at a dose rate of 185.77 rads/min. The dose varied from 10 rads to 5,000 rads. Growth kinetics were analyzed for irradiated and nonirradiated treatments during a period of 20–24 weeks. The dose range between 600–2,500 rads induced compact callus at high frequencies as compared with the controls that were friable. Irradiation also markedly reduced the rate of growth and differentiation of callus tissues. After a period of 12 weeks of growth, the control cultures completely differentiated into vegetative buds, whereas irradiated callus exposed to 800–2,500 rads continued to grow but remained in the callus form. The tissues irradiated in the range of 1,000–2,000 rads, and controls were subcultured on the media without growth hormones and inositol. Under these conditions, the control cultures did not grow, whereas irradiated callus tissues grew well. Further, using a bioassay system high concentrations of cytokinins were detected in the callus tissues. The ability of irradiated callus to produce endogenous cytokinins and to grow autonomously is indicative of its tumorous nature. To our knowledge, this is the first report on the induction of plant tumors in callus tissues by gamma irradiation.

K. N. Pandey and P. S. Sabharwal, T. H. Morgan School of Biological Sciences, University of Kentucky, Lexington, Kentucky 40506.

A. SIEGEL, V. HARI, AND K. KOLACZ

A Remarkably High Rate of Tobacco Mosaic Virus-Specific Protein Synthesis in Infected Protoplasts

Tobacco mesophyll protoplasts were either infected with tobacco mosaic virus or mock infected and then incubated in the light of 25°C in a medium containing only the salts of Murashige and Skoog. Samples containing 500,000 protoplasts in 1.5 ml were sequentially pulsed for two hours, starting one hr after inoculation, with 40 μCi ^{3}H-leucine and 300 μg chloramphenicol. The proteins labeled in 0.1 of each sample (50,000 protoplasts) during each 2 hr time period were analyzed by SDS-PAGE and fluorography on a 12–20 percent slab gel. It was found, in agreement with others, that infected protoplasts synthesize three unique proteins with nominal molecular weights of 160,000, 135,000 and 17,500 daltons, the smallest being the viral capsid protein. Synthesis of these proteins was first detected during the 5–7 hour pulse period, during which time the 135,000 protein was synthesized more rapidly than the other two. During all other time periods, except for 7–9 hrs, the viral-specific proteins incorporated label at a rate inversely proportional to their size. The rate of capsid protein synthesis increased rapidly until during the final period analyzed, 44–46 hrs, it accounted for 67 percent of the protein-synthetic activity of the protoplast. The actual increase in rate of capsid protein synthesis was actually greater than this because it was determined that synthesis of virus-specific proteins was in addition to, rather than at the expense of, synthesis of all of the other protoplast proteins. On this basis, the rate of synthesis of the 135,000 protein was constant from 11–13 hrs to the end of the experiment at 46 hrs. Analysis of the soluble fraction of disrupted 2 hr ^{3}H-leucine-pulsed protoplasts by non-SDS PAGE on a 3–20 percent low

A. Siegel, Biology Department, Wayne State University, Detroit, Michigan 48202.

cross-linked slab gel revealed that only a fraction of the newly synthesized capsid protein had been incorporated into virus particles.

D. G. DAVIS, R. H. HODGSON,
K. E. DUSBABEK, AND B. L. HOFFER

Metabolism of the Herbicide Diphenamid (*N, N*-Dimethyl-2,2-Diphenylacetamide) in Cell Suspensions of Soybean (*Glycine max* Merr.)

The fate of the herbicide diphenamid (*N, N*-dimethyl-2,2-diphenylacetamide), was determined in cell suspensions of soybean, (*Glycine max* Merr. "Wilkin"). (^{14}C-carbonyl)-diphenamide was added as an acetone solution to aseptic cultures to give final concentrations of 2 to 3 μM diphenamid and 0.5 percent (v/v) acetone. Neither diphenamid nor acetone were phytotoxic at these concentrations. Diphenamid was present in the cultures during three growth phases of the cell cultures: early log phase (3 to 7 d), log phase (7 to 14 d), and stationary phase (14 to 18 d). The ^{14}C-labeled products were identified tentatively by thin-layer chromatographic comparison with reference compounds. The major metabolic products appeared to be the same as those found in intact plants; the *N*-hydroxy methyl; monomethyl; and demethylated analogs of diphenamid; and two polar metabolites (0.9 percent to 25 percent of the applied ^{14}C) that were probably two glucoside conjugates, one with an acidic group attached to the glucose. A minor $CHCl_3^-$ soluble metabolite appeared to be different from those previously identified in intact plants. Although cell cultures cannot be expected to behave exactly like intact plants or to be used to replace intact plants, the results on diphenamid metabolism in soybean cell suspensions indicate that they can be used to obtain reliable information in the fate of agricultural chemicals in plants.

D. G. Davis, Metabolism and Radiation Laboratory, U. S. Department of Agriculture, Agricultural Research Service, Fargo, North Dakota 58102.

K. W. HUGHES

Isolation of a Herbicide-Resistant Line of Soybean Cells

The herbicide paraquat is most toxic to photosynthesizing plants. There is, however, a toxic effect on plants grown in the dark. Toxicity in the light has been ascribed to a free radical formation via the photosynthetic pathway. The dark reaction is unknown. Callus induction and growth of soybean cotyledon tissue was studied at levels of paraquat from 10^{-8}M to 10^{-3}M. Callus induction and growth is completely inhibited at 10^{-3}M paraquat and severely restricted at 10^{-4}M in both the light and the dark. In the light, callus growth is stimulated at low molarities above the control level, but this response is highly variable. There is no significant difference between plant variation for callus induction in the light, but there is a highly significant difference in the dark. This confirms findings by other investigators that the light and dark reactions are different. A resistant line of soybean cells has been isolated by selection at 10^{-3}M paraquat and maintained on 10^{-4}M paraquat. The resistant line grows normally at 10^{-3}M paraquat and retains its resistance after several months growth on medium without paraquat. Studies indicate that the resistance has a genetic basis. Studies concerning the physiological basis of paraquat resistance are in progress.

K. W. Hughes, Department of Botany, University of Tennessee, Knoxville, Tennessee 37916.

Organogenesis from Callus Cultures of Orchardgrass

Because of a high degree of heterozygosity and self-incompatibility, outstanding clones of "cool season" grass species, such as orchardgrass, must be propagated vegetatively to retain a specific genotype. Since these species do not produce stolons or rhizomes, the number of individuals that may be propagated at one time from a single plant is limited. Cloning at the cellular or tissue level could be of great importance in the improvement of these species from the standpoint of greatly increasing the number of individuals and decreasing the time for clonal propagation. Thus, the objective of the present research was to develop technology for the induction of callus and plantlet regeneration in orchardgrass (*Dactylis glomerata* L.).

Calli were initiated from either mature embryos or whole caryopses of orchardgrass (*Dactylis glomerata* L., cv. Boone) on a modified Schenk-Hildebrandt (SH) medium containing 2.15 mg per liter of kinetin and 15 mg per liter of 2,4-dichlorophenoxyacetic acid (2,4-D). The calli were transferred twice at two- to three-week intervals and maintained on an SH medium containing 5 mg/1 of 2,4-D. Small leaves developed on the third subculture when the 2,4-D concentration was lowered to 1 mg/1. These leaves then developed into shoots. Roots were also initiated; but root growth proceeded more rapidly when the young plantlets were transferred to an SH medium without hormones and in which the inorganic salt concentration was reduced by one-half. After sufficient root and shoot growth, seedlings could easily be transferred to small pots and the plants grown to maturity.

B. V. Conger and J. V. Carabia, Department of Plant and Soil Science, University of Tennessee, Knoxville, Tennessee 37901.

R. A. MAZUR AND J. X. HARTMANN

Freezing of Plant Cells and Protoplasts

The survival rate of cultured plant cells and protoplasts was assessed after freezing to –196°C in cryoprotectant solutions. Two-day-old cells and enzymatically prepared protoplasts of *Bromus inermis* Leyss and *Daucus carota* L. were frozen in a four-step protocol (+4°C, –20°C, –60°C and –196°C). Glycerol and dimethyl sulfoxide (DMSO) were compared at three concentrations against a protoplast culture medium for their ability to protect against damage by freezing. The results showed that carrot protoplasts survived better (68 percent viability) than cells (38 percent) when 0.7 M glycerol was utilized as a cryoprotectant. When DMSO was utilized, the survival rates were nearly equivalent but significantly lower than with glycerol (44 percent-cells; 41 percent-protoplasts). Similar results were obtained with bromegrass cells and protoplasts. Viability was assessed by trypan blue dye exclusion. An equilibration period prior to freezing was essential for survival of plant protoplasts. The role of the cell wall in freezing will be discussed.

R. A. Mazur and J. X. Hartmann, Department of Biological Sciences, Florida Atlantic University, Boca Raton, Florida 33431.

Index